Applications of Space Techniques on the Natural Hazards in the MENA Region

Mashael M. Al Saud
Editor

Applications of Space Techniques on the Natural Hazards in the MENA Region

Editor
Mashael M. Al Saud
Space Research Institute
King Abdel Aziz City for Science and
Technology (KACST)
Riyadh, Saudi Arabia

ISBN 978-3-030-88873-2 ISBN 978-3-030-88874-9 (eBook)
https://doi.org/10.1007/978-3-030-88874-9

This Springer imprint is published by the registered company Springer Nature Switzerland AG
The registered company address is: Gewerbestrasse 11, 6330 Cham, Switzerland

Preface

Natural hazards, and the resulting catastrophism, are the result of naturally occurring processes that have been operating throughout Earth's history. They became widespread physical processes that often strike many regions of the globe. They usually occur in places with frequent natural risk, but they sometimes suddenly occur in regions which have never witnessed such disastrous processes. Lately, natural hazards have attracted more attention, because of threats from naturally occurring events, and thus leaving negative impact on humans and the environment severely destructed; therefore, there arise consequences in many communities, notably those with weak risk controls as well as poor societies. The majority of natural hazards implies floods, ocean storms, landslides, forest fire, earthquakes, volcanoes, and other aspects of natural disasters. Many of these hazards are characterized for defined regions, and even their impact can be predicted in those regions.

Recently, natural hazards have been exacerbated with obvious increase in their impact, and this is attributed to several factors, including mainly the changes in the climatic conditions, notably the extreme climatic events, as well as the urban expansion which increases the exposure to more damages and loss in human life. Over the last two decades (2000–2020), there were more than 7400 major recorded disaster events claiming more than a million lives, affecting more than a billion people, and resulting in more than three trillion US dollars in global economic losses, where a large part of these events is attributed to the natural disasters.

Natural hazards are well pronounced in the Middle East and North Africa (MENA) Region. Over the past four decades, these disastrous events have affected more than 40 million people in the MENA Region and cost their economies about US$ 20 billion. In the MENA Region, there is rapid urbanization, which increases the exposure of people and economic assets to disaster events. Besides, there are many projects applied to reduce and mitigate the impact of the hazards. These projects have been based on scientific studies done to identify the main factors behind natural hazards, and to apply assessment and then to propose the best measures required. Thus, many methods and models have been adopted to determine the vulnerable zones for natural disasters. However, the applied methodologies differ between the applied studies, notably when diverse tools are used.

As reliable and effective tools for studying natural hazards, remote sensing techniques have been lately raised and proved their reliability in natural risk management approaches. These techniques are being rapidly developed with the success of technological industries; therefore, they enable capturing observations for Earth's surface from different altitudes and at different time periods. Thus, a miscellany of observatory satellite images and other space techniques are processed to study natural hazards, including their vulnerability and assessment, and propose risk controls. The advancement of space techniques in the field of natural hazards is worth mentioning, and they must be highlighted as major tools for risk analysis.

This motivated the author to develop this book document on the applications of space techniques in the study of natural hazards in the MENA Region, notably this region is considered as a global hotspot for natural disasters. Furthermore, there are many outstanding scientists from the MENA Region who work on the applications of space techniques in different themes, with a special emphasis on natural risk analysis. Therefore, the assembly of several topics obtained by these scientists, the case of this book, will introduce a distinguished volume that highlights the issue of natural disasters.

Due to its significance, this book is a first of its type on the use of space techniques on natural hazards. It is composed of 26 chapters where topics (i.e., types of natural risks) are crossed with different geographic zones from the MENA Region. It includes three main parts, starting with the introductory part on application of space techniques on the entire MENA Region, then the second part that illustrates different case studies applied also to the whole MENA Region, and, finally, the third part that reveals case studies from different countries in the region, where it has been classified into different themes, including comprehensive assessment of natural hazards, floods, drought and desertification, landslides and subsidence, and different aspects of natural hazards.

The author is aiming to introduce this book for a spectrum of audiences including scientific researchers and academics, experts and decision makers, university and research centers, as well as national institutes concerned with natural risk management.

Riyadh, Saudi Arabia Mashael M. Al Saud

Contents

Part III Case Studies: Space Techniques and Natural Hazards in the MENA Region

About the Editor

Mashael M. Al Saud is a research scientist at the Space Research Institute, King Abdel Aziz City for Science and Technology (KACST), Kingdom of Saudi Arabia. Prof. Al Saud is specialized in applied geomorphology and geodesy. She has been working in several disciplines where space techniques and geo-information systems were usually utilized. The author has been involved in research studies and projects on watershed management, applied geomorphology and surficial processes, natural hazards, water resources assessment and exploration, geophysical prospecting, and many other topics where satellite images are used to monitor Earth' surface.

Prof. Al Saud is a member of numerous national and international scientific societies, and has been granted a number of national and international awards and honorariums. She has remarkable number of published scientific works including books, publications in international peer-reviewed journals, and regional and international conferences and symposia.

Prof. Al Saud is continuously producing research projects, technical repots, and thematic maps for different regions of the Kingdom of Saudi Arabia, and always aiming to add scientific inputs for the research and development. Recently, Prof. Al Saud started to integrate studies on the regional scale, such as in this book which covers the entire Middle East and North African Region.

Part I
Introduction

Chapter 1
Space Techniques for Earth Observation

Mashael M. Al Saud

Abstract Features on Earth's surface remained undiscovered for long time and the interlinkage between these features was undefined until human was initiated conventional tools to figure out all these feature in one view from above. Thus, space techniques have been created with simple launching of sensors and cameras at heights to observe all objects on Earth's surface. This has been developed from low-flight by airplanes to space shuttles that can turn around the Earth along define orbital rotation. Therefore, the era of Remote Sensing has been existed since early 1990s, and tens thousands of these space shuttles, represented mainly by satellites, are rotating and capturing data on different themes required to be investigated. There are numerous of Land Observatory satellites retrieving daily images from hundreds of kilometers above Earth's surface and then transmitting these images for to receiving station which became widespread. Recently the use of space techniques in studying Earth's features has been rapidly developed, and the remote sensing techniques and its complimentary geo-information systems are new tools implied in many institutions and they became a primary tool for the management of Earth's resources. These tools have a wide spectrum of applications, on terrestrial and marine environments, including mainly the monitoring approaches, resources assessment, change detection, natural hazards assessment. This chapter will demonstrate a detailed discussion on these techniques with a special emphasis on satellite images and their specifications.

Keywords Satellite image · Polar orbit · Spatial resolution · Geo-spatial data · GIS

1.1 Concepts and Definitions

Natural resources, environment, infrastructure as well as human are usually under natural risks which is resulted in severe damages and destruction of major components on Earth's surface. Traditional methods for the analysis of these risks has been

M. M. Al Saud (✉)
Space Research Institute, King Abdel Aziz City for Science and Technology (KACST), Riyadh, Saudi Arabia

M. M. Al Saud (ed.), *Applications of Space Techniques on the Natural Hazards in the MENA Region*, https://doi.org/10.1007/978-3-030-88874-9_1

initiated since few centuries when urbanism has taken place over wide geographic areas and the number of natural risks have been exacerbated.

Primarily, maps were utilized along with field surveys and investigations and accompanied with laboratory tests. Thus, people has long identified the importance and value of maps in several topics notably those belong to surficial processes on Earth's surface. In fact, the history of mapping can be referred to some thousands of years ago, and this remained until the last few decades when maps production was following traditional methods of cartography. These methods were primitive whether in terms of the acquirement of data and information on terrain surfaces and the existing processes or in the ways by which these maps were drawn.

Nowadays, mapping is represented mainly by topographic maps (i.e. base maps) or charts, which are characterized by large-scale dimensions and variety of quantitative representation of vertical and horizontal features of terrain surface that are illustrated as contour lines (i.e. virtual lines with equal elevations). These maps were primarily extracted from aerial photos that are acquired by aircraft or by any low-flight shuttles. Aerial photos, which have a miscellany of specifications and are interpreted by stereoscopic analysis, represent the beginning of Remote Sensing era when the first recorded photograph was taken from an airplane by Wilbur Wright in 1909. This led to the development of science of photogrammetry which enables applying analysis and measurements for the Earth's surface.

Recently, stereoscopic photographs and satellite images can produce topographic maps with high accuracy, and then enabling the extraction of thematic information in order to apply several calculations and analysis on the topographic maps. Therefore, satellite remote sensing has been used for topographic mapping purposes since the launch of ERTS-1 in 1972 (Dixon-Gough, 1994).

Recently, the use of space techniques has become a common tool in many regions and they are applied for different applications, including natural resources and surficial processes. In a broad sense, these techniques tackle the processing of digital (i.e. electronically produced) satellite images and remotely sensed products, which are significant instruments used to draw maps, identify, analyze and measure the observable terrain features that are reflecting several natural and anthropogenic processes (Al Saud, 2020).

The following are significant definitions belong to space techniques including their specifications and applications:

– Remote Sensing (RS): It is recently described as a science of exploring and monitoring Earth's components from space. It is therefore, the process for detecting and monitoring the physical characteristics of the objects on Earth's surface without physical contact with these objects. RS depends mainly on measuring the reflected and emitted radiation on Earth's at a distance (typically from satellite or aircraft).

GIS: The Geographic Information System is computer-based system of hardware and software capable to display and manage information about geographic localities, analyze spatial relationships, and model spatial processes.

- Temporal resolution: The frequency at which images are captured over the same location on the earth's surface.
- Sensors: A device (e.g. camera) mounted on satellite to collect data by detecting the energy that is reflected from Earth.
- Pixel: is a physical point in a raster image, or the smallest controllable element of a picture represented on the screen.
- Digital number (DN): It is a variable assigned to a pixel where it represents a binary integer in the range of 0–255 (i.e. a byte). Thus, a single pixel may have several digital number variables corresponding to different bands registered.
- Raster image: Is a pixel-based image, called also a "bitmap", where the latter is a grid of individual pixels that collectively compose an image.
- Vector image: This refers to vector graphics are based on mathematical formulas that define geometric primitives such as polygons, lines, curves, circles and rectangles.
- Radar: Acronym for radio detection and ranging. A device or system that detects surface features on the earth by bouncing radio waves off them and measuring the energy reflected back.
- Space shuttle: *A space* Transportation System, partially reusable rocket-launched vehicle designed to span into orbit around Earth. It often carry sensors, human or cargo.

1.2 Space Techniques

There are different types of space techniques where sensors are mounted of platforms or shuttles forming satellites that travel in space and follow define orbital turns. Up to date, there are more than sixty thousand satellites have been lunched to the space around the Earth, and many of them are now non-functional.

Thus, there are the communication and land (or Earth) observatory satellite images where the first is concerned in the communication purposes (e.g. radio, TVs, GPS, etc.), while the observatory satellites are acquiring images to the Earth's surface including the marine and terrestrial environments. For the latter, it has been assigned as the science of "Remote Sensing"which has a wide spectrum of applications and it is being rapidly developed.

Satellites follow two orbiting traverses; these are either polar or geostationary orbiting. Hence, meteorological satellites and most of the communication satellites follow the geostationary orbit, while land observatory satellites travel along the polar orbit (Fig. 1.1).

Therefore, Remote Sensing is the process of geo-spatial data acquisition from satellites, aircraft, and lately drones, etc. therefore, satellite images, aerial photos, thermal images, Digital Elevation Models (DEMs), etc. are produced. The acquired geo-spatial data is derived from the electro-magnetic radiation (EMR) which is reflected and emitted from the objects, with different physical characteristics, on Earth's surface. Even though, the flight characteristics of the platforms are

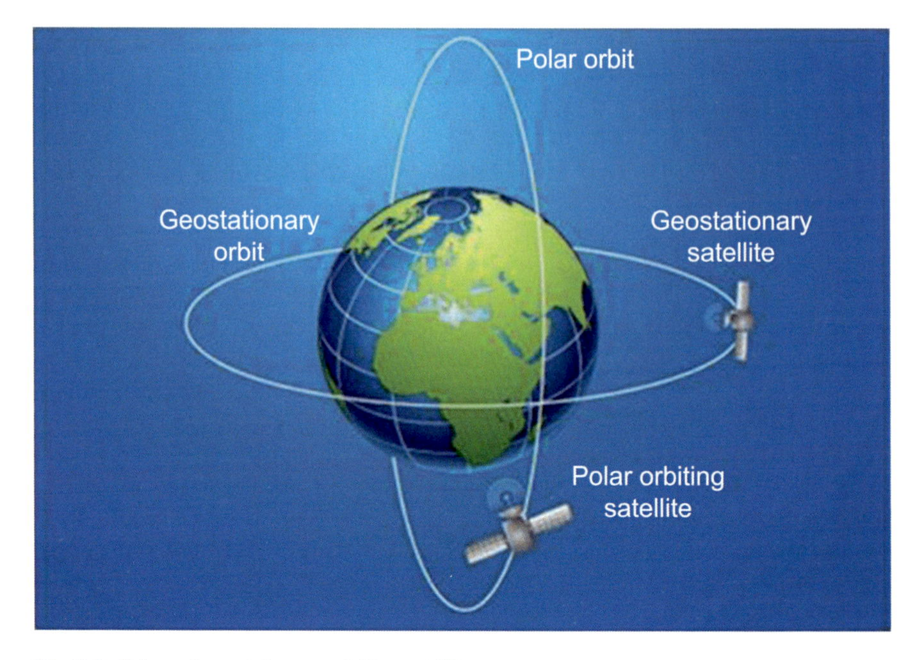

Fig. 1.1 Polar and geostationary orbiting satellites

significant for the resolution of the acquired geos-spatial data, yet sensors on satellites are the most significant in this respect.

This electromagnetic energy, as electric and magnetic disturbance, propagates in media (e.g. atmosphere) and spans a broad spectrum from very long radio waves to very short gamma rays. The electromagnetic spectrum is classified by wavelength that ranging from 10 to 10 mm to cosmic rays up to 1010 mm, and the broadcast wavelengths, which extend from 0.30 to 15 mm.

Two types of remote sensing sensors are known to acquire the electromagnetic radiation. They are the passive (optical remote sensing) and the active (microwave, or radar) sensors. Passive sensors detect the naturally radiation reflected or emitted by the objects on Earth's surface, whereas active sensors transmit their own energy for illumination which is directed toward the target to be identified, and then it capture the reflected radiation reflected from that target.

Remote sensing can be classified with respect to the type of radiation captured by sensors and the ranges wavelengths, as follows:

- Visible and Reflective Infrared RS,
- Thermal Infrared RS,
- Microwave (or radar) RS.

Sensors on platforms have different optical and spectral specifications, one of them is the altitude above Earth's surface, as well as the time needed to complete a turn around the Earth. The most known platforms are: 1) satellites (altitude: 500–900 km), 2) space shuttle (altitude: 185–575 km), 3) high-latitude flying aircraft (altitude: 10–12 km), 4) moderate-latitude flying aircraft (altitude: 1.5–3.5 km) and 5) low-latitude flying aircraft (with altitude below 1 km).

The first lunched remote sensing satellites, namely Sputnik, 1957; Explorer, 1958 and Corona, 1960 have been widely used and many innovative techniques were developed since then. Thus, remote sensing has a very wide range of applications in many different fields including (but not limited) water resources, marine environment, coastal zones management, agricultural, natural hazards, change detection and monitoring, marine pollution, etc.

Satellite remote sensing is a wide topic with many applications, including solid earth science, physical oceanography, land/ocean biology, cryosphere science, atmospheric science and near-earth space science. There is a vast array of passive and active sensors currently in orbit around the Earth providing a wealth of remote sensing products that are transformed into information using a variety of techniques and tools. Learning the entire field would require studying perhaps a dozen books on the topic. Physical principles of remote sensing: third edition does not attempt to cover all these tools and applications but instead exposes the basic physics and chemistry underlying remote sensing methods (Sandwell, 2013).

1.3 Satellite Images

First aerial photograph was simply practiced from a balloon flew over Paris in 1858. However, the produced aerial photograph was no longer exist, and then the earliest surviving aerial photograph, which was taken also from a balloon and entitled "Boston", was taken in 1860. Therefore, photography and photogrammetry have been rapidly developed, and then widely used notably for the military purposes during World Wars I and II.

Photography, as the beginning of remote sensing, proved to be a significant tool for visual interpretation and the production of analog maps. However, the development of satellite platforms during 1950s, the associated need to telemeter imagery in digital observation, and the need for consistent digital images have given the chance for the development of sensors as a fundamental format for the capture of remotely sensed data.

Nowadays, several satellite systems are in operation in collecting images that is subsequently distributed to users. Each type of satellite data offers specific characteristics that make it more or less appropriate for a particular application (Eastman, 2001).

Usually, identifying the specifications of a satellite image is significant, they are often determined as a first step forward while determining the tools of analysis in studies and projects. In this regard, many specifications characterize satellite images. We illustrate the most significant one as follows (according to Al Saud, 2020):

1. Spatial resolution:

The pixel size recorded in a raster images is described as "spatial resolution". Where the pixel is the smallest controllable element of a picture represented on the screen. Hence, the pixel size enables distinguishing features, and then the pixel in satellite images controls the clarification and discrimination of objects on Earth's surface. In this regard, the compound remote-sensing system is significant, including lens antennae, display, exposure, processing and many other specifications to achieve well recognizable image.

Images with low-resolution are characterized by large pixel size where they do not show objects as clearly as in high-resolution images. Hence, satellite images, from different sensors, have different spatial resolution (example Fig. 1.2) that ranges from tens of centimeters (e.g. Geo-Eye, 41 cm; Quick-bird, 61 cm) up to few of kilometers (e.g. NOAA, AVHRR, 4 km).

2. Spectral resolution:

Spectral resolution is defined as the band range or band width reported by the sensor. Thus, it is the ability of a sensor to highlight fine *wavelength* intervals. Therefore, the finer the *spectral resolution,* the narrower the *wavelength* range for a particular channel or band. For example, the satellite image of Landsat sensor has seven bands, some of them are in the infrared spectrum, where the spectral resolution ranges between 0.7 and 2.1 μm. Besides, the satellite image of Aster sensor has 14 spectral bands where 3 of them are in the visible range and 11 in the infrared range.

3. Revisit time:

This is also the return rime, or it is often described as the temporal resolution of the image where it is attributed to the temporal frequency of sampling by repeat imaging. In other words, it is defined as the time needed by the satellite to comeback over the same point on Earth's surface. The revisit time turn depends on the orbitography of the platform or satellite on which the sensor is mounted. In addition, most of the Earth observation satellites having quasi-polar orbits, thus the frequency of revisits is also depending on the latitude of the area of study. Therefore, it is important for remote sensing element, notably in mapping and monitoring approaches.

The temporal resolution (or revisit time) on the satellite image does not belong to its spatial or spectral resolution. For example, Landsat (30 m) and Aster (15 m) require 16 days to make one turn around the Earth, while the revisit time of Spot-6 (6 m) is 26 days.

4. Swath width:

Swath width describes the area imaged on the Earth's surface. Thus, it is the areal coverage that sensors can viewed and imaged, and then capture settled image size, which is called as "image scene". Hence, in aerial photogrammetry, swath width mainly depends on flight altitude type; nevertheless, the altitude has no essential effect in swath width for images retrieved by satellites which it depend mainly on the specifications of the sensor (Al Saud, 2020).

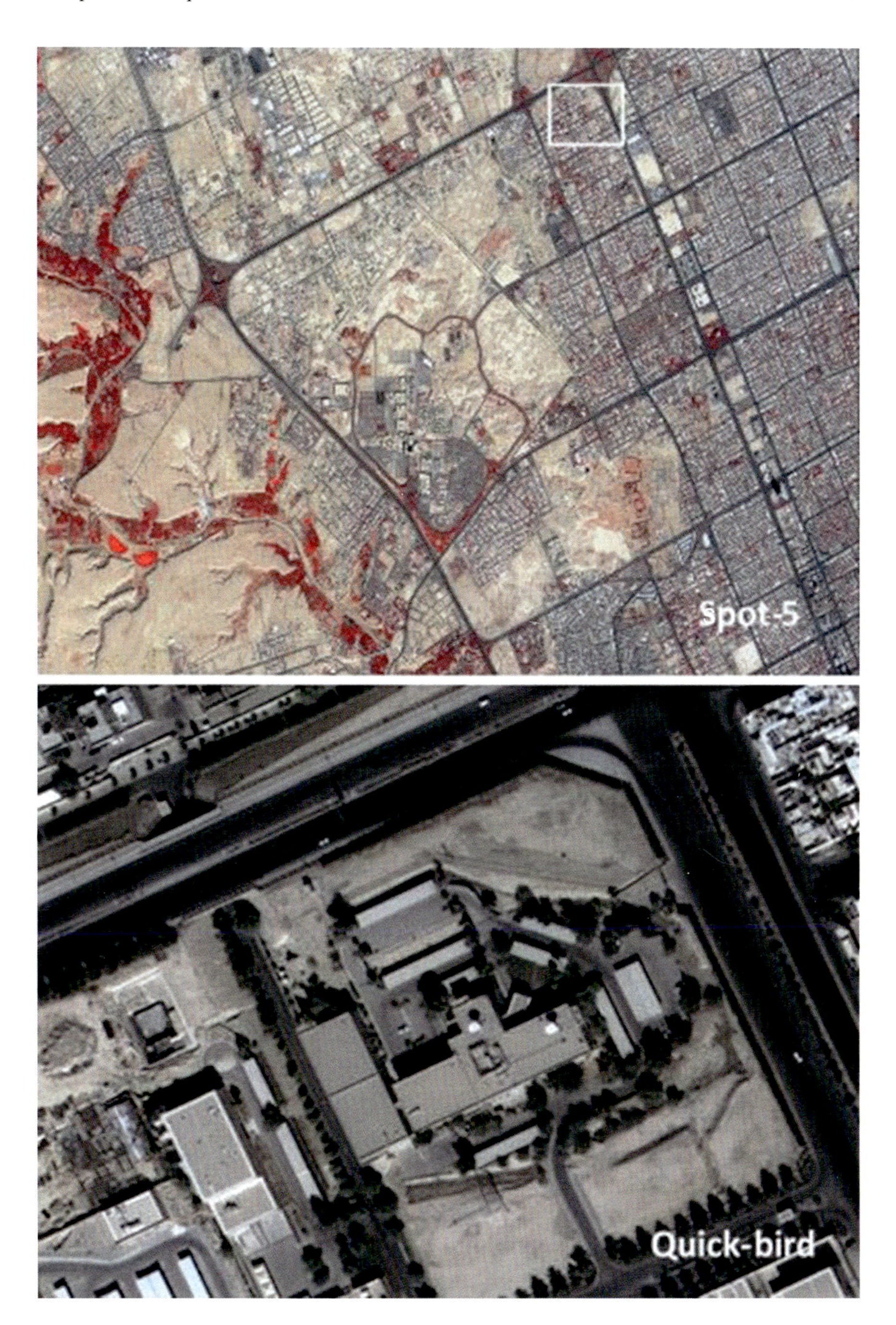

Fig. 1.2 Satellite images with different spatial resolution (Spot-5 with 5 m and Quick-bird with 0.61 m; the Quick-bird image is indicated in the white frame above in the Spot image)

Table 1.1 Example of commonly used land observatory satellite images.

Satellite	No of Bands	Spatial resolution	Revisit time	Swath width (km)
Worldview-4	6	0.31m	1.7 days	13.1 × 13.1
Geo-eye	5	0.50 m	2.8 days	15.2 × 15.2
Kompsat-6	6	0.50 m	28 days	30 × 30
Quick-bird	5	0.61 m	1–3 days	16.5 × 16.5
IKONOS	5	0.82 m	3 days	11.3 × 11.3
Rapid-eye	5	5 m	5.5 days	77
Sentinel-1A	13	5 m	6 days	250
SPOT-7	4	1.5 m	26 days	60 × 60
Aster	14	15 VNIR, 30 m SWIR, 90 m TIR.	16 days	60 × 60
IRS 1D	4	23 m	5 days	70 × 142
Landsat 7 ETM +	8	30 m, 120 m thermal, 15 m pan	16 days	183 × 183
Landsat 8 OLI-TIRS	14	30 m, 100 m thermal, 15 m pan	16 days	183 × 183
MODIS	36	250 m, 500 m, 1 km, 2 km, 4 km	Twice/day	2030 × 1354
AVHRR	4	1.1 km	Twice/day	3000

Even though, it is not often the case, yet many satellite images with big swath width are found with low spatial resolution; in spite that swath width and image resolution have no relationship. For example, a Sentinel-1 image has a swath width of 80 × 80 km, while it is 60 × 60 km for Aster and Spot-7 satellite images.

In this regard, image specifications, especially those acquired by satellites, are significant prerequisite for the selection of an image to be analyzed as shown in Table 1.1.

1.4 Satellite Image Processing

The available digital raster images of spectral reflectance data is resulted from the solid state multispectral scanners and other raster input devices. The advantage of having these digital data implies enabling to apply computer analysis techniques to the image data-a field of study called Digital Image Processing (Eastman, 2001). Thus, the efficiency of satellite images is mainly based on their optical and spectral specifications which are significant in the assessment and study of different terrain themes including mainly monitoring approaches, natural hazards assessment, change detection, etc.

For example, observing Earth's surface from space using by Ikonos satellite images enables clearly distinguishing objects with approximately 0.82 × 0.82 m area, and this is more likely to observe these objects from about100 m altitude. This virtually means that mankind can fly over any area and look down and distinguish all objects exceed 0.67 m^2, but this observation can be each 3 days only. Thus, different type of satellite images are used for several purposes to attain dates overlapping, and utilizing from the spatial resolution of one image to another, etc. (Al Saud, 2020).

This is why is must be made clear the selection of appropriate satellite images before starting the processing of these images. This includes, in a broad sense, the spatial and spectral resolution, swath width and the revisit time. In addition, the knowledge on and expertise to analyze these images using proper computerized devices is significant. Moreover, there are many other specifications to be considered while selecting satellite images, such as the spectral and electronic properties, number of bands, their wavelength and spectrum ranges (e.g. thermal band, micro-wave, optical band, etc.).

Usually, the availability of proper satellite images is a constraint. This makes it necessary for many countries to establish space center where they can retrieve a considerable number of satellite image for research and commercial application. In this regard, the Kingdom of Saudi Arabia is a typical example in the MENA Region where the Space Research Institute has been established at the King Abdulaziz City for Science and Technology (KACST).

The digital (electronic) raw data and the appropriate software for image processing are the main two components to be ready in order to start the analysis of satellite images. In this regard, there are several software types used, the most commonly ones are:

- ENVI Image Analysis: Produced by: *IBM*. Colorado, USA.
- ERDAS Imagine: Produced by: *Lucia*, Georgia, USA.
- PCI Geomatics: Developer of *Geomatica*, Toronto, Canada.
- ILWIS: Produced by *ITC*, Enschede, Netherlands.

Each software has define characteristics and follows different manners of use. Thus, the processing of satellite images implies the pre-processing steps and image analysis and classification. They can be concluded as follows:

1. Pre-processing: This represents the first step applied to prepare the images for further processing and classification. Therefore, satellite images can be analyzed after applying the following processes using the software.
   - Image sub-setting, which is the process applied to extract the area of interest (AOI) from the entire images scene. It is applied to have the lowest size and then to avoid the slow-down the work and any demanding on computing resources.
   - Atmospheric correction, is a process applied for noise removal and to clarify true surface reflectance where it is done by removing atmospheric effects from satellite images.
   - Geometric correction, which is the correction of noise and sun-angle which usually result in images displacement due to the altitude of shuttle-bearing sensor. Thus, registration is applied using “rubber sheet“transformation which warps the image on defined points.

- Geo-referencing, is a significant step to rectify the image for the assignment of geographic location, scale, and alignment to a file, and it is performed on raster and vector data to interrelate internal coordinate system to associate objects with locations in physical space.
- Mosaicking, is an opposite process for sub-setting, but in this case, multiple images on different separate scenes are interlinked together in order to have a unified scene for the AOI.

2. Image analysis and classification (processing): This is applied after the image has been prepared following the previous pre-processing steps. It is applied in order to recognize the existing objects on Earth's surface, calculating dimensions, monitoring changes and features, etc. This will follow two major methods:

   - Image enhancement and detection, where several digital and spectral applications are applied in the used software in order to reach the clearest observation and to distinguish objects as much as the resolution of the images allows. These applications are mostly tentative and their use is dependent on the knowledge of the analyzer.

The known applications of images enhancement are the: color slicing, edge detection, directional filtering, enhancement, interactive stretching, contrasting and sharpness. In addition, band combination which is also performed for the single band and multi-band where different band are ordered in different ways unless reaching the suitable observation.

- Image classification, "Classifier"on the software is used and it involves grouping the image pixel values into indicative categories. There are many types of image classification, and the most significant ones are the following: (1) Unsupervised classification by which pixels are grouped into "clusters" based on their properties, and the each cluster can be tentatively attributed to land class, (2) Supervised classification where the analyzer selects representative classes of land, and therefore the software utilizes these "training sites" and applies them to the entire image, (3) Object-based classification where pixels are grouped into representative vector shapes with size and geometry. It is not therefore similar to the supervised and unsupervised classification which are pixel-based (GIS Geography, 2020).

1.5 Geographic Information System (GIS)

Geographic Information System, also called "Geo-information System" is a computer-based system applied for the extraction, storing, drawing and the display of the geo-spatial data. It enables to expose different types of geo-spatial data on one display, such as forests, urban areas, streams, streets, pipelines, water bodies, etc. The use of GIS performs the easily observe, analyze, and understand patterns and relationships (NGS, 2020).

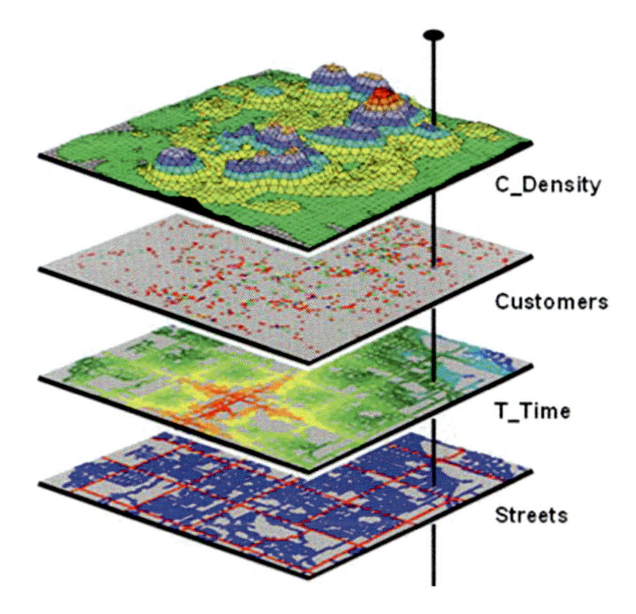

Fig. 1.3 Example showing the integration of different themes in the GIS system

The GIS technology performs applications on digital information, such as the digitization of the geo-spatial data, where hard copy maps or survey plans are transferred into digital data by using computer-aided design and geo-referencing capabilities. Therefore, GIS is used for mapping the locations, quantities and even qualities, densities, finding the inside and nearby features, change detection and many other applications of geo-spatial data illustration. It is therefore, applied in data modeling to overlap different digital maps (as layers) where each map has a define theme, and this enables integration of different themes into a unified figure (example in Fig. 1.3).

The increase in the use and sophistication of these systems has led to a new academic interest which has resulted in a vigorous and expanding research community (Clarke, 1986). Lately, the techniques of the geo-information system have been involved in the management planes and instrumentation in many institutions (e.g. authorities, research centers, universities, etc.), then these techniques became a primary tool for data management.

The manipulation of geo-spatial data in the GIS systems is performed using different types of software, which have been well developed recently and usually accompanied with Remote Sensing tools. Hence, ESRI (Environmental System Research Institute, Redlands, USA) is the major used software where it implies *Arc-GIS,* as the principal Geo-information system tool extended on ITS computers, and is installed in UNIX and Networked PC devices. Thus, *Arc-GIS* has three main digital components:

Arc-GIS, as a software, is utilized to generate, display and analyze geo-spatial data. It includes three digital components:

- *Arc-Map* which enables visualizing spatial data, performing spatial analysis and drawing maps.
- *Arc-Catalog* is a tool for browsing and exploring spatial data, as well as viewing a creating metadata and managing spatial data
- *Arc-Toolbox* is an interface for accessing the data conversion and analysis function the come from *Arc-GIS*.

References

Al Saud, M. (2020). *Sustainable land management for NEOM region* (p. 220). Springer Publisher.
Clarke, K. (1986). Advances in geographic information systems. *Computers, Environment and Urban Systems., 10*(3–4), 175–184.
Dixon-Gough, R. W. (1994). Geographical information management: the way forward for remote sensing. *Geodetical Info Magazine, 8*(8), 68–74.
Eastman, J. (2001). *Guide to GIS and image processing. Vo l. IDRISI 32 release 2* (p. 172). Clark University.
GIS Geography. (2020). *Image classification techniques in remote sensing*. Available at: https://gisgeography.com/image-classification-techniques-remote-sensing/
NGS (National Geographic Society). (2020). *GIS (Geographic Information System)*. Resource Library. Encyclopedic Entry. Available at: https://www.nationalgeographic.org/encyclopedia/
Sandwell, D. (2013). Physical principles of remote sensing: third edition. *Geophysical Journal International, 195*(3). https://doi.org/10.1093/gji/ggt314

Chapter 2
Natural Hazards in the MENA Region

Mashael M. Al Saud

Abstract Natural hazards occupy a significant geo-environmental problem in many regions worldwide. It is rarely a year goes by without a severe regional disastrous event influencing human life, urban structure and the environment. The impact of natural hazards has been dramatically increased due to many physical and man-made factors. There are several aspects of disastrous natural processes where some of them occur as flash events, other with considerable time periods to touch its impact. The assessment, monitoring and identification of natural hazards remain a challenge, notably in the absence of sufficient tools for analysis and weak knowledge and skills to tackle the subject matter. Recently, the development of space techniques played a major role in the analysis of natural risk including predications, change detection, and pre/post-conflict assessment, as well as the diagnoses of the influencing factors. Natural hazards are well pronounced in the MENA Region where several physical components, including the meteorology and exposures, makes it vulnerable for natural risks. Lately, the use of remotely sensed data and tools has been adopted in the MENA Region in order to investigate these disastrous events. Thus, many studies and projects proved to be very helpful in the reduction of risk impact. This chapter introduces a comprehensive knowledge of the remote sensing applications and the natural hazards occur in the MENA Region. Therefore, case studies from different countries and for diverse aspects of natural hazards, will be illustrated.

Keywords Climate · Arid region · Mass movement · Floods · Risk reduction

2.1 MENA Region

It is alternatively named the WANA (West Asia and North Africa). The MENA acronym is usually used in academia, military planning, disaster relief, media planning as a broadcast region, and business writing (WEF, 2010). The MENA Region includes.

M. M. Al Saud (✉)
Space Research Institute, King Abdel Aziz City for Science and Technology (KACST), Riyadh, Saudi Arabia

M. M. Al Saud (ed.), *Applications of Space Techniques on the Natural Hazards in the MENA Region*, https://doi.org/10.1007/978-3-030-88874-9_2

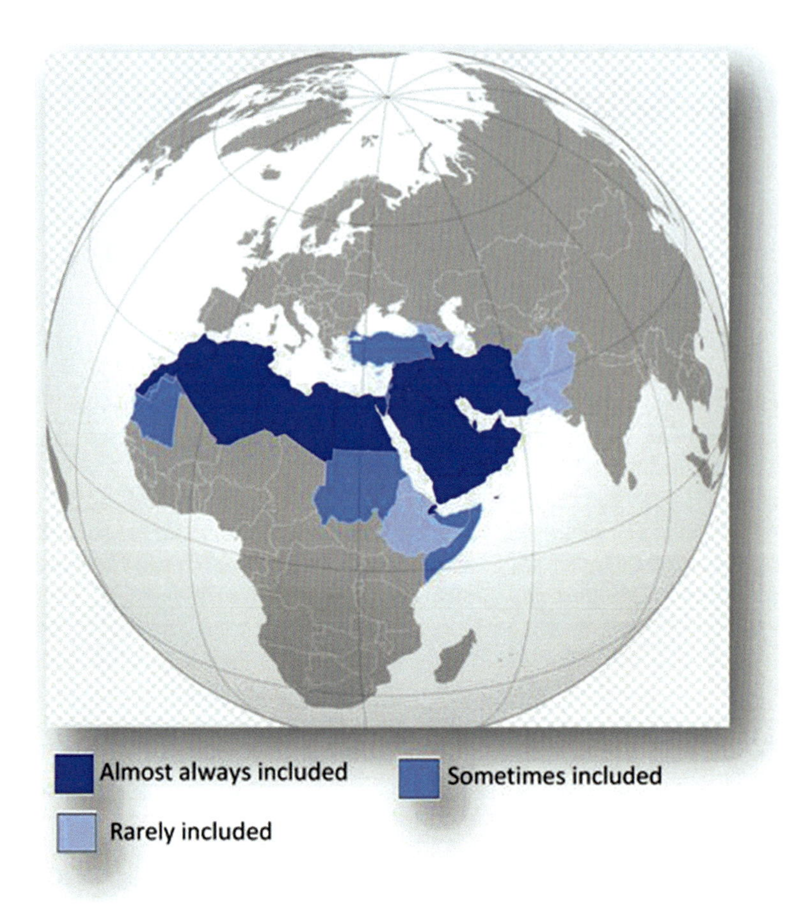

Fig. 2.1 MENA Region, and the belonging of the countries to MENA Region. (Perthes, 2004)

According to the World Bank geographic classification (2003), there are 21 countries or territories constitute the Middle East and North Africa (MENA) Region. These are:

- 6 Gulf Cooperation Council (GCC) members (Bahrain, Kuwait, Oman, Qatar, Saudi Arabia, and United Arab Emirates),
- 15 other countries or territories: Algeria, Djibouti, the Arab Republic of Egypt, Iraq, the Islamic Republic of Iran, Israel, Jordan, Lebanon, Libya, Malta, Morocco, the Republic of Yemen, the Syrian Arab Republic, Tunisia, and West Bank and Gaza.

There is sometimes a contradictory in the identifying the MENA countries, where some of them are not often included. In this regards, a map was prepared and showing the belonging of the countries to MENA Region (Fig. 2.1).

However, as for January 2021, the World Bank website groups the same set of 21 countries/territories as MENA: Algeria, Bahrain, Djibouti, Egypt, Iran, Iraq,

Israel (Occupied Palestinian Territory), Jordan, Kuwait, Lebanon, Libya, Malta, Morocco, Oman, Qatar, Saudi Arabia, Syria, Tunisia, United Arab Emirates, West Bank and Gaza and Yemen. However, many countries are also well known to belong to the MENA Region, such as Turkey and Greece which compose a part of the Middle East and then the MENA Region.

The largest number of these countries belong to the Arab Region, and they share numerous similarities including mainly the cultural, economic and environmental similarities across the countries; for example, most of these countries have almost geographic attributes, and thus similar climatic conditions and even natural disasters.

The surface area of the MENA Region is about five million km^2, and it contains about 6% of the World's population (7.7 billion), and about the same as the population of the European Union. Thus, it extends between the following geographic coordinates:

12°57`30" E & 63°09`15" E and 41°57`00" N & 12°30`12" N.

The MENA Region encompasses abundant human and natural resources, accounts for a large share of the World's petroleum production and exports, and it enjoys on average a reasonable standard of living. Within this general characterization, countries vary substantially in resources, economic and geographical area, population, and standards of living. Meanwhile, intra-regional interaction is weak, being restricted mainly to labor flows, with limited trade in goods and the related services (IMF, 2020).

McKee et al. (2017) reported key messages about the demographic and economic prospects of the MENA Region. These are:

1. The population in the MENA Region will be double that of Europe by 2100, and then will exceed1billion people.
2. By 2100, the population in the MENA Region will exceed the China's population.
3. The demographic trend in the MENA Region is towards a gradually ageing society.
4. The majority of MENA countries are resource-poor and labor-abundant, and the economies are insufficiently diversified.
5. Apart from exploiting oil and gas resources to industrial development holds great opportunities.
6. MENA economies need greater investment in STEM subject skills (science, technology, engineering and mathematics) to prepare the new generation for labor market demands.
7. The demographic and economic perspective of MENA is challenging; however, there is also a "demographic dividend" if policies are steered towards economic inclusion of the new generation.

It can be said that the MENA region is composed of a heterogeneous group of countries ranging from the high-income (oil-exporting countries of the GCC), to middle and lower middle income countries, in addition to the least developed countries such as Sudan and Yemen (Table 2.1). While, the MENA Region is one

Table 2.1 Contextual indicators for the Middle East and North Africa, 2014

MENA country	GPD ($)	Agricultural land (% of total land area)	Arable lands (% of total land area)	Renewable freshwater resources (Billion m^3)	Freshwater withdrawals (Billion m^3)
Qatar	86.9	6	1	0.06	0.44
UAE	44.5	5	0	0.15	4.0
Kuwait	42.9	9	1	0.0	0.9
Bahrain	24.9	11	2	0.004	0.35
Saudi Arabia	24.5	81	2	2.0	24
Oman	20.5	5	0	1.4	1.32
Cyprus	19.6	12.2	–	0.32	23.4
Greece	14.5	48	–	0.2	6.9
Turkey	9.1	50	26	112	120.3
Lebanon	8.5	64	13	4.8	1.3
Iraq	6.7	21	12	35	66
Libya	5.6	9	1	0.7	5.8
Iran	5.5	28	9	129	93
Algeria	5.4	17	3	11	8.0
Tunisia	4.2	65	19	4.0	3.0
Jordan	4.0	12	3	0.7	0.9
Egypt	3.3	4	3	2.0	78
Morrocco	3.2	69	18	29	10
Palestine	2.9	50	11	0.81	0.42
Sudan	2.1	29	8	4.0	27
Syria	2.0	76	25	7.0	17
Yemen	1.6	45	2	2.0	4.0
Mauritania	1.3	39	0.5	0.4	1.4

Adapted after World Bank (2018); UNCTAD (2018); FAO (2018)

of the largest global net food importing regions, and thus it often faces uncertainties on both the supply side and the demand side.

The high and growing dependence on international markets for key staple food products makes is one of the main concern in the MENA Region. This has been resulted in elaborating policies which appear strikingly inappropriate with respect to the existing resources. For example, while MENA is one of the most land and water constrained regions of the world, it has the lowest water tariffs in the world and it heavily subsidizes water consumption at about 2% of its GDP. As a result, the productivity of water use is about half the world average (World Bank, 2018).

The interlinkage and cooperation between the MENA countries is less than it with countries from outside the region. This includes several sectors and activities. However, there are a number of characteristics shared between these countries in spite of the existed heterogeneity between MENA countries. Hence, the growth in the region has underperformed, with GDP per capita growing at about 1.6% per year between 2001 and 2016 (Table 2.1).

Of the total land area of the MENA Region, only one-third is agricultural land (i.e. cropland, agricultural activities, pastures, etc.), while only 5% is arable lands (cropland) (Table 2.1). The rest of the land is either urban or dry desert. Due to the dry climate, about 40% of cropped area in the region requires irrigation (FAO, 2018).

From the natural point of view, the MENA Region has its geographic location that makes it with almost similar physical conditions. It is characterized by semi-arid to arid climate, with humidity in some mountainous regions. The average annual rainfall rate ranges between <10 mm to >1500 mm, while the average annual temperature ranges between 16 °C and 32 °C. In addition, MENA Region is well known by its climatic extremes, notably the extreme heat waves, sand storms and torrential rainfall.

Recently, MENA Region is witnessing changes in the climatic conditions which is moving towards torrential rainfall and increased temperature rate, and this in turn affects tremendous sectors including mainly water, agriculture and energy sectors. The existing climatic conditions and the natural setting including geology and geomorphology play a major role in creating natural hazards in the region.

2.2 Natural Hazards and Their Identification

Natural hazards is crucial geo-environmental issue in many regions, notably where there is high of hazardous event that result in harmful consequences on communities under lurking risk. Another way of conceptualizing natural hazards is the coexistence of humans in a normal environment that may threaten their safety, property at any time. Thus, the concept "natural hazard" is brought to mind when the word "disaster" is mentioned (Al Saud, 2018).

Natural hazards have different definitions describing; thus they are described as the elements of the physical environment, harmful to man and caused by forces extraneous to him (Burton et al., 1978). Whereas, the Asian Disaster Preparedness Center (ADPC, 2000) described a natural hazard as a threat, and a future source of danger. While (CBSE, 2006) defined natural hazard as a dangerous condition or event, that threat or have the potential for causing injury to life or damage to property or the environment. Therefore, natural hazards have the potential to cause harm to:

- People (decease, injury, disease, pandemic and stress),
- Human activity (trading, economic, life style, educational etc.),
- Property (property damage, economic loss),
- Environment (loss in fauna and flora, pollution, loss of comforts).

Natural hazards originate from many sources with different aspects of physical systems including mainly the atmospheric, hydrologic, oceanographic and tectonic systems where the impacts are catastrophic in many cases. Therefore, many several concepts were on natural hazards (or natural disastrous events) were demonstrated, but all concepts imply the risk, loss and damaging terms. The resulted damages are often harmful and in many instances, they result in hundreds of deaths and cost several billions of dollars, disruption of commerce, and destruction of homes and infrastructure (Al Saud, 2018).

The physical processes initiating natural hazards are often interlinked. For example, transportation of surficial materials (i.e. erosion) is caused by floods and torrents where the latter two processes are interlinked with climatic conditions. In fact, there are types of natural hazards which result in one aspect of impact, such as drought which is causes only famine, but the case of earthquakes and floods and are different. Also, types and degree of impact of natural hazards differ between regions, and they are mainly induced by the existing natural influencers and partially by the anthropogenic ones.

The event recurrence (i.e. return period) of natural hazards is mainstreamed in the management plans and environmental policies. Thus, the event recurrence is significant on human time-scale. For example, there are seven-year flood, or decade earthquake. This reflected by the statistical records of how often a hazard event of a define intensity will occur. Thus, the recurrence of hazard is a functioned by its frequency (Al Saud, 2020).

There are many types of hazards, which can be natural/ or man-made (Table 2.2). Hence, in many instances both can be combined and then result severe damages. Thus, human might directly or indirectly trigging the disastrous processes. This is well known due to many negative activities, such as explosions, excavation, forest fire, deforestation, industries, etc. (Table 2.2).

There are different tools and methods used to monitor and assess natural hazards. They include conventional and advanced tools and techniques. These can be summarized as follows:

1. Natural Hazards on Geologic Maps

Geologic maps are one of the most indicative maps used for natural hazards assessment. They can support the ability of investigator to locate natural resources and many other significant terrain features, notably the surficial processes and thus natural hazards; therefore, they enable identify on hazardous events (GSA, 2008).

Geological maps can be used to recognize indicative geologic-related clues and signatures belong to natural hazards. This is dependent on the accuracy of the available maps, their scale, and on the skill of interpreters. It is well known that there are several elements (or clues) of recognition for natural hazards and the related signatures that can be obtained from geologic maps.

According to Al Saud (2015), the success of recognition implies a number of the elements existed on the map, notably: 1) the distribution of rock lithologies and their contacts in the vertical and horizontal aspects, 2) lithological characteristics and the rock rigidity, porosity and permeability, compaction, etc. 3) inclination of bedding planes, and abrupt dip changes between different rock masses, 3) rock deformations and their dimensions and orientation and 4) existence of other geologic structures (e.g. folds, flexures, plunging rocks, etc.).

Table 2.2 Types of natural and man-made hazards. (CBSE, 2006)

Type of natural hazards	
Geological hazards	Earthquakes Tsunami Volcanics Landslides Dam burst Mine collapse
Hydrologic and climate hazards	Tropical Cyclone Cloudburst Tornado and Hurricane Landslide Floods Heat & Cold waves Drought Snow Avalanche Hailstorm Sea erosion
Environmental Hazards	Environmental pollutions Desertification Deforestation Pest Infection.
Biological	Human / Animal Epidemics Food poisoning Pest attacks Weapons of mass Destruction
Chemical, Industrial and Nuclear accidents	Chemical disasters Oil spills/Fire Industrial disasters Nuclear
Accident related	Boat / Road / Train/ air crash/ rural and urban fires, bomb, serial bomb blasts Building collapse Electric accidents Festival related disasters Mine flooding Forest fires

2. Natural Hazards on Topographic Maps

Topographic maps are significant products revealing signatures on terrain surface which reflect factors exist from Earth interior. This implies mainly slopes, depressions and drainages. Therefore, the perturbation in the behavior of these signatures provides evidence on the vulnerability to anomalous natural processes, including the natural risk (Al Saud, 2015). This makes the topographic maps as significant as geological maps for risk assessment and vulnerability.

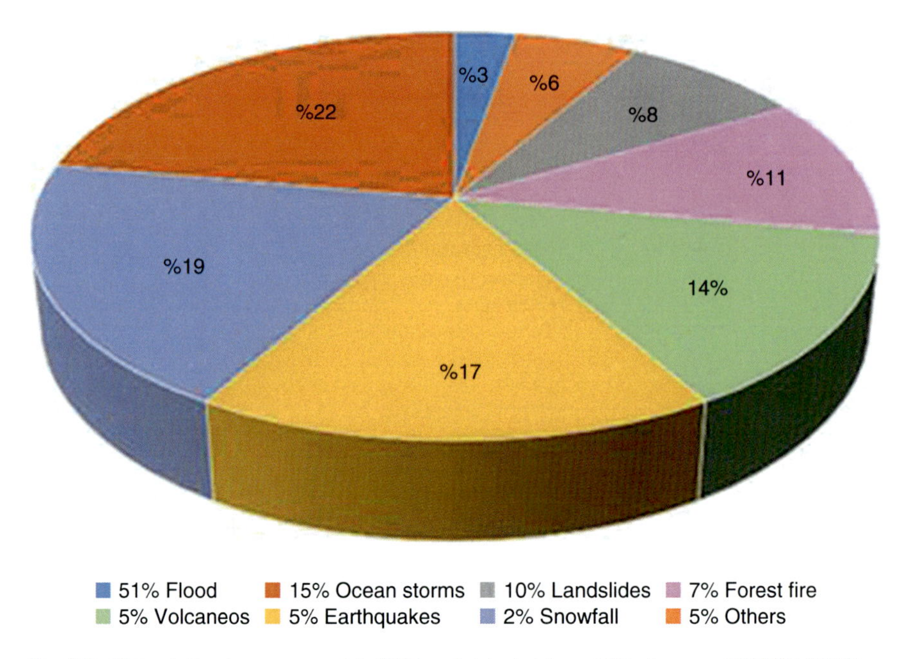

Fig. 2.2 Natural disastrous occurred in 2014 and detected by satellite images. (ICSMD, 2014)

3. Natural Hazards on Satellite Images

Recently, most studies on natural hazards and surficial processes are applied using satellite image processing. This was described in details in Chap. 1. These images are useful for monitoring and assessment for all types of natural disasters including of forest fires, floods, sea level rise, mass movement, weather extremes, desertification, erosion and many more. As an advantage of using satellite technology can be found in natural disasters implies providing advance warning for specific hazardous events (Kohiyama and Yamazaki 2005). They assist monitoring and the quick evaluation of the damage, and therefore support the decision-making process in the preparedness implementations. Thus, satellite and airborne imagery can offer an efficient contribution in studying natural risks notably when they are integrated with the geographical information systems (GIS).

Data and information retrieved by ICSMD (2014), where satellite images and remote sensing products are adopted, proved to be creditable approaches for monitoring natural hazards. Thus, there are t 19 major disastrous events occurred in Asia, 8 in South America, 5 in Africa, 4 in Europe, 2 in Oceania, 2 in the Caribbean and 1 in Central America. It was obvious that the most recurrent natural hazard types being floods (51%) and ocean storms (15%) while solid earth-related hazards represented 10% (Fig. 2.2).

4. Filed Survey for Natural Hazards

All tools used for natural risk identification, predication, monitoring and assessment are integrated with field survey and in-situ verification. This provides accurate data and information about the studied areas, as well as they are complimentary for the retrieved thematic maps from satellite images or from other tools.

There are several types of field surveys, and these surveys are applied at different stages during the study of a natural hazard. According to Al Saud (2015), the application of field survey is usually scheduled as follows:

1. Prerequisite reconnaissance: It is applied for selected sites before the treatment of the supplementary tools (e.g. satellite imageries).
2. Joint survey: Sometimes there is need to apply field verification during/or along with the investigation which is carried out by the supplementary tools.
3. Post-investigation survey: It is often carried out after producing maps when data and information become ready.

There are many types of tools used during field surveys, and the selection of these tools depends on the type and purpose of the study and availability of these tools; in addition, the natural setting of the investigated area is significant. Thus, the tools encompass different specifications and characteristics. They can be simple (e.g. GPS, measuring tapes, sampling kits, etc.), or they can be sophisticated ones (e.g. radio-spectrometer, surveying devices, Lidar, etc.).

5. Early Warning System

Early warning systems (EWSs) have been lately adopted in many programs in order to assess geophysical and biological hazards, complex socio-political emergencies, industrial hazards and many other related risks (Basher, 2006). Lately, the increased number of natural hazards in many regions and the resulted harmful damages makes it necessary to establish integrated monitoring (i.e. predication and post-assessment) systems. In this regard, EWSs are built and based on Geographic Data Guide (GDG) which represents an electronic system for monitoring the geographic areas, and thus to identify their behaviours/resilience according to different physical and man-made conditions. The established EWSs can identify areas and localities with geographic attributes which are benefit to predict the natural risk. Hence, EWSs are usually linked to operational and controlling rooms which receive and deliver data and information to the concerned institutes and governmental bodies as well as to the inhabitants.

2.3 Natural Hazards in MENA Region

Over the past four decades, disasters have severely affected more than 40 million people in the Middle East and North Africa (MENA) Region and have cost their economies about US$ 20 billion. Between 2000 and 2005 alone, there are more than

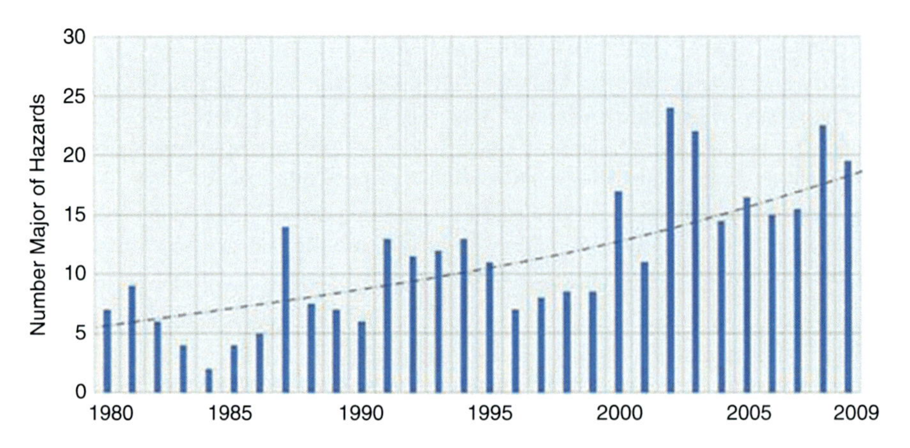

Fig. 2.3 Increased trend of disasters in the MENA Region. (WB, 2014)

120 disasters have caused an average of US$ one billion per year in damages and losses. In the MENA Region, the interlinkage between different natural disasters, rapid urbanization, water scarcity, and climate change has emerged as a serious challenge for policy and planning (WB, 2014).

In the MENA Region, there is rapid urbanization which increases the exposure of people and economic assets to disaster events. The population size already accounts for 62% of total population and is anticipated to double in the next three decades (UN-HABITAT, 2012).While the number of disasters around the World has almost doubled since the 1980s, the average number of natural disasters in the MENA Region has almost tripled over the same period (EM-DAT, 2011).

The management of disaster risks is utmost complex development challenges in MENA Region. With special emphasis, floods, earthquakes, and drought represent the most significant challenges to growth and stability in the region. Globally, the number of disasters has almost doubled since the 1980s in the MENA Region (Fig. 2.3), where the average number of natural disasters has almost tripled over the same period. Therefore, approximately 40 million people have been affected by over 350 natural disasters between 1981 and 2010 (EM-DAT, 2011).

The impact of urbanization development is especially significant in the coastal areas of the MENA Region, where the largest cities and economies are located, with approximately 60 million people (about 17% of MENA's total population).

As it always happening, the most frequent natural hazards in the MENA Region are floods, earthquakes, storms, and drought. Over the last four decades years, floods have been the most recurrent natural disasters reported by EM-DAT who recorded that at least flood 300 events (53% of the total number of natural disasters). While, earthquakes account for 24%, 21% for and storms and 10% for drought. However, the time factor of is different between disastrous events, thus they can be as a flash events, such as earthquakes and mass movements (e.g. seconds), relatively rapid such as flood (e.g. hours), and creeping like drought and sea level rise (Al Saud, 2018).

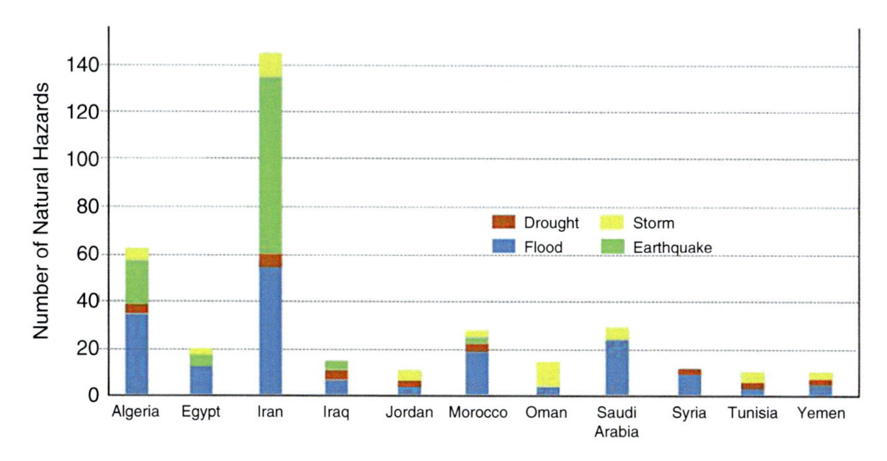

Fig. 2.4 Comparison of total number and type of natural hazards in the MENA Region, 1980–2010. (EM-DAT, 2011)

This increase in the number of natural disasters is due to an increase in exposure and vulnerability across the region In addition to many other reasons including lack to infrastructure, urban Sprawl, poor risk management and lack to early warning systems and monitoring tools, etc. The 2011 Global Assessment Report on Disaster Risk Reduction reported that: although global flood mortality risk has been on the decrease since 2000; nevertheless, it is still increasing in the MENA Region (Fig. 2.4). The percentage of GDP exposed to floods has tripled in the region from 1970–79 to 2000–2009 (UNISDR, 2019).

For example, in Yemen, cumulative loss in real income over the 5-year period following the 2008 floods in Hadramout was estimated to about 180% of pre-flood regional agricultural value added. This large increase can be attributed to the growing concentration of assets at risk, particularly in urban areas, and insufficient structural and nonstructural mitigation measures. While for Saudi Arabia, floods have become a yearly disasters in many regions of the Kingdom and the majority of these floods is due to the chaotic urban expansion along valley courses, as well as the unmaintained infrastructures. A good example is the floods of Jeddah city and the surroundings in 2009 and 2011 where a number of people were deceased and the damages in the infrastructure and environment were severe enough (Al Saud, 2010a, b, 2015).

Projections suggested that the economies and livelihoods of the MENA Region will be the second most affected by climate change (WB, 2018). This merely will be reflected on the large number of natural hazards, notably the changing rainfall patterns. Therefore, governments across the region are increasingly seeking integrated disaster risk management (DRM) services. Increasing awareness of DRM issues has brought about progressive policy shifts, as seen in the creation of a number of DRM-specific institutions and investments in programs around the region. These programs, include mainly the monitoring systems and Early Warning Systems (e.g. remote sensing-based stations) and national and city-level risk assessments. In 2010, for instance, the Islamic Conference of Environment Ministers adopted the Islamic Strategy for Disaster Risk Reduction and Management, which aimed at establishing integrated and comprehensive DRM structures and policies across Islamic countries (WB, 2014).

On the national and regional level, there are many initiatives and programs elaborated in order to monitor and reduce the impact of disasters. In this regard, thanks to the Global Facility for Disaster Reduction and Recovery (GFDRR), who initiated a number of country-level programs since 2007, such as those obtained in Morocco and Yemen to empower the disaster resilience. These programs include activities aimed at enhancing data and information availability on natural disaster, developing strategies and policies for risk assessment and reduction, building capacities in DRM through training programs at the national and local levels, and creating state-led, post-disaster recovery and reconstruction programs (WB, 2014).

The integration with the formal entries was emphasized. Thus, in Algeria, Djibouti, Jordan, Lebanon, Morocco, Kingdom of Saudi Arabia and Yemen, donor-funded hazard risk assessments have been achieved; flood-recovery projects have been launched in Djibouti and Yemen; and inter-ministerial steering committees on DRM have been established under umbrella the prime ministers in Djibouti, Lebanon, Morocco and Yemen, and under the Ministry of Interior in Algeria.

The number of efforts done on DRM in the MENA Region have not yet been matched with the integrated approaches required to manage disaster risks effectively. In 2012, the region launched a 10-year Arab Strategy for Disaster Risk Reduction, including the reduction of climate change impacts and disaster losses through the identification of strategic priorities; the improvement of coordination mechanisms; and monitoring at the national, regional, and local levels (WB, 2014).

The impacts of climate change is well pronounced in the MENA Region. Hence, the year 2010 was the hottest in the region since the 1800s, and therefore, 19 countries set new national temperature rise. Thus, temperature has been increased by 0.2–0.3 °C per decade. In addition, climatic models and projections presumed that sea-level rise, as well as the frequency and intensity of natural hazards will likely increase in the MENA Region.

These unfavorable climatic conditions affected many geographic localities, and then the MENA's population, notably that the 100 million poor who have limited resources to adapt. According to the climate projections of the Fourth Assessment Report of the Intergovernmental Panel on Climate Change, 38 temperatures in MENA Region will increase by 3 to 4 °C by the end of the century, which is a significant 1.5% faster than the global average.

The dimensional impact on the socioeconomic due to climate change is not well determined, but it well touched in many parts of the MENA Region. For example, studies showed that the long-term implications of climate change over the next four decades are likely to lead to a cumulative decrease in household incomes estimated at about US$ 1.8 billion (7% of GDP) and US$ 5.7 billion (24% of GDP) in Tunisia and Yemen; respectively.

Based on the MENA countries' Hyogo Framework for Action (HFA), a number of pillars were adopted. They were given that most governments have been reporting on their DRM progress in achieving the HFA objectives. Nevertheless, the World Bank is also considering the principles illustrated by the GFDRR Pillars in the Sendai Report in 2012, which focus more on the DRM tools available rather than the outcomes (WB, 2014). These adopted pillars are:

1. Institutional capacity building, education, and knowledge sharing.
2. Risk identification, including risk information, integration of DRM with the climate change agenda and hazards and climatic data.
3. Risk reduction which starts from disaster response to risk reduction and building codes.
4. Risk financing and transfer including budgets and risk financing
5. Response, recovery, and reconstruction which implies institutional arrangements and World Bank instruments to assist MENA countries in emergency response.

References

ADPC (Asian Disaster Preparedness Center). (2000). *Capacity building in Asia using information technology applications*. Concepts of hazards, disasters and hazard assessment. Course material. Module 2. Available at: http://www.adpc.net/casita/course-materials/Mod-2-Hazards.pdf

Al Saud, M. (2010a). Assessment of flood hazard of Jeddah area 2009, Saudi Arabia. *Journal of Water Research and Protection (JWARP), 2*, 839–847.

Al Saud, M. (2010b). Application geo-informatics techniques in the study of floods and torrents in Jeddah in 2009. *Arab Journal of Geographic Information Systems, III*(1) (Arabic edition).

Al Saud, M. (2015). *Flood control management for the city and surroundings of Jeddah, Saudi Arabia* (p. 177). Springer. ISBN13: 978-94-017-9660-6.

Al Saud, M. (2018). *Using space techniques and GIS to identify vulnerable areas to natural hazards along the Jeddah-Rabigh Region, Saudi Arabia* (p. 306). Nova Science Publisher Inc. ISBN13: 978-15-361-33134.

Al Saud, M. (2020). *Sustainable land management for NEOM region* (220 pp). Springer Publisher.

Basher, R. (2006). Global early warning systems for natural hazards: Systematic and people-centred. *Philosophical Transactions, 364*, 2167–2182.

Burton, I., Kates, R. W., & White, G. F. (1978). *The environment as hazard*. Oxford University Press.

CBSE (Central Board of Secondary Education). (2006). *Natural hazards and disaster management*. A Supplementary Textbook in Geography for Class XI. 51 pp.

EM-DAT. (2011). *International Disaster Database, Université Catholique de Louvain, Brussels*. Available at: https://www.emdat.be

Food and Agriculture Organization of the UN (FAO). (2018). *Aquastat main database*. http://www.fao.org/nr/water/aquastat/data/query/index.html?lang=en

GSA (Geological Society of America). (2008). Available at: http://geoscociety.org/positions/pos3_mapping.pdf

IMF (International Monetary Fund). (2020). *Growth and stability in the Middle East and North Africa*. Available at: https://www.imf.org/external/pubs/ft/mena/04econ.htm

Kohiyama, M., & Yamazaki, F. (2005). Image fluctuation model for damage detection using middle-resolution satellite imagery. *International Journal of Remote Sensing, 26*(24), 5603–5627.

McKee, M., Keulertz, M., Habibi, N., Mulligan, M. Woertz, E. (2017). *Demographic and economic material factors in the Mena Region*. Technical Report of an EU Project entitled: Middle East and North Africa Regional Architecture. 43 pp.

Perthes, V. (2004). America's Greater Middle East and Europe: Key issues for dialogue. *Middle East Policy Council, XI*(3), 85–97.
UNCTAD. (2018). *UNCTAD Stat*. http://unctadstat.unctad.org/wds/ReportFolders/reportFolders.aspx?sCS_ChosenLang=en
UN-HABITAT. (2012). *The state of Arab Cities 2012: Challenges of urban transition, 2012*. Available at: www.unhabitat.org.
UNISDR. (2019). *Global assessment report on disaster risk reduction*. Available at: https://digitallibrary.un.org/record/3825375?ln=en
WB. (2014). *Natural disasters in the Middle East and North Africa: A regional overview*. The International Bank for Reconstruction and Development /the World Bank. 114pp.
WB. (2018). *World development indicators, World Bank, Washington, D.C.* http://databank.worldbank.org/data/reports.aspx?source=world-development-indicators.
World Economic Forum. (2010). *The Middle East and North Africa*. Marrakesh, Morocco, 26–28 October, 2010.

Part II
Applications of Space Techniques for the MENA Region

Chapter 3
Open Data and Tools for Multispectral Satellite Analysis of Desert Sand Dunes Migration: Case Studies in the MENA Region

Daniele Oxoli and Maria Antonia Brovelli

Abstract The recent availability of medium resolution free and open multispectral satellite imagery, with short revisiting time, opens newsworthy opportunities to the monitoring of natural events, like the sand dunes migration that is of high interest in desert areas such as the MENA Region. In parallel, open tools such as algorithms, software, and web platforms have also emerged by facilitating imagery exploitation. The chapter investigates the availability and characteristics of the modern open multispectral satellite imagery and assesses the suitability of employing it for the analysis of desert sand dunes migration through the design of an ad-hoc data analysis framework. The technological requirements and challenges in the management of a large volume of satellite images and the benefits derived from the systematic use of cutting-edge GIS and remote sensing free and open-source software technologies are discussed. A preliminary application of the conceived analysis framework on two study areas in Egypt and the United Arab Emirates is presented. A discussion on lessons learned and both possible improvements and uses of the presented framework concludes the chapter.

Keywords Multispectral imagery · Sand dunes migration · Change detection · Open data · Free and open-source software

3.1 Introduction

The emerging availability of medium resolution, global coverage, and multitemporal imagery – captured by modern satellite missions – has empowered a number of applications focusing on the detection of changes occurring on the Earth's surface.

On the one hand, the increased interest in multitemporal imagery analysis is mainly triggered by the rising number of satellites with short revisiting time that

D. Oxoli (✉) · M. A. Brovelli
Politecnico di Milano, GEOlab, Milan, Italy
e-mail: daniele.oxoli@polimi.it; maria.brovelli@polimi.it

M. M. Al Saud (ed.), *Applications of Space Techniques on the Natural Hazards in the MENA Region*, https://doi.org/10.1007/978-3-030-88874-9_3

enables the acquisition of either long time series or frequent bi-temporal images (Ghamisi et al., 2019). On the other hand, the open distribution policies of both archive data as well as new observations – adopted by some of the principal Earth Observation (EO) programs – enable outstanding opportunities for the user community to grow and, in turn, facilitate the spread of case studies exploiting multitemporal imagery (Belward & Skøien, 2015). The best-known examples of the above are the US National Aeronautics and Space Administration (NASA) Landsat (US National Aeronautics and Space Administration, 2021) and the European Space Agency (ESA) Sentinel (European Space Agency, 2021b) missions which fully embrace the open data principles. Together with the availability of open satellite imagery, powerful free and open-source software tools for processing remote sensing data have been also provided during the last decade by allowing for a wide spectrum of applications to potentially any user interested in thereof. Nowadays, such software tools are often integrated into Geographic Information Systems (GIS) platforms thus facilitating both direct extraction of geographic information from remote sensing data as well as its combination with contextual information from other sources to carry out broader analyses (Ghamisi et al., 2019). In this context, meaningful outcomes have been achieved for the monitoring of large-scale environmental phenomena including e.g. urban sprawl (Wendl et al., 2018), deforestation (Finer et al., 2018), and polar ice degradation (Runge & Grosse, 2019).

However, the spatial and spectral resolution – nowadays provided by the open multispectral satellite imagery – also allows for applications at a finer scale. In view of the above, an experimental analysis framework to investigate sand dunes migration in desert areas by means of multitemporal analysis of open multispectral imagery is presented in this chapter. The proposed application is critical to a number of construction engineering operations that require, for example, planning of mitigation measures to prevent undesired windblown sand interactions with infrastructures and other features in the desert environment (Oxoli et al., 2020). The interest in applying open satellite imagery for sand mitigation planning is justified by the significantly higher costs of the traditional field surveys that are often inadequate to understand sand dynamics on a large scale and to identify locations where such sand mitigation measures are to be applied. A number of authors in the literature have exploited the multitemporal analysis of multispectral imagery for assessing dunes movement (Hugenholtz et al., 2012). However, most of these studies focused on single dune fields, where only a few target dunes were tracked, see e.g. (Al-Mutiry et al., 2016; Michel et al., 2018; Hassoup, 2019). Indeed, the proposed analysis framework aims at identifying movements at a pixel level but remaining suitable to wide scenes analysis across multiple dune fields as well as desert land cover textures.

The proposed analysis framework leverages open multispectral satellite imagery, such as the one provided by Landsat 8 and the Sentinel-2 missions. The computational work is carried out exclusively through the use of free and open-source GIS and remote sensing software and encompasses multiple steps including imagery acquisition and preprocessing, temporal stacks generation, dunes migration analysis, and results synthesis. Basically, dunes migration is analysed by looking at changes that occurred within a stack of multispectral imagery repetitively acquired by a multispectral satellite camera over the same study area at different times. Atmospherically corrected imagery with an equal set of spectral bands is spatially co-registered and then classified to extract raster masks distinguishing target dune pixels from the other background land cover classes (build-up, crops, etc.). Dunes migration is estimated through local cross-correlation analysis applied in sequence to each pair of masks in the multitemporal stack. Movements direction and magnitude are assigned to each pixel of the resulting maps. Global observations of principal dunes movements occurred in the study area can be then summarized and presented by using e.g. a windrose graph.

The procedure is finally tested on two study areas located in the Middle East and North Africa (MENA) region, namely in Egypt and the United Arab Emirates (UAE) deserts. Preliminary results show a marginal agreement of dunes migration patterns if compared with the dominant wind directions in the areas that were here considered as the principal driving forces of dune dynamics (Lancaster, 2013). Besides early numerical results, which require robust validation, operations on data that are pointed out by the framework provide users with a pattern that can be adapted to most multispectral satellite imagery-based change detection studies. This is supported also by the exclusive use of open data and free and open-source software thus providing large rooms to empower, replicate and improve the proposed framework. The chapter continues as follows. Section 3.2 describes state-of-art open multispectral imagery and related software tools. In Sect. 3.3, a general analysis framework to perform dunes migration analysis is outlined. Section 3.4 presents applications of the framework into two case studies in the MENA region. Conclusions and possible improvements to the work are discussed in Sect. 3.5.

3.2 Open Satellite Imagery and Software Tools

Free and open access to satellite imagery is a result of the open data policies embraced by some of the leading national and international EO programs. Examining the history, NASA and the US Geological Survey (USGS) paved the way to satellite open data sharing with the free release of Landsat data archive in the Year 2008 (Wulder et al., 2012). ESA and the Copernicus Programme of the European Union (European Commission, 2021) followed with the free distribution of Sentinel data starting from the Year 2012. Focusing on open multispectral medium resolution imagery, best available data sources are currently represented by the NASA Landsat 8 and the ESA Sentinel-2 missions. Accordingly, these two data sources were considered in the development of the proposed framework for dunes migration analysis.

The Landsat 8 is an Earth observation satellite launched in 2013 within the Landsat program, a joint NASA – USGS program providing the longest continuous space-based record of Earth's land. The Landsat 8 satellite provides 11-bands multispectral global-coverage imagery at a spatial resolution of up to 30 m by offering a revisit time of 16 days (US National Aeronautics and Space Administration, 2021). USGS policy assures all users the free-of-charge, open and full access to the data. The ESA Sentinel-2 data are provided by the Copernicus Programme which mission is to monitor and forecast the state of the environment by means of satellite Earth observations. Copernicus policy assures all users the free-of-charge, open and full access to the data. The Sentinel-2 mission consists of a constellation of two polar-orbiting satellites placed in the same sun-synchronous orbit launched respectively in 2015 and 2017 providing 13-bands multispectral global-coverage imagery at a spatial resolution of up to 10 m and offering a revisit time up to 5 days (European Space Agency, 2021b). Further than open data policies, a pivotal component of the Landsat 8 and Sentinel-2 missions is the technical and logistical support offered by their belonging EO programs in making both newly acquired and archive data accessible online. The systematic distribution of the imagery to the final user by means of online data portals and infrastructures is among the assets that have boosted the recognition of these missions as a cornerstone for scientific advances in land remote sensing. Information on the official online data portals of the Landsat 8 and Sentinel-2 missions are reported in Table 3.1.

Besides the availability and accessibility of quality multispectral imagery, a critical aspect for any remote sensing application is the need for robust software tools enabling consistent numerical processing of imagery. Typical operations accomplished by remote sensing software tools include orthorectification, radiometric correction, spatial and spectral co-registration, bands combination, and classification in support to further statistical or numerical processing of remote sensing-derived products (Bunting et al., 2014). A number of software and libraries has been made available which are fueling applications of remote sensing in manifold scientific domains. Nowadays, most of these tools are available as free and open-source software or free license software and provide cutting-edge functionalities that are comparable, and often superior, to the one offered by proprietary solutions (Brovelli et al., 2017). Furthermore, remote sensing software and libraries are often fully integrated into popular GIS platforms thus enabling processing, visualization and combination of remote sensing-derived information with contextual geospatial data in a one-stop shop fashion. A selection of software tools to accomplish most of the multispectral imagery processing operations are suggested in Table 3.1. The selection is not exhaustive and refers to the software toolkit used to implement the framework for dunes migration analysis outlined in the following section. Many other remote sensing software tools are available, a fairly complete list is available e.g. in (Bunting et al., 2014).

Table 3.1 Free and open source software toolkit used for dunes migration analysis with open multispectral satellite imagery

Software	Purpose	Description
Copernicus Open Access Hub	Data exploration and download	Official online data portal of the Copernicus program of the European Commission including Sentinel-2 imagery (Copernicus Programme, 2021)
USGS Earth Explorer	Data exploration and download	Official online data portal of the NASA missions, including Landsat 8 imagery, maintained by the USGS (US Geological Survey, 2021)
*Google Earth Engine©	Data exploration, processing and visualization	Web-platform for cloud-based processing of remote sensing data on large scales (Google, 2021)
SNAP	Data processing and visualization	Sentinel's application platform providing a suite of functionalities to handle and processing Sentinel data (European Space Agency, 2021a)
GRASS GIS	Data processing and visualization	Free and open source cross-platform GIS software suite used for geospatial data management and analysis, image processing, graphics and maps production, spatial modeling, and visualization (GRASS Development Team, 2018)
GDAL	Data processing	Cross-platform library providing a variety of utilities for geospatial data handling and processing (GDAL/OGR contributors, 2021)
JavaScript	Programming language	High-level, interpreted programming language most well-known as the scripting language for Web applications
Python	Programming language	High-level, interpreted, general purpose programming language offering large support to data analysis and integration with most of GIS software

*Free but not open-source

3.3 A Framework for Dunes Migration Analysis

Detection of changes occurring on Earth's surface has become an established practice in remote sensing science since the availability of multitemporal and spatially coherent multispectral images has constantly increased during the last decades.

Time-series analysis of multispectral satellite images is generally employed in land surface change detection including urban area expansion, deforestation and polar ice monitoring, among others. Sand dunes migration can be investigated using techniques developed in the above context. In this section, a general analysis framework to accomplish such is outlined. The summary workflow of the proposed analysis framework is depicted in Fig. 3.1. Details are presented in the following sections.

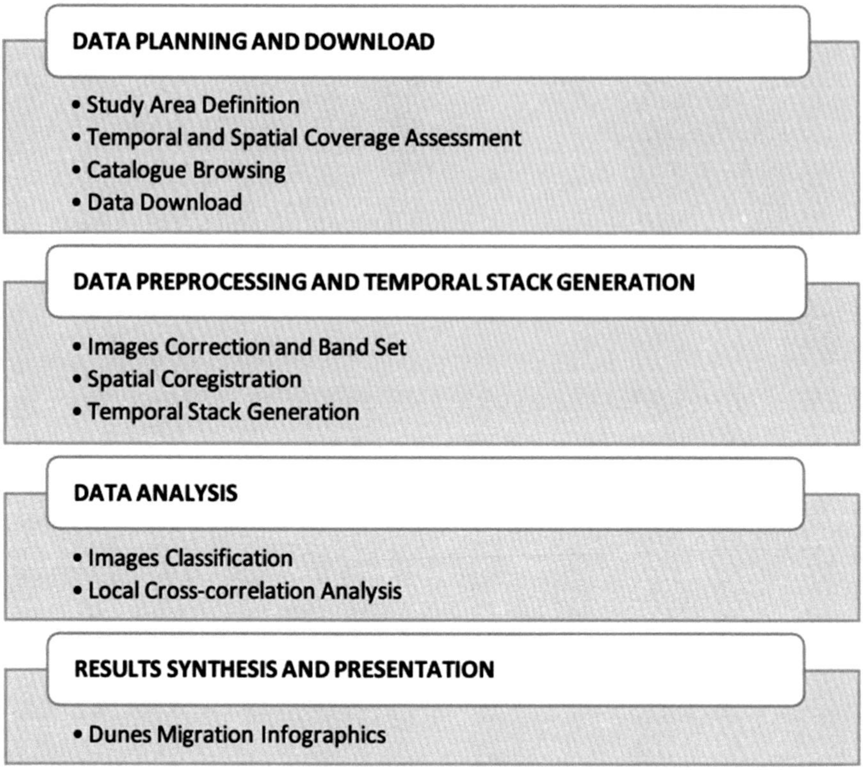

DATA PLANNING AND DOWNLOAD

- Study Area Definition
- Temporal and Spatial Coverage Assessment
- Catalogue Browsing
- Data Download

DATA PREPROCESSING AND TEMPORAL STACK GENERATION

- Images Correction and Band Set
- Spatial Coregistration
- Temporal Stack Generation

DATA ANALYSIS

- Images Classification
- Local Cross-correlation Analysis

RESULTS SYNTHESIS AND PRESENTATION

- Dunes Migration Infographics

Fig. 3.1 Proposed analysis workflow for dunes migration analysis using multispectral satellite imagery

3.3.1 Data Planning and Download

Prior to any satellite remote sensing data analysis, preliminary steps have to be considered to gather a consistent dataset and avoid the processing of unnecessary data. This is especially true for multitemporal analyses, such as change detection, that are generally data demanding. Typically, multispectral satellite image files require the allocation of large computer memory and computing resources to be analysed. Therefore, a thorough data planning step is always advised.

After the definition of the study area extent and the analysis time period, the most suitable data source has to be identified and both quality and temporal/spatial images coverage for the defined study area and period have to be assessed. Multispectral satellite images are generally distributed as *tiles*. Each tile corresponds to an image acquisition at a specific time and the tile location and extend coincide to a reference grid cell called *granule*. The reference grid is fixed and specific for each satellite mission considered. For example, Landsat 8 granules extend is 185x185 km^2 and approximately one tile every 16 days is available, according to the revisit time of the satellite. Sentinel-2 granules have an extent of about 100 × 100 km^2 and approximately one tile every 5 days is available. The study area may be adjusted to minimize the number of granules required for its complete coverage. This expedient may significantly reduce the amount of image files to be handle during the analysis.

The temporal availability of tiles on the study area has then to be assessed to ensure the selected data source being able to support multitemporal analyses with its actual frequency of images in the considered time period. In fact, it may happen than the actual acquisition frequency is lower than the theoretical one due to technical faults either of the satellite or the archive infrastructure. Furthermore, in dealing with optical multispectral remote sensing, data availability is often affected by the could cover which may cause one or more tiles being not actually serviceable for the analysis.

The data planning step aims to address the above questions and it can be achieved by querying archives of the provides using their online data portal (see e.g. Table 3.1 for Landsat 8 and Sentinel-2). An alternative and powerful tool that can be adopted in data planning is Google Earth Engine©. This freely accessible Web-platform allows querying extensive remote sensing data archives, including the Landsat 8 and the Sentinel-2 archives. Querying operations are accomplished programmatically using JavaScript-base scripting (Google, 2021). Summary statistics for tiles availability and quality for a user-defined area and time period can be computed and displayed using graphical widgets directly on the Web interface of the platform, as shown in Fig. 3.2. According to the information obtained in the data planning step, the list of tiles to be downloaded for starting the analysis can be defined. The

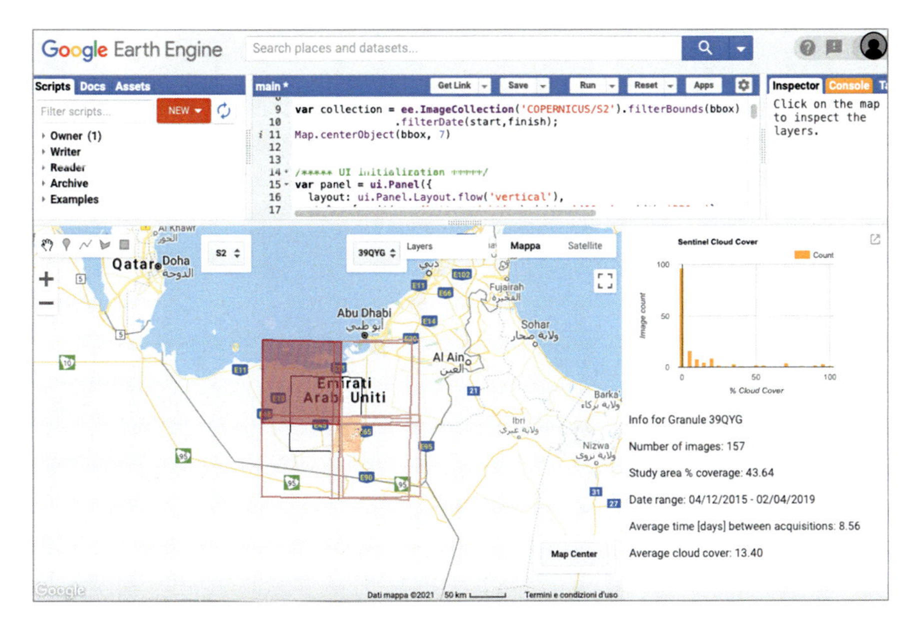

Fig. 3.2 Summary statistics generation for a sample Sentinel-2 granule (pink canvas) using Google Earth Engine© for a user-defined study area (black canvas)

download can be thus performed again through the online data portal of the data providers.

3.3.2 Data Preprocessing and Temporal Stack Generation

Once tiles have been downloaded, they have to be preprocessed to obtain a multi-temporal and spatially co-registered stack composed of radiometrically calibrated images with an equal set of spectral bands so that pixels in the stacked bands represent always the same land portion without atmospheric noises or other artefacts (Tewkesbury et al., 2015). These operations are key to any change detection technique, including dunes migration analysis. Changes are investigated in terms of alterations of pixel characteristics in time and therefore even partially inaccurate geometric and/or radiometric corrections would inevitably bias the results.

Atmospheric correction aims at removing atmospheric effects on the reflectance values of images that vary between each image acquisition time and location, thus enhancing images comparability in the multitemporal stack. A number of methods have been proposed in the scientific literature to perform such correction (Vermote & Kotchenova, 2008). Satellite data providers often distribute preprocessed tiles on which atmospheric correction has been already applied (i.e. Level-2 data). This is the case, for example, of Landsat 8 which provides on-demand services for downloading every tile in the catalogue at processing Level 2 (surface reflectance). A Sentinel-2 Level-2 data catalogue is also available even if not the whole tiles archive is currently accessible at this processing level. However, atmospheric correction algorithms are integrated into most remote sensing software packages, including for example GRASS GIS (GRASS Development Team, 2018) and SNAP (European Space Agency, 2021a) (see Table 3.1), and can be eventually applied by the user on the downloaded tiles.

A second step to reduce the analysis data volume is to remove unnecessary spectral bands form each tile. Indeed, not all bands are useful for land remote sensing applications. In the case of Landsat 8, thermal, aerosol, panchromatic, and cirrus bands are sensed by the satellite sensors for different purposes (US Geological Survey, 2021) and can be therefore removed. The same applies to the Sentinel-2 for which aerosol, water vapour, and cirrus bands are generally not considered in applications such as the one proposed in this chapter. A fuller treatment of multi-spectral satellite image processing can be found e.g. in (Ose et al., 2016).

Finally, the remaining atmospherically correct bands have to be spatially matched on the study area extent with an equal pixel resolution. This procedure is often referred to as spatial co-registration and guarantees each bands pixel representing exactly the same land portion through the time in the stack. The stacking operation consists merging, cropping and resampling the bands on a common grid overlapping the study area. Bands of each tile can be then grouped and the reference acquisition time, inherited from tile metadata, can be stored within the resulting raster file name.

A series of multispectral scenes with time references is thus obtained and it represents the multitemporal stack to be employed in change detection analyses. The stacking operation can be performed using functionalities available in many remote sensing software, including GRASS GIS and SNAP. Recursive operations on tiles can be also automatized thanks to the extended support of the aforementioned software to Python scripting. Most of the input/output operations and bands manipulation can be further optimized directly invoking in those scripts auxiliary libraries such as GDAL (GDAL/OGR contributors, 2021).

3.3.3 Data Analysis

The proposed change detection analysis to investigate dunes migrations considers iteratively a couple of scenes in the stack acquired at two different times. The set of scenes composing the stack is sorted according to the time interval and the time step (Δt) at which dunes migration is intended to be assessed. From now on, we will use the term *master* to identify the initial scene acquired at time t, and the term *slave* to identify the consecutive scene acquired at time $t + \Delta t$. A schematic of the proposed data analysis is reported in Fig. 3.3.

The initial operation for the master and slave scenes comparison is to match the band histograms of the slave scene with the ones of the master scene. This operation improves comparability by balancing colour contrasts between the master and the slave scenes which are captured by the same satellite camera but in different times and, likely, under different exposure conditions (Shapira et al., 2013). Support to histogram matching operations is provided e.g. by GRASS GIS or it can be executed programmatically using image processing libraries such as the Scikit-image Python library (Van der Walt et al., 2014).

Matched scenes are then classified by exploiting all spectral bands to extract raster masks distinguishing target sand pixels from the other background land cover classes (e.g. build-up, crops, etc.), that are to be excluded from the dunes migrations analysis. In general terms, classification techniques for multispectral remote sensing images can be of two types (Lu & Weng, 2007), as follows.

- Supervised classification: The grouping operations for pixels with common characteristics is performed by comparing their spectral characteristics with the one of user-identified subsets of pixels on the image, generally called *training samples*, which are representative of specific land cover classes. Training samples are selected based on the experience of the user and must adequately represent the spectral characteristics of each class in the image to be classified. In principle, each image of a temporal stack requires the definition of dedicated training samples unless no significant local changes in the land cover are expected.
- Unsupervised classification: The grouping operations for pixels with common characteristics are based on an automated statistical processing without the user providing training samples describing land cover classes. The user must then

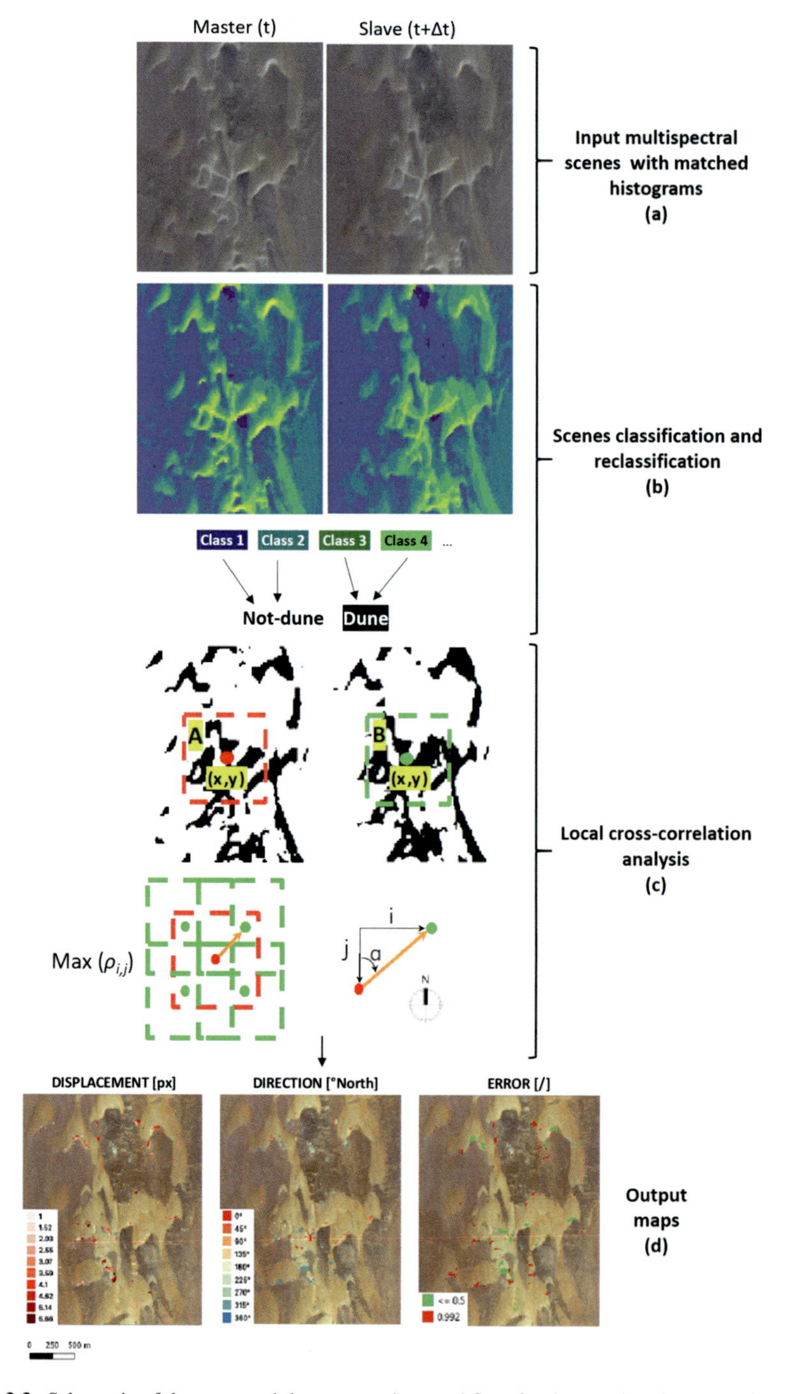

Fig. 3.3 Schematic of the proposed data processing workflow for dunes migration analysis

have substantial knowledge of the area being classified when automatically defined classes have to be related to actual features on the ground.

While supervised techniques generally allow achieving higher classification accuracies than unsupervised techniques, they require higher operational costs due to the manual provision of training samples that has to be performed by the user. Additional details on image classification methods can be found e.g. in (Lu & Weng, 2007; Ose et al., 2016). In the case of the sand dunes, significant changes are expected among different scenes leading to the need for modifying training samples for each scene in the input stack. Thereby, supervised classification techniques will not be considered further in the proposed framework. The reason is the excessive manual intervention that would be necessary to process a large number of scenes, such as the one that may be included in image stacks covering long study periods. Nevertheless, supervised classification techniques may be considered to provide change detection methods with more accurate inputs. Support to both supervised and unsupervised classification is provided by most of remote sensing software, including GRASS GIS and SNAP.

In principle, multispectral imagery-based change detection can be also performed on the original spectral bands, bypassing the classification (Ghamisi et al., 2019). However, classification is suggested in this framework to simplify the information on the land cover carried by the original multispectral scenes, and to better focus detection of changes on dunes only (Tewkesbury et al., 2015) (see Fig. 3.3b). In particular, binary raster masks were used in the preliminary applications of the proposed framework to render the distinction between dunes and all the other back- ground land cover classes (Oxoli et al., 2020). Binary raster masks are obtained through re-classification of the classified images by manually labelling the detected pixel sub-classes into dune or not-dune classes.

Once masks from both master and slave scenes are obtained through the classification, dunes migration analysis is performed by detecting and describing shifts, respectively between the master and the slave masks pixels, which are estimated by means of local cross-correlation analysis (You et al., 2017). Shifts are detected for each pixel by using a moving window (i.e. a subset of neighbouring pixels). Two homologous windows of equal shape are defined respectively on the master and the slave masks. At each iteration, the slave mask window is shuffled in all directions around its central pixel. For each simulated shift, a cross-correlation coefficient ρ is computed with respect to the master mask window as in Eq. 3.1 (You et al., 2017); where A is the set of pixels in the master moving window and B is the homologous on the slave mask. x and y define the position of the window central pixel in the mask grid, whereas i and j are the pixel shifts imposed to the central pixel of B respectively in the horizontal and vertical directions. Cov and Var are the covariance and variance operators.

$$\rho_{i,j} = \frac{Cov\left[A_{(x,y)}, B_{(x+i,y+j)}\right]}{\sqrt{Var\left[A_{(x,y)}\right] Var\left[B_{(x+i,y+j)}\right]}} \quad (3.1)$$

The features of the (i, j) vector (i.e. directions and displacements) resulting from the shift that maximize ρ are stored into output raster maps (which have equal extension and resolution of the input scenes) in the position of the master mask window central pixel (see Fig. 3.3c). The moving window is then centred on another pixel and the procedure is repeated such as the above information is computed for each pixel in the masks. Outputs of the local cross-correlation analysis mainly consist of three raster maps for each master/slave comparison (Fig. 3.3d). Outputs include a map with shift directions (azimuths) and a map with displacements (measured in pixels unit). In addition, a map of normalized root mean square cross-correlation errors (Fienup, 1997) or, alternatively, of the ρ values associated with the shifts (You et al., 2017) can be also computed. Both the former and the latter can be used as metrics of the reliability of each detected shift and arbitrarily adopted as a filter to remove weakly correlated shift from the results (see Sect. 3.4). A temporal series of directions and displacements is thus obtained from the procedure applied iteratively to each pair of masks in the multitemporal stack.

The local cross-correlation analysis can be fully performed by coupling GRASS GIS with the image matching functions of the Scikit-image library through Python scripting.

It is worth to notice that the proposed change detection procedure is experimental and requires users a substantial knowledge of the outlined processing algorithms and software tools to be implemented. While the suggested data planning and preprocessing steps have general validity in most optical remote sensing data work, the processing operation suggested is mostly specific for dunes migration analysis and cognate applications. A number of parameters, which may significantly affect results, have to be set or tuned to perform the analysis. In particular, the following parameters needs to be spelled out prior to the processing algorithms application according to the expectations of the analyst.

- Master/slave relationship: Changes can be computed with respect either to the initial scene in the stack only, such as for the couples (t; t + Δt), (t; t + 2Δt), etc., or by using a consecutive master and slave couple at each comparison, such as (t; t + Δt), (t + Δt; t + 2Δt), etc. In the first case, the master scene is fixed and output shifts are cumulative in time, while in the second case the master is moving in time so net shifts that occurred within the time step Δt are detected.
- Time step: The time step between master and slave scenes also influences results. For example, the adoption of long time steps (e.g. one year) on a fixed master scene concentrates the analysis on baseline or principal dunes displacements. The use of shorter time steps and a moving master should be preferred when the dynamic of periodical or seasonal displacements (Hassoup, 2019) is the focus of the analysis.
- Moving window: The moving window size should reflect somehow the spatial configuration of the land cover phenomena under investigations (e.g. average

dune extension, maximum expected dunes displacement, etc.) to improve results (You et al., 2017).

Validation procedures should also be considered in the framework to assess both reliability of classification and change detection outputs. Finally, the selection of the error or, alternatively, of the ρ value threshold to remove weakly correlated shift from the results should be also supported by sensitivity analyses.

3.3.4 Results Synthesis and Presentation

The ultimate goal of the proposed framework is to provide a synoptic view of dunes migration patterns. To that end, a visual comparison of output maps aided by GIS software may be misleading due to the amount of information (i.e. multitemporal pixel-wise displacement features) to be considered. To that end, the use of summarizing graphs is here suggested. In particular, the use of windrose diagrams or polar plots is advised to create snapshots of the dominant movement directions and displacements. The windrose diagram allows displaying oriented histograms of cumulative shifts (in percentage) that are read directly from the temporal series of output maps, obtained by the local cross-correlation analysis. Results can be filtered by removing from the diagram all those shifts that present a cross-correlation error higher – or alternative a ρ lower – than an arbitrary threshold. Detected shifts from the output maps are then included in the diagram as shown in Fig. 3.4. During the synthesis of the results, displacements can be multiplied by the scenes pixel resolution to display them in meters. The creation of windrose diagrams can be accomplished, for example, by exploiting Python data plotting functionalities such as the Windrose library (Roubeyrie & Celles, 2018).

3.4 Preliminary Applications in the MENA Region

The proposed framework for dunes migration analysis was preliminary tested on two real case-study in the MENA region (Oxoli et al., 2020). The selected study areas (see Fig. 3.5) are located in the UAE desert (6000 Km^2) and in the Egypt desert (2000 Km^2). The UAE study area is mostly characterized by wide linear crescentic Quaternary dunes and Quaternary aeolian sand sheet composed by continental sands and gravels. The Egypt study area includes widespread development of barchans dunes. Additional information on the geomorphology of the two deserts can be found e.g. in (Lancaster, 2013).

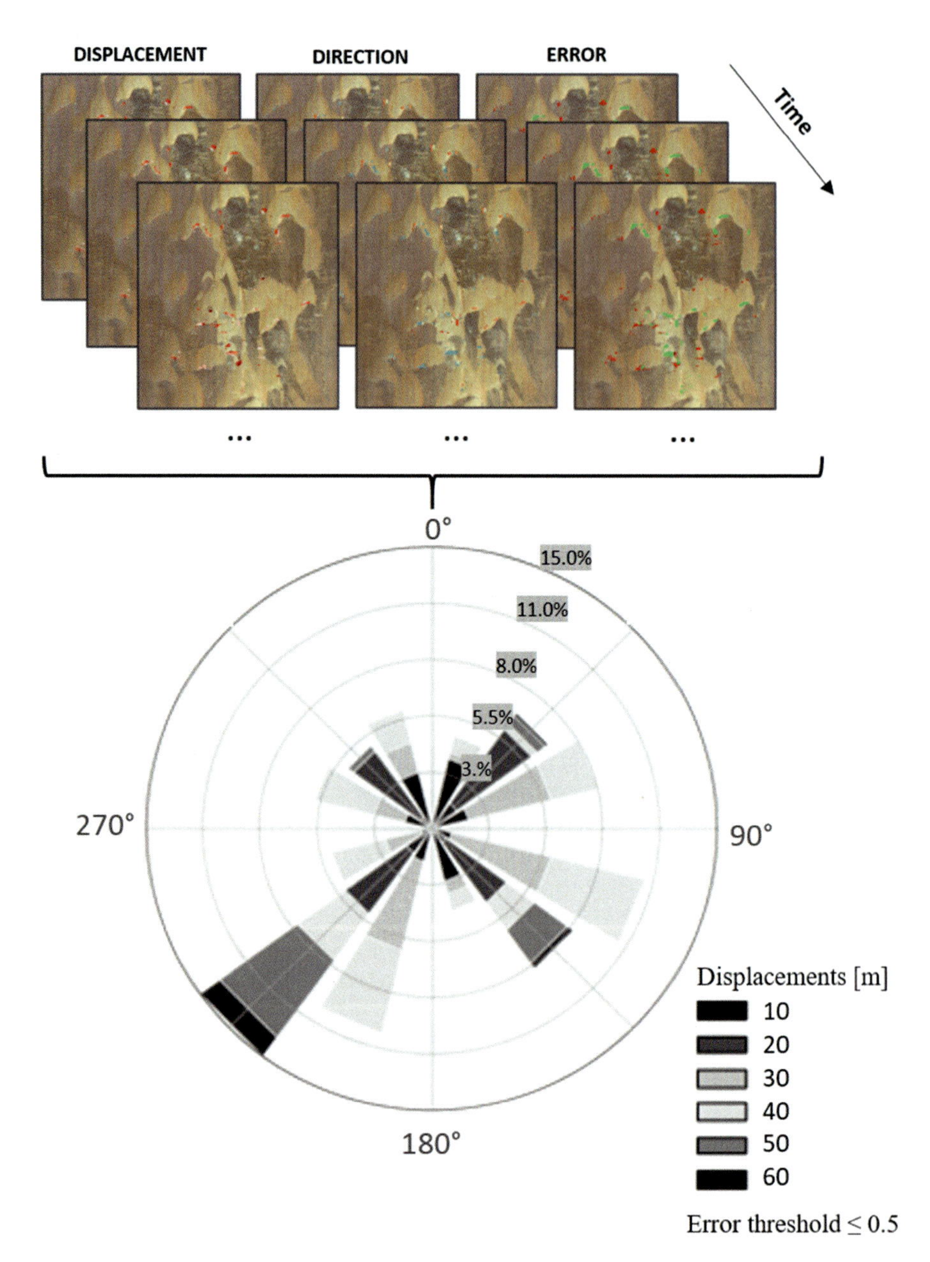

Fig. 3.4 Schematic of dunes migration pattern synthesis by means of a windrose diagram

3.4.1 Experiment Setting

The preliminary test considered both Landsat 8 and Sentinel-2 images acquired in the periods May 2013 – November 2015 (Landsat 8) and December 2015 –

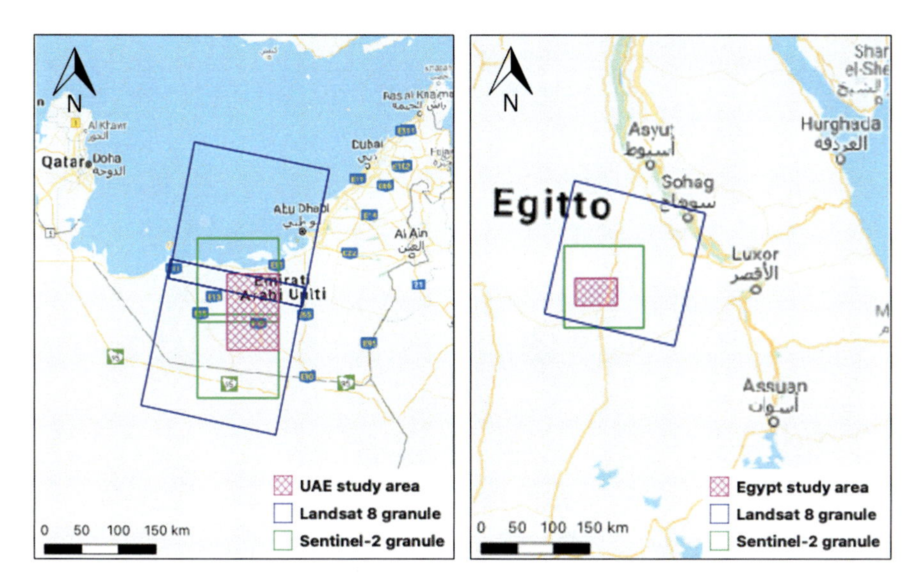

Fig. 3.5 Study areas (pink pattern boxes) in UAE (right) and Egypt (left) with Landsat 8 (blue boxes) and Sentinel-2 (green boxes) granules coverage. (Basemap: ©2020 Google)

December 2018 (Sentinel-2). Reference studies in the literature asserted that dunes migration patterns require a long time observation baseline to be characterized (Hugenholtz et al., 2012). Therefore, the choice of the period allowed investigating the longest time series of observations – at the time of the analysis – without duplicated data, according to the first passage of the satellites over the study areas. This was supported also by the availability of reference dominant winds directions (see Fig. 3.7). Dominant winds directions were here assumed as the main driving force of dunes migration (Hassoup, 2019) and used for early validation of results.

The data planning performed for this test revealed that the average acquisition time over the two areas was approximately 16 days for the Landsat 8 and 7 days for the Sentinel-2. A further check on cloud cover pointed out the possibility of retrieving from the catalogues at least one scene per month for both areas with limited or absent clouds disturbance. Nevertheless, a time step of one year was finally used in this preliminary analysis with the aim of limiting data volumes and address the detection of principal dunes migration patterns rather than periodical or seasonal displacements (see Sect. 3.3.3).

Images were downloaded according to the granules overlapping the study areas (see Fig. 3.5) by using the Web data portals of the providers (see Table 3.1). Landsat 8 images were available from the catalogue as Level-2 data (surface reflectance). Sentinel-2 catalogue offered instead only Level-1C data (top-of-atmosphere reflectance) for both areas in the considered time period. Atmospheric correction was applied to Sentinel-2 tiles using the dedicated GRASS GIS functionalities. The stacking operation was performed as reported in Sect. 3.3.2 and two separate stacks for the Landsat 8 and the Sentinel-2 were created.

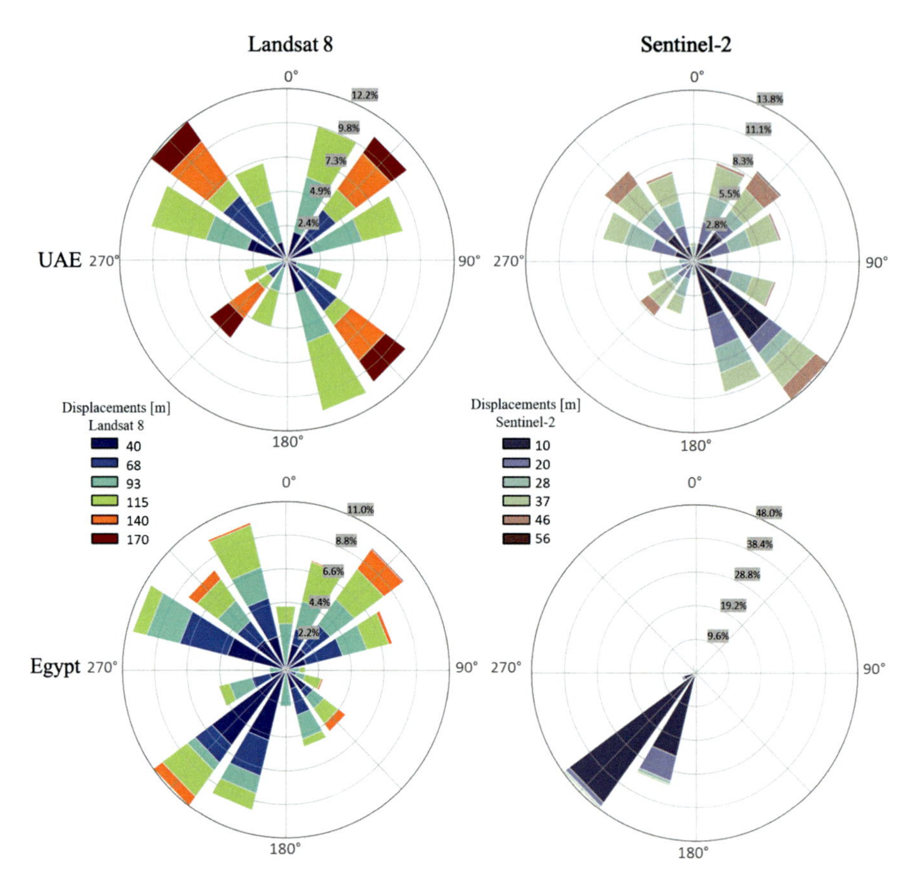

Fig. 3.6 Results synthesis for the preliminary application of the presented framework for dune migration analysis on the UAE and Egypt study areas. Only shifts with an associated error lower than 0.5 were considered in the graphs

To sum up, the local cross-correlation analysis to infer dunes migration for the proposed case studies used a time step of one year between images for the periods 2013 – 2015 for Landsat 8, and 2016 – 2018 for Sentinel-2. September was selected as reference month. A fixed master approach with a moving window of 15 x 15 pixels was adopted (see Sect. 3.3.3). The synthesis of results, provided through windrose graphs, is included in Fig. 3.6. Dominant winds directions that were here considered for qualitative validation of results are reported in Fig. 3.7.

3.4.2 Preliminary Results Discussion

Preliminary results show a marginal agreement of dunes shifts directions with the dominant winds (see Figs. 3.6 and 3.7) (Oxoli et al., 2020). Displacements appear

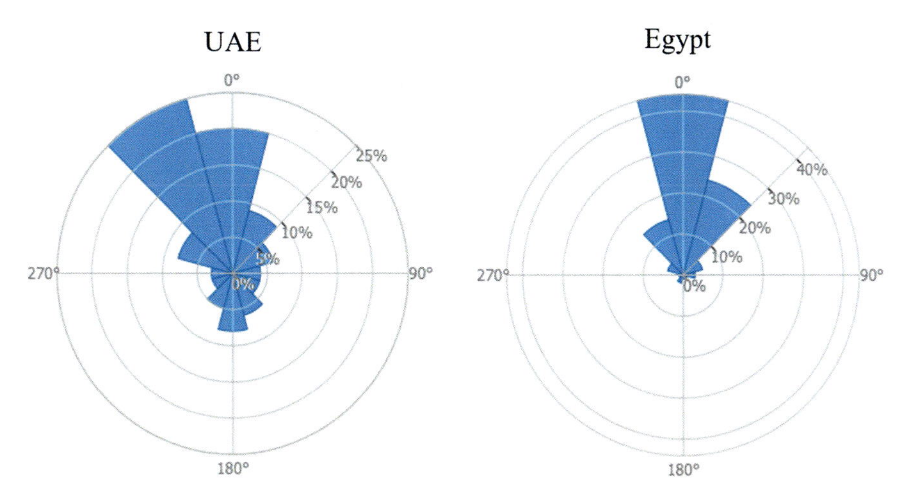

Fig. 3.7 Average of wind blow directions in the considered study areas between 2008 and 2018 (wind data: GWA 3.0 ©2020 (Technical University of Denmark – DTU, 2021))

moderately overestimated according to reference data in the literature, see e.g. (Lancaster, 2013; Hassoup, 2019) where the estimated net movement rate for dunes in the studies areas varies from dozens of meters (Egypt) to few meters per year (UAE). According to the above, Sentinel-2 imagery (10 m resolution) outperformed the Landsat 8 (30 m resolution), suggesting the pixel resolution being critical for dunes migration studies. In fact, the minimum shift that can be detected is equal to or larger than the pixel resolution. Therefore, movements of a few meters cannot be sensed by the procedure because of the native resolution of input data, to which the minimum mappable change is limited.

In this experiment, Sentinel-2 imagery procured more clustered results to the dominant winds blow directions than the Landsat 8 by enforcing the above consideration on the resolution. However, many shifts were registered also in opposite or different directions than the dominant winds. It is known from the literature that dunes generally show heterogeneous patterns both in time ad space due to local major forcings or obstacles such as wind transport capacity, terrain morphology, vegetation, and sand supply (Hugenholtz et al., 2012). Furthermore, dune mobility might not always imply migration but simply changes in dunes morphology (Lancaster, 2013). All these features require primarily consideration in order to exploit dunes migration patterns identified through the presented analysis for operational purposes.

Additional sources of error may then derive from the analysis parameters setting and quality of input data. In particular, the satellite imagery georeferencing offered by the data providers has a root mean square error for the positional accuracy that ranges between 10 m (Sentinel-2) to 12 m (Landsat 8). These errors may manifest as uncontrolled shifts among stacked images that are not due to the actual dynamics of the dunes and biasing the change detection. Next is the classification step. It is intuitive that the more accurate is the classification the better is the detection of shifts and therefore the selection of the classification procedure is key to provide the analysis with the best compromise between accuracy and applicability on a large images stack. Lastly, a procedure for tuning parameters for the local cross-correlation analysis, such as time steps, master/slave relationship and moving windows size, should be considered to improve results. Validation of results through quantitative comparison with ground truth information – intended as field data or remote sensing products with higher resolution and accuracy than the target satellite imagery – is finally advised (Hugenholtz et al., 2012).

Issues and improvements to the analysis discussed above were not directly tackled in this early stage of work. Nonetheless, these provide an exhaustive list of directions for future developments as well as recommendations for proper use of the proposed framework.

3.5 Conclusions

This chapter presented a framework for desert sand dunes migration analysis that combines multispectral and multitemporal open satellite imagery with free and open-source software. A preliminary application on two study areas in the MENA region was provided to outline both potentials and limitation of the proposed framework as well as to discuss improvements and recommendations thereof. The computational steps were described according to both data and software requirements. Preliminary results showed marginal agreements in terms of detected dune shifts with dominant winds blow directions for the considered study areas in Egypt and UAE deserts.

Although being experimental, the framework promises to produce valuable results for investigating sand dunes migration. Resulting maps and summary graphs may provide support to sand mitigation planning by supplying synoptic views of dunes dynamics in a timely and cost-effective manner. The monitoring of sands dunes activity is critical for communities and ecosystems living alongside arid regions which are forced to co-evolve with such dynamic environments. Despite being a natural process, changes in dunes and sand masses migration patterns are also affected by climate changes and human interactions with desert environments (Yizhaq et al., 2009). In this context, the development of new techniques for simulating and monitoring dunes migration represents a pressing concern to be addressed.

To that end, the proposed framework leverages exclusively free and open-source GIS and remote sensing software and open satellite imagery. This is of paramount importance to provide the analysis with the potential to be empowered, replicated and improved. Furthermore, cross-correlation of imagery was originally proposed mainly to track movements of the atmosphere, oceans, and glacier fringes by leaving large room for future research on applications to inland remote sensing such as land cover displacements (You et al., 2017).

Finally, the suitability of the modern open satellite imagery for highly resolved environmental analysis in space and time opens new opportunities to unpin both the economic and scientific potential of this data, while adding value to the open data policies and motivating investments in satellite Earth Observations platforms and techniques development.

References

Al-Mutiry, M., Hermas, E., Al-Ghamdi, K., & Al-Awaji, H. (2016). Estimation of dune migration rates North Riyadh city, ksa, using spot 4 panchromatic images. *Journal of African Earth Sciences, 124*, 258–269.

Belward, A. S., & Skøien, J. O. (2015). Who launched what, when and why; trends in global land-cover observation capacity from civilian earth observation satellites. *ISPRS Journal of Photogrammetry and Remote Sensing, 103*, 115–128.

Brovelli, M. A., Minghini, M., Moreno-Sanchez, R., & Oliveira, R. (2017). Free and open source software for geospatial applications (foss4g) to support future earth. *International Journal of Digital Earth, 10*(4), 386–404.

Bunting, P., Clewley, D., Lucas, R. M., & Gillingham, S. (2014). The remote sensing and gis software library (rsgislib). *Computers & Geosciences, 62*, 216–226.

Copernicus Programme. (2021). *Open access hub*. https://scihub.copernicus.eu, (21 February 2021).

European Commission. (2021). *Copernicus programme*. https://www.copernicus.eu.copernicus.eu, (21 February 2021).

European Space Agency. (2021a). *Sentinel application platform (SNAP)*. https://step.esa.int/main/toolboxes/snap, (21 February 2021).

European Space Agency. (2021b) *Sentinel missions*. https://sentinel.esa.int/web/sentinel, (21 February 2021).

Fienup, J. R. (1997). Invariant error metrics for image reconstruction. *Applied Optics, 36*(32), 8352–8357.

Finer, M., Novoa, S., Weisse, M. J., Petersen, R., Mascaro, J., Souto, T., Stearns, F., & Martinez, R. G. (2018). Combating deforestation: From satellite to intervention. *Science, 360*(6395), 1303–1305.

GDAL/OGR contributors. (2021). GDAL/OGR geospatial data abstraction software library. *Open Source Geospatial Foundation*. URL https://gdal.org

Ghamisi, P., Rasti, B., Yokoya, N., Wang, Q., Hofle, B., Bruzzone, L., Bovolo, F., Chi, M., Anders, K., Gloaguen, R., et al. (2019). Multisource and multitemporal data fusion in remote sensing: A comprehensive review of the state of the art. *IEEE Geoscience and Remote Sensing Magazine, 7*(1), 6–39.

Google. (2021). *Google earth engine*. https://earthengine.google.com, (21 February 2021).

GRASS Development Team. (2018) *Geographic resources analysis support system (GRASS GIS) Software 7.4*. Open Source Geospatial Foundation, URL https://grass.osgeo.org

Hassoup, A. (2019). Sand dunes hazard assessment in el-kharga oasis, Egypt. *Geophysical Research Abstracts, 21*.

Hugenholtz, C. H., Levin, N., Barchyn, T. E., & Baddock, M. C. (2012). Remote sensing and spatial analysis of aeolian sand dunes: A review and outlook. *Earth Science Reviews, 111*(3-4), 319–334.

Lancaster, N. (2013). *Geomorphology of desert dunes*. Routledge.

Lu, D., & Weng, Q. (2007). A survey of image classification methods and techniques for improving classification performance. *International Journal of Remote Sensing, 28*(5), 823–870.

Michel, S., Avouac, J. P., Ayoub, F., Ewing, R. C., Vriend, N., & Heggy, E. (2018). Comparing dune migration measured from remote sensing with sand flux prediction based on weather data and model, a test case in Qatar. *Earth and Planetary Science Letters, 497*, 12–21.
Ose, K., Corpetti, T., & Demagistri, L. (2016). Multispectral satellite image processing. In *Optical remote sensing of land surface* (pp. 57–124). Elsevier.
Oxoli, D., Brovelli, M., Frizzi, D., & Martinati, S. (2020). Detection of land cover displacements through time-series analysis of multispectral satellite imagery: Application to desert sand dunes. *The International Archives of Photogrammetry, Remote Sensing and Spatial Information Sciences, 43*, 739–744.
Roubeyrie, L., & Celles, S. (2018). Windrose: A python matplotlib, numpy library to manage wind and pollution data, draw windrose. *Journal of Open Source Software, 3*(29), 268.
Runge, A., & Grosse, G. (2019). Comparing spectral characteristics of landsat-8 and sentinel-2 same-day data for arctic-boreal regions. *Remote Sensing, 11*(14), 1730.
Shapira, D., Avidan, S., & Hel-Or, Y. (2013). Multiple histogram matching. In *2013 IEEE international conference on image processing* (pp. 2269–2273). IEEE.
Technical University of Denmark – DTU. (2021). Global Wind Atlas. https://globalwindatlas.info, (21 February 2021).
Tewkesbury, A. P., Comber, A. J., Tate, N. J., Lamb, A., & Fisher, P. F. (2015). A critical synthesis of remotely sensed optical image change detection techniques. *Remote Sensing of Environment, 160*, 1–14.
US Geological Survey. (2021). *EARTHEXPLORER*. https://earthexplorer.usgs.gov, (21 February 2021).
US National Aeronautics and Space Administration. (2021). *Landsat sciences*. https://landsat.gsfc.nasa.gov, (21 February 2021).
Van der Walt, S., Schönberger, J. L., Nunez-Iglesias, J., Boulogne, F., Warner, J. D., Yager, N., Gouillart, E., & Yu, T. (2014). Scikit-image: Image processing in python. *PeerJ, 2*, e453.
Vermote, E. F., & Kotchenova, S. (2008). Atmospheric correction for the monitoring of land surfaces. *Journal of Geophysical Research: Atmospheres, 113*(D23).
Wendl, C., Le Bris, A., Chehata, N., Puissant, A., & Postadjian, T. (2018). Decision fusion of spot6 and multitemporal sentinel2 images for urban area detection. In *IGARSS 2018-2018 IEEE international geoscience and remote sensing symposium* (pp. 1734–1737). IEEE.
Wulder, M. A., Masek, J. G., Cohen, W. B., Loveland, T. R., & Woodcock, C. E. (2012). Opening the archive: How free data has enabled the science and monitoring promise of landsat. *Remote Sensing of Environment, 122*, 2–10.
Yizhaq, H., Ashkenazy, Y., & Tsoar, H. (2009). Sand dune dynamics and climate change: A modeling approach. *Journal of Geophysical Research: Earth Surface, 114*(F1).
You, M., Filippi, A. M., Güneralp, I., & Güneralp, B. (2017). What is the direction of land change? A new approach to land-change analysis. *Remote Sensing, 9*(8), 850.

Chapter 4
How Severe is Water Stress in the MENA Region? Insights from GRACE and GRACE-FO Satellites and Global Hydrological Modeling

Ashraf Rateb, Bridget R. Scanlon, and Sarah Fakhreddine

Abstract Freshwater scarcity in the Middle East and North Africa (MENA) is increasingly exacerbated by rapid population growth demands and climate change and currently impacts ~0.6 billion people in the region. In this chapter, we revisited the trends in terrestrial water storage (TWS) over the last 18 years between 2002 and 2020 using observations of the Gravity Recovery and Climate Experiment and its Follow On (GRACE-FO) missions. We evaluated the interdecadal TWS trends in the MENA region against the variability of climate-driven TWS between 1901 and 2020 derived from GRACE, GRACE-FO, and natural simulation of the global WaterGAP hydrological model. Climate-driven TWS represents TWS anomalies that are only forced by non-anthropogenic stressors and vary from annual cycle to centennial variations. These TWS patterns were derived using the cyclostationary empirical orthogonal functions method over grid and MENA's polygon scales. The interdecadal trend of TWS in the MENA region shows that the entire MENA region lost about ~760 Gigaton (Gt) between 2002 and 2020, equivalent to ~2.6x the annual rate of ice loss from Greenland or ~ 2 mm of global sea-level increase. Depletion is more severe in the Middle East (e.g., Iran, Saudi Arabia) than in North Africa, except for Tunisia. Current GRACE-GRACE (FO) TWS depletion trends in MENA exceed past climate variability magnitude by at least a factor of 50 (considering GRACE period only), especially in northern Saudi Arabia, southern and eastern Iran, western Iraq, Egypt, Libya, and Algeria. These regions are characterized by a hyper-arid climate, an absence of groundwater recharge, overexploitation of surface water and groundwater resources. Sustainable surface and groundwater management is more urgent than ever to meet increasing demands. Interpreting GRACE trends relative to the magnitude and variability of climate-driven interannual and

---

A. Rateb (✉) · B. R. Scanlon · S. Fakhreddine
Bureau of Economic Geology, University of Texas at Austin, Austin, TX, USA
e-mail: ashraf.rateb@beg.utexas.edu

M. M. Al Saud (ed.), *Applications of Space Techniques on the Natural Hazards in the MENA Region*, https://doi.org/10.1007/978-3-030-88874-9_4

interdecadal variations helps to evaluate the reliability and forecast skills of current TWS trends and highlights the role of the anthropogenic activity in draining MENA's water resources.

Thousands have lived without love, not one without water. W. H. Auden (1907–1973)

4.1 Introduction

During the past decades, water scarcity has become an increasingly serious challenge globally due to the widening gap between global demands and limited supplies. Stress on freshwater resources has more than doubled since the 1960s, with 25% of the world's population, across 17 countries, facing an "extremely high" level of water shortages as defined by an 80% imbalance between supply and demand. Of these 17 countries, 11 are located in the Middle East and North Africa (MENA) (WRI: Aqueduct Water Risk Atlas, 2019). (Fig. 4.1). The water crisis in MENA is primarily a product of the hyper-arid environment (high temperature and high evaporation throughout most of the year), and it is worsened by overexploitation of rivers and aquifers. Water severity in MENA is unevenly distributed, ranging from highly severe in the Gulf Cooperation Countries (GCC), Jordan, Yemen, and

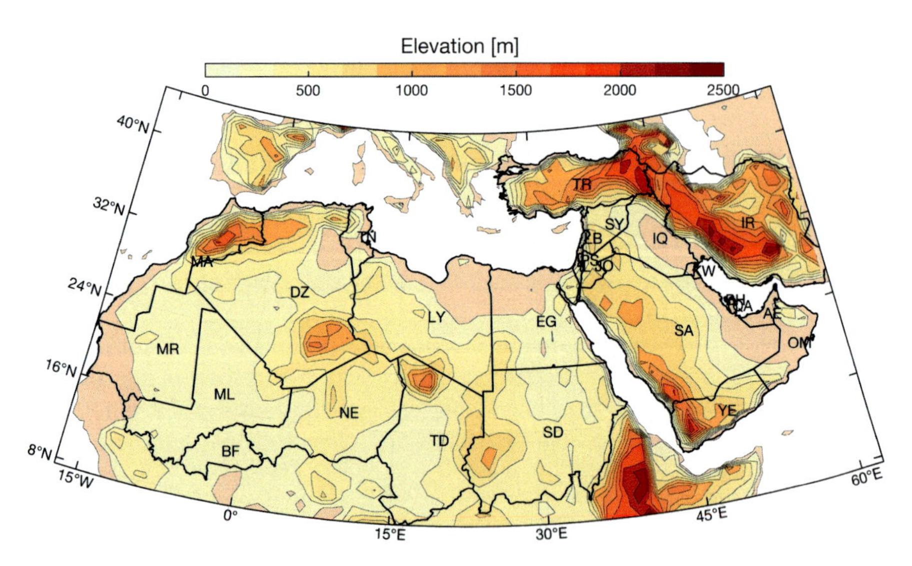

Fig. 4.1 Location of the study area that includes the 26 countries in the Middle East and North Africa; Algeria (DZ); Chad (TD); Libya (LY); Niger (NE); Tunisia (TN); Egypt (EG); Iraq (IQ); Israel (IL); Jordan (JO); Lebanon (LB); Palestinian Territory (PS); Sudan (SD); Syria (SY); Turkey (TR); Bahrain (BH); Iran (IR); Kuwait (KW); Oman (OM): Qatar (QA; Saudi Arabia (SA); United Arab Emirates (AE); and Yemen (YE). Contour lines represent topography elevations. (Amante & Eakins, 2009)

Libya to less severe in countries with high recharge (e.g., Morocco), high groundwater reserves (e.g., Lebanon and Iran) or large rivers (e.g., Iraq) (Bank, 2017).

Ample evidence has accumulated for the role of anthropogenic activity in driving current water scarcity in the MENA. However, it is unclear how this role compares with climate change magnitudes. This chapter outlines the current staus of TWS depletion in MENA relative to the historical magnitude of the climate variability between 1901 and 2020. We highlight the hot spots of TWS depletion in MENA, where current water abstractions are out of balance relative to supplies and greatly exceed observed and simulated impacts of climate variability.

Since 2002, the Gravity Recovery and Climate Experiment (GRACE) mission has resulted in a paradigm shift in understanding the water cycle and the interplay between climate change and human interference (Tapley et al., 2004; Landerer et al., 2020). GRACE provides unprecedented spatiotemporal information on the water cycle that enables monitoring changes in TWS from monthly to interdecadal scales and from a resolution of ~330 km to a global scale. GRACE mission includes twin satellites launched by the National Aeronautics and Space Administration (NASA) and the German Aerospace Center (DLR) in March 2002 to map temporal changes of the Earth's gravity field that can vary the distance between the two satellites at an elevation of ~470 km (Wahr et al., 1998). The GRACE mission was decommissioned in October 2017, and a new mission, named GRACE Follow On (FO), was launched in May 2018 to continue the same legacy (Landerer et al., 2020). GRACE was used to monitor trends in TWS (Rodell et al., 2018), mapping groundwater storage (GWS) depletions and assessing the role of climate and human impacts (Frappart & Ramillien, 2018; Russo & Lall, 2017; Xiao et al., 2015; Kelley et al., 2015), and mapping drought and flood occurrences and frequencies (Reager et al., 2014; Chen et al., 2010, 2018). GRACE-TWS represents all land reservoirs from surface to the Moho including (groundwater (GW), surface water (SW), soil moistures, snow water, and ice). GRACE has been validated against in-situ groundwater level observations and showed good performance when sufficient in-situ data were available (Rateb et al., 2020a). GRACE has been used to validate and enhance the performance of global hydrological models in simulating trends and annual components of TWS (Scanlon et al., 2018, 2019; Eicker et al., 2014; Döll et al., 2014) and trends in GWS in regional modeling (Rateb et al., 2020a).

In MENA region, early studies using GRACE show TWS depletion primarily caused by GW depletion in Iran, Saudi Arabia, and North (N) Africa. During the 2007–2009 drought in the Tigris Euphrates Basin (TEB), TWS declined by a total of 143 Gt, with 60% of the depletion arguably attributed to GW pumping for irrigation (Xiao et al., 2015). Other studies attributed the large depletion predominantly to SW decline during the 2007–2009 drought (Longuevergne et al., 2013) and during the past four decades (Rateb et al., 2020b). Rateb et al. (2020b) investigated the water balance and drought occurrences in the TEB using GRACE/GRACE-FO, hydrological models, and in-situ observations between 1979 and 2020. They concluded that severe droughts have occurred at decadal timescales in the basin since the 1980s. Exceptional droughts are documented in 1998–2000 and 2007–2009, and mild to moderate droughts in 1983–1984, 1989–1992, 2011–2013, and 2018. SW declined the most and human impacts on SW and GW amplified TWS depletion during these drought periods by at least 50% (Rateb et al., 2020b). TWS declined by ~80 Gt during the 2007–2009 drought, mainly from the surface reservoir. GW depletion contributed only 25%–30% of the TWS depletion (Rateb et al., 2020b). The severe drought of 2007–2009 in the ME led to land degradation, desertification (Albarakat et al., 2018), land subsidence (Rateb & Kuo, 2019), and mass migration in Iraq and Syria (Kelley et al., 2015). Recent floods in 2019 and 2020 have resulted in TWS recovery of ~144 Gt by July 2020 to the levels prior to 2006 (Rateb et al., 2020b). However, by the time of this writing in February/March 2021, most of Turkey is subjected to drought, with reservoirs declining to their lowest level within the past 15 years (Observatory, 2021). Saudi Arabia experiences an annual TWS loss of 10.5 Gt/year. in the northwestern part of the country (Rodell et al., 2018), and an annual GWS loss of 6 to 5 Gt/year in north (Joodaki et al., 2014). TWS depletion in the north stabilized after 2016 when the Saudi government ended their three decades of domestic wheat production program (USDA, 2016). In N Africa, TWS depletion in the Nubian aquifer and northern western Saharan aquifers was about 50 Gt and 30 Gt between 2003 and 2016 (Frappart, 2020). Using recent GRACE-FO data, Ahmed (2020) estimated that GW withdrawal and recharge could be balanced if GW withdrawal decreased by 1.5 Gt/year and 2.7 Gt/year in the Nubian and northwestern Sahara aquifers, respectively.

This chapter updates the status of TWS trends in the MENA region and further evaluates the reliability of the TWS trend over the past 18 year. (2002–2020) relative to observed and simulated climate-driven TWS variability between 1901 and 2020. Climate-driven TWS represents TWS variations driven by climate only without human intervention and includes annual cycle and low-frequency components (e.g., interannual to centennial) that may be derived by climate teleconnections (e.g., El Niño–Southern Oscillation and Atlantic Multi-decadal Oscillation). The study area includes 26 countries in MENA regions (Fig. 4.1).

4.2 Data and Methods

4.2.1 GRACE Satellites and Global Hydrological Modeling

GRACE data described in this chapter represent TWS anomalies from GRACE/ GRACE-FO based on the two mascons solutions from the University of Texas at Austin, Center for Space Research Mascon (UTCSR-M) and NASA Jet Propulsion Laboratory Mascon (JPL-M) for the period 2002 and 2020. Macson solutions have a higher signal-to-noise ratio relative to traditional spherical harmonics solutions (Scanlon et al., 2016) and higher spatial resolution (~200 km at the equator and 100 km at the poles) (Rodell et al., 2018). CSR mascons were built using the Tikhonov regularization technique based on a geodetic grid (Save et al., 2016). JPL mascons use altimetry satellite observations and geophysical models to constrain the global mass flux anomalies (Watkins et al., 2015). The following corrections were applied to both mascons. 1) Estimates of C_{20} and C_{30} for GRACE after August 2016 and GRACE-FO were replaced by the satellite laser ranging estimates (Loomis et al., 2019), and C_{21} was corrected (Swenson et al., 2008; Sun et al., 2016). The global glacial isostatic adjustment correction has been applied (Peltier et al., 2018). JPL mascons have a native GRACE resolution of 3° but are downscaled to 1° and resampled to 0.5° using the community land model (CLM). The gain factor based on the difference between filtered and unfiltered CLM-TWS is applied to the JPL grid to strengthen the weakened signal under 3°. CSR mascons have a 1° resolution but are downscaled to 0.25° to better resolve leakage between the oceans and land. Physical interpretation of GRACE and GRACE-FO should be inferred at the missions' native resolution, even with downscaling and resampling the solutions. Downscaling GRACE data to spatial resolutions finer than 3° introduces a dependency of the neighboring grid cells and does not reflect a real mass change at the downscaled resolution.

We incorporated the climate-driven simulation (natural, excluding human intervention) of the WGHM model. WGHM includes five water use models (irrigation, household, livestock manufacturing, and cooling hydropower plants) (Döll et al., 2003). The model version used in this study is forced by outputs of WATCH-Forcing-Data-ERA-Interim (WFDEI) GPCC (Weedon et al., 2014). It has a spatial resolution of 0.5° × 0.5° (55-km at the equator) and monthly temporal resolution between 1901 and 2016. For additional information on WGHM-2.2, we refer to (Doell et al., 2014; Schmied et al., 2014; Müller Schmied et al., 2016).

4.2.2 *Cyclostationary Empirical Orthogonal Functions (CSEOF)*

To derive the climate-driven TWS in MENA, we applied cyclostationary empirical orthogonal functions (CSEOF) (Kim & North, 1997; Kim et al., 1996) to TWS variability from WGHM model. In CSEOF, spatiotemporal data are decomposed into a number of interpretable components of principal component times series (PCTs) and loading functions representing the spatial patterns of the PCTs. CSEOF differs from other decomposition methods (e.g., EOF) or empirical mode decompositions by being time-dependent, periodic, and can be applied to a single time series (1-D) or cube data (3-D). Implementing CSEOF requires knowing a nested period at which temporal components and their spatial patterns are common and can be derived. We used 12-month as the default climate cycle at which different climate variables yield a periodic change. We applied CSEOF on the 1-D time series (grid-scale) and for the MENA region polygon detrended TWS, which allows a detailed analysis of the time series. The derived components are independent with different variances (cyclostationary) (Fig. 4.2). The first derived component represents interannual to centennial variations of TWS; the climate-driven trends. The second principal component of the CSEOF decomposition represents the modulated

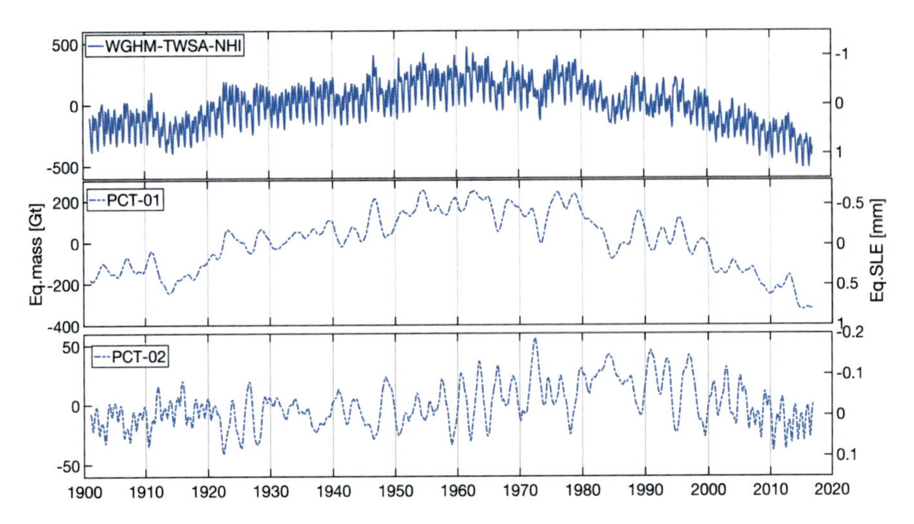

Fig. 4.2 (**Upper panel**) WGHM-TWS with no human intervention in the MENA region expressed as a mass movement in gigatons and equivalent global sea level in mm. (**Middle panel**) The first principal component time series derived after decomposition using CSEOF technique. PCT01 represents the low-frequency TWS component —the interannual to centennial climate-driven TWS. (**Lower panel**) PCT02 represents the modulated seasonal cycle of the climate-driven TWS

annual cycle (MAC). MAC is a better representation of the annual cycle and differs from the stationary seasonal cycle by being time-dependent with different mean and variance.

4.3 Current TWS Depletion Trends and Variability of Past Climate-Driven TWS

The magnitude of the TWS depletion in the ME is higher than that in N Africa (Fig. 4.3). In the ME, TWS depletion is more severe in Iran, south Turkey, and northern Saudi Arabia, with rates ranging from -20 to -10 mm/year. Similar depletion is only found in Tunisia in N Africa. A moderate-severe trend in TWS depletion prevailed in most of Turkey, southern Saudi Arabia, and most of N Africa, with rate ranges from -10 to -3 mm/year. Stable trends with a range of ± 3 mm/year, were found in west Yemen, northwest Egypt, and most of Mauritania and Morocco. Rising trends with a range (5–15 mm/year) are mainly found in N Africa, south the Sahara regions in Sudan, Chad, Niger, and Mali. Overall, the MENA region experienced an annual loss of water mass of -42 ± 2 Gt/year. Water mass loss between 2002 and 2020 totaled ~ -760 Gt—equivalent to ~2.6x the annual rate of Greenland ice loss or ~ 2 mm increase of the global sea level, considering the areas of the global oceans (361×10^6 km^2) (Fig. 4.3).

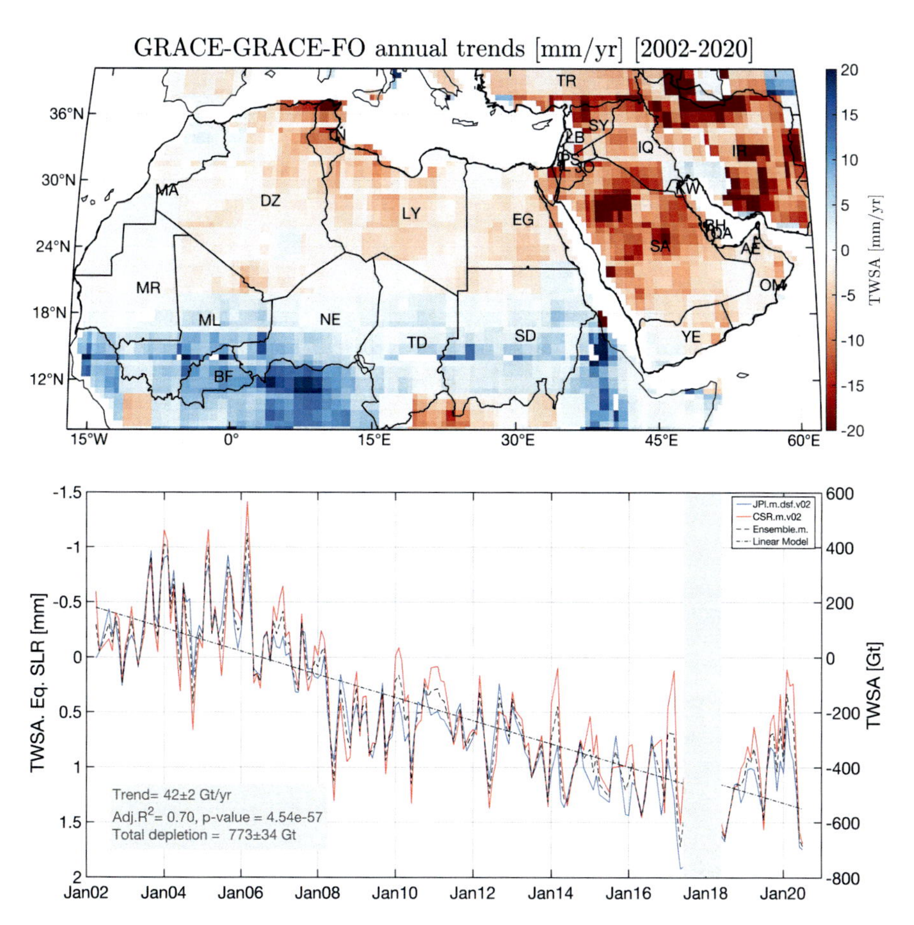

Fig. 4.3 (**Upper panel**) Spatial patterns in annual TWS trends based on GRACE JPL mascons expressed in mm. (**Lower panel**) Temporal patterns in TWSA trends in the MENA region from CSR and JPL mascons, and their ensemble mean, expressed in equivalent volume (Gt) and global sea-level rise (mm). Note, the difference in TWSA Eq.SLR signs show the loss of land water corresponds to global sea-level rise.

Climate driven TWS represents the simulated TWS driven only by climate forcing (e.g., precipitation and temperature). Variability in climate driven TWS is $\leq$10 mm throughout most of the region. The standard deviation (SD) is higher in southern Turkey, Northern Iran, western coasts of the Persian Gulf, Nile River, northwestern Africa (coastal), and south of Sahara (e.g., Chad, Mali, and Sudan). These regions with higher SD either receive substantial annual precipitation or (e.g., Chad, Mali, northern Iran) or represent a surface water body (e.g., Nile river) (Fig. 4.4).

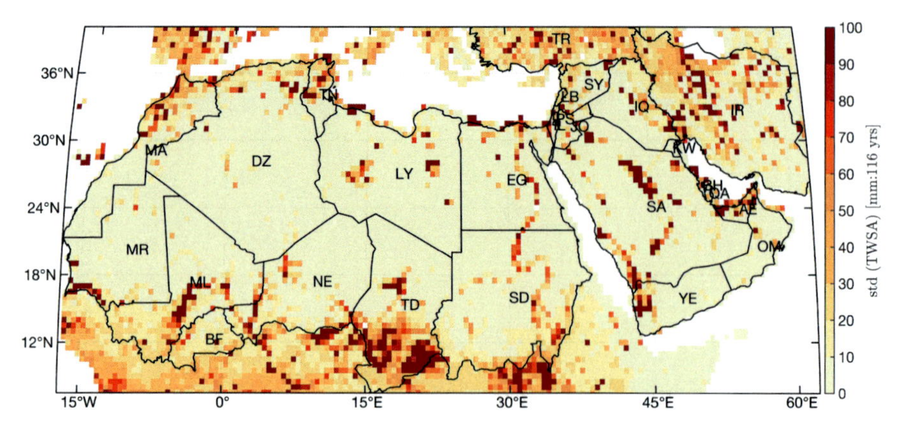

Fig. 4.4 Standard deviations in climate-driven WGHM-TWS extracted by CSEOF (Fig. 4.2)

4.4 Severity of Total Water Storage Depletion

Interdecadal trends in GRACE-TWS are driven by climate variability, change, and human intervention. While the long-term variability in WGHM – TWS are climate-driven only (Fig. 4.4). The interdecadal trend to interannual variability ratio (TIVR), based on GRACE TWS for the period (2002 to 2020), or the WGHM – TWS represents the severity for the current trends relative to historical climate variability. Grid cells with TIVR ratios that exceed ± 3 represent cells with severe or exceptional TWS depletion that exceed the magnitude of climate variability impacts by ± 3 standard deviations. The severe-exceptional TWS depletion mostly occurred in the central basin in Iran, the Arabian Aquifer system in Saudi Arabia, and Nubian Aquifer sandstone in N Africa (Fig. 4.5). In these regions, the TIVR ratio is $50\times$ the magnitude of the interannual variations between 2002 and 2020 and $150\times$ the magnitude of MAC, and interannual-centennial variations of TWS between 1901 and 2016. The 100-150x factor could be uncertain given the uncertainty in the climate forcing data (e.g., precipitation, temperature) in the WGHM model; however, it doesn't change the higher relative deviation of the climate-driven variability of TWS relative from the current interdecadal GRACE TWS. The climate in these regions is drier than in the MENA, and the regions receive <200 mm/year. annual precipitation (Fig. 4.5) with no single surface water body. GW storage variability constitutes most of the TWS variability and was overexploited for irrigation (>80%). The regions with surface water (e.g., Nile in N Africa and TEB in ME) have relatively less severe trends and smaller TIVR ratios. In these regions, SW is the primary source of irrigation and use in Egypt and Iraq, and GWS provides a buffer during dry periods. However, given the high water demands and extending these rivers beyond the borders, the resources are subjected to climate impacts and control by the upstream countries (e.g., Ethiopia on the Nile and Turkey on TEB). Such control diminished flow in the Tigris and Euphrates rivers (Kim et al., 1996), with Turkey singly controlling the resource of the two rivers by building dams and

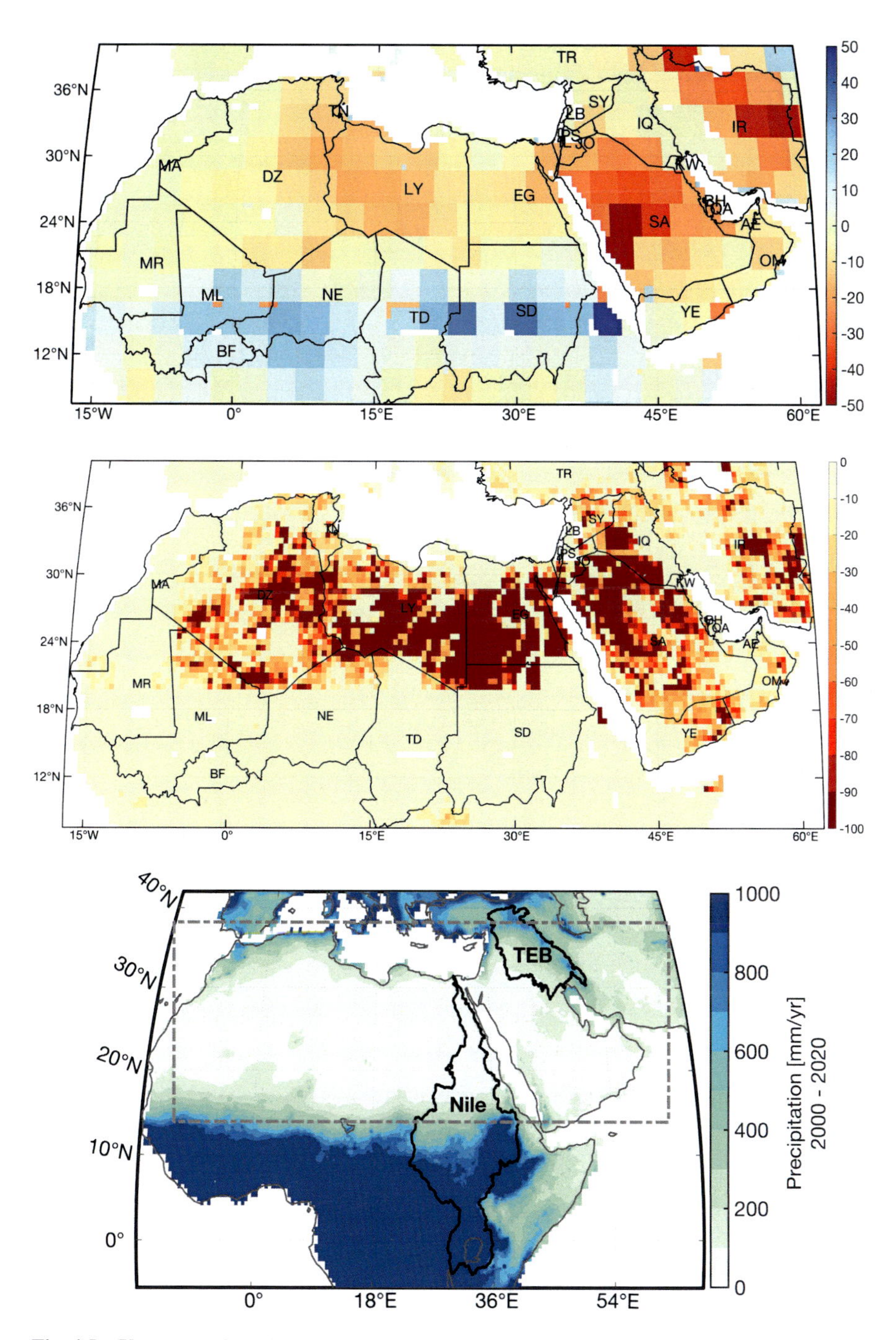

Fig. 4.5 (**Upper panel**) Ratio of interdecadal trends to the variability of climate-driven TWS based on JPL mascons for the 18 year. [2002–2020]. (**Middle panel**) Ratio of interdecadal trends to variability in climate-driven TWS based on WGHM model over 116 year. [1901–2016]. Ratios that exceed ±3 are considered severe or exceptional trends in TWS depletion. (**Lower panel**) Mean annual precipitation in the MENA regions for the period 2002–2020 derived from GPM-Imerge data

expanding irrigation projects (e.g., the southeastern Great Anatolian Project). Ethiopia is also building the Grand Ethiopian Renaissance Dam without reaching an agreement with the riparian countries (Sudan and Egypt), which could impact the annual flow in the Nile, especially during dry years. As a result, the situation is likely to worsen in the future without transboundary cooperation in these countries.

Other regions with lower interdecadal trends and TIVRs that may reflect the window of climatic variability impacts include the upper streams in the TEB, Turkey, Coastal regions in Oman, Yemen, and N Africa. These regions have higher annual precipitation (200–400 mm/year), which replenishes neighboring aquifers (Fig. 4.5). The exceptional wetness in the MENA is only found south of the Sahara, where the mean annual precipitation exceeds 400 mm/year.

Water use in MENA is mainly based on nonrenewable GW resources. At the lowest level of unsustainable use is Saudi Arabia and Libya, where abstraction from the Nubian aquifer (Libya) and the Arabian aquifer sustains (Saudi Arabia) water demand at 75% and 80%, respectively. Water use in other countries (e.g., Oman, Iran, and Djibouti) is unsustainable in terms of GW. Countries with sustainable use are coastal countries with use based on surface water (e.g., Morocco), on a high GW reserve (e.g., Lebanon), or relying heavily on desalination (e.g., Kuwait) (Bank, 2017).

4.5 Future Considerations for Water Security in MENA

Water scarcity in the MENA region is spatially variable and exceeds impacts of centennial climate variability by factors of 100–150×. Regions with high interdecadal rates and negative TIVRs represent regions with water resource deficits driven chiefly by anthropogenic activity. Rapid growth in human population, economic devolvement, and poor water management increased pressure on the MENA freshwater resources. More sustainable water management requires regional cooperation towards mitigation and plans to secure future water resources. The following highlights some measures ranging from common sense to cutting-edge ideas that could be adopted to reduce water insecurity in MENA.

- **Increase irrigation efficiency:** Irrigation in MENA is not fully optimized and primarily based on flood irrigation, with 60–92% of the water used for irrigation (Bank, 2017). Given the limited resources, adopting better irrigation technologies (e.g., sprinkle and drip irrigation) and plants seeds that require less water can reduce water waste and food loss and increase water availability.
- **Artificially recharge the aquifers:** In MENA, relocating the surface water in areas where the surface water is available (e.g., Nile and TEB, Jordanian rivers) to recharge surrounding aquifers could help alleviate temporal disparities in water availability between wet (floods) and dry (drought) climate cycles. In the GCC and Israel, aquifer recharge practices include recharge dams, aquifer storage transfer and recovery, recharge ponds, soil aquifer treatment, rooftop rainwater

harvesting, and Karez/Ain systems which use underground tunnels to recharge rainfall-runoff (Parimalarenganayaki, 2020). Expanding these projects during episodic events can even out impacts of rainfall variability and increase resilience in the MENA region.

- **Reuse water after treatment, for irrigation, or to replenish the aquifers**. Only 27% of wastewater is being treated and reused, with 43% of wastewater reuse for irrigation purposes in the MENA (Qadir et al., 2010). Water scarcity can be alleviated by the reuse of treated wastewater. Water reuse can also help improve crop production. One of the prominent examples of water treatment in MENA is the Dan Region Reclamation Project (Shafdan), south of Tel Aviv, Israel. Shafdan operates the largest wastewater treatment plant in Europe and the MENA with a treated volume greater than 140 million m^3 per year and was established in 1977 (Maliva, 2020). The secondary treated wastewater is recharged and further treated using SAT by infiltrating 120–140 M m^3/year. over 110 hectares with a retention time of 6–12 months in the aquifer (Cikurel et al., 2012).
- **Desalinization:** Desalinization is a lifeblood in MENA, GCC (e.g., Qatar, Kuwait) rely almost totally on desalinated water (>90%), UAE (70%), and Saudi Arabia (60%). For example, UAE operates the largest desalinated seawater plant globally with a stored volume of ~10 Mm^3 /year. It is expected to recover potable water for direct use (Stuyfzand et al., 2017). The full project was completed in 2017 and includes ~300 injection wells to store 26 Mm^3 of supply of desalinated water which is sufficient to supply the Abu Dhabi Emirates with emergency water for three months (Dawoud, 2020). Desalination is energy-intensive and may not be suitable for other poor countries in the MENA without access to vast energy resources (e.g., Yemen, Libya), besides its carbon footprint and chemical disposal. However, using other renewable energy sources for desalinization plants (e.g., solar power) should provide an alternative to long-term desalination after the oil and gas era. Also, involving the private sector in desalination contracts should help other countries —given the high prices and costs in building desalination plants.
- **Transboundary cooperation and holistic management:** Most of MENA‘s freshwater resources are transboundary (e.g., Nile, TEB, Jordan river basins, Arabian, and Nubian aquifers). Overexploiting these resources jeopardizes the ecosystem, fueling social unrest and sparking conflicts between countries. Current water scarcity is an existential threat to MENA with unknown future consequences imposed by climate change. Tackling this shared threat requires regional cooperation to manage the shared resources among countries and integrated approaches to water management during climate extremes (drought, floods) to maximize the shared benefits and minimize the economic costs and social conflicts.
- **Public outreach and economic policies:** Coping with the current water crisis in MENA requires changing individual behavior and lifestyle norms. Many of the alternative solutions for water scarcity in MENA face obstacles due to social unacceptance. (e.g., water reuse) or high prices of water treatment. Most of

MENA lacks water economic policies; water prices are issued at national levels, and prices are paid regardless of use or considering the available resources. Lowering the prices of treated wastewater for irrigation and raising pricing of fresh SW/GW resources will help protect and maintain the limited freshwater resources, adapting improved irrigation practices and reducing pollution.

MENA has a population growth rate of 1.6%, exceeding the global rate of 1.1%. These rates widen the water demand-supply gap as >70% of the finite freshwater is used for agriculture production. Also, high population growth rates exacerbate climate change impacts and drive mass migration. Policymakers need an effective strategy and key policies to reduce population growth rates to decrease stresses on water resources and food production.

References

Ahmed, M. (2020). Sustainable management scenarios for northern Africa's fossil aquifer systems. *Journal of Hydrology, 589*, 125196.

Albarakat, R., Lakshmi, V., & Tucker, C. J. R. S. (2018). Using satellite remote sensing to study the impact of climate and anthropogenic changes in the Mesopotamian marshlands, Iraq. *Remote Sensing, 10*(10), 1524.

Amante C, Eakins BW. ETOPO1 arc-minute global relief model: Procedures, data sources and analysis. 2009.

Bank, W. (2017). *Beyond scarcity: Water security in the Middle East and North Africa*. The World Bank.

Chen, J. L., Wilson, C. R., & Tapley, B. D. (2010). The 2009 exceptional Amazon flood and interannual terrestrial water storage change observed by GRACE. *Water Resource Research, 46*(12). https://doi.org/10.1029/2010wr009383

Chen, X., Jiang, J., & Li, H. (2018). Drought and flood monitoring of the Liao River basin in Northeast China using extended GRACE data. *Remote Sens-Basel, 10*(8), 1168. https://doi.org/10.3390/rs10081168

Cikurel, H., Guttman, J., & Aharoni, A. (2012). Managed aquifer recharge for agricultural reuse in Shafdan, Israel. In *Water reclamation technologies for safe managed aquifer recharge* (pp. 83–102). IWA Publishing.

Dawoud, M. A. (2020). Strategic water reserve using aquifer recharge with desalinated water in Abu Dhabi Emirate. *Desalination and Water Treatment, 176*, 123–130.

Doell, P., Mueller Schmied, H., Schuh, C., Portmann, F. T., & Eicker, A. (2014). Global-scale assessment of groundwater depletion and related groundwater abstractions: Combining hydrological modeling with information from well observations and GRACE satellites. *Water Resources Research, 50*(7), 5698–5720.

Döll, P., Kaspar, F., & Lehner, B. (2003). A global hydrological model for deriving water availability indicators: Model tuning and validation. *Journal of Hydrology, 270*(1–2), 105–134.

Döll, P., Fritsche, M., Eicker, A., & Müller, S. H. (2014). Seasonal water storage variations as impacted by water abstractions: Comparing the output of a global hydrological model with GRACE and GPS observations. *Surveys in Geophysics, 35*(6), 1311–1331. https://doi.org/10.1007/s10712-014-9282-2

Eicker, A., Schumacher, M., Kusche, J., Döll, P., & Schmied, H. M. (2014). Calibration/data assimilation approach for integrating GRACE data into the WaterGAP global hydrology model (WGHM) using an ensemble Kalman filter: First results. *Surveys in Geophysics, 35*(6), 1285–1309. https://doi.org/10.1007/s10712-014-9309-8

Frappart, F. (2020). Groundwater storage changes in the major North African transboundary aquifer systems during the GRACE Era (2003–2016). *Water, 12*(10), 2669.

Frappart, F., & Ramillien, G. (2018). Monitoring groundwater storage changes using the gravity recovery and climate experiment (GRACE) satellite Mission: A review. *Remote Sens-Basel, 10*(6), 829. https://doi.org/10.3390/rs10060829

Issa, I., Al-Ansari, N., Sherwany, G., & Knutsson, S. (2013). Trends and future challenges of water resources in the Tigris–Euphrates Rivers basin in Iraq. *Hydrology and Earth System Sciences Discussions, 10*(12), 14617–14644.

Joodaki, G., Wahr, J., & Swenson, S. (2014). Estimating the human contribution to groundwater depletion in the Middle East, from GRACE data, land surface models, and well observations. *Water Resources Research, 50*(3), 2679–2692. https://doi.org/10.1002/2013wr014633

Kelley, C. P., Mohtadi, S., Cane, M. A., Seager, R., & Kushnir, Y. (2015). Climate change in the Fertile Crescent and implications of the recent Syrian drought. *Proceedings of the National Academy of Sciences of the United States of America, 112*(11), 3241–3246. https://doi.org/10.1073/pnas.1421533112

Kim, K.-Y., & North, G. R. (1997). EOFs of harmonizable cyclostationary processes. *Journal of the Atmospheric Sciences, 54*(19), 2416–2427.

Kim, K.-Y., North, G. R., & Huang, J. (1996). EOFs of one-dimensional cyclostationary time series: Computations, examples, and stochastic modeling. *Journal of Atmospheric Sciences, 53*(7), 1007–1017.

Landerer, F. W., Flechtner, F. M., Save, H., Webb, F. H., Bandikova, T., Bertiger, W. I., et al. (2020). Extending the global mass change data record: GRACE Follow-On instrument and science data performance. *Geophysical Research Letters, 47*(12), e2020GL088306.

Longuevergne, L., Wilson, C. R., Scanlon, B. R., & Cretaux, J. F. (2013). GRACE water storage estimates for the Middle East and other regions with significant reservoir and lake storage. *Hydrology and Earth System Sciences, 17*, 4817–4830., doi:105194/hess-17-4817-2013, 2013.

Loomis, B., Rachlin, K., & Luthcke, S. (2019). Improved earth oblateness rate reveals increased ice sheet losses and mass-driven sea level rise. *Geophysical Research Letters, 46*(12), 6910–6917.

Maliva RG. Anthropogenic aquifer recharge and water quality. Anthropogenic aquifer recharge. Springer; 2020. p. 133–164.

Müller Schmied, H., Adam, L., Eisner, S., Fink, G., Flörke, M., Kim, H., et al. (2016). Impact of climate forcing uncertainty and human water use on global and continental water balance components. *Proceedings of the International Association of Hydrological Sciences, 374*, 53–62.

Observatory NE: Turkey Experiences Intense Drought. (2021). Accessed.

Parimalarenganayaki, S. (2020). Managed aquifer Recharge in the Gulf countries: A review and selection criteria. *Arabian Journal for Science and Engineering*, 1–15.

Peltier, R. W., Argus, D. F., & Drummond, R. (2018). Comment on "an assessment of the ICE-6G_C (VM5a) glacial isostatic adjustment model" by Purcell et al. *Journal of Geophysical Research: Solid Earth, 123*(2), 2019–2028. https://doi.org/10.1002/2016jb013844

Qadir, M., Bahri, A., Sato, T., & Al-Karadsheh, E. (2010). Wastewater production, treatment, and irrigation in Middle East and North Africa. *Irrigation and Drainage Systems, 24*(1), 37–51.

Rateb, A., & Kuo, C. Y. (2019). Quantifying vertical deformation in the Tigris-Euphrates Basin due to the groundwater abstraction: Insights from GRACE and Sentinel-1 satellites. *Water, 11*(8), 1658. https://doi.org/10.3390/w11081658

Rateb, A., Scanlon, B. R., Pool, D. R., Sun, A., Zhang, Z., Chen, J., et al. (2020a). Comparison of groundwater storage changes from GRACE satellites with monitoring and modeling of major U.S. aquifers. *Water Resource Research, 56*(12), e2020WR027556. https://doi.org/10.1029/2020WR027556

Rateb, A., Scanlon, B. R., & Chung-Yen, K. (2020b). Multi-decadal assessment of water resources in the Tigris-Euphrates Basin using satellites, hydrological modeling, and in-situ data science of the total environment. *The Science of the Total Environment*.

Reager, J. T., Thomas, B. F., & Famiglietti, J. S. (2014). River basin flood potential inferred using GRACE gravity observations at several months lead time. *Nature Geoscience, 7*(8), 588–592. https://doi.org/10.1038/ngeo2203

Rodell, M., Famiglietti, J. S., Wiese, D. N., Reager, J. T., Beaudoing, H. K., Landerer, F. W., et al. (2018). Emerging trends in global freshwater availability. *Nature, 557*(7707), 651–659. https://doi.org/10.1038/s41586-018-0123-1

Russo, T. A., & Lall, U. (2017). Depletion and response of deep groundwater to climate-induced pumping variability. *Nature Geoscience, 10*(2), 105–108. https://doi.org/10.1038/ngeo2883

Save, H., Bettadpur, S., & Tapley, B. D. (2016). High-resolution CSR GRACE RL05 mascons. *Journal of Geophysical Research: Solid Earth, 121*(10), 7547–7569. https://doi.org/10.1002/2016jb013007

Scanlon, B. R., Zhang, Z., Save, H., Wiese, D. N., Landerer, F. W., Long, D., et al. (2016). Global evaluation of new GRACE mascon products for hydrologic applications. *Water Resources Research, 52*(12), 9412–9429. https://doi.org/10.1002/2016wr019494

Scanlon, B. R., Zhang, Z., Save, H., Sun, A. Y., Muller Schmied, H., van Beek, L. P. H., et al. (2018). Global models underestimate large decadal declining and rising water storage trends relative to GRACE satellite data. *Proceedings of the National Academy of Sciences of the United States of America, 115*(6), E1080–E10E9. https://doi.org/10.1073/pnas.1704665115

Scanlon, B., Zhang, Z., Rateb, A., Sun, A., Wiese, D., Save, H., et al. (2019). Tracking seasonal fluctuations in land water storage using global models and GRACE satellites. *Geophysical Research Letters, 46*(10), 5254–5264.

Schmied, H. M., Eisner, S., Franz, D., Wattenbach, M., Portmann, F. T., Flörke, M., et al. (2014). Sensitivity of simulated global-scale freshwater fluxes and storages to input data, hydrological model structure, human water use and calibration. *Hydrology and Earth System Sciences, 18*(9), 3511–3538.

Stuyfzand, P. J., Smidt, E., Zuurbier, K. G., Hartog, N., & Dawoud, M. A. (2017). Observations and prediction of recovered quality of desalinated seawater in the strategic ASR project in Liwa, Abu Dhabi. *Water., 9*(3), 177.

Sun, Y., Riva, R., & Ditmar, P. (2016). Optimizing estimates of annual variations and trends in geocenter motion and J2 from a combination of GRACE data and geophysical models. *Journal of Geophysical Research: Solid Earth, 121*(11), 8352–8370.

Swenson, S., Chambers, D., & Wahr, J. (2008). Estimating geocenter variations from a combination of GRACE and ocean model output. *Journal of Geophysical Research: Solid Earth, 113*(B8). https://doi.org/10.1029/2007jb005338

Tapley, B. D., Bettadpur, S., Watkins, M., & Reigber, C. (2004). The gravity recovery and climate experiment: Mission overview and early results. *Geophysical Research Letters, 31*(9). https://doi.org/10.1029/2004gl019920

USDA. (n.d.). *Grain and Feed Annual Saudi Arabia*. https://apps.fas.usda.gov/newgainapi/api/report/downloadreportbyfilename?filename=Grain%20and%20Feed%20Annual_Riyadh_Saudi%20Arabia_3-14-2016.pdf2016. pp. 18.

Voss, K. A., Famiglietti, J. S., Lo, M., Linage, C., Rodell, M., & Swenson, S. C. (2013). Groundwater depletion in the Middle East from GRACE with implications for transboundary water management in the Tigris-Euphrates-Western Iran region. *Water Resources Research, 49*(2), 904–914. https://doi.org/10.1002/wrcr.20078

Wahr, J., Molenaar, M., & Bryan, F. (1998). Time variability of the Earth's gravity field: Hydrological and oceanic effects and their possible detection using GRACE. *Journal of Geophysical Research: Solid Earth, 103*(B12), 30205–30229. https://doi.org/10.1029/98jb02844

Watkins, M. M., Wiese, D. N., Yuan, D.-N., Boening, C., & Landerer, F. W. (2015). Improved methods for observing Earth's time variable mass distribution with GRACE using spherical cap mascons. *Journal of Geophysical Research: Solid Earth, 120*(4), 2648–2671. https://doi.org/10.1002/2014jb011547

Weedon, G. P., Balsamo, G., Bellouin, N., Gomes, S., Best, M. J., & Viterbo, P. (2014). The WFDEI meteorological forcing data set: WATCH forcing data methodology applied to ERA-interim reanalysis data. *Water Resources Research, 50*(9), 7505–7514.

WRI: Aqueduct Water Risk Atlas. (2019). Accessed 2021.

Xiao, R., He, X., Zhang, Y., Ferreira, V., & Chang, L. (2015). Monitoring groundwater variations from satellite Gravimetry and hydrological models: A comparison with in-situ measurements in the Mid-Atlantic region of the United States. *Remote Sens-Basel, 7*(1), 686–703. https://doi.org/10.3390/rs70100686

Zekri, S. (2020). *Water policies in MENA countries*. Springer.

Chapter 5
The Application of Remote Sensing on the Studies of Mean Sea Level Rise in the Arabian Gulf

Nada Abdulraheem Siddig, Abdullah Mohammed Al-Subhi, and Mohammed Ali Alsaafani

Abstract Mean sea level (MSL) trend at the Arabian Gulf has been estimated based on hourly tide gauge (TG) data of seven stations at the west of the gulf (1979–2008) and multi-missions satellite altimetry monthly mean (1993–2018). Analysis exposes that MSL is rising due to global warming. Altimetry data reveals a global rising trend by about 2.8 ± 0.4 mm/year while for the Arabian Gulf, trend estimation shows higher rate by about 3.6 ± 0.4 mm/year. This value almost in agreement with previous trend estimations for the gulf by many researchers and trend values in adjacent seas such as the Red Sea and Gulf of Aden. Based on TG hourly values, sea level trend is also showing a rising trend at all stations with variable rates. For example, at Mina Salman the trend value is about 3.4 ± 0.98 mm/year which agrees with the above estimate from the altimetry data followed by values from Arrabiyah Island station; 2.4 ± 0.66 mm/year. However, not all stations reflect the same MSL trend rising rates; for example, Ras Tanura recorded the lowest value of trend followed by Jubail station by about 0.7 ± 0.31 mm/year and 1.6 ± 0.71 mm/year respectively.

Keywords Remote sensing · Mean sea level rise · Arabian gulf

5.1 Introduction

Marine studies gained countless benefits from remote sensing in the past four decades. To illustrate, these benefits include but not limited; coastal applications programs such as tracking sediment and erosion prevention. Furthermore, another advantage is to ocean applications where ocean dynamics can be tracked and modelled. This includes ocean circulation, climate studies, and tide and sea level fluctuations. Moreover, additional influence lies on the hazard assessments such as

N. A. Siddig (✉) · A. M. Al-Subhi · M. A. Alsaafani
Department of Marine Physics, Faculty of Marine Sciences, King Abdulaziz University, Jeddah, Saudi Arabia
e-mail: nsiddique@stu.kau.edu.sa

M. M. Al Saud (ed.), *Applications of Space Techniques on the Natural Hazards in the MENA Region*, https://doi.org/10.1007/978-3-030-88874-9_5

oil pollution. Of course, the natural marine resources have earned well assets from remote sensing such as coral reef health monitoring programs. One of the strong points to oceanographers is the advent of satellite altimetry with the Skylab in the mid-1970s. Satellite altimetry has accomplished revolutionary bounce in the advancement of ocean studies and those related to the climate. Climate changes, as a result of global warming, has raised the sea level in a dramatic rhythm affecting millions of lives in the global coastal areas. Sea level rise has become a concerning issue to almost all countries due to the increasing threat to their coastal areas and hence development. Satellite altimetry precisely measures the topography of the global sea surface in a short revisit time.

It is well known that oceans are the dynamic of the earth's climate. The most significant roles are played by the Pacific Ocean since it represents more than one-third of the surface of our globe and nearly half the area of its oceans. However, the ocean and seas get affected by climate change in two ways; direct and indirect. The direct effect is by the increasing of the water volume by melting of snow and ice and consequently, the sea level rises. When the ocean got more heat, its water column expands and hence, the sea level rises; this is an indirect effect of climate change and for oceanographers it is known as Steric Sea Level. For this reason, the projection of sea level rise is becoming the most concern for many countries and agencies. Before the era of satellites sea level data was collected merely from tide gauges distributed around almost all coasts. Lately these data were plugged in into hydrodynamic statistical models to figure out the future sea level rise projections. However, these models have so many shortage and limitations due to the absence of data in some parts of the globe. After the satellite altimetry data gained its wide coverage and reliability starting early 90s with TOPEX/Poseidon (T/P) revolutionary mission both climate and sea level rise studies got a lot of strength and scientific significant.

Satellite altimeter is an active microwave sensor measures the time took a radar pulse to hit the sea surface and reflects back to the receiving altimeter antenna in order to estimate SSH (Fig. 5.1). Thus, the SSH equals to the satellite orbit (O) which represents its height above the earth's ellipsoid after subtracting the satellite corrected range (R) which represents its height, above the sea surface:

$$\mathrm{SSH} = \mathrm{O} - \mathrm{R}_{\mathrm{corrected}}$$

Like many other satellites remote sensing, altimeters face difficulties and challenges. To illustrate, the troposphere and ionosphere distort microwave signals and attenuate their quality and hence required some technical corrections. The other difficulties arise due to complicated conditions of coastal areas, including their bottom topography, water dynamics and land proximity (Taqi et al., 2020). These costal difficulties prevent the extracting of SLA information directly from the microwave waveform within the first 50 km of the coasts (Taqi et al., 2017). For the above reasons, satellite altimetry data must go through many corrections for the atmospheric attenuations and land proximities. However, the land proximities for semi-enclosed seas and gulfs remains one of the big challenges that always prevent the full

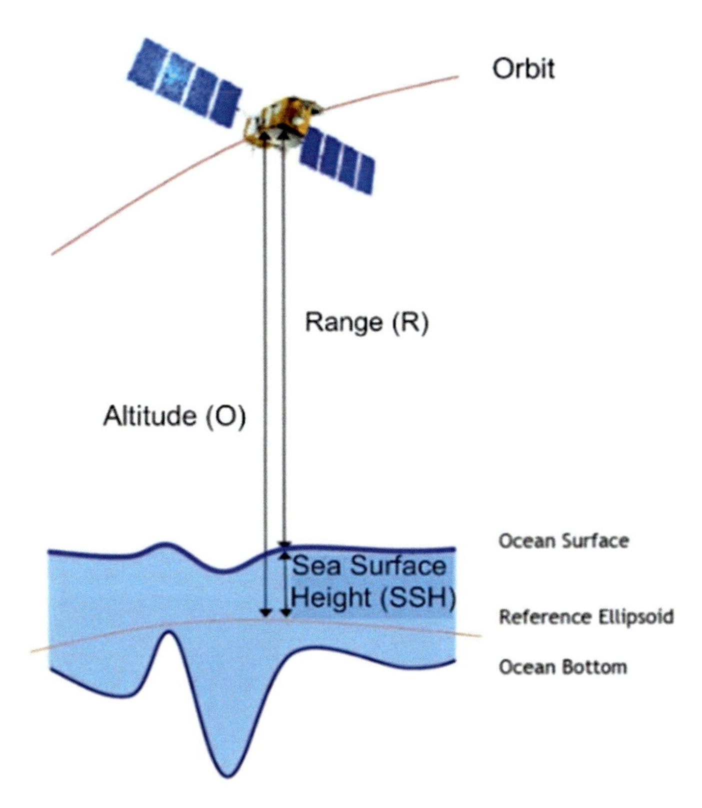

Fig. 5.1 Satellite altimetry estimates Sea Surface Height (SSH) by measuring travelling time of radar pulse to sea surface and back to antenna. Modified version of the image from the Archiving, Validation, and Interpretation of Satellite Oceanographic Data (AVISO) at: http://www.aviso.altimetry.fr

use of altimeter data in these areas until recently when these issues have been fully resolved. (Taqi et al., 2020) have successfully improved SLA precision using a technique called the improved "Fourier series model (FSM01)" method in the Red Sea which in turn can be applied in any coastal areas around the world.

The Arabian Gulf (AG), is a semi-enclosed marginal sea covered a total area of about 239 × 103 km^2 with an average depth of about 36 m (Emery, 1956). Coastal areas in the northwest and the west are shallow. The average length of the AG is 990 km (Fig. 5.2). The main water exchange between the AG and the Indian Ocean is through the Strait of Hormuz. The wind (Shamal) blowing from north and northwest in the AG, that blow through winter and summer, is characterized by strong wind speeds during winter due to high atmospheric pressure disturbances and by a relatively lower intensity during summer (Perrone, 1979). The wind speed at the coast reaches as high as 15 ms^{-1} (Reynolds, 1993). The annual evaporation over the

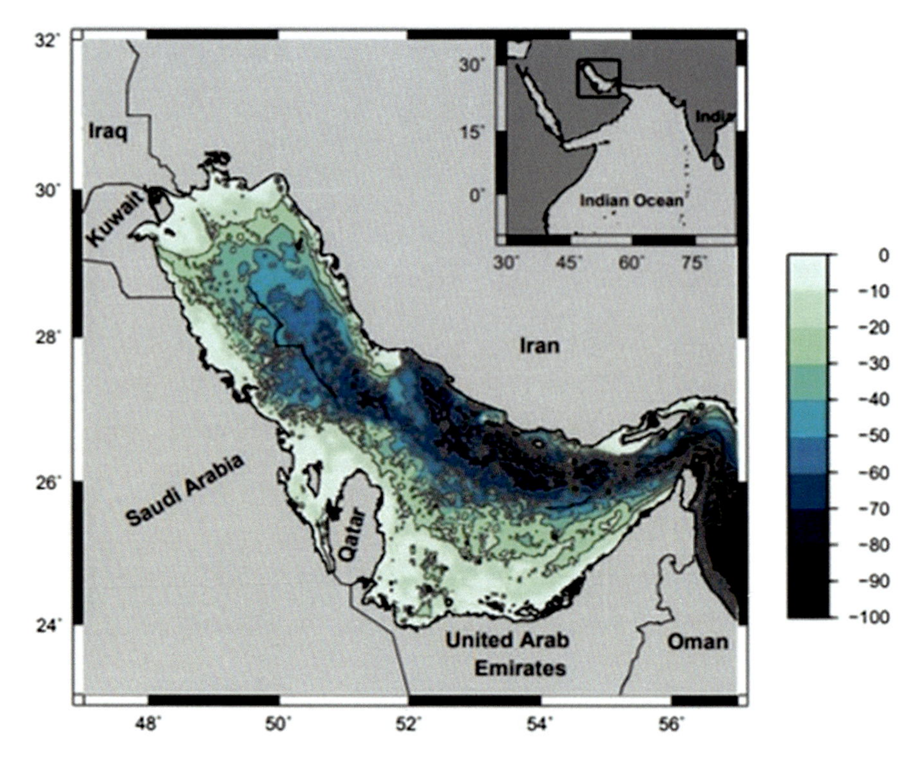

Fig. 5.2 A map of the study area shows the name and location of the stations

AG is about 2 m/year (Ahmad & Sultan, 1991; Hastenrath & Lamb, 1979; Meshal & Hassan, 1986; Privett, 1959; Xue & Eltahir, 2015), while fresh water input by precipitation is ~0.15 m /year (Johns et al., 2003). The main source of fresh water was occurred mostly in the northern end of the AG, through the Shatt Al Arab river by convergences the Euphrates, Tigris and Karun rivers in it. However, the discharge of rivers is very small compared to evaporation.

Tides in the AG are complex, and the major tide is varying in nature from being semi-diurnal, diurnal, and mixed type (Reynolds, 1993). Semi diurnal tides have two amphidromical points in the north-west and south, while the diurnal tide has one amphidromical point in the center of the AG, near the Kingdom of Bahrain. It also shows that the primary constituents are M2, S2, K1, and O1 (Najafi, 1997). The tidal propagation in the AG basin is counterclockwise from the Iranian coast north to the Saudi Arabia coast south.

The sea level variations in the west and northwest coasts of the AG has been the focus of the researchers (Al-Subhi, 2010; Alothman et al., 2014; El Din, 1990; Khalilabadi & Mansouri, 2013; Reynolds, 1993; Sultan et al., 2000) calculated the meteorological effects causing (up to 75%) the seasonal signals of mean sea level in the AG, out of which the atmospheric pressure is contributed by 62% and wind stress by 12%.

Since 1992, high quality satellite altimeters (TOPEX/Poseidon, ERS-2, GFO, Jason-1, Envisat, Jason-2 and Jason-3) lead to accurate estimates of the sea level rise in global measurements. Gornitz (1995) estimated an increasing trend of the global sea level non-satellite records to be 1–2 mm/year. Meanwhile, satellite altimetry data shows an increase around 3 mm/year (Antonov et al., 2005; Bindoff et al., 2007). However, (Church et al., 2008) found that the sea level trend from both TG and satellite altimetry data show sea level is rising by more than 3 mm/year. In the Arabian Gulf (AG), Sultan et al. (2000) found that the sea level trend at Ras Tanura during 1980 and 1994 is rising by 1.70 mm/year. Moreover, (Alothman & Ayhan, 2010) analysed sea level data in 13 stations in the north-western coast of the AG and found a relative rise of about 1.96 mm/year by correcting the vertical land motion. Later study conducted by (Alothman et al., 2014) in the same part of the AG found a trend of 2.20 mm/year and after correcting this value for vertical land motion the trend became 1.50 mm/year.

Strait of Hormuz, with great economic importance (industry, commerce, and oil) and marine life. AG is an important area being an extension of the Indian Ocean across the.

The aim of this study is to investigate sea level trend from seven stations on the west coast of the AG from 1979 to 2008. These data serve as a ground truth and validation for all satellite altimetric sea level data available from (1993–2018). It is important to mention that most of the results presented in this chapter have been published in (Siddig et al., 2019) and re-presented here with the permission of the publisher.

5.2 Data and Methods

The National Oceanic and Atmospheric Administration (NOAA) was the main source of the satellite altimetry data which extracted for the period between 1993 and 2018 from official website https://www.star.nesdis.noaa.gov/socd/lsa/SeaLevelRise/LSA_SLR_timeseries.php. The TG data of six selected stations along the Arabian Gulf, Saudi Arabian coast was obtained from Saudi Aramco Company. While the Permanent Service for Mean Sea Level (PSMSL) https://www.psmsl.org/data/obtaining/map.html was used for the data of the Mina Salman, Kingdom of Bahrain. Details of these stations, including the data duration, names and coordinate is listed in (Table 5.1). The study was done within the longest recorded period of 29 years conducted for Mina Salman, Jubail and Ras Tanura stations and the shortest recorded period in this study is 9 years for the Abu Ali Pier station.

In order to extract residual sea level, which is caused by forces other than tide mainly the meteorological forces, from the observed TG records, the World Tide MATLAB Software (WTWC) (Boon, 2004), has been used. The software applies a selective, least squares harmonic analysis to identify the tidal constituents and predict tides and tidal currents using a total of 35 tidal constituents.

Table 5.1 The location of TG stations in the western coast of the Arabian Gulf

NO	Station	Longitude	Latitude	Period
1	Murjan Island	49.63°	28.45°	1986–2008
2	Arrabiyah Island	50.17°	27.77°	1985–2000
3	Abu Ali pier	49.68°	27.31°	2000–2008
4	Jubail	49.91°	26.86°	1980–2008
5	Ras Tanura	50.16°	26.64°	1980–2008
6	Qurayyah pier	50.11°	25.88°	1980–2000
7	Mina Salman	50.61°	26.20°	1979–2007

Table 5.2 The highest astronomical tide (HAT) and lowest astronomical tide (LAT) values for each station along with the period length of data recorded

Station	HAT(m)	LAT(m)	Data recording time (day)
Murjan Island	0.56	−0.44	8187
Arrabiyah Island	0.92	−0.67	5556
Abu Ali pier	0.88	−0.40	3110
Jubail	1.00	−1.01	10,074
Ras Tanura	1.13	−1.13	10,440
Qurayyah pier	0.15	−0.19	7649
Min Salman	1.02	−1.10	10,535

By mathematical application to analysis the predict tidal, we obtain the equation.

$$h(t) = h_0 + \sum_{j=1}^{m} f_j H_J \cos\left(\omega_j t + u_j - k_j^*\right) \tag{5.1}$$

where, t = time in serial hours;

h(t) = predicted water level at t;
h0 = mean water level;
fj = lunar node factor for jth constituent;
Hj = mean amplitude for jth constituent over 18.6-year lunar node cycle;
ωj = frequency of jth constituent;
uj = nodal phase for jth constituent;
κj* = phase of jth constituent for the period origin is utilized (midnight beginning December 31, 1899) and m = number of constituents.

For purely solar constituents, fj = 1 and uj = 0.

The monthly mean sea level elevations were determined by analysing the long-term TG data as illustrated in (Table 5.2). The length of the recorded data varies from station to station with some short gaps in a few stations while at Min Salman, the station has seven-year gaps in 1981 and 1998–2003.

In order to estimate the linear trend from the residual sea level data, first the seasonal effect has to be eliminated through calculation of the monthly averaged residual sea level. Statistical testing is carried out on the time series for the trend significance prior to fitting the linear model using Mann-Kendall method, which tests whether to reject the null hypothesis (H0, no trend) or the alternative hypothesis (Ha, if trend is present), is to be accepted.

The trend is fitted using the Least-Square Line (LSL) method that is often used in the approximation of the general pattern of time series over the selected period (Crum, 1925; Hoshmand, 1997).

The linear equation in general

$$y = a + bx(i) \tag{5.2}$$

Where the a and b, can be expressed as follows:

$$a = \frac{y}{c} \quad \text{where } c = \text{length of data} \tag{5.3}$$

$$x_m = \sum_{i=1}^{N} x_i - \widetilde{x} \tag{5.4}$$

The sum of square coefficient of the element x

$$X^2 = \sum_{i=1}^{N} x_{m(i)}^2 \tag{5.5}$$

Write the sum of y

$$y = \sum_{i=1}^{N} y_i \tag{5.6}$$

We may rewrite these equations as

$$b = \frac{XY}{X^2} \tag{5.7}$$

In order to estimate the value of the linear trend, Eq. (5.3) and (5.7) are substituted in Eq. (5.2).

The standard error is calculated by dividing the standard deviation from the mean as

$$se = \frac{\sqrt{\frac{1}{N-1} \sum_{i=1}^{N} (x_i - x_m)^2}}{\sqrt{N}} \tag{5.8}$$

The XLSTAT software (http://www.xlstat.com/en/) is used for calculating the probability value (p-value), and hypothesis testing. The Mann-Kendall test was applied to assess the significance of mean sea level trend, as follows.

$$S = \sum_{k=1}^{n-1} \sum_{j=k+1}^{n} sgn\left(X_j - X_k\right) \tag{5.9}$$

Where, the data collected over time = X1, X2, X3.......Xn.

(X_j -X_k) is the difference between current values and all previous values, where j > k, that takes on the values 1, 0, or − 1.

For the satellite altimetry data, the sea level linear trend was determined from the merged sea level from all altimeter data. The data filtered to approximately 10-days' time interval of 1° x 1° grids with seasonal signals removed. The final stage involved determination of the monthly mean sea level for Arabian Gulf area as well as that of the global coverage which lies on the latitudes 66°S to 66°N.

5.3 Results and Discussion

As shown in Table 5.2, the HAT and LAT indicate that there are high variations of the tidal range between the stations over the study period. To illustrate, the highest range is noted in the station Min Salman, Ras Tanura and Jubail at 2.00 m and the lowest range is Qurayyah pier at about 0.34 m. the recorded TG time series of all stations are plotted in (Fig. 5.3). From the figure it is clear that Jubail and Ras Tanura station has the longest continuous recorded data while the lowest record is in Abu Ali pier.

5.3.1 *The Linear Trend Analysis*

The linear trends for AG and global oceans have been estimated from the entire satellite altimetry monthly mean time series as shown in (Figs. 5.4 and 5.5). The linear trend for AG and global oceans is 3.6 ± 0.4 mm/year and 2.8 ± 0.4 mm/year respectively with the significant values of P = 0.0001. The TG monthly mean sea level show significant positive trends with P value <0.05, for all stations except at Mina Salman and Arrabiyah Island. The inconsistencies in these two stations are due to large data gaps in the time series.

Figure 5.6 show the monthly mean sea level with fitted trend at all the TG station. Mina Salman, however, has a discontinuity in the record for about 6 years. The trend analysis for the uninterrupted period (1982–1997) shows a significant trend with P value = 0.0001. Similarly, at Arrabiyah Island after removing the periods with severe gaps, the trend analysis (during 1990–2000) shows significant positive

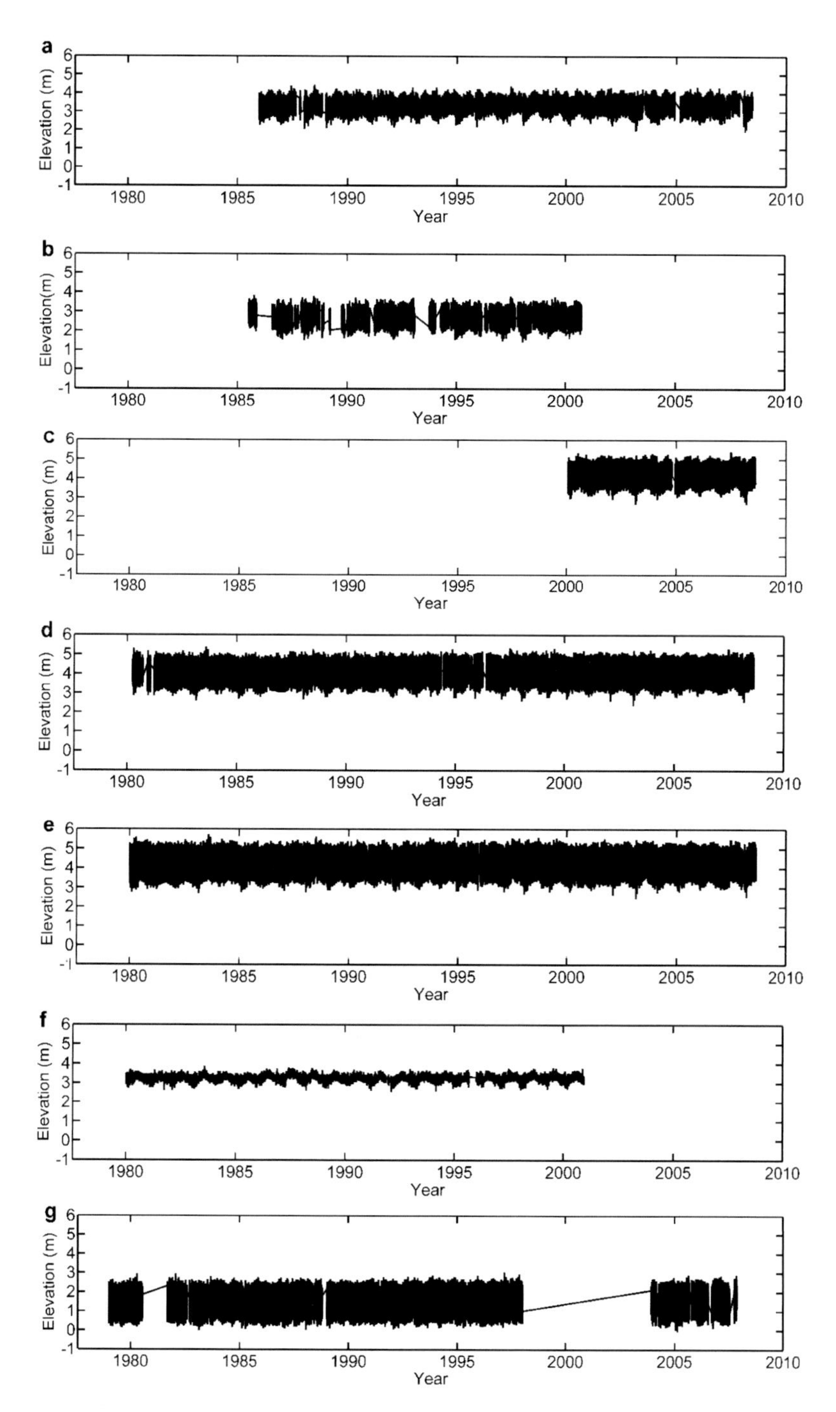

Fig. 5.3 Time series of the records for seven stations. (**a**) Murjan Island, (**b**) Arrabiyah Island, (**c**) Abu Ali, (**d**) Jubail, (**e**) Ras Tanura, (**f**) Qurayyah, (**g**) Mina Salman.

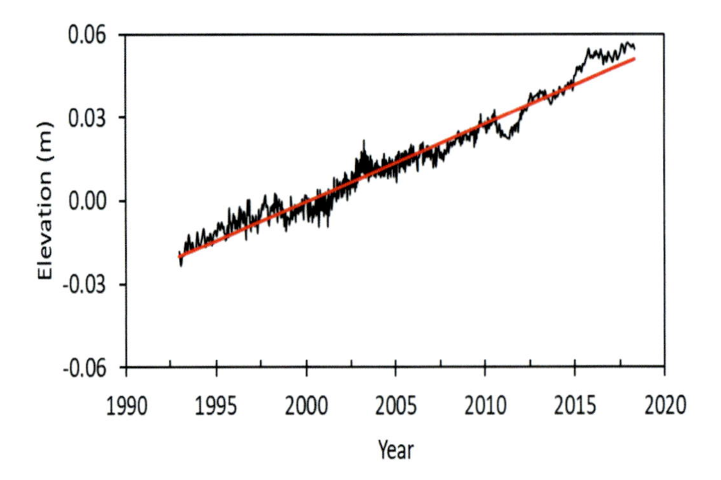

Fig. 5.4 Global oceans mean sea level trend 2.8 mm/year, from multi-mission satellite altimetry

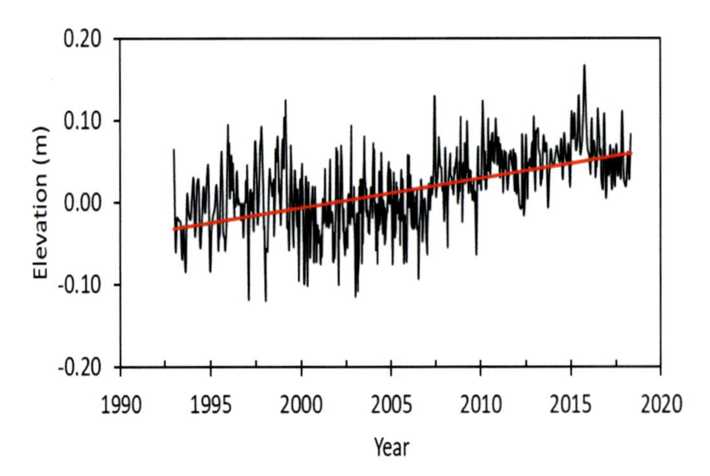

Fig. 5.5 Arabian Gulf mean sea level trend 3.6 mm/year, from multi-mission satellite altimetry

trend. The highest trend of the monthly sea level can be noted at the stations Abu Ali Pier and Mina Salman with the values being 3.4 ± 0.98 mm/year and 3.1 ± 0.7 mm/year respectively. While at Ras Tanura and Jubail stations the trend values is 0.7 ± 0.31 mm/year and 1.6 ± 0.71 mm/year which considered the lowest trends among all station. The other stations recorded their trends as illustrated:

- Qurayyah Pier station =2.2 ± 0.82 *mm/yrar*
- Arrabiyah Island= 2.4 ± 0.66 *mm/year*
- Murjan Island station= 2.4 ± 0.94 *mm/year*

Table 5.3 list all the estimated trends in this study as well as in the previous studies for an inter-comparison of values, however, the data duration may vary among the

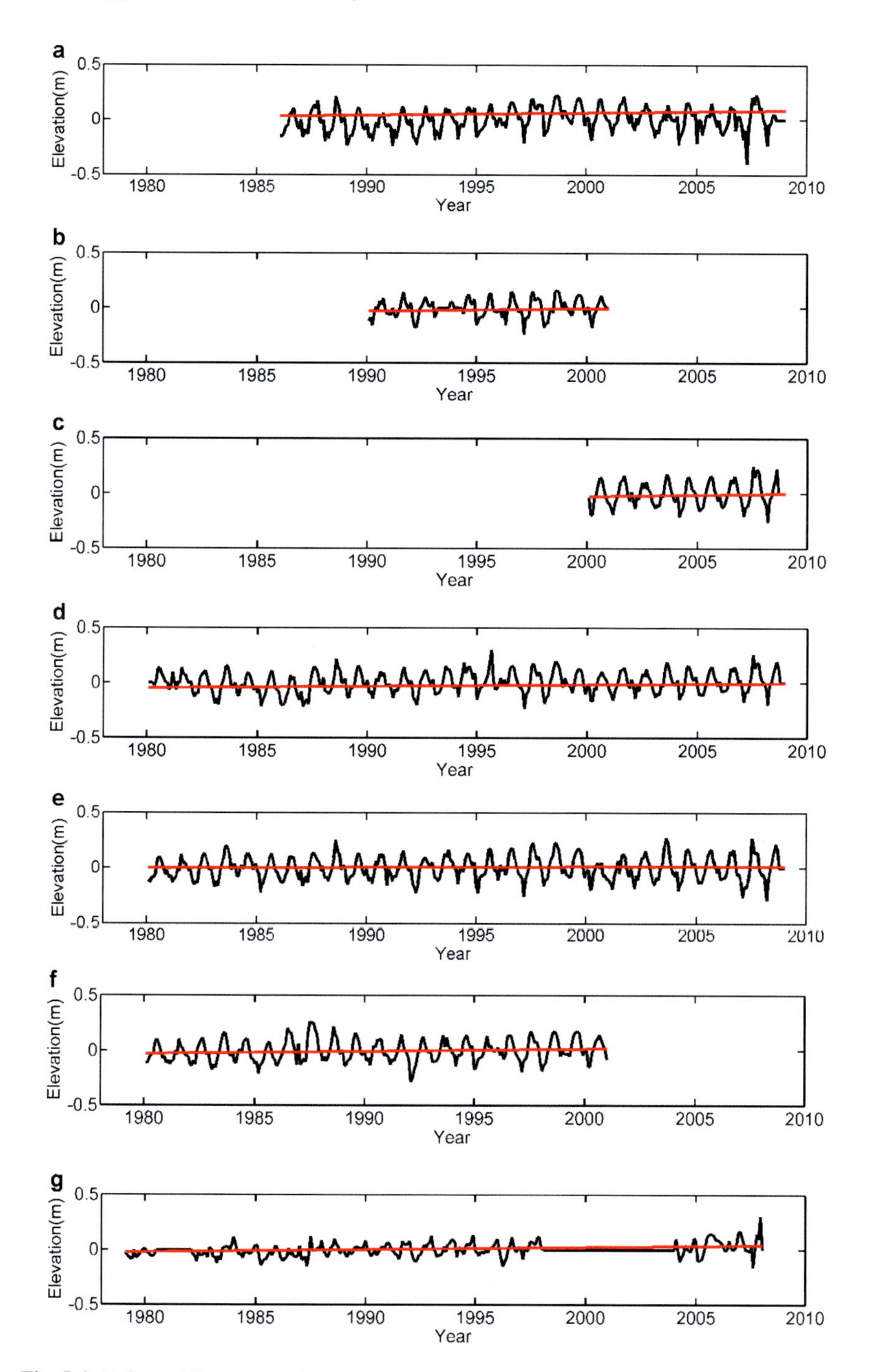

Fig. 5.6 Estimated linear trend for TG stations. (**a**) Murjan Island, (**b**) Arrabiyah Island, (**c**) Abu Ali, (**d**) Jubail, (**e**) Ras Tanura, (**f**) Qurayyah, (**g**) Mina Salman

Table 5.3 Estimated mean sea level trend rates for north-western Arabian Gulf area as compared with the previous estimates

Stations	This study			Hosseinibalam et al., 2007		Alothman & Ayhan, 2010		Alothman et al., 2014	
	Estimate trend mm/year	P-value	Period	Estimate trend mm/year	Period	Estimate trend mm/year	Period	Estimate trend mm/year	Period
Murjan Island	2.4 ± 0.94	0.0001	1986–2008	9.75 ± 0.15	1990–2000	9.37 ± 2.02	1986–2001	7.05 ± 1.17	1986–2001
Arrabiyah Island	2.4 ± 0.66	0.032	1990–2000			−4.15 ± 3.52	1985–1998	−0.33 ± 0.18	1985–1998
Abu Ali pier	3.1 ± 0.70	0.000	2000–2008	4.5 ± 0.04	1990–2000	1.74 ± 1.14	1980–2001	1.18 ± 0.63	1980–2001
Jubail	1.6 ± 0.71	0.002	1980–2008						
Ras Tanura	0.7 ± 0.31	0.015	1980–2008	0.84 ± 0.03	1990–2000	1.85 ± 1.05	1980–2001	0.74 ± 1.11	1980–2001
Qurayyah pier	2.2 ± 0.84	0.001	1980–2000			3.29 ± 1.35	1980–1998	2 ± 0.99	1980–1998
Min Salman	3.4 ± 0.98	0.0001	1979–2007			3.22 ± 0.58	1979–2007	2.97 ± 0.51	1979–2007

studies. In the present study, the estimated trend at Murjan Island station is in the same range with that of the rest of stations in AG, even though it is much less than the previous estimates of (Alothman et al., 2014; Alothman & Ayhan, 2010; Hosseinibalam et al., 2007). There are clear decadal signals in our estimates with 23 years of data (Fig. 5.6). This may explain the reason for the inconsistency in the trend estimation between the present study and previous studies, where the short data duration in the previous studies (11–15 years) may overestimate the trend values within the decadal signal. For confirmation, the trends were re-estimated in same periods of the previous studies and the results show similarly high values (Figure not shown). At Abu Ali Pier, the estimated trend is 3.1 mm/year, which in between the available estimates of the previous studies (Alothman et al., 2014; Alothman & Ayhan, 2010; Hosseinibalam et al., 2007) (Table 5.3). At Qurayyah Pier, Ras Tanura and Mina Salman, the estimated trends agrees with that of Alothman et al. (2014) for all stations and with Hosseinibalam et al. (2007) for Ras Tanura and with Alothman and Ayhan (2010) for Mina Salman. At Arrabiyah Island, the analysis shows increasing trend, which is in contradictory with previous estimate, where they reported a decreasing trend (Alothman et al., 2014; Alothman & Ayhan, 2010). Alothman et al. (2014) related that decrease to human activities in that area and the existence of oil platforms near the station. It is clear from our findings that the gaps in data records significantly affect the estimated trends. A rough estimate of trends by incorporating data with gaps of the same period of that of previous studies, leads to a decreasing trend in this region. In present study, period contains gap is excluded and only the data of minimal discontinuity is used (1990 to 2000) in the trend estimation. The mean trend value for all the stations is $\approx$ 2.3 mm/year in the AG.

5.4 Conclusion

In this study, seven TG stations on the west coast of the AG have been analyzed. The highest tidal range is recorded at Ras Tanura with 2.26 m and the minimum tidal range is seen at southern coastal station (Qurayyah Pier) with 0.34 m (Table 5.2). Based on satellite data for the period from 1993 to 2018, the trend was estimated for global oceans and for AG region with about 2.8 + 0.4 mm/year, and about 3.6 $\pm$ 0.4 mm/year respectively. The trend of the AG is higher than that of the global ocean, which is expected in semi-enclosed seas and regional gulfs. The monthly mean residual sea level is used for the estimation of the linear trend at all stations. Mina Salman station show the highest trend value with about 3.4 $\pm$ 0.98 mm/year, while at Abu Ali Pier the estimated trend is about 3.1 $\pm$ 0.7 mm/year. The estimated trend at Arrabiyah Island, Murjan Island and Qurayyah Pier stations show similar values with about 2.4 $\pm$ 0.66 mm/year, 2.4 $\pm$ 0.94 mm/year, and 2.2 $\pm$ 0.84 mm/year respectively. Lower trends have been estimated at Jubail and Ras Tanura stations with about 1.6 $\pm$ 0.71 mm/year and 0.7 $\pm$ 0.31 mm/year respectively. The average linear trend for all seven stations is about 2.3 mm/year. The present study shows the trend estimates for Jubail station from 29 years (1980–2008) for the first time. Similarly, at Arrabiyah Island the present study shows positive trend, which agrees with all other stations, while all previous studies show negative trend at the same station. The main reason for the negative trends in the previous studies was due to inclusion of high variability data (the period from 1985 to 1989) having lots of gaps. At Murjan Island, the longer duration data (23 years) produced good trend estimates, which agrees with that of the other stations. The previous studies show very high trend at this station, which is mainly due shorter data record they analysed. The increasing trends in the AG indicate that the gulf is responding positively to the global warming phenomenon. However, this response will have its impact on several environmental issues such as coral reef which already experience high vulnerability of bleaching. On the other hand, coastal communities will be affected with sea level rising problem due to global warming.

Acknowledgments The authors would like to thank ARAMCO for providing the data. Also, recognitions for Permanent Service for Mean Sea Level (PSMSL) data, available at URL: http://www.psmsl.org/data/obtaining/map.html, and (NOAA) National Oceanic and Atmospheric Administration data, available at URL: https://www.star.nesdis.noaa.gov/sod/lsa/SeaLevelRise/LSA_SLR_maps.php

References

Ahmad, F., & Sultan, S. A. R. (1991). Annual mean surface heat fluxes in the arabian gulf and the net heat transport through the strait of hormuz. *Atmosphere – Ocean, 29*, 54–61. https://doi.org/10.1080/07055900.1991.9649392

Alothman, A. O., & Ayhan, M. E. (2010). *Detection of sea level rise within the Arabian Gulf using space based GNSS measurements and insitu tide gauge data: Preliminary results*. CHANGE, Impacts, Vulnerability & Adaptation. Environment Agency-Abu Dhabi.

Alothman, A. O., Bos, M. S., Fernandes, R. M. S., & Ayhan, M. E. (2014). Sea level rise in the north-western part of the Arabian Gulf. *Journal of Geodynamics, 81*, 105–110. https://doi.org/10.1016/j.jog.2014.09.002

Al-Subhi, A. M. (2010). Tide and sea level characteristics at Juaymah, west coast of the Arabian gulf. *Marine Scienes, 21*, 133–149. https://doi.org/10.4197/Mar.21-1.8

Antonov, J. I., Levitus, S., & Boyer, T. P. (2005). Thermosteric Sea level rise, 1955-2003. *Geophysical Research Letters, 32*, 1–4. https://doi.org/10.1029/2005GL023112

Bindoff, N. L., Willebrand, J., Artale, V., Cazenave, A., Gregory, J. M., Gulev, S., Hanawa, K., Le Quéré, C., Levitus, S., Nojiri, Y., & others. (2007). *Observations: oceanic climate change and sea level, in: Climate Change 2007: The physical science basis: Contribution of Working Group I to the fourth assessment report of the intergovernmental panel on climate change* (p. 996). Cambridge University Press.

Boon, J. (2004). *Secrets of the tide: Tide and tidal current analysis and applications, storm surges and sea level trends (Marine Science)* (p. 224). Horwood Publishing Limited. https://doi.org/10.1016/C2013-0-18114-7

Church, J. A., White, N. J., Aarup, T., Wilson, W. S., Woodworth, P. L., Domingues, C. M., Hunter, J. R., & Lambeck, K. (2008). Understanding global sea levels: Past, present and future. *Sustainability Science, 3*, 9–22. https://doi.org/10.1007/s11625-008-0042-4

Crum, W. L. (1925). The least squares criterion for trend lines. *Journal of the American Statistical Association, 20*, 211–222.

El Din, S. H. S. (1990). *Sea level variation along the western coast of the Arabian Gulf.* The International Hydrographic Review.
Emery, K. O. (1956). Sediments and water of Persian gulf. *AAPG Bulletin, 40*, 2354–2383. https://doi.org/10.1306/5CEAE595-16BB-11D7-8645000102C1865D
Gornitz, V. (1995). Monitoring sea level changes. *Climatic Change, 31*, 515–544. https://doi.org/10.1007/BF01095160
Hastenrath, S., & Lamb, P. J. (1979). *Climatic atlas of the Indian Ocean. Part II: The oceanic heat budget.* The University of Wisconsin Press.
Hoshmand, R. (1997). *Statistical methods for environmental and agricultural sciences* (2nd ed.). CRC press. https://doi.org/10.1201/9780203738573
Hosseinibalam, F., Hassanzadeh, S., & Kiasatpour, A. (2007). Interannual variability and seasonal contribution of thermal expansion to sea level in the Persian Gulf. *Deep-Sea Research Part I: Oceanographic Research Papers, 54*, 1474–1485. https://doi.org/10.1016/j.dsr.2007.05.005
Johns, W. E., Yao, F., & Olson, D. B. (2003). Observations of seasonal exchange through the Straits of Hormuz and the inferred heat and freshwater budgets of the Persian Gulf. *Journal of Geophysical Research, 108*. https://doi.org/10.1029/2003JC001881
Khalilabadi, M. R., & Mansouri, D. (2013). Effect of super cyclone "GONU" on sea level variation along Iranian coastlines. *Indian Journal of Marine Sciences, 42*, 470–475.
Meshal, A. H., & Hassan, H. M. (1986). Evaporation from the coastal water of the central part of the Gulf. *Arab Gulf Journal of Scientific Research, 4*, 649–655.
Najafi, H. S. (1997). Modelling tides in the Persian Gulf using dynamic nesting. *University of Adelaide*. https://doi.org/10.1142/9789814350730_0003
Perrone, T. J. (1979). *Winter Shamal in the Persian Gulf, Naval environmental prediction research facility,Monterey*. California, Technical Report, IR-79-06.
Privett, D. W. (1959). Monthly charts of evaporation from the N. Indian Ocean (including the Red Sea and the Persian Gulf). *Quarterly Journal of the Royal Meteorological Society, 85*, 424–428. https://doi.org/10.1002/qj.49708536614
Reynolds, R. M. (1993). Physical oceanography of the Gulf, Strait of Hormuz, and the Gulf of Oman-Results from the Mt Mitchell expedition. *Marine Pollution Bulletin, 27*, 35–59. https://doi.org/10.1016/0025-326X(93)90007-7
Siddig, N. A., Al-Subhi, A. M., & Alsaafani, M. A. (2019). Tide and mean sea level trend in the west coast of the Arabian Gulf from tide gauges and multi-missions satellite altimeter. *Oceanologia, 61*, 401–411. https://doi.org/10.1016/j.oceano.2019.05.003
Sultan, S. A. R., Moamar, M. O., El-Ghribi, N. M., & Williams, R. (2000). Sea level changes along the Saudi coast of the Arabian Gulf. *Indian Journal of Marine Sciences, 29*, 191–200.
Taqi, A. M., Al-Subhi, A. M., & Alsaafani, M. A. (2017). Extension of satellite altimetry Jason-2 sea level anomalies towards the Red Sea coast using polynomial harmonic techniques. *Marine Geodesy, 40*, 315–328. https://doi.org/10.1080/01490419.2017.1333549
Taqi, A. M., Al-Subhi, A. M., Alsaafani, M. A., & Abdulla, C. P. (2020). Improving sea level anomaly precision from satellite altimetry using parameter correction in the Red sea. *Remote Sensing*, 12. https://doi.org/10.3390/rs12050764
Xue, P., & Eltahir, E. A. B. (2015). Estimation of the heat and water budgets of the Persian (Arabian) gulf using a regional climate model. *Journal of Climate, 28*, 5041–5062. https://doi.org/10.1175/JCLI-D-14-00189.1

Chapter 6
Assessment of Vulnerability, Risk, and Adaptation of MENA Region to SLR by Remote Sensing and GIS

M. E. El Raey

Abstract An assessment of sea-level rise to the vulnerability of impacts, based on the UNFCCC national communications carried out by satellites for the countries: Egypt, Libya, Saudi Arabia, Jordan, Tunisia Algeria, and Morocco is carried out to illustrate conditions after 30 years from now, based on comparison of scenarios We will introduce some outlines of research based on vulnerability measurements and mapping, and resilience. The measurements of the vulnerability of each country are determined by the changes in temperature, precipitation, socioeconomic parameters, and increase of hazards due to climate changes that occur. The risk associated with climatic changes is also discussed, and ground-based adaptation needed are explored for each country. Development of institutional capabilities for ICZM and upgrading awareness are highly recommended for adaptation in the long run. Periodic nourishment of Alexandria and Rosetta beaches, detached breakwaters for Alexandria, Port-Said, and dune fixation are the recommended for no regrets management policy. A program for raising resilience and awareness is necessary for all.

6.1 Origin and Causes of Climate Change

The Earth's climate has changed in the past for a variety of reasons and several times, for example changing the relationship between land and water (continents and oceans) and changing the intensity of radiation or solar energy, and change in Earth orbit and volcanic eruptions. For example, average global temperatures were higher than today, where water was estimated to be above its current level by several milli- meters (IPCC, 2012). Temperatures also dropped by about 5 Celsius Degrees in the ice ages, which were about 100,000 years old.

Scientists estimated that about 20,000 years ago, the sea level was about 120 meters below its current level because the water was trapped in polar ice sheets. The last 8000 years, which include most of the recorded human history, have been

M. E. El Raey (✉)
University of Alexandria, Alexandria, Egypt
e-mail: melraey@Alexu.edu.eg

M. M. Al Saud (ed.), *Applications of Space Techniques on the Natural Hazards in the MENA Region*, https://doi.org/10.1007/978-3-030-88874-9_6

relatively stable enabling stability, supporting agriculture, and forming human settlements as a result leading to population growth.

6.1.1 Natural Causes

Throughout Earth's history, the climate has changed globally and locally in nearly all periods. Climate change has many natural causes such as solar activity, ocean currents, volcanic eruptions, meteorites, eccentricity, and tectonic movements among others.

6.1.2 Solar Activity

The change in solar radiation was the main driver of climate change over geological time, but its role in the current climate change on Earth is very small and unnoticeable (Hegerl & Zwiers, 2007). Since 1978, solar radiation has been measured by satellites with high accuracy, these measurements indicate that total solar radiation fluctuates every 11 years in the so-called solar cycle, but it has not increased since 1978, which compatible with the argument of Hegerl, 2007. However, estimations indicate that solar radiation and volcanic activity had a very weak effect in the three decades following 1978 and that these solar and volcanic activities could explain only the periods of warmth and cold that occurred between 1000 and 1900 (Hegerl & Zwiers, 2007). There is no doubt that the only logical relationship to solar activity is that of weather, not climate, as changes in solar radiation and sunspots can make the winter cooler, or warmer, with very small changes in global averages.

6.1.3 Ocean Currents

The oceans, which cover about 71% of the Earth's planet, absorb about twice the amount of solar radiation from the Earth's atmosphere and surface. The oceans are a key component of the climate system, they have a more active and dynamic role in determining the climate system through interacting with continents by carrying vast amounts of heat and redistribute it to the surrounding landmasses (Rahmstorf et al., 2018). Evaporation in the ocean waters increases air humidity, forming rain and storms that then move to large areas of the earth.

Therefore, it was argued that the oceans play an important role in the distribution of unequal solar radiation on the Earth's surface by global climate currents. Without currents, regional temperatures will be more extreme and very hot at the equator and very cold at the poles (Rahmstorf et al., 2018).

6.1.4 Volcanic Eruptions

Volcanic eruptions can affect climate patterns for years due to the flow of certain gases from volcanoes and ash staying in the atmosphere for a relatively long time. The largest portion of gases released into the atmosphere is water vapor. Other gases include carbon dioxide (CO_2), sulfur dioxide (SO_2), hydrochloric acid (HCl), hydrogen fluoride (HF), hydrogen sulfide (H_2S), carbon monoxide (CO), hydrogen gas (H2), NH3, and methane (CH4) (Lockwood et al., 2011).

Sulfuric acid is one of the most important gases; it can remain in the atmosphere for up to 3 or 4 years in the stratosphere. Large explosions can affect the radiation balance of the Earth because volcanic clouds absorb ground radiation, dispersing a large amount of incoming solar radiation, the so-called "radiative effect" that can last from 2 to 3 years after volcanic eruptions.

Gases and dust particles in the atmosphere during the volcano affect the climate, where most of the particles emitted by volcanoes cool the planet by blocking incoming solar radiation. The cooling effect can last for months to years depending on the characteristics of the explosion. Volcanoes have also caused global warming over millions of years during times in the history of the Earth where massive amounts of volcanoes occurred, leading to the release of greenhouse gases into the atmosphere (Lockwood et al., 2011). Although volcanoes are located only in specific places on Earth, their impact is global because of atmospheric circulation patterns. The volcano releases particles of dust, ash, sulfur dioxide, and greenhouse gases such as water vapor and carbon dioxide, these dust and ash particles are small and light so they can stay in the stratosphere for several months preventing sunlight and causing cooling on large areas of the earth.

For greenhouse gases, volcanoes are known to release greenhouse gases such as water vapor and carbon dioxide. However, these amounts, which are emitted into the atmosphere from volcanic eruptions, do not considerably change the global quantities of these gases in the atmosphere. However, there have been some times in the history of the Earth where intense volcanic eruptions have increased the amount of carbon dioxide in the atmosphere significantly, causing global warming.

Monsoon systems interact sensitively with radiation disturbances such as volcanic eruptions and summer monsoon winds mainly driven by stronger heating of subtropical land masses compared to the surrounding oceans. The land blocks react more rapidly with volcanic cooling, leading to a slowdown in monsoons. Understanding monsoons is also very important for future climate. The response of monsoon systems to changes in greenhouse gases, aerosols, spin-off, ice cover, and other factors is not well known.

6.1.5 Meteorites

The effects of many large meteorites are observed in geological records of the Earth, and depending on their size can have significant impacts on the larger climate of the volcanoes. One example of the great influence of meteorites is the Yucatan peninsula in Mexico called the Chicxulub crater, which is believed to have been formed 65 million years ago. The diameter of the crater is more than 180 km. It is believed that the meteorite that caused the crater was at least 10 km in diameter. Many scientists have attributed the extinction of dinosaurs to such an event, causing debris to be thrown into the atmosphere, preventing sunlight from reaching the Earth's surface. All plants and animals are believed to have been affected by an action like this. The rich layer of Iridium sediments found in the geological record throughout the world is strong evidence for this reason (National Research Council, 2012).

6.1.6 Eccentricity

Eccentricity is the change in the shape of the earth's orbit around the sun. There is about a 3% difference between the time when we are closer to the sun and the time when we are farther from the sun. The tilt angle of the Earth's axis varies slowly so that it takes about 41,000 years to shift between 22.1 degrees to 24.5 degrees and back again. This is lower than the current angle we are at right now 23.45 degrees, which means less seasonal differences between the north and south hemispheres, while the larger angle means more seasonal variations, for example, warmer summers and colder winter. Current researches show that Earth's tendency is currently declining and will reach its lowest values around the world within 10,000 years. This trend will tend to make the winter warmer and the summer cooler (IPCC, 2012).

6.1.7 Tectonic Movements

The continental land plates move all the time. As they were at one time combined in only one continent, and broke apart and formed the continents and oceans among them. As continents move, climatic types change accordingly, as ocean currents are expected to shift warm bodies to cooler regions and so on.

Panel movements determine the shapes and sizes of continents and ocean basins that have a significant impact on the climate. When all continents join only one continent, as it was 225 million years ago, most of the Earth's surface is far from the oceans and dominated by the continental climate. But when continents are as separate as today, ocean currents are more able to distribute heat because of proximity to oceans, leading to a less extreme global and regional climate.

The movement of plate tectonics also causes geological activity that in turn affects the climate like volcanoes that was mentioned above.

6.1.8 Human Causes

Natural sources of carbon dioxide are estimated to be more than 20 times like humans, but over long periods up to several years, nature can rebalance them too much through their natural banks, for example through photosynthesis by plants or marine plankton. Because of this natural equilibrium, the level of carbon dioxide remains between 260 and 280 ppm, as was the case in the 10,000 years that mediated the period from the end of the last ice age and the beginning of the industrial age.

Global warming caused by human activities has played a significant role in many physical and biological systems, such as sea-level rise, which is expected to increase in the future, as well as increased frequency and intensity of some extreme weather events and loss of biological diversity in agricultural production. Many studies refer to the issue of global warming to human activities due to increased concentrations of greenhouse gases in the atmosphere, especially carbon dioxide, which is the largest contributor. Several studies have shown that there is a direct relationship between measured anthropogenic emissions of greenhouse gases, high greenhouse gases in the atmosphere, and high temperatures. The results also indicate that the rates of increase in each of them are unprecedented.

There is a direct correlation between population growth, high greenhouse gases, and temperature. Computer simulations indicate that man-made greenhouse gas emissions alone are sufficient to cause the observed rise in temperature of the atmosphere and other terrestrial systems (oceans, continents, and ice).

There is no natural cause that can account for the current rapid rise in temperature and carbon dioxide, which is about 20,000 more than normal, due to human factors.

6.1.9 Land Use Changes

When people change the way they use the land, they inadvertently change the climate. The most obvious example is the effect of urban heat island, a phenomenon that makes urban areas hotter than the surrounding rural areas during the day, and especially during the night. The change of green areas or forests to farms or urban areas is also a significant change in the water balance (transpiration) and heat as well as the amount of absolute carbon dioxide.

6.1.10 Transportation

The transport sector includes the movement of people and goods by cars, trucks, trains, ships, aircraft, and other vehicles, to arr the majority of greenhouse gas emissions, particularly carbon dioxide, are derived from the combustion of oil-based products such as gasoline. Some relatively small quantities are also emitted from this sector, such as methane and nitrous oxide during fuel combustion. Besides, a small number of hydro-fluorocarbons (HFCs) emitted from air conditioners used in cars as well as refrigerated transport or refrigerators are included.

Global tourism is closely related to climate change as was confirmed by The United Nations; it is one of the major contributors to the increase in concentrations of greenhouse gases in the atmosphere. It also represents about 50% of traffic movements in the world. The rapid expansion of air traffic contributes about 2.5% of the production of carbon dioxide, and the tourism rate is expected to increase from 564 million in 1996 to 1.6 billion by 2020, adding much to the problem unless steps to reduce emissions from international travelers and to increase the transportation and welfare they need (Becken, 2007).

6.1.11 Industry

The industrial sector produces commodities and raw materials that we use in our daily lives. Greenhouse gases emitted from industrial production are divided into two categories: direct emissions produced during the industry itself, and other indirect emissions produced off-site.

Direct emissions are generated from fuel combustion methods to generate electricity through chemical reactions and leakage from the same processes as well as industrial equipment. Most direct emissions come from fossil fuel consumption for power generation. As well as the leaks of natural gas and oil systems and the use of fuel in production processes and chemical reactions during the production of chemicals, iron, steel, and cement. Indirect emissions are generated by the burning of fossil fuels in power plants to generate electricity, which is used by the industrial plant to operate industrial buildings and machinery.

6.1.12 The Construction Sectors

The construction sector is one of the largest energy-consuming sectors with an estimated 30% to 40% of total GHG emissions. Energy consumption in this sector has increased over the past years due to the large increase in new buildings.

The developed countries differ from the developing countries in this sector, where in the first they plan well before the construction of any more buildings while in the second the construction is random, which affects the rest of the other sectors as in the drainage, water, electricity and other important sectors, those buildings cause consuming more energy.

6.1.13 Energy Supply

The energy supply sector is the largest contributor to global greenhouse gas emissions, with energy contributing relatively little to greenhouse gas emissions, especially in developed countries. They include, in large part, the electricity sector through generation, transmission, and distribution of electricity. Carbon dioxide (CO_2) accounts for the vast majority of greenhouse gas emissions from this sector, but small amounts of methane and nitrous oxide are also emitted.

To sum up, as was described the increase in human activities has led to an increase in greenhouse gases in the atmosphere (especially carbon dioxide), causing the recently observed global warming. If greenhouse gases continue to grow at the same speed, it is expected that temperatures will rise very dramatically on the Earth's surface it could reach around 4 °C above mid-nineteenth century temperatures. It does not mean that if the emissions are reduced quickly enough the heat will stop rising but there is a chance that it does not exceed 2 °C, which may reduce the destructive effects and makes the ability to adapt better.

It was observed that the concentration of carbon dioxide increased from 280 ppm before 1800 to 396 ppm in 2013 and this was known through many measurements by scientists (Trofimenko, 2011). Scientists have pointed out that for thousands of years and until about 200 years ago, the carbon cycle was in a fairly stable equilibrium. Since the nineteenth century, anthropogenic emissions of carbon dioxide from burning fossil fuels, manufacturing, and deforestation have affected this equilibrium.

Over the past two centuries, the increase in the burning of fossil fuels has been closely linked to global growth in energy use and economic activity. It has also increased significantly from 2000 to 2010 (NASA and GISS, 2013).

6.1.14 Uncertainty

In the context of climate change risk assessment, uncertainty arises because, although we may already be confident that the climate is changing, we do not know precisely the magnitude of these changes or their associated effects. For example, in some areas, it may not be clear whether rainfall will increase or decrease. It is also an important fact for decision-makers as they are in a very sensitive position but cannot make a large percentage of the exact point or threshold at which the climate will change at a certain level concerning their area of control. Uncertainty is a broad range of possible outcomes and complexity makes it impossible to define a set of probabilities (Rueter, 2013).

6.1.15 Error Versus Uncertainty

It is important not to confuse the terms "Error" and "uncertainty"because they are very different. The error is the difference between the measured value and the real value of the thing being measured. Uncertainty is estimating the doubt about the outcome of the measurement. For example, the error can be corrected by making corrections from calibration. The error that its value is not well known is uncertainty. Uncertainty describes the novelty and accuracy of measurement in any field (Rueter, 2013).

In climate change, there are several uncertainties, such as in the greenhouse gas emissions/concentration scenario, model configuration, and bias and downscaling uncertainty.

The development of a set of global-scale climate change projections for the twenty-first century as a response to increased greenhouse gas emissions is necessary to assess the impacts of global warming from greenhouse gases and to develop appropriate adjustment and mitigation strategies. In the development of these projections, all the causes of climate change that have already been taken into account, both natural and human, must be developed with their uncertainty in mind. Uncertainty is influenced by our incomplete knowledge of these causes and processes and their precise description (Quiggin, 2008).

Over the past decade, research interest has shifted significantly to quantitative assessment and representation of uncertainties in climate change projections for use in impact and risk analysis studies. It is important to note that the term "uncertainty"has a generally negative connotation, which means that uncertainty is associated with our weak knowledge of the problem and should therefore be reduced as much as possible by further research.

6.1.16 Uncertainties in Climate Change Projections

As was mentioned above, the climate could be changed as a result of several causes, both human and natural. Among the major anthropogenic factors are atmospheric greenhouse gases, tropospheric aerosols caused by pollutant emissions, and changes in land use, for example, the greenhouse gases affect the climate by absorbing infrared radiation emitted from the Earth's surface and aerosol can absorb the dispersion of solar and infrared radiation, as well as the significant impact of land-use change.

The problem lies in the accuracy and validity of future projections as the first step in the series of steps to be followed to produce an appropriate projection of climate change globally and to generate scenarios for GHG emissions and aerosols based on future socio-economic and technological development assumptions, which are difficult to predict in the next century, resulting in different emission paths, making them a high uncertainty. Land-use change is also difficult to predict as human behavior and urbanization are difficult to project, adding to the vulnerability of these models (Giorgi, 2010).

In addition to the main natural effects seen in climate simulations, such as in volcanic eruptions, for example, they can lead to the injection of small particles in the stratosphere, where they can remain for several months to years as mentioned above, causing a change in the amount of solar radiation reaching the Earth.

Considering that natural phenomena such as volcanoes, for example, or the number of sunspots is an unexpected and cannot be project accurately, climate models have a large error rate that makes it difficult to rely on to make critical decisions that lives of communities depend on (Giorgi, 2010).

Since emissions scenarios are the first step in the sources of uncertainty and due to their extreme ambiguity and unpredictability, whether natural or human during the next century or even the next few years, this makes uncertainty unavoidable and will not be eliminated, even with developing a great range of potential social and economic paths in the future. The uncertainty about climate change will remain very high.

This may be why IPCC has developed a series of scenarios that have moved from low to high emission levels, intending to cover the entire range of future developments. IPCC has not linked any probabilities to these scenarios, making them all plausible (Giorgi, 2010).

The next stage is the introduction of these greenhouse gas (GHG) scenarios into the global biological and geochemical models for the corresponding GHG concentration scenarios. These biogeochemical models are affected by uncertainty due to poor knowledge of the biogeochemical cycles and the use of approximate representation. Therefore, until the science of biogeochemical cycles improves, uncertainty in these models will remain unreliable and should be reduced.

Experts are taking greenhouse gas concentration scenarios and introducing them into special systems for producing climate projections for the twenty-first century. In other words, climate projections are a sensory experience of the climate system's response to certain increases in greenhouse gas levels. Future climate predictions do not make any assumptions about the potential for volcanoes or future changes in solar activity, as they are very difficult and maybe impossible, which mean that climate models assume that the natural causes are stable in the future, that adds another factor of the uncertainty that is extremely difficult to assess.

Generally, science makes predictions about how a system will be followed and then measured seriously. The problem for ecologists is to try to focus on making testable predictions about the real environment we live in to understand natural processes so that we can either respond or control future events. The natural system, in contrast to experimental systems, is full of uncertainties caused by all possible types of interacting factors, and so uncertainty enters into daily routines in the environmental field.

The important and convincing message that has been revealed in the work of (Rueter, 2013) is that we cannot just study a problem and gather a lot of information to make a good and appropriate decision on something. There are some cases where the uncertainty is very high where it is impossible to predict the results with any degree of certainty.

The same reference has identified three types of unknowns: **Risk**, a potential estimate of the likelihood of an event occurring or being exposed. If, for example, we calculate the risk or potential damage from exposure, we can calculate how much money and effort we have to spend to control this risk. **Uncertainty** is a wide range of potential results and complexity that makes it impossible to determine a set of probabilities. For example, we can create a set of scenarios and use them to describe the different paths that may occur in the future, but we do not have a specific way of knowing the future that will happen. Finally, **Indeterminacy** where there is a set of information that we are not able to know. The author asserts the reason here that sometimes we concentrate our energy and resources to address a problem which results in a very large set of results that may include a set of surprises and often surprises.

The surprise is a change in the system which is qualitatively different from what we expected. For example, the response of nature to overfishing was not as predicted and logical "lack of fish," but the surprise was that excessive fishing led to an increase in the number of jellyfish in a very large amount (Rueter, 2013).

These problems in future projections do not detract from the work of model developers, but they should develop also some ideas about how to live with them as they are inevitable problems and affect the success of measuring accurate climate predictions in the future. In this sense, this study will not review or rely on any future climate models or specific projections in the study area. Rather, the focus will be on the vulnerability and responsiveness.

6.1.17 Vulnerability Assessment

Is to determine the extent to which the community is affected by the risks as identified and evaluated in the above points, as the good profile of the community identifies the sites and places of the people in the community and the important facilities on the maps, and the risk assessment places the risks on those maps, enabling the vulnerability assessment by comparing areas where risks overlap with important people and facilities, allowing the assessment of possible losses in dollars, and prioritizing the most serious risks.

Hazard vulnerability assessments describe who is threatened (hazard identification) and what is being exposed to (potential loss, injury, damage, negative effects on livelihoods), and the effects of such exposure. In other words, the goal is not only to identify the risk factors (who and what is vulnerable) but also the driving forces that are vulnerable in a particular place. These assessments can be qualitative in their approach or quantitative in estimating populations at risk and rating vulnerability. Risks can be single or multiple and can range from localized site-specific analyzes to more regional ones (Cutter et al., 2009).

Climate-related Hazards impact of climate change on hazards is hard to quantify because, under a constant climate, it is generally an order of magnitude below the natural existing uncertainty. Conversely, vulnerability is not well understood and few rigorous methods are available to quantify this factor (Gilard, 2016).

6.1.18 Risk Assessment

The quality of risk assessments is critical to making informed and good risk management decisions. It is important to understand risk-related vulnerabilities and to make it clear to decision-makers and civil society as a whole, to bridge the gap between academics, decision-makers, and the public (Sørensen & Jebens, 2015).

Risk assessment may include quantitative and qualitative techniques and information to describe the nature of risks. Specific techniques are particularly useful in conditions such as climate change, where there is uncertainty about probability and consequences. Regardless of the sources of uncertainty, the initial assessment process can provide a comprehensive and rigorous means of prioritizing the risks of climate change. After identifying or determining the types of risks that threaten each sector, further planning is needed to reduce uncertainties (Cutter et al., 2009).

6.1.19 Risk Reduction

There is a range of ways to reduce risk, begins first with identifying the risk, and is called "**Risk Assessment**" by this method ways to minimize risk exposure is developed and called "**Risk management**". It is important to keep in mind that although risk management measures exist, they may not be sufficient to protect those who are exposed to risk. To be effective, it must demonstrate how to avoid risk to all concerned in many ways for example raising awareness and repeating important messages whenever possible using the various media and this step called "**Risk Communications**". So, dealing with risk implies risk assessment, management, and communication efforts.There is a rule when dealing with risk (the more you are exposed, the greater the risk).

6.1.20 Vulnerability

There is great interest in the concept of vulnerability, where there are many definitions in the literature, derived from different conceptual models and frameworks.

Vulnerability can be defined as "the characteristics of a person or group and their situation that influence their capacity to expect, cope with, resist and recover from the impact of a natural hazard" (Wisner et al., 2004). This means that the person's vulnerability is determined by influencing one or more elements that are at risk to some extent. These elements vary according to economic, cultural, and social differences and may be tangible or intangible, including life risks and changes in livelihoods and property. There are vulnerabilities at all levels of global, national, and local and can have an effect known as the snowball effect; for example, the global economy can affect the local one. It is mostly dynamic and depends on time and can change annually or even every hour. Vulnerability is highly dependent on community actions and can increase or decrease as a result (Sørensen & Jebens, 2015).

6.1.21 Understanding Vulnerability

Vulnerability is a term that has been mentioned frequently in the literature on disaster management. However, as stated in (Yasir, 2009) the definition of this term is still controversial, with each scientific field-shaping it differently. It was mentioned in the same reference that the first who used this term were engineers in the application of the physical structures, and the term was then expanded over the years by sociologists to include the social, economic, political, and institutional aspects. More than 25 different definitions have been identified in the current literature reflecting its multifaceted nature.

Concerning natural hazards, vulnerability can be defined as an inadequate capacity to overcome disasters and their impacts. Although this term was introduced in disaster studies in the 1970s, it gained importance in the 1980s and has been defined in this period as a threat factor limiting the ability of society to absorb and recover from a dangerous and harmful event (Yasir, 2009).

In the 1990s, the concept of vulnerability was further changed as people were able to respond and protect themselves from disasters. Here the definition changes to become the characteristics of people or society that determine their capacity to expect, cope with, resist and recover from the impact of natural hazards.

In addition to these socio-economic processes, (Dow & Downing, 1995) add other demographic aspects, such as population ages and dependency, which are the conditions under which the characteristics that relate primarily to the physical, social, economic, and environmental factors are determined (Yasir, 2009). Others (Wisner et al., 2004) show that instead of simply taking care of these factors, there is a greater need to focus on the problems that arise from their interaction together. This approach transforms people from being mere passive recipients into active players by increasing their tendency to change.

Therefore, vulnerability is a multilayered and multidimensional perspective and must be addressed from all sides to determine it correctly. Some other studies have used a more integrated approach to determining vulnerability as a combination of potential exposure and social response within a specific geographic range (Yasir, 2009).

6.1.22 Vulnerability to Climate Change

Vulnerability varies widely across communities, sectors, and regions, and is addressed in a variety of ways by different specializations. Studies have begun to address the term "vulnerability"very early, and the literature on vulnerability has grown enormously over the past few years. It was defined by (Timmerman, 1981) as "The degree to which a system acts adversely to the occurrence of a hazardous event. The degree and quality of the adverse reaction are conditioned by a system's resilience (a measure of the system's capacity to absorb and recover from the event)" and in 1982 the definition of UNDRO (United Nations Disaster Relief Organization) was presented "Vulnerability is the degree of loss to a given element or set of elements at risk resulting from the occurrence of a natural phenomenon of a given magnitude."

Then some other definitions addressed the social perspective such as Susman et al. 1984 "Vulnerability is the degree to which different classes of society are differentially at risk", while Dow (1992) defined vulnerability as "the differential capacity of groups and individuals to deal with hazards based on their positions within physical and social worlds." Cutter (1993) Vulnerability is the likelihood that an individual or group will be exposed to and adversely affected by a hazard. It is the interaction of the hazards of place (risk and mitigation) with the social profile of communities (Lynn et al., 2011).

Blaikie et al. (1994) argues that vulnerability is "the characteristics of a person or group in terms of their capacity to anticipate, cope with, resist and recover from the impact of a natural hazard. It involves a combination of factors that determine the degree to which someone's life and livelihood are put at risk by a discrete and identifiable event in nature or society" (Blaikie et al., 2003). And in the same year Bohle et al. (2003) defined vulnerability from a social perspective as "an aggregate measure of human welfare that integrates environmental, social, economic and political exposure to a range of potentially harmful perturbations. Vulnerability is a multi-layered and multidimensional social space defined by the determinate, political, economic and institutional capabilities of people in specific places at specific times."

Later in 1994 Cannon related the vulnerability again to the risk and described it as "a measure of the degree and type of exposure to the risk generated by different societies concerning hazards. Vulnerability is the characteristic of individuals and groups of people who inhabit a given natural, social and economic space, within which they are differentiated according to their varying position in society into more or less vulnerable individuals and groups."

The definition of UNEP (1999) was more comprehensive as they described the vulnerability as a function of sensitivity to present climatic variability, the risk of adverse future climate change, and the capacity to adapt. The extent to which climate change may damage or harm a system; vulnerability is a function of not only the systems' sensitivity but also its ability to adapt to new climatic conditions. While UNISDR (2008) identified it as the conditions determined by physical, social, economic, and environmental factors or processes, which increase the susceptibility of a community to the impact of hazards (Lewis, 2010).

From these different definitions, it could be concluded that vulnerability can refer to the sensitive systems, both natural and human, as the study area is located on the coastline, making it vulnerable to sea-level rise, as well as it is a low lying land and high densely populated area, and what multiple the vulnerability to climate change the agricultural land that secures the livelihood of local communities is vulnerable to salinization and sea-level rise as well as both floods and flash floods and also prone to urban encroachment leaving the community highly vulnerable to environmental changes. That is, vulnerability is the weakness of natural and human systems and their inability to deal with imminent or potential danger, i.e., the inability of natural systems to adapt quickly enough to changes and to cope with the negative effects.

The inability of communities to recognize the danger in the first place and to deal with it during its occurrence or after it ends; in other words, preparedness, resilience, and response at different stages greatly affect vulnerability, and based on the community's ability to carry out these previous steps, societies can be divided into more or less vulnerable.

The most vulnerable are those who have fewer options and those who are subject to restrictions such as related to gender, social class, physical disability, lack of education and work, ill people, and others who do not make their own decisions.

The inability of political systems represented in decision-makers and how they predict, deal with, and recover from a natural event can turn this event into a major disaster. In other words, failure to deal with a natural normal event transforms it into a natural disaster, and when vulnerability interacts with a hazard a disaster results.

Several major weaknesses in the study area are associated with many climate-sensitive systems, including, for example, food supplies represented in vulnerable agricultural land, weak infrastructure, health, water resources, and sensitive coastal systems among others.

Therefore, environmental, social, economic, and political considerations are the basis for evaluating the vulnerability of any society, and neglecting any part of them has a very significant impact on vulnerability assessment.

6.1.23 Social Vulnerability

Social vulnerability refers to the resilience of communities when faced with external pressures on human beings, health, and welfare, as in the case of natural and human disasters and disease outbreaks. Reducing social vulnerability can reduce human suffering and economic losses.

Some types of social vulnerability may be due to lack of awareness, wrong decision making, land misuse, abuse of natural resources, poverty, lack of control, disability, etc. There are millions of people who are somehow vulnerable to natural disasters and climate-related events, as they have no choice about where they live. They are subject to policies and laws which may not be democratic or external pressures without awareness of what is happening and without understanding their rights as citizens. The most vulnerable occurs as a result of exploitation, greed, marginalization, and abuse by some stakeholders putting the vulnerable population in more danger(Kelman & Lewis, 2010).

Social, political, and economic forces are pushing poor and marginalized communities to the most vulnerable areas around the world, causing more poverty and poor environments and the places, making these places very vulnerable to disasters. In Egypt, for example, those who live in flood plains are the poor who are driven by poverty to find anywhere without caring about their environmental status. Those who live in weak or partially destroyed buildings are the poor who do not have enough money to renew and preserve them.

Areas that are highly vulnerable to many climate-related events (e.g. flash floods) occupied by the poor and marginalized groups of citizens, as well as slums that are not serviced by good infrastructure, where the most vulnerable groups of society are at risk. Due to their social and economic situation, any environmental event turns into an environmental disaster in those areas at risk.

The livelihoods of the population are the source of their vulnerability, so the salinization of agricultural land on the northern coast of the Nile Delta as well as the salinization of freshwater resources puts people in these areas under greater pressure, making them more vulnerable to climate change.

Recent climate events on the northern coast of the Nile Delta in the last 10 years as well as the expected rise in sea level, the relatively high rainfall, the inability of the government to deal with these events, and the citizen's lack of awareness adding more vulnerability, especially shortly. Extreme events are expected to increase, and if the status of decision-makers and citizens will still stable, there will be no other option than to turn these events into real disasters. Some researchers have called this type of change "creeping environmental change" (Glantz, 1994; Kelman, 2009).

To reduce the risk of disasters, it is necessary to remove the root causes of vulnerability. Therefore, the analysis of weakness in the study area has to be rationalized and detailed, by reaching the poor and marginalized population, stakeholders, and some decision-makers to measure and test the vulnerability indicators in the real world.

This study will focus particularly on a poor and random area with a focus on all aspects of vulnerability to disasters related to climate as possible and try to enumerate all indicators of vulnerability in a sample of Egyptian society in some detail.

6.1.24 Vulnerability Process

The "vulnerability process" refers to the values, ideas, behaviors, and actions that have led to characteristics such as fragility, weakness, exposure, and susceptibility and that can perpetuate or absolve these issues (Kelman & Lewis, 2010).

6.1.25 Vulnerability and Poverty

Disaster literature has linked vulnerability to marginalization and poverty, and research has shown that vulnerability and poverty interact closely with one another. This is because poverty increases the inability to access resources that can enable people to adapt or become more resilient, and this group of people lacks quality education and appropriate work which makes them more vulnerable to disasters, which makes vulnerability a component of poverty (Yasir, 2009).

While (Wisner et al., 2004) suggest that vulnerability is a more complex concept because it combines both external and internal characteristics of the population, while poverty tends only to focus more on the personal needs of individuals. In the same vein (Yasir, 2009) suggests from quoting from the work of (Blaikie et al., 2003) that vulnerability is the potential for future loss of risk, while poverty is a measure of the current situation.

As Chambers et al. noted that anti-poverty programs focused on improving people's income while programs to combat or deal with vulnerability tend to enhance security and reduce potential losses to natural disasters.

6.1.26 Vulnerability Risk

There is a great need to review the basic concepts that place natural hazards in the context of vulnerability. To assess the impact of risks on individuals or communities, it is necessary to distinguish between 'hazard', 'risk', and 'vulnerability'components in disaster analysis. The risk can be defined in this case as a severe geophysical event that is likely to be the cause of the disaster. Thus risk can be defined as the probability of loss resulting from the interaction of a certain level of hazard and vulnerability (Yasir, 2009) (Fig. 6.1).

(Wisner et al., 2004) schematize this relationship as follows:

$$\text{R (Risk to Disaster)} = \text{H (level of hazard)} \times \text{V (degree of vulnerability)}\ 9$$

Based on the above equation, (Alexander, 2000) put the total risk as follows (as was mentioned in the study of (Yasir, 2009))

$$\text{Total Risk} = (\Sigma \text{ elements at risk}) \times (\text{hazard and vulnerability})$$

From the above, it is clear that total risk is a complex production of its various constituent elements (population, societies, infrastructure, economic activities, services, etc.) which are exposed to the threat of a hazard by staying in a position of vulnerability (Yasir, 2009).

(Alexander, 2000) argues that although vulnerability can be estimated in the absence of risk, it cannot quantify without including risk in the aggregate equation.

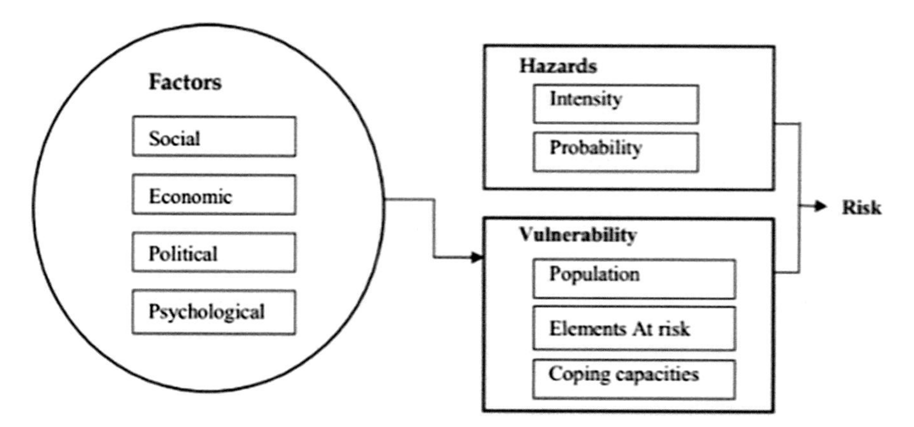

Fig. 6.1 Relationship between Hazard, Vulnerability and Risk. (Adopted from (Yasir, 2009) cited from Frantzova et al. (2008))

Intensive Risk is the risk associated with the exposure of a very large number of people and economic activities to serious and intense events and can lead to potentially catastrophic effects involving high rates of mortality and asset loss. Intensive risk is mainly a characteristic of large cities or high population areas that are not only exposed to extreme hazards such as strong earthquakes or heavy floods but also have very high levels of vulnerability to these hazards.

Extensive Risk is the widespread risk associated with the exposure of populations in each locality to frequent or persistent low- or moderate-intensity conditions, often of a high local nature, which can lead to devastating cumulative effects. High risk is a feature of rural areas and urban margins where communities are exposed to recurrent floods or landslides. This type of risk is also associated with poverty, urbanization, and environmental degradation.

Residual Risks are those risks that remain unmanaged until there are effective disaster risk reduction measures, which must therefore maintain emergency response and recovery capabilities. Residual risk means the continued need to develop and support the effective capacity of emergency services, preparedness, response, and recovery as well as social and economic policies as in safety nets and risk transfer mechanisms.

Risk Reduction there is a range of ways to reduce risk, begins first with identifying the risk and is called "**Risk Assessment**" by this method ways to minimize risk exposure is developed and called "**Risk management**". It is important to keep in mind that although risk management measures exist, they may not be sufficient to protect those who are exposed to risk. To be effective, it must demonstrate how to avoid risk to all concerned in many ways for example raising awareness and repeating important messages whenever possible using the various media and this step called "**Risk Communications**". So, dealing with risk implies risk assessment, management, and communication efforts. There is a rule when dealing with risk (the more you are exposed, the greater the risk).

Prospect Disaster Risk Management is management activities that address and seek to avoid the development of new and growing disaster risks (Thinda, 2009). This concept focuses on addressing risks that may develop in the future if risk reduction policies are not put in place, rather than waiting to deal with the risks that occur.

6.1.27 Vulnerability Analysis

The inconsistency between different vulnerability definitions makes it difficult to perform a vulnerability analysis correctly. The current methodologies for wide-scale vulnerability analysis have recognized the need to develop a more comprehensive approach to shift attention from existing methodologies to functional frameworks that focus more on processes rather than on the fixed boundaries of the social system (Yasir, 2009).

Several studies have analyzed the vulnerability concept where they have been concerned about the importance of the ability of people and society to protect themselves from disaster. Thus, capacity could be understood as the counterpart to vulnerability. The capacity of the society and population includes the willingness and presence of responsible local governments and leaderships and awareness among members of society among other things. By building capacities, the level of risk will move to an acceptable level where vulnerability decreases (Sørensen & Jebens, 2015).

(Wisner et al., 2004) identifies four main methods for evaluating vulnerability, each with its effects. The first approach takes the demographic perspective to see vulnerability as a societal situation where people are considered to be vulnerable. The second approach uses taxonomic behavior using physical, economic, social aspects of vulnerability. A third approach is a situational approach; it takes a multi-dimensional perspective as a dynamic rather than a static concept. The basic premise of this approach is that disasters are not seen as exceptional events but rather as an extension of the problems of daily life where vulnerability is a period that extends normal circumstances to exceptional situations. However, although this model is specifically designed to deal with organized complexity, it is limited because of its geographical peculiarities and cannot generalize results from the context of disasters.

The fourth approach is seen as contextual and proactive by (Wisner et al., 2004), who takes a participatory view of vulnerability analysis where community members participate as active agents of change rather than passive recipients of aid. With the involvement of the community in the actual evaluation process, the contextual and proactive approach goes beyond the scope of vulnerability to the empowerment of the society that forms the core of a democratic society, yet its success remains limited by the participatory capacities of the members of society and the long time frame (Yasir, 2009).

Together, these approaches suggest that vulnerability analysis needs to adopt a more comprehensive approach that requires a shift in the conceptualization of vulnerability from static to dynamic, and thus moving away from demographics and classifications to situational approaches.

The fourth case provides an additional benefit to the analysis of community capacity and adaptability, yet requires community participation that will be taken into account later.

6.1.28 Vulnerability Assessment

A vulnerability assessment is an analysis of the expected impacts and risks of an area or system and of a community or capacity to adapt to it. For vulnerability to be properly evaluated, sensitivity, susceptibility, and adaptability must first be identified (Byrne, 2014).

6.1.29 Vulnerability Measurement

The development of a tool for measuring vulnerability across all disciplines is difficult given the ambiguity of the definition of vulnerability from the beginning as well as the constant change of dynamic nature and changing scales of analysis (temporal and spatial). However with advances in the science of vulnerability in all disciplines and building on the motivation that empirical evidence-based information is needed to support planning pressures and vulnerability, especially for disasters, a range of approaches have been developed in this regard (Cutter et al., 2009).

A key indicator of leading economic indicators is a set of 10 economic variables used to estimate future economic activity, which has proved to be very strong in predicting economic recessions over the past 50 years. Economic, social, and environmental indicators have been started since the 1940s but have become more pronounced in the 1960s and 1970s, followed by environmental indicators. While the 1990s saw more emphasis on the development of environmental sustainability indicators as well as vulnerability.

However, there are some complexities when using indicators, as in the complexities that are often within a system of a given variable or set of variables, which has a significant impact on what is measured and how. This is clearly illustrated by issues such as social networks, trust in government, institutional capacity, and disaster preparedness, which are difficult to measure quantitatively. Besides, the availability of data presents another problem for both hazard and population parameters, and this in itself may hinder the selection of input variables (Cutter et al., 2009).

In this respect, the vulnerability can guide policy development on national and sub-national vulnerability and act as a means of measuring progress towards this specific goal (Birkmann, 2008).

6.1.30 *Resilience*

Resilience is not only a term for hazards or natural disasters (Timmerman, 1981). There are many examples of this concept in many areas, such as material science, engineering, psychotherapy, social sciences, biology, ecology, and many other sciences that use this term differently from one another.

From an environmental perspective, resilience was treated from the point of view that the risk was as much about normal community activities as it was about the extreme event itself, and it still works to this point of view. However, there is still a need to unify the traditional bilateral or doubled vision that separates social life from nature so that linking it enables understanding of vulnerability and resilience. Resilience itself is a natural and long-term social activity.

The IPCC definition of resilience (2012) puts it within the framework of the social or ecological capacity of the system to absorb disturbances while maintaining the same infrastructure, standard work methods, self-regulation, and adaptation. While UNISDR (**United Nations International Strategy for Disaster Reduction**) introduced another definition in 2008 that demonstrated resilience as the capacity of the system or the potentially vulnerable community to adapt, through resistance or change to reach and maintain an acceptable level of performance, which is determined by the degree to which social systems can organize themselves to increase capacity to learn from past disasters to ensure better protection from future disasters and to improve risk reduction measures as much as possible (Kelman & Lewis, 2010).

6.1.31 *Pressure and Release Model*

This model is a valuable tool for identifying disaster risk reduction measures by analyzing and understanding the root causes of a potential event. It explores the relationship between hazards and vulnerability and looks at the link between **root causes**, **dynamic pressure,** and insecure or **unsafe conditions**. This progress can be used in the vulnerabilities to describe and explain the interrelationships between different vulnerabilities. The root causes, dynamic pressure, and unsafe conditions by the model are three layers of social processes that lead to vulnerability. Root causes also give rise to several dynamic pressures that in turn explain how unsafe conditions have begun (Thinda, 2009).

In turn, this model allows for the assessment of effectiveness, the identification of capacities required, and the flexibility required to mitigate disasters. Root causes are often the result of very long-term actions and are implemented in society to the extent that does not change.

6.1.32 Previous Experiences

Past experiences could shorten the way to reach a suitable solution to deal with the vulnerability of the northern coast's community in Egypt. Not all studies in this field are useful to be applied in the study area, but we should choose what suits our society, our economic circumstances, and other considerations that make every society Unique in dealing with disasters.

Some studies (Sørensen and Jebens, 2015) show that until recently, developed countries were not interested in valuing their vulnerability, because as they thought, their society was well aware of disasters and they could deal with all kinds of disasters easily. But Hurricane Katrina was one of the most striking examples which revealed the error of this belief. Thus, the analysis of vulnerabilities in developed countries can reveal unexpected results, and all developed countries are now taking this action. So, by selecting the best results of their work it could save time and effort as well as allowing developing countries to develop new measures of their own.

The work of (Sørensen and Jebens, 2015) summed up the problems that should be addressed in Denmark to reduce vulnerability to climate change as follows: **First**, there are no political and financial linkages between DRM (Disaster Risk Management) as well as reducing the preparedness budgets, which suggest the presence of problems in political management, **Second**, the need to close the gap between the tools and maps in the academic community and the needs of decision-makers and the public; **Third,** Lack of knowledge and awareness of risk and risk reduction among the public; **Fourth,** lack of multidisciplinary work, especially between natural and social sciences.

The study also focused on qualitative approaches to identify various social vulnerabilities to improve risk management and advise on how to proceed with disaster risk management and adaptation to climate change.

The same work has shown that although Denmark is a developed country with both technical and financial solutions, only a few are thinking about the future, as well as a lack of public confidence in decision-makers. The results of the study showed that participants from civil society made it clear that they were not sure whether municipalities and emergency management can protect them at the time of disasters. The results also show that the past has shown that *a disaster is required before society implements risk reduction measures*, possibly because of a lack of awareness and knowledge of the real problem.

6.1.33 Vulnerability Mapping

A vulnerability map can determine the exact location where people, natural resources, or property may be in danger due to a potential disaster that may result in death, injury, pollution, or any other destruction of any kind. This type of map is created along with the various other types that present the risks in one way or another. The vulnerability map can indicate areas of housing in which a large population lives, suffers from randomization, or that the buildings are old and are not qualified to deal with natural events, such as in the case of floods. Also, identifying the commercial, tourist, and residential areas that may be affected in the event, which gives the possibility of developing a plan to deal with it (Edwards et al., 2007).

The best way to map vulnerability is to use GIS technology where all possible data can be placed to feed this program, such as land survey data, satellite imagery, land boundary maps, road maps, topographic maps, or other maps used for other related purposes.

6.1.34 Benefits of vulnerability Mapping

Vulnerability maps can illustrate the places at risk, the type of hazard, and the individuals exposed to it, which means a full understanding of the risks and vulnerabilities so that decision-makers can know what is needed to protect these areas and allow them to identify mitigation measures to prevent or minimize loss of life, injury, and environmental consequences. A major benefit of this type of map is the potential for use in all stages of disaster management: prevention, mitigation, preparedness, operations and relief, recovery, and lessons learned. For example, in the prevention phase, vulnerability maps can be used to avoid high-risk areas when developing residential, commercial, or industrial areas. It also tells technicians about where infrastructure can be affected in a disaster. A vulnerability map can contain evacuation routes for eviction in situations where this is required (Edwards et al., 2007).

These maps can be used to assess the disaster, how to deal with the disaster during and after it, evaluate the results, take lessons learned to correct the errors that were present before the actual event, as well as the ability to clarify the extent of damage, assess the quality of emergency management, to help for recovery.

6.1.35 Products of Vulnerability Mapping

Vulnerability maps exist in a large variety, having various content, various scales, and fulfilling different objectives. The basic vulnerability maps and few applied maps are (World Meteorological Organization, 2013).

Detailed descriptions and examples for each vulnerability mapping product are shown below.

- **Preliminary vulnerability map**

It is the map that indicates on a small scale the general or potential risks of some areas, and it is considered to be the first step to more detailed mapping and is therefore considered important (World Meteorological Organization, 2013).

The main objective of these preliminary maps is to identify the risks (eg, floods, flash floods, heavy precipitation, cold or heatwaves, and other hazards that are likely to occur in the study area) and determine the outer limits of the potential event.

By superimposing the potentially affected areas with land-use maps or other layers in the GIS database like densely populated places; "hot spots"can be identified in the community under study. This type of map is therefore referred to as simplified risk maps. This map can be used for general purposes or even for detailed assessments. These maps are also important in that knowledge of the type of potential risk is closely linked to making decisions on the future development of the region.

- **Event map**

This type of map shows the locations and extent of disasters in different ways. It is based on events that occurred in the past, both near and far. Event data are collected to feed this type of map through a variety of sources. The importance of this type of map lies in the use of past events to raise awareness among local populations and thus reduce potential losses. Specialists and decision-makers also use this kind of mapping to respond to emergencies. After extreme events, this kind of map is referenced to conclude to reduce the likelihood of recurrence of the same event in the areas where it occurred and to avoid future damage to the least possible extent.

- **Hazard map**

This map shows some detailed data about the expected event as it is in the range, speed, and expected direction of the event. For example, in the case of floods, there is a set of key elements that must be incorporated into the risk map for a particular probability of occurrence, as in flood extent (areas covered by water), flow velocity (m / s), water depth (m); as well as other parameters such as flood propagation (km / h), flood depth * velocity (m * m/s; as an indicator for the degree of hazard) (World Meteorological Organization, 2013).

Hazard layers are superposed with the topographic map of the area, DEM, and the geomorphological map to show the hazard in high accuracy, and to determine where it may be happening. This type of mapping provides basic information for the development of technical guidance on various issues related to disaster management, assisting various stakeholders including local governments to make right and correct decisions for disaster management, developing a risk mitigation plan, and developing comprehensive risk management plans; these maps are also important in emergency management.

- **Vulnerability map**

This map shows the potential harm to people with priority given to a particular group of people (such as the elderly, the disabled, and others), assets, infrastructure, and economic activities at risk either directly or indirectly, and information is presented either quantitatively or qualitatively through different indicators.

Vulnerability maps provide the basis for risk maps that are needed for contingency planning. These maps show the potential consequences of an event on human activity. These maps are important to ensure proper planning for the future away from potential danger. The database for this type of the map must be regularly and continuously updated for variable vulnerability parameters as they are changing over time.

- **Risk map**

These maps include potential hazards with the vulnerabilities of current or potential economic activities at risk. These kinds of maps are an update of the hazard maps and vulnerability maps, which show the average damage per unit of space, often expressed in terms of monetary (the potential loss per unit area and time), which is necessary for the economic assessment.

Risk maps are an assessment tool, thus supporting prioritization for risk reduction. It is also very important for land use planning, where future development planners are concerned with maps that present potential risks in the future. These maps can also identify and see the consequences of past mistakes (World Meteorological Organization, 2013).

- **Zoning map**

This type of mapping can be considered as an important type of planning, as it illustrates existing risks and classifies them (low, medium, high). This map is based on a hazard map as well as monitoring land use in a particular area.

- **Emergency map**

This map is based on hazard, vulnerability, and risk maps according to its purpose. Emergency response requires precise settings because response time is a limiting factor.

Emergency preparedness plans are also possible scenarios that can develop during the event, including worst-case scenarios. The following elements are very closely associated with each other; they are warning, emergency planning, and rescue operations. Forecasting and warning are key elements of risk management to avoid loss of life.

Emergency maps can identify the area or locations where the hazard is expected to occur and indicate fast and safe routes for evacuation. Emergency maps are developed on a needs basis; however, the real importance of this map lies in the ability to implement it.,

- **End-user map**

(Hazard zoning, emergency, insurance, etc.): There is a variety of end-user maps. They are all deduced in one or the other way from the base maps. The end-user maps address different planning issues and are developed according to the need. Here only end-user maps (hazard zoning and emergency map) (World Meteorological Organization, 2013).

6.2 Geographic Information Systems

There are multiple definitions of Geographic Information Systems (GIS) perhaps the most common definition is the one of the National Center of Geographic Information and Analysis-a GIS that defined it as "a system of hardware, software and procedures to facilitate the management, manipulation, analysis, modeling, representation, and display of georeferenced data to solve complex problems relating to planning and management of resources" (Hulbutta, 2009).

GIS is considered an essential tool for resource planning and management. GIS allows the user to analyze spatial information, modify data, map, and display results. It is used in a wide range of applications, such as risk management, identification of risk areas, and other multiple applications.

The power of GIS comes from the ability to create interactive queries and connect different information in a spatial context and reach a conclusion about this relationship. New information can also be created and presented in different ways to suit specific purposes. The information consists of several different layers containing a specific geographic reference so that they all can be linked together (Fig. 6.2).

Each layer consists of one type of information and together forms a database containing information, that information can answer the various questions in mind. Making the GIS a good tool for different types of analysis, for example, the consequences of specific climatic events in a given area can be assessed and an appropriate plan of action developed (ESRI, 2013).

6.2.1 Nature of Geographical Data

The map is the most common form where geographic data is represented, and geographic features include four main components: its geographic position, attributes, spatial relationships, and time. Geographical data are essentially a form of spatial data. The GIS requires the use of a common coordinate system for all data sets that will be used together.

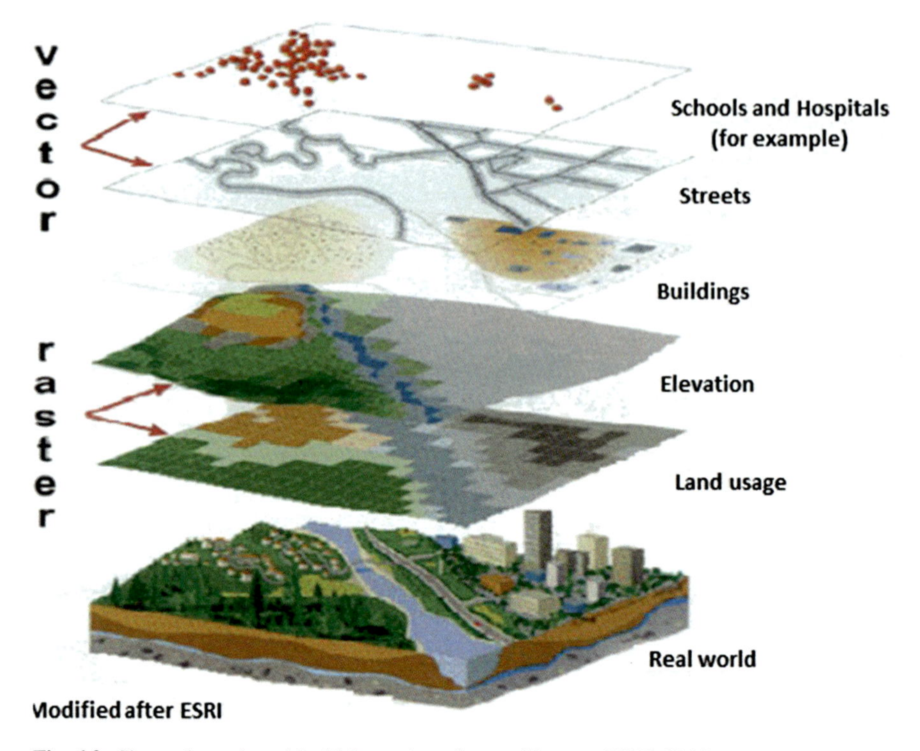

Fig. 6.2 Shows the real world which consists of several layers. (ESRI, 2013)

6.2.2 *Data Entry*

Both spatial and attribute data are entered into a computer system by different input devices like scanners, digitizers, keyboards, mice, etc. For entering spatial data scanner, digitizer, the mouse is used, while the attribute data representing in reports and tables among other things enter my keyboards. Since the data are derived from a different set of sources, they have a different set of scales, projections, reference systems, and so on. So there is a great need to standardize the database according to common standards.

6.2.3 *Building GIS*

The use of Geographic Information Systems (GIS) is a useful tool for various and diverse environmental studies, a particularly powerful system for studying climate change and sea-level rise with spatial representation and analysis, where it is easy to explore the problem and develop a strategy to reduce it.

Steps to build GIS begin with data collection and processing and finally the output results. In the case of **data collection**, data for the construction of geographic information systems include topographic maps, socio-economic data, statistical data, scheduled data, and field surveys conducted to verify the accuracy of the data obtained (Hulbutta, 2009).

As for **data processing**, it passes through a set of stages that begin with the **input stage**. Maps are the main spatial input of GIS through different methods, either through digital conversion or by importing data from other computer information systems. Then, **the editing stage** is the correction of the error that may result from the previous stage (input stage) as it involves many types of errors, so this step is very important to correct and complete errors and add more preliminary data.

This is followed by the **data management** phase where data is stored and retrieved from the database. The methods used to implement these functions affect the efficiency of the systems tool for all operations with different data. Then the **Data Manipulation and Analysis phase**. This phase identifies information that can be generated by GIS. Through this phase, it is possible to do several things, including relate multiple data files by shared key fields and treat the resulting collection as a unit for all tabular processing functions, including data entry, analysis, and report generation. Calculate automatically values of new or existing fields with arithmetic expressions or table lookup in related files. Modify the database by adding and removing files or fields. Finally summarize user-specified fields (ESRI, 2013).

The final form of **output** can be either in the form of maps, reports, or tables. The desired format is determined according to users' needs and therefore user participation is important in determining output requirements. The output is the procedure by which information from the GIS is presented in a form suitable to the user.

The interdependence of ecosystems and human impact on the environment is a complex challenge for governments and individuals, as well as for environmentalists and specialists from all disciplines. GIS technology is used to support and communicate information to decision-makers and the public. GIS allows the integration and analysis of multiple layers of location-based data, including environmental measurements, which means GIS is a powerful software that allows the linking of an unlimited amount of information to a particular geographic location. The importance of geographic information systems is to ensure accurate reporting while improving data collection, improving decision-making processes, increasing productivity with simplified work processes, providing better analysis of data and multiple display options, and developing predictive scenarios for several different studies (Blaikie et al., 2003).

It is useful to have a good idea of the actual needs of the designated information before starting to compile and input data, as this will help prioritize spatial data. So, the next lines explore vulnerability mapping as it is the main point in this work.

6.3 Egypt

6.3.1 Background

Egypt lies between Latitude 22° and 32°, and the country's maximum distances are 1024 km from north to south, and 1240 km from east to west. Egypt is bordered by the Mediterranean Sea to the north, by Sudan to the south, by the Red Sea, Palestine, and Israel to the east, and by Libya to the west. The total area of Egypt is 1,001,450 km2, with a land area of 995,450 km2 and a coastline of 3500 km on the Mediterranean and the Red Sea. The surface level extremes range from 133 m below sea level in the Western Desert to 2629 m above sea level in the Sinai Peninsula.

The general climate of Egypt is dry, hot, and desertic, with a mild winter season with rain over the coastal areas, and a hot and dry summer season. Data collected by the Egyptian Meteorological Authority and local universities for the period 1961–2000 indicate that there is a general trend towards warming of the air temperature, with increases in the number of hazy days, the misty days, the turbidity of the atmosphere, frequency of sand storms and hot days. At the time of the last census (2006), the total Egyptian population amounted to 76.5 million, with an average growth rate of about 2.3% per year. In 2006, the GDP and the GNP amounted to about 107 billion US$ and 113 billion US$ respectively. Egypt is categorized as a lower middle-income country, with its GNP per capita being 1556 US$ in 2006.

A survey of the detailed quantitative assessment of the vulnerability of the Nile delta coast of Egypt to the impacts of sea-level rise is presented. GIS and remote-sensing techniques are used together with ground-based surveys to assess the vulnerability of the most important economic and historic centers along the coasts, the cities of Alexandria, Rosetta to the west, and the city Port-Said to the east.

Results of the analysis of satellite images and GIS (El Raey et al., 1997) indicate that, in these cities alone, over two million people will have to abandon their homes, 214,000 jobs, and over $ 35.0 billion in land value, property, and tourism income may also be lost due to an SLR of 50 cm. The loss of the world-famous historic, cultural and archeological sites is unaccountable for. The vulnerability of other low lands in Egypt, outside these cities remains to be assessed.

Development of institutional capabilities for ICZM and upgrading awareness are highly recommended for adaptation in the long run. Periodic nourishment of Alexandria and Rosetta beaches, detached breakwaters, and dune fixation are recommended for no regret management measures. A UNDP project is already in progress.

The vulnerability of various resources of Egypt to the impacts of climate change has been recognized for a long time (Sestini, 1991). In particular, quantitative vulnerability assessment of water resources, agricultural resources, and coastal zone resources has been investigated thoroughly. The coastal zone of Egypt,c in particular, is most vulnerable to the impacts of climate changes.

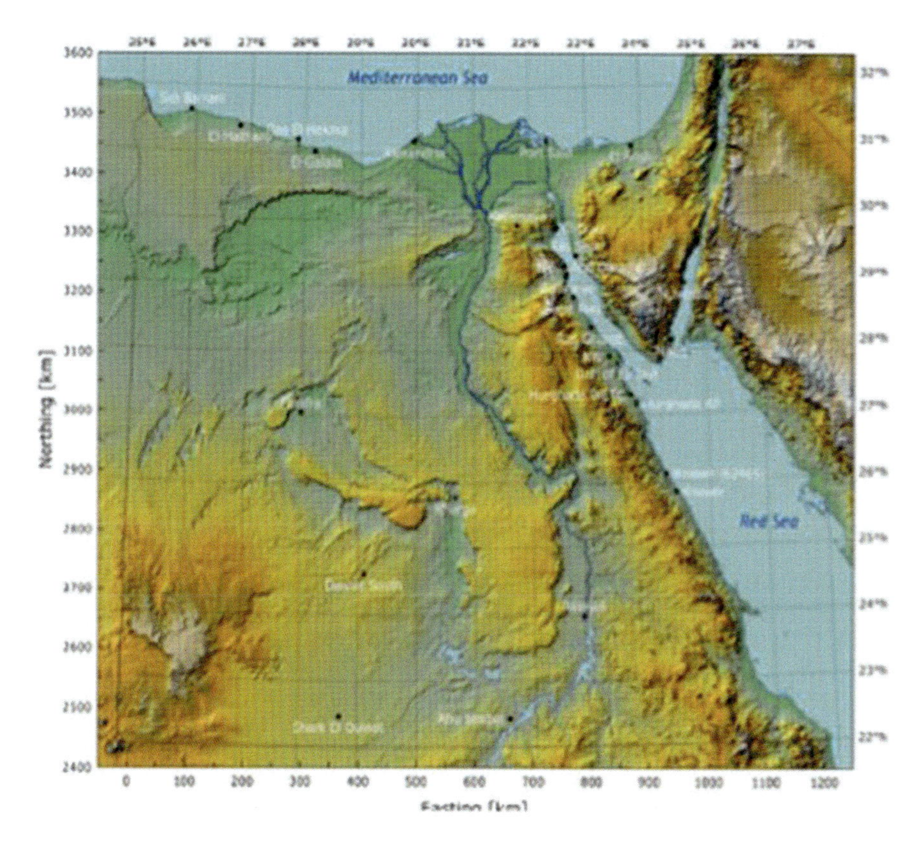

Fig. 6.3 Topography of Egypt

change, not only because of the impact of sea-level rise but also because of the impacts on water, agriculture (food security) and tourism resources, and human settlements. A framework of an action plan has recently been advanced by Egyptian authorities.

6The shoreline of Egypt extends for more than 3500 km along the Mediterranean Sea and the Red Sea coasts, (Fig. 6.3). The Nile delta coast under `consideration, constitutes only 250 km and hosts several highly populated cities, such as Alexandria, Port-Said, Rosetta, and Damietta. These cities also encompass a large portion of the Egyptian industrial and economic sectors. Also, the Nile delta coastal zone includes a large portion of the most fertile low land of Egypt. As a result of the large human activities in these cities, it suffers from some major problems including population pressure, interference of land use, pollution, waterlogging, and lack of institutional capabilities for integrated management.

The objective of this work is to survey the results of a detailed quantitative assessment of the vulnerability of important cities of Egypt to the Mediterranean Sea; namely Alexandria, Rosetta, and Port-Said.

Geographic information systems, remote sensing, and modeling techniques together with ground-based surveys were used to assess potential impacts on each sector and to evaluate socioeconomic losses. The stepwise vulnerability assessment procedure suggested by IPCC has been followed to identify and quantify potential risks of each environmental sector for each district of these cities. A summary of the main results is presented below.

6.3.2 Alexandria City

Alexandria city is located to the west of the Rosetta branch of the River Nile and is famous for its beaches, historic and archeological sites. It is the second-largest city in Egypt with a population of about 4.0 million. It hosts the largest harbor in the country and about 40% of the Egyptian industrial activities. During summer, the city receives over a million tourists. The extension of the city to the south is delayed by the existence of a large water body, Lake Maryut, south of the city. The level of water in Lake Maryut is kept at 2.8 m below SL by continuous pumping of lake water, into the Mediterranean.

To assess the impacts of SLR on Alexandria city, a multi-band high-resolution LANDSAT image (TM, September 1995) of the city is classified to identify and map land-use classes. A geographical information system is built and upgraded in ARC/INFO environment and included layers of

- City district's boundaries
- Topographic maps,
- Land cover from satellite image classification,
- Land use classes,
- Population distribution and employment of each district

Figure 6.4 shows a part of the classified image for Alexandria city (Fig. 6.5).

A scenario of sea-level rise of the city of Alexandria of 0..25, 0.5, and 1.0 m over the next century is assumed, taking land subsidence (2.0 mmyr^{-1}), into consideration. The percentage of the population and land-use areas at risk for each scenario level are identified and quantified by GIS analysis. Table 6.1 shows results of the risk of inundation due to each scenario, if no action is taken, or 'business as usual. These results, together with statistical ground-based employment data are extrapolated to assess the potential loss of employment for each sector. This is also presented in Table 6.2.

Results of the analysis of satellite imagery and the GIS (El Raey et al., 1997, 1999; EL Raey & Atricia, 2011) indicate that for an SLR of 0.5 m if no action is taken, an area of about 30% of the city will be lost due to inundation. Over 2 million people will have to be moved away 195,000 jobs will be lost and an economic loss of land and properties of over $30 billion is expected over the next century. Figure 6.3 shows estimates of losses for each sector of each district in the city of Alexandria.

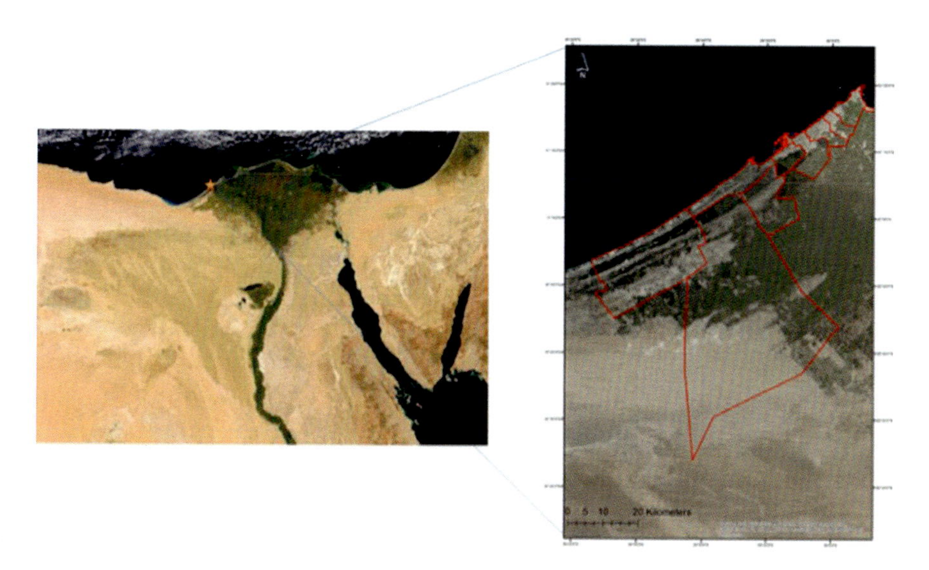

Fig. 6.4 Egypt Nile delta with Alexandria to the west and Port Said to the east

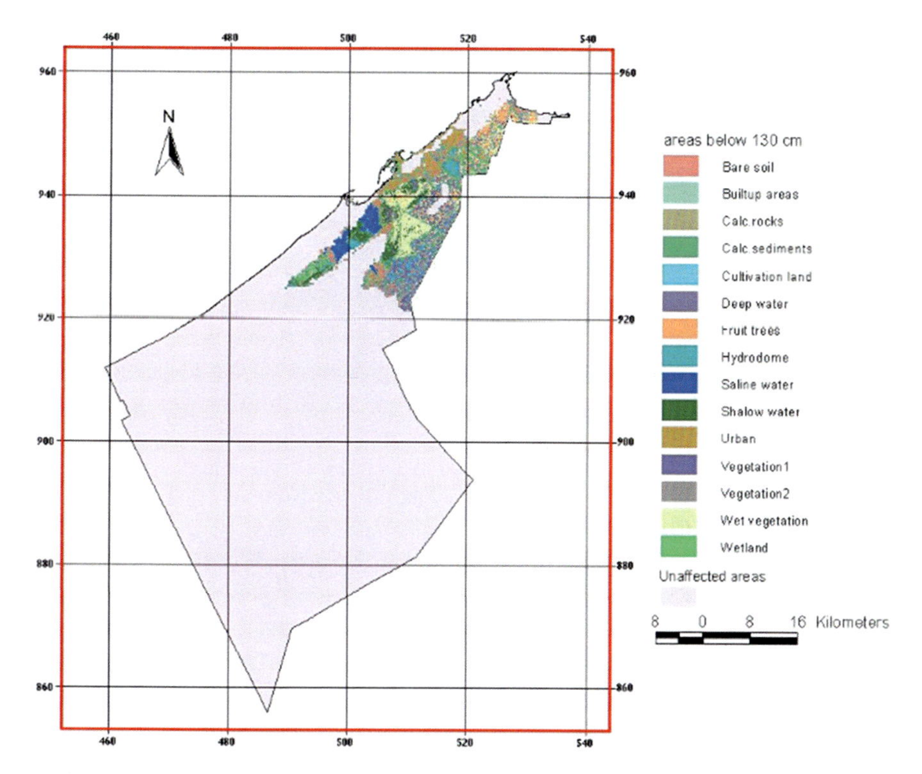

Fig. 6.5 Alexandria Governorate and districts classified image

Table 6.1 Percentage areas, populations, and land use above each elevation contour

Sector/elevation (m)	0.0	*SLR* = 0.25	*SLR* = 0.5	*SLR* = 1.0
Population	54.9	40.1	33.1	24.0
Beaches	98.7	89.1	52.2	36.0
Residential	73.8	72.5	60.7	48.0
Industrial	46.1	43.9	34.1	27.8
Services	54.1	44.8	24.1	17.8
Tourism	72.0	69.0	51.0	38.0
Restricted area	80.0	79.0	75.0	73.0
Urban	62.0	56.0	44.0	33.0
Vegetation	45.0	41.0	37.0	25.0
Wetland	53.0	51.0	42.0	2.0
Bare soil	85.0	76.0	71.0	69.0

Table 6.2 Population expected to be displaced and loss of employment in each sector due to SLR scenarios in Alexandria Governorate

Year sector	2010, SLR = 18 cm	2025, SLR = 30 cm	2050, SLR = 50 cm
Area loss (km^2)	114	190	317
Population displaced × 1000	252	545	1512
Employment loss			
a-agriculture	1370	3205	8812
b-tourism	5737	12,323	33,919
c-industry	24,400	54,936	151,200
Total loss of employment	32,507	70,465	195,443

Given the severe losses in Alexandria due to sea-level rise, and the huge development in progress, and the frame of 'no regrets policy', it is suggested that both a strategic plan and a short-term plan, have to be adopted. An integrated coastal zone management approach, including building provincial institutional capacity, upgrading awareness, and expanding the GIS and decision-making process, must be adopted. On the short-term approach, it is concluded that periodic beach nourishment is the cheapest alternative available, against direct inundation. However, the impact of saltwater intrusion impact on water resources, soil salinization, and building structures has to be further investigated and mitigated (Fig. 6.6).

Because of the severe losses in Egypt due to sea-level rise, and the huge development in progress, and in the frame of 'no regrets policy', it is suggested that both a strategic plan and a short-term plan, have to be adopted. Integrated coastal zone management (ICZM) approaches, including building provincial institutional capacity, upgrading awareness, and expanding the GIS and decision-making process, must be adopted. On the short-term approach, it is concluded that periodic beach nourishment is the cheapest alternative available, against direct inundation. However, the impact of saltwater intrusion impact on water resources, soil salinization, and building structures has to be further investigated and mitigated.

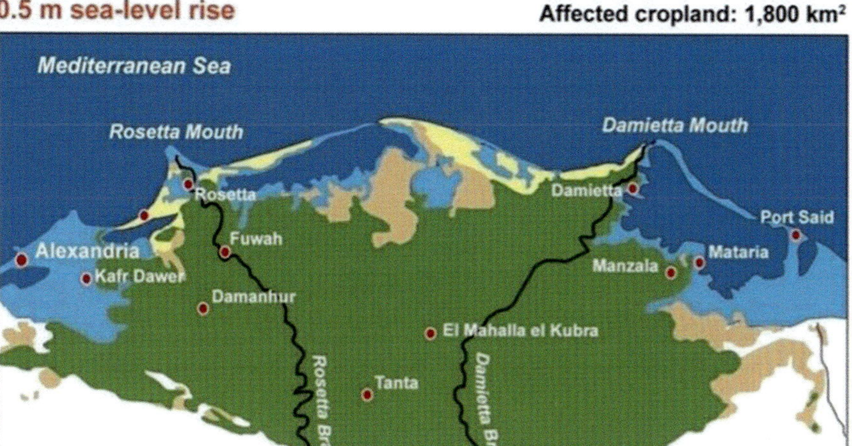

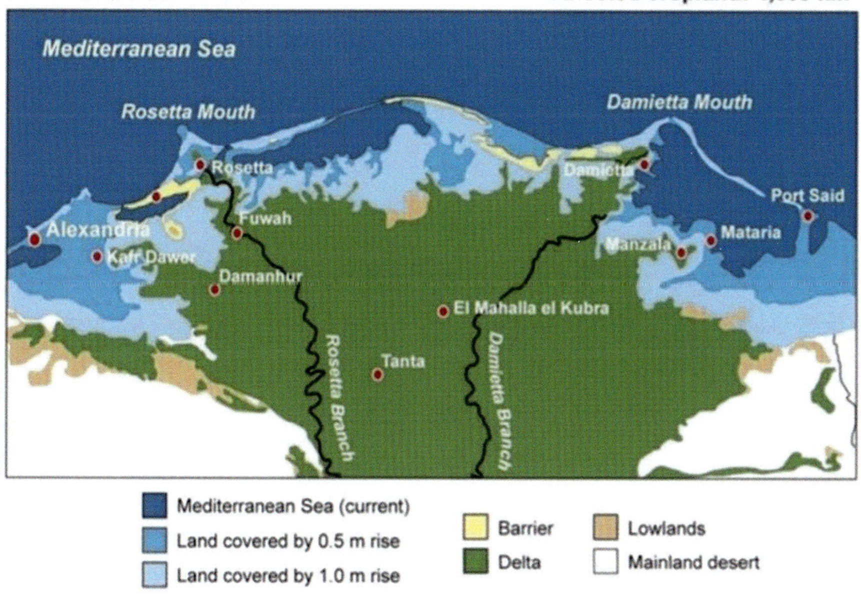

AR FitzGerald DM, et al. 2008.
Annu. Rev. Earth Planet. Sci. 36:601–47.

Fig. 6.6 The assessment of sea-level rise in the Nile delta, of Egypt. (According to FitzGerald et al., 2008)

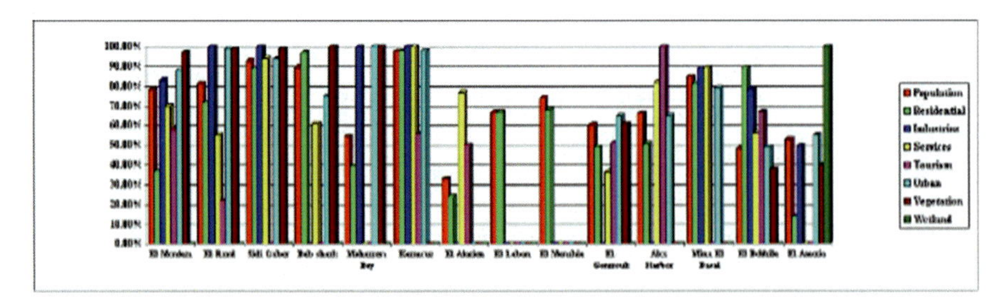

Fig. 6.7 Percentage vulnerability to SLR of Alexandria different localities of 1.0 m

6.3.3 *Rosetta City and Vicinity*

Rosetta city is a well-known Pharaonic and Islamic city located in the Rosetta region near the intersection of the Rosetta branch of the River Nile with the Mediterranean, east of Alexandria.

It has a population of 2.5 million people. Excessive erosional rates have been observed near Rosetta promontory, due to cessation of sediments after building the High Dam on the River Nile about 1000 km to the south. The region surrounding the city is well known for its water-logging and water-bogging problems.

To assess quantitatively possible impacts of sea-level rise, a high-resolution GIS of the city and surrounding area is built. The GIS includes layers of:

1. Political boundaries
2. Topographic maps
3. Land use (urban. Vegetation, historic,. . .)
4. Socioeconomic maps.

Detailed ground surveys were carried out to update and verify available maps. Impacts were evaluated based on land cover/land-use losses for every 20 cm of sea-level rise (SLR). Figure 6.4 shows a series of GIS plots illustrating vulnerable and safe areas for each 20.0 cm of SLR. Results were extended to estimate employment losses and economic losses for a 50 cm SLR scenario based on extrapolated scenarios. Estimates showed that about 1/3 of the employment power in the city will be affected and a loss of about $2.9 billion in land and property is expected, over the next century. The loss of historic and archeological sites is again unaccounted for. Table 6.3 presents the results of the impacts on each employment sector.

The government has built a massive sea wall near the tip of the promontory, as a protective measure against erosional problems, and excessive salinization. However, recent observations indicate that this massive hard structure is seriously challenged by coastal erosion. Beach nourishment may be a possible solution. Plans for the development of the area should seriously consider the ICZM approach in which decision-making is based on a detailed GIS suitability analysis.

Table 6.3 Results of the impacts on each employment sector

A sector of economic activities	Actual Employment × (10^3)	Employment Expected loss (jobs)
Agriculture and fishing	18.5	4633
Conversion industries	4.5	502
Municipal utilities	0.139	78
(electricity, gas, and water supplies)		
Construction and buildings	1.355	433
Commercial	2.877	1640
Transportation	2.191	1248
Community services	6.275	3576
Total	35.936	
		(33.7% of all employment)

6.3.4 *Port-Said City*

Port-Said city is located on the Mediterranean Sea to the east of the Damietta branch of the River Nile, at the entrance exit of the Suez Canal. In addition to its strategic position, it is the second-largest tourist and trade center of Egypt on the Mediterranean. Lake Manzala, the largest of Nile delta lakes is located just to the west of the city and receives a sizable amount of effluent pollution from various sources on the eastern region of the delta. A field survey has concluded that the Mediterranean coast of the city hosts the most important economic and tourist areas of the city, in addition to its relatively low elevation. The impact of SLR on this part of the city is carried out in detail (Fig. 6.8 and Table 6.4).

6.4 Saudi Arabia

6.4.1 *Background*

The Kingdom of Saudi Arabia comprises about four-fifths (80%) of the Arabian Peninsula, occupies approximately 2,250,000 km^2 area, and bordered on the west by the Red Sea; on the east by the Arabian Gulf, Bahrain, Qatar, and the United Arab Emirates; on the north by Jordan, Iraq, and Kuwait; and on the south by the Sultanate of Oman and Yemen.

Saudi Arabia‘s Red Sea coast on the west stretches to approximately 1760 km, while its eastern coast on the Gulf covers 650 km, including 35 km^2 of mangroves and 1480 km^2 of coral reefs. The country has an arid climate with an average annual rainfall of 70.5 mm. Almost two-thirds of the country is arid steppe and mountains with peaks as high as 3000 m, and most of the remainder is sand desert (Fig. 6.9).

Fig. 6.8 A satellite figure of Port Said and Port Fouad

Table 6.4 An assessment of conditions at the Nile Delta (Frehy etal 2021) has been carried out recently on Alexandria and found that the city is hugly vulnerable. The government in Cairo has taken steps to carry out integrated coastal zone management (ICZM) and have asked UNDP Cairo to carry out risk assessment. Estimation of expected economic loss in land cover Rosetta – (1995–1996 prices)

Land cover	Lost area (m2)	Square meter price 1990, in the US ($)	Total estimated cost in the US ($) (Millions)
Coast bare land	3,581,740	200	716
Marine bushes	4,504,750	150	675
Palm cultivation	5,790,260	200	1158
Bare land	27,508	250	7
Urban clusters	1,251,204	300	375
Total	15,155,462	–	2931

Saudi Arabia consists of a variety of habitats such as sandy and rocky deserts, mountains, valleys, meadows, salt-pans ('sabkhas'), lava-areas, etc. It includes most types of terrain which can be generally divided into two distinct groups of rocks; the Arabian shield and the Arabian Platform. Seventy-eight percent of the population is concentrated in the urban areas (Saudi, 2005).

Coral Reefs represent the most significant habitat found along the Saudi shores (both Red Sea and Arabian Gulf). Coral reefs play an important role in the coastal

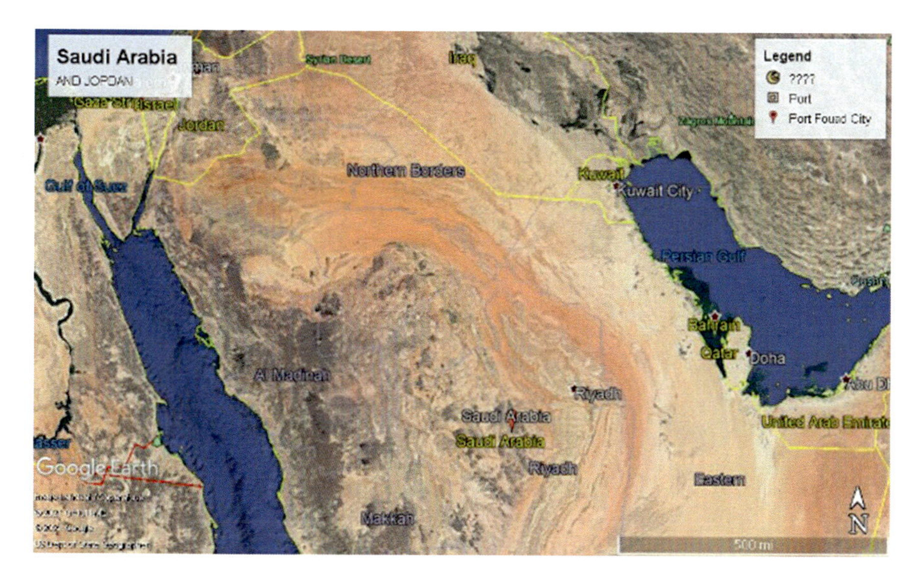

Fig. 6.9 Saudi Arabia and Jordan to the North

ecosystem. These reefs as well as the Mangrove forests form the basic framework of tropical habitats and provide shelter and food for a wide array of marine life. The highest coral diversity occurs in the central Saudi Arabian Red Sea area. Coral reef harbors a longstanding and important artisan fishery.

6.4.2 Vulnerability to Sea Level Rise

While mangroves are found scattered along much of the Red Sea coast, the major concentration is in the southern red sea where factors such as increased sediments create an environment more conducive to their development. Agricultural development, properly planned and managed, could be beneficial to certain coastal habitats such as mangroves. Mangroves have a variety of values: they provide food in the form of detritus, shelter for numerous organisms (such as mollusks, crabs, shrimps, and fish) fodder for camels, and goats, and fuel for human use. Mangroves are also important nesting sites for several species of birds.

The development of coastal recreational facilities and coastal villages in the Ras Hatiba area north of Jeddah and shrimps aquaculture along the southern Red Sea coast have contributed to the decline of Saudi Arabia‘s coastal mangroves.

Coastal cities of Saudi Arabia extend along the Red Sea coast as well as the Arabian Gulf coast. Four coastal cities have been selected as the most vulnerable cities to Accelerated Sea-level Rise (ASLR) along the Red Sea namely Jeddah, Rabigh, Yanbu, and Gizan). The selection of these cities was based on the population growth, socio-economic activities, and historical and cultural importance to the

Fig. 6.10 The Saudi Arabian Cities on the Red Sea

Kingdom. Again, on the Arabian Gulf cities such as El Khafji, Al Jubail, Al Dhahran and Al Khobar (Fig. 6.10) could be considered vulnerable.

The following potential impacts were identified in a study on Saudi SLR impacts (Saudi, 2005).

An increase in sea level rise will increase the intrusion of saline water from both the Arabian Gulf and the Red Sea into coastal aquifers, which will potentially affect the freshwater supply in coastal zones. In cases of flooding in coastal areas, saltwater will further intrude into aquifers. This intrusion will increase the demand for 88 h water from other sources, mainly Desalination Plants. At the same time, sea-level rise will increase the saltwater intrusion of estuaries, potentially benefiting marine fish at the expense of freshwater ecosystems.

Groundwater levels in these areas might also be affected by the intrusion of saline water. The groundwater level itself and the soil structure determine the potential for the intrusion of saline waters. Managed areas with a reduction in groundwater level because of drainage are more vulnerable to intrusion.

Sehat and Qateef are the main agricultural cities along the Arabia Gulf and Gizan along the Red Sea. These cities could be impacted by Accelerated Sea Level Rise. Recently increase in soil salinity has been observed in some of these coastal cities. This increment has impacted the production of cultivated products. It is suspected that saltwater intrusion may be one of the factors impacting agricultural activities.

One of the most significant impacts of sea-level rise is the acceleration of coastal erosion as well as the inundation of mangroves, wetlands, and coral reefs. The rich biodiversity of the wetlands in Saudi Arabia is seriously threatened by the loss of wetlands due to sea-level rise. The effect of sea-level rise will depend on the type of mangrove forest. These mangrove forests may either keep pace with the rising sea level rise or may be submerged. Large-scale changes in species composition and zoning in mangrove forests are also expected due to changes in sedimentation and organic accumulation, nature of coastal profile, and species interaction.

An additional threat of Accelerated Sea Level Rise affecting the Saudi Arabian coasts will come from an exacerbation of sandy beach erosion. As the beach is lost, fixed structures nearby are increasingly exposed to the direct impact of storm waves, and will ultimately be damaged or destroyed unless expensive protective measures are taken. It has long been speculated that the underlying rate of long-term sandy beach erosion is two orders of magnitude greater than the rate of rising sea level. Therefore, any significant increase in sea level has direct consequences for coastal inhabitants.

Results from the studies on various aspects of the impacts and possible responses to sea-level rise on the Saudi Arabian coasts indicate that a sizable proportion of the Arabian Gulf and the Red Sea will be affected by a combination of inundation and erosion, with consequent loss of developed properties including industrial, recreational and residential areas.

No detailed socioeconomic study of accelerated sea-level rise has been carried out yet in Saudi Arabia. However, it has been estimated that 20% of Saudi Arabian coastal areas have been subject to development, 130 km along the Arabian Gulf coasts and 352 km along the Red Sea coasts. A conservative scenario of 1% annual coastal development was applied on the Arabian Gulf and the Red Sea coasts. This scenario was applied to the coastal erosion model to estimate the area of sandy beaches that may demolish as a result of sea-level rise.

Considering the annual coastal development in the Kingdom is 1% and the IPCC Sea Level Rise projection Scenarios towards the year 2100 and by applying Bruun model to estimate the high-risk areas subjected to coastal erosion along the Arabian Gulf, it was found that:

- For the Low Sea Level Rise Scenario (LSLRS) of 0.2 m rise, 401 hectares of sandy beaches are estimated to be lost by the year 2100.
- For the Medium Sea Level Rise Scenario (MSLRS) of 0.49 m rise, 984 hectares of sandy beaches are estimated to be lost by the year 2100, and
- For the High Sea Level Rise Scenario (HSLRS) of 0.86 m rise1, 726 hectares of sandy beaches are estimated to be lost by the year 2100.

6.4.3 Institutional and Practical Adaptation Measures (El Raey, 2008)

It is well recognized that the following adaptation measures at least are needed:

1. An integrated institutional structure must be developed. A Regional Circulation Model (RCM) has to be developed and a strong institutional monitoring system has to be established.
2. A monitoring system of tide gauges and systematic observations of the coastal zone with provisions for land subsidence must be established.
3. An early Warning Center for climate change, which must be able to predict flash flood and other phenomena, must be established

6.5 Libya

6.5.1 Background

Officially the Great Socialist People's Libyan Arab Jamahiriya is located in North Africa. Bordering the Mediterranean Sea to the north, Libya lies between Egypt to the east, Sudan to the southeast, Chad and Niger to the south, and Algeria and Tunisia to the west. With an area of almost 1,800,000 km2, Libya is the fourth largest country in Africa by area. The capital, Tripoli, is home to 1.7 million of Libya's 5.7 million people. The three traditional parts of the country are Tripolitania, Fezzan, and Cyrenaica. Libya has the highest HDI in Africa and the fourth-highest GDP (PPP) per capita in Africa as of 2009, behind Seychelles, Equatorial Guinea, and Gabon. These are largely due to its large petroleum reserves and low population (Fig. 6.11).

Libya has a small population residing in a large land area. Population density is about 50 persons per km^2 in the two northern regions of Tripolitania and Cyrenaica but falls to less than one person per km^2 elsewhere. Ninety percent of the people live in less than 10% of the area, primarily along the coast. About 88% of the population is urban, mostly concentrated in the two largest cities, Tripoli and Benghazi. With

Fig. 6.11 EGYPT and Eastern Libya

the longest Mediterranean coastline among African nations, Libya's mostly virgin beaches are an important social gathering place.

6.5.2 *Vulnerability to Sea Level Rise*

Regardless of the inland areas that lie below the sea level, most of the coastal area at the southern part of Sert Bay is vulnerable to direct inundation and/or saltwater intrusion as seen by comparison to the simulation of sea-level rise of 1 m. Besides, many of the coastal cities such as Benghazi, Libya's second-largest city are considered vulnerable to sea-level rise and potential impacts of extreme storm events. Saltwater intrusion on already scarce groundwater resources may also be damaging to important water resources.

Sert Bay as Seen Today a Sert Bay in Libya After 1.0 Sea-Level Rise (Fig. 6.12)

Practical and Institutional Adaptation Measures

- Building up institutional and human capacity for monitoring coastal parameters including land subsidence, developing databases and modeling are necessary prerequisites for adaptation
- Development of proactive planning and integrated coastal zone management is also necessary
- Carrying out a complete vulnerability survey of the coastal zone and associated protection measures are necessary
- A program of upgrading awareness and resilience of the community must start for the decision-makers and population at large.

Fig. 6.12 a & b Sert Bay in Libya in case the sea rises by 50 cm. Notice the formation of a lake on the eastern side of the bay

6.6 Morocco

6.6.1 Background

The Moroccan Atlantic coast is the most important area for the national economy, taking into account its demographic and economic weight. The Atlantic coast of Morocco is divided into:

61% of the urban population of the large cities
80% of permanent manpower of industries

53% of the tourist capacity
92% of the foreign trade.

However, this situation is changing with the realization of the new port Tangier Med and the economic development it brings to the Mediterranean coast of Morocco (Etude V&A, 2006). The total coastal population represented more than 50% of the population of Morocco, increasing on average by 2.77% per annum, whereas the total population of Morocco increased only by 2.5% per annum on average for the same period. The densely populated coast is subject to major pressures from human development and this is only projected to increase in the future. The urban population of the coastal areas did not cease growing since the beginning of the century. The density of the population reaches 162 habitants/km ^{2} between Kenitra and Casablanca, compared to the 93 inhabitants/km ^{2} on the Mediterranean coast (Etude V&A, 2006).

77% of Morocco's industrial activities are located in the coastal areas and 98% of Moroccan foreign trade relies on shipping as a transportation method. Therefore, the economic importance of coastal areas is imperative. Also increasing popularity of beach tourism highlights the importance of coastal areas to Morocco's economy. Beach holiday was the primary motivator for both national and international tourism in Morocco, and the coastal areas boast more than 50% of accommodation capacity, the most dynamic centers being Agadir and Casablanca (Etude V&A, 2006).

6.6.2 Vulnerability to Sea Level Rise

The coasts of Morocco already strongly weakened by human activities would be confronted with major socio-economic and environmental difficulties if no adaptation measures or vulnerability studies are undertaken. A 2002 study by the Ministry of Environment identified two coastal zones as being most vulnerable: The coast of Saïdia and the coast of the bay of Tangier.

Fig 20 through Fig. 21 shows the variation of land use and elevation along the coastal strip of the Mediterranean coast of Morocco. The figures demonstrate the vulnerability of the dense urban areas and tourist units to potential impacts of sea-level rise and saltwater intrusion. Figure 6.12 presents a simulation of the land to be flooded in case of SLR of 2 m FIGDigDigital (Fig. 6.13 and 6.14).

Bio-Geophysical Impacts of SLR Impacts of sea-level rise on coastal areas are numerous and varied but the most significant are generally and in the case of Morocco: the low-lying coastal flooding, coastal erosion, and salinization of estuaries and coastal aquifers

- **Phenomenon of flooding**: The first threat is that of flooding of evergreen coastal marine waters which are currently poorly or partially emerged, as the shores of deltaic plains, salt marshes, mangroves, coral reefs, etc. There seem to have

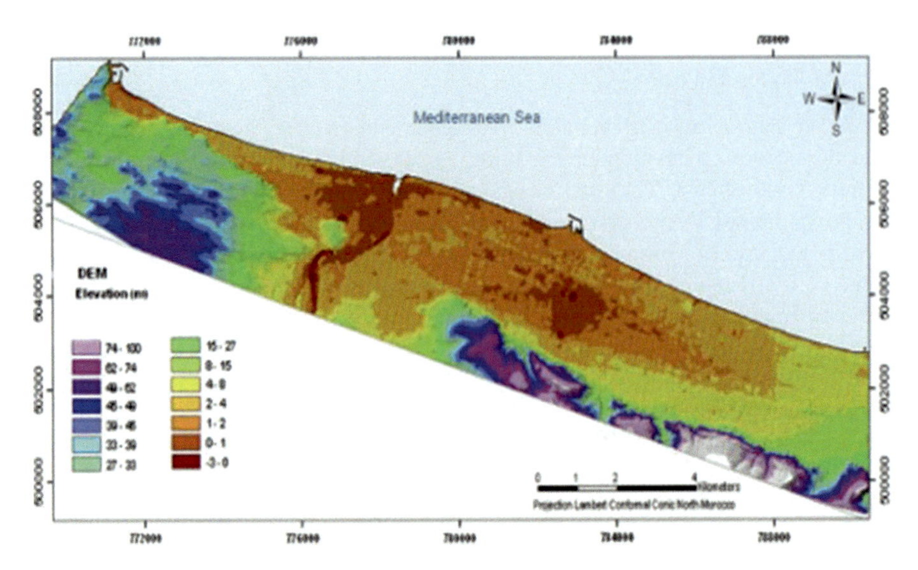

Fig. 6.13 Digital Elevation Model DEMof Eastern Mediterranean coast of Morocco (Snoussi et al., 2008)

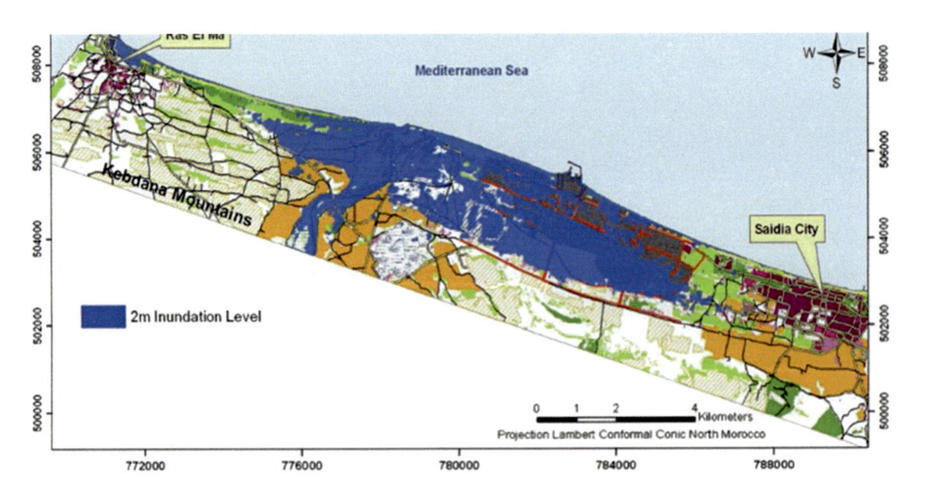

Fig. 6.14 The coastal area of Eastern Mediterrnean illustrating inundation level (Snoussi et al., 2008)

increased the intensity and wave heights over the past three decades, and that the waves of storms were more frequent with the expected climate CC.

In Morocco, the importance of the coastline (3500 km), is marked by a large number of environments marginal-coastal and coastal wetlands such as lagoons, estuaries, berries or less closed beaches, coastal islets, and predisposes. These areas generally have a low topography and high vulnerability to flooding by seawater.

Coastal wetlands of Morocco are renowned for their ecological values, as well as goods and services they provide to local populations, who long have developed know-how traditional use of natural resources. The flooding of these wetlands could ultimately damage their ecological, social, and economic conditions and force people who depend on their activities (including agriculture and livestock) to new conditions by their conversion to fishermen, aqua-culturalists, and others. The barrier beaches could also readily admit a cross-cut larger volume of seawater, which could be detrimental to the vegetation of halophilic salt marshes, which will be subject to a longer duration of submergence and higher salinity, for all species sensitive to salinity and water column, for salt marshes which could be used for grazing, and the protection works (breakwaters, jetties, groins, etc..)

- **Coastal Erosion**:

In Morocco, two-thirds of the beaches are eroding, particularly among the cliffs Jorf Lasfar and Oualidia recede seven major coastal environments of the Atlantic are total or partial closure and port facilities have increased volumes dredging of 15% during the last 5 years. It is often difficult to attribute these changes to a single cause. While the share of anthropogenic activities is undeniable, these findings could provide tangible evidence of a rise in sea level combined with strong swells storms becoming more frequent.

6.6.3 *The Phenomenon of Salinization*

The phenomenon of salinization (Increasing the salt content of freshwater by seawater intrusion) is likely to affect estuaries and coastal aquifers. Indeed, the SLR may lead to an increase in the depth of estuaries but especially a greater penetration upstream intrusion of saltwater. Increased salinity of surface water will no doubt have an impact on the fauna and flora. It seems that disturbances of this kind should be more sensitive in areas with a low tidal range.

Concerning fresh groundwater, if the sea level increases, the separation between the continental freshwater and saltwater marine will move sideways to the ground, and level piezometric groundwater will be enhanced. Thus, this will result in a reduction in the volume of fresh groundwater through salt wedge intrusion. The hydrogeology of low areas, often composed of alluvial sedimentary permeable soil can be changed. The aquifers at risk of a rise of the same order as that of sea level would have a considerable impact on vegetation, and even at ultra-high elevations.

Many studies have shown salinity more or less advanced into coastal aquifers. This has been attributed to several anthropogenic factors (including over-pumping, the return of irrigation water, etc.) which are the main cause of the decreased levels of freshwater as a result of the invasion of the salt wedge. However, no study has explicitly linked salinity to sea level rise due to CC. Thc penetration of the salt tide further upstream is felt in several estuaries, without series of temporal data can confirm. Many living species may disappear because they cannot adapt to changes in salinitY.

Socio-Economic Impacts of SLR

- **Agriculture:** The intrusion of saltwater into groundwater can affect the quality of products and yields significantly.
- **Water resources**: As the sea rises, the fresh groundwater and surface water can be displaced by saltwater, which can have significant adverse impacts on the drinking water supply. Following CC, prospective studies have shown that in 2020 the water cut by 10–15%. The SLR would therefore exacerbate the reduction in coastal areas.
- **Fisheries and Aquaculture**: SLR could affect coastal activities such as (fishing, coastal lagoon, the harvest of shellfish and seaweed, etc.). Besides, aquaculture activities that are related to physical and chemical conditions (salinity, chemistry, temperature, oxygenation, etc.) specific husbandry should adapt to changes associated with SLR, the increased temperature, and salinity.
- **Tourism**: Impacts of CC and the SLR on tourism will affect quality and availability of water resources, erosion of beaches, and loss /degradation of coastal infrastructure. A very powerful example is that of the Bay of Tangier, which represented the first station national tourism in the years 1970–1980. It has fallen sharply in the years 1990, mainly because of the chronic degradation of its coastline. Thus, Tangier has lost 53% of its international nights, resulting in a decrease in tourism revenues ($ 20 M / year), income crafts (25%), and tourist transportation (40%) (Snoussi et al., 2008). SLR would have catastrophic effects if nothing is done to rehabilitate and protect the bay.
- Thermal power plants installed on the coast, refining plants, and deposits industrial centers and roads along the coast, are particularly vulnerable to SLR plus swells from storms or surges.
- **The sanitation sector**: In the big coastal cities, were particularly systems of sewage and stormwater that would be threatened by SLR? Coastal stations of water treatment may be damaged and no longer fulfill their function, which will be detrimental to water quality, and consequently to the health of populations.
- **Forests**: Forests on the coast would be affected by the marine invasion and salinization waters of the aquifer.

In conclusion:

1. A change in sea level, even a few inches may, in different segments of coastline, cause a significant withdrawal of the shore either by erosion or flooding;
2. The intrusion of seawater can lead to forms of degradation by salinization in extensive grounds operated by coastal agriculture; Early warning systems of extreme events such as heatwaves, flash floods, and dust storms must be established
3. Policies and measures should be developed based on model studies and participation of stakeholders
4. Upgrading awareness programs of Stakeholders should also be carried out

The main coastal port structures, harbor pools, and sanitation are also vulnerable to rising sea levels. Rising sea levels will not only impact the environment but also different sectors of the economy including in particular tourism, and will require interventions (protection, rehabilitation) which are not always easy or even possible sometimes. Hence the interest is to give priority to the issue of rising sea levels in future decisions for the management of coastal environments.

studies and research to identify vulnerabilities to rising sea levels across Morocco have remained very limited and data that may help in realizing such studies are unavailable. Also, it is urgent to establish a research program integrated with sampling measurements and modeling of the tidal effect of the elevation sea level on coasts and in the image of what has been done in neighboring countries.

Practical and Institutional Adaptation Measures
Based on previous considerations, it is suggested that the following adaptation measures are necessary:

1. Projects for vulnerable sectors (such as the coastline, forest, or precarious human establishments) have to be identified and protection measures worked out.
2. Establishing a strong institutional capacity for monitoring, building a geographic database, modeling, and assessment
3. Adopting proactive planning and integrated coastal zone management approaches for development along the coast
4. Upgrading awareness of decision-makers and stakeholders of the potential impacts of sea-level rise on various aspects of development

But, it is clear that the Moroccan economy, which is still caught up in the problems of development and struggles against poverty, cannot withstand the costs of such projects without sacrificing the major components of its social and economic development programs (education, health, basic infrastructures, rural development, etc.).

6.7 Algeria

Due to its geographical position and climatic characteristics, Algeria is highly vulnerable to climate change. Even a small temperature rise would lead to various socio-economic problems that hinder the development of the country. The models predict that rainfall events are less frequent but more intense, while droughts are more common and longer. The spatial and temporal distribution of rainfall will also change. The analysis of climate data from 1931 to 1990 in northern Algeria reveals a rise in temperature of 0.5 °C would reach an increase of 1 °C by 2020.

A temperature rises of 2 °C is expected by 2050. The decrease of water resources, declining agricultural yields, encroaching desert, the challenge of planning, and the energy consumption for air conditioning are only the initial impacts to which Algeria must find answers supportable economically and socially. Thus, although the contribution of Algeria to global warming is minimal (less than 0.5% of global GHG emissions), the country is very vulnerable and should integrate adaptation into its development policy. We present in this study an analysis of the current situation about support sustainable development and climate change issues, the footprint of Algeria, trends in emissions of CO_2 in Algeria, mitigation and adaptation strategy Algeria, national climate plan, and especially the impact of the new national plan for promoting renewable energy adopted in 2011 and expects to produce 40% of electricity needs from solar. Avenues of consideration that can mitigate the impacts induced by medium-term climate change will also be presented. Due to its geographical position and climatic characteristics, Algeria is highly vulnerable to climate change. Even a small rise in temperature would lead to various socio-economic problems that hinder the development of the country.

The models predict that rainfall events are less frequent but more intense, while droughts are more common and longer. The spatial and temporal distribution of rainfall will also change. The analysis of climate data from 1931 to 1990 in northern Algeria reveals a rise in temperature of 0.5 °C would reach an increase of 1 °C by 2020. A temperature rises of 2 °C is expected by 2050. The decrease of water resources, declining agricultural yields, encroaching desert, the challenge of planning, and the energy consumption for air conditioning are only the initial impacts to which Algeria must find answers supportable economically and socially.

Thus, although the contribution of Algeria to global warming is minimal (less than 0.5% of global GHG emissions), the country is very vulnerable and should integrate adaptation into its development policy. We present in this study an analysis of the current situation concerning support sustainable development and climate change issues, the footprint of Algeria, trends in emissions of CO_2 in Algeria, mitigation and adaptation strategy Algeria, national climate plan, and especially the impact of the new national plan for promoting renewable energy adopted in 2011 and expects to produce 40% of electricity needs from solar. Avenues of consideration that can mitigate the impacts induced by Energy Procedia (2013) 1286–1294 1876–6102 © 2013.

6.8 Tunisia

6.8.1 Background

Based on updated climate projections from the latest IPCC report, this chapter presents the.

vulnerability of different sectors to climate change (water resources, agriculture, and ecosystems, coast and fisheries, tourism, health,

der), by describing the potential impacts of climate change on each sector and their capacity to adapt. An overview of the main ongoing or scheduled adaptation initiatives is also provided, followed by a presentation of the priority axes and strategies of adaptation to climate change in Tunisia.

For this third national communication, the outputs of a set of models used in the latest IPCC report(AR5) with two representative concentration pathways (RCP 4.5 and 8.5) are analyzed to provide climate change projections for Tunisia at both the 2050 and 2100 horizons. These data come from the simulations of all EURO-CORDEX models at a 12.5 km resolution.

6.9 Synthesis of Climate Projections for Tunisia

Past Climate Trends in Tunisia In Tunisia, over the 1978–2012 period, we observe a significant upward trend in annual maximum, average, and minimum temperatures of around 2.1 ° C, with regional disparities.

Observations show a slight upward (statistically insignificant) trend for the total annual rainfall over this same period of observed data. Nevertheless, the disparities between seasons and the interannual variability are very strong. The beginning of the 1980s is known for a succession of several dry years in Tunisia, influencing the general trend of the curves Regarding meteorological and climatic extremes, a change in the frequency of thermal and rainfall extremes is observed over the past period, based on the observed data from the meteorological station S of the Tunisian National Institute of Meteorology (INM).

2050 and 2100 Climate Projections Under RCP 4.5 and RCP 8.5 Scenarios Projections show an increase in annual temperature at the 2050 and 2100 horizons for both scenarios .his increase ranges between 1 ° C and 1.8 ° C by 2050 on average for the set of studied models and between 2 ° C and 3 ° C at the end of the century, with the RCP 4.5 scenario. For tender the RCP 4.5 scenario, the seasonal rainfall projections show a much more pronounced.

variability by 2100 with significant rainfall decreases in summer in southern Tunisia (– 35%) and a slight increase in rainfall in autumn in the northwest of the country (+ 5%). Under the RCP 8.5 scenario, the seasonal rainfall projections show very pronounced decreases in precipitation (−35%) in winter in southern Tunisia and spring in northern Tunisia, which differs from the scenario RCP 4.5.

Regarding climate extremes, Tunisia could experience more frequent and longer heatwaves by.

2100 under the scenario RCP 8.5. On the other hand, the cold waves would decrease as well as the episodes of extreme rainfall.

6.9.1 Water Resources

Among the 36 billion of m3 of rainwater that Tunisia receives each year, 16.3 billion of m3 (45%).

can be mobilized. 4.8 billion m3 (13%) constitute the annual potential for blue water which can be used to meet socio-economic needs6. The main permanent watercourse is the Medjerda River, which has its sources in Algeria and on which is set the Sidi Salem dam, the largest one in Tunisia. 11.5 billion m3 (32%) infiltrates into the soil and constitutes water reserves for rainfed agriculture forests and rangelands. The remaining 19.7 billion m3 (55%) evaporate, are stored in the wetlands, or flow towards the sea.

An increase in the intensity and frequency of dry periods, combined with the increase in temperature, should reduce soil moisture and surface and underground water stocks. These impacts of climate change are likely to be worsened by the increasing water needs, notably for human use, but more particularly for agriculture, given the increase in evapotranspiration and the decrease in soil humidity Groundwater forms 44.5% of Tunisia's water potential, with 226 shallow water tables and 340 deep aquifers. Most of the groundwater comes from deep aquifers in the south, among which the largest resources are non-renewable fossil groundwater (610 Mm3 / year are non-renewable, coastline, the overall decrease of water resources due to climate change could be significant. In particular, we could assist in a drying up of water sources, which constitute the main resources in some rural areas of the country.

Among a set of about 215 water tables in Tunisia, almost a quarter is located in the coastal area. These coastal aquifers store about 290 Mm3, i.e. 40% of the groundwater potential and almost 6% of the total water resources that can be mobilized in the country. Seawater intrusion into the coastal waters tables will contribute to their progressive salinization, especially since many of these aquifers already show signs of degradation (salinization, overexploitation). Sea level rise could be responsible for the loss of 220 Mm3 of water resources, i.e. about 30% of the total groundwater potential and 75% of the phreatic resources.

Water use conflicts are already observed and are intensifying in Tunisia, especially during drought periods. Rural areas relying on springs for drinking will be greatly affected, given the drying up of b.

6.9.2 *Major Initiatives for Adaptation*

Given the fact that Tunisia is already affected by water scarcity in some regions and that the rate of conventional water resources mobilization reaches 92%, various adaptation measures have already been implemented. These measures include investments for a greater mobilization of unconventional water. For instance, the seawater desalination plant in Djerba will be operational in 2018 and the desalination plants in Sfax, Zarrat, and Sousse should be operational by 2020. These investments should allow securing the supply of drinking water until 2030. The potential reuse of wastewater, estimated at 300 Mm^3, is also an ongoing adaptation measure in Tunisia.

The possibility of transferring a part of the surplus water from the extreme north, estimated at 400 Mm3 / year, to the center of the country in the area of Kairouan, is currently under study. This would be done through the reinforcement of infrastructures: connection of dams, doubling of transfer lines, the building of new storage facilities. A national program for artificial groundwater recharge, by surface water or treated wastewater, is also initiated. However, the volume of water injected underground is dependent on the availability of surface water, which has decreased from 66.2 Min 1996 to 30.52 Mm3 in 2015.

Water and soil conservation is part of a new strategy integrating the impact of climate change on the national territory. Adaptation measures are also initiated in the agricultural sector to reduce the water demand, such as the use of conservation agriculture9. A national water-saving strategy in the agricultural sector implemented since 1995 led to the equipment of more than irrigated perimeters with water-saving techniques, which reduced the water demand in some of these areas.

However, the effectiveness of this strategy is limited by the obsolescence of the facilities (65% of the meters are and ecosystems, Tunisian agriculture is of crucial importance because of its contribution to national food security.

Tunisia has more than ten million hectares of agricultural land representing 62% of the country's total area, with 32% of the total area covered by annual and permanent crops. Livestock production, with more than 400,000 breeders, is dominated by a flock of sheep, cattle, goats, and poultry. The agriculture and agri-food sector accounted for 11.5% of the GDP in 2012. The agricultural sectovides permanent income for 470,000 farmers, contributing to the stability of the rural population, which represents 35% of the country's population.

Climate change impacts on the sector 30%. Globally, the consequences of climate change would be economically negative for the most vulnerable populations, notably for women, of whom 32.3% live in rural areas and who constitute more than 70% of active jobs in agriculture and forestry. Poverty is likely to intensify and to affect small farmers whose agricultural activities, already economically unprofitable, may disappear under the impact of climate change. As a consequence of the expected decline in suitable areas for agriculture, and considering constant yields, agricultural GDP would decline by 5–10% in 203.

Many ecosystems would also be strongly impacted by climate change. An increase in forest fires is expected, which currently already reduce the forest area by 1200 ha each year, leading to the gradation and loss of biodiversity. Alfa ecosystems and rangelands are already under heavy anthropogenic pressure like overgrazing. A study conducted in 2014 in the governorate of Medenine13estimated that the reduction of natural fodder resources would be around 23% and 26% respectively.

In 2020 and 2050, threatening the sustainability of pastoralism. Wetlands would also be threatened by increased salinization and eutrophication. The oasis ecosystems are already highly vulnerable due to their strong dependency on water resources. For the Tunisian oasis zone, average warming of 1.9 ° C is expected by 2030 and 2.7 ° C by 2050, with decreasing rainfalls by 9% in 2030 and 17%in 2050, and an increase in evapotranspiration of 8% by 2030 and 14% by 205014.

Major Initiatives for Adaptation Initiatives for climate change adaptation in agriculture are generally multi-scalar and more developed since the revolution. Meteorological and climatic monitoring has been widely developed to anticipate extreme events that can cause severe damages and losses on agricultural yields.

Looking for more efficiency in water use within irrigated perimeters, Tunisia began its National Program on Water Saving in 1995. In 2015, notably, t response to climate change impacts already obese rved, localized irrigation (drip irrigation) is applied to 46% of the total area of irrigated perimeters, compared to 30% for sprinkler irrigation and 24% for gravity irrigation15. Several other adaptation measures are implemented in the agricultural sector to reduce the water demand, such as the use of conservation agriculture, both at a local and regional scale. In terms of traditional agriculture, one of the initiatives for adaptation is to educate farmers and encourage them to use ancestral know-how.

Each governorate, depending on the types of agriculture developed within its geographical area, has also planned adaptation to climate change by strengthening participative Deba t es between managers and farmers and/or inhabitants of rural areas for example. Capacity building of dedicated national services is also a priority. Several strategies have been recently developed to protect.

Tunisian ecosystems from the impacts of climate change. Many strategies focus on sensitive oasis ecosystems. Moreover, in 2015, a study formulated strategic directions and developed a plan for the adaptation to climate change. of biodiversity, in all its components. However, the implementation of the advocated actions and recommendations of all these studies and strategies remains constrained.

The coast has a total length of 2290 km, distributed as follows: 1280 km of continental coastline, 450 km of island coastline, and 560 km of lagoons. The three main types of fishing encountered in Tunisia are trawling lamp fishing and inshore fishing. The main marine production area..is located in the governorates of Sousse, Monastir, and Nabeul, while the production of continental aquaculture, is mainly located in the governorate of Béja. Fishery production has steadily increased since the 1990s, reaching a total production of 118,000 tons in 2012, with an annual growth rate of 2.6% between 1996 and 2012. The production of the year 2016 was estimated at 126,528 tons, for a total amount of 828 million dinars.

For this third national communication, the outputs of a set of models used in the latest IPCC report (AR5) with two representative concentration pathways (RCP 4.5 and 8.5) are analyzed to provide climate change projections for Tunisia at both the 2050 and 2100 horizons. These data come from the simulations of all EURO-CORDEX models at a 12.5 km resolution. 1. Synthesis of climate projections for Tunisia ~Past climate trends in Tunisia, over the 1978–2012 period, we observe a significant upward trend in annual maximum, average, and minimum temperatures of around 2.1 ° C, with regional disparities. Observations show a slight upward (statistically insignificant) trend for the total annual rainfall over this same period of observed data. Nevertheless, the disparities between seasons and the interannual variability are very strong. The beginning of the 1980s is known for a succession of several dry years in Tunisia, influencing the general trend of the curves.

Regarding meteorological and climatic extremes, a change in the frequency of thermal and rainfall extremes is observed over the past period, based on the observed data from the meteorological stations of the Tunisian National Institute of Meteorology (INM) Projections show an increase in annual temperature at the 2050 and 2100 horizons for both scenarios. This increase ranges between 1 ° C and 1.8 ° C by 2050 on average for the set of studied models and between 2 ° C and 3 ° C at the end of the century, with the RCP 4.5 scenario. For the RCP 8.5 scenario, this increase ranges between 2 ° C and 2.3 ° C by 2050 on average for the set of studied models and between 4.1 ° C and 5.2 ° C at the end of the century. The coast in the North and the East of Tunisia would warm up less quickly than the West and the extreme South.

6.10 Jordan

6.10.1 Background

Jordan lies in the north of the Saudi Arabian Kingdom, It has one city on the eastern side of the Gulf of Aqaba called Aqaba city, having a coastal zone of 27 km along the Gulf

Currently, there is intensive exploitation of underground resources8, which provide around 81% of the water needs of the irrigated sector. The decrease in available stocks would be more pronounced in the north of the country which concentrates 80% of the resources already mobilized, and in the center of the country where groundwater is the main resource for agriculture and drinking water.

As Jordan is one of the most vulnerable countries to the risks of climate change the country is undergoing a rapid and effective process of enhancing its institutional and policy-relevant framework for addressing climate change challenges. It is the first country in the Middle East to develop a national climate change policy in 2013.

Jordan has three distinct ecological systems: (i) Jordan Valley which forms a narrow strip located below the mean sea level and has warm winters and hot summers with irrigation mainly practiced in this area; (ii) the western highlands where rainfall is relatively high and climate is typical of Mediterranean areas; and (iii) the arid and semiarid inland to the east (estimated to cover over 80% of the total area), known as the "Badia", where the annual rainfall is below 50 mm. Badia is an Arabic word describing the open rangeland where Bedouins (nomads) live and practice seasonal grazing and browsing. According to the IUCN Red List of 2006, Jordan has 47 globally threatened species. Of the 83 mammal species existing in Jordan, 12 are considered globally threatened. As for birds, there are 15 globally threatened species in Jordan. Around 2500 (Fig. 6.15).

Based on long historical data obtained from Jordan Metrology Department (JMD), climatic variables are changing significantly at both the national and station level, indicating that climate change is becoming more apparent. Both the Mann-

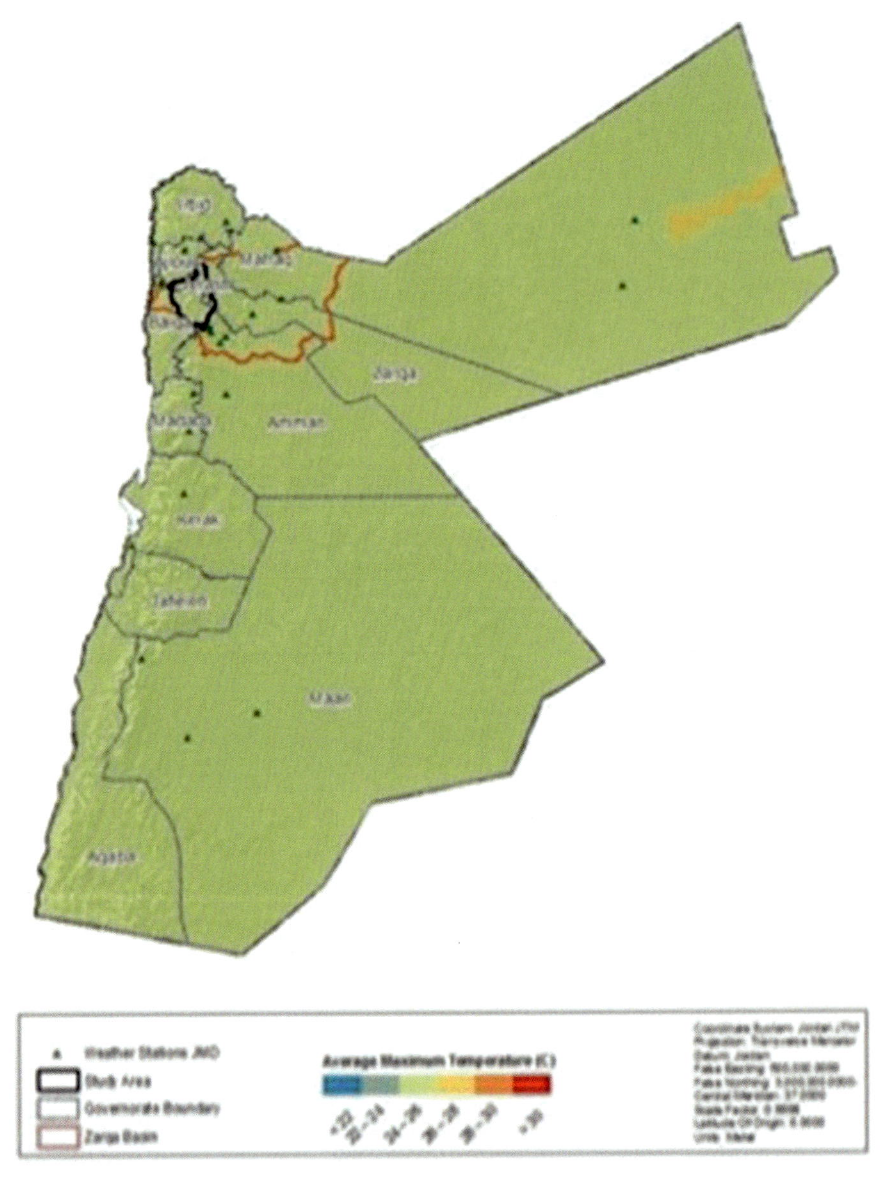

Fig. 6.15 Average mean Temperature in Jordan

Kendall rank trend test and linear regression trends indicate that the annual precipitation tends to decrease significantly with time at a rate of 1.2 mm per year. Simultaneously, the mean, maximum and minimum air temperature tends to increase significantly by 0.02, 0.01, and 0.03 °C/year, respectively.

On the other hand, the relative humidity tends to increase significantly by an average of 0.08%/year, while class A-pan evaporation seems to have non-realistic estimations of decreasing significantly by 0.088 mm/year. The number of days of dust storms tends to decrease significantly by 0.09 days/ year and 0.06 days/year for visibility less than 1 km and 5 km. Besides, the historic data tested on both annual and monthly basis indicated that precipitation reduction is highly significant during the whole rainy season except for January. Similarly, during the dry seasons of June, July, and August, the precipitation has tended to increase over time, although this increase is considered negligible in its quantity as indicated by the magnitude of the slope. Interpolated spatial maps show the locations of these changes to be more apparent at both northern and southern parts. Dynamic downscaling for this study was achieved using the Africa CORDEX domain, in which 43 grid points with 50 km resolution were crossed throughout the country. Nine different GCM coupled with two RCMs for two RCPs (4.5 and 8.5) were used to assess future projections as compared to reference historic data (1980–2010).

Three-time horizons were selected; 2020–2050, 2040–2070, and 2070–2100. Climatic indices were extracted, processed, and debiased using delta and quantile-quantile scientific techniques. The selected climatic variables represent precipitation, mean temperature, maximum temperature, minimum temperature, wind speed and direction, relative humidity, class A evaporation, drought indices at 3 and 6 months basis, number of consecutive dry days, number of heavy rainfall days, and snow depth. The suggested reference model that was close to the median from all 9 models was "SMHI – NCC-NorESM-LR" a combination of the Norwegian Earth System Model as global climate model, and the Swedish SMHI regional climate model.

This model was further used to further to interpolate the climatic indices at 1 km resolution using combined statistical projections at the station level (Delta method) and geostatistical interpolation using the digital elevation model (DEM).

Based on the definitions of exposure consisting of likelihood, geographic magnitude, and confidence, the IPCC definitions were used. The qualitative measures (e.g. rare, unlikely, possible, likely, and extremely likely were based on the probability of occurrence per year, while the confidence scale was based on the processing of a multi-model ensemble (i.e. Very high, high, medium, low, and very low confidence). The projections' results agree with previous work of Second National Communication (SNC) to UNFCCC and are consistent with IPCC-AR5. For the year 2085, the two RCPs extremely likely predicted rise in mean temperature for all of the country, up +2.1 °C [+1.7 to +3.1 °C] for RCP 4.5, and + 4 °C [3.8–5.1 °C] for RCP 8. The increase was predicted to be homogeneous for the RCP 4.5, and stronger for the Eastern and the Southern regions for RCP 8.5.

Future dynamic projections predict extremely likely warmer summer compared to other seasons. Compared to the SNC that used CMIP3 results, multi-ensembles projections of CMIP5 results coupled with regional climate models in CORDEX give a more consistent trend to a likely drier climate. In 2070–2100, the cumulated precipitation could likely decrease by 15% [−6% to −25%] in RCP 4.5, by – 21% [−9% to −35%] in RCP 8.5. The decrease would be more marked in the western part of the country. It is more likely to have drier autumn and winter as compared to spring, with a median value of precipitation decrease reaching −35% in autumn in 2070–2100. Also, the dynamic projections predict more extremely likely heat waves where the analysis of summer temperature, monthly values, and the inter-annual variability reveal that some thresholds could be exceeded especially for a summer month where the average maximum temperature for the whole country could exceed 42–44 °C.

Drought events were likely predicted as indicated by the two indices of consecutive dry days and SPI. The maximum number of consecutive dry days would likely increase in the reference model of more than 30 days for 2070–2100 20 Jordan's Third National Communication on Climate Change Executive Summary period. The SPI indicates more frequent droughts with a 3 to 4 years lag. In contrast to drought, annual values still show possible heavy rainy years at the end of the century. More intense droughts would be (partly) compensated by rainy years, in a context of a general decrease of precipitation. Potential evaporation would also likely increase. However, the occurrence of snow would be unlikely to decrease.

Finally, the future projections unlikely to predict no trend for winds, where maximum wind speed does not evolve significantly. Climate exposure, risks, sensitivity, impacts, and adaptive capacity:

Concluding Remarks

1. All seven countries are vulnerable to climatic changes one way or another. The following management and policy suggestions are in place:

   (a) Institutional setup
   (b) Proactive Planning
   (c) Monitoring system and Research capabilities
   (d) Identification and assessment of options for adaptation
   (e) Upgrading resilience and awareness
   (f) Regional ....follow-up of implementation
   (g) Adoption in the educational system

6.11 Summary of Action Plan

6.11.1 Institutional Capacity

1. MENA countries must develop a regional center or organization for climate change. The center should collect and analyze data, build a geographic database and establish monitoring systems for indicators of concern to all Arab countries. It should be responsible for carrying out and enforcing strategies, plans, policies, and measures for proper adaptation
2. MENA countries must introduce concepts of ICZM and disaster reduction in the educational system and must develop the institutional capacity for integrated coastal zone management and build up capacity and follow up in these directions

3. MENA countries must develop the institutional capacity for risk reduction by adopting Early Warning Systems of flash floods, storm surges, and heatwaves.
4. MENA region countries must develop transparent data and information systems to allow early warning of problems

Awareness

1. All MENA region countries must develop awareness programs for upgrading the resilience of vulnerable communities, population, stakeholders, and investors
2. MENA Region countries must Work to create new job opportunities in safe areas and exercise environmental law enforcement of regulations such as SEA, EIA

Monitoring

1. It is necessary to monitor and assess land subsidence especially in coastal areas of excessive urban loads and excessive rates of oil and/or water extraction

References

Alexander, D. (2000). Vulnerability to landslides. In T. Glade, M. Anderson, & M. Crozier (Eds.), *Landslide hazard and risk* (pp. 175–198). John Wiley & Sons.

Becken, S. (2007). *Climate change and tourism – advances in knowledge and practice*. People Research Centre, Lincoln University.

Birkmann, J. (2008). *Measuring vulnerability to natural hazards*. UNU-Press.

Blaikie, P., Cannon, T., Davis, I., et al. (1994). *At risk: natural hazards, people's vulnerability and disasters*. Routledge.

Blaikie, P., Wisner, B., Cannon, T. & Davis, I. (2003). *At-risk: natural hazards, people's vulnerability, and disasters*.

Bohle, H.-G., Downing, T. & Watts, M. (2003). Climate change and social vulnerability: the sociology and geography of food insecurity. *Global Environmental Change, 4*, 37–48.

Byrne, T. R.. (2014). *Household adaptive capacity and current vulnerability to future climate change in Rural Nicaragua*. M.Sc., University of Lethbridge.

Cutter, S. (1993). Vulnerability to environmental hazards. *Progress in Human Geography, 20*, 529–539.

Cutter, S. L., Emrich, C. T., Webb, J. J., & Morath, D. (2009). *Social vulnerability to climate variability hazards: a review of the literature*. Hazards and Vulnerability Research Institute. Department of Geography. The UniversConclusionsCcC.

Dow, K. (1992). Exploring differences in our common future(s): the meaning of vulnerability to global environmental change. *Geoforum, 23*, 417–436.

Dow, K. & Downing, T. (1995). Vulnerability research: where things stand. *Human Dimensions Quarterly, 1*, 3–5.

Edwards, J., Gustafsson, M., & Näslund-Landenmark, B. (2007). Handbook for vulnerability mapping. In *Disaster reduction through awareness, preparedness, and prevention mechanisms in coastal settlements in Asia*. Swedish Rescue Services Agency.

El Raey, M. E. (1997). *Vulnerability assessment of the coastal zone of the Nile Delta, in the impact of sea-level rise; ocean and coastal manag* (pp. 371–329).

El Raey, M. E. (2008). *Impact of sea-level rise on the Arab region*. Published Report.

El Raey, M. E., & Atricia, K. (2011). *Mbote* (pp. 773–788). Springer.

El Raey, M., Fouda, Y., & Nasr, S. (1997). GIS assessment of the vulnerability of Rosetta, area to the impacts of sea-level rise. *Environmental Monitoring and Assessment.*
El Raey, M., Frihy, O., Nasr, S., & Dewidar, K. H. (1999). Vulnerability assessment of sea- level rise on port said governorate. *Environmental Monitoring and Assessment, 56*(2), 118–128.
ESRI. (2013). *Principals of remote sensing.*
Etkin, D. (1999). Risk transference and related trends: driving forces towards more mega-disasters. *Environmental Hazards, 1*, 69.
FitzGerald, G., Fenster, M., Argow, B., & Buynevich, I. (2008). Coastal Impacts Dueto Sea-Level Rise. *Annual review on earth planetary science, 36*, 601–647
Frantzova et al. (2008). Creative Commons Attribution 4.0 International.
Gilbert, C. (1995). Studying disaster: A review of the main conceptual tools. *International Journal of Mass Emergencies and Disasters, 13*, 231–240.
Gillard, O. (2016). *Hazards, vulnerability, and risk* (pp. 19–29).
Giorgi, F. (2010). Uncertainties in climate change projections, from the global to the regional scale. *EPJ Web of Conferences, 9*, 115–129.
Glantz, M. (1994). *Creeping environmental phenomena and societal responses to them.* Workshop report. National center for atmospheric research. Boulder, USA, 274 pp.
Harrod, R. P., & Martin, D. L. (2014). *Bioarchaeology of climate change and violence ethical considerations*. Springer.
Hegerl, G. C., & Zwiers, F. W. (2007). Understanding and attributing climate change. In *Climate change 2007: The physical science basis. Contribution of working Group I to the fourth assessment report of the intergovernmental panel on climate change*. Cambridge University Press.
Hulbutta. (2009). GIS analysis of global impacts from sea level rise. *Photogrammetric Engineering & Remote Sensing, 75.*
IPCC. (2012). *Managing the risks of extreme events and disasters to advance climate change adaptation, A special report of working Groups I and II of the intergovernmental panel on climate change*. Cambridge University Press.
Kelman, I. (2009). Understanding vulnerability to understand disasters. *Center for International Climate and Environmental Research* - Oslo (CICERO).
Kelman, I., & Lewis, J. (2010). Places, people, and perpetuity: community capacities in ecologies of catastrophe. *ACME: An International E-Journal for Critical Geographies, 9*, 191–220.
Lewis, D. (2010). *Emergency management 101: "The questions you should be asking in your community"*. Utah League of Cities and Towns, Annual Convention.
Lockwood, J. P., Hazlett, R. W., & Wiley-Blackwell. (2011). Volcanoes: Global Perspectives. *Pure and Applied Geophysics, 168*, 1871–1872.
Lynn, K., Mackendrick, K. & Donoghue, E. M. 2011. *Social vulnerability and climate change: synthesis of literature*. United States Department of Agriculture Forest Service. Pacific Northwest Research Station.
Moss, R. H., Brenkert, A. L. & Malone, E. L. (2001). *Vulnerability to climate change: A quantitative approach* 88
NASA & GISS. (2013). *Global climate change.*
National Research Council, N. (2012). Climate change, evidence, impacts, and choices. *National Academy of Sciences.*
Quiggin, J. (2008). Uncertainty and climate change policy. *Economic Analysis and Policy, 38*, 203–210.
Rahmstorf, S., Caesar, L., Robinson, A., Saba, V., & Feulner, G. (2018). An observed fingerprint of a weakening Atlantic Ocean overturning circulation. *EGU General Assembly, 20.*
Rueter, J. (2013). *Environmental sciences and management program.* Chapter 9 – Risk and Uncertainty.
Scheer, D., Benighaus, C., Benighaus, L., Renn, O., Gold, S., Roder, B., & Bol, G. F. (2014). The distinction between risk and hazard: understanding and use in stakeholder communication. *Risk Analysis, 34*, 1270–1285.

Sestini, G. (1991). *The implication of climate changer to the Nile Delta*. Report for the Nile DeltaeportWC/14Nairobi, Kenya, UNEP.
Snoussi, M., Ouceni, T. & Nazgin, S. S. (2008). *Precision.*
Sørensen, C. S. & Jebens, M. (2015). How can awareness in civil society and governance be raised? Reducing risks from coastal hazards. *Proceedings of the 24th New South Wales coastal conference.*
Thinda, T. K. A. (2009). Community-based Hazard and vulnerability assessment: a case study in Lusaka informal settlement, City of Tshwane. In *A mini-dissertation submitted in partial fulfillment of the requirements for the degree of masters in disaster risk management*. The University of the Free State
Timmerman, P. (1981). Vulnerability. *Resilience, and the Collapse of Society Environmental Monograph, Institute of Environmental Studies, 1.*
Trofimenko, N. (2011). *Climate change: current issues*. Kiel Institute for the World Economy.
UNEP (United Nations Environment Programme). (1999). *Chemical, human, environmental, and ecological risk assessment*. Training module No. 3, 67 pp.
UNFCCC. (2011). *Framework convention on climate change, water, and climate change impacts and adaptation strategies*
UNISDR. (2008). Hyogo framework for 2005–2015: Building the resilience of the nations and communities to disasters. Available at: www.unisdr.org/wcdr/intergover/official-docs/Hyogoframework-action-english.pdf.
Wisner, B., Blaikie, P., Cannon, T., & Davis, I. (2004). *At-risk: natural hazards, people's vulnerability, and disasters* (2nd ed.). Routledge.
World Meteorological Organization, W (2013). *Integrated flood management tools series 20*
Yasir, A. (2009). *The political economy of disaster vulnerability: a case study of Pakistan earthquake 2005*. London School of Economics & Political Science.

Part III
Case Studies: Space Techniques and Natural Hazards in the MENA Region

Chapter 7
Satellite Images for Modeling Terrain Instability in Saudi Arabia (Jeddah-Rabigh)

Mashael M. Al Saud

Abstract Natural hazards became the foremost geo-environmental issue in several regions worldwide. Rarely a month goes by without a disastrous event that impacts urbanism and environment. The influence of natural hazards is being increased due to the development of many physical and man-made influencers. These hazards imply many types and occur with different dimensions. Due to its complicated geology and geomorphology, the Kingdom of Saudi Arabia encompasses a miscellany of disasters of the natural origin, and terrain instability is one of the influencing types where it occupies tremendous localities including urban and rural ones. It usually impacts many communities and plays a role in the socioeconomic changes. The increased number of events related to terrain instability makes it necessary to elaborate relevant studies. Therefore, mapping instable terrain has become urgent and must be accounted in many projects. The use of satellite images proved to be significant tool for modeling different factors controlling the existence of terrain instability. This chapter will illustrate these factors, as thematic layers, which have been extracted and manipulated by satellite images processing and GIS systems. Results show that about 89% of the area between Jeddah and Rabigh is under terrain instability risk (ranges from high to very high risk).

Keywords Mass movement · Damages · Slope · DEM · Spot images

7.1 Types of Terrain Instability

Terrain instability is a geo-environmental issue, especially where urbanism is increasing. It is usually interlinked with the term "Mass Movement" which expresses the movement of different terrain materials (rock and soil) by the effect of physical forces (e.g. water. Slope, wind, etc.) and in some cases by the influence of human activities such as: excavation, construction, etc. Thus, mass movement follows

---

M. M. Al Saud (✉)
Space Research Institute, King Abdel Aziz City for Science and Technology (KACST), Riyadh, Saudi Arabia

M. M. Al Saud (ed.), *Applications of Space Techniques on the Natural Hazards in the MENA Region*, https://doi.org/10.1007/978-3-030-88874-9_7

diverse mechanisms and resulted severe damages and destruction, which might be harmful to human and nature. Thus, it is necessary to investigate the stability of terrain where new construction projects, transportation systems, planned settlements, etc. will take place.

Slopes are most hazardous terrain surfaces, notably when unconsolidated materials are located. It is a widespread risk occurs in many localities of the Kingdom of Saudi Arabia. In this regard, several studies and projects are applied on natural risk, aiming to reduce and mitigate its impact. This is well pronounced on slopes, certainly along the descents between the Arabian Shield and the Red Sea coast. Therefore, many rock deformations exist such as escarpments, faulting, rock toppling, etc. and then create slope instability and rock fall and sliding, especially after rain storms (Al Saud, 2018).

Three major types of the instability of terrain exist the Kingdom of Saudi Arabia where they are viewed from the hazardous point of view. The first two are interlinked with materials mobility along slopes (i.e. landslides and rock falls); while the third one is often with limited and local movement and then represents either land subsidence or uplift.

7.1.1 Landslides

Landslides represent one of the well-known natural hazards occur in many regions where they might be at small and large scale. Thus, the degree of damage results from landslides is mainly controlled by its dimension and the mechanism of movement. However, a large landslide are always resulted in severe damages in the environment and human and it is considered as a risky threat and this is exactly the case in the Kingdom of Saudi Arabia.

Landslides are described as rocks, soil, or debris flows (sliding) by the impact of gravity along slopes. They represent the movement of surficial materials along terrain where a direct contact between these materials and terrain remain from the beginning of movement until these materials settle down. While, the duration of landslide movement ranges from very slow to very rapid, and it exists on any terrain given the suitable conditions of soil, moisture, and the angle of slope. Integral to the natural process of the earth's surface geology, landslides serve to redistribute soil and sediments in a process that can be in sudden collapses or in slow gradual slides (Al Saud, 2018).

There are many factors controlling the occurrence of landslides. These factors implies the geologic, morphological, meteorological and man-made origin. Thus, some factors can act separately in creating terrain instability, while sometime two or more of them act together. Hence, landslides might be triggered by torrential rain, floods, earthquake and other causes obtained by human intervention, such as excavation, terrain cutting and filling, etc.

There are many features characterize the type of landslides, with a special emphasis to: rotational, translational, debris flow and avalanche, earth flow, creeping

Table 7.1 Major types of landslides (Varnes, 1978)

Type of landslide	Bed rock	Engineering soils	
		Predominantly coarse	Predominantly fine
Rotational	Rock slide	Debris slide	Earth slide
Translational			
Lateral spread	Rock spread	Debris spread	Earth spread
Flows	Rock flow	Debris flow	Earth flow
	Creep	Soil creep	
Complex	Combination of two or more types of movement		

Fig. 7.1 Rocks susceptible to roll towards the road. (Al Saud, 2018)

and lateral spread. According to Varnes (1978), landslides were interlinked to their rock type and the engineering soils as in Table 7.1.

7.1.2 Rock Falls

Another aspect of mass movement, rock fall is a common risky phenomenon. However, rock fall (or sometimes called mass wasting) differs from landslide in that it is attributed to the movement of materials on air, whereas a landslide moves (or intermittently) on terrain surface. These can be as rock crumpling, toppling and rolling (example in Fig. 7.1). Thus, the movement mechanism of surface materials belongs to rock fall is characterized by fast process.

For rock fall, factors controlling the mechanism of movement are almost similar to those for landslides, but the type of surface materials in this case are mainly of hard and consolidated rocks with different size ranging between gravel and debris to rock boulders and masses.

Normally, free rock fall typically happens along slopes steeper than 76°, bouncing on slopes between 45 and 76°, and rolling on slopes less than 45°. However, a rock may alternate between the three modes during its downslope movement, because slopes are often irregular (Al Saud, 2018).

The dimension of event resulted from fallen rocks, as mass wasting, is mainly governed by thickness of **rock beds**, joint sets and the existing fractures systems which act in the fragmentation of rocks, and then detach the fragmented and collapsed rocks long the slope. In addition, the length of fissures and the volume of fallen rocks, tend to follow power law or fractal distributions, meaning that their numbers decrease exponentially as fracture length and rock fall volume increases.

7.1.3 Subsidence and Uplifts

This type of terrain instability is well known in many regions, likewise in the Kingdom of Saudi Arabia (Fig. 7.2). It is often has a minimal movement of surface materials if compared with landslides and rock fall. This type of terrain instability includes the pull-down (i.e. subsidence) or the pull-up (i.e. uplift) of the terrain surface. These movements are usually accompanied with ground fissuring and cracking whether in soil or in rocks, and the associated subsidence or uplift, which occur in the middle part of the subsided or uplifted locality.

The resulting fissures associated with the subsidence or uplift of the ground surface, often ranging from a small-scale cracks to large faults; where they begin as small traces that expand later on due to the external factors such as erosion and

Fig. 7.2 Land subsidence near Jeddah area

rainfall. Sometimes this type of fissures are caused by the change in the soil nature, especially along the sides of the basins due to the presence of rugged rock topography underneath these soil layers (Budetta, 2004). Moreover, ground cracks can evenly begin at great depths below the surface, as a result of horizontal movement in the aquifers due to excessive withdrawal of the ground water from the fragmented-rock reservoirs, also due to loess soil, and earthquakes. These fissures and subsidence could cause many problems in the urbanized and agricultural areas and induce damages to the infrastructure. (Holzer, 1984).

Under arid conditions such as in desert areas, the shortage of groundwater resources leads to excessive pumping that may aggravate land subsidence incidents. This is the case in many areas in the Kingdom of Saudi Arabia they suffer from groundwater overexploitation resulting in land subsidence and fissuring which also lead to damage of agricultural areas due to loss of water and sometimes injuries from the hidden fissures. (Bankher, 1996).

Terrain uplift is also a widespread phenomenon, but it has different mechanism than subsidence in term of movement direction, and it is mainly originated as a result of tectonic activities and the related rocks deformation. Furthermore, earthquakes can cause ground movement through cracking which may span for long distances. In addition, delta sediments that are saturated with water and Sabkhas are the suitable localities that could be affected by subsidence and fissures or faulting due to earthquakes. Examples on faulting and fissuring, that lead to subsidence and uplifts in Saudi Arabia and attributed to the earthquake and volcanic activities, occurred in Harrat Lunayyir (east to Younba City), where the movement of magma led to seismic activity, resulting from the rise and subsidence in the terrain surface, causing numerous cracks and faults. (Bankher, 1996).

7.2 Factors Influencing Terrain Instability

There are several physical and anthropogenic factors controlling the stability of terrain. However, these factors differ from one region to another depending on the existing natural and man-made characteristics and intervention which is triggering terrain instability. According to Al Saud (2018), these factors can be summarized as follows:

1. Geology: This includes the characteristics of rock lithology and the existence of weak and weathered, sheared and fissured surficial materials, plus rock layer discontinuity and many other features of rock deformation and lithological friability.
2. Geomorphology: This encompasses tremendous morphological features and processes that act in disintegrating and weakening the terrain material, such as: deposition loading slope, thawing, freezing alluvial and glacier erosion.

   Meteorology: Meteorological impact on terrain instability and other natural hazards is well pronounced, where it plays from the accumulated and running

water due to torrential rainfall for example, as well as the increased temperature and its impact on the shrinking of the surficial materials, and then losing their consolidation. In addition, there are several meteorological parameters that may play a role in terrain instability, such as wind, heat waves, freezing, etc.; as well as the abrupt changes in climatic conditions has a significant role.
3. Human interference: In many cases human works (e.g. excavation, dumping, mining, artificial vibration, etc.) act in the enhancement of ground fragility and then accelerating mass movements from their original place. This interference differ between regions and over different time periods.

Classification of factors influencing terrain stability are often different between researchers, but all accorded that these factors are resulted in damages which might be severe in some cases. For example, an acute tectonic seismicity can result the movement of rock and soil masses where other factor, such the geomorphology including mainly the slope, can also control the mechanism of any movement. Moreover, climatic extremes results detaching of surface materials and then enhance transportation processes.

According to the physical setting of the Kingdom of Saudi Arabia, factors affecting the occurrence of terrain instability are many, even though they differ between the Saudi regions. However, each region encompasses a number factors that might not exist in other regions. Thus, the study area is located between Jeddah and Rabigh (19,723 km^2), is a representative region where it is characterized by mountain ridges and coastal plain along the western Saudi coast. Therefore, the most influencing factors on terrain instability in the area of study were primarily determined. This has been done using many tools including satellite images analysis and field surveys, as well as, records and datasets from previous studies. In this regards, the use of space techniques was the most significant tool in this study notably in identifying instable localities. In addition, geological maps were used as supportive tool. Therefore, the following factors were adopted for investigating instable terrain in the area of study (Al Saud, 2018).

1. Slope: The inclination of topographic surfaces is has high gravitational impact on moving objects on these surfaces. Thus, when slope angle exceeds, soil and rock masses will move either slowly or suddenly following different mechanisms along the slope surface. In this regard, there are many elements controlling the timing of slope failure and its mechanism, and this includes, in addition to the slope angle, hardness of the objects located along the slope and surface roughness.

More than one parameter play together in moving terrain materials along slopes. However, the angle of friction can be separately able to move surface materials since it represents the maximum angle for slope-tolerant holding objects. Hence, the area of study is known by the flat terrain surface at the coastal plain, besides a sloping terrain at the adjacent mountain chains located to the east, and thus, tremendous slopes exist.

Al Saud (2012, 2018), classified terrain instability, as viewed from the morphological setting and field observations in the study area, as follows:

- Sudden movement and hazardous location at high slopes,
- Rapid movement and instable terrain at slight high slopes,
- Instable terrain exists when more than one parameter exists at moderate slopes,
- Difficult movement along terrain (almost stable) at slight low slopes,
- No remarkable movement at very low slopes.

2. Lithological characteristics: This lithological feature is one of the major factors that affecting terrain stability. It reflects the consolidation of rocks either as hard rock lithology or soft ones. These characteristics are controlled by the physical properties of rocks, such as the porosity, permeability, grain sorting, etc.

Normally, soft lithologies are susceptible to move than hard ones, but the latter might exist if more factors act. For example, hard limestone rocks are often stable, but when fractures occur among, surfaces of weakness will be developed between the bedding planes and result terrain instability.The area between Jeddah and Rabigh is known by numerous types of rock lithologies. This occupies the three major rock types (i.e. sedimentary, igneous and metamorphic) with diverse features of rock stratification and deformation. Therefore, five classes were considered for the lithological characterization with respect to terrain instability. This classification has been viewed from rock rigidity. According to Al Saud (2018), these classes are:

- All igneous and metamorphic rocks – very stable
- Carbonates facies – slightly stable
- Sandstone and other clastic rocks – non-compacted
- Alluvial and talus deposits and saline muds – instable
- Aeolian sediments and dunes – movable.

3. Fractures: This type of rock deformation is utmost significant in terrain instability and occurs at different scale and shapes. They can be as fault alignments with long distances or as local fissure systems with few centimetres. It is often considered as the primary factor in rock instability, since fractures increase rock fragmentation and friability. However, this factor is also related to/or originated from acting physical forces and specifically the tectonic activities. Therefore, terrain surface with dens fracture systems, including fissures and joints, are often instable.

   The area between Jeddah and Rabigh is characterized by dense fractures due to the complicated tectonic setting, especially it is located in the proximity of the Arabian Shield which is almost surrounded the eastern side of the study area. Therefore, the obtained classification of fractures according to terrain instability depended mainly on the number of fracture density (a function of number of fractures in a define area). Hence, zones with dense fractures are considered as unstable and vice versa. Based on this concept; however, five classes of fracture density were adopted in this study.

Landforms: A landform is characterized by the belonging physical attributes such as slope, elevation, orientation, rock and soil type, soil type, and rock exposure. This sometimes represented by land cover which describes a number of terrain features and topographic aspects. Thus, landforms in the area of study constitute a number of terrain surfaces and orientations. Therefore, the physical properties of the mentioned landforms govern the stability of terrain, and more certainly lithological characteristics and slope degree, which are in direct relationship with soil and rock types in the area of study. In this regards, Al Saud (2018), classified landforms in the area of study as: slopping surfaces as the most vulnerable to instability, and this should be integrated with lithological characteristics where soft rocks and soil exist. Besides, flat surfaces with hard terrain materials would be much stable.

7.3 Tools for Analysis

Perhaps the identification of factors influencing terrain instability should be primarily determined, and then they should be digitally mapped for the entire area between Jeddah and Rabigh. This requires using different tools and techniques for data retrieve, extraction and manipulation. This implies advanced techniques to elaborate and analyze geospatial data required.

The availability of tools required is a significant element. They also need skilled expertise. Table 7.2 shows the tools and techniques used for investigating the influencing factors on different types of terrain instability in the area of study.

The concept behind the applied method includes mainly the preparation of data on the influencing factors, and each factor was treated separately (Fig. 7.3). Hence, the major factors influencing terrain instability (i.e. slope, lithology, fractures and landforms) were treated. Each factor has a number of elements reflecting the effect on terrain instability. Some elements act towards increasing terrain instability while other elements work in opposite direction, and this must be restrictedly identified. Thus, each factors will be classified into 5 classes according to terrain instability, and therefore, these classes will be systematically integrated in the GIS system (Arcmap) in order to establish the final terrain instability map (Fig. 7.3).

The classes of terrain instability indicate relative ranking of the likelihood of terrain instability that occurs after any natural or anthropogenic event. However, they do not indicate of the level of the expected impact of terrain instability or potential damage. The 5-class terrain instability classification, as shown in this study, is to flag on the potential risk areas. Hence, it should not be understood as an onsite description for terrain stability field assessments.

Table 7.2 Major tools used for studying terrain stability (Al Saud, 2018)

Theme	Tool	Details	Data and information
Documentation	Previous studies and records	Documents and indicative maps	Recognize the available and obtained data and information to support knowledge.
Thematic maps	Geologic maps	Scale 1:250000, Skiba et al., 1986	Identify the geographic distribution of different lithologies and structures.
	Landforms maps	Land resources map, scale 1: 500.000, Ministry of Agriculture and Water, 1986	Determining the geographic distribution of different landforms.
Fracture systems	Satellite images	Geo-Eye-1: 5 bands; 46 cm resolution	Direct identification of observable instable terrain features (landslides, etc.).
		Aster: 14 bands, 15 m resolution (visible bands); 90 m resolution (thermal bands);	Linear thermal anomalies as indicator for fractures systems (i.e. faults).
		Landsat 7 ETM^{+}: 7 bands, 30 m resolution (visible bands); 120 m resolution (thermal bands);	Identify distinguished geologic features needed (e.g. ring structure, domes, etc.).
Slope	SRTM	Radar-based data; 30 m resolution	To extract the digital elevation model (DEM) and slopes
Data manipulation	Software	ERDAS-Imagine-11 (*Leica product*)	Satellite image processing and classification.
		PCI's Geomatica-10.2 (*Geomatica product*)	
		Arc-GIS-10.2 (*Esri product*)	Digital data manipulation and drawing
Filed surveys	Variety of field devices and instruments (e.g. incline-meter, GPS, laser-meter, etc.)		Field verification and inspection.

7.3.1 Slope

Usually, slope maps are generated from topographic maps depending on contour lines. Then, slope calculation can be obtained from these maps when they are digitally produced. Recently, the generation of slope maps is obtained using the geographic information system (GIS) which can be elaborated to build three-dimensional models for any geographic area. Thus, the representation of terrain topography needs require three-dimensions (z, y, x), which is known as digital elevation model (DEM). The applications of DEM became dominant in several applications, specifically in identifying terrain components, such as slope, sunlight exposure, depressions, drainages, etc. (Al Saud, 2012).

Establishing DEMs needs calculating the elevation points whether from digital contour lines or from stereoscopic satellite images (e.g. Spot images). Therefore, Triangulated Irregulated Network (TIN) would be primarily constructed, and they

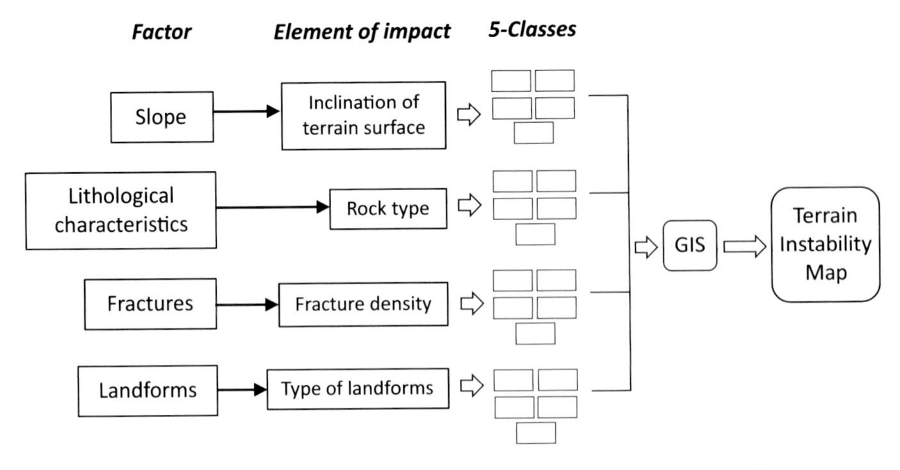

Fig. 7.3 Scheme of elements used in the applied method for mapping terrain stability in Jeddah-Rabigh Region

Table 7.3 Slope classes according to terrain instability in Jeddah-Rabigh Region

Class	Slope description	Terrain instability[a]	Slope Degree[a]	Area (km^2)	%[b]
I	High slope	Moveable	> 35°	3168	16
II	Slightly high slope	Instable	35–25°	2668	13.5
III	Moderately sloping	Non-compacted	25–15°	3072	15.5
IV	Slight low slope	Slightly stable	15–5°	4526	23
V	Low slope	Very stable	<5°	6289	32

[a]According to Fig. 7.4
[b]Percentage of the total studied area

represents digital data structure used in GIS for surface attributes of the natural land surfaces. This will be made up of irregularly distributed nodes and lines with three dimensional coordinates which are arranged in a network of non-overlapping triangles. Therefore, TINs are obtained from mass points, break-lines, and polygons, where they become nodes in the network, and represent primary input into a TIN in order to determine the overall shape of the surface (Al Saud, 2012).

Shuttle Radar Topography Mission (SRTM) was used in this study, to generate DEM. SRTM is a radar-based remote sensing product introduced by NASA. This product is originally made publicly available at a three-arc-second pixel size (1/1200 of a degree of latitude and longitude) with 90 m spatial resolution. In some studies, 90 m pixels are large enough to display the required resolution from SRTM DEM; however, it is now revealed by the much smaller 30 m pixels.

Digital elevation model (DEM) of the study area was generated, and consequently, terrain slopes were extracted and categorized into five classes to illustrate aspects in the area of concern, which will be a used factor while mapping terrain instability. These classes were illustrated in Table 7.3 and mapped Fig. 7.4.

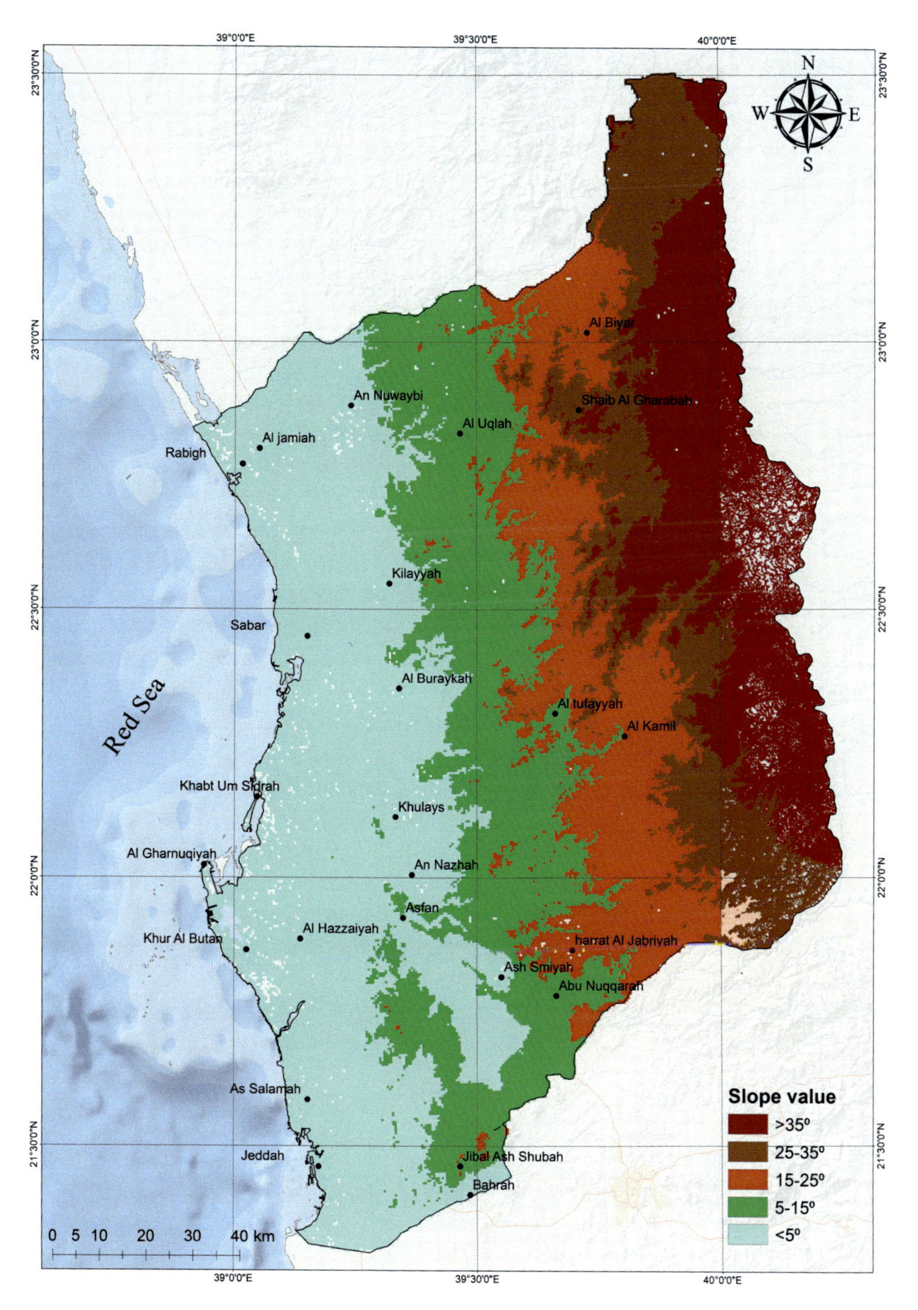

Fig. 7.4 Slope maps of Jeddah-Rabigh Region. (Al Saud, 2018)

7.3.2 *Lithology*

The geological rock formations in the area between Jeddah and Rabigh were studied and chronologically classified based on geological map done by Al Saud (2018). There 44 geological formations are exposed in the study area and extending from the coast of the Red Sea to the mountains to the east. The oldest geological age of these rock formations is the Precambrian, and they belong to 13 rock groups, suites and complexes.

Almost all rock types with diverse lithologies are exposed in the study area, including the sedimentary, igneous and metamorphic. There is a dominancy for carbonate rocks, silicates, clastic rocks, alluvial and colluvial deposits, intrusive and extrusive igneous rock and lavas, as well as varieties of metamorphosed rocks. The distribution of these lithologies is controlled by the geologic structures, as well as by the geomorphological features, resulting in different bending aspects, unconformable bedding planes, dikes and irregular rock bodies from lava and the igneous and metamorphosed rocks.

Rock formations in the study area have different degrees of impact on terrain instability which is controlled mainly by lithology including the physical and mineralogical characteristics. Therefore, an empirical classification (i.e. categorization of rock types) for the exposed rock formations, was adopted regardless of their age and location. Emphases were on the lithological characteristics and then how to respond to terrain instability in terms of water retention, porosity, permeability, hardness, and many other physical and mineralogical rock properties (Al Saud, 2018).

Many classifications were followed on rock lithologies according to terrain instability, such the classification done by Church (1983), Hungr (1984) and Chatwin, et al. (1994). However, these properties are different according to rock characteristics with respect to the movement in the terrain surface. This can be attributed to the different characteristics of areas where these classification were applied. In this study, the classification of lithology-terrain instability (5-classes categorization) was obtained as lithological categorization for lithologies that match similar impact on terrain instability (Table 7.4).

The classified lithologies were mapped to illustrate their geographic distribution and their respond to terrain instability and mass movement (Fig. 7.5). A systematic

Table 7.4 Lithological classes according to terrain instability in Jeddah-Rabigh Region

Class	Major lithology	Area (km^2)	%	Terrain stability**
I	Alluvial and colluvial deposits, dunes, Sabkha, aeolian deposits	6049	30.5	Moveable
II	Conglomerate, sand, silt, tuff, marl, clay, argillaceous rocks	2386	12	Instable
III	Lava and volcanic ash, limestone	5532	28	Non-compacted
IV	Amphibole, ultramafic rocks, tonalite,	2735	14	Slightly stable
V	Basalt, rhyolite, andesite, granite, quartzite, gabbro, gneiss, Syenite	3021	15.5	Very stable

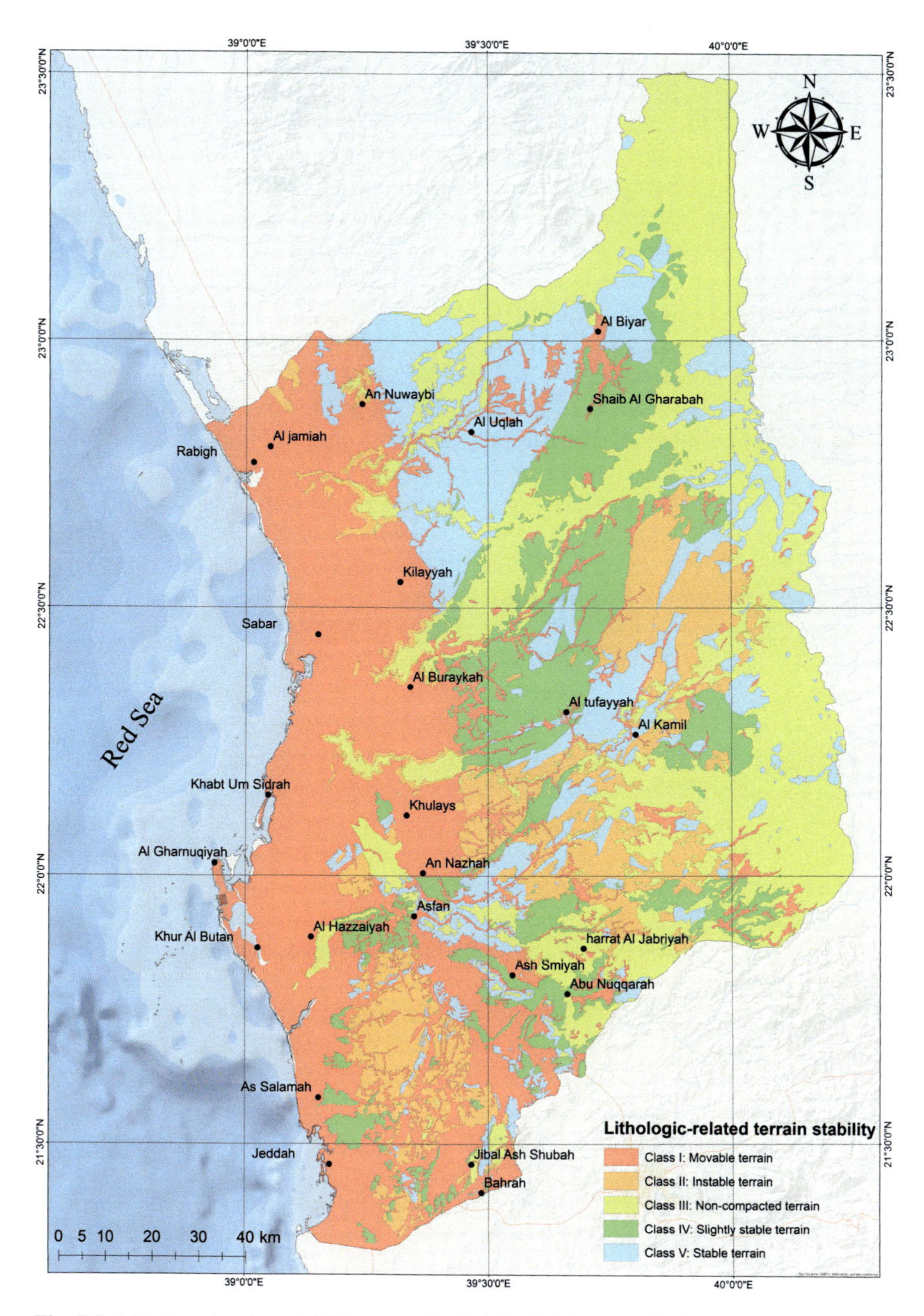

Fig. 7.5 Lithology-terrain instability map of Jeddah-Rabigh Region. (Al Saud, 2018)

lithological merging was followed for the lithologies with similar impact on terrain instability, whether these lithologies are attributed to the same rock formation or not.

7.3.3 *Fractures*

Large-scale fractures (e.g. faults) and significant factors in terrain instability. These are represented by large alignments of rock deformations. These faults resulted largely in creating other (secondary) fractures such as fissures and joints. Whereas, fissures and joints often exist in very local exposure and it could not be well identified in large-scale areal assessment.

Identification of large-scale fault alignments is successfully obtained by using satellite images, which enable identifying linear geological features on these images, the so-called "Lineaments". Therefore, space techniques are good tools for the identification of linear features as observed on satellite images. They appear to be tremendous on hard rock bodies and negligible on soft terrain (Al Saud, 2008).

Advanced Space-borne Thermal Emission and Reflection Radiometer (ASTER) satellite images were processed in this study. Aster L1A Reconstructed Unprocessed Instrument Data V003. ASTER is an advanced multi-spectral sensor with electromagnetic spectral region ranges from visible to thermal infrared, and it occupies 14 spectral bands. The spatial resolution of these images is 15 m with a visible bands, while they have 60 km x 60 km swath width, and thus 15 image scenes were used to cover the entire study area.

ERDAS Imagine-11, and PCI's Geomatica-10.2 software were used for image processing. These types of software encompass several digital advantages for image processing. Therefore, a number of steps have been followed starting by the preprocessing step with a series of consequent operations, including mainly: 1) atmospheric correction or normalization, image registration, geometric and radiometric corrections; 2) linking of the required image scenes, which is also called "Images Mosaicking". This is necessary in order to obtain a unique scene from all images required to cover the entire area of interest, and to facilitate further image treatment approaches.

Lineament identification from satellite images depends mainly on the thermal differentiation where thermal anomalies are the target features to be identified. These features represent the linear alignment of the existing fractures. These fractures, with dominant soil and detrital sediments and thus higher water retention (i.e. wetness) than the surrounding, will show cooler temperature than the rest terrain body.

Thermal bands (i.e. band number 10–14) in ASTER images, as attributed to Thermal Infrared (TIR), were processed. They are characterized by 90 m spatial resolution. In addition to the use of TIR bands, other spectral and digital advantage on ASTER images were utilized such as: color slicing and contouring, filtering, edge detection, band combination, enhancement, etc. Consequently, the lineament map was produced for the area between Jeddah and Rabigh.

There are 3189 lineaments detected in the area of study where they have different lengths. Many of the identified lineaments exist for long distance exceeding several tens of kilometers, while small-length lineaments also exist and they were found as geographic clusters.

The identified linear features are unevenly distributed in the study area, and thus their density was different between different localities. Therefore, a density map for lineaments is necessary to illustrate their concentration. In this regard, areas with dense lineaments (fractures) are characterized weak terrain stability and vice versa (Greenbaum, 1985; Al Saud, 2008).

The density of Lineaments can be elaborated using several approaches, but the most creditable one is that accounts the total number of lineaments with as specific area (geographic frame with known area). This concept is also followed by many researchers such as Gustafsson (1994) and Teeuw (1995). Hence lineaments density (L_D) can be expressed by the following equation:

$$L_D = \frac{\sum Ln}{A}$$

Where Σ Ln is the total number of lineaments, and A the define area where the counted lineament located in. In this study a lineament density map was produced with five classes (Fig. 7.6 and Table 7.5).

The generation of this map followed Sliding Windows approach, in which the area of study was classified into frames of 5 km x 5 km, and then the number of lineament was counted in each frame. Consequently, the resulting numeric values were again plotted in the middle of each frame. The resulting values were used once more to build up the contours map using Arc-GIS software (Fig. 7.6).

The average lineament density for the area between Jeddah and Rabigh was calculated as 4 lineaments per 25 km^2 (Al Saud, 2018). The description of the five classes of lineaments density was illustrated in Table 7.5.

7.3.4 Landforms

The influence of landform on terrain instability and mass movement implies several elements, with a special emphasis to the rigidity of terrain materials and structure. Hence, these physical characteristic are often deduced from the obtained landform maps. In this study, landforms were adopted from the Land Resources Maps of Kingdom of Saudi Arabia, which was obtained by the Ministry of Agriculture and Water (1986). For the area between Jeddah and Rabigh, there are 13 landform. The landforms with almost similar impact on terrain instability were digitally combined, using Arc-GIS, and then categorize into 5 classes (Fig. 7.7 and Table 7.6).

The obtained landforms map shows that 30% of the region between Jeddah and Rabigh is characterized suitability terrain instability.

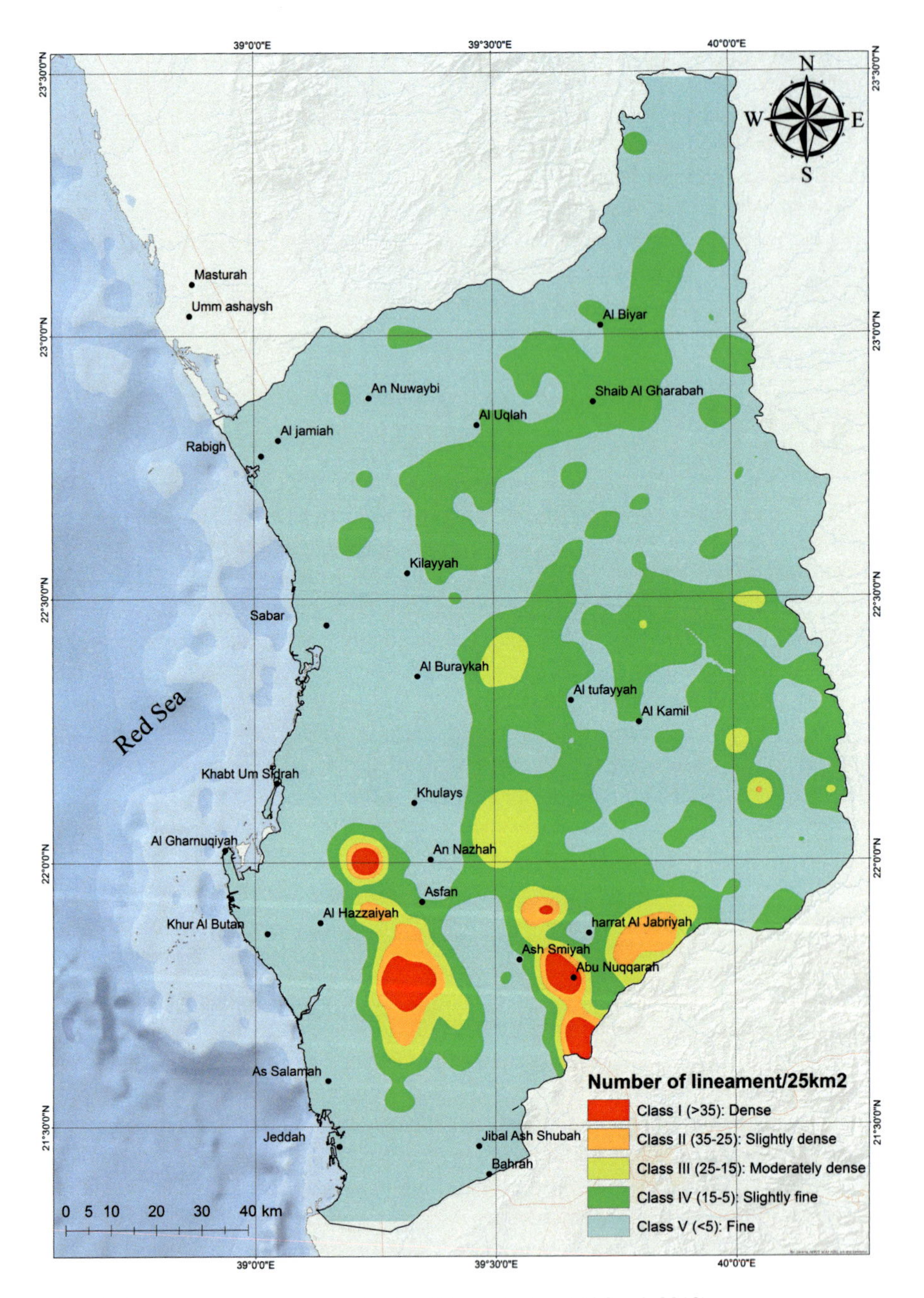

Fig. 7.6 Lineaments density map of Jeddah-Rabigh Region. (Al Saud, 2018)

Table 7.5 Classification of lineaments density in Jeddah-Rabigh Region

Class	Lineament density (lineaments/25km^2)	Description	Terrain instability
I	>35	Dense	Moveable
II	35–25	Slightly dense	Instable
III	25–15	Moderately dense	Non-compacted
IV	15–5	Slightly fine	Slightly stable
V	<5	Fine	Very stable

7.4 Data Manipulation

The geospatial data for the main influencing factors on terrain instability were identified and digitally illustrated using Arc-GIS. The GIS layers of the influencing data were plotted with their 5 classes for each. In order to create a map for the terrain instability in the area of interest; therefore, all these factors must be integrated together in one digital unit. In other word, all identified factors and their elements must act together to localize the geographic areas with different characteristics towards terrain instability (i.e. hotspots). Nevertheless, not all these factor have the same impact on the existence of terrain instability. Hence, some factors have larger influencing than others, and this must be considered during the digital manipulation of these factors together. Therefore, each factor must be attributed to specific degree of impact, which has been described as "weight".

For example, the impact of landforms aspect on terrain stability is not similar as that of fractures, and so on for the rest factors and their elements. The determined weights of each factors and the belonging elements will be used during data integration in the Arc-GIS.

The following steps must be followed for data manipulation:

1. Factors to be prepared as digital maps, and each factor will be considered as a GIS layer with five classes (elements).
2. The obtained classes will be ordered from the highest to lowest impact. Thus, Class I must represent the most effective (i.e. unstable) terrain instability to Class V as the lowest impact (i.e. very stable). Also, there will be a defining for the weights of influence for each factor.
3. Producing the digital maps of all factors after considering their weights. This can be done using ESRI's Arc-GIS (i.e. Arc-View) software by "superimposing" different GIS layers together in the GIS system.
4. Each weight will attributed to a numeric percentage value, where the total for all factors must be 100%.
5. The five classes of each factor have also specific impact on terrain instability. For example, Class I in slope factor differs from Class I in lithology factor. Therefore, a "rate" has been dedicated for each class.

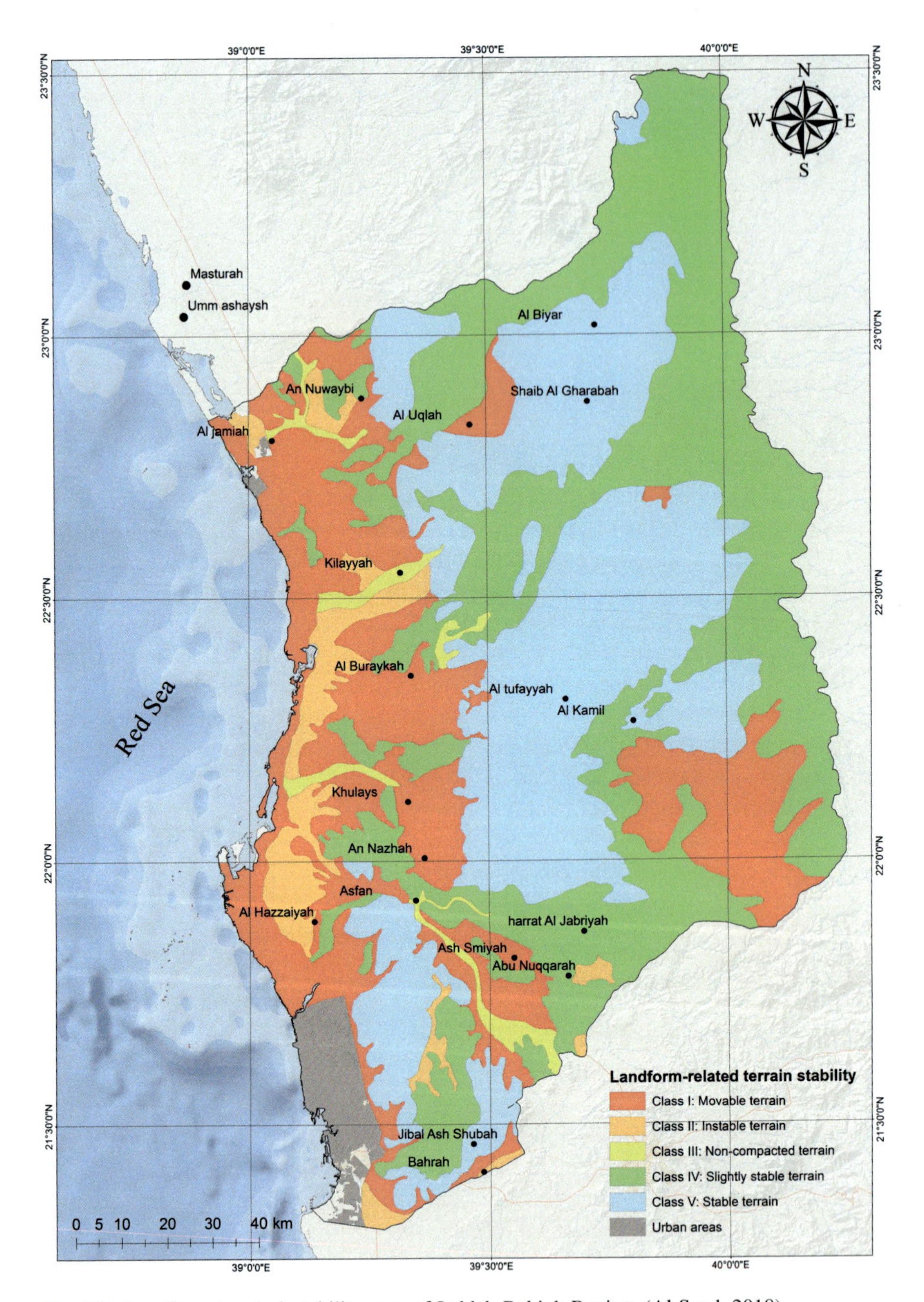

Fig. 7.7 Landform-terrain instability map of Jeddah-Rabigh Region. (Al Saud, 2018)

Table 7.6 Landform classes according to terrain instability in Jeddah-Rabigh Region

Class	Land form aspect	Area (km^2)	%	Terrain instability
I	Coastal plain, tidal flats, alluvial fans and foot slopes, degraded plain/active slopes	4750	24	Moveable
II	Alluvial plain, Pediplan with deep soils	983	5	Instable
III	Pediplan with shallow soils, Wadi deposits	461	2.5	Non-compacted
IV	Hills and rock outcrops, lava fields and volcanic hills	7180	36.5	Slightly stable
V	Mountains	6349	32	Very stable

The adoption of values for weights and rates were determined based on the following, according to Al Saud (2018):

1. Field experience and observations in field studies and the related ground investigations.
2. The assessment of the impact of factors depended on by many studies obtained by Edet et al. (1998); Robinson, et al. 1999); Das (2000) and Shaban et al. (2006).

Based on the above points and considering Al Saud (2010), the adopted factors in this study were given the following weights: slope (35%), lithology (30%), fractures (20%) and landform (15%). Table 7.7 shows weights of factor (F_w) and rate (F_r) and Effectiveness (E_f) for the influencing factors on terrain instability.

Rates are accounted with 100 as the maximum value 0 for the minimum value, thus, rates of the five classes will illustrated as follows: 100–80, 80–60, 60–40, 40–20 and 20–0. Therefore, the average of rating for each class will be 90, 70, 50, 30 and 10 for classes from I to V; respectively (Table 7.7).

To calculate the degree of effectiveness (E_f) for each factor depending on weight (F_w) and rate (F_r). Hence, each weight will be multiplied by the rate (F_w x F_r). For example, the weight of lithology is 30%, if multiplied by class II of the rate, which is 70%; therefore, the degree of effectiveness will be as follows:

$$E_f = F_w \times F_r = 30/100 \times 70 = 21$$

The sum of effectiveness for all classes will be 250 as shown in Table 7.8. Therefore, for each class the net effectiveness (En) is calculated by dividing the factor effectiveness (E_f) on the Sum of effectiveness, which equals 250 (Table 7.8). For example, E_n for Class II in the lithology was calculated as follows: $E_n = E_f /\sum$ effectiveness = 21/250 x 100 = 8.4%.

The E_n for each classes will be digitally converted in the GIS system, and they will represent the elements of the systematic integration of factors in order to produce the terrain instability map.

Table 7.7 Weights and rates of effectiveness on terrain stability (Al Saud, 2018)

Class/Factor	I	II	III	IV	V	
Slope (°)						
$\mathbf{F_w}$	35%					
$\mathbf{F_r}$	90	70	50	30	10	
$\mathbf{E_f}$	31.5	24.5	17.5	10.5	3.5	87.5
Lithology						
$\mathbf{F_w}$	30%					
$\mathbf{F_r}$	90	70	50	30	10	
$\mathbf{E_f}$	27	21	15	9	3	75
Fractures (lineament/25km^2)						
$\mathbf{F_w}$	20%					
$\mathbf{F_r}$	90	70	50	30	10	
$\mathbf{E_f}$	18	14	10	6	2	50
Landform						
$\mathbf{F_w}$	15%					
$\mathbf{F_r}$	90	70	50	30	10	
$\mathbf{E_f}$	13.5	10.5	7.5	4.5	1.5	37.5
Sum of effectiveness						**250**

Table 7.8 Net effectiveness of classes composing influencing factors in terrain instability (Al Saud, 2018)

Class/Factor	I	II	III	IV	V
Slope					
$\mathbf{E_n}$	12.6	9.8	7	4.2	1.4
Lithology					
$\mathbf{E_n}$	10.8	8.4	6	3.6	1.2
Fractures					
$\mathbf{E_n}$	7.2	5.6	4	2.4	0.8
Landforms					
$\mathbf{E_n}$	5.4	4.2	3	1.8	0.6

7.5 Results

The obtained map, with five main categories, evidences the degree of influence on terrain instability and the related mass movements (Fig. 7.8).

This maps shows about 90% of the area between Jeddah and Rabigh is under risk of terrain instability (ranges from high to very high risk). Therefore, from the obtained map, the following dimensions were calculated using GIS application:

- Very high risk (movable) = 8308 km^2 (42% of the studies area).
- High risk (instable) = 9383 km^2 (48% of the studies area).
- Moderate risk (non-compacted) = 1118 km^2 (6% of the studies area).
- Low risk (slightly stable) = 568 km^2 (2.5% of the studies area)
- No risk (very stable) = 346 km^2 (1.5% of the studies area).

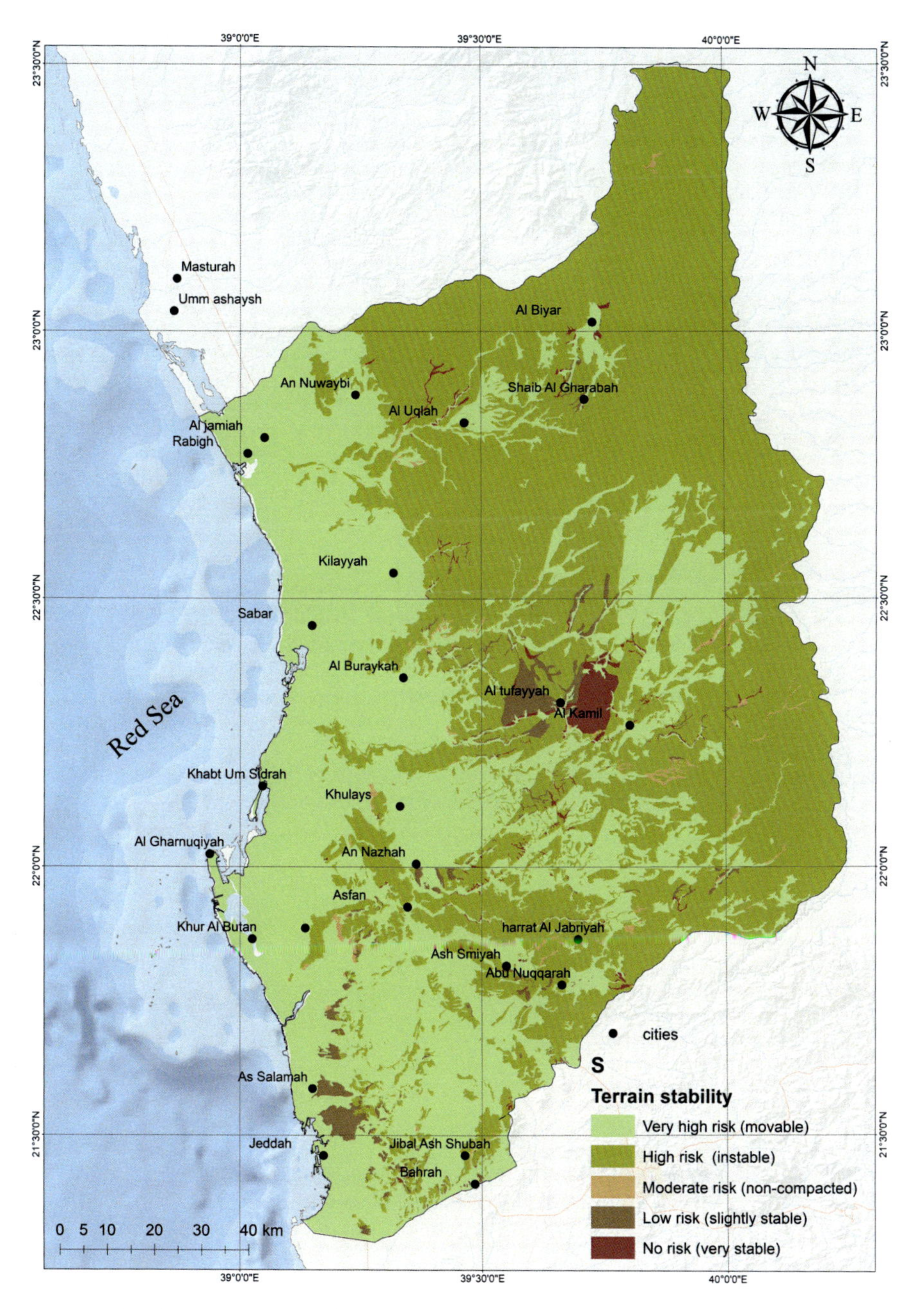

Fig. 7.8 Terrain stability map of Jeddah-Rabigh Region. (Al Saud, 2018)

References

Al Saud, M. (2008). Using ASTER images to analyse geologic linear features in Wadi Aurnah Basin, Western Saudis Arabia. *The Open Remote Sensing Journal, 2008*(1), 7–16.

Al Saud, M. (2010). Mapping potential areas for groundwater storage in Wadi Aurnah Basin, western Arabian Peninsula, using remote sensing and GIS techniques. *Hydrogeology Journal, 18*(6), 1481–1495.

Al Saud, M. (2012). Use of remote sensing and GIS to analyze drainage system in flood occurrence, Jeddah – Western Saudi Coast. In *Book on "Drainage systems"*. InTech Publishing. https://doi.org/10.5772/34008

Al Saud, M. (2018). *Using space techniques and GIS to identify vulnerable areas to natural hazards along the Jeddah-Rabigh Region, Saudi Arabia* (p. 306). Nova Science Publisher Inc. ISBN13: 978-15-361-33134.

Bankher, K. (1996). *Engineering geological evaluation of earth fissures in Wadi Al-Yutamah South Al-Madinah Al-Munawwarah.* M.Sc thesis, FES, Jeddah.

Budetta, P. (2004). Assessment of rock fall risk along roads. *Natural Hazards and Earth System Sciences, 4*, 71–81.

Chatwin, S. C., Howes, D. E., Schwab, J. W., & Swanston, D. N. (1994). *A guide for management of landslide-prone terrain in the Pacific Northwest* (Handb. No. 18) (2nd ed.). B.C. Ministry of Forests, Land Manage.

Church, M. (1983). *Concepts of sediment transfer and transport on the Queen Charlotte Islands.* FFIP Working Paper 2/83, Fish Forestry Interaction Program, B.C. Ministry of Forests, B.C. Ministry of Environment, Canada Department of fisheries and Oceans.

Das, D. (2000) *GIS application in hydrogeological studies.* Available at www.gisdevelopment.net

Edet, A., Okereke, S., Teme, C., & Esu, O. (1998). Application of remote sensing data to groundwater exploration: A case study of the Cross River State, southeastern Nigeria. *Hydrogeology Journal Springer-Verlag, 6*(3), 394–404.

Greenbaum, D. (1985). Review of remote sensing applications to groundwater exploration in basement and regolith. *British Geological Survey Report OD, 85*(8), 36p.

Gustafsson, P. (1994). SPOT satellite data for exploration of fractured aquifers in a semi-arid area in Botswana. *Hydrogeology Journal 1994, 2*(20), 9–18.

Holzer, T. L. (1984). Ground failure induced by ground-water withdrawal from unconsolidated sediment. In T. L. Holzer (Ed.), *Man-induced Land Subsidence* (pp. 67–105). Reviews in Engineering Geology, VI, the Geological Society America.

Hungr, O., Morgan, G. C., & Kellerhals, R. (1984). Quantitative analysis of debris torrent hazards for design of remedial measures. *Canadian Geotechnical Journal, 21*, 663–677.

Ministry of Agriculture and Water. (1986). *Land resources map of the Kingdom of Saudi Arabia* (33pp). Land management Department.

Robinson, C., El-Baz, F., & Singhory, V. (1999). Subsurface imaging by RADARSAT: Comparison with Landsat TM data and implications for groundwater in the Selima area, northwestern Sudan. *Remote Sensing Abstracts, 25*(3), 45–76.

Skiba, W., Gilboy, C., & Smith , J. (1986). Geological map of Rabigh Quadrangle sheet 22 D, Ministry of Petroleum and Mineral Resources. Kingdom of Saudi Arabia.

Shaban, A., Khawlie, M., & Abdallah, C. (2006). Use of remote sensing and GIS to determine recharge potential zones: The case of occidental Lebanon. *Hydrogeology Journal, 14*(4), 433–443.

Teeuw, M. (1995). Groundwater exploration using remote sensing and a low-cost geographic information system. *Hydrogeology Journal 1995, 3*(3), 21–30.

Varnes, D. J. (1978). Slope movement types and processes. In R. L. Schuster & R. J. Krizek (Eds.), *Special report 176: Landslides: Analysis and control* (pp. 11–33). Transportation and Road Research Board, National Academy of Science.

Chapter 8
Remote Sensing Studies on Monitoring Natural Hazards Over Cultural Heritage Sites in Cyprus

Athos Agapiou and Vasiliki Lysandrou

Abstract This chapter presents examples of remote sensing studies for monitoring natural hazards related to ancient monuments and archaeological sites in Cyprus. Through these studies, the use of Earth Observation, and specifically the contribution of the European Copernicus Programme, is highlighted. Most of them have been carried out during the last years, within the framework of funded research projects. The various case studies presented in this chapter underscore Earth Observation's mingling with other remote sensing techniques (both middle range and terrestrial) and geoinformatics towards inclusive monitoring of cultural heritage and prevention against possible hazards.

The chapter unfolds in two parts: The first part introduces an overview of the potential contribution of Earth Observation to the Cultural Heritage Disaster Risk Management (DRM) cycle, with specific focus on the Eastern Mediterranean basin. The DRM includes six consecutive steps that require various inputs, including the context, threats and monitoring phases. For each step, the role of Earth Observation sensors and their related products are discussed.

The second part of this chapter focuses on studies dealing with natural hazards in Cyprus using optical and radar datasets. These studies include the following: (a) soil erosion by water, (b) vegetation growth, and (c) detection of surface displacements, in sites with archaeological interest.

The chapter ends with a comprehensive risk assessment report of various hazards (both natural and anthropogenic) using the Analytical Hierarchy Process (AHP) method. This assessment concerns the western part of Cyprus, namely the Paphos District, where more than 200 monuments are found.

A. Agapiou (✉) · V. Lysandrou
Department of Civil Engineering & Geomatics, Cyprus University of Technology, Limassol, Cyprus

Eratosthenes Centre of Excellence, Limassol, Cyprus
e-mail: athos.agapiou@cut.ac.cy; vasiliki.lysandrou@cut.ac.cy

M. M. Al Saud (ed.), *Applications of Space Techniques on the Natural Hazards in the MENA Region*, https://doi.org/10.1007/978-3-030-88874-9_8

8.1 Introduction

This chapter aims to present various authors' studies performed in the last years, through research activities, dealing with natural hazards over cultural heritage sites. The chapter has a specific focus on the Eastern Mediterranean basin and Cyprus, as well as remote sensing data, and Earth Observation sensors. At the beginning of the chapter, an overview of the potential contribution of Earth Observation to the Cultural Heritage Disaster Risk Management (DRM) cycle is presented. The DRM cycle and the role of the Earth Observation sensors are emphasised in the next section. The synthesis and assessment from various hazards (both natural and anthropogenic) using the Analytical Hierarchy Process (AHP) methodology are then presented.

Hazards can have a negative impact on cultural heritage, while a combination of hazards may trigger other secondary ones. As identified by International Council for Science (ICSU) and the World Meteorological Organization (WMO), and adopted by UNESCO, the most common categories of hazards are the following: meteorological, hydrological, geological, astrophysical, biological, and climate change (UNESCO, 2010).

In the literature, various terms have been used to study the hazards' phenomena. *Disaster* is defined by the United Nations International Strategy for Disaster Reduction as "a severe disruption of the functioning of a community or a society causing widespread human, material, economic or environmental losses which exceeds the ability of the affected community or society to cope using its resources" (United Nations, 2009). This definition was extended by other international organisations, dealing with the management of cultural heritage sites, in order to include disaster impacts not only on people and properties but also on the cultural heritage values of the World Heritage property (UNESCO, 2010).

Risk is defined as "the chance of something happening that will have an impact upon objectives" (Emergency Management Australia, 2000), while the United Nations (2009) refers to *risk* as to the combination of the probability of an event and its negative consequences. Moreover, *hazard* is defined as "a dangerous phenomenon, substance, human activity or condition that may cause loss of life, injury or other health impacts, property damage, loss of livelihoods and services, social and economic disruption, or environmental damage or any phenomenon, substance or situation" (United Nations, 2009). Hazard can eventually cause disruption or damage to different infrastructures and services, people, property, and environment (Abarquez and Murshed, 2004). Finally, *vulnerability* refers to "the characteristics and circumstances of a community, system or asset that make it susceptible to the damaging effects of a hazard" (United Nations, 2009). Based on these terms, vulnerability is an intrinsic characteristic of an asset, independent of its exposure. Consequently, *disaster risk* is described as the result of hazard and vulnerability (UNESCO, 2010).

Earth Observation plays an essential role in monitoring cultural heritage sites' purposes against various anthropogenic or natural hazards. The existing literature

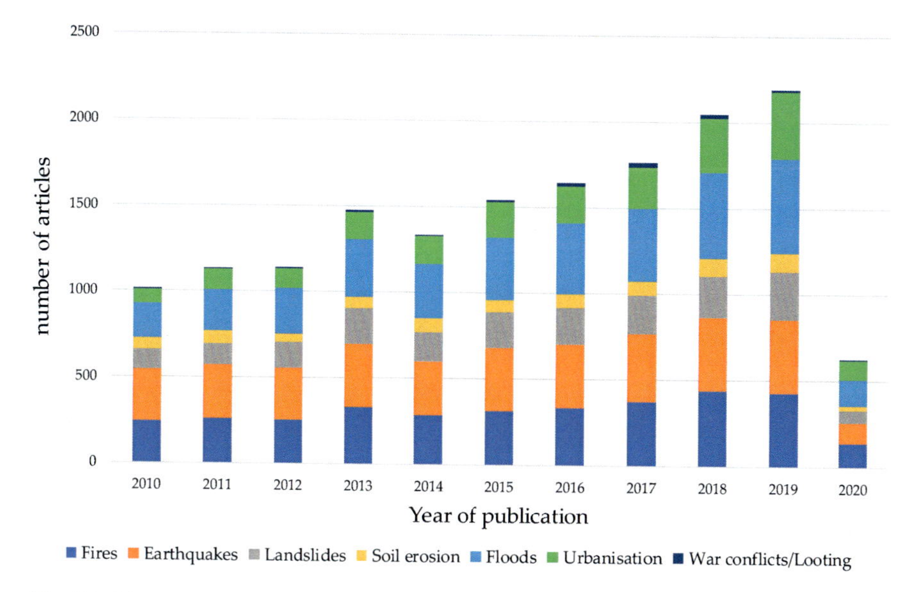

Fig. 8.1 The total number of published articles related to various hazards with the support of satellite and space-based observations for the last decade (2010–2020). (Source: Agapiou et al., 2020b)

indicates a steady increase in scientific studies dealing with this topic (Fig. 8.1). This is mainly due to satellite sensors' advantages, such as performing systematic measurements remotely covering large areas. The increasing number of new sensors has also released new opportunities to support cultural heritage sites' systematic monitoring. For instance, the European Copernicus Sentinel-2 constellation provides optical images with a revisit time of 5 days at the equator, while these data are freely distributed and open access (Li & Roy, 2017). Other initiatives and services like the Copernicus Emergency Management Services, (Bosco et al., 2021), the International Charter Space and Major Disasters (2021), the United Nations Platform for Space-based Information for Disaster Management and Emergency Response (UN-SPIDER) (2021) or Group on Earth Observation (GEO) (2021), can provide support to local authorities after significant disastrous situations.

8.2 Disaster Risk Management (DRM) Cycle and Earth Observation Contribution

The Disaster Risk Management (DRM) cycle proposed by the International Strategy for Disaster Reduction and other international organisations and committees (Unesco, 2010, ICCROM 2016, ISO 31000, 2018) comprises six steps, briefly outlined below. All steps are interlinked between them, and hence any assumptions and ambiguities impact the full implementation of the DRM cycle. The design and conceptualisation of a DRM cycle plan is a synergistic effort of several qualified parties to obtain specific local value information.

Disaster Risk Management (DRM) Cycle Steps *Step I*: Understanding the context; *Step II*: Identifying risks; *Step III*: Analysing risks; *Step IV*: Evaluating risks; *Step V*: Treating risks; *Step VI*: Monitoring phase.

The first step includes collecting relevant information about various aspects of a cultural heritage site. The information can include details related to the physical environment and administrative, legal, political, socio-cultural, and economic aspects of a site. Ste II of the DRM cycle comprises identifying relevant natural and human-made risks that can potentially threaten cultural heritage. Then, at step III, the possibility of a threat is calculated, and the expected impact of all risks is estimated. Then, at step IV, the hierarchy and the classification of all potential risks is performed. At the following step, relevant effective measures can be planned by local stakeholders to eliminate or minimise the negative impact of the identified risks. Finally, the monitoring phase, includes a periodic update of all information

Following this brief introduction regarding the DRM cycle, a recent study (Agapiou et al., 2020b) attempted to link these steps of the DRM cycle with Earth Observation potentials. Therefore, for each step, likely synergies between Earth Observation sensors' existing capacity and cultural heritage management needs were investigated.

Regarding step I (context) and step II (identification of risks), remotely sensed sensors might be used for documentation and cartographic purposes. Diachronic observations over the site can be achieved through satellite and aerial observations providing time-series land use maps. This can support a better understanding of the archaeological site's potential changes. Beyond the existing high-resolution satellite multispectral sensors, archival satellite and aerial datasets can be used to map landscapes before modern development and changes (Hritz, 2013; Agapiou et al., 2016a; Ur, 2016; Lysandrou and Agapiou, 2020; Casana, 2020). Other existing geo-datasets like risk maps for geohazards and maps produced after processing satellite-based information (such as geo-datasets related to soil erosion by water) are available for Europe through specific platforms (Panagos et al., 2012).

Step III, concerns the risk assessment/risk analysis, which can be achieved from satellite images and related products, as can be seen in Solari et al., 2020; Tapete et al., 2016; Pastonchi et al., 2018, for floods, landslides, looting etc. Based on all previous steps, the overall risk is estimated in step IV. During step V, actions to prevent and/or limit the identified risks' overall damage occur. Here the role of Earth Observation sensors can be supportive for cases like illegal actions over archaeological sites. Even though satellite-based observations cannot prevent illegal actions on the ground, identifying looted areas can be considered a critical step towards increasing awareness for potential illegal trafficking and the protection of cultural assets.

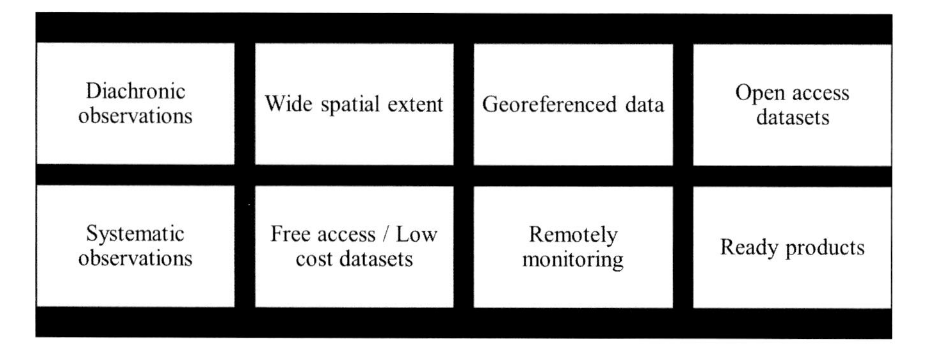

Fig. 8.2 Favourable characteristics of satellite observation datasets for cultural heritage's DRM. (After Agapiou et al., 2020b)

Finally, at step VI, the role of remotely sensed data can be fundamental for systematic monitoring of archaeological sites. The spatial extent of satellite images and the continuous observation over archaeological sites can be easily achieved using remote sensing techniques. For instance, the Copernicus satellite sensors' high temporal revisiting time (5-days) is ideal for the systematic observation since they can continuously provide new data, even of areas that are not physically accessible. Figure 8.2 summarises the primary characteristics of satellite sensors, which can support DRM cycles.

Therefore, the role of satellite observations and connected remotely sensed sensors are multiple. Examples from this role can be seen in Fig. 8.3. Despite that the list is not exhaustive, it indicates satellite observations' potentials toward implementing the DRM cycle.

8.3 Examples of Remote Sensing Studies for Monitoring Natural Hazards Over Cultural Heritage Sites in Cyprus

In this section, various studies regarding natural hazards over heritage sites in Cyprus using Earth Observation and other remotely sensed data are presented. Natural hazards include the impact of soil erosion by water, vegetation growth and dynamics, and the detection of surface displacements.

8.3.1 Soil Erosion

Soil erosion by water is a natural phenomenon that involves the detachment of soil material rainfall and the flow traction (Erosion by Water, 2021). Models regarding

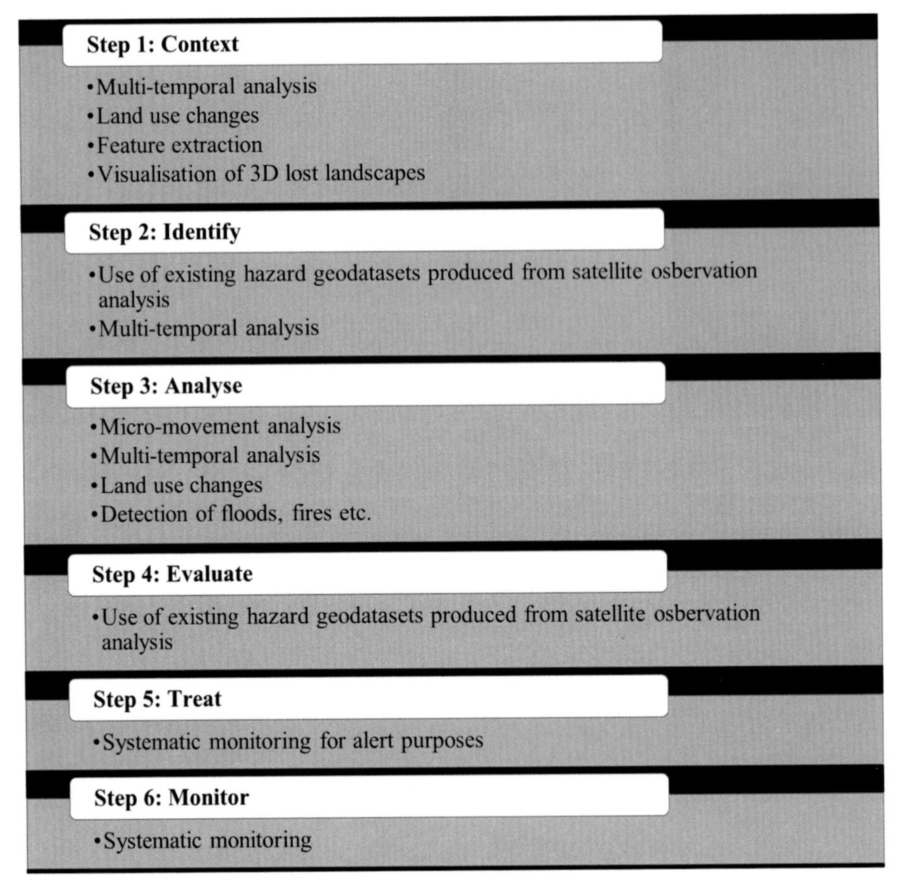

Fig. 8.3 Indicative key satellite observation processing chains beneficial for the various steps of a DRM cycle intended for cultural heritage. (After Agapiou et al., 2020b)

the study of soil erosion have been investigated in the past by Panagos (2015), Panagos and Katsoyannis (2019), Pena et al. (2020) and Chandramohan et al. (2015); for more related literature, see Agapiou et al. (2020a).

These studies have used various models to better estimate and predict soil loss (Quinton, 2011). The use of empirical models like the Revised Universal Soil Loss Equation (RUSLE) has been widely used due to reasonably accurate estimations and its capacity to be implemented, using a Geographical Information System (GIS) and satellite datasets (Borrelli et al., 2013). RUSLE model, developed by Renard et al. (1997), uses five factors: multiplying each other. These factors are the rainfall erosivity factor (R), the soil erodibility factor (K), the slope length and steepness factor (LS), the cover management factor (C), and the conservation practice factor (P) for the estimation of the average annual soil loss (A) (for a review regarding this model see Ghosal and Das Bhattacharya, 2020). Through the Panagos et al. (2012) study, RUSLE datasets have become accessible on a European level. European and global soil erosion maps and geodata sets, like those of Panagos et al. (2012), can be accessed by the European Soil Data Centre (ESDAC).

Soil erosion by water remains one of the most important natural hazards that are threatening archaeological sites. Both soil loss and soil deposition can alter an area's archaeological context, transferring, for instance, ceramics from one location to another. Archaeological context includes not only the standing monuments but also subsurface archaeological remains. For the lastest, our knowledge is limited for several sites, making their protection very difficult. While some heritage management methods using ground-based strategies have been reported in the past (see Luo et al., 2019), these have a limited spatial extent. Therefore, evaluating the risk of subsurface archaeological remains from soil erosion over large areas is peculiarly difficult. In a recent article, Agapiou et al. (2020a) used Kibblewhite et al. (2015) datasets to develop a sub-surface archaeological proxy map at a European level. In their work, Kibblewhite et al. (2015) have categorised European soil according to how the various archaeological materials can be affected by the pertinent soil type, following a standard taxonomic classification. The study from Agapiou et al. (2020a) has integrated these datasets to provide for the first European estimation of subsurface archaeological exposure due to soil loss. Simultaneously, the results are biased on the models' uncertainty and assumptions and the datasets used—the analysis aimed to pave the way to implement extensive-scale studies related to subsurface archaeological materials threats. The overall results are shown in Fig. 8.4. The analysis indicated that 75% of the area is characterised as a low threat due to soil erosion, with soil loss of less than 5 t/ha per year. In comparison, 13% and 12% are characterised with moderate (soil loss between 5–10 t/ha per year) and high-risk (soil loss more than 10 t/ha per year) level.

In detail, as shown from Fig. 8.4, four different subsurface materials, namely the metals, bones, organics, and stratigraphy evidence, have been investigated against soil loss. The preservation state of each type of stratigraphic material is mapped as poor (red), fair (yellow), and good (green). Areas with no data available are visualised with white colour in the background. The majority of the area for all types of archaeological material is considered low-threat regardless of their soils' preservation capacity. However, regarding the moderate- and high-threat level areas, fluctuations can be observed. South Europe, including the case study of Cyprus and the Mediterranean basin, is exposed to higher soil-loss threats compared to northern countries. A country-level statistic based on the findings of Fig. 8.4 was also implemented. These are shown in Figs. 8.5, 8.6, 8.7 and 8.8. Figure 8.5 indicates the mean values per country level for metals concerning the level of soil-loss threats, while Figs. 8.6, 8.7 and 8.8 indicate the same results for bones, organics, and stratigraphic evidence, respectively. For all Figs. 8.5, 8.6, 8.7 and 8.8, values close to 1 indicate the low preservation status, values close to 2 indicate fair conditions of preservation, and values close to 3 good preservation status.

The primary outcomes of the previous study were the following: (1) Most European countries can be considered low threat areas for all types of subsurface materials; (2) Northern European countries can be considered low threat areas in contrast to Mediterranean countries, which are characterised with moderate and

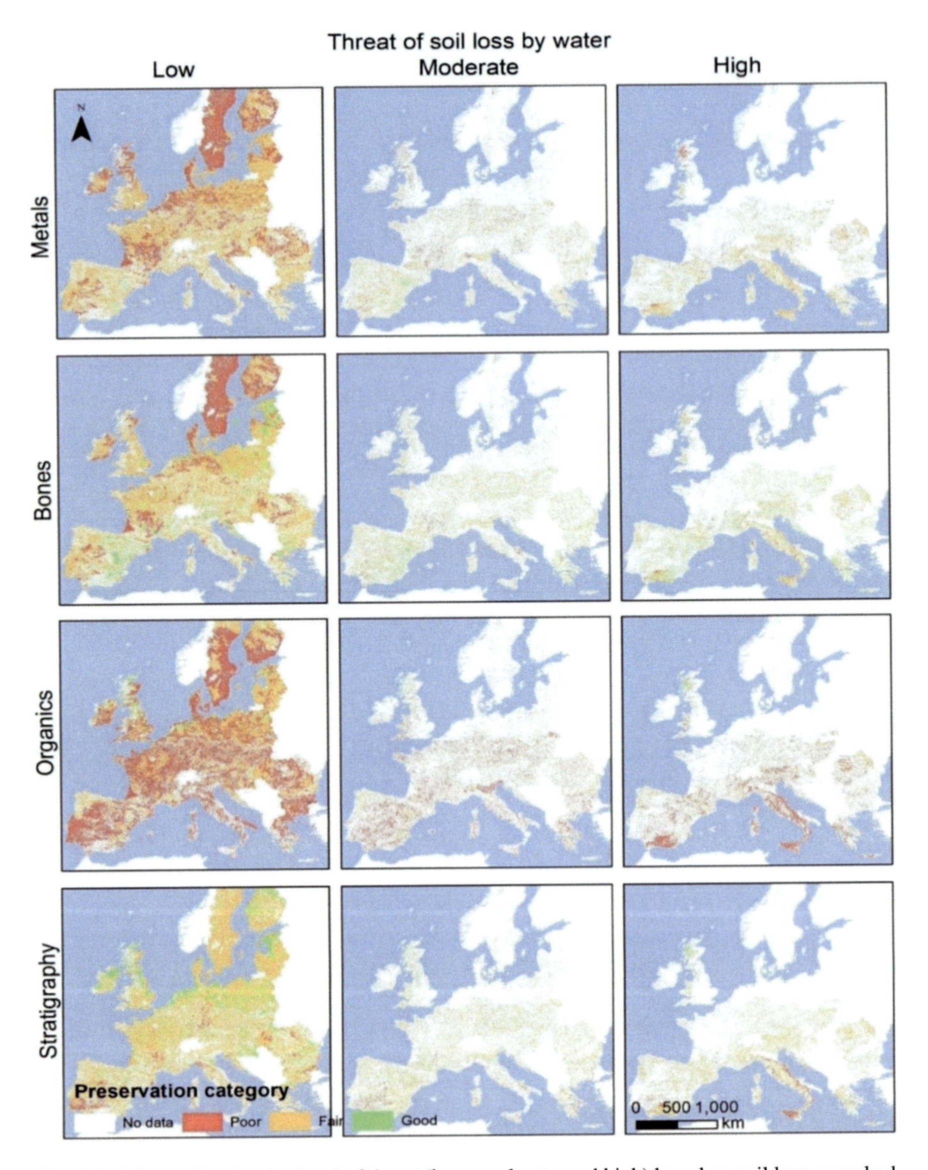

Fig. 8.4 Maps indicating the level of threat (low, moderate, and high) based on soil loss provoked by water activity (soil erosion) for each of the four different subsurface materials (metals, bones, organics, and stratigraphy evidence): The preservation state is also indicated (red for poor, yellow for fair, and green for good). (Source: Agapiou et al., 2020a)

high-risk levels; (3) Areas characterised as low threat from soil erosion present approximately 10% of adequate preservation capacity, which varies depending on the type of the material; (4) Similar patterns on a European scale for all types of materials are reported for areas characterised with moderate and high risk from soil loss.

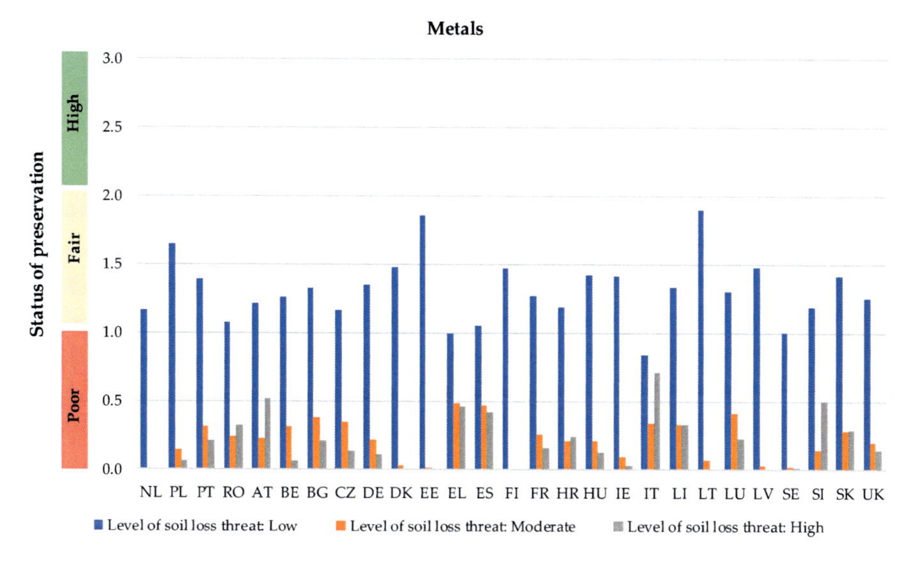

Fig. 8.5 Mean values per country level for metals per level of soil-loss threats: Values close to 1 indicate the poor preservation status, values close to 2 indicate fair conditions of preservation, and values close to 3 good preservation status. (Source: Agapiou et al., 2020a)

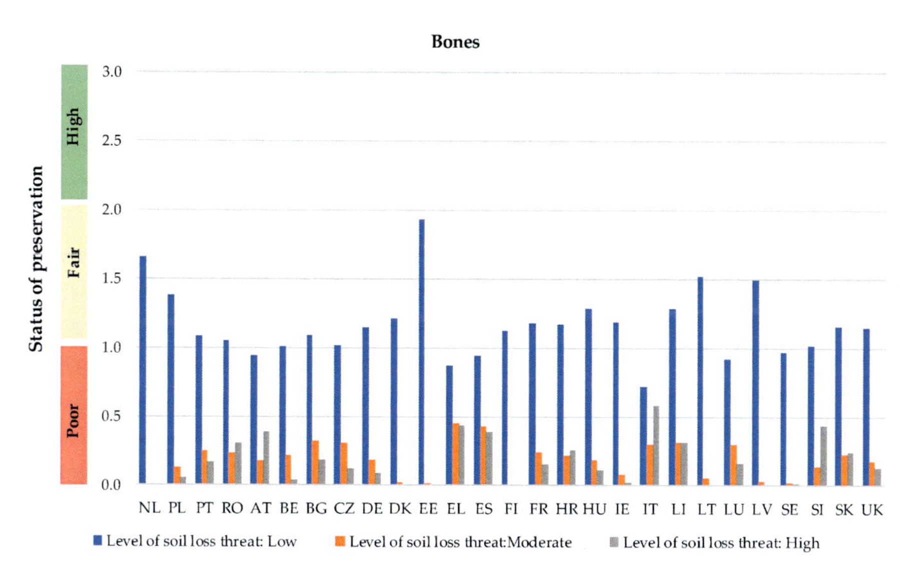

Fig. 8.6 Mean values per country level for bones per level of soil-loss threats: Values close to 1 indicate the poor preservation status, values close to 2 indicate fair conditions of preservation, and values close to 3 good preservation status. (Source: Agapiou et al., 2020a)

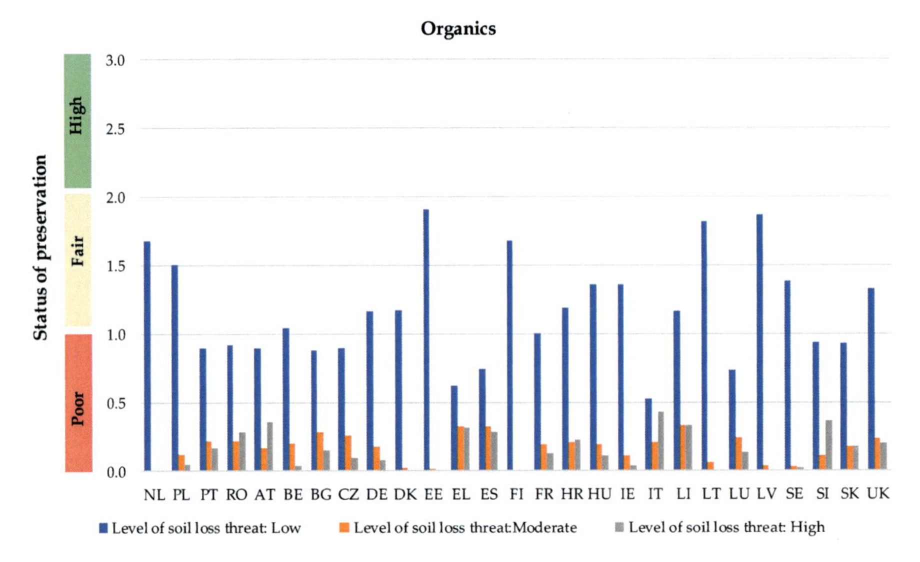

Fig. 8.7 Mean values per country level for organics per level of soil-loss threats: Values close to 1 indicate the poor preservation status, values close to 2 indicate fair conditions of preservation, and values close to 3 good preservation status. (Source: Agapiou et al., 2020a)

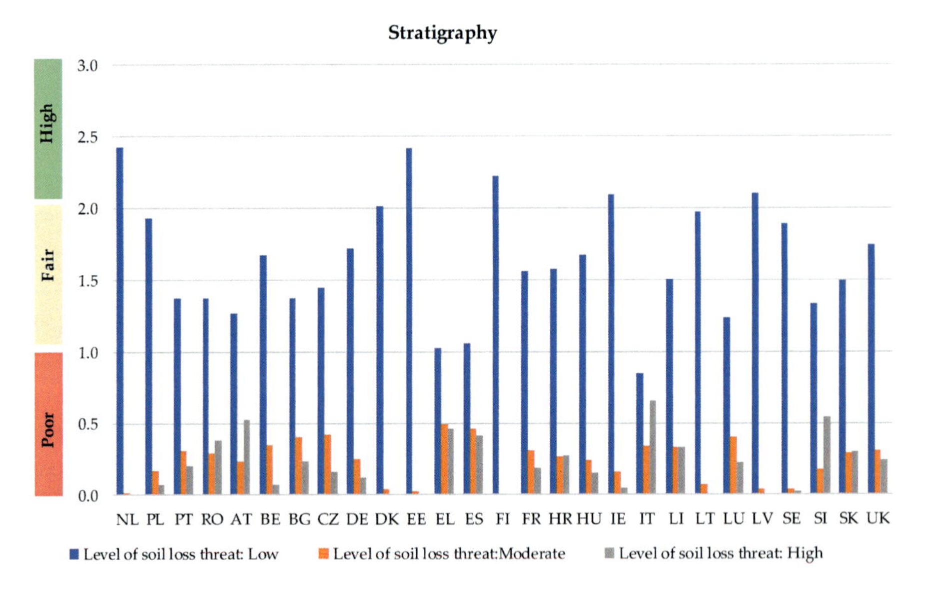

Fig. 8.8 Mean values per country level for bones per level of soil-loss threats: Values close to 1 indicate the poor preservation status, values close to 2 indicate fair conditions of preservation, and values close to 3 good preservation status. (Source: Agapiou et al., 2020a)

8.3.2 Vegetation Growth and Its Dynamics

Monitoring vegetation dynamics and long-term temporal changes of vegetation cover is of great importance for assessing the risk level of a natural or anthropogenic hazard. Vegetation plays a critical role in several hazards, like the soil loss mentioned earlier. The increase or decrease of vegetation cover through vegetation indices has been used in the past as an indicator for land-use change and urbanisation sprawl. At the same time, it can be an indication of agricultural pressure near archaeological sites' surroundings. The extraction of vegetation cover has been systematically investigated in the literature using optical satellite sensors. However, the use of radar vegetation indices is limited, while their combination is even rarer.

Several studies have demonstrated the benefits of satellite-based monitoring, providing comprehensive and systematic coverage over archaeological sites (Luo et al., 2019; Agapiou and Lysandrou, 2015). Open and freely distributed optical and radar satellite images are available from the European Copernicus Programme (2021). The Sentinel sensors, with a high-temporal revisit time, medium resolution satellite images can be downloaded through specialised big data cloud platforms such as the Sentinel Hub. At the same time, radiometric and geometric corrections can be applied.

To evaluate the overall performance of the synergistic use of optical and radar vegetation indices from the Sentinel-1 and Sentinel-2 sensors, Agapiou (2020) has used these datasets over the archaeological site of "Nea Paphos" in Cyprus. The study has also used other open access services, namely the crowdsourced OpenStreetMap initiative. In detail, optical and radar Sentinel datasets, acquired over the archaeological site of "Nea Paphos" have been used, while Sentinel ready products from the Sentinel Hub service and crowdsourced vector geodata available at the OpenStreetMap service have been explored. Finally, compressed red-green-blue (RGB) high-resolution optical data from the Google Earth platform for validation purposes were used (Fig. 8.9).

From the Sentinel Hub service, radar and optical Sentinel images were retrieved, and the Normalised Difference Vegetation Index (NDVI) and the Radar Vegetation Index (RVI) were processed using Eqs. 1 and 2:

$$\mathrm{NDVI} = (\rho_{\mathrm{NIR}} - \rho_{\mathrm{RED}})/(\rho_{\mathrm{NIR}} + \rho_{\mathrm{RED}}), \quad (8.1)$$

$$\mathrm{RVI} = (\mathrm{VV}/(\mathrm{VV} + \mathrm{VH}))^{0.5}\ (4\ \mathrm{VH})/(\mathrm{VV} + \mathrm{VH}), \quad (8.2)$$

Where ρ_{NIR} and ρ_{RED} refer to the reflectance values (%) of the near-infrared and red bands of the optical Sentinel-2 sensor (band 8 and band 4), while the VV and VH refer to the polarisation bands of the Sentinel-1 sensor, implemented by a custom script available with the Sentinel-Hub services (2021).

Based on the optical and radar vegetation indices, the proportion of vegetation was then retrieved. In our study, two different models have been applied for both optical and radar datasets:

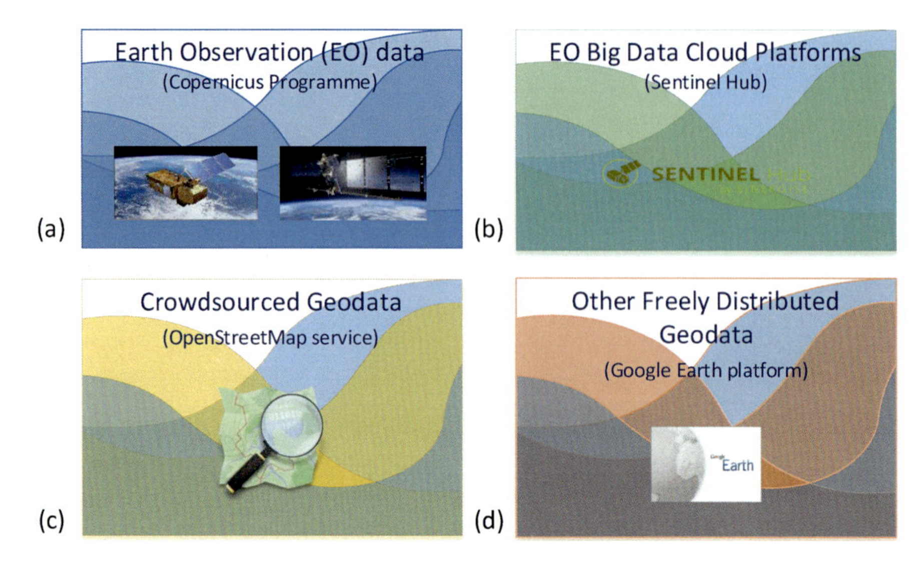

Fig. 8.9 A schematic representation of the four "layers" of information used the study of Agapiou (2020): the Earth Observation Sentinel-1 and Sentinel-2 images (top left), the Sentinel Hub, an Earth Observation big data cloud platform (top right), crowdsourced geodata from OpenStreetMap (bottom left) and the Google Earth platform (bottom right)

$$\mathrm{Pv}_1 - \mathrm{radar} = (\mathrm{RVI} - \mathrm{RVI\ non-veg.})/(\mathrm{RVI\ veg} - \mathrm{RVI\ non-veg.}), \tag{8.3}$$

$$\mathrm{Pv}_1 - \mathrm{optical} = (\mathrm{NDVI} - \mathrm{NDVI\ non-veg.})/(\mathrm{NDVI\ veg} - \mathrm{NDVI\ non-veg.}), \tag{8.4}$$

$$\mathrm{Pv}_2 - \mathrm{radar} = [(\mathrm{RVI} - \mathrm{RVI\ min})/(\mathrm{RVI\ max} - \mathrm{RVI\ min})]^{0.5} \tag{8.5}$$

$$\mathrm{Pv}_2 - \mathrm{optical} = [(\mathrm{NDVI} - \mathrm{NDVI\ min})/(\mathrm{NDVI\ max} - \mathrm{NDVI\ min})]^{0.5} \tag{8.6}$$

Where vegetation index veg (NDVI veg and RVI veg) and non-vegetation index (NDVI non-veg. and RVI non-veg.) represent the vegetated and non-vegetated pixels of the considered index, respectively, vegetation index max (NDVI max and RVI max) and vegetation index min (NDVI min and RVI min) represent the maximum and minimum histogram value of the vegetation image.

To investigate pottential correlation between the NDVI and the RVI indices, a regression analysis was carried out. As shown in Fig. 8.10, no specific pattern between the two indices can be extracted since there was a high variance. This is also aligned with the previous findings indicating that optical and radar indices do not produce similar findings.

In the light of the above, periodic monthly RVI and NDVI indices covering May 2019 to May 2020 were extracted from the Sentinel Hub service. RVI results over the "Nea Paphos" archaeological site are shown in Fig. 8.11-left, while Fig. 8.11-right shows the NDVI index results (whereas a – l in both figures refers to months starting from May 2019). Higher RVI values that could correspond to

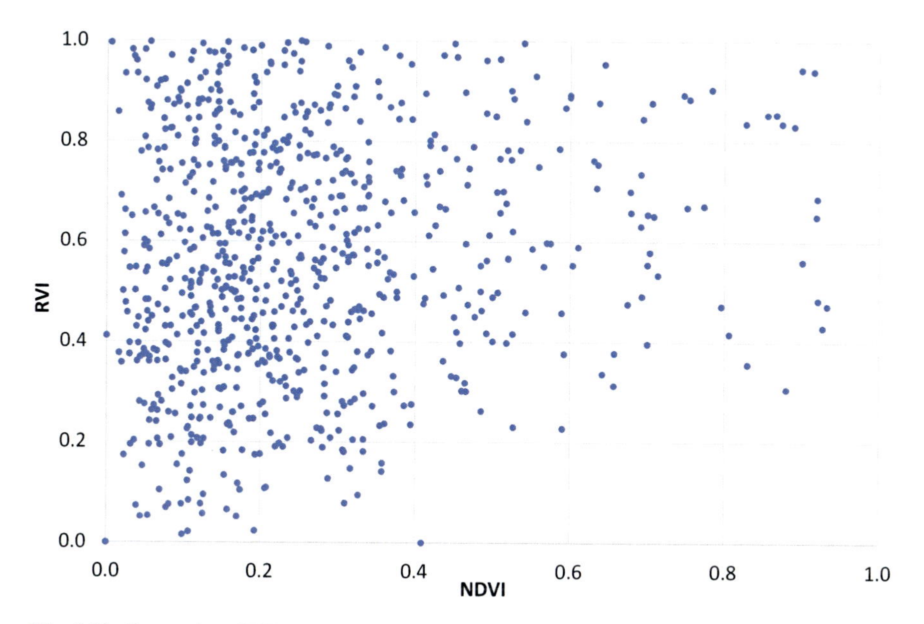

Fig. 8.10 Scatterplot of NDVI and RVI values over 1000 random points in the case study area

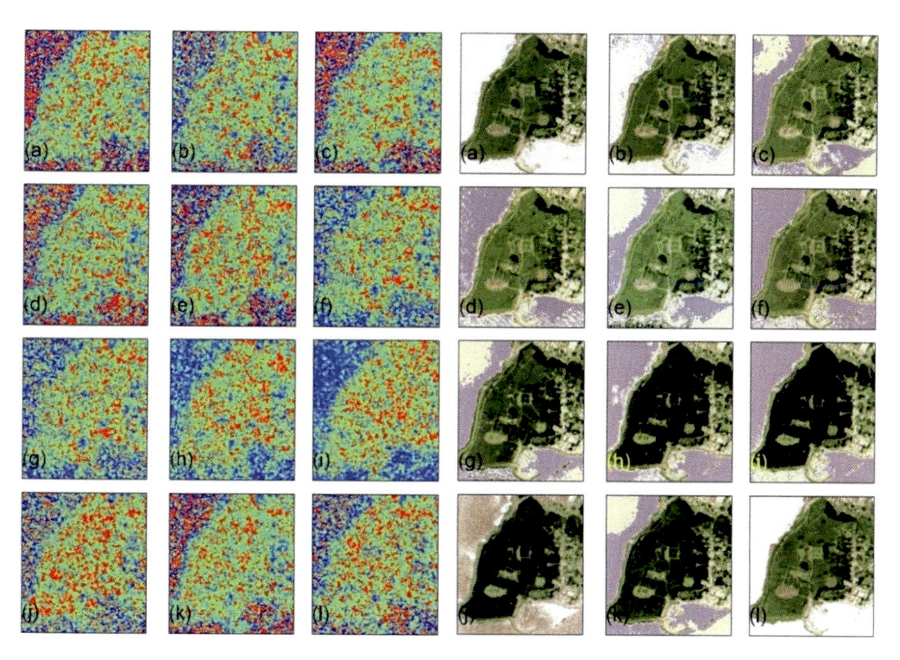

Fig. 8.11 Monthly RVI (left) and NDVI results (right) (**a**: May 2019, **b**: June 2019, **c**: July 2019, **d**: Aug. 2019, **e**: Sept. 2019, **f**: Oct. 2019, **g**: Nov. 2019, **h**: Dec. 2019, **i**: Jan. 2020, **j**: Feb. 2020, **k**: March 2020 and **l**: April 2020). (Source: Agapiou, 2020)

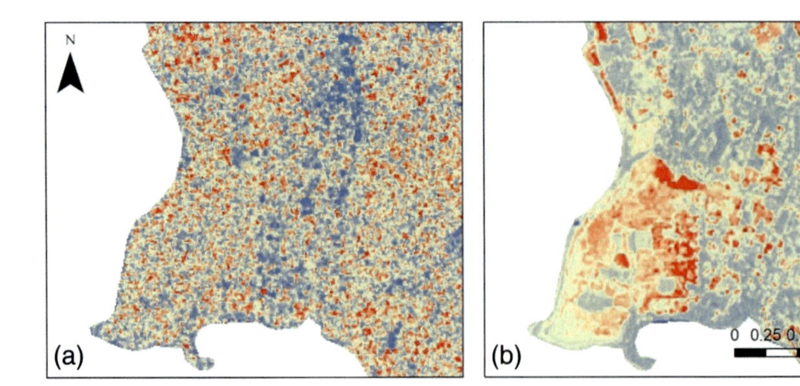

Fig. 8.12 (**a**) The difference in the proportion of vegetation cover as estimated from the RVI index is based on Eqs. 8.3 and 8.5 and (**b**) the difference in the proportion of vegetation cover as estimated from the NDVI index based on Eqs. 8.4 and 8.6. Higher difference values are indicated with red colour, while lower values with blue colour. (Source: Agapiou, 2020)

vegetation growth are highlighted with red colour in Fig. 8.11-left, while the dark green colour in Fig. 8.11-right shows vegetation at the optical products. The results show that Sentinel-2 images using the NDVI index can depict the vegetation's phenological changes over the "Nea Paphos" archaeological site throughout the year. In contrast, the interpretation of the RVI index is still problematic (Fig. 8.11-left). Nevertheless, an increase in vegetation (red colour) is evident during the months Dec. 2019 to Feb. 2020 (Fig. 8.11-left, h–j), which is also visible to optical products as well (Fig. 8.11-right, h–j).

Then, the vegetation proportion (see Eqs. 8.3, 8.4, 8.5, and 8.6) have been estimated for each type of sensors. These results were compared over the archaeological site of "Nea Paphos". Figure 8.12 a shows the difference between the OpenStreetMap (Eq. 8.3) and image statistics (Eq. 8.5) for the RVI index, while Fig. 8.12b the difference between the proportion of vegetation cover using the NDVI index (Eqs. 8.4 and 8.6). Higher differences are highlighted with red colour, while lower differences are estimated with blue colour.

The findings described above show that the NDVI and the RVI indices did not provide comparable results. Optical indices like the NDVI can be interpreted more easily in contrast to the RVI results. However, in some cases, such as the findings of Fig. 8.11, suggest that radar products can be used as an alternative to optical data for detecting patterns (e.g. vegetation growth) in specific areas of interest.

The study of Agapiou (2020) proposed a framework, whereas both proportions of vegetation indices derived from Sentinel-1 and -2 sensors can be used by multiplying the RVI x NDVI datasets. This outcome can be used with the VV and VH polarisations of Sentinel-1 to create a new pseudocolour composite. Radar datasets can depict urban areas, which can enhance the difference between vegetated and urban areas. An example of such a new composite is shown in Fig. 8.13. The NDVI proportion was estimated over the archaeological site of "Nea Paphos", using Sentinel-2 spectral bands 4 and 8. In contrast, the RVI vegetation proportion was

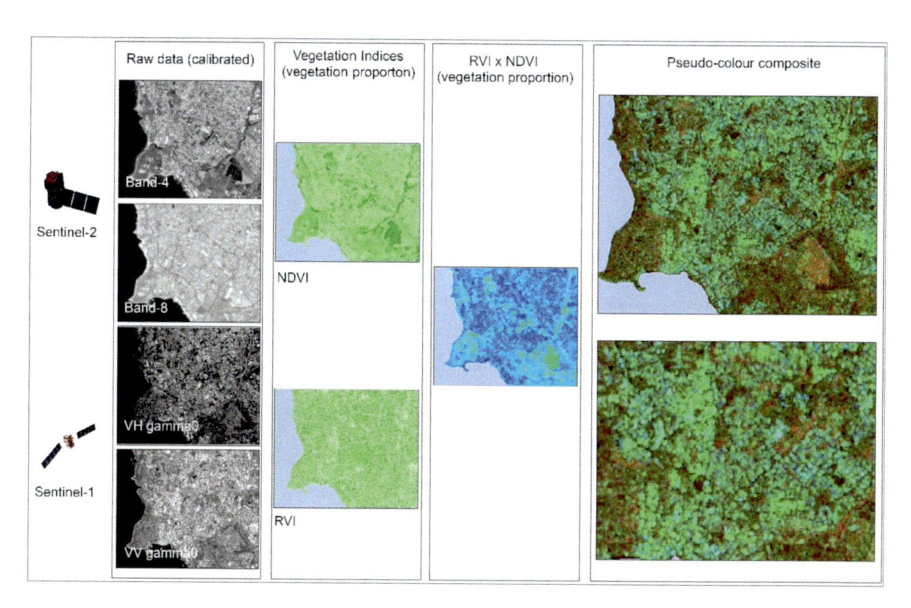

Fig. 8.13 New pseudo colour composite on the right-left of the Figure integrating the RVI x NDVI vegetation proportion index and the VV / VH polarisations from Sentinel 1. Vegetated areas are shown with red colour, while buildings with green colour. On the left side, the overall procedure (as explained in the previous sections) for estimating the RVI x NDVI index. (Source: Agapiou, 2020)

estimated using the VV and VH polarisations of the radar Sentinel-1. The combination of the two new products generates the new RVI x NDVI vegetation proportion index, which can be combined with the VV and the VH polarisations to highlight vegetated areas (see red colour areas under the pseudocolour composite of Fig. 8.13) and buildings (see with green colour areas under the pseudocolour composite of Fig. 8.13).

8.3.3 Surface Displacements

Earth observation may support the disaster risk management cycle, as this is understood for cultural heritage in multiple ways. At the same time, related technologies can change the traditional way of processing earth observation's data, specifically, from desktop analysis to cloud-based, for instance, the use of Google Earth Engine (GEE, 2021) for archaeological and heritage management studies (Orengo et al., 2020; Agapiou, 2017). This section presents an example of the Hybrid Pluggable Processing Pipeline (HyP3) cloud-based system, operated by the Alaska Satellite Facility (ASF), a related new cloud platform. To our knowledge, this platform has not been hitherto used for heritage management (Agapiou & Lysandrou, 2020).

One of the most significant earthquakes that hit Cyprus in recent years was a 5.6 magnitude scale seismic event on 15th April 2015, at 08:25 UTC, and it was strongly felt throughout the country. The earthquake's epicenter was estimated at 8 km NW of Paphos (western Cyprus), with a depth of 27.62 km. This earthquake remains the biggest in Cyprus -until today- from the launch of the Sentinel-1 sensors in 2014.

In this study, two pairs of Sentinel-1 images were used in ascending (south pole towards the north pole) and descending orbit (north pole towards the south pole). For each pair, an image before and after the event was elaborated. InSAR deformation analysis was executed through the HyP3 platform. In particular, the InSAR GAMMA algorithm was used. The methodology describing the GAMMA software for InSAR analysis using Sentinel images comprises eleven (11) steps as described in Agapiou and Lysandrou (2020).

A vertical displacement map was generated from the HyP3 platform under the assumption that the interferometric phase is related solely to the topography of the area. Values were given in meters, with positive values indicating uplift and negative values indicating subsidence. The area of the earthquake's epicentre, the Paphos town, hosts significant archaeological sites and monuments, some of them listed as UNESCO World Heritage.

The InSAR analysis resulted in small ground displacements in this area, both from the images taken in ascending orbit and the seismic network. Figure 8.14a indicates the results from the unwrapped interferogram, while Fig. 8.14b shows the results from the vertical displacement analysis. Fig. 8.14c shows the coherence values based on the pair of Sentinel-1 images used in the ascending orbit. Areas with low coherence values (less than 70%, Fig. 8.14c) were excluded from the analysis.

A critical finding of that study was that the satellite datasets were processed in less than 1-hour for each orbit, significantly minimising the computational time compared to traditional desktop analysis. The use of ARD products produced from cloud-based platforms like the HyP3 is significant for heritage management. They can provide displacement information over large areas in a short time. However, as in almost all earth observation processing chains, these results require ground verifications from ground stations.

8.4 Risk Maps Using Various Hazards

As earlier stated, several hazards can affect archaeological assets, both individual monuments and entire sites. An Analytical Hierarchy Process (AHP) method, a multi-criteria decision-making method based on comparing concepts (alternatives) in pairs, can be implemented to address individual and unique characteristics of monuments and sites, creating small clusters. AHP is a straightforward approach,

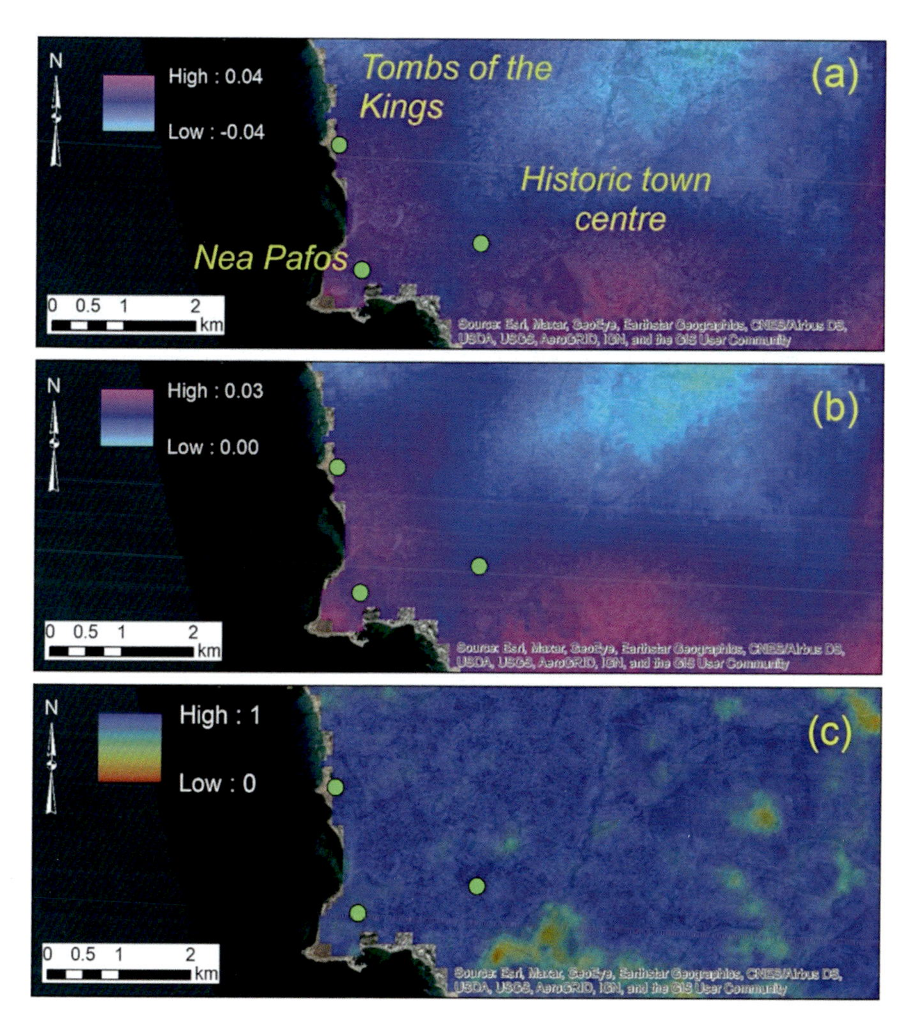

Fig. 8.14 (**a**) Unwrapped interferogram; (**b**) Vertical displacements; (c) Coherence map, enveloping archaeological sites of the area. (Source: Agapiou & Lysandrou, 2020)

widely applied to help decision-making mainly when several conflicting criteria are simultaneously occurring. Saaty (1977) has proposed AHP in the 1970s.

The study carried out by Agapiou et al. (2016b) was focused on the Paphos district in western Cyprus, an area that abounds in antiquities. More than 150 declared Ancient Monuments of First (Ancient Monuments on State Land) and Second Schedule (Ancient Monuments on Private Land) protected by the Antiquities Law had been mapped with high accuracy (Fig. 8.15). The authors have conducted previous research in this area to create a common geo-database of all monuments, estimating hazards, and produce risk maps from remote sensing data

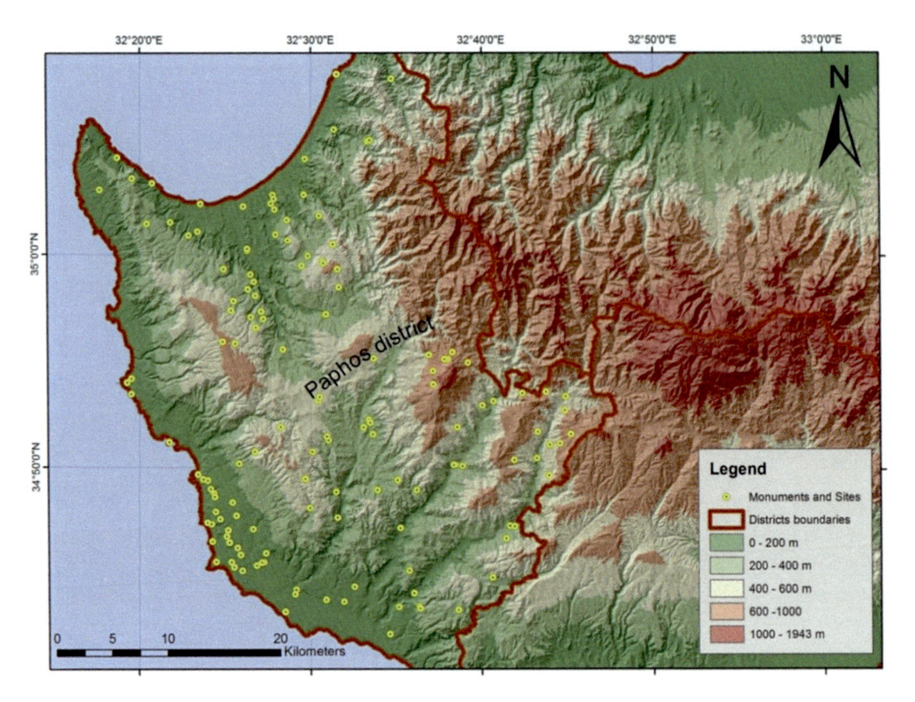

Fig. 8.15 Protected monuments and sites in Paphos district, Cyprus. (Source: Agapiou et al., 2016b)

(Agapiou et al., 2013). These monuments were clustered into five groups that shared similar geomorphological characteristics.

A variety of remote sensing datasets were used to map natural and anthropogenic hazards over this area. The list included low, medium and high-resolution satellite images like the MODIS, Landsat and QuickBird, along with ready satellite products, like the ASTER Global Digital Elevation Model, ASTER GDEM. Initially, each hazard was analysed. Then, the overall risk was estimated based on the AHP methodology. A series of risk maps were created relative to anthropogenic (urban sprawl, modern road network, fires) and environmental (erosion, salinity, neotectonic activity) hazards that affect the archaeological sites in the Paphos district. The resulting risk maps for each hazard are illustrated in Fig. 8.16 (for further information, see Agapiou et al., 2015).

For each one of the five different clusters, a separate AHP was implemented. Table 8.1 shows the weight factors for each group of monuments and each hazard. The highest weight for each class is highlighted in the table. The weights might vary significantly for each hazard, depending on the importance of each of the five classes. This difference of weights recorded for the same hazards in the different classes is normal since each group of monuments (class) faces dissimilar proportions since these are correlated to the site's location, amongst others.

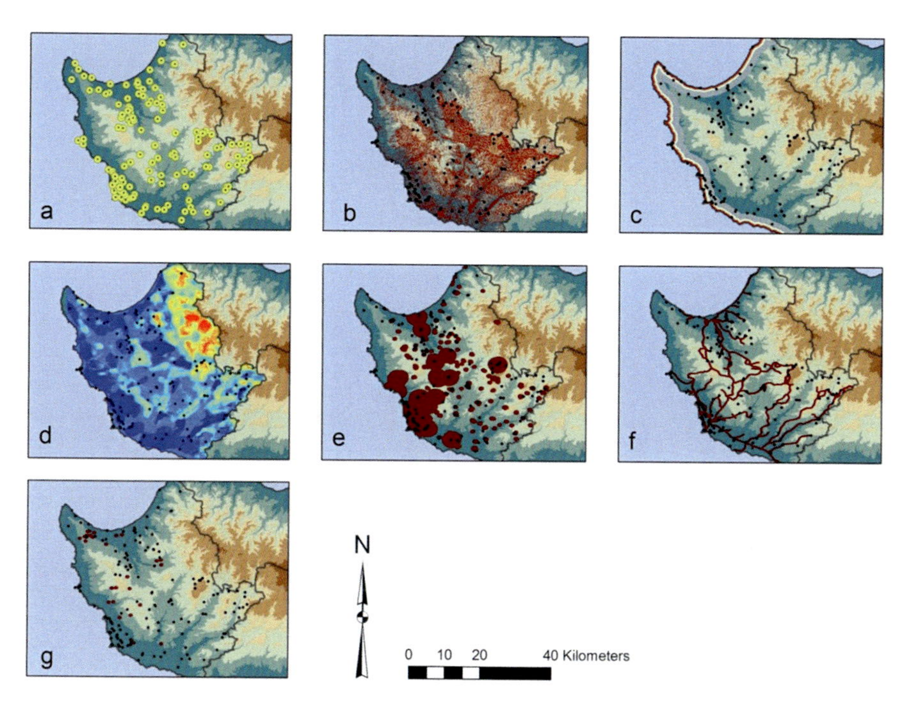

Fig. 8.16 Map indicating the different anthropogenic and natural hazards over the Paphos District. (**a**) Archaeological sites of the Paphos district; (**b**) Erosion map: areas, where the soil loss is greater than the mean value soil loss of the whole district, are indicated with red colour; (**c**) Salinity map: areas close to the sea are indicated with red; (**d**) Tectonic Activity: high and very high hazard area are indicated with red while the very low hazard is indicated with blue colour; (**e**) Urban expansion indicated with red colour; (**f**) Road network proximity (250 m) indicated with red colour which is linked with potential future urban expansion; and (**g**) Fires map observed during the period 2010-2013 indicated with red colour. (Revised map from Agapiou et al., 2015)

Using the weights shown in Table 8.1, the following equations for overall risk hazard for each group of monuments was estimated:

$$\text{Risk Hazard for Group 1} = 0.2500 * \text{F1} + 0.0213 * \text{F2} + 0.1302 * \text{F3} + 0.2747 * \text{F4} + 0.0555 * \text{F5} + 0.2683 * \text{F6} \quad (8.7)$$

$$\text{Risk Hazard for Group 2} = 0.3928 * \text{F1} + 0.0174 * \text{F2} + 0.0679 * \text{F3} + 0.1209 * \text{F4} + 0.1142 * \text{F5} + 0.2868 * \text{F6} \quad (8.8)$$

$$\text{Risk Hazard for Group 3} = 0.0906 * \text{F1} + 0.0207 * \text{F2} + 0.2061 * \text{F3} + 0.2390 * \text{F4} + 0.1468 * \text{F5} + 0.2968 * \text{F6} \quad (8.9)$$

$$\text{Risk Hazard for Group 4} = 0.1436 * \text{F1} + 0.1251 * \text{F2} + 0.2339 * \text{F3} + 0.4114 * \text{F4} + 0.0257 * \text{F5} + 0.0603 * \text{F6} \quad (8.10)$$

Table 8.1 AHP factors for the five different classes (groups) of monuments

AHP weight factors for Group 1									
	Factors	F1	F2	F3	F4	F5	F6	Total Sum	Normalised Weights
1	Tectonic	1	1/9	5	3	9	3	21.11	0.2500
2	Salinity	1/9	1	1/7	1/9	1/3	1/9	1.800	0.0213
3	Road Network	1/3	7	1	1/3	3	1/3	11.00	0.1302
4	Urban areas	1/5	9	3	1	7	3	23.20	0.2747
5	Soil erosion	1/9	3	1/3	1/7	1	1/9	4.690	0.0555
6	Fires	1/3	9	3	1/3	9	1	22.66	0.2683
								84.46	**1**

AHP weight factors for Group 2									
	Factors	F1	F2	F3	F4	F5	F6	Total Sum	Normalised Weights
1	Tectonic	1	9	9	9	7	5	40.00	0.3928
2	Salinity	1/9	1	1/5	1/5	1/7	1/9	1.77	0.0174
3	Road Network	1/9	5	1	1/3	1/3	1/7	6.92	0.0679
4	Urban areas	1/9	5	3	1	3	1/5	12.31	0.1209
5	Soil erosion	1/7	7	3	1/3	1	1/7	11.62	0.1142
6	Fires	1/5	9	7	5	7	1	29.20	0.2868
								101.82	**1**

AHP weight factors for Group 3									
	Factors	F1	F2	F3	F4	F5	F6	Total Sum	Normalised Weights
1	Tectonic	1	5	1/3	1/3	1/3	1/3	7.33	0.0906
2	Salinity	1/5	1	1/9	1/9	1/7	1/9	1.67	0.0207
3	Road Network	3	9	1	1/3	3	1/3	16.67	0.2061
4	Urban areas	3	9	3	1	3	1/3	19.33	0.2390
5	Soil erosion	3	7	1/3	1/3	1	1/5	11.87	0.1468
6	Fires	3	9	3	3	5	1	24.00	0.2968
								80.87	**1**

AHP weight factors for Group 4									
	Factors	F1	F2	F3	F4	F5	F6	Total Sum	Normalised Weights
1	Tectonic	1	1/3	1/3	1/5	7	3	11.87	0.1436
2	Salinity	3	1	1/5	1/7	3	3	10.34	0.1251
3	Road Network	3	5	1	1/3	5	5	19.33	0.2339
4	Urban areas	5	7	3	1	9	9	34.00	0.4114
5	Soil erosion	1/7	1/3	1/5	1/9	1	1/3	2.12	0.0257
6	Fires	1/3	1/3	1/5	1/9	3	1	4.98	0.0603
								82.64	**1**

AHP weight factors for Group 5									
	Factors	F1	F2	F3	F4	F5	F6	Total Sum	Normalised Weights
1	Tectonic	1	9	7	3	3	9	32.00	0.3289
2	Salinity	1/9	1	1/7	1/9	1/5	1/5	1.76	0.0181
3	Road Network	1/7	7	1	1/3	3	3	14.47	0.1487
4	Urban areas	1/3	9	3	1	5	5	23.33	0.2398
5	Soil erosion	1/3	5	1/3	1/5	1	3	9.86	0.1014
6	Fires	9	5	1/3	1/5	1/3	1	15.86	0.1630
								97.28	**1**

Source: Agapiou et al. (2016b)

$$\text{Risk Hazard for Group 5} = 0.3289 * \text{F1} + 0.0181 * \text{F2} + 0.1487 * \text{F3} + 0.2398 * \text{F4} + 0.1014 * \text{F5} + 0.1630 * \text{F6} \quad (8.11)$$

Where F1 to F6 stand for the different hazards (Tectonic; Salinity; Road Network; Urban areas; Soil erosion and Fires respectively), based on Table 8.1, the normalised weights for each risk have been added to the attribute table of the monuments in a GIS environment. Then, interpolation was carried out in a GIS environment, based on the Inverse Distance Weight (IDW) algorithm. The results from the interpolation of the weight factors are presented in Fig. 8.17. It should be noticed that in comparison with traditional AHP methodology, a single value would be allocated for all monuments in the Paphos district.

Then the overall risk hazard map was produced by multiplying the weight factor and the hazard:

$$\text{Overall Risk} = \text{Weight 1} * \text{F1} + \text{Weight 2} * \text{F2} + \text{Weight 3} * \text{F3} + \text{Weight 4} * \text{F4} + \text{Weight 5} * \text{F5} + \text{Weight 6} * \text{F6} \quad (8.12)$$

Figure 8.18 shows the overall risk hazard map, where the five main categories are classified using natural breaks values. These categories are: (1) very low hazard, (2) low hazard; (3) medium hazard; (4) high hazard; (5) very high hazard. The areas under the indication of very high hazard, are located in Paphos town and immediate environs, where significant archaeological areas lie.

8.5 Conclusions

This chapter presented examples from earth observation studies related to the monitoring of archaeological/cultural heritage sites over Cyprus. The chapter was based on published work, a result of recent research, while it also delivered new concepts and applications using cloud-based earth observation platforms.

Section 3.1 reports the threat of subsurface archaeological remains from soil erosion, exploiting existing geo-data. The results showed that although most of the archaeological sites are characterised as "low threat" areas, significant differences between regions (north and south Europe) can be seen. Section 3.2 focused on the caption of vegetation dynamics using integrated optical and radar sensors. Vegetation is a critical factor for several hazards like urbanisation and soil loss. A pseudocolour composite was retrieved over urban and vegetated areas by applying the NDVI and the RVI indices. Further studies are needed in this direction. Moreover, InSAR analysis through the HyP3 platform has been presented.

Estimation of threats over extensive areas, like the case study of Paphos District, was carried out using the AHP methodology. More than 150 monuments of this area were grouped into five classes based on specific characteristics. The overall risk was

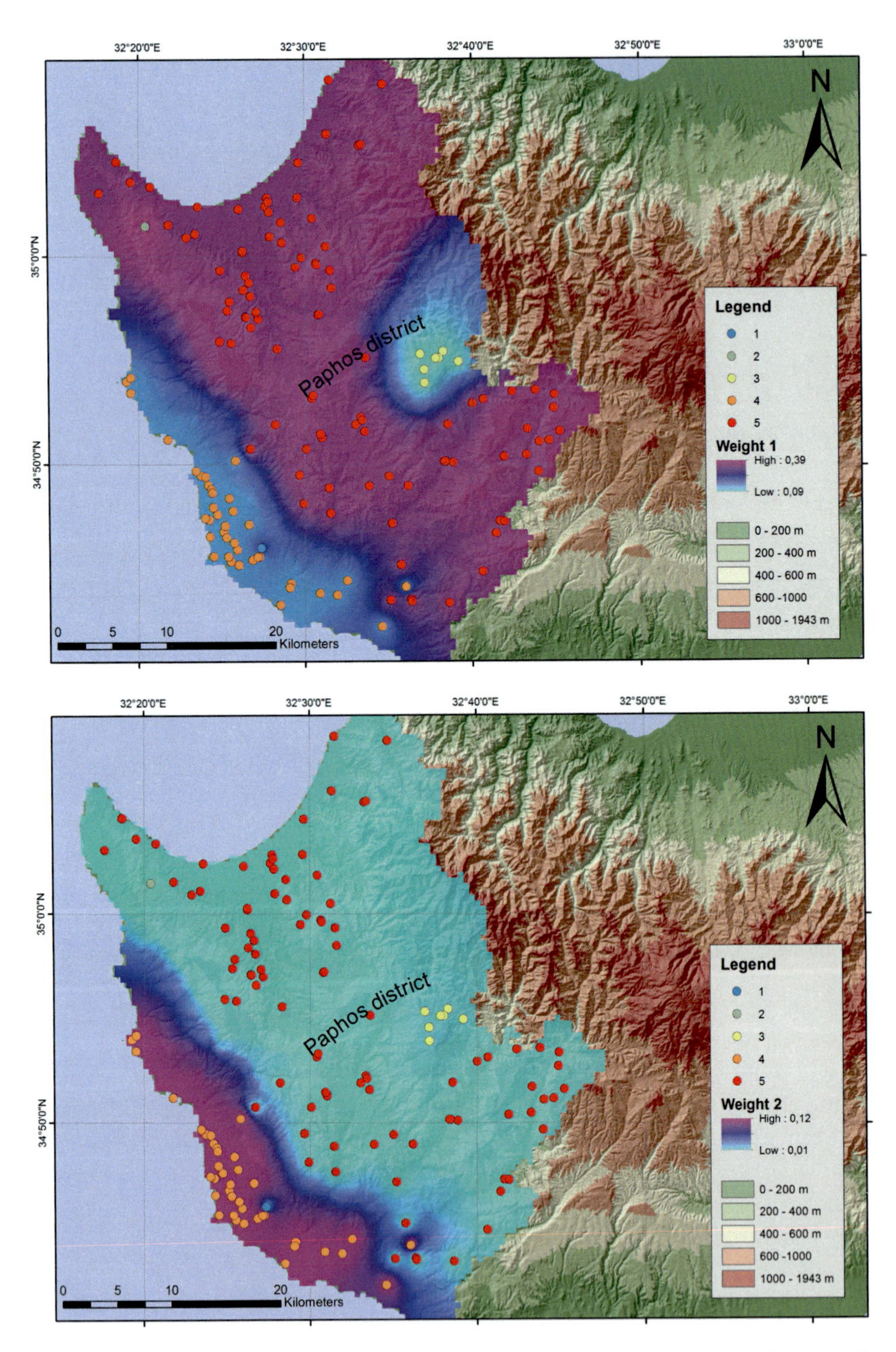

Fig. 8.17 Interpolation of the different normalisation weight for each hazard (Weight 1 to 6: Tectonic; Salinity; Road Network; Urban areas; Soil erosion and Fires respectively). (Source: Agapiou et al., 2016b)

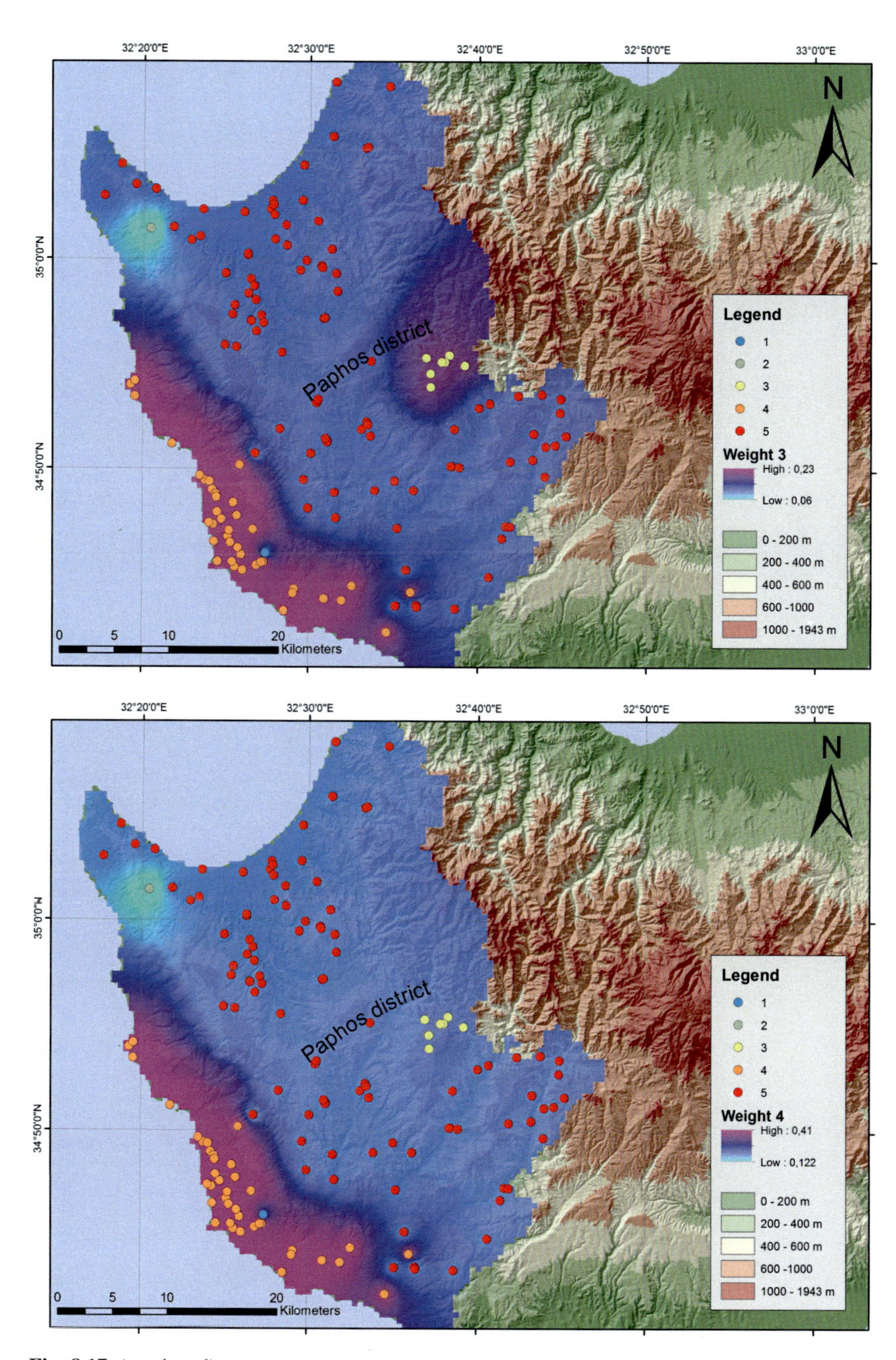

Fig. 8.17 (continued)

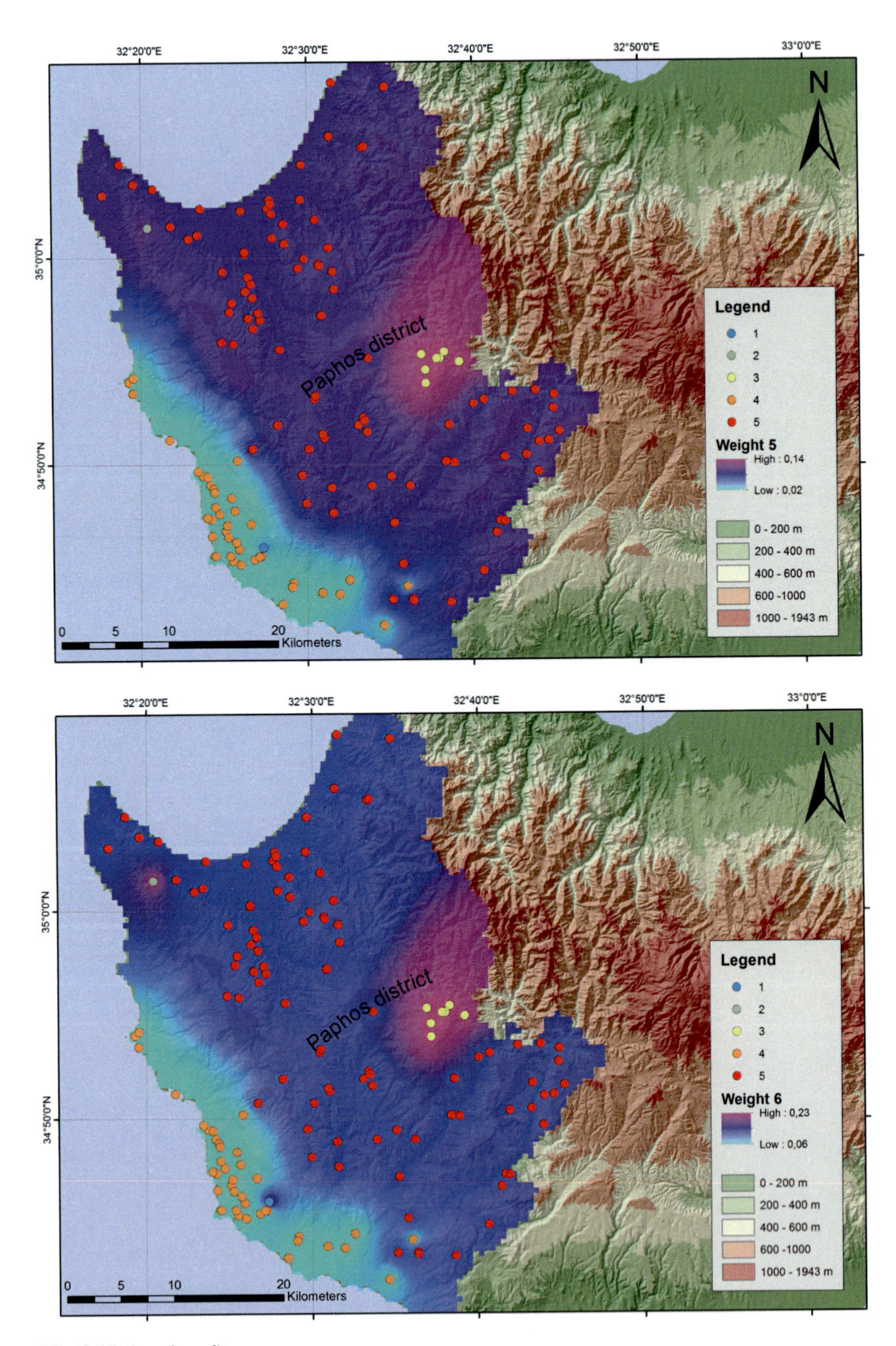

Fig. 8.17 (continued)

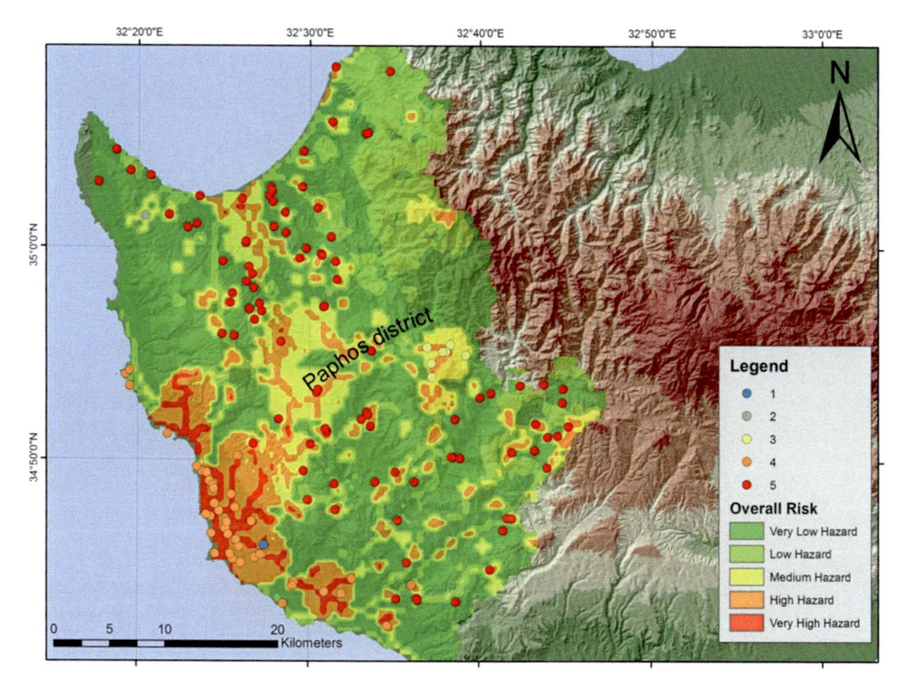

Fig. 8.18 Overall risk hazard map for the Paphos district, based on the clustering of the sites. (Source: Agapiou et al., 2016b)

estimated for each group, and afterwards, the overall risk prioritisation was carried out.

A key to the quality assurance of the applications above, methodologies and tools, is multidisciplinary collaboration. Future trends in the domain of earth observation indicate that technological changes will affect how space-based monitoring and observation are performed. This will be primarily based on cloud-platforms while the use of ready products will become more frequently used by the end-users and local stakeholders.

Acknowledgements This paper is submitted under the NAVIGATOR project. The project is being co-funded by the Republic of Cyprus and the Structural Funds of the European Union in Cyprus under the Research and Innovation Foundation grant agreement EXCELLENCE /0918/0052 (Copernicus Earth Observation Big Data for Cultural Heritage).

References

Abarquez, I., & Murshed, Z. (2004). *Field practitioners' handbook, community-based disaster risk management*. Asian Disaster Preparedness Centre.

Agapiou, A. (2017). Remote sensing heritage in a petabyte-scale: Satellite data and heritage earth Engine© applications. *International Journal of Digital Earth, 10*, 85–102.

Agapiou, A. (2020). Estimating proportion of vegetation cover at the vicinity of archaeological sites using Sentinel-1 and -2 data, supplemented by Crowdsourced OpenStreet Map Geodata. *Applied Sciences, 10*, 4764. https://doi.org/10.3390/app10144764

Agapiou, A., & Lysandrou, V. (2015). Remote Sensing Archaeology: Tracking and mapping evolution in scientific literature from 1999–2015. *Journal of Archaeological Science: Reports, 4*, 192–200.

Agapiou, A., & Lysandrou, V. (2020). Detecting displacements within archaeological sites in Cyprus after a 5.6 magnitude scale earthquake event through the Hybrid Pluggable Processing Pipeline (HyP3) cloud-based system and Sentinel-1 Interferometric Synthetic Aperture Radar (InSAR) Analysis. *IEEE Journal of Selected Topics in Applied Earth Observations and Remote Sensing, 13*, 6115–6123. https://doi.org/10.1109/JSTARS.2020.3028272

Agapiou, A., Nisantzi, A., Lysandrou, V., Mammouri, R., Alexakis, D. D., Themistocleous, K., Sarris, A., & Hadjimitsis, D. G. (2013). Mapping air pollution using earth observation techniques for cultural heritage. Proceedings SPIE 8795. *First International Conference on Remote Sensing and Geoinformation of the Environment (RSCy2013) 87950K*. https://doi.org/10.1117/12.2028234.

Agapiou, A., Lysandrou, V., Alexakis, D. D., Themistocleous, K., Cuca, B., Sarris, A., Argyrou, N., & Hadjimitsis, D. G. (2015). Cultural heritage management and monitoring using remote sensing data and GIS: the case study of Paphos area, Cyprus. *CEUS Computers Environment and Urban Systems, 54*, 230–239. https://doi.org/10.1016/j.compenvurbsys.2015.09.003

Agapiou, A., Alexakis, D. D., Sarris, A., & Hadjimitsis, D. G. (2016a). Colour to greyscale pixels: re-seeing greyscale archived aerial photographs and declassified satellite CORONA images based on image fusion techniques. *Archaeological Prospection, 23*, 231–241. https://doi.org/10.1002/arp.1536

Agapiou, A., Lysandrou, V., Themistocleous, K., et al. (2016b). Risk assessment of cultural heritage sites clusters using satellite imagery and GIS: the case study of Paphos District, Cyprus. *Natural Hazards, 83*, 5–20. https://doi.org/10.1007/s11069-016-2211-6

Agapiou, A., Lysandrou, V., & Hadjimitsis, D. G. (2020a). A European-scale investigation of soil erosion threat to subsurface archaeological remains. *Remote Sensing, 12*, 675. https://doi.org/10.3390/rs12040675

Agapiou, A., Lysandrou, V., & Hadjimitsis, D. G. (2020b). Earth observation contribution to cultural heritage disaster risk management: Case study of Eastern Mediterranean open air archaeological monuments and sites. *Remote Sensing, 12*, 1330. https://doi.org/10.3390/rs12081330

Borrelli, P., Robinson, D. A., & Fleischer, L. R. (2013). An assessment of the global impact of 21st century land use change on soil erosion. *Nature Communications, 8*. https://doi.org/10.1038/s41467-017-02142-7

Bosco, C., de Rigo, D., Dewitte, O., Poesen, J. (2021). *Copernicus emergency management services*. https://emergency.copernicus.eu. Accessed on 22 Feb 2021.

Casana, J. (2020). Global-scale archaeological prospection using CORONA satellite imagery: Automated, crowd-sourced, and expert-led approaches. *Journal of Field Archaeology, 45*(Suppl 1), S89–S100. https://doi.org/10.1080/00934690.2020.1713285

Chandramohan, T., Venkatesh, B., & Balchand, A. N. (2015). Evaluation of three soil erosion models for small watersheds. *Aquatic Procedia, 4*, 1227–1234. https://doi.org/10.1016/j.aqpro.2015.02.156

Emergency Management Australia. (2000). *Emergency risk management – Applications guide* (Australian emergency manuals series). www.ema.gov.au

Erosion by Water, Joint Research Centre European Soil Data Centre (ESDAC). (2021). Available online: https://esdac.jrc.ec.europa.eu/themes/erosion. Accessed on 22 Feb 2021.

European Copernicus Programme. (2021). https://www.copernicus.eu/en. Accessed on 22 Feb 2021.

Ghosal, K., & Das Bhattacharya, S. (2020). A Review of RUSLE Model. *Journal of the Indian Society of Remote Sensing*. https://doi.org/10.1007/s12524-019-01097-0

Google Earth Engine (GEE). (2021). https://earthengine.google.com. Accessed on 22 Feb 2021.

Group on Earth Observations (GEO). (2021). https://www.earthobservations.org/geoss_wp.php. Accessed on 22 Feb 2021.

Hritz, C. (2013). A malarial-ridden swamp: using Google Earth Pro and Corona to access the southern Balikh valley, Syria. *Journal of Archaeological Science, 40*(4), 1975–1987.

ICCROM. (2016). *A Guide to Risk Management of Cultural Heritage 2016*. Government of Canada, Canadian Conservation Institute.

International Charter Space and Major Disasters. (2021). https://disasterscharter.org/web/guest/home;jsessionid=25223736C25BDBEEF6EBF9BC7CA46695.jvm1. Accessed on 22 Feb 2021.

ISO 31000. (2018). *Risk management – Guidelines*. https://www.iso.org/standard/65694.html. Accessed on 22 Feb 2021.

Kibblewhite, M., Tóth, G., & Hermann, T. (2015). Predicting the preservation of cultural artefacts and buried materials in soil. *Science of the Total Environment, 529*, 249–263.

Li, J., & Roy, D. P. (2017). A global analysis of Sentinel-2A, Sentinel-2B and Landsat-8 data revisit intervals and implications for terrestrial monitoring. *Remote Sensing, 9*, 902.

Luo, L., Wang, X., Guo, H., Lasaponara, R., Zong, X., Masini, N., Wang, G., Shi, P., Khatteli, H., Chen, F., et al. (2019). Airborne and spaceborne remote sensing for archaeological and cultural heritage applications: A review of the century (1907–2017). *Remote Sensing of Environment, 232*, 111280.

Lysandrou, V., & Agapiou, A. (2020). The role of archival aerial photography in shaping our understanding of the Funerary Landscape of Hellenistic and Roman Cyprus. *Open Archaeology, 6*(1), 417–433. https://doi.org/10.1515/opar-2020-0117

Orengo, A. H., Conesa, C. F., Garcia-Molsosa, A., Lobo, A., Green, A. S., Madella, M., & Petrie, C. A. (2020). Automated detection of archaeological mounds using machine-learning classification of multisensory and multitemporal satellite data. *Proceedings of the National Academy of Sciences Aug, 117*(31), 18240–18250. https://doi.org/10.1073/pnas.2005583117

Panagos, P. (2015). Modelling soil erosion at European scale: Towards harmonisation and reproducibility. *Natural Hazards and Earth System Sciences, 15*, 225–245. https://doi.org/10.5194/nhess-15-225-2015

Panagos, P., & Katsoyiannis, A. (2019). Soil erosion modelling: The new challenges as the result of policy developments in Europe. *Environmental Research, 172*, 470–474. https://doi.org/10.1016/j.envres.2019.02.043

Panagos, P., Van Liedekerke, M., Jones, A., & Montanarella, L. (2012). European Soil Data Centre: Response to European policy support and public data requirements. *Land Use Policy, 29*(2), 329–338. https://doi.org/10.1016/j.landusepol.2011.07.003

Pastonchi, L., Barra, A., Monserrat, O., Luzi, G., Solari, L., & Tofani, V. (2018). Satellite data to improve the knowledge of geohazards in world heritage sites. *Remote Sensing, 10*, 992.

Pena, B. S., Abreu, M. M., Magalhães, R. M., & Cortez, N. (2020). Water erosion aspects of land degradation neutrality to landscape planning tools at national scale. *Geoderma, 363*, 114093. https://doi.org/10.1016/j.geoderma.2019.114093

Quinton, J. N. (2011). Soil Erosion Modeling. In J. Gliński, J. Horabik, & J. Lipiec (Eds.), *Encyclopedia of agrophysics; encyclopedia of earth sciences series*. Springer.

Renard, K. G., Foster, G. R., Weesies, G. A., McCool, D. K., & Yoder, D. C. (1997). *Predicting soil erosion by water: A guide to conservation planning with the Revised Universal Soil Loss Equation (RUSLE)*. US Department of Agriculture, Agricultural Research Service.

Saaty, T. L. (1977). A scaling method for priorities in hierarchical structures. *Journal of Mathematical Psychology, 15*(3), 234–281.

Sentinel-Hub services, custom scripts. (2021). https://custom-scripts.sentinel-hub.com/sentinel-1/radar_vegetation_index. Accessed on 22 Feb 2021.

Solari, L., Bianchini, S., Franceschini, R., Barra, A., Monserrat, O., Thuegaz, P., Bertolo, D., Crosetto, M., & Catani, F. (2020). Satellite interferometric data for landslide intensity evaluation in mountainous regions. *International Journal of Applied Earth Observation and Geoinformation, 87*.

Tapete, D., Cigna, F., & Donoghue, N. M. D. (2016). Looting marks' in space-borne SAR imagery: Measuring rates of archaeological looting in Apamea (Syria) with TerraSAR-X staring spotlight. *Remote Sensing of Environment, 178*, 42–58.

UNESCO (Paris), et al. (2010). *Managing disaster risks for world heritage. World heritage resource manual* (p. 69). United nations educational, scientific and cultural organisation (UNESCO).

United Nations. (2009). *UNISDR terminology on disaster risk reduction* (p. 35). United Nations International Strategy for Disaster Reduction (UNISDR) 2009.

United Nations Platform for Space-based Information for Disaster Management and Emergency Response (UN-SPIDER). (2021). https://www.unoosa.org/oosa/en/ourwork/un-spider/index.html. Accessed on 22 Feb 2021.

Ur, J. A. (2016). Middle Eastern archaeology from the air and space. *In Situ Fall*, 1–4.

Chapter 9
Remote Sensing and GIS Application for Natural Hazards Assessment of the Mauritanian Coastal Zone

Abdoul Jelil Niang

Abstract Coastal zones are dynamic and vulnerable spaces subject to the influences of coastal processes which are exacerbated by the human activities and extreme meteorological events associated to climate change. The Mauritanian coastal zone located in the western part is the most vital in the country. It is home to the two largest cities and over a third of the country's population and concentrates the most important economic activities. The Mauritanian coastline is vulnerable to various hazards including marine submersion, coastal erosion, desertification and dunes movement and flooding by rainfall. Climate change is therefore a major challenge for this area. The shoreline change which is one of the main factors in the modification of the morphology of the coast, will be analyzed from 1986 to 2021, using Digital Shoreline Analysis System (DSAS) software to calculate the rates of changes. The results show an accretional trend of the shoreline, but the sectors most threatened by coastal erosion and marine submersion are located near the port facilities.The risk of flooding by rainwater mainly concerns the city of Nouakchott; as Nouadhibou is in a coastal desert, precipitation is rare there. The Coastal geomorphology is generally in the form of a low plain occupied in several places by sabkhas and whose altitude at the level of certain sectors (such as Nouakchott) is below sea level from which it is protected by a thin and fragile dune ridge which can be crossed in some places during strong storms. Port facilities and other human activities at the coastline contribute to exacerbating the vulnerability of the coastal area. This study therefore focuses on the use of diachronic and multisource remote sensing data and GIS application for coastal multi-risk hazard assessment. The risks linked to climate change will also be analyzed from different outlook for assessing the degree of vulnerability and adaptive capacity of Mauritania face the various types of threats.

Keywords Costal hazards · Shoreline change · DSAS · Climate change · Marine submersion · Floods · Dunes encroachment · Mauritania

A. J. Niang (✉)
Geography Department, Umm Al-Qura University, Makkah, Saudi Arabia
e-mail: anniang@uqu.edu.sa

M. M. Al Saud (ed.), *Applications of Space Techniques on the Natural Hazards in the MENA Region*, https://doi.org/10.1007/978-3-030-88874-9_9

9.1 Introduction

Coastal zones are dynamic and vulnerable spaces subject to both the influences of coastal processes exacerbated by human activities concentrated there and extreme meteorological events associated with climate change and sea-level rise. Thus, they appear to be extremely sensitive sectors, characterized by rapid population increases for strategic, economic, or recreational reasons and the installation of various infrastructures causing territorial changes, which can be a source of sometimes irreversible environmental degradation.

The Mauritanian coastal zone is vulnerable to various hazards (e.g., erosion, flooding, desertification) and human interventions. The consequences of climate change will exert increasing pressure on this part of the territory, which is home to a constantly increasing population, inevitably leading to increased risk of natural disasters. This coastline, which has previously experienced marine transgressions in which the sea invaded hundreds of kilometers of the continent (Barusseau et al., 1998; Giresse et al., 2000), is under the threat of marine submersion favored by coastal erosion, which is one of the main factors in the modification of the morphology of the coastline. The Intergovernmental Panel on Climate Change (IPCC, 2019) has predicted that the impacts of climate change will result in a sea-level rise of between 0.29 m and 1.1 m.

With the increasing sea level and the topography of shores (characterized by low-lying areas), coastal areas are considerably vulnerable to the risks of coastal erosion, marine incursions, floods, and extreme climatic events (Goussard & Ducrocq, 2014; Tragaki et al., 2018). Among the dramatic consequences of erosion, we can highlight the loss of land with great economic, social, and environmental value and infrastructures as well as the degradation of maritime and coastal biodiversity by the inundation of wetlands (Senhoury et al., 2016; Tragaki et al., 2018). According to many researchers, numerous coastal and island areas in African countries are threatened by sea-level rise and marine submersion or, increasingly, frequent flooding, particularly in the case of large populated cities along coasts (Hinkel et al., 2012; Goussard & Ducrocq, 2014; IPCC, 2019; Alves et al., 2020). Nouakchott (the capital of Mauritania), because of its topography, is considered the most vulnerable and exposed area to coastal hazards in West Africa (Alves et al., 2020).

This is why many studies interested in coastal erosion have focused on the region of Nouakchott (Ould Mohameden, 1995; Courel et al., 1998; Senhoury, 2000; Trebossen, 2002; Barry, 2003; Wu, 2003; Ould Sidi Cheikh et al., 2007; Niang, 2014; Tragaki et al., 2018; Alves et al., 2020). The other parts of the coast have been the subject of a few studies focusing on erosion. Among these, the study of Littaye and Ould Ahmed (2018) was interested in the dynamics of the area along the northwestern Mauritanian coast, particularly the Banc d'Arguin area. In this context, we have targeted the entire coastline because any marine incursion in the south could go as far as the capital through the depression of the Aftout es Saheli (see Fig. 9.1). The topography and location of the city also make it more vulnerable to flooding by

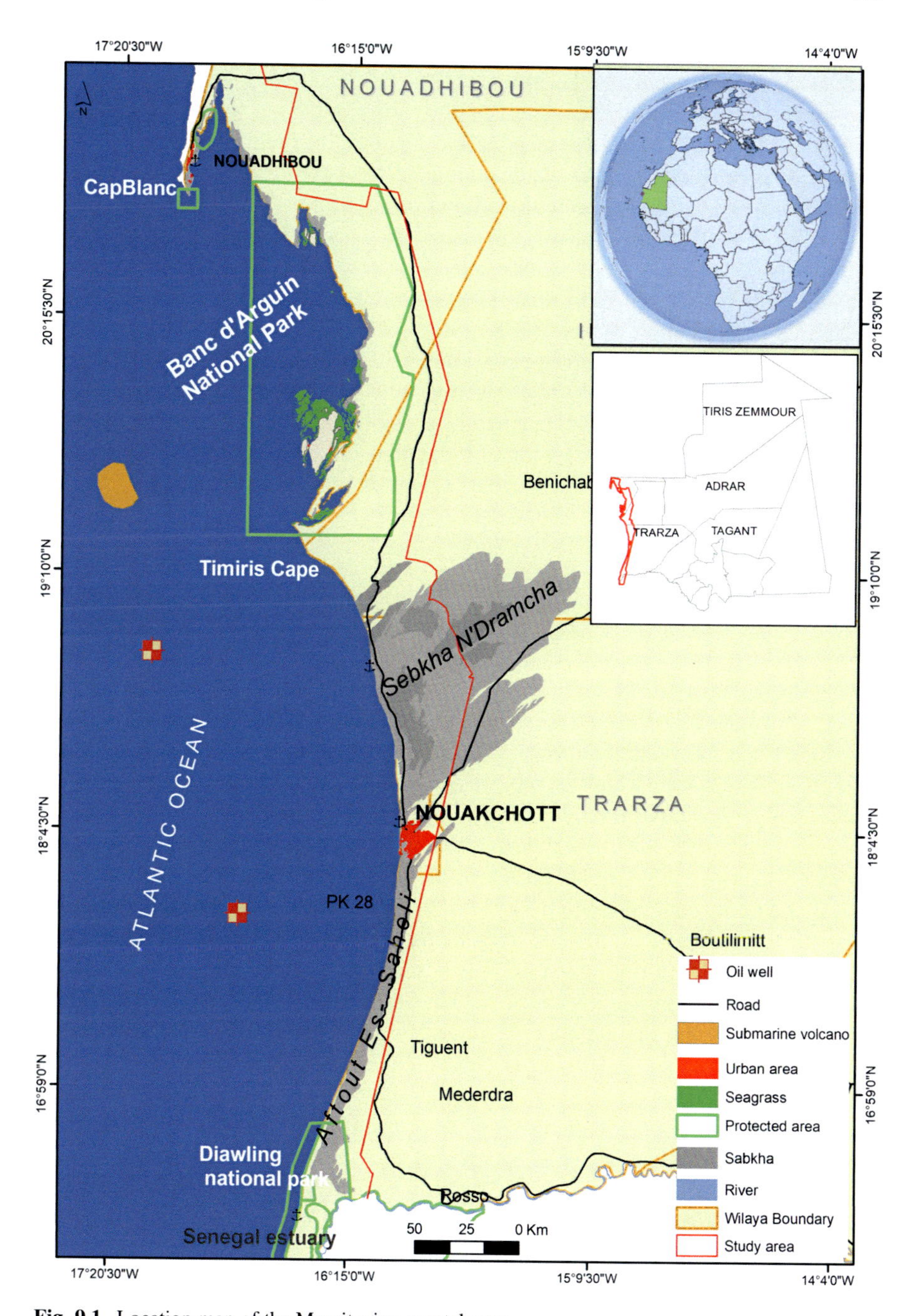

Fig. 9.1 Location map of the Mauritanian coastal zone

rainwater. The coast of Mauritania is located between the western end of the large aeolian sand dune systems of the Sahara and the Atlantic Ocean, which also exposes it to processes of desertification.

Thus, the identification, assessment, and mapping of highly vulnerable coastal areas is of great importance to the development of coastal management plans, risk mitigation measures, and appropriate adaptation solutions.

The main aim of this study is to assess the coastal hazards along the coast shoreline of Mauritania. The evolution and spatiotemporal dynamics of the shoreline between 1986 and 2021 will be analyzed using remote-sensing satellite images and Geographic Information System (GIS) techniques. The detection and analysis of extracted shorelines from remote-sensing data will be important for gaining a better understanding of the morphological dynamics of coastal areas and the main risks to which they are exposed. Thus, evaluating possible and real risks to the coastline requires the monitoring and mapping of the spatial and temporal changes of various hazards for the sustainable disaster management of coastal areas.

9.2 Study Area

The Mauritanian coastline stretches along the Atlantic coast of Africa from the Cap Blanc peninsula (21 ° 10’N) to the Senegal River estuary (16 ° N) for nearly 800 km. The coast can be subdivided into two large groups separated by Cape Timiris (Fig. 9.1). The south, which is very slightly curved, is characterized by a linear, arid, sandy, and windy beach. It is based on a low coastline (5–10 m) that limits the sea and protects vast salty depressions (Sebkha Ndramcha from +3 to −3 m, Aftout es Saheli from −1 to +1 m height above mean sea level). The shoreline generally runs in the north–south direction. To the right of Nouakchott, located at around 18 ° 07 north latitude and 16 ° 01 west longitude, the coastline shows a slight concavity with a very large radius of curvature. The north is rocky, with steep sandstone cliffs 10–20 m high and numerous caves, sabkhas, and sandy areas. It presents a very sinuous outline with the presence of numerous islands and bays. This coastline is entirely part of the Senegalese–Mauritanian sedimentary basin.

The coastal area is home to the two main cities of the country and more than a third of the country’s population as well as four important port sites that constitute poles of attraction where important economic activities are concentrated.

The coast of Mauritania is marked by strong aridity, being subject to a sub-arid climate in the southwest and an arid climate further north. The wind blows for much of the year, moving sand and dust. The harmattan is a hot, dry winter wind that blows from the east and northeast, carrying with it huge volumes of sand and dust that end up in the Atlantic. A large dune complex was set up over nearly 500 km along the Atlantic coast of Mauritania during the arid phases of the Quaternary. During the Ogolian period—between 25 ka and 15 ka (Lancaster et al., 2002)—considerable amounts of sand were accumulated or reworked over the entire study area (Fig. 9.2).

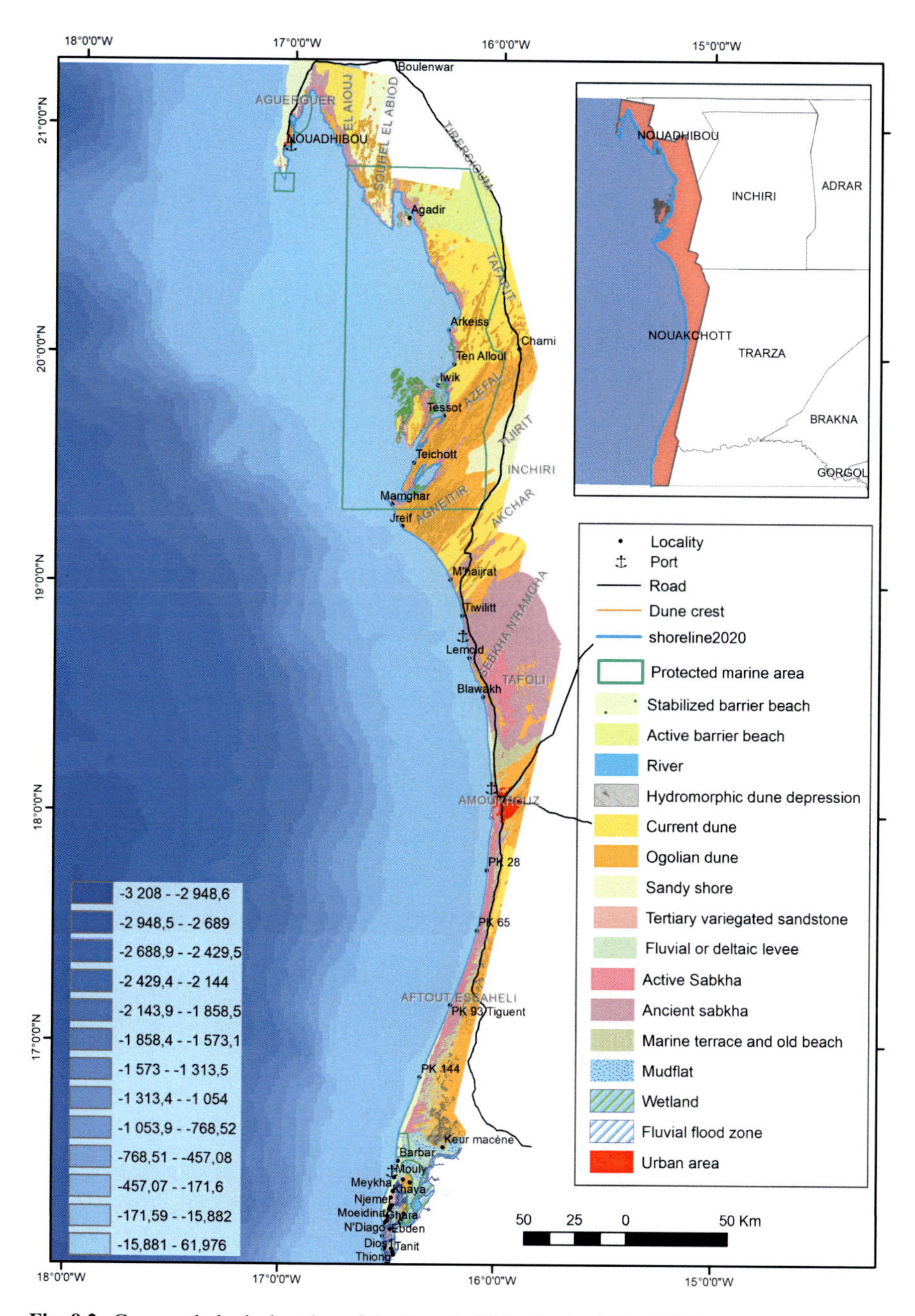

Fig. 9.2 Geomorphological setting of the littoral of Mauritania. (After MPEM, 2004, modified)

Fig. 9.3 Illustrations of geomorphological setting of coastline: (**a**) rocky cliff coast in of Cap Blanc peninsula, (**b**) Coast in shoals (mudflats) with seagrass beds and small isolated beaches in Banc d'Arguin zone, (**c**), low and sandy coast in the South of Cap Timiris. (Source: author. 2011)

The study area presents a relatively varied geomorphology (Fig. 9.2). North of Cape Timiris, in the region of Nouadhibou, the rocky cliff coast is at the level of the Cap Blanc peninsula, with steep sandstone cliffs 10–20 m high, it is gravel-sandy in Baie de Levrier. In the Banc d'Arguin region up to Cape Timiris, the coast is essentially low, with seagrass beds and small isolated beaches. South of Cape Timiris, the coast is low and sandy (see Fig. 9.3).

The mainland is geomorphologically characterized by the presence of the Aftout es Saheli depression. It is a narrow plain that stretches over 200 km (Hebrard, 1973) between the coastal dunes and the continental dunes of Erg Trarza, from Chott Boul in the south—the former mouth of the Senegal River (64 km north of Saint-Louis)—

to Nouakchott. This 3–5 km wide coastal plain can be inundated by the waters of the Senegal River during exceptional floods, such as that of 1950 (Duchemin, 1951).

The sabkha of Aftout es Saheli constitutes a low zone, the altitude of which is lower than the sea level. The water table is everywhere sub-flush, and its level is in direct relation with that of the ocean. There are other sabkhas along the coast, including the great Sebkha Ndramcha and others that occupy the interdune corridors.

Different dune ergs occupy most of the mainland. Lancaster et al. (2002) identified three dune-trend classes of linear dunes of different ages: northeast large linear ridges, north–northeast moderate-sized linear dunes, and north small linear dunes. These dune formations are sometimes reworked on the surface, forming transverse dunes.

9.3 Materials and Methods

9.3.1 Data Used

Multisource and multitemporal remote-sensing and climatic data were used to assess coastal hazards along the coast of Mauritania. To monitor shoreline changes, six satellite images (Operational Land Imager [OLI]) and six thematic mappers (TMs) taken by Landsat were used to cover the coast of Mauritania in 1986 and 2021. The images were acquired from the US Geological Survey‘s (USGS) EarthExplorer website (http://earthexplorer.usgs.gov) and were already orthorectified. Other multitemporal images were used to analyze the processes of desertification along the coastline and flooding by rainwater, namely a CORONA image acquired in 1965, a SPOT-5 image from 2006, and two Sentinel-2B images from 2020. The details of the images are noted in Table 9.1.

9.3.2 Automated Shoreline Detection

To monitor shoreline dynamics, six Landsat satellite images from both 1986 and 2021 were used. Pre-processing steps included radiometric calibration and atmospheric correction using ENVI 5.3. The Modified Normalized Difference Water Index (MNDWI) was utilized to delineate the coastline for change detection. This algorithm was chosen because it performs consistently well irrespective of the land cover type (Sunder et al., 2017) and is better at suppressing errors from built-up land as well as vegetation and soil noises compared to NDWI (Feyisa et al., 2014). The index is calculated based on the following formula: MNDWI = (Green − SWIR) / (Green + SWIR). Positive values indicate water, while negative values indicate land. A land/water threshold was manually applied to classify the images into two classes: land and water. Suitable land/water thresholds for each index were determined through trial and error and comparison to reference maps generated using visual

Table 9.1 Details of the image data set

Sensor	Acquisition date	Path/row	Resolution (m)	Nb of bands	Cloud cover
Landsat 8 OLI	03/20/2021	206/45	30	11	0
Landsat 8 OLI	03/20/2021	206/46	30	11	0
Landsat 8 OLI	03/20/2021	206/47	30	11	0
Landsat 8 OLI	03/13/2021	205/47	30	11	0
Landsat 8 OLI	03/13/2021	205/48	30	11	0
Landsat 8 OLI	03/13/2021	205/49	30	11	0
Landsat 5 TM	10/30/1986	206/45	30	7	0
Landsat 5 TM	10/30/1986	206/46	30	7	0
Landsat 5 TM	10/23/1986	205/48	30	7	0
Landsat 5 TM	10/23/1986	205/49	30	7	0
Landsat 5 TM	02/25/1986	205/47	30	7	0
Landsat 5 TM	11/15/1986	206/47	30	7	0
Sentinel-2B	09/29/2020	T28QCE	10	12	0
Sentinel-2B	09/29/2020	T28QCF	10	12	0
SPOT-5 XS	02/21/2006	–	10	4	0
CORONA KHB-4	12/26/1965	–	6	1	0

interpretation. The coastline was finally extracted by converting the water/land binary image into a vector format with ArcGIS software.

Various types of uncertainty sources are often identified by researchers and have been divided into two categories (Hapke et al., 2011). Those related to georeferencing and digitalization can be limited because Landsat images are orthorectified and shorelines are extracted automatically. The errors retained are associated with image resolution and correspond to pixel precision (30 m).

9.3.3 Calculation and Interpretation of Shoreline Change Rates

The rates of change of the shoreline were calculated using two statistic methods among these available in the Digital Shoreline Analysis System (DSAS) application. The DSAS is a freely available tool designed to work with ESRI ArcGIS software. It was used for the statistical analysis of shoreline changes between 1986 and 2021. For this study, the DSAS version 5 software developed by the USGS (Himmelstoss et al., 2018) was used to generate orthogonal transects.

spaced 100 m apart along the coast of Mauritania. Two techniques were utilized to understand the general trend of shoreline movement. The end point rate (EPR) highlighted the trend for each transect between 1986 and 2021. The calculated rates represented the shore position differences between the 2 years divided by the time elapsed between the two shorelines. In this study, the time that elapsed between the two dates was about 35 years. The net shoreline movement (NSM) was used to calculate the total distance change from 1986 to 2021. An analysis of the kinematics of the shoreline was performed along the Mauritanian coast from 1986 to 2021, mainly to identify the sections of the coast subject to erosion that are threatened by marine submersion.

9.3.4 Monitoring Risks Linked to Desertification and Flooding by Rainfall

The risks associated with desertification were analyzed using diachronic images to follow the movements of dunes in different periods since 1965 on CORONA aerial photographs and satellite images. Spectral indices and different filters were applied to identify the evolution of dune ridges. Field photos were used to observe the problems caused by dune movements in certain areas.

The surface water flooding zones were defined through various satellite images and field photographs taken during various missions. Floods caused by rainwater mainly concern the city of Nouakchott, as the city of Nouadhibou—which is located in a coastal desert—receives little precipitation.

9.4 Results and Discussion

9.4.1 Shoreline Change Analysis and Implications for the Mauritanian Coastal Area

Erosion and accretion were calculated using two techniques. The EPR was used to compute changes in terms of rates or areas between the two shorelines by comparing erosion and accretion in the study area.

The dynamics of the shorelines over the period 1986–2021 (35 years) were calculated using the EPR and NSM methods. The results illustrated in Fig. 9.4 show the overall shoreline change rates calculated from the analysis. Positive values indicate an aggradation of the shoreline, while negative values are related to erosional rates. This figure also illustrates the locations of accretion and erosion areas along the coast. Over the period of 35 years, the NSM and EPR overall averages indicate an accretional trend of the shoreline, with an average distance of 43.87 m (1.29 m/year).

In addition, the overall transect averages show that the Mauritanian coastline is mainly subject to an accretion; 59.44% of all transects were accretional and 35.44% of them were statistically significant, while 40.56% of all transects were erosional but only 11.73% of them were statistically significant. The maximum accretion

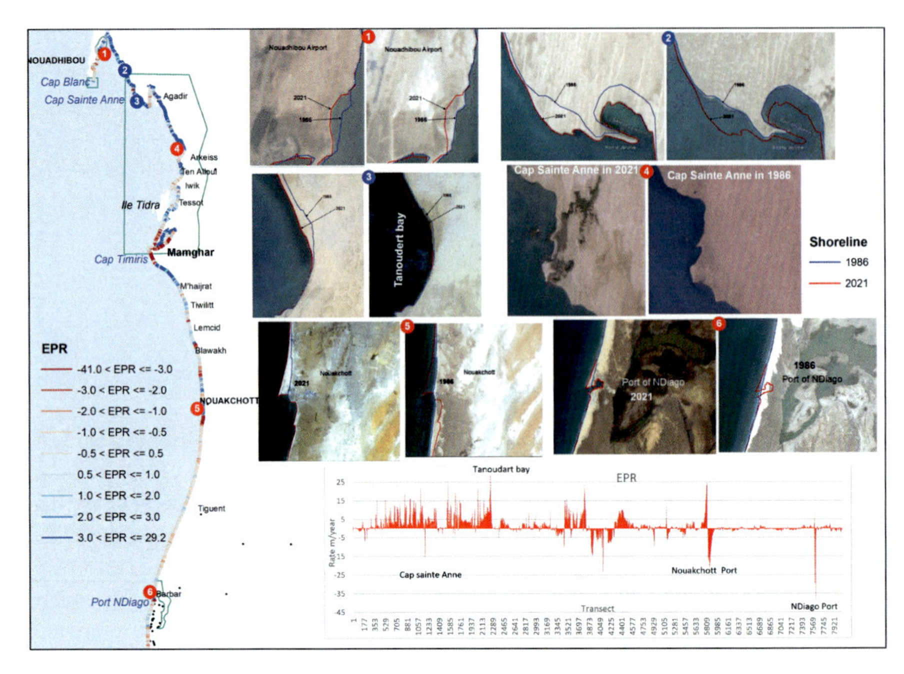

Fig. 9.4 End point change rates for all transects in m/year for period 1986–2021. The transects with negatives values indicate erosion and the positive values show accretion rates. The inserted illustrations indicate the evolution of the coastline at the place numbered on the map, the numbers in blue show the sectors of maximum accretion while the numbers in red show the maximum erosion according to the division of the coastline. The graph in red indicates the EPR change rate for the entire coastline between 1986 and 20121. (Author own Figure)

distance was found to be 990.54 m (29.13 m/year based on EPR) in the Tanoudert bay in the National Banc d'Arguin Park, with an average distance of 43.87 m and an average rate of 1.29 m/year. The maximum erosion distance was found to be −1392.55 m (−40.95 m/year based on EPR) in the port of Ndiago, where construction began in 2016 in the west part of the Mauritanian coast.

On the basis of different geomorphological units of the coastline, there was a difference in the evolution of the change rate of the shoreline position. In the north part of the coastline, in the peninsula of Cap Blanc and the region of Nouadhibou, the maximum accretion rate reached 18.74 m/year in Pointe Jerome at the Baie des Pelicans, while the maximum erosion rate was −7.95 m/year at the airport of Nouadhibou (see Fig. 9.4). In the Banc d'Arguin area and in the north of Cape Timiris, the maximum erosion rate recorded was −23.9 m/year at Cap Sainte-Anne, while the maximum accretion rate was 29.13 m/year in the Tanoudert bay (Fig. 9.4). In the south part of the littoral, the maximum erosion rate on the entire shoreline (−40.95 m/year) was noted at the Ndiago port, while the maximum progradation rate (24.65 m/year) was at the port of Nouakchott, which was built during the 1980s.

9.4.2 Impact and Drivers of Spatiotemporal Shoreline Changes

The evolution of the shoreline measured over different periods from 1986 to 2021 shows that the dynamics of the shoreline are controlled by natural and anthropogenic drivers. Many studies conducted in different geographic areas have concluded that human activities are the main driving force behind shoreline changes (Amrouni et al., 2019; Tian et al., 2020). Mauritania is no exception. The phenomenon of coastal erosion was observed over the entire Mauritanian coastline. Coastal risks have increased due to widespread changes in coastal infrastructures, given the topography of the coast. The sectors most threatened by coastal erosion and marine submersion are located near the port facilities (de Boer et al., 2019). Part of the urban expansion of Nouakchott and Nouadhibou has come at the expense of the coastal area. The most significant changes have occurred in the ports of Nouakchott and Ndiago.

According to Oppenheimer et al. (2019), anthropogenic drivers have played an important role in increasing low-lying coastal zones' exposure and vulnerability to sea-level rise (e.g. demographic, coastal infrastructures and land use changes). Thus, the southern part of the coast, which is the lowest lying, is the most vulnerable to the combined consequences of coastal erosion and climatic changes.

9.4.3 Littoral Erosion and Risk of Incursion or Marine Submersion

Marine submersion is undoubtedly the most serious risk in the coastal zone in the context of climate change, particularly the southern part, which has the lowest-lying sectors characterized by linear sand barriers. Remember that this area has already experienced several marine transgressions in the past (Giresse et al., 2000).

- **The north part of the coast**

At Cap Blanc, very active erosion was observed. Landslides threaten the infrastructures located there (the lighthouse in particular). We should also mention the fall (in 2007) of a chapel and a cross located on Cape Point, which can be seen in the photos below (Fig. 9.5). This active erosion threatens the existence of the caves, which constitute the refuges of the monk seals. It also risks destroying this natural barrier that protects the port of Nouadhibou against storms.

The city of Nouadhibou is under the threat of coastal flood hazards from the sea due to the presence of bays and lagoons to the north and south of the airport, as can be seen in Fig. 9.6. The erosion is also felt in the Baie de l'Etoile, where there is an undermining of the base of the building of the sport fishing center.

In Cap Sainte-Anne, the breaking of the dune ridge has led to the formation of a permanent lagoon that currently covers a large area, as we can see in Fig. 9.4.

Fig. 9.5 Field photos shows Infrastructures destroyed by erosion (top right chapel and church – photo found on Google, other rubble photos and scree. (Source, Author, 2010)

At Banc d'Arguin, several islands of sedimentary origin are threatened by erosion. Sidi Cheikh et al. (2009)—based on a multitemporal study using satellite imagery—showed significant morphodynamic changes, particularly at the level of the islands of Nair and Niroumi, where certain areas of the dune ridge became sebkhas between 1987 and 2001. The submersion of parts of the islands destroys the nesting areas of certain species of birds and vegetation, which risks affecting biodiversity.

- Nouakchott and the south part of the coast

The city of Nouakchott, a large part of which is at a level close to or even lower than that of the sea, is separated from it only by a thin coastal strip that has been weakened in several places (Niang, 2014). This situation was worsened by the construction of the port of Nouakchott beginning in 1984, which led to a spectacular modification of the coastline (Fig. 9.4). Climate change will increase the threat natural disasters presents to the city. Nouakchott has been identified as the most exposed area to coastal hazards (de Boer et al., 2019; Alves et al., 2020).

While the port is by far the most significant cause of the spectacular changes in the Nouakchott coastline, other anthropogenic elements have also contributed considerably to disturbing this coastal system. Sand withdrawals from coastal dunes have made a significant impact on the city's short-term risk of marine submersion. Their presence at the top of the dune bordering on constructions (fish markets, hotels, factories, etc.) disrupts the natural functioning of the beach by creating

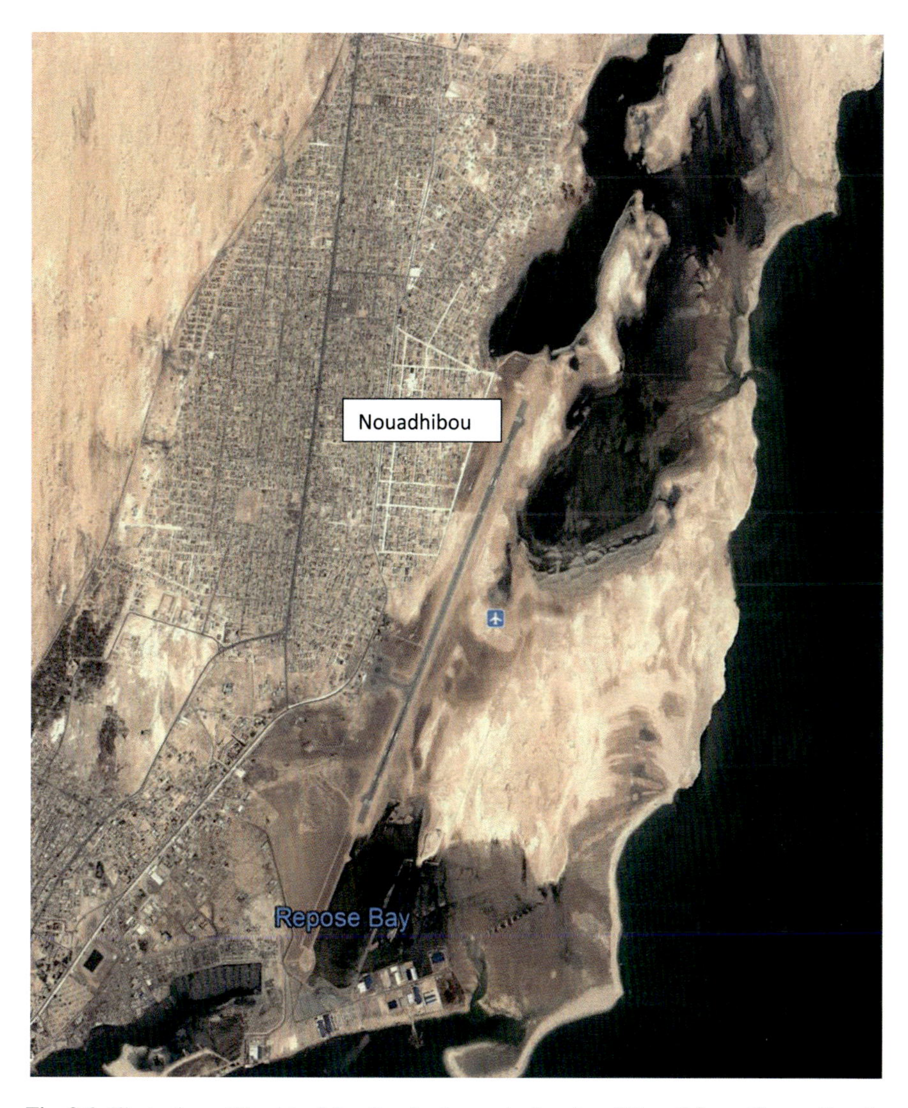

Fig. 9.6 Illustration of the risk of flooding by the sea in the city of Nouadhibou. (Source Google Earth)

obstacles to dynamic exchanges between the dune and it. The use of vegetation—already hit by the drought of recent decades—as firewood, the presence of large concentrations of animals to supply the capital with fresh milk, and the movement of vehicles between the beach and the depression of the Aftout through the dune rim for seaside activities are other weakening elements that have played a part.

The city of Nouakchott is located—for the most part—between 0 and + 4 m Institut Géographique National (IGN) (geodetic and altimetric reference points calculated by National Geographic Institute of France where 0 is the Atlantic sea level for Mauritania), with a full sea level of +2.16 m IGN (IRC-Consultant and Saint Martin Paysage, 2004). The extreme fragility of the coastal barrier that protects the city from the sea, the uncontrolled exploitation of this coastline, and the development of unsuitable infrastructures have made the coastal barrier's protection extremely vulnerable and exposed a large part of the city to the risk of flooding or, more exactly, real submersion. These submersions have typically hit the wharf area/ fishing market beach. In August 1992, the storm waves crossed the cordon and dug a breach of about 50 m north of the El Ahmedi hotel. As a result, the authorities decided to ban sand sampling in the cordon between the port and the fishermen's beach, a ban that is respected very little. During the night of December 14, 1997, surges accompanied by strong swells magnified by a sea breeze during high tide caused several points of rupture between the El Ahmedi hotel and the Sabah hotel. The damage was significant at the fishermen's beach, where there were several victims. On December 18, 1999, an episode of storms resulted in a new flooding of the lower sectors of the sebkha where a whole new district of Nouakchott had developed, forcing the municipality to rapidly displace all the inhabitants of the area. In early March 2006 and late 2009, the tide submerged a good strip of land, threatening the port infrastructure of the capital and causing the destruction of dozens of boats.

The analysis of the topographic surveys of Nouakchott measured by GRESARC (2006) shows that approximately a third of the built-up areas of the city are submersible (GRESARC, 2006). Figure 9.7 shows the sectors most vulnerable to the risk of marine submersion. This map shows the areas subject to the risk of submersion in the city of Nouakchott in the event of large breaches formed in the dune ridge in an exceptional hydrometeorological context.

Currently, two zones weakening the littoral and inducing the risk of marine submersion have presented themselves.

The first zone is located between the industrial area of the wharf and the fish market (see Fig. 9.7). The dune ridge presents two critical sectors (Fig. 9.7) likely to evolve into breaches in the event of a storm from southwest to northeast associated with a high sea level. This risk is particularly high due to the low altimetry and the narrowness of the cordon at this level. Although these critical sectors still partially reach an altitude of more than 1 m, the state of the dune ridge at this level must be the subject of special attention.

The second zone is located immediately to the south of the port and is following the retreat of the shoreline (which reached 300 m between 1995 and 2006) (GRESARC, 2006), resulting in the total erosion of the dune ridge (see Fig. 9.8). The current altimetry of the bordering lands would also allow an intrusion of marine waters in an exceptional context. Figure 9.9 shows some manifestations of the erosion north and south of the port of Nouakchott.

In this context, and with the absence of internal relief making it possible to contain the water (except, perhaps, the raised roads crossing the zones of sebkhas), the extension of the submersion within the low zone located behind the littoral fringe would be very fast, regardless of the location of the breaking point (north or south of the port).

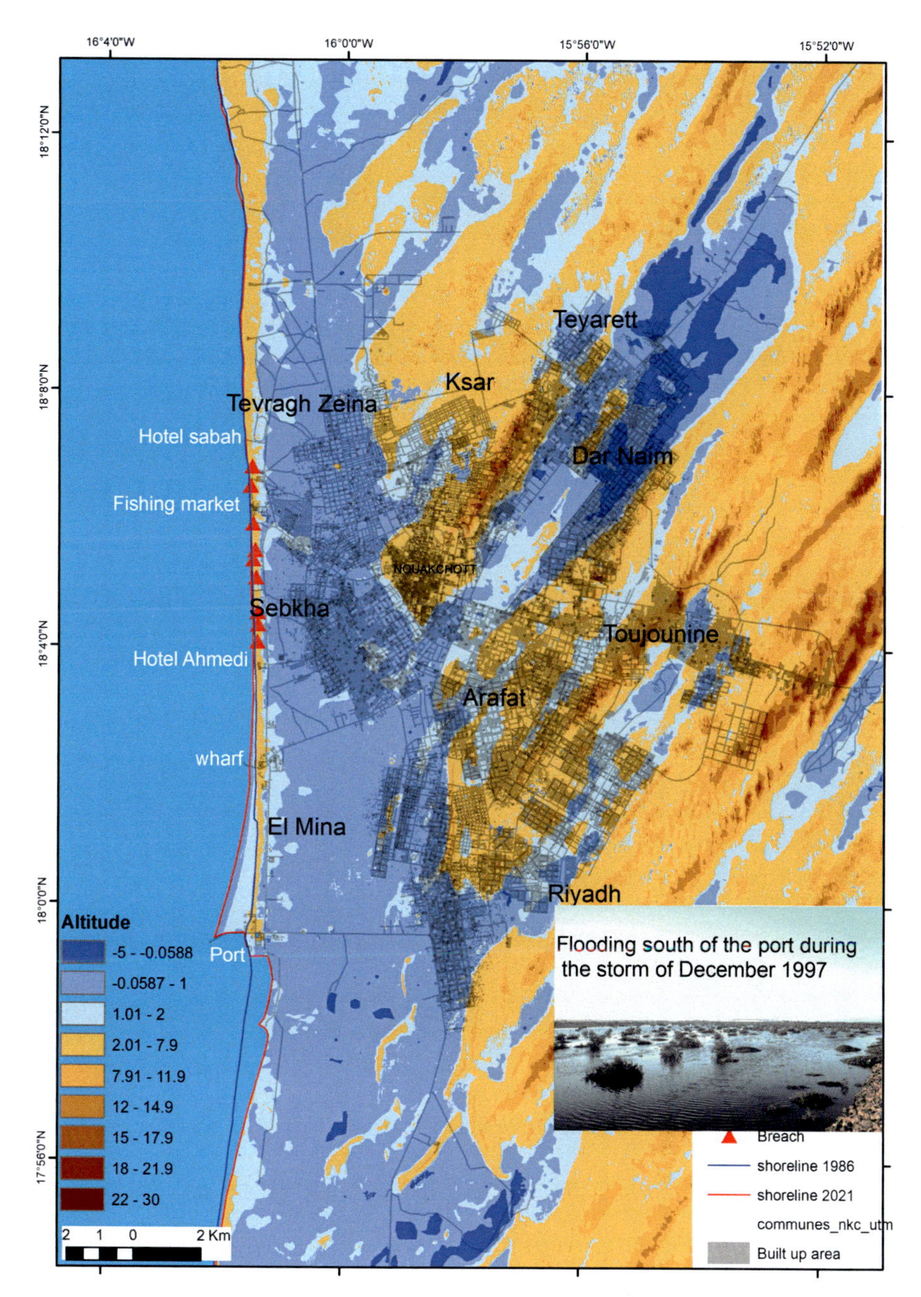

Fig. 9.7 Submersion risk map of Nouakchott for differents sea level values (author own map using various data sources)

Fig. 9.8 Some manifestations of the erosion at Nouakchott (**a**) erosion of buildings on the coastal dune north of the port, (**b**) degradation of the protective dike south of the port, (**c**) Subsidence and bypass of the end of the epi in the southern part of the port. (Source: author, 2011)

From a socioeconomic point of view, the inhabited areas most threatened by submersion are mainly the districts of Sebkha, El Mina-Nord, and Riyadh, as well as the western part of Tévragh-Zeina (see Fig. 9.7). It must also be noted that the peripheral districts are gaining more and more on the coastal area, further increasing coastal risks. The submersion hazard map shows that a large part of the city is at an altitude of around 1 m. However, if we consider the maximum level of the tide (which can reach 2 m), the foreseeable rise in the mean sea level, and exceptional events, the sea level could exceed 3 m. This would be enough for it to cross several breaches located on the dune cordon and the south of the port and thus invade the

Fig. 9.9 Fixing the littoral dune by Typha mats (aquatic vegetation) (**a**) 2004 in IRC and Saint-Martin Paysage, 2004; (**b**) in 2010, source: Author, 2010), (**c** and **d**) fixation of dunes by wattle. (Source: author 2010)

city, with disastrous consequences. According to Senhoury et al. (2016), the economic losses in case of flooding by the sea might reach the equivalent of seven billion U.S. dollars.

In response to this threat, and to reduce the effects of climate change, various projects-including African Strategy for Fight against Coastal Erosion (SALEC), Adaptation to Climate and Coastal Change (ACCC), and, recently, Coastal Cities Climate Change Adaptation (ACCVC)—were selected to, among other things, (i) integrate coastal erosion control in the framework for a strategic vision for the development of coastal areas, (ii) find financing and build capacities for the fight against coastal erosion, (iii) reduce the vulnerability of countries and develop effective resistance mechanisms to cushion the effects of coastal erosion, and (iv) analyze all of the problems linked to the development of the coast to provide lasting solutions.

It is in this context that the ACCC project, which ended in 2011, selected a pilot site in Nouakchott to improve the protection of the dune cordon over an area that covers 50 ha. The ACCVC project launched at the end of 2012 will ensure the continuation of this work, using biological fixation of dune cordon with local plants adapted to the marine and coastal environment, such as *Tamarix aphylla*, *Nitraria retusa*, and *Atriplex halimus*.

On the coast of Nouakchott, developments aimed at combating coastal erosion have been conducted. These mainly involve fixing and raising the dune cordon via wattle systems. Mechanical fixings using Typha mats (*Typha australis*) were put in place in 2006 (IRC-Consultant and Saint-Martin Paysage, 2004) between the fishing market and the wharf as well as to the north of the fishermen's beach. This system has had some success, as we can verify with the photo taken in 2010 that the dune has been raised at this location. The examples of fasteners we present here are located in the area between the fishermen's port and the wharf (Fig. 9.9).

In view of all of these measures, we can see that the public authorities are aware of the threat of submersion and of the latent danger posed by increasing human pressure on coastal environments. In this sense, concrete measures have been taken at the legislative, institutional, and practical levels. These measures are all likely to fight coastal erosion and mitigate the effects of climate change, but their implementation shows disparities between rules and practices.

Considering the various manifestations of coastal erosion and the infrastructures built on the coast, we perceive either a lack of knowledge of the ongoing processes or an unconsciousness of certain actors of the Mauritanian coastal environment. If the establishment of port infrastructures is a necessity for the economic development of the country, particular attention must be paid to the dramatic consequences that this may entail (in particular, the risk of the capital's submersion), and drastic measures must be taken.

In addition, at the legislative and regulatory levels, we notice a mismatch between the texts and their application. Article 28 of the Coastal Law stipulates that the occupation and use of coastal land must make it possible to preserve areas that are remarkable or necessary to maintain natural balances. However, we note the establishment of several infrastructures (hotels in particular) that not only destroy this balance but are built for commercial purposes.

The ban on permanent constructions on the coastal dune cordon or at a distance of 500 from it, the flood-prone areas, and the sebkhas set out in Article 38 is not respected in any of these sectors. The BMCI residential area built by SOCOGIM (national company of construction and real estate management called now ISKAN) are in some cases located less than 300 m from the beach.

In the southern part of the coast, especially around the village of Ndiago, there is also the threat of marine incursion. In 2009, a strong storm caused a crossing of the littoral cordon and the destruction of certain constructions of fishermen at the edge of the sea (Fig. 9.10). Let us recall that any incursion in the southern part of the littoral could affect the city of Nouakchott, which can be flooded from the Aftout es Saheli. The recent construction of the port of Ndiago, which may have the same effects as the port of Nouakchott, could increase the vulnerability of this part to marine submersion. This could also result in a threat to the biodiversity of the Diawling park very close to the port.

Fig. 9.10 Illustration of marine submersion risk (**a**) and littoral erosion (**b**) in the south part of Mauritanian coast in 2009. (Source: author, 2010)

9.4.4 Risk of Flooding by Rainwater

The risk of flooding by rainwater mainly concerns the city of Nouakchott; as Nouadhibou is in a coastal desert, precipitation is rare there.

Flooding by rainwater, or the problem of "stagnant water," is linked to the conjunction of three essential factors: precipitation in the context of climate change, the level of the water table, and the absence of a water network that provides functional sanitation.

The increase in precipitation as a result of climate change and the rise in the water table contribute to increasing the risk of flooding, already favored by unchecked urbanization. The massive influx of populations following the rural exodus of the 1970s has led to an increase in construction in flood-prone areas.

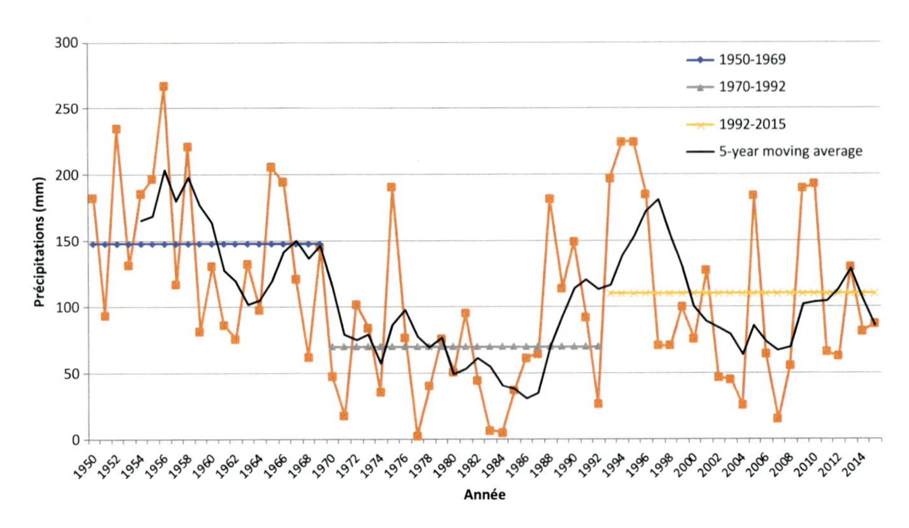

Fig. 9.11 Evolution of annual precipitation in Nouakchott from 1950 to 2015. (Source: National Weather Office)

The essential characteristic of these precipitations is their variability in importance, rhythm, and spatiotemporal distribution. In the Saharo–Sahelian zone, the analysis of a long series of rainfall data shows a succession of three distinct phases identified by Pettitt's homogeneity test (see Fig. 9.11).

The first period was the so-called wet years (from 1950 to 1969), when the highest rainfall totals were recorded. This was a phase of heavy rainfall, with the average annual accumulation reaching 150 mm.

The second period was that of the great drought of the 70s and 80s, when the driest years were recorded. The recorded quantities rarely reached 70 mm. However, this period was marked by very strong population growth in and urbanization of Nouakchott. Entire neighborhoods emerged in areas that are prone to flooding if heavy rainfall returns.

The third is the current period, which is marked by an improvement in annual rainfall accumulations from the beginning of the 1990s.

Note, however, that the rains that cause the current floods have a return period of less than 6 years according to Ould Sidi Cheikh et al. (2007). After performing statistical analyses on daily precipitation, they note that the rains responsible for floods are neither abnormal nor exceptional, as they were more frequent during the so-called wet period.

It is therefore evident that improving current rainfall conditions will imply increasingly frequent rainfall.

In addition, if no wastewater evacuation system is built while the volumes of water used continue to increase, especially with the establishment of a supply from the Senegal River in 2010, some neighborhoods will be condemned. This is the case in certain sectors of Socogim PS, where the junction between rainwater and groundwater forces people to abandon their homes (as illustrated in Fig. 9.12). In these

Fig. 9.12 Flooded urban areas (**a**) sebkha in 2005, (**b**) SOCOGIM beach 2010, (**c**) SOCOGIM PS, 2009, (**d**) Carrefour BMD downtown, 2009; (**e**) sebkha september 2013 from google earth and enhaut.org. (Source photos: author, 2010)

neighborhoods, loss of human life has been reported due to electrocution problems. Some neighborhoods have been emptied of their population during the rainy season.

We note that, apart from the fact that it constitutes a receptacle for marine water and rain, the water table threatens the city of Nouakchott by outcropping and makes the housing conditions precarious even outside the rainy season. Some areas keep stagnant water more than 6 months after the rains, while others keep it all year round. They have turned into lagoons and have been abandoned by their populations. In some places, schools, mosques, and roads totally occupied by water become impassable for several weeks or even months. In the city center, rainwater from certain arteries or crossroads is evacuated by tankers.

In the commune of Sebkha, the construction of paved roads has had the unexpected effect of trapping and pushing back water inside the houses. The Mauritanian authorities put in place an interim solution that consisted of evacuating water to coastal basins by means of motor pumps. The problem was that this took too long, as the water also had to be sucked out of the slick.

The risk of flooding is aggravated by the topography of the land, the presence of sabkha soils, and the rising groundwater level. Despite the recent installation of a drainage system, the problem remains unresolved.

This situation can lead to public health problems because the permanent water table is in direct contact with accumulations of garbage and is enriched with domestic wastewater from septic tanks, resulting in the proliferation of mosquitoes and numerous pathologies in certain neighborhoods. The city presents ecological conditions that favor the development of urban pathology (Sy et al., 2011; Ahmed-Salem et al., 2017).

9.4.5 Desertification Risk

In the part of the coast located north of the Sebkha Ndramcha (see Figs. 9.1 and 9.2), the dune system often is in direct contact with the Atlantic Ocean and participates in the accretion noted on the coastline. In the south, the dunes are often separated from the ocean by the depression of the Aftout es Saheli. In both parts, desertification constitutes a permanent threat, and the phenomenon of silting up affects towns and infrastructures along the coastal zone. The roads connecting the cities must be constantly cleared as they are often invaded by sands.

Most of the dune formations of Mauritania were formed during the Ogolian period, between 20 ka and 14 ka (Nicholson & Flohn, 1980; Nguer & Rognon, 1989; Lancaster et al., 2002). A large dune complex has been set up over nearly 500 km along the Atlantic coast of Mauritania. During the ogolian period considerable amounts of sand were accumulated or reworked over the entire Mauritanian coast. The post-Ogolian dunes are characterized by their vigorous relief. They also have a northeast–southwest orientation, like the Ogolian dunes, and they form regular alignments (Barbey, 1989). These formations experience climatic pulsations that lead to accumulations, fixations, or erosions.

Since 3 ka, there has been a trend toward aridity. The reactivation of the dunes is then linked to drought situations, which can be amplified by human actions. Currently, with the drought that has raged since the early 1970s, we are witnessing an upsurge in wind phenomena (favored by the degradation of the plant cover, the rainfall deficit, human pressure, etc.) imposed on the landscapes of the area. The wind has become a major morphodynamic agent in the studied sector.

The current dunes are the result of past droughts. Along the coast, there are large fields of ergs. Before the 1960s, the vegetation in this area was weak but quite visible, particularly in the CORONA satellite photographs of 1965. The end of the 1960s marked the beginning of a long period of vegetation degradation in the northwest, southwest, and east of Nouakchott. The current dunes are aggravating factors in the silting up of the cities of Nouakchott and Nouadhibou due to their mobility. According to Mainguet et al. (2008), the maximum degradation threshold was reached in the early 1990s. Even if the improvement in rainfall conditions from 1994 onward has contributed to slowing down environmental degradation, Nouakchott has recorded a number of days of sand wind higher than the national average (Niang, 2014).

To better understand the mobility of the sands, we conducted a diachronic study by comparing the aerial photographs. The analysis of the aerial photographs of 1963 shows that the 1960s were situated in a period. This manifested itself on the morphogenetic level by the fixation of aeolian deposits. We can notice from these aerial shots that the dune model inherited from the arid phases of the Quaternary is still fixed. The absence of sharp ridges and bare dune sets allows us to confirm this. In addition, traces of runoff can also be detected on the sides of the dunes, and vegetation, although sparse, covers the Ogolian dunes. With regard to human settlements, it can be noted that the city of Nouakchott did not experience a great spatial extension, and humans had less of a hold on the environment.

An examination of the aerial photographs from 1980 highlights that they bear the marks of the great drought that has raged throughout the Sahel since 1968. This climatic deterioration is distinguished above all by its intensity, extent, and duration. Annual precipitation is for the most part deficient and irregular. This has caused an extension of desertification processes such as the destruction of the vegetation cover, the worsening of wind processes, and the reworking of dunes.

The impacts of these restrictive environmental conditions are perceptible in the aerial photographs of 1980 in several aspects. The first signs of the remodeling of the sands and re-movement of the formerly fixed dunes are the appearance of sharp ridges that can be clearly seen in the series of photos. We note the presence of nebkhas at the foot of most obstacles, including fences and walls. We can also mention the presence of wind deposits at inter-dunes and depressions, their advance toward certain infrastructures (such as the airport runway), and the depletion or even disappearance of plant cover.

This persistent drought also has negative consequences for humans. Human pressure accentuates environmental degradation. Here, the rural exodus has had the effect of expanding the city, the population of which has increased sharply, hence the occupation of certain areas very sensitive to silting up.

The context of the aerial photographs of 1991 is that of the exacerbation of aeolian processes and of environmental degradation conditions. The length of the drought has ended up creating arid conditions in the Sahelian zone. Environmental

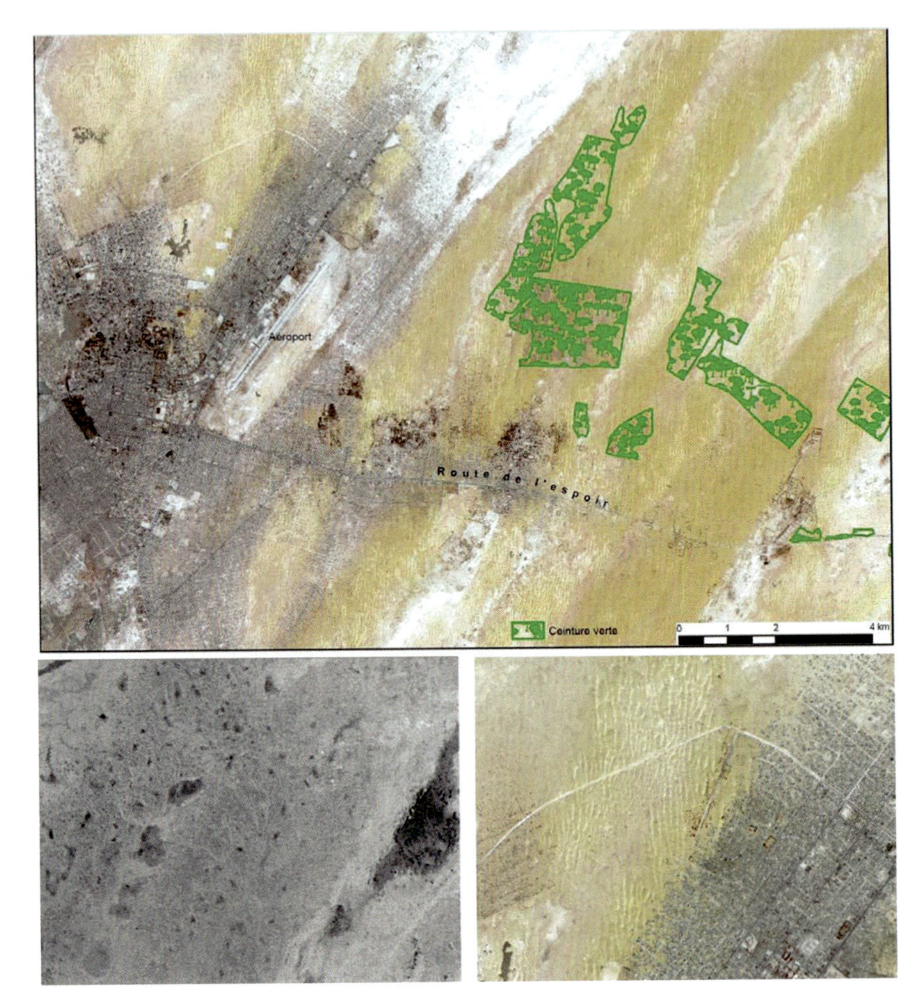

Fig. 9.13 Dunes movement around Nouakchott in 2006, re-mobilization of dune sands between 1965 (Corona satellite at the bottom left) and 2006 (SPOT image at the bottom right)

degradation is so intense that even the return of normal rainfall would require a longer or shorter response time from the physical environment.

Signs of the exacerbation of desertification phenomena and wind turbines are predominant and prevail over the entire area studied (Figs. 9.13 and 9.14). This is evidenced by the generalization of living dunes and their very significant advance toward infrastructures. The sand deposits are becoming very busy. Attempts to fix the dunes have not generated the expected results, and living dunes have been found in the study area. The winds and lithometeors imprint morphological modifications on the dune shapes (Niang, 2014).

Fig. 9.14 Illustrations of different aspects of desertification. Dune massifs in contact with the Atlantic Ocean in the Banc d'Arguin area. Dune movement around the city of Nouakchott. (Sources: a https://geoimage.cnes.fr/sites/default/files/drupal/202101/image/arguin01-blecoquierre.jpgb- http://liftoff.msfc.nasa.gov/ c Enhaut.org)

It is under these conditions that the public authorities, aware of the problem of silting up, have undertaken a vast campaign to plant trees around Nouakchott (see Fig. 9.13) to create a green belt in order to fight against the advance of the dunes.

9.5 Conclusion

This study focused on the coastal hazards weighing on the coast of Mauritania in view of the spectacular evolution of the coastline due particularly to anthropogenic activities and climatic changes. The erosion and coastal flooding that threaten the Mauritanian coast seriously jeopardize the livelihoods and security of the growing populations living in this area. Climate change, the increasingly strong extreme climatic events that accompany them, and the rise of the sea level constitute serious risks, the social and economic consequences of which can be very dramatic and catastrophic for this vital part of the country. At the level of Nouakchott and Nouadhibou, several sectors are affected by the impact of floods and the worsening of erosion in coastal areas, with considerable economic losses. The erosion resulting from port facilities has reached worrying proportions.

The Mauritanian coast, by its geomorphological setting, is particularly vulnerable to climate change, which constitutes a major challenge to its development and even its existence. Despite the adaptation measures taken, some of which aim to protect the coastline that defends this area against marine submersion, it is clear that the solutions proposed are still insufficient.

The results of shoreline change analysis show that over the period of 35 years, the NSM and EPR overall averages indicate an accretional trend of the shoreline, with an average distance of 43.87 m (1.29 m/year). The sectors most threatened by coastal erosion and marine submersion are located near the port facilities. Part of the urban expansion of Nouakchott and Nouadhibou has come at the expense of the coastal area. The most significant changes have occurred in the ports of Nouakchott and Ndiago. Thus, the southern part of the coast, which is the lowest lying, is the most vulnerable to the combined consequences of coastal erosion and climatic changes. The threat of the marine incursion weighing on the coast is closely linked to the spectacular modification of the coastline induced by the construction of the port of Nouakchott. Efforts are mainly concentrated on the dune ridge located north of the port, while the southern part where it has completely disappeared is only protected by a 2 m high dike. However, even if the sea were to cross the cordon to the south, the water would reach the city by the Aftout es Saheli depression. In the northern part, we can underline that the fixing and raising of the cordon are not very compatible with the existing constructions on the top of the dune.

The risk of flooding by rainwater mainly concerns the city of Nouakchott; as Nouadhibou is in a coastal desert, precipitation is rare there. Flooding by rainwater, or the problem of "stagnant water," is linked to the conjunction of three essential factors: precipitation in the context of climate change, the level of the water table, and the absence of a water network that provides functional sanitation. The risk of flooding is aggravated by the topography of the land, the presence of sabkha soils, and the rising groundwater level. Despite the recent installation of a drainage system, the problem remains unresolved. The risk of flooding from rainwater and the rise in the water table, which is very much felt by some residents of Nouakchott, is reinforced by poor urban development. The flooded districts sometimes belong to the urban periphery, but others are spaces developed by the national real estate company. If a sanitation and rainwater collection network is not put in place very quickly, entire neighborhoods will be uninhabitable after each rainy season.

Anarchic urbanization also implies a degradation of urban infrastructures by sand, stagnant water, and salt, rendering certain neighborhoods uninhabitable and causing significant economic and health damage. Mitigating the adverse effects of climate change would undoubtedly require more appropriate measures, which require much more resources.

Signs of the exacerbation of desertification phenomena and wind turbines are predominant and prevail over the entire area studied. This is evidenced by the generalization of living dunes and their very significant advance toward infrastructures.

In addition, oil reserves of commercial interest have been identified off Mauritania, This activity represents an additional pressure on fishery resources and a risk of pollution.

References

Ahmed-Salem, M., Leduc, C., Marlin, C., Wagué, O., & Sidi Cheikh, M. A. (2017). Impacts of climate change and anthropization on groundwater resources in the Nouakchott urban area (coastal Mauritania). *Comptes Rendus Geoscience, 349*(6–7), 280–289. https://doi.org/10.1016/j.crte.2017.09.011

Alves, B., Angnuureng, D. B., Morand, P., et al. (2020). A review on coastal erosion and flooding risks and best management practices in West Africa: What has been done and should be done. *Journal of Coastal Conservation, 24*, 38. https://doi.org/10.1007/s11852-020-00755-7

Amrouni, O., Hzami, A., & Heggy, E. (2019). Photogrammetric assessment of shoreline retreat in North Africa: Anthropogenic and natural drivers. *ISPRS Journal of Photogrammetry and Remote Sensing, 157*, 73–92.

Barbey, C. (1989). Etude chronologique de la sédimentation éolienne dans le Sud-ouest de la Mauritanie et dans le nord du Sénégal. *Bulletin de la Société Géologique de France, 5*(1), 21–24.

Barry, S. (2003). Contribution à l'étude géomorphologique de la côte mauritanienne : cas de Nouakchott et ses environs. DES en géographie, Université Cheikh Anta Diop de Dakar, Département de géographie, 116p.

Barusseau, J. P., Ba, M., Descamps, C., Diop, E. S., Diouf, B., Kane, A., Saos, J. L., & Soumare, A. (1998). Morphological and sedimentological changes in the Senegal River estuary after the constuction of the Diama dam. *Journal of African Earth Sciences, 26*(2), 317–326.

Courel, M. F., Rudant, J. P., Leterrier, E. & Tulliez, G. (1998). Apport de l'imagerie R.S.O. en milieu aride: cas de la région de Nouakchott en Mauritanie; 7è journées scientifiques de l'AUPELF UREF, Montréal, nov. 97 : 225–232.

de Boer, W., Mao, Y., Hagenaars, G., de Vries, S., Slinger, J., & Vellinga, T. (2019). Mapping the Sandy Beach evolution around seaports at the scale of the African continent. *Journal of Marine Science and Engineering, 2019*(7), 151. https://doi.org/10.3390/jmse7050151

Duchemin, J. (1951). L'inondation de l'Aftout es-Saheli et du poste de Nouakchott (Mauritanie, Trarza occidental). *Bulletin IFAN, B, XIII*(4), 1303–1305.

Feyisa, G. L., Meilby, H., Fensholt, R., & Proud, S. R. (2014). Automated Water Extraction Index: a new technique for surface water mapping using Landsat imagery. *Remote Sensing of Environment, 140*, 23–35.

Giresse, P., Barusseau, J. P., Causse, C., & Diouf, B. (2000). Successions of sea-level changes during the Pleistocene in Mauritania and Senegal distinguished by sedimentary facies study and U/Th dating. *Marine Geology, 170*(1), 123–139.

Goussard, J. J., & Ducrocq, M. (2014). West African coastal area: Challenges and outlook. In S. Diop et al. (Eds.), *The Land/Ocean interactions in the coastal zone of West and Central Africa*. Estuaries of the World, Springer. https://doi.org/10.1007/978-3-319-06388-1_2

GRESARC. (2006). Cartographie des risques d'inondation de Nouakchott. Rapport d'études, 39 p.

Hapke, C. J., Himmelstoss, E. A., Kratzmann, M. G., et al. (2011). *National assessment of shoreline change: historical shoreline change along the New England and Mid-Atlantic coasts*. U.-S. Geological Survey Open-File Report 2010–1118; 2011. p. 57.

Hebrard, L. (1973). Contribution à l'étude géologique du quaternaire du littoral mauritanien entre Nouakchott et Nouadhibou. Participation à l'étude des désertifications du Sahara. Thèse de l'Université de Lyon. 483 p.

Himmelstoss, E. A., Henderson, R. E., Kratzmann, M. G. et al. (2018). Digital shoreline analysis system (DSAS) version 5.0 user guide: U.S. Geological Survey Open-File Report 2018–1179; 2018, p. 110. https://doi.org/10.3133/ofr

Hinkel, J., Brown, S., Exner, L., et al. (2012). Sea-level rise impacts on Africa and the effects of mitigation and adaptation: An application of DIVA. *Regional Environmental Change, 12*, 207–224. https://doi.org/10.1007/s10113-011-0249-2

IPCC. (2019). *Climate change and land: an IPCC special report on climate change, desertification, land degradation, sustainable land management, food security, and greenhouse gas fluxes in terrestrial ecosystems* (P.R. Shukla, J. Skea, E. Calvo Buendia, V. Masson-Delmotte, H.-O. Pörtner, D. C. Roberts, P. Zhai, R. Slade, S. Connors, R. van Diemen, M. Ferrat, E. Haughey, S. Luz, S. Neogi, M. Pathak, J. Petzold, J. Portugal Pereira, P. Vyas, E. Huntley, K. Kissick, M. Belkacemi, J. Malley, (eds.)). In press.

IRC-Consultant, Saint Martin Paysage. (2004). Etude de l'environnement aux abords de Nouakchott. Programme de Développement Urbain de Nouakchott – PDU. Nouakchott. 189 p.

Lancaster, N., Kocurek, G., Singhvi, A., Pandey, V., Deynoux, M., Ghienne, J. F., & Lo, K. (2002). Late Pleistocene and Holocene dune activity and wind regimes in the western Sahara Desert of Mauritania. *Geology, 30*(11), 991–994.

Littaye, A., & Ould Ahmed, S. C. (2018). The dynamics of the coastal land scapes over the last decades: Wind drivers for change along the North Western Mauritanian coast. *Journal of Earth Science and Climatic Change, 9*, 450. https://doi.org/10.4172/2157-7617.1000450

Mainguet, M., Dumay, F., Ould Elhacen, M. L., Georges, J. C. (2008). Changement de l'état de surface des ergs au nord de Nouakchott (1954–2000). Conséquences sur la désertification et l'ensablement de la capitale. Géomorphologie : relief, processus, environnement, 2008, n° 3 : 143–152.

MPEM. (2004). Le littoral mauritanien, un patrimoine national, une ouverture sur le monde. Nouakchott (ISBN : 2-9514914-5-X).

Nguer, M., & Rognon, P. (1989). Homogénéité des caractères sédimentologiques des sables ogoliens entre Nouakchott (Mauritanie) et Mbour (Sénégal). *Géodynamique, 4*(2), 19–133.

Niang, A. J. (2014). La résilience aux changements climatiques : cas de la ville de Nou kchott. (Resilience against climate change : Case of Nouakchott City). *Geo-Eco-Trop, 38*, 155–168.

Nicholson, S. E., & Flohn, H. (1980). African environmental and climatic changes and the general atmospheric circulation in late Pleistocene and Holocene. *Climate Change, 2*, 313–348.

Oppenheimer, M., Glavovic, B. C., Hinkel, J., van de Wal, R., Magnan, A. K., Abd-Elgawad, A., Cai, R., Cifuentes-Jara, M., DeConto, R. M., Ghosh, T., Hay, J., Isla, F., Marzeion, B., Meyssignac, B., & Sebesvari, Z. (2019). Sea level rise and implications for low-lying islands, coasts and communities. In H. O. Pörtner, D. C. Roberts, V. Masson-Delmotte, P. Zhai, M. Tignor, E. Poloczanska, K. Mintenbeck, A. Alegría, M. Nicolai, A. Okem, J. Petzold, B. Rama, & N. M. Weyer (Eds.), *IPCC special report on the ocean and cryosphere in a changing climate*. In press.

Ould Mohameden, A. (1995). Aménagement et évolution du littoral, Apports de la télédétection et de la modélisation mathématique: cas du port de Nouakchott (Mauritanie). Thèse. Université de Nice, 149 p.

Ould Sidi Cheikh, M. A., Ozer, P., & Ozer, A. (2007). Risques d'inondations dans la ville de Nouakchott (Mauritanie). *Geo-Eco-Trop, 31*, 19–42.

Senhoury, A. E. (2000). Influence d'un ouvrage portuaire sur l'équilibre d'un littoral soumis à un fort transit sédimentaire, l'exemple du port de Nouakchott. Thèse de Doctorat. Université de Caen., 162p.

Senhoury, A., Niang, A., Diouf, B., Yve-Franços, T. (2016). Managing Flood Risks Using Nature-Based Solutions in Nouakchott, Mauritania. In *Book edit by Springer: ecosystem-based disaster risk reduction and adaptation in practice* (pp. 435–455).

Sidi Cheikh, M. A., Antonio, A. A. & Yelli, D. Y. (2009). Le Banc d'Arguin face aux effets des – changements climatiques. 8ème congrès Maghrébin des sciences de la Mer 7-9 octobre 2009.

Sunder, S., Ramsankaran, R., & Ramakrishnan, B. (2017). Inter-comparison of remote sensing sensing-based shoreline mapping techniques at different coastal stretches of India. *Environmental Monitoring and Assessment, 189*, 290. https://doi.org/10.1007/s10661-017-5996-1

Sy, I., Koita, I. Traoré, M., Keita, D. M., Lo, B.,Tanner, M. & Cissé, G. (2011). Vulnérabilité sanitaire et environnementale dans les quartiers défavorisés de Nouakchott (Mauritanie): analyse des conditions d'émergence et de développement de maladies en milieu urbain sahélien. Vertigo Volume 11.

Tian, H., Xu, K., Goes, J. I., Liu, Q., HDR, G., & Yang, M. (2020). Shoreline changes along the coast of Mainland China—Time to pause and reflect? *ISPRS International Journal of Geo-Information, 9*(10), 572. https://doi.org/10.3390/ijgi9100572

Tragaki, A., Gallousi, C., & Karymbalis, E. (2018). Coastal Hazard vulnerability assessment based on geomorphic, oceanographic and demographic parameters: The case of the Peloponnese (Southern Greece). *Land, 7*, 56. https://doi.org/10.3390/land7020056

Trebossen, H. (2002). Apport des images RADAR à Synthèse d'Ouverture à la cartographie marin*e*. Thèse à l'Université de Marne la Vallée. Spécialité : Sciences de l'Information Géographique. 161p.

Wu, C. (2003). Application de la géomatique au suivi de la dynamique environnementale en zones arides. Exemples de la région de Nouakchott en Mauritanie, du Ningxia nord et du Shaanxi nord en Chine du nord-ouest. Université de Paris 1-Pantheon-Sorbonne et de l'Ecole Pratique des Hautes Etudes, 217p.

Chapter 10
Flooding Hazard Assessment Considering Climate Change in the Coastal Areas of Algeria Based on a Remote Sensing and GIS Data Base

Barbara Theilen-Willige and Rachid Mansouri

Abstract Evaluations of satellite data, geophysic, bathymetric and meteorologic data as well as digital elevation data, help to identify critical coastal areas in N-Algeria exposed to flooding due to flash floods, storm surge, meteo-tsunamis or tsunami waves. Data mining is the prerequisite for flooding hazard preparedness. It is aiming at visualizing critical areas and providing information about damage in case of emergency due to flooding hazards as fast as possible, as the civil protection units need this information for their management. The actual inventory of land use and infrastructure (bridges, railroads, roads, river embankments, etc.), industrial facilities and the structure of settlements and cities is an important issue for the hazard assessment and damage loss estimation.

Sea level rise due to climate change has to be considered when dealing with the detection of areas susceptible to flooding. Evaluations of optical satellite data such as Landsat 8 and Sentinel 2 and radar data (Sentinel 1, ALOS PALSAR) contribute to a better understanding of the development of currents at the coast and their interactions with the coastal morphology. A systematic inventory of coast segments showing traces of flooding in the past is presented as well as the monitoring of coast-near lowlands, lakes and ponds. Digital elevation model (DEM) including bathymetric data help to identify areas prone to flooding.

Keywords Remote sensing · GIS · Flooding hazard · Tsunamis · Sturm surge · Flash floods · N-Algeria

B. Theilen-Willige (✉)
TU Berlin, Berlin, Germany
e-mail: barbara.theilen-willige@campus.tu-berlin.de

R. Mansouri
Department of Civil Engineering and Hydraulics, Laboratory LGCH, 8 Mai 1945 Guelma University, Guelma, Algeria
e-mail: mansouri.rachid@univ-guelma.dz

M. M. Al Saud (ed.), *Applications of Space Techniques on the Natural Hazards in the MENA Region*, https://doi.org/10.1007/978-3-030-88874-9_10

10.1 Introduction

The Algerian coast has been prone to hazardous tsunamis as Northern Algeria is located at the boundary between the African and European plates. The convergence rate between those tectonic plates is in the range of order of 4–7 mm per year (Amir & Theilen-Willige, 2017). The compressive motion results in the occurrence of tsunamogenic earthquakes in Western Mediterranean, in particular, the Alboran Sea separating the Ibero area in Spain and Maghreb in Morocco and Western part of Algeria (Fig. 10.1). On October 1790, a devastating earthquake hit Oran city in North Western of Algeria (Io = IX – X, MSK Scale). Historical spanish documents reports that two thousand people died in North Africa (Lopez Marinas & Salord, 1990). The seismic crisis consisted of a series of foreshocks reported during September and beginning of October 1790. The main shock occurred at 01:15 AM on the ninth of October and aftershocks are reported until February 1790. The earthquake was felt as far as 200 km from Oran to Almeria and Carthagena in southern Spain. A tsunami was generated just a few minutes after the main shock. In the Algerian coast, the sea withdrawal was about 200 m. In Spain, the sea penetrated inland by 50 m nearby Almeria. In the harbour of Carthagena the sea rose by 1.8 m (Lopez Marinas & Salord, 1990). Tsunami water waves in the Oran, Arzew and

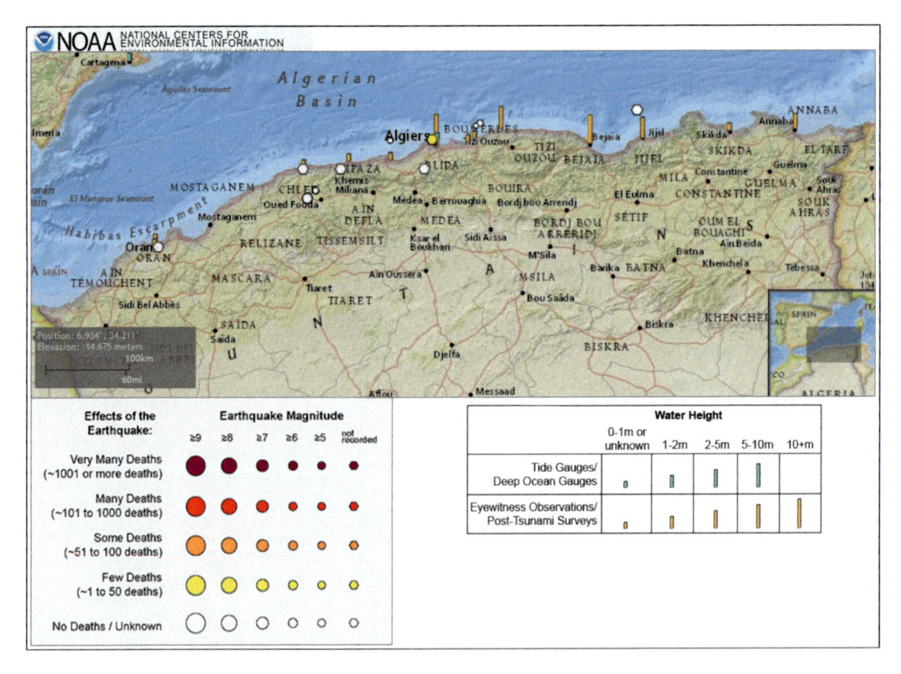

Fig. 10.1 Tsunami events along the Algerian coast
NOAA Web Map https://maps.ngdc.noaa.gov/viewers/hazards/?layers=0

Almeria Bay were trapped along the coasts for a long time. The tsunami propagation in the Alboran Sea lasted about an hour (Amir et al., 2012; Amir, 2014).

The Djidjelli Earthquake of August 21 and 22, 1856 triggered sea waves of 2–3 m high (observed at Djidjelli) that flooded the eastern Algerian coast several times. At Bougie (Bejaia) and Philippeville (Skikda), small towns located eighty kilometers west and east of Djidjelli, Ambraseys & Vogt (1988) and Soloviev et al. (2000) reported that the sea rose from about 5 m, flooding the shore five to six times (Yelles-Chaouche et al., 2009).

During the instrumental period, two tsunamigenic events are evidenced as they were the first recorded by geophysical instruments. The first one occurred after the destructive El Asnam event of October 10, 1980 (Ms: 7.3). Although located at a distance of about 60 km from the coast, the earthquake triggered a submarine landslide inducing a weak tsunami recorded by several tide gauge stations of southeastern Spain (Papadopoulos & Fokaefs, 2005). The second tsunami, the more recent one, is the tsunami of Boumerdes of 21.05.2003. This event, one of the most important in the western Mediterranean region within the last century, was generated by an earthquake of magnitude Mw 6.8 that occurred on the offshore reverse fault of Zemmouri (Yelles et al., 2004; Alasset et al., 2006). This thrust fault, with a length of about 50–55 km, is assumed to outcrop near the seafloor at about 10–15 km from the shoreline (Deverchere et al., 2005). Effects of this tsunami were felt in the entire westernnMediterranean region and especially along the Balearic coasts (Alasset et al., 2006). The Boumerdes tsunami demonstrated the high potential of the Algerian margin for tsunami generation (Yelles-Chaouche et al., 2009), see Fig. 10.1.

The main cause of tsunami generation in the Mediterranean Sea is tectonic activity associated with strong earthquakes. However, tsunami waves are also generated by landslides. Most of them were caused by subaerial landslides or marine slides induced mainly by earthquakes and less frequently by volcanic eruptions. For the vast majority of volcanic events the actual generation mechanism of destructive water waves is a volcanic slope failure (Gusiakov, 2020). Several types of landslides cause tsunamis with significant heights in near-source coasts that attenuate rapidly due to frequency dispersion (Papadopoulos et al., 2007). Due to the concentrated large wave heights, such tsunamis may result in catastrophic consequences. The magnitude 6.8 earthquake (21.05.2003) located near Boumerdes (central Algerian coast) triggered large turbidity currents responsible for 29 submarine cable breaks at the foot of the continental slope over 150 km from west to east (Cattaneo et al., 2012).

When dealing with the flooding hazard potential along the Algerian coasts it is necessary to consider meteotsunamis as well, see as an example Fig. 10.2. Meteotsunamis caused by extreme weather events and storm surge can affect coastel areas with increasing intensity due to the effects of climate change. Meteotsunamis are much less energetic than seismic tsunamis and that is why they are occurring local, similar to landslide tsunamis. Destructive meteotsunamis are always the result of a

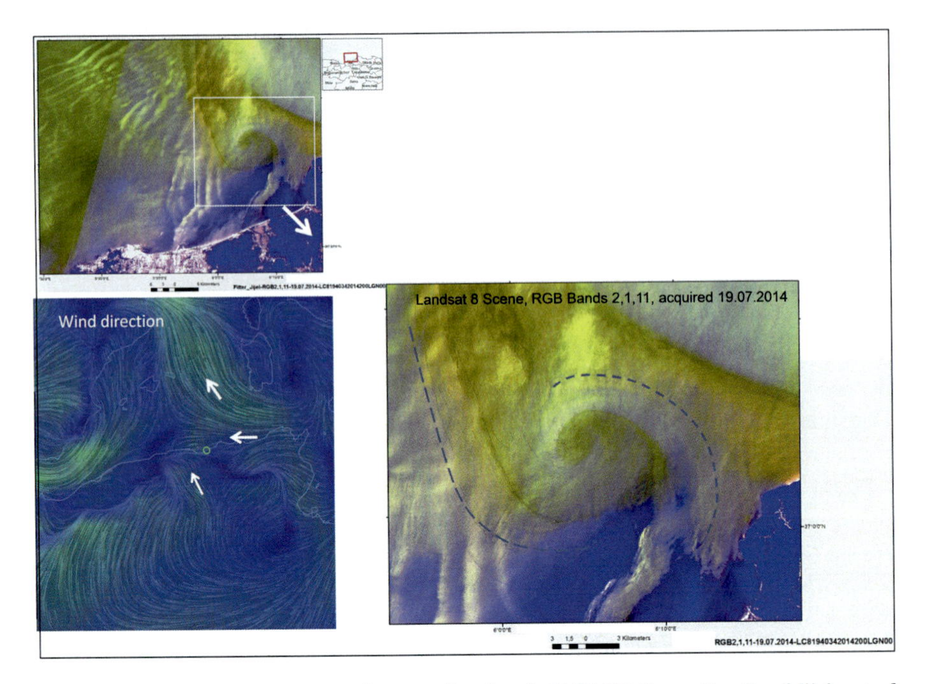

Fig. 10.2 Surface water current development (Landsat 8, 19.07.2014) near the city of Jijel east of Algiers
There were no clouds, fog or mist at the acquisition time. The wave pattern visible on the image is related to the water surface current pattern at the acquisition time (correlated with data from https://earth.nullschool.net)

combination of several resonant factors (atmospheric gravity waves, pressure jumps, frontal passages, squalls, etc.). The low probability of such a combination is the main reason why major meteotsunamis are infrequent and observed only at some specific locations (Monserrat et al., 2006).

As extreme weather events will increase due to climate change the occurrence of extreme sea level oscillations associated with meteotsunamis and storm surge will be an important issue at the Algerian coast.

The sea level rise related to climate change has to be taken into account as well when dealing with the detection of areas prone to potential tsunami flooding. Climate models project a Global mean sea level (GMSL) rise during the twenty-first century that will likely be in the range of 0.29–0.59 m for a low emissions scenario and 0.61–1.10 m for a high one. GMSL projections that include the possibility of faster disintegration of the polar ice sheets predict a rise of up to 2.4 m in 2100 and up to 15 m in 2300 (European Environment Agency, 2021).

How this will affect the tidal range has still to be investigated.

These sea level rise trends have to be considered when dealing with flooding hazard preparedness in coastal areas This means that tsunami flooding modeling has to consider and to include these sea level changes when calculating the potential indundation extent.

10.1.1 The Role of Remote Sensing and GIS

When catastrophic earthquake and tsunami hazards happen and affect cities, settlements and infrastructure, immediate and efficient actions are required which ensure the minimization of the damage and loss of human lives. Proper mitigation of damages following disastrous events highly depends on the available information and the quick and proper assessment of the situation. Responding local and national authorities should be provided in advance with information and maps where the highest damages due to unfavourable, local site conditions in case of stronger flooding events can be assumed. The better a pre-existing reference database of an area at risk is prepared and elaborated, the better a crisis-management can react in case of hazards and related secondary effects. The potential of social and economic losses due to those events is increasing. Therefore, information of geodynamic processes is a basic need for the long-term safety of cities, settlements, infrastructure and industrial facilities. The assessment of potential hazard prone areas is fundamental for planning purposes and risk preparedness, especially with regard to supervision and maintenance of settlements, infrastructure, industrial facilities and of extended lifelines (Amir & Theilen-Willige, 2017).

The ability to undertake the assessment, monitoring and modeling can be improved to a considerable extent through the current advances in remote sensing and GIS technology. Causal or critical environmental factors influencing the disposition to be affected by flooding hazards and the potential damage intensity can be analysed interactively in a GIS database. The interactions and dependencies between different causal and preparatory factors can be visualized and weighted step by step in this GIS environment.

The detection of coastal areas prone to flooding due to torrential rains causing flash floods can be carried out with the support of remote sensing and GIS tools. Flooding can be one of the manageable of natural disasters, if flood prone areas are identified and suitable flood mitigation strategies are implemented. A practical way of identifying flood prone areas is offered by the application of mathematical models, which consider complex hydrological and hydraulic processes of these areas.

Coastal area and river basin management poses big challenges, especially in developing countries, because of lack of continuous data, particularly streamflow data, that require automatic recording instruments to acquire. Remote sensing technology can augment the conventional methods to a great extent in rainfall-runoff studies. The role of remote sensing in runoff calculation is generally to provide a source of input data or it is used as an aid for estimating equation coefficients and model parameters. For example, one of the applications of remote sensing data in hydrologic models is focused on the determination of both, urban and rural land use for estimating runoff coefficients.

10.1.2 Goals of This Study

The aim of this research is focused on a contribution to a coastal hazard geoinformation system with interdisciplinary dynamic content, enabling the communication between local authorities, public organizations and universities. It combines and uses actual and previous efforts by adding new researches, focusing on remote sensing, and new techniques in Geographic Information Systems (GIS). This work could be part of a basic risk assessment presenting information for strategic planning on where potential problems may occur in the infrastructure in the coastal areas of N- Algeria.

This study aims to contribute to the:

- mapping of traces of earlier flooding events on the different satellite images
- influence of different wind directions on current development in the coastal ares and the interaction of currents with the coastal morphology
- delineation of areas prone to flooding due to their morphometric disposition
- evaluation of bathymetric data to analyse the influence of the sea bottom topography on tsunami wave development, especially in the coast-near area
- contribution to a disaster managment system.

10.2 Materials and Methods

The interdisciplinary approach used in the scope of this research comprises remote sensing data, geological, geophysical and topographic data and GIS methods. Earthquake and tsunami data were provided by the European-Mediterranean Seismological Centre (EMSC), International Seismological Centre (ISC), US Geological Survey (USGS) and National Oceanic and Atmospheric Administration (NOAA). Satellite imageries and Digital Elevation Model (DEM) data were used for generating a GIS data base and combined with different geodata and other thematic maps. Satellite data such as Sentinel 1 – C-Band, Synthetic Aperture Radar (SAR) and optical Sentinel 2 images, Advanced Spaceborne Thermal Emission and Reflection Radiometer (ASTER) and Landsat optical data (Landsat TM and Landsat 8, the Operational Land Imager -OLI). Digital Elevation Model (DEM) data gained from the Shuttle Radar Topography Mission (SRTM), ASTER DEM data and Advanced Land Observing Satellite-1 (ALOS), Phased Array type L-band Synthetic Aperture Radar (PALSAR) mission/Japan Aerospace Exploration Agency (JAXA), were downloaded from open-sources such as USGS/Earth Explorer, Sentinel Hub/ESA, Alaska Satellite Facility (ASF) and Google Earth. The data were processed using geoinformation systems ArcGIS from ESRI and QGIS. Shapefiles from Algeria were downloaded from the Geofabrik's download server.

Meteorologic data such as about wind directions and wind speeds were downloaded from the 'power regional data access tool' of the 'POWER Data Access Viewer' provided by NASA as well as from earth.nullschool.net (Beccario, 2020).

10.2.1 *Digital Image processing of Different Optical and Radar Satellite Data*

ENVI software from Harris Geospatial Solutions and the Sentinel Application Platform (SNAP) provided by ESA were used for the digital image processing of the optical Landsat 5 and 8, Sentinel 2 and ASTER data. SNAP provided as well the tools for the processing of radar data. The evaluation of Sentinel 1 A and B radar images requires geometric correction and calibration.

The different steps of digital image processing used in this research are described in the following text and Fig. 10.3. Digital image processing of LANDSAT 5 Thematic Mapper and Landsat 8 – The Operational Land Imager (OLI) data was carried out by merging different Red Green Blue (RGB) band combinations with the panchromatic Band 8 to pan-sharpen the images. The Red, Green, Blue (RGB)-Principle is reviewed briefly: Three images from different optical satellite bands to be used as end-members in a triplet are projected, one image through one primary color each, one image is coded in blue, the second in green and the third in red.

For the detection of water currents along the Algerian coast various image-processing procedures were tested and the results combined with available meteorological (wind, temperature, etc.) information. Evaluations of LANDSAT, ASTER and Sentinel 2 satellite imageries acquired in time series support a better knowledge of water current dynamics, influenced among other factors (wind direction and speed, upwelling etc.) by the coastal morphology.

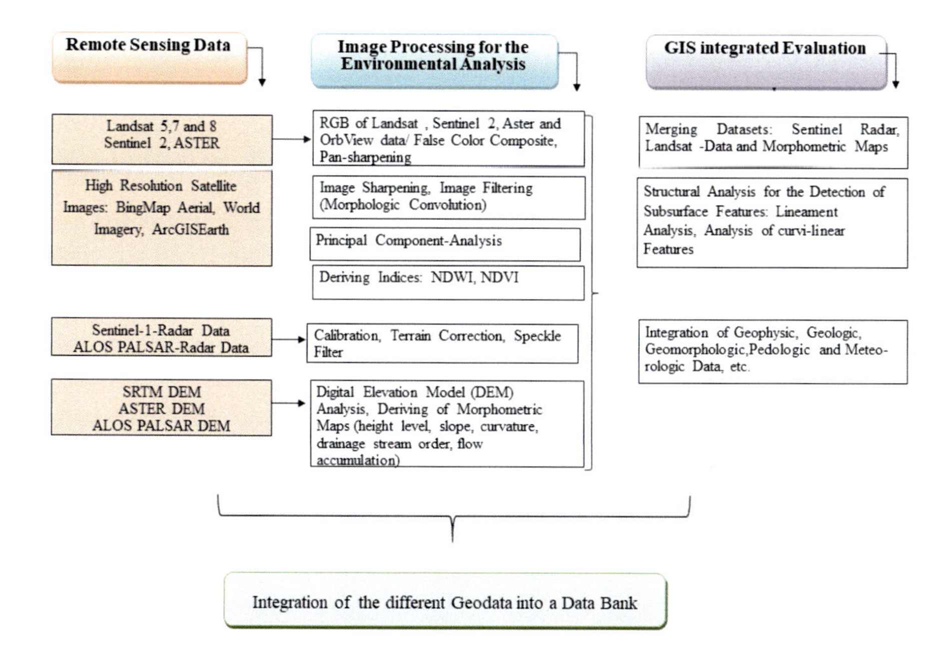

Fig. 10.3 Workflow of the digital data processing

A systematic correlation of the streaming pattern visible on satellite data with wind data, especially wind directions, will contribute to a better understanding of which coastal segments might be prone more to flooding in case of high energetic flood waves from specific directions.

In the scope of this study, the comparative analysis of optical satellite data and the Sentinel 1 radar data was carried out, in order to derive wave pattern and current information. The evaluation of Sentinel 1 A and B radar images requires geometric correction and calibration. The processing of the Sentinel 1 radar data was carried out using the SNAP software of ESA. The illumination geometry of the radar signals in relation to the wind direction plays an important role for the visibility of the water surface properties. Therefore, radar data from descending and ascending orbits with different illumination geometries were used.

10.2.2 Evaluation of Digital Elevation Model (DEM) Data

Digital elevation data help to identify and categorize the different geomorphologic units, their size and arrangement. SRTM, ASTER GDEM (30 m spatial resolution) and Advanced Land Observing Satellite (ALOS), Phased Array type L-band Synthetic Aperture Radar (PALSAR) Digital Elevation Model (12.5 m resolution) were obtained from open sources such as USGS, EarthExplorer and Alaska Satellite Facility (ASF) used with GIS to evaluate terrain features. Terrain features can be described and categorized into simple topographic relief elements or units by parameterizing DEMs such as height levels, slope gradients, and terrain curvature. From DEM (Digital Elevation Model) data derived morphometric maps (slope gradient maps, drainage, etc.) were combined with geologic information in a GIS data base. In the scope of this study DEM data were mainly used to derive information of the lowest and flattest areas prone to flooding in the coastal reas of Algeria

Additionally, the bathymetric DEM data provided by General Bathymetric Chart of the Oceans (GEBCO), International Hydrographic Organization and the Intergovernmental Oceanographic Commission of UNESCO) and EMODnet Bathymetry Consortium (2020) were integrated into the research in order to combine the information of the sea bottom topography with earthquake, tsunami and submarine landslide data (Fig. 10.4).

The integration of different morphometric factors in a GIS environment using weighting procedures plays an important role in the GIS application in the frame of this study. The basic pre-requisite for the use of weighting tools of GIS is the determination of weights and rating values representing the relative importance of factors and their categories. The weighted overlay method takes into consideration the relative importance of the parameters and the classes belonging to each parameter (ESRI, online support in ArcGIS). The application of a weight-linear-

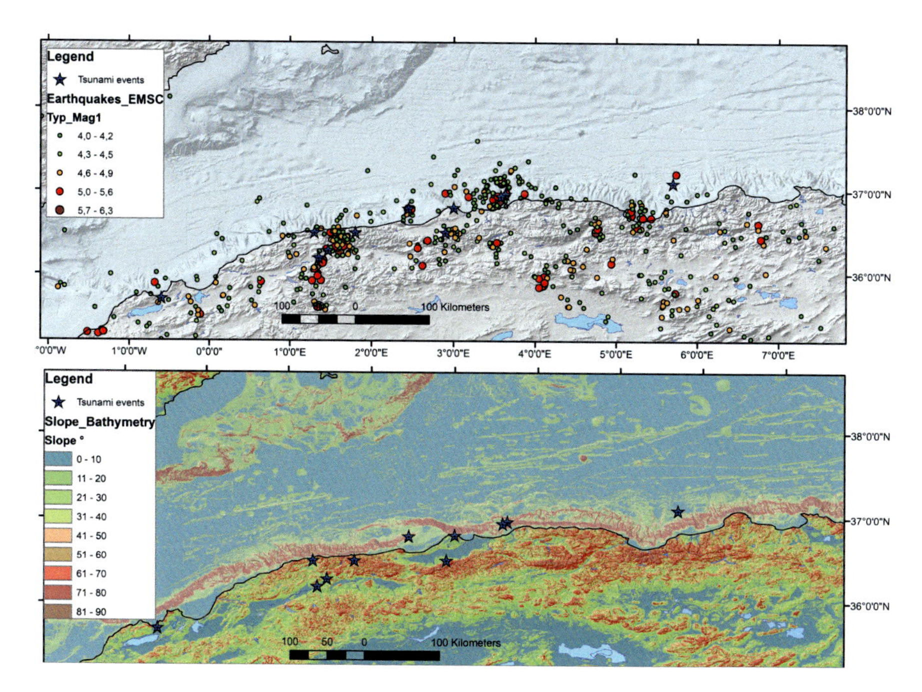

Fig. 10.4 Earthquakes and tsunami events (data sources, EMSC, USGS, ISC, NOAA) Bathymetric data: GEBCO and EMODnet Bathymetry Consortium

combination in susceptibility assessment has been identified as a semi-quantitative method, involving both expert evaluation and the idea of ranking and weighting factors. The efficacy of the weighted overlay-method lies in the fact that human judgments can be incorporated in the analysis. The weights and ratings are determined using the expert's subjective knowledge. The method starts by assigning an arbitrary weight to the most important criterion (highest percentage), as well as to the least important attribute according to the relative importance of parameters (Fig. 10.5).

The sum over all the causal factors/layers that can be included into GIS provides some information about the susceptibility to flooding and the extent of inundation. This susceptibility is calculated by adding every layer, as described below, to a weighted influence and summing all layers. After weighing (in %) the factors according to their probable influence, susceptibility maps can be elaborated, where those areas are considered as being more susceptible to flooding, where "negative" causal factors occur aggregated and are interfering with each other (Fig. 10.5).

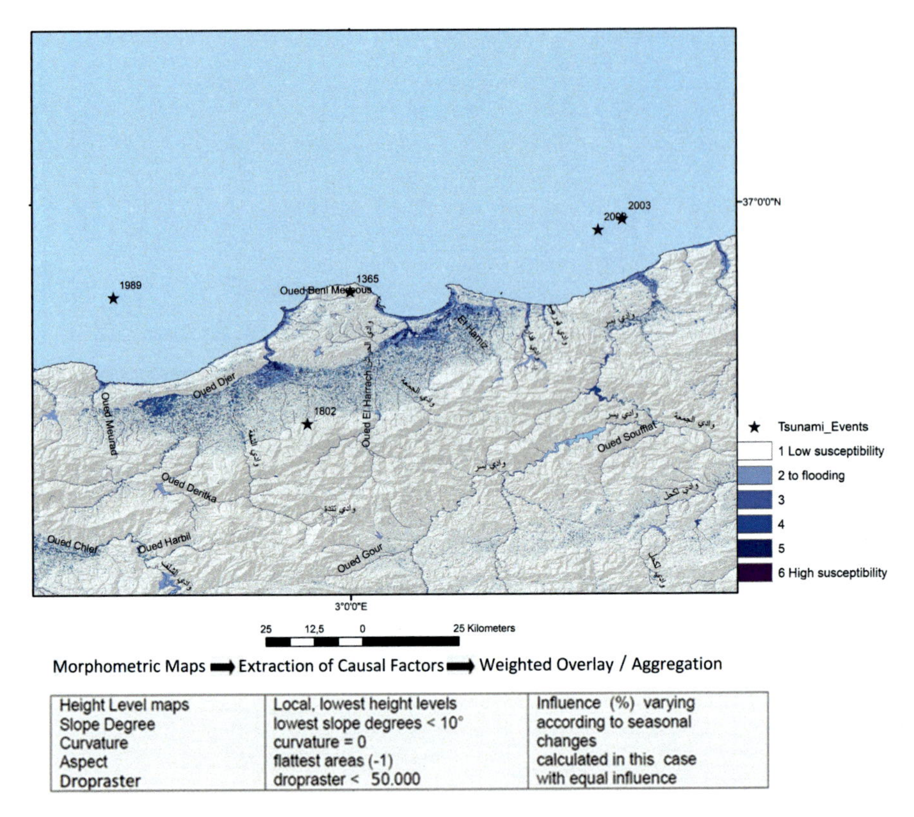

Height Level maps Slope Degree Curvature Aspect Dropraster	Local, lowest height levels lowest slope degrees < 10° curvature = 0 flattest areas (-1) dropraster < 50.000	Influence (%) varying according to seasonal changes calculated in this case with equal influence

Fig. 10.5 Weighted overlay workflow in ArcGIS to derive information about areas susceptible to flooding due to their morphometric disposition
By extracting first morphometric factors like slope degrees below 10° from the slope gradient map, by deriving from the drop raster the areas with values below 50.000 units using the hydrology-tools in ArcGIS (flow direction), by extracting from curvature maps areas with terrain curvature = 0 and from the aspect map the flat areas (−1), the flattest areas are highlighted. In the weighted overlay procedure these selected morphometric data were summarized, merged with different percentages of influence and represented in a map. The flattest areas with lowest slope degrees and curvatures are classified by values from 0 to 7, thus, enhancing and accentuating the visibility of of areas ptone to flooding

10.3 Results

The direction and angle of incoming, high energetic flood waves will have a significant influence on the water current development, on the interaction with the submarine topography, on the formation of the coastal morphology and on wave dynamics (amplification, interfering). Thus, the wave angles might have an impact on the potential flooding extent in the affected coastal areas. River mouths and their oriention forming an entrance for potential incoming tsunami waves have to be monitored.

The different satellite data and their combined evaluations offer many of the needed information.

10.3.1 Evaluations of Optical Satellite Data

Optical satellite data of the Algerian coast reveal information of coastal currents (forming coherent water masses in motion) that are found in the region between the coastline and the edge of the continental shelf (Gelfenbaum, 2005).

The exposition to flooding of the coastal segments depends as well on their oriention to the main wave direction and energy.

The next figure shows as an example the water current situation near Jijel on 20.02.2020 (Fig. 10.6). The main wind and water current flow direction was directed towards E-NE (US National Weather Service). Smaller longshore currents can be observed parallel along the arc-shaped coastline. The effects of bottom friction on coastal trapped waves become visible. Every tongue of land or peninsula is causing further currents, depending on the direction of the main wave front.

The sea surface streaming pattern as visible on the Landsat-scenes (only the upper centimeters of the water surface) is mainly influenced by the wind situation at the

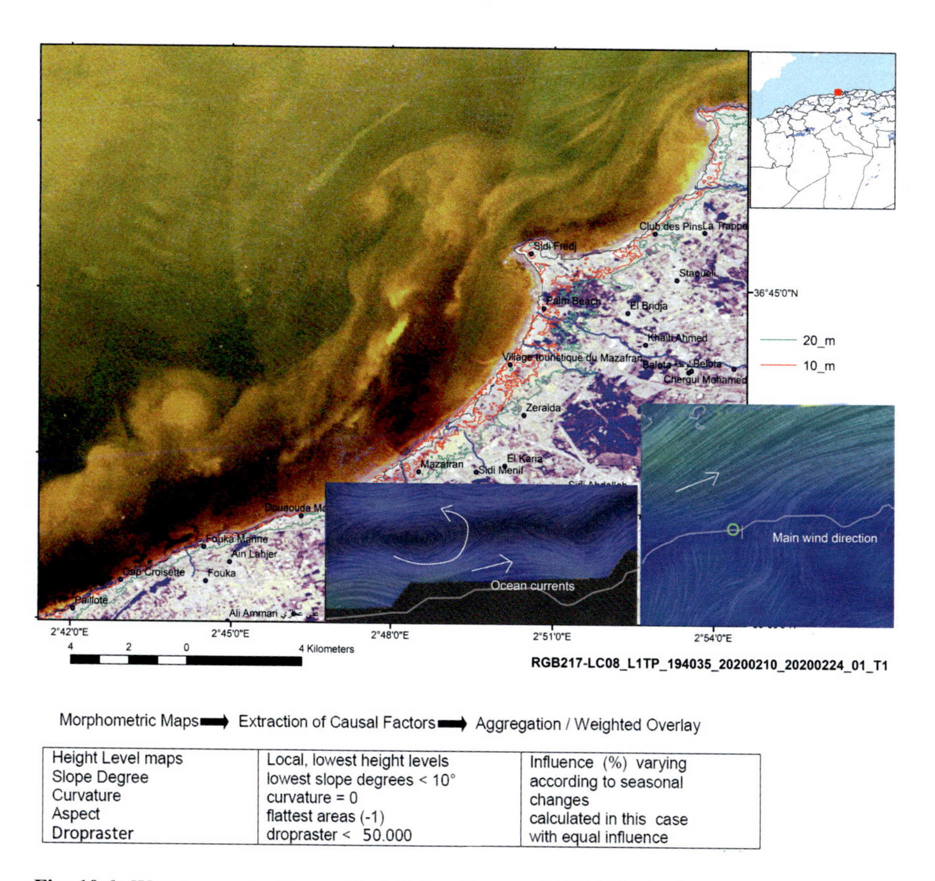

Height Level maps Slope Degree Curvature Aspect Dropraster	Local, lowest height levels lowest slope degrees < 10° curvature = 0 flattest areas (-1) dropraster < 50.000	Influence (%) varying according to seasonal changes calculated in this case with equal influence

Fig. 10.6 Water current pattern and wind directions on 10.02.2020 in the Bay of Jijel showing coastal segments exposed to stronger current activity during westwinds (wind and ocean current data from earth: a global map of wind, weather, and ocean conditions, https://earth.nullschool.net/)

acquisition time (wind direction, wind speed), further on by the tidal situation and the input of river water.

Upwelling and downwelling occurs along coastlines, depending on the main wind directions. However, these up-and downwelling processes cannot be monitored directly by remote sensing due to the low penetration depth of the sensors only up to several centimeters. Thermal bands of Landsat 8 and ASTER help to identify current patterns that might be related to these currents.

Satellite images can be used to derive information on water currents and coastal sediment flow dynamics after earthquakes and tsunami events: As the magnitude Boumerde- 6.8 earthquake (21.05.2003) struck at 17:44 local time (USGS, 2003), submarine mass movements and higher sediment discharge of the rivers into the sea had an influence on the water dynamics (Fig. 10.7).

Tsunami-induced morphological changes on coasts prone to inundation and sediment transport changes are controlled by the specific tsunami event and sediment characteristics,

The evaluations of Landsat data reveal that especial semicircular bays are affected by circular currents such as the Bay of Oran, Arzew, Algiers or Bejaia. In case of a stronger tsunami event circular currents could occur in these semicircular bays and, thus, amplify the intensities by interfering and superimposing of wave energy.

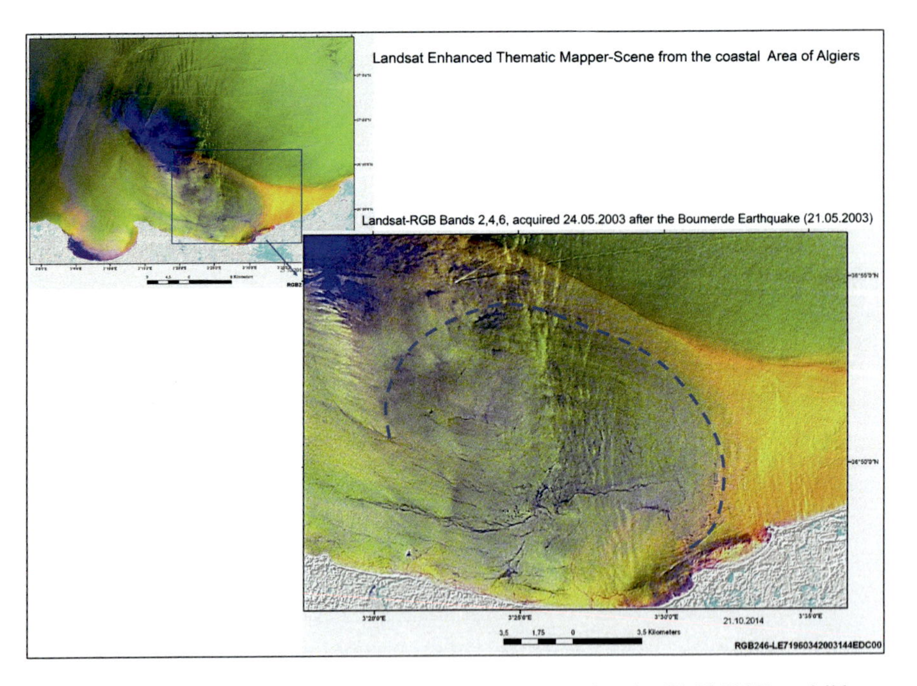

Fig. 10.7 Sediment input (dark-yellow) after the Boumerdes earthquake (21.05.2003) as visible on a Landsat satellite scene
The magnitude 6.8 earthquake struck at 17:44 local time (USGS, 2003), wreaking extensive damage throughout five provinces. Submarine mass movements and higher sediment discharge of the rivers into the sea had an influence on the water dynamics

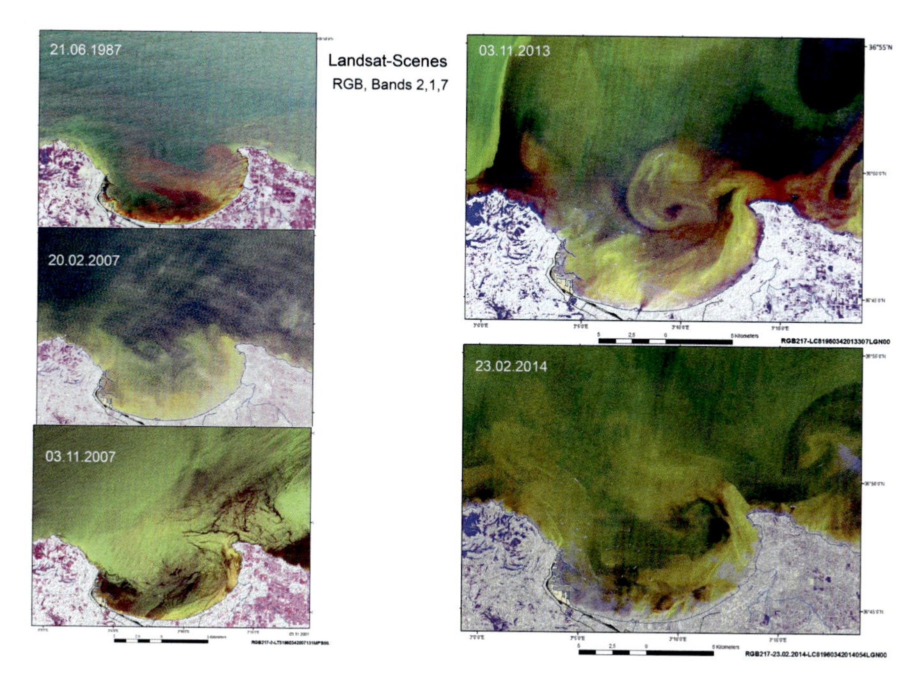

Fig. 10.8 Development of water currents depending on the specific meteorologic conditions and tide situations at the acquisition time of the Landsat data from the Bay of Algiers

The development of the semicircular shaped Bay of Algiers could be explained by such circular streaming dynamics with the consequence of erosional effects as well. The sea surface streaming pattern as visible on the Landsat- and RapidEye-scenes of the Bay of Algiers (visualizing the situation only of the upper centimeters) is mainly influenced by the wind situation at the acquisition time, further on by the tidal situation and the input of river water. This is revealed very clearly by a Landsat time series (Fig. 10.8).

In case of stronger tsunami events with several tsunami wave fronts circular currents might occur in the Bay of Algiers as well and, thus, amplify the intensities by interfering and superimposing of incoming waves. The direction and angle of incoming, high energetic flood waves will have a great influence on the currents and dynamics.

In the next figures different wind situations are presented combined with the current situation at the same time. Figure 10.9 shows the west-wind, Fig. 10.10 the eastwind and Figs. 10.11a and 10.11b the north-wind situation.

The current pattern according to the specific wind directions and wind properties and the interactions with the coastal morphology help to understand which coastal segments might be more affected in case of high energetic waves directed towards the coast. In case of north-winds larger currents could be observed in the semicircle bays along the Algerian coast such as in the Bay of Algiers (Fig. 10.11b).

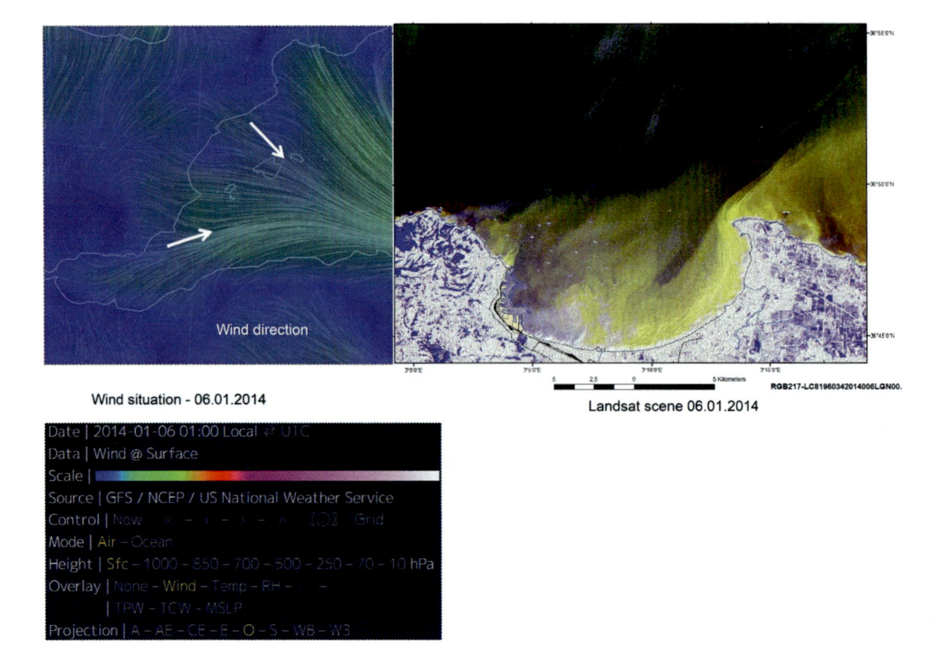

Fig. 10.9 Landsat scene (RGB, Bands 2,1,7, 06.01.2014) showing a west-wind situation
In case of high energetic waves coming from western direction the coast segments oriented more towards these waves would be prone to a higher flooding exposure. Wind data: http://earth.nullschool.net/#2014/02/23/1800Z/wind/surface/level/orthographic=-355.92,38.50,3000

10.3.2 Evaluations of Radar Data

Sentinel 1 synthetic aperture radar (SAR) has been used to detect wave information in a large coverage of the coastal areas of Algeria. Wave parameters such as significant wave height and mean wave period, can be usually obtained from SAR-derived wave spectra (Pleskachevsky et al., 2019; Shao et al., 2016). The evaluation of wave spectra needs a good understanding of complicated SAR wave imaging mechanisms. The C-band VV- and HV polarization Sentinel-1 SAR images are an additional tool whenever cloud cover is a hindrance, provided that the wind properties allow the detection of currents, see Fig. 10.12.

The differences in brightness between pixels in the radar image, marked by changes in the gray scale and backscatter intensity due to surface roughness changes contribute to the detection of wave properties. Dark-blue image tones of the colour-coded radar scene (Fig. 10.12) are associated with low waves because the incident radar signals were largely reflected from their "radar-smooth" surfaces in a mirror-like fashion away from the satellite antenna. High wave areas appear in lighter tones (here in blue-green) as their more radar-rough surfaces generate a diffuse and stronger signal return/radar backscatter. Coast near currents become clearly visible.

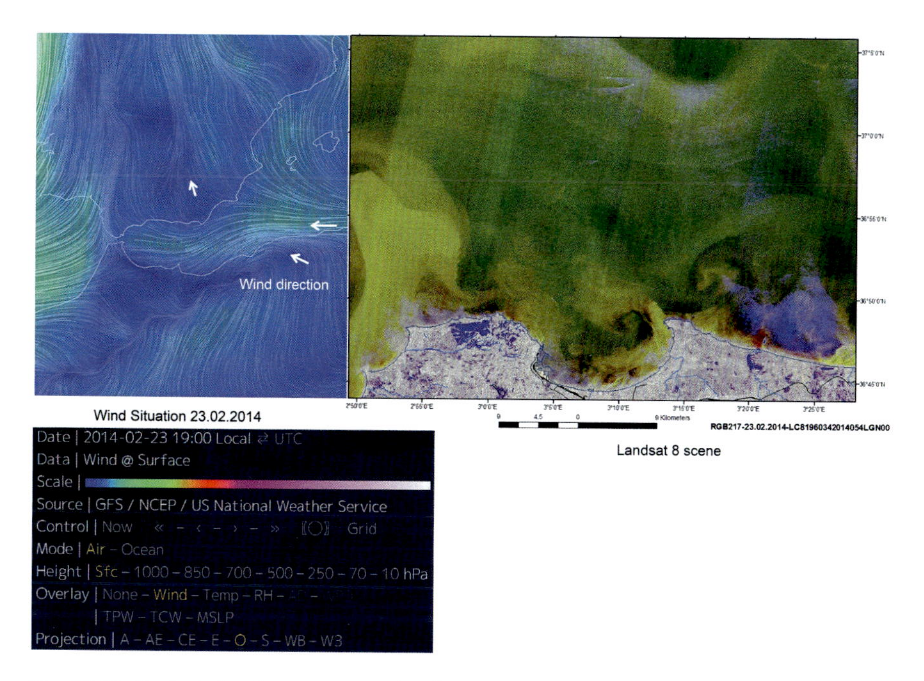

Fig. 10.10 East-wind and water current situation at the coast of Algiers on 23.02.2014
Wind direction data: http://earth.nullschool.net/#2014/02/23/1800Z/wind/surface/level/orthographic=5.07,36.54,3000

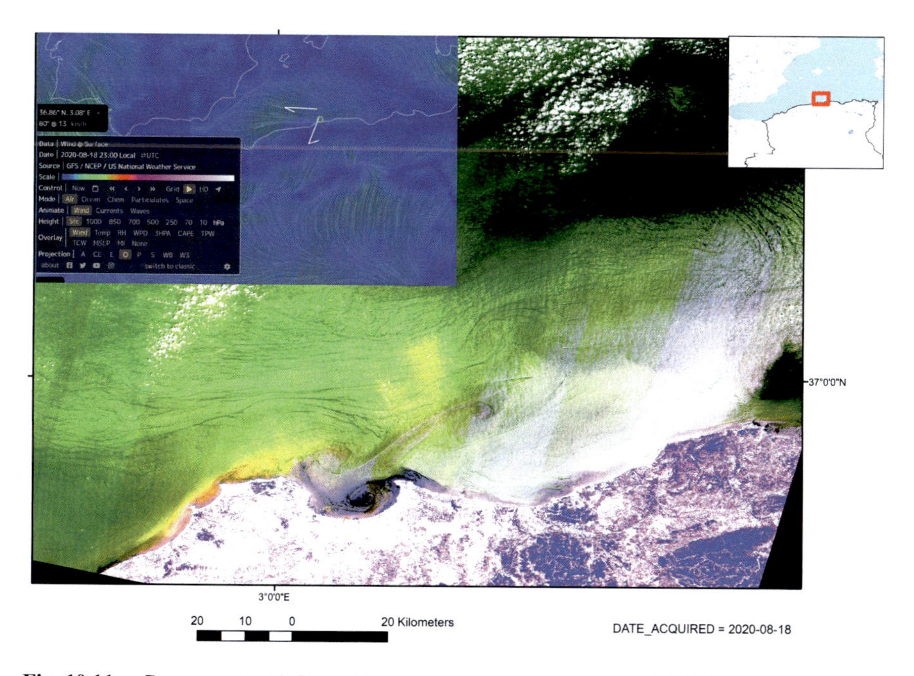

Fig. 10.11a Current pattern influenced by east to northeast-winds on 18.08.2020

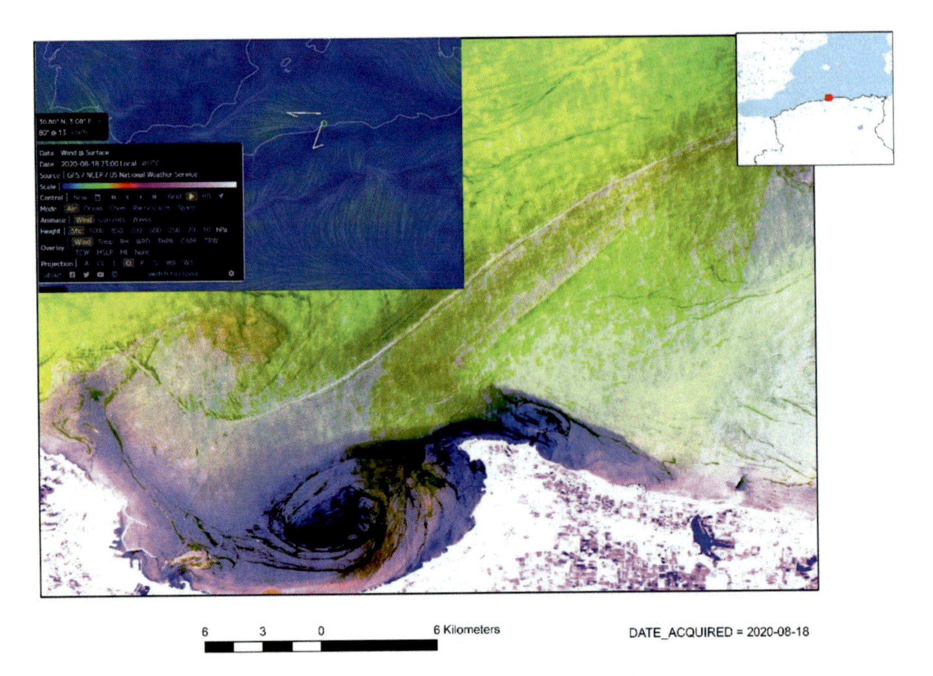

Fig. 10.11b Amplification of Fig. 10.11a: Northern wind direction causing circular coastal currents in the semicircular Bay of Algiers

10.3.3 Evaluations of DEM Data in Combination with Other Satellite Data

A flooding susceptibility map and a potential tsunami hazard map of coastal areas, that predicts the probable locations of possible future inundation and tsunami occurrences, is required which takes into consideration as well the potential morphodynamic consequences of these events at the coasts such as, abrasion, sedimentation and landslides. Undercutting the slopes at the coasts by abrasion and erosion those flood waves initiate a high potential of slope failure.

DEM data help to identify those areas that are most likely to be flooded in case of extreme weather constellations by flash floods or in case of tsunami events which coastal areas might be affected due to their morphometric disposition.

The amount of the sea level rise is still in discussion and in research. In a worst case scenario it will comprise about several meters as mentioned before. Therefore, flooding susceptibility maps of coastal areas should include the future sea level rise. Based on the current free available DEM data the areas below 10 m height level are delineated to show which areas might be prone in future to flooding, wether by sea level rise, sturm surge or by flash floods.

As river mouths are forming an entrance for intruding water waves from the sea a systematic assessment of larger river mouths prone to flooding will contribute to hazard preparedness.

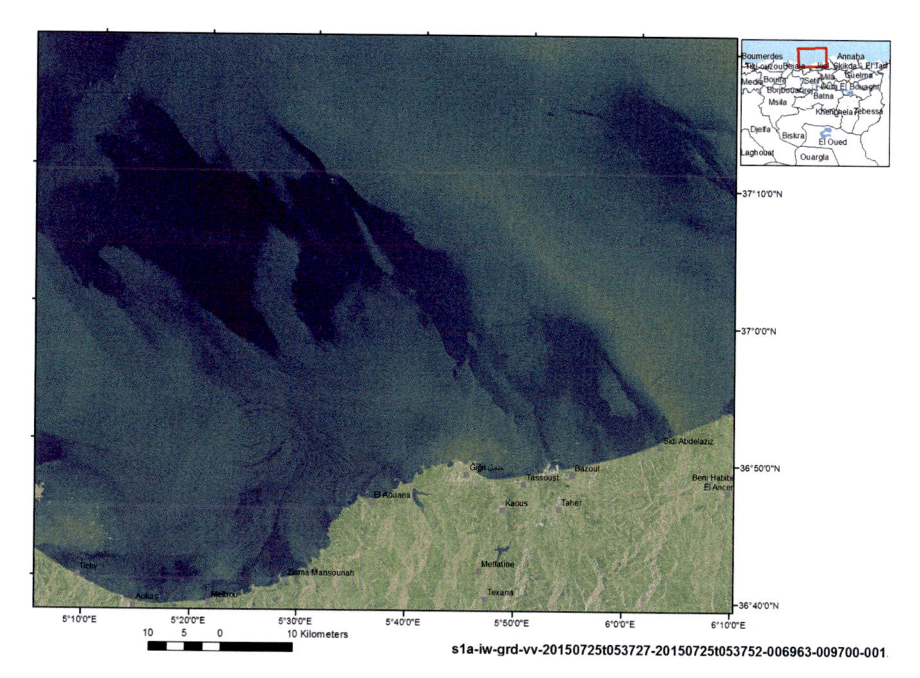

Fig. 10.12 Sentinel-1- Satellite Radar Scene
The dark-blue areas correspond to sea-surfaces with radar-smooth backscatter (mirror-like reflection) related to lower wave heights, the light blue-green areas are related to stronger radar reflection due to higher waves

Further on, the inventory of traces of former sea water intrusion is another important step towards the delineation of coastal segments prone to flooding. Figure 10.13 presents the area of Mersat El Hadjadi (Theilen-Willige, 2006). On the SRTM based morphometric maps and the corresponding Landsat scene traces of former flooding are evident. In case of storm surge or tsunami events it is important to know where the backwater areas are situated which might remain longer flooded after sea waves intrusions and return flow.

Tsunami inundation and damage was not uniform along the Algerian coast in the past. Evaluations of satellite data reveal that coast segments with SSW-NNE to SW-NE orientation have been affected in the past by strong, high energetic sea waves more than W-E oriented coast segments. Most of the latter are steep, cliff coast. The SW-NE orientated sections can be find often within the eastern part of larger semi-circular bay segments.

The affected coast segments that could be identified are shown in Fig. 10.14. Some more detailed views are presented in the Figs.10.15 and 10.16. Due to the coverage by settlements and cities in some sections traces of sea water intrusions cannot be observed on satellite images.

Wave traces of parallel, arrow-shaped debris walls, partly modified by eolian overprint, like in the area of Qued Lekbir in the east of the city of Skikda, are visible

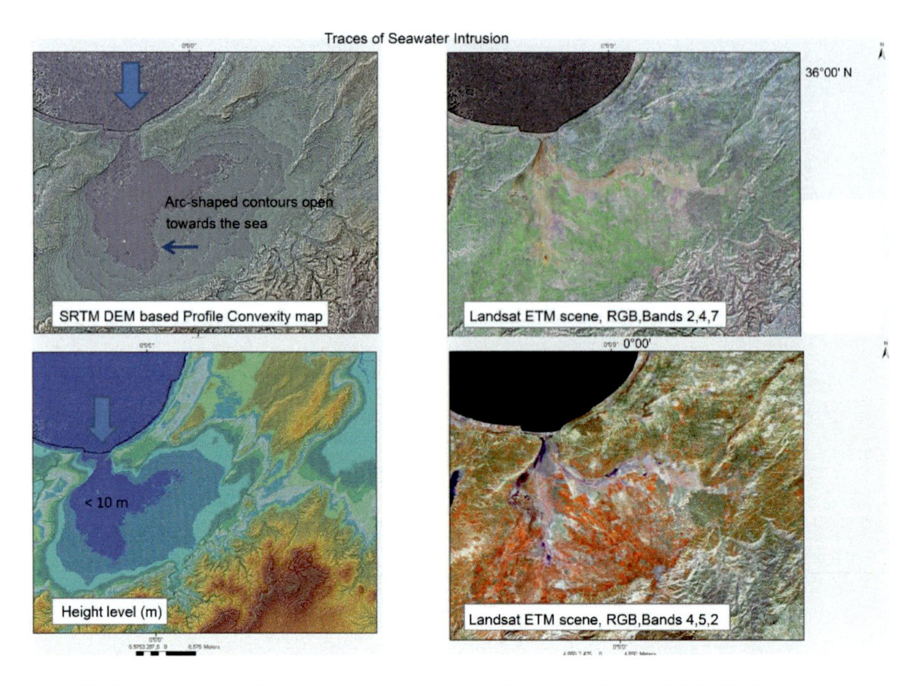

Fig. 10.13 River mouth forming an entrance for flood waves at Mersat El Hadjadj

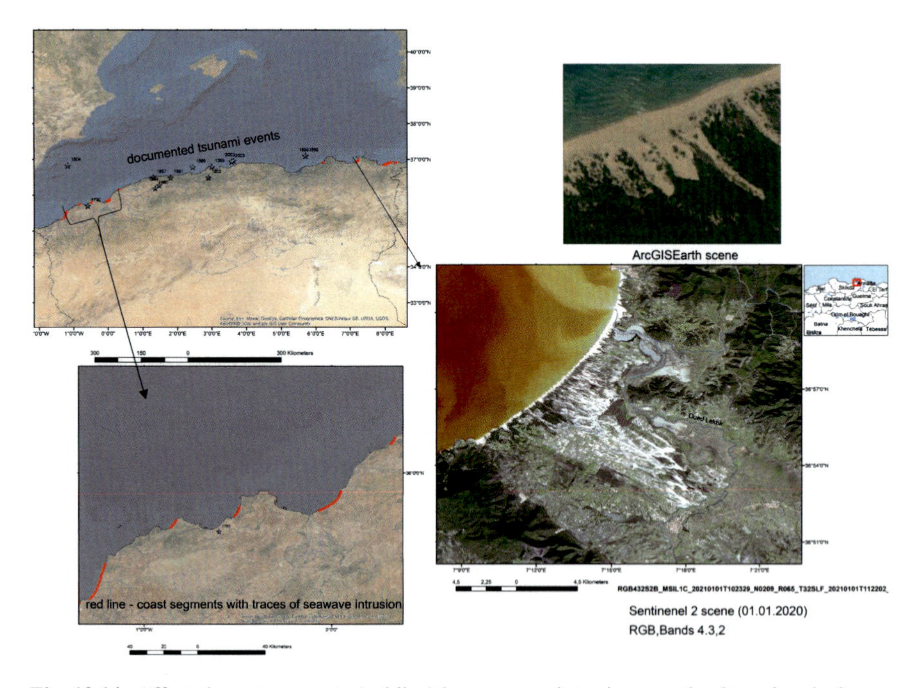

Fig. 10.14 Affected coast segments (red line) by sea wave intrusions, partly along river beds

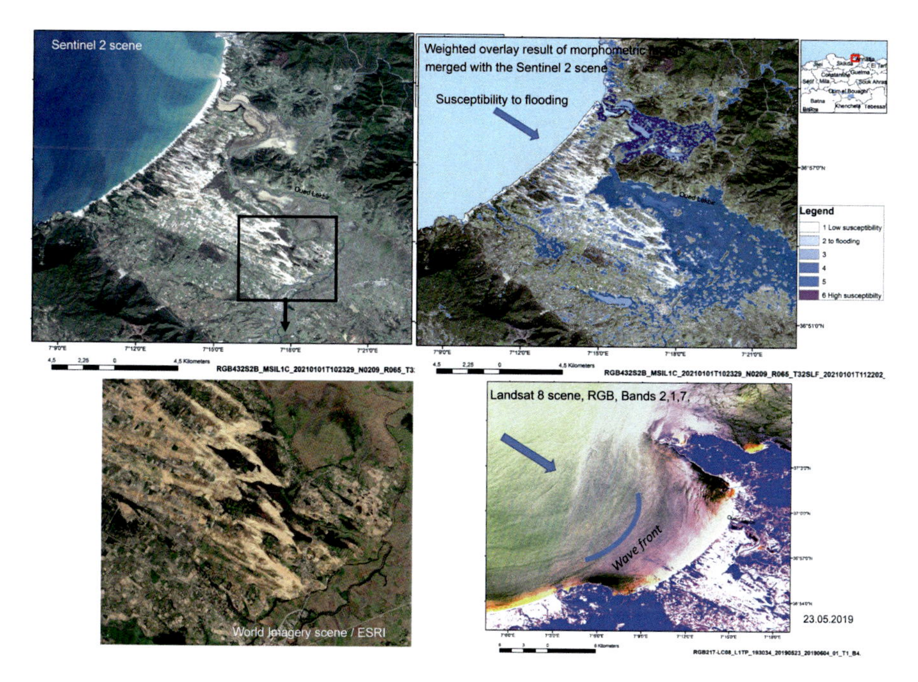

Fig. 10.15 Traces of high energetic flood waves visible on a Sentinel 2 (RGB, Bands 4,3,2) scene and on a Landsat 8 scene (RGB, Bands 2,1,7), modified by eolian activity from the Qued Lebkir east of Skikda, NE-Algeria
The weighted overlay of morphometric factors influencing the susceptibility to flooding (lowest and flattest areas) provide information of areas that might be affected by flooding due to their morphometric properties

at the Algerian coast where sea waves intruded several km landward (Fig. 10.14). In case of the Qued Lekbir traces of high energetic flood waves are clearly visible on satellite images with arrow-like shaped, parallel debris-walls, intersecting each other, directed towards SE. When merging the satellite images (Sentinel 2 and Landsat 8) with ASTER DEM data, the areas susceptible to flooding can be visualized. Areas with height levels below 10 m clearly show traces of previous flooding events like residual ponds. Further on, areas below 20 m are prone to flash floods in case of high precipitations as well.

The return flow (backwash) after tsunami or storm surge flooding has to be monitored. In general, the return current flows into the low-elevated areas or into river beds and channels, sometimes partly remaining there for days before discharging back into the sea, leading to the salinization of groundwater and of soils.

A systematic inventory of backwater areas, of coast-near lakes and ponds and their seasonal water level and volume changes as a prerequisite for flooding preparedness can be carried out as well with the support of remote sensing. With regard to climate change the monitoring of those seasonal changes (water volume, outline, etc.) gets an increasing importance. When high energetic flood waves intrude landwards during a humid season with high water levels in the lakes and ponds the extent

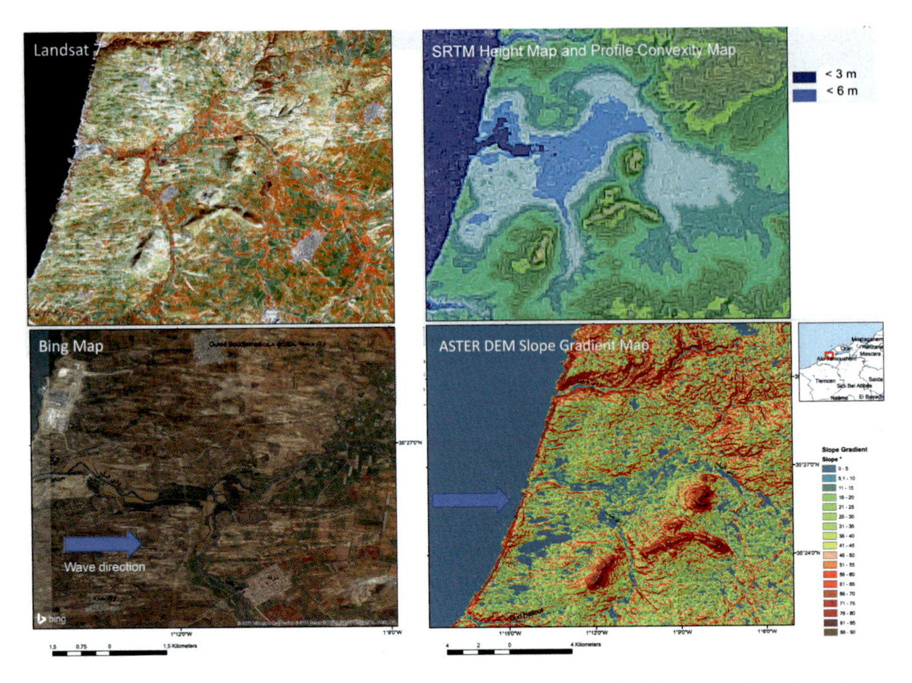

Fig. 10.16 Traces of flood waves more than 12 km landward visible on satellite data of the Algerian west coast, the sandy material later affected by eolian activity (erosion and sedimentation)

of flooded areas will be larger than during a dry season with low water levels or dried-up streams and lakes. This is demonstrated by the example from the coastal area in the east of the city Annaba in NE-Algeria (Fig. 10.17). Whereas the lowlands are nearly dried during the summer, they are partly flooded during the humid season in wintertime. The resulting map of the weighted overlay approach (merging and summarizing the morphometric factors influencing the disposition to flooding) shows in dark-blue colours those areas with the highest susceptibility to flooding in case of flash floods or tsunami flooding.

Height level maps indicating those areas below 10 m height level help to identify areas at risk such as in the next figures from Algiers. Merging height level data with actual satellite images show the infrastructure prone to potential flooding (Figs. 10.18a and 10.18b).

The weighted overlay approach in ArcGIS can contribute to the detection of those lowlands, that after the return-flow could remain longer time flooded than the environment. The result of a weighted overlay of morphometric factors in ArcGIS influencing the susceptibility to flooding is shown in the next figure of the Bay of Algiers (Fig. 10.19). Even in longer distances from the shoreline flooding might be possible affecting the southern city area of Algiers.

GEBCO bathymetric data were used to derive hillshade and flow accumulation maps (Fig. 10.20, image above) to localize canyons along the coast. The Algerian

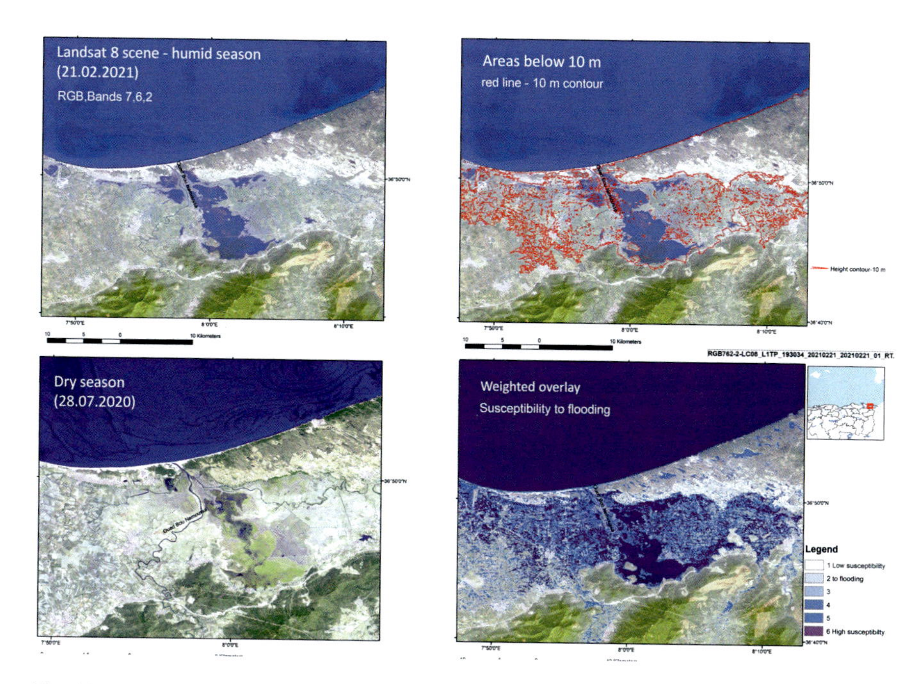

Fig. 10.17 Seasonal variability of the flooding extent influencing the additional flooding in case of tsunami events

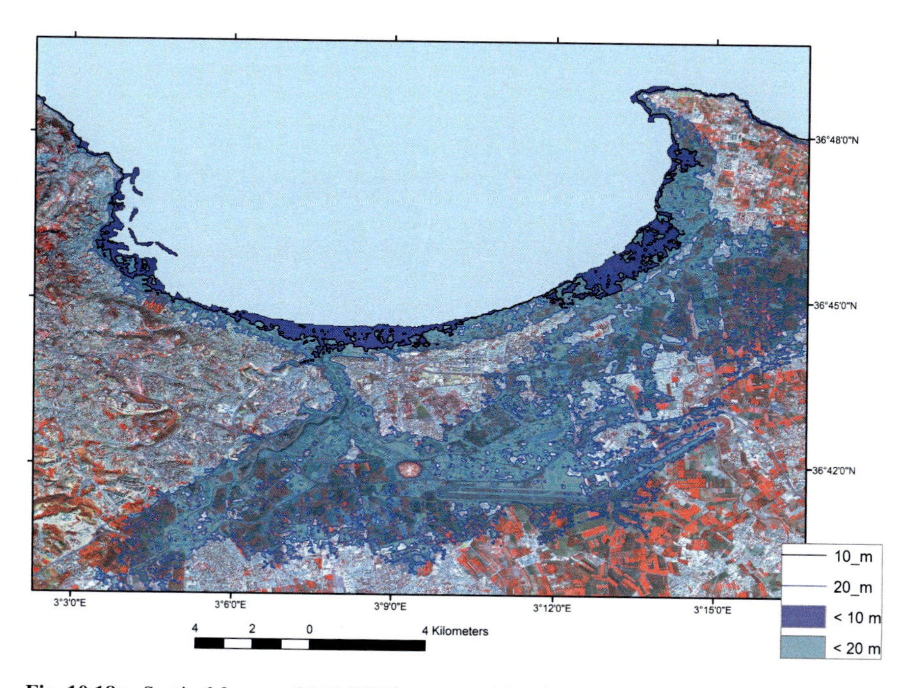

Fig. 10.18a Sentinel 2 scene (31.10.2020) merged with ASTER DEM data below 10 m and 20 m height level

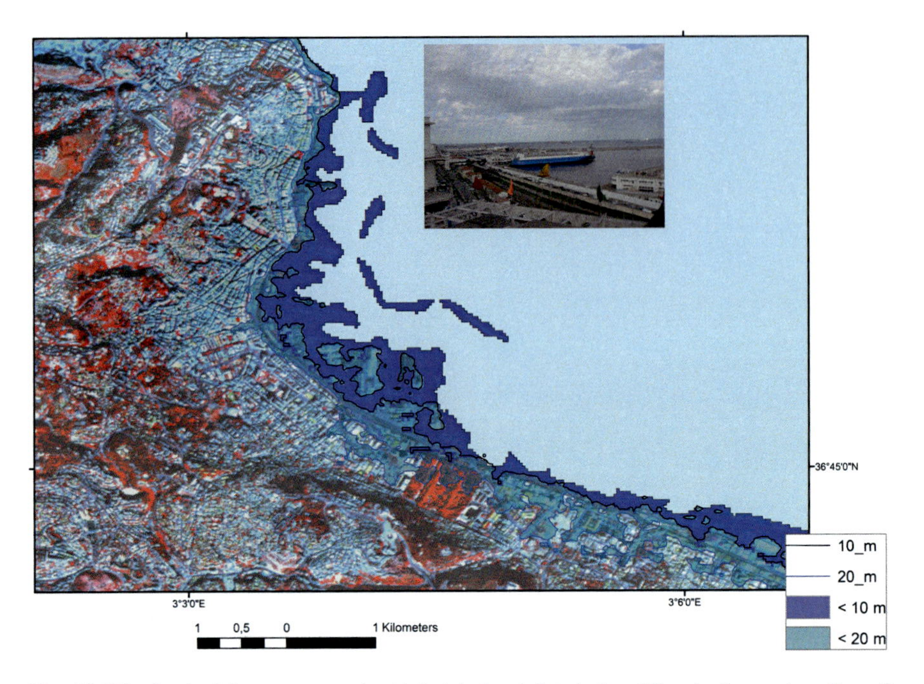

Fig. 10.18b Sentinel 2-scene merged with height level data below 10 m in the western Bay of Algiers (image of the harbour: Theilen-Willige, taken 2014)

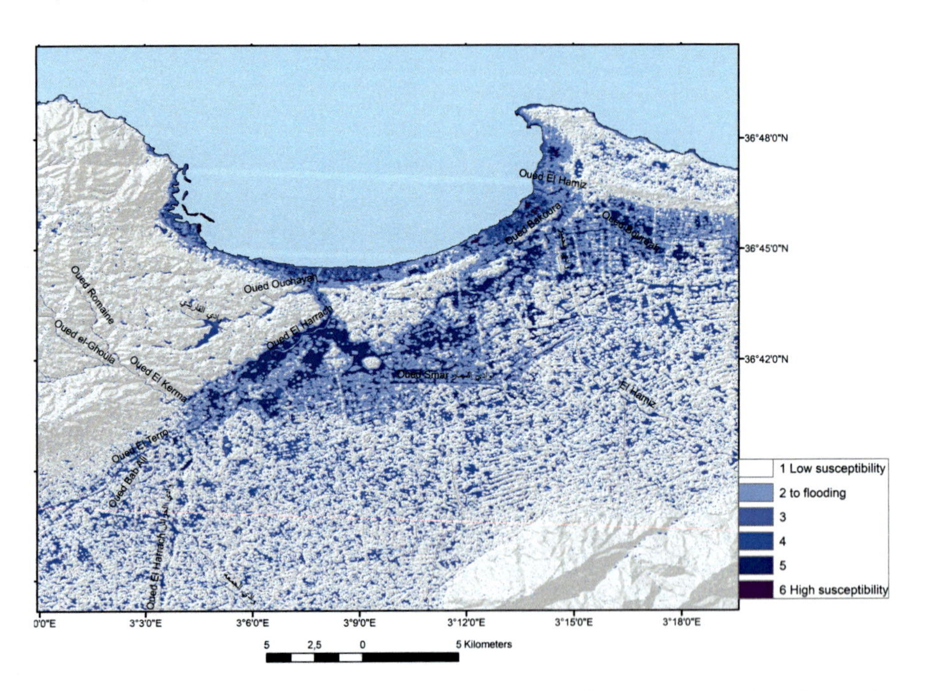

Fig. 10.19 Weighted overlay of morphometric factors to visualize areas susceptible to flooding in case of flash floods and tsunami waves in the coast-near part in the area of Algiers based on ASTER DEM data (30 spatial resolution)

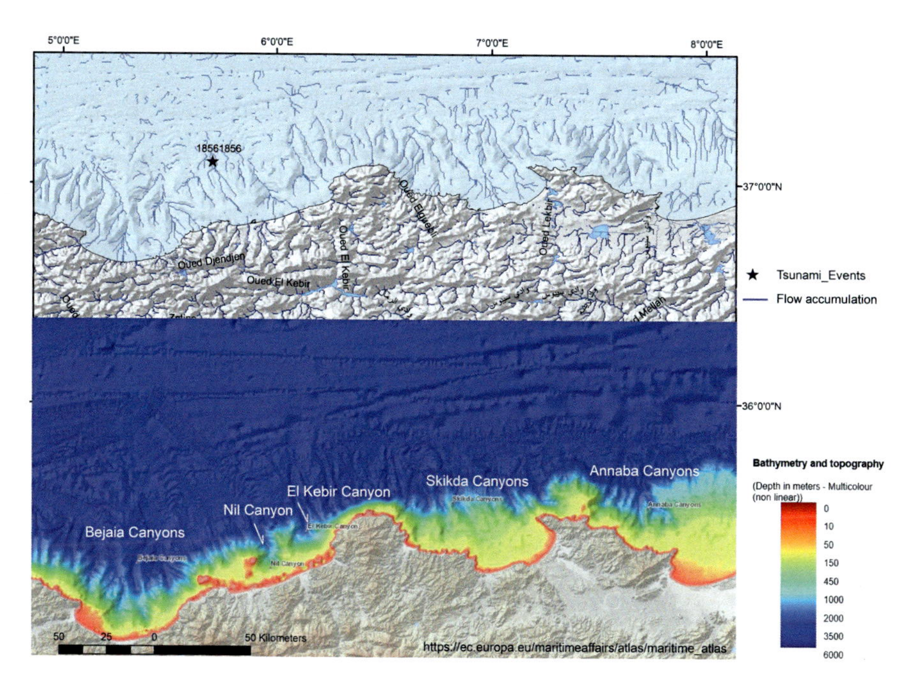

Fig. 10.20 The role of canyons along the Algerian coast in case of high energetic flood waves from northern direction by focusing wave energy
The blue lines in the map above are derived from the flow accumulation calculation in ArcGIS based on GEBCO bathymetric data. The lower map was provided by the European Atlas of the Seas/European Commission

continental shelf is relatively narrow, 5–10 km wide on average (Cattaneo et al., 2012). The continental slope along the Algerian coast is irregular, with intermediate flats of various width and a rather sinuous slope breaks (Domzig et al., 2006). The morphology of the continental slope of central Algeria presents abrupt scarps that at least partly correspond to active tectonic structures.

In case of significant tsunami occurrence from northern direction these canyons oriented perpendicular towards the coast could focus wave energy within these canyons which might lead in a worst case scenario to a higher flooding extent along the affected coast segments. Deep submarine canyons can lead to the amplification of waves on both sides of the canyons. The width length, and depth of the canyons play a role as well as the width of the shelf area. The effect of the parallel, NW-SE, to N-S-oriented arrangement of the canyons perpendicular to the coast has to be taken into account as well.

Tsunami wave propagating through submarine canyons has been studied by Jinadasa (2008) and Ioualalen et al. (2007), who analyzed the effect of 2004 Indian Ocean Tsunami in Sri Lanka and Bangladesh, respectively. The former found that the maximum inundation occurred just in front of the submarine canyons. In a similar manner, the latter found that the waves were amplified near the coast

(Aranguiz & Shibayama, 2013). In addition tsunami wave heights could be further enhanced due to overlapping of incoming waves with reflected waves from the coastal area. Tidal effects are a potential further source to increase tsunami heights up to a few centimeters (Jinadasa, 2008). Thus, the effect of the tide when investigating tsunami hazard, particularly, in coasts where tidal variations are significant has to be considered.

The next figures present examples of coast segments where submarine valleys are oriented in the same direction as the rivers along the coast (Figs. 10.21, 10.22 and 10.23). River mouths of rivers with the same orientation as the canyons along the coast might be exposed to focused wave energy due to funneling effects. In case of high energetic waves such as tsunami waves or landslide-tsunamis from northwestern and northern direction, the river mouths, especially those with the same orientation as of submarine valleys and canyons, could form an entrance for the flood waves and, thus to extended flooded areas.

The behavior of tsunami waves propagating over submarine canyons along the Algerian coast could also be influenced by abrupt changes in bathymetry, such as those caused by the steep continental slope with height differences of about 3000 m, see Fig. 10.21.

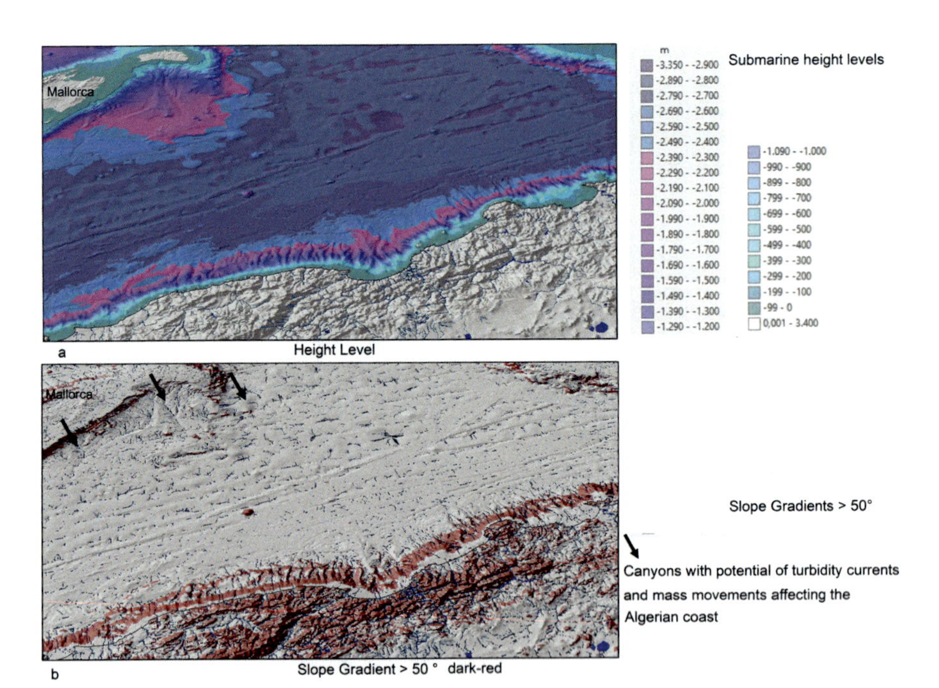

Fig. 10.21 3D perspective view of the Algerian coast based on GEBCO bathymetric data
(**a**) Height level
(**b**) Slope gradients > 50° combined with flow accumulation calculations to visualize potential valley and canyon sites prone to submarine turbidity currents

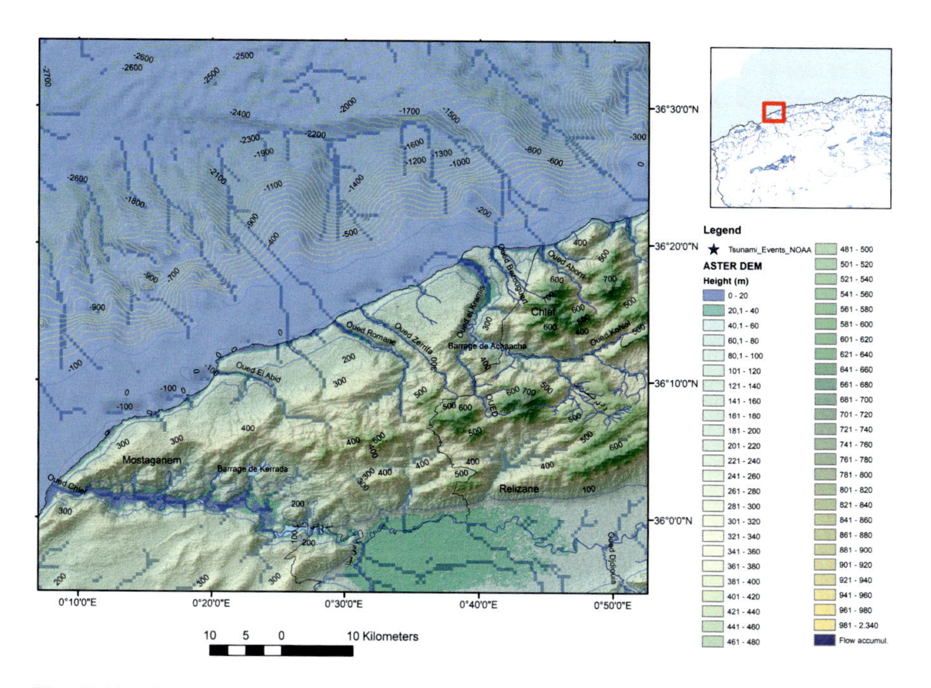

Fig. 10.22 River mouths and oriention of the rivers with the same NW-SE orientation of canyons along the coast

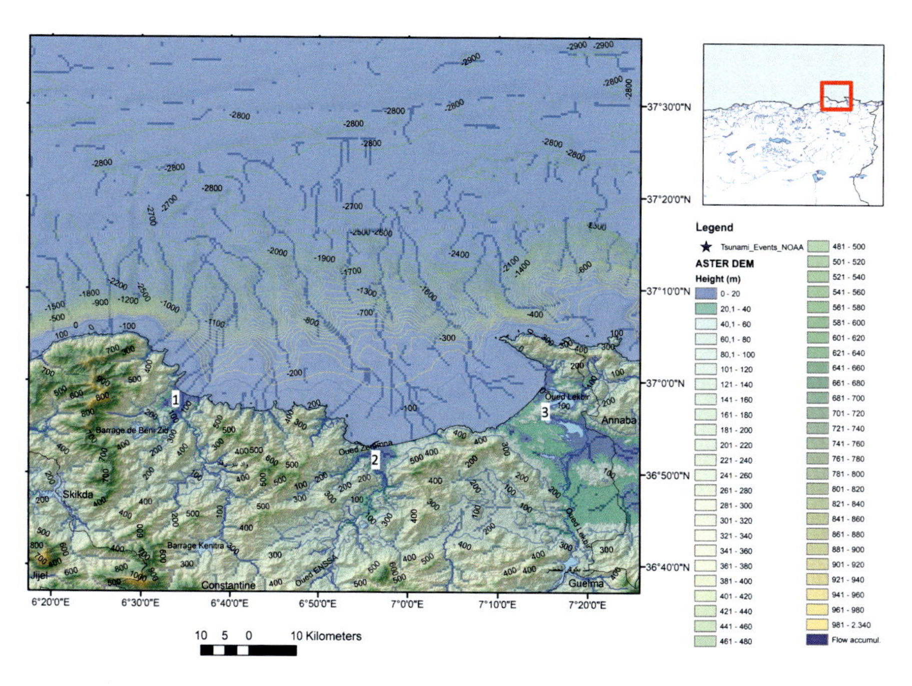

Fig. 10.23 River mouths and oriention of the rivers with the same NNW-SSE orientation of canyons along the coast of Collo and Skikda with exposure of Qued Elguebli (1) in the western part, Qued Zeramna (2) in the central part and Qued Lekbir (3) in the eastern part of the scene

Whenever the sources and main directions of tsunami waves or meteotsunamis are known, a warning should be given especially to those settlements within larger river mouth situated in areas below 10 m height levelt hat are exposed to the main wave fronts. The height levels have to be adapted and modified, however, in future according to sea level rise changes.

10.4 Conclusions

The input of remote sensing and GIS can be considered only as a small part of the whole "mosaic" of flooding susceptibility research approaches. Nevertheless, it offers a low-cost to no-cost approach (as the used DEM and satellite data are free), that can be used in any area, providing a first basic data stock for emergency preparedness by including susceptibility-to-flooding maps.

Summarizing factors influencing flooding susceptibility such as relatively low height levels (<10 m), terrain curvature (values calculated in ArcMap = 0, corresponding to flat terrain), slope gradients <10°, drop raster <50.000, and high flow accumulation values by using the weighted-overlay tools in ArcGIS, help to detect areas with higher flooding susceptibility due to their geomorphologic disposition. Due to the aggregation of causal, morphometric factors, those areas can be visualized. This approach is suited to obtain a first basic overview on susceptible areas according to a standardized approach. This might be of interest especially for countries with low financial resources wherein such maps are still unavailable.

The impact of storm surge or tsunami waves on the coast depends on geomorphological settings of the coast as well as the shelf (Abdul Rasheed et al., 2006). Variation in run-up along the coastline depends on the topography of the coastline, near-shore bathymetry, beach slope, coastal orientation, direction of the arriving wave etc. Variations in the near shore bathymetry and bottom topography were be studied to understand the relationship with the amplification of tsunami height development. The evaluations of bathymetric data reveal the steep submarine slopes in front of the Algerian coast and the submarine valleys and canyons perpendicular to the coast line. Depending on the source of high energetic waves and wave angle a focusing effect of the canyons on wave energy is possible.

Satellite images contribute to a better understanding of the interactions between the coastal morphology on the development of water currents, of course depending on meteorologic and tidal conditions at the acquisition time of the images.

Based on DEM and different satellite data, as well as infrastructural data those settlements and cities situated in lowlands and river mouth areas can be identified and integrated into a disaster management system (Fig. 10.24). When creating buffer zones with a 2 km-radius along the river mouth areas and intersecting these buffer zones with known settlements and cities within theses buffer zones the potentially affected settlements can be identified. In case of emergency help could be better focused, coordinated and organized according to the size of the population.

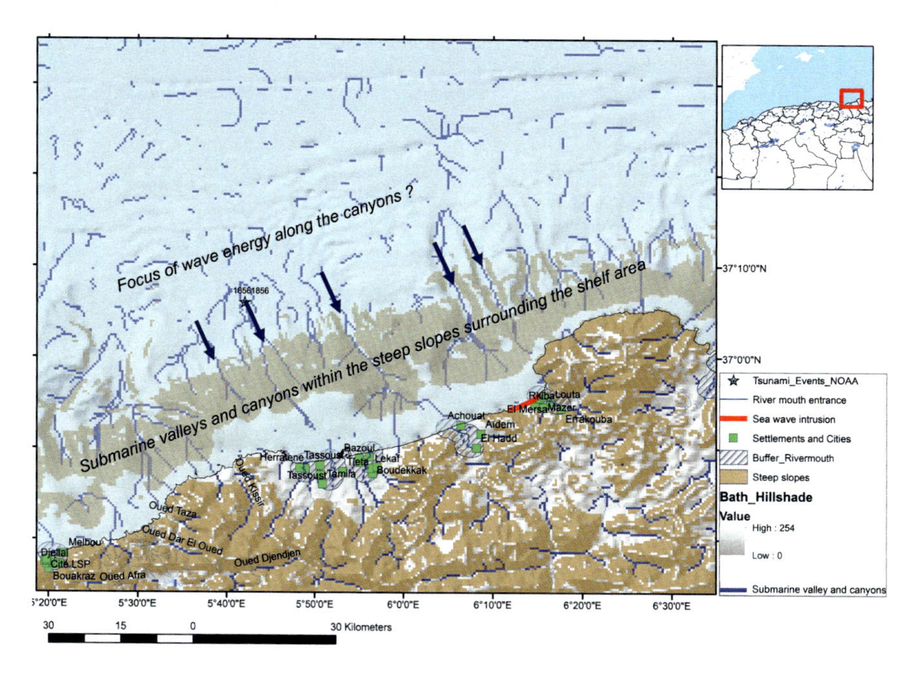

Fig. 10.24 Cities and settlements situated within a 2 km-radius-buffer zone along river mouth areas of rivers oriented in the same direction as the submarine canyons and valleys (see arrows)

It must be added that storm surges or meteo-waves, although generally not as destructive as a major tsunami-waves, can be comparatively more frequent. Therefore, inundation maps indicating the extent of the coastal area that could be affected by potential events of both tsunami waves, storm surges and flash floods should an input in any data base dealing with flooding hazards. The influence of sea-level rise due to climatic change has to be considered as well.

Acknowledgements Ms.Verena Willige, Munich is kindly acknowledged for her helpful discussions and input of ideas.

References

Abdul Rasheed, K. A., Kesava Das, V., Revichandran, C., Vijayan, P. R., & Thottam, T. J. (2006). Tsunami impacts on morphology of beaches along South Kerala coast, west coast of India. *Science of Tsunami Hazards, 24*(1), 24–34.

Alaska Satellite Facility (ASF). https://search.asf.alaska.edu/#/. Accessed 20 Nov 2020.

Alasset, P.-J., Hebert, H., Maouche, S., Calbini, V., & Meghraoui, M. (2006). The Tsunami induced by the 2003 Zemmouri earthquake (Mw=6.9): Modelling and results. *Geophysical Journal International, 166*, 213–226.

Ambraseys, N. N., & Vogt, J. (1988). Material for the investigation of the seismicity of the region of Algiers. *European Earthquake Engineering, 3*, 16–29.
Amir, L. (2014). Tsunami Hazard assessment in the Alboran Sea for the Western coast of Algeria. *Journal of Shipping and Ocean Engineering, 4*, 43–51.
Amir, L. A., & Theilen-Willige, B. (2017). Coastal risk and water flow analysis in Eastern Algeria (Western Mediterranean). *Universal Journal of Geoscience, 5*(4), 99–111. https://doi.org/10.13189/ujg.2017.050403
Amir, L., Cisternas, A., Vigneresse, J.-L., Dudley, W., & Mc Adoo, B. (2012). Algeria's vulnerability to tsunamis from near-field seismic sources. *Science of Tsunami Hazards, Journal of Tsunami Society International, 31*(1), 82–98.
Aranguiz, R., & Shibayama, T. (2013). Effect of submarine Canyons on Tsunami wave propagation: A case study oft he Biobio Canyon, Chile. *Coastal Engineering Journal, 55*(4), 1350016. https://doi.org/10.1142/S0578563413500162. (23 pages).
Beccario, C. (2020). *Earth – A visualization of global weather conditions forecast by supercomputers updated every three hours*. https://earth.nullschool.net/
Cattaneo, A., Babonneau, N., Ratzov, G., Dan-Unterseh, G., Yelles, K., Bracène, R., Mercier de Lépinay, B., Boudiaf, A., & Déverchère., J. (2012). Searching for the seafloor signature of the 21 May 2003 Boumerdes earthquake offshore central Algeria. *Natural Hazards and Earth System Sciences, 12*, 2159–2172. www.nat-hazards-earth-syst-sci.net/12/2159/2012/. https://doi.org/10.5194/nhess-12-2159-2012
Deverchere, J., Yelles, K., Domzig, A., Mercier de Lepinay, B., Bouillin, J.P., Gaulier, V., Bracene, R., Calais, E., Savoye, B., Kherroubi, A., Le Roy, P., Pauc, H., & Dan, G. (2005). Active thrust faulting offshore Boumerdes, Algeria, and its relation to the 2003 Mw 6.9 earthquake. *Geophysical Research Letters, 32*. https://doi.org/10.1029/2004GL021646
Domzig, A., Yelles, K., Le Roy, C., Déverchère, J., Bouillin, J.-P., Bracène, R., Mercier de Lépinay, B., Le Roy, P., Calais, E., Kherroubi, A., Gaullier, V., Savoye, B., & Pauc, H. (2006). Searching for the Africa–Eurasia Miocene boundary offshore western Algeria (MARADJA'03 cruise). *Comptes rendus geoscience, 338*, 80–91. https://doi.org/10.1016/j.crte.2005.11.009
EMODnet Bathymetry Consortium. (2020). EMODnet Digital Bathymetry (DTM). https://doi.org/10.12770/bb6a87dd-e579-4036-abe1-e649cea9881a
European Environment Agency. https://www.eea.europa.eu/data-and-maps/indicators/sea-level-rise-7/assessment. Accessed 05 Jan 2021.
European Space Agency (ESA). https://scihub.copernicus.eu. Accessed 15 Feb 2021.
Gelfenbaum, G. (2005). Coastal currents. In M. L. Schwartz (Ed.), *Encyclopedia of coastal science* (Encyclopedia of Earth Science Series). Springer. https://doi.org/10.1007/1-4020-3880-1_78
General Bathymetric Chart of the Oceans (GEBCO). https://www.gebco.net/https://www.gebco.net/data_and_products/gebco_web_services/web_map_service/mapserv? Accessed 17 Nov 2020.
Gusiakov, V. K. (2020). Global occurrence of large tsunamis and tsunami-like waves within the last 120 years (1900–2019). *Pure and Applied Geophysics, 177*, 1261–1266. https://doi.org/10.1007/s00024-020-02437-9
Ioualalen, M., Pelynovsky, E., Asavanant, J., Lipikorn, R., & Deschamps, A. (2007). On the weak impact of the 26 Indian Ocean tsunami on the Bangladesh coast. *Natural Hazards and Earth System Sciences, 7*, 141–147.
Jinadasa, S. U. P. (2008). *Interaction of tsunami wave propagation with coastal bathymetry and geomorphology: A case study in Sri Lanka*. General Bathymetric Chart of the Oceans (GEBCO) Science Day 2008, Japan Coast Guard, Tokyo, Japan.
Lopez Marinas, J. M., & Salord, R. (1990). *El periodo sismico oranes de 1790 a la luz de la documentación de los archivos espanoles*. I.G.N. Publication, 6, 64 pages. Madrid.
Monserrat, S., Vilibic, I., & Rabinovich, A. B. (2006). Meteotsunamis: Atmospherically induced destructive ocean waves in the tsunami frequency band. *Natural Hazards and Earth System Sciences, 6*, 1035–1051. www.nat-hazards-earth-syst-sci.net/6/1035/2006/.

Papadopoulos, G. A., & Fokaefs, A. (2005). Strong tsunamis in the Mediterranean Sea: A re-evaluation. *ISET Journal of Earthquake Technology, 463*(42, 4), 159–170.
Papadopoulos, G. A., Daskalaki, E., & Fokaefs, A. (2007). Tsunamis generated by coastal and submarine landslides in the Mediterranean Sea. In V. Lykousis, D. Sakellariou, & J. Locat (Eds.), *Submarine mass movements and their consequences* (pp. 415–422). Springer. https://doi.org/10.1007/978-1-4020-6512-5_43
Pleskachevsky, A., Jacobsen, S., Tings, B., & Schwarz, E. (2019). Estimation of sea state from Sentinel-1 synthetic aperture radar imagery for maritime situation awareness. *International Journal of Remote Sensing, 40*(11), 4104–4142. https://doi.org/10.1080/01431161.2018.1558377
Shao, W., Zhang, Z., Li, X., & Li, H. (2016). Ocean wave parameters retrieval from Sentinel-1 SAR imagery. *Remote Sensing, 8*, 707. https://doi.org/10.3390/rs8090707
Soloviev, S. L., Solovieva, O. N., Go, C. N., Kim, K. S., & Shchetnikov, N. A. (2000). *Tsunamis in the Mediterranean Sea, 2000 B.C.–2000 A.D. Advances in natural and technological hazards research* (p. 260). Kluwer Academic Publishers.
Theilen-Willige, B. (2006). Emergency planning in northern Algeria based on remote sensing data in respect of Tsunami Hazard preparedness. *Science of Tsunami Hazards, 25*(1), 3–18.
US Geological Survey (USGS)/Earth Explorer. https://earthexplorer.usgs.gov/. Accessed 12 Dec 2020.
Yelles-Chaouche, A., Roger, J., Déverchère, J., Bracène, R., Domzig, A., Herbert, H., & Kherroubi, A. (2009). The 1856 Tsunami of Djidjelli (Eastern Algeria): Seismotectonics, modelling and hazard implications for the Algerian Coast. *Pure and Applied Geophysics, 166*, 283–300. Basel: Birkhäuser Verlag, 0033–4553/09/010283–18, https://doi.org/10.1007/s00024-008-0433-6
Yelles, A. K., Lammali, K., Mahsas, A., Calais, E., & Briole, P. (2004). Coseismic deformation of the 21st May 2003, Mw = 6.8 Boumerdes earthquake, Algeria, from GPS measurements. *Geophysical Research Letters, 31*. L13610, https://doi.org/10.1029/2004GL019884

Chapter 11
Geographical Information Systems (GIS) and Multi-criteria Analysis Approach for flood Risk Mapping: Case of Kasserine Region, Tunisia

Salwa Saidi, Walid Dachraoui, and Belgacem Jarray

Abstract The floods have showed an increasing in recent years in Tunisia and especially in Kasserine region, in the central part of Tunisia. This research aims to develop updated and accurate flood risk map in Kasserine region. The vulnerability and risk maps use geographical information systems (GIS) and multi-criteria analysis with the application of Analytical Hierarchy Process (AHP) methods to estimate weights for each parameter that contribute to flood risk.

The flood risk map is obtained by superposition of vulnerability and hazard maps. It shows that risk is very high in the North of the study area and high in the South, and weak in the center of the study area. The most vulnerable areas are those in the northern part because the slope is very low and the density of the hydrographic network is high with the presence of impermeable urban areas with very low permeability. Socio-economic hazard mapping was carried out based on land use and spontaneous habitat. The high hazard zone corresponds to urban areas and the spontaneous settlements. In fact, according to the risk map, these same areas present high risks. What makes these areas require management measures in order to protect it against flooding during exceptional floods.

Keywords GIS · AHP · Vulnerability · Flooding risk · Management

S. Saidi (✉)
Faculty of Sciences of Tunis, University of Tunis El Manar, Tunis, Tunisia

Water, Energy and Environment Laboratory, ENI-Sfax, Sfax, Tunisia
e-mail: salwa.saidi@fst.utm.tn

W. Dachraoui
Faculty of Sciences of Tunis, University of Tunis El Manar, Tunis, Tunisia

B. Jarray
General Directorate of Dams and Hydraulic works, Ministry of Agriculture, Tunis, Tunisia

M. M. Al Saud (ed.), *Applications of Space Techniques on the Natural Hazards in the MENA Region*, https://doi.org/10.1007/978-3-030-88874-9_11

11.1 Introduction

Floods are the most common and devastating phenomena that affecting both developed and developing countries around the world. The situation has become alarming due to the increasing of urbanization, the global warming and the lack of adequate water resources managements. Particularly, in Tunisia over the last 50 years, floods have caused a huge damage and serious socio-economic risks. In fact, between 1959 and 2015 (floods of 1959, 1962, 1969, 1973, 1979,1980, 1982, 1990, 1995, 2003, 2007, 2011, 2015), at least 815 people were killed by floods, thousands were injured, and 76,300 people were dislocated because of habitat destruction (AUGT, 2015; Saidi et al., 2018).

Assessment of flood risk is considered a pre -hazard management and planning activity since it is able to elaborate a zonation in function of the susceptibility of areas to flooding, as evidenced by previous studies (Rahmati et al., 2016; Ajim Ali et al., 2020).

Therefore, the use of remote sensing (RS) and Geographical Information System (GIS) tools have been conducted by many researchers, for flood risk mapping (Bates, 2004; Sanyal & Lu, 2004; Pradhan, 2009; White et al., 2010; Bates, 2004; Haq et al., 2012; Strobl et al., 2012; Patel & Srivastava, 2013; Jaafari et al., 2014; Tehrany et al., 2014, 2015; Wanders et al., 2014; Rahmati et al., 2016; Das, 2019).

Since this is a complex phenomenon, the flood risk implicates many factors and the combination of GIS and multicriteria analysis has proven its efficacity. There are various methods of multi-criteria decision analysis and many researchers have applied them in natural hazard and flood risk mapping and modeling (Ghanbarpour et al., 2013; Papaioannou et al., 2015; Rahmati et al., 2016; Costache et al., 2020; Ajim Ali et al., 2020). The Analytic Hierarchy Process (AHP) introduced by Saaty (1977) is one of the most common multi- criteria decision methods, and it has been widely applied to solve decision-making problems related flood risk (Saidi et al., 2017, 2018; Sedghiyan et al., 2021; Sutradhar et al., 2021; Nitin et al., 2021).

So, the main objectives of this paper are

- to spatially analyze the vulnerability of communities to surface water flooding
- to determine the relation between the spatial distribution of vulnerability and hazard in order to investigate the risk to communities.
- To demonstrate the relation between vulnerability, hazard and risk maps and the parameters influencing the flood risk
- To calculate weights of all parameters influencing the flood risk unsing Multicriteria analysis via Analytical Hierarchic Process (AHP)

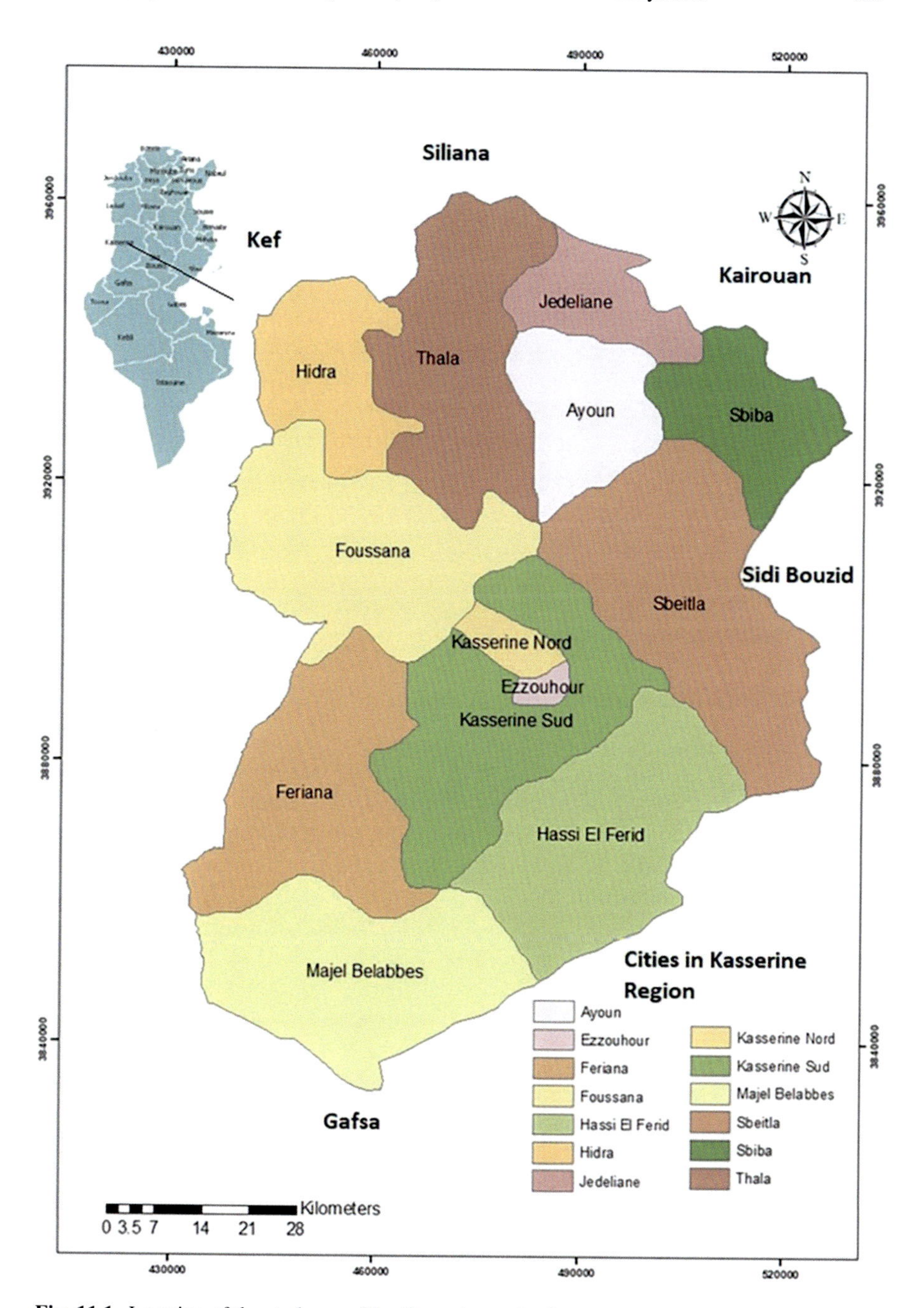

Fig. 11.1 Location of the study area "the Kasserine region"

11.2 Study Area Characteristics

The Kasserine region is located in the Center-West of the country, along the Algerian border (200 kilometers) (Fig. 11.1). It is bounded by the regions of El Kef and Siliana in the North, Sidi Bouzid in the East and Gafsa in the South. The Kasserine region has an arid to semi-arid climate. The pedology of Kasserine is characterized by the presence of calcified soil, alluvium, Humus soil and regenerating soil. The Kasserine region is characterized by a dense hydrographic network fed by 4 main wadis, in particular the most famous wadi which is Oued Lahtab. The area of the Kasserine region is estimated at 8,260,090 km^2, or 5.19% of the total area of the country. The region of Kasserine presents the highest point topographically of Tunisia Djebel Chaambi at 1544 m (CRDA Kasserine, 2018).

The Kasserine region is characterized by a stratigraphic series varying from Triassic sandstones and evaporites, through to Quaternary siliceous deposits (Hassen et al., 2016).

11.3 Methodology

11.3.1 Flood Risk Mapping

According to EU Directive (COM, 2006) for flood management, "flood risk"is defined as the likelihood of a flood event together with the actual damage to human health and life, the environment and economic activity associated with that flood event. In this context the flood risk is determined in function of the relationship of hazard (H) and vulnerability (V).

$$R = (H) * (V)$$

11.3.1.1 Flood vulnerability Mapping

Based on the literature review, the flood vulnerability represents a very important component of the flood risk. The factors chosen for the vulnerability mapping are: the density of the hydrographic network, the permeability and the slope. The vulnerability map is obtained by weighted superposition of these three parameters.

The weights are assessed using the Analytical Hierarchic Process (AHP) method by the use of Intelligent Decision software.

11.3.1.2 Hazard Assessment

The American Planning Association defines a hazard as "the process of defining and describing a hazard, including its physical characteristics, magnitude and severity, probability and frequency, causative factors, and locations or areas affected." (HIRA, 2017).

The hazard map in our context is the weighted superposition of three parameters: land use, rainfall intensity and spontaneous habitat. The calculation of the weights is done using the same methodology used for the vulnerability map. The process of calculating weights using the AHP method and the IDS software will be detailed later.

11.3.1.3 Process of Weights Calculation

The major drawback of multicriteria methods is the subjectivity of the determination of the rating scale and the weighting coefficients (Saidi et al., 2011). In this study, the analytic hierarchy process (AHP), as developed by Saaty (1977) is adopted to weight assessment in Intelligent Decision System software. In fact, the hierarchical process consists of the object layer, the index layer and the sub index layer. The object layer is the flood risk whose index is (R). The index layer includes two categories: The vulnerability and the Hazard, whose indexes are labeled respectively (V) and (H). The sub index layer includes all parameters of vulnerability and hazard indexes.

The hierarchical process is one of the expert systems that deal with imprecision of results by affecting a weighting to each criteria or parameter (Kim et al., 2012) and by reducing complex decision to a series of one-on-one comparisons, and then synthesizes the final results (Ting & Cho, 2008).

11.4 Results and Discussion

11.4.1 Vulnerability Assessment

In order to delineate the vulnerability to flood degree, the most important parameters are mapped: Slope, the hydrographic network density and the permeability.

After establishing the thematic maps of the different parameters were ranked. Hence, the ranked layers are defined relying on their relative importance to control the flood vulnerability according to this classification: very low (1), low (2), medium (3), high (4), and very high (5).

The application shows that the permeability map shows 3 different permeability classes: In the North West there is a high permeability which gives this zone a low risk. On the other hand, in the center of the Kasserine region and in the east, there is an average to low permeability and therefore likely to be the most inundated (Fig. 11.2).

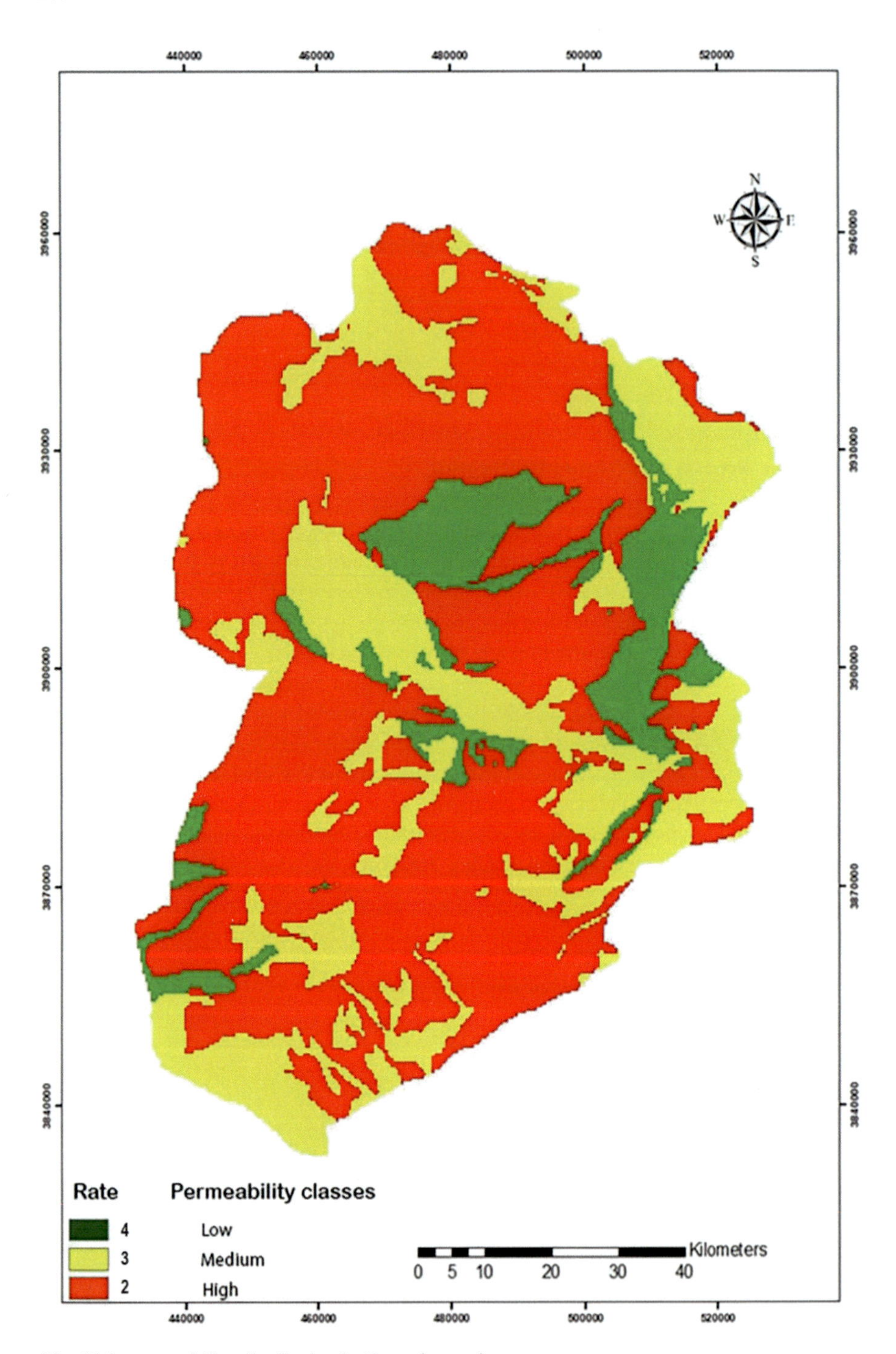

Fig. 11.2 permeability distribution in Kasserine region

The slope map reveals that the areas where the slope is low are considered the most risked areas for flooding and which occupy the majority of the study area, a score of 5 will be assigned to these areas. However, areas with a high slope are the least risky and subsequently a score of 2 will be assigned to these areas (Fig. 11.3).

The hydrographic network density map shows 4 classes: very low, low, medium and high density (Fig. 11.4).

The vulnerability map shows that high vulnerability occupies the northern part and corresponds to zones with low permeability (urban zone) (Fig. 11.5)

- A moderate vulnerability occupies the central part of the region which coincides with areas where the permeability is very high and the presence of vegetation can also reduce the flood risk. In these sectors the topography marked with average spreading (average slope) plays a key role in the evacuation of rainwater to areas with flat topography.
- low vulnerability zones are located in the East where the permeability is high and the slope is steep.

11.4.2 Hazard Assessment

The parameters used in hazard assessment are: Rain intensity, Land use and anarchic and unplanned urban zones (Figs. 11.6, 11.7 and 11.8). The hazard map, resulting of weighted superposition of these three parameters, shows (Fig. 11.9): -The northern and southern parts of the region are characterized by a high hazard index corresponding to a high rainfall intensity and the presence of anarchic habitats. These areas have no sewerage and rainwater drainage network. As for the West, the West and the North East, they are marked by a low and medium hazard because of the low rainfall and the presence of high-density forests.

11.4.3 Flood Risk

The present flood risk map of the study area reflecting the influences of different parameters/criteria. The influence of each parameter affecting floods is expressed by a weight and which is illustrated in the Table 11.1.

The flood risk map, result of multiplying the hazard by the vulnerability maps reveals that areas at risk of flooding are in the north (Fig. 11.10). This due to high vulnerability and moderate and high Hazard in the Kasserine region. It corresponds to flat areas with high density of hydrographic network and unplanned urban areas. This is revealing the influence of physical characteristics on flood risk.

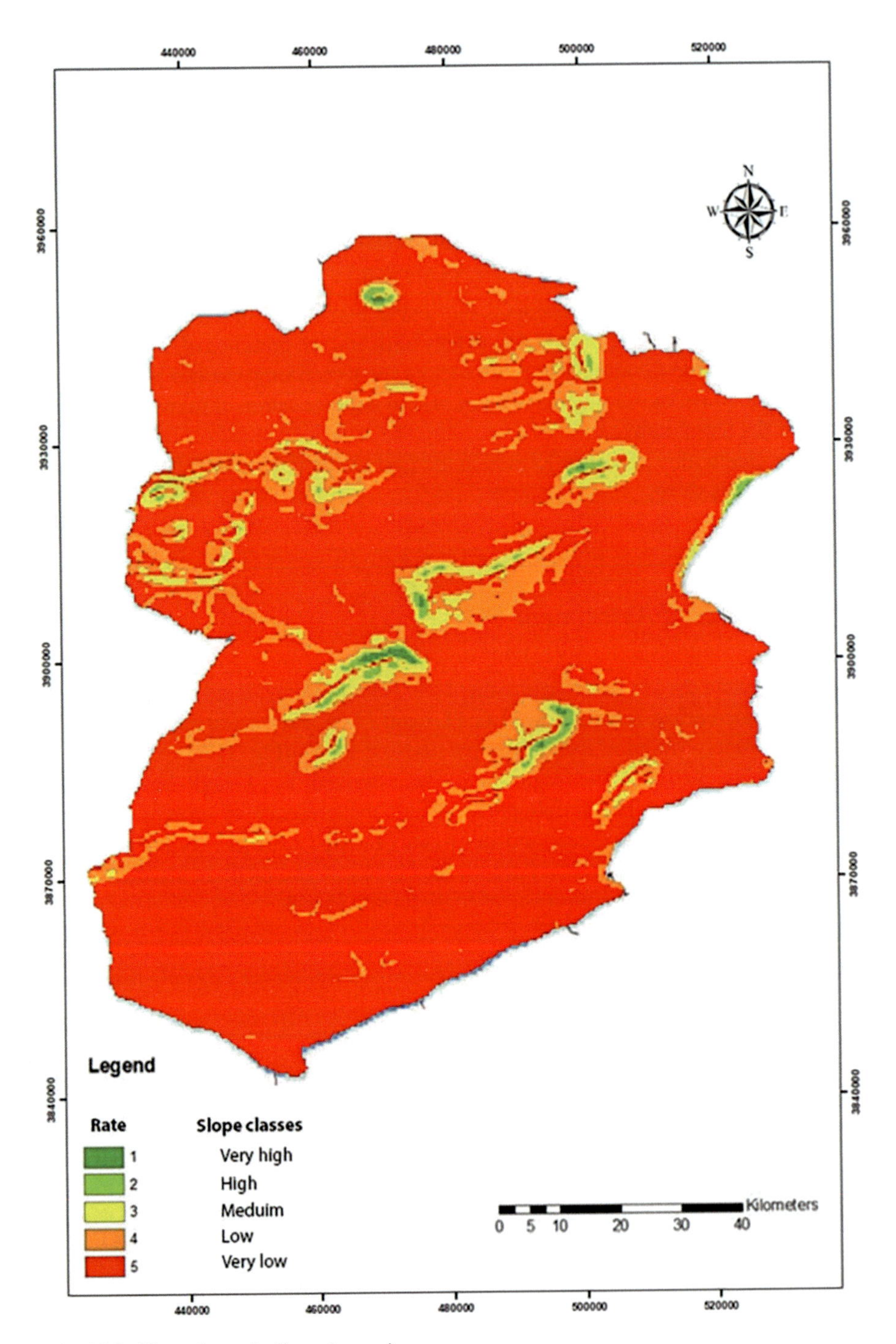

Fig. 11.3 Slope classes in Kasserine region

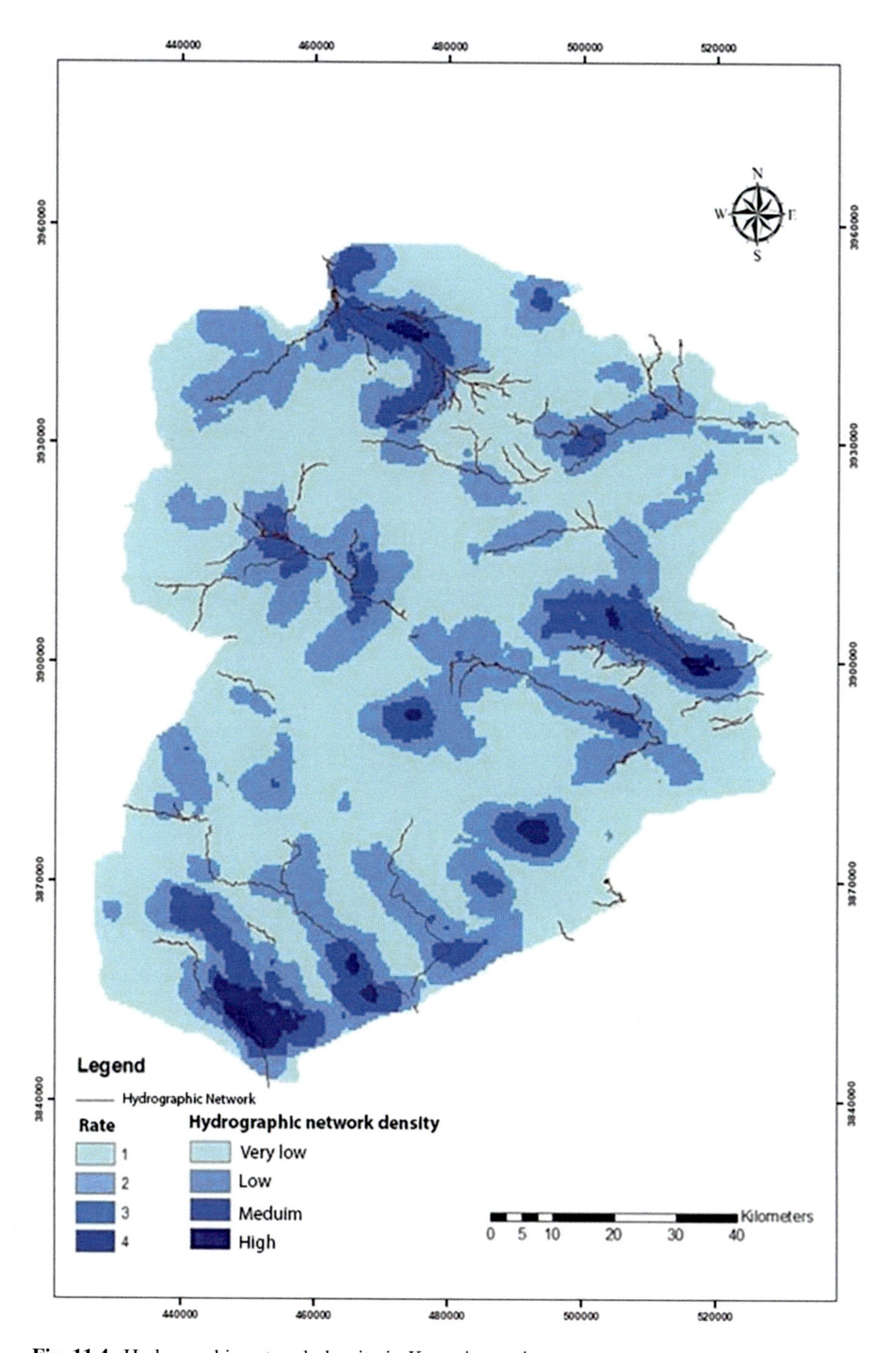

Fig. 11.4 Hydrographic network density in Kasserine region

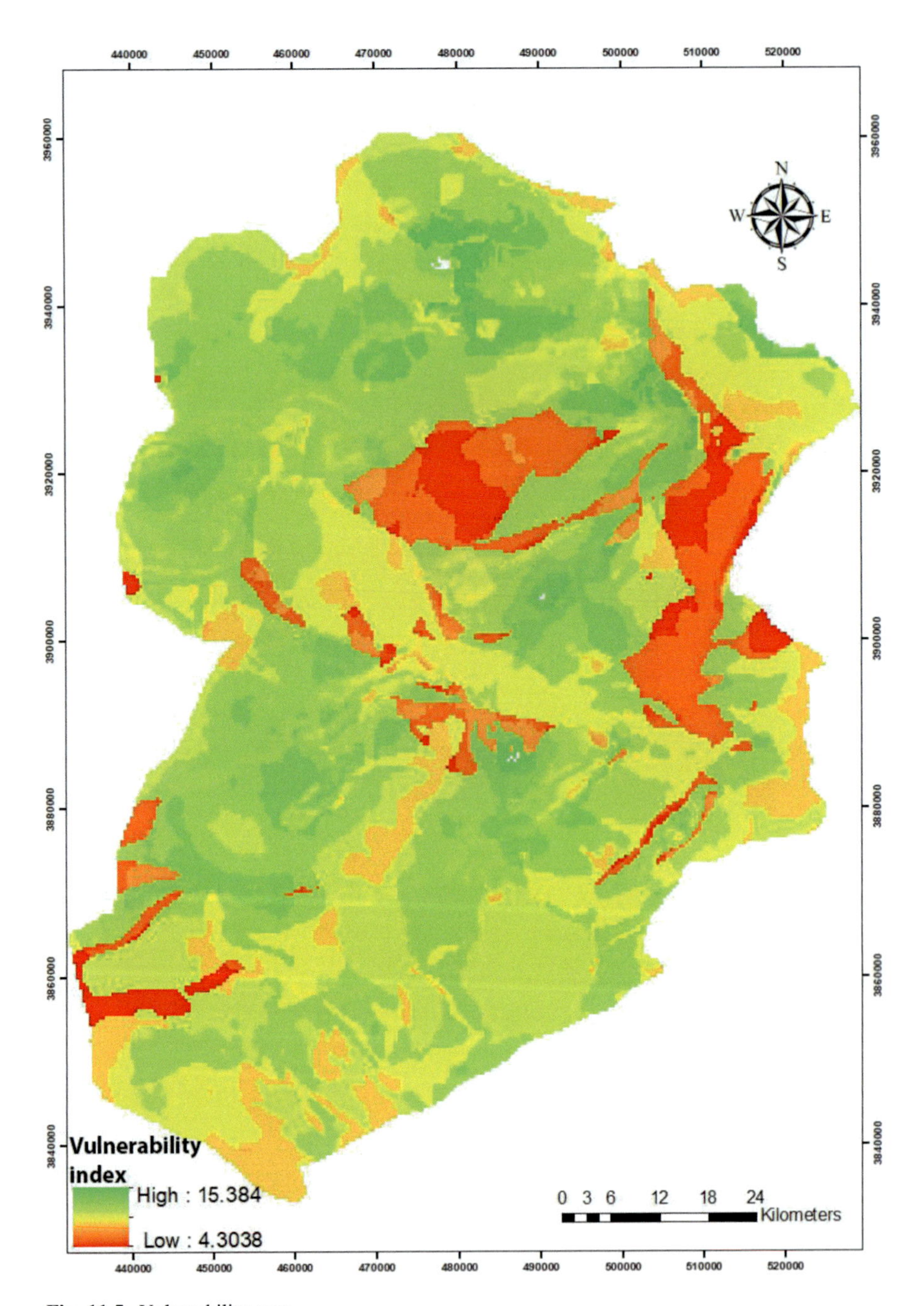

Fig. 11.5 Vulnerability map

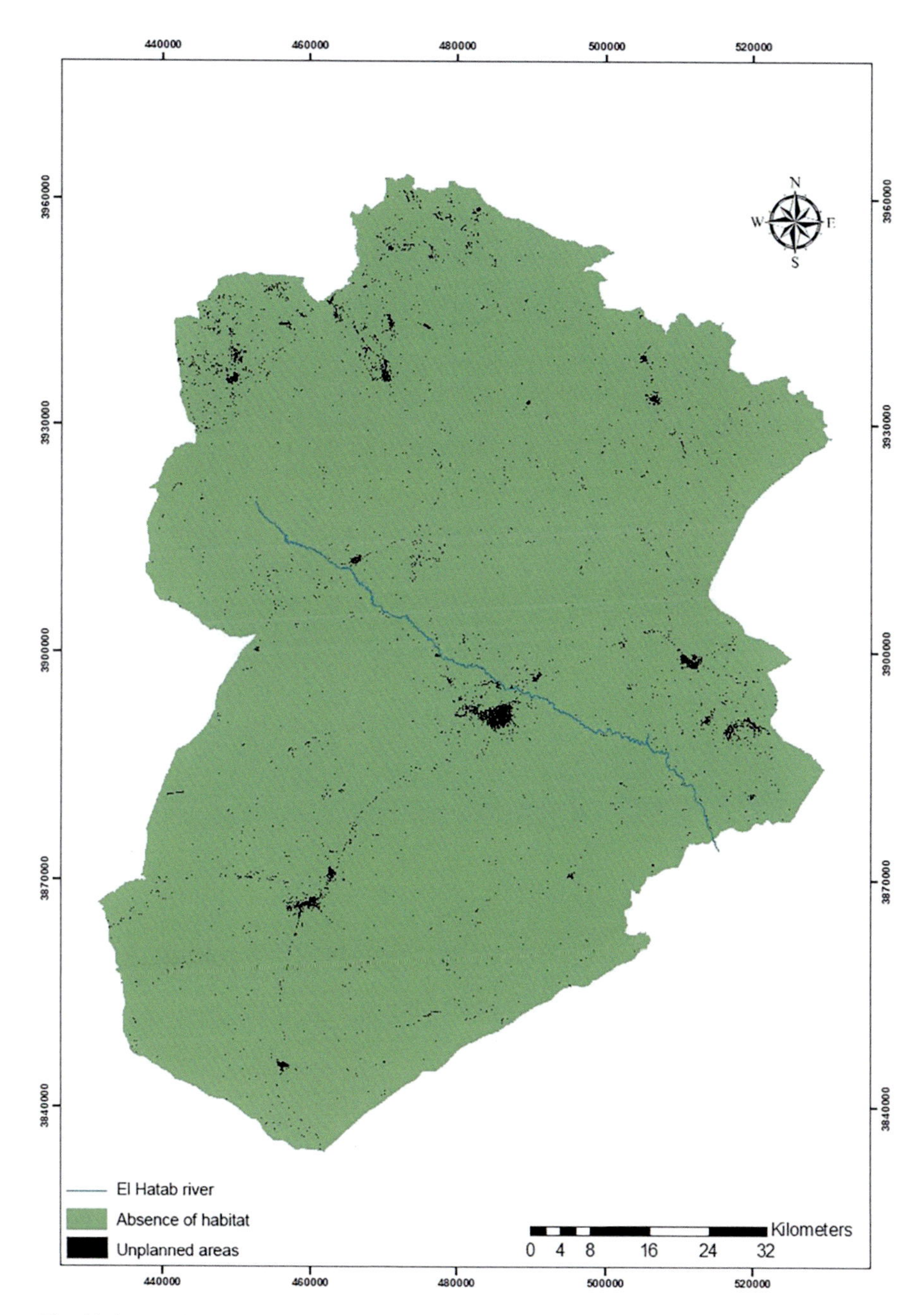

Fig. 11.6 Location of the unplanned urban areas in Kasserine region

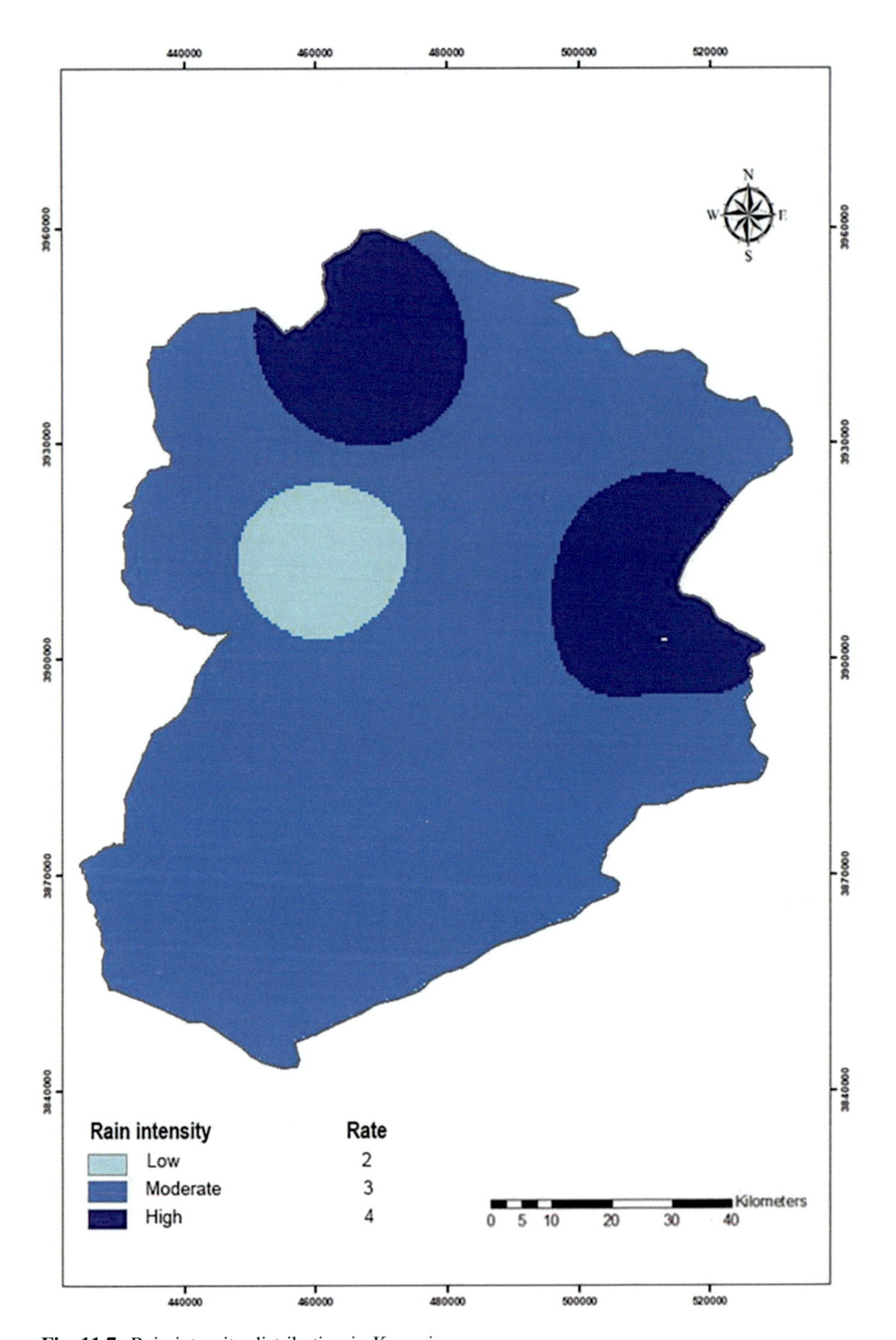

Fig. 11.7 Rain intensity distribution in Kasserine

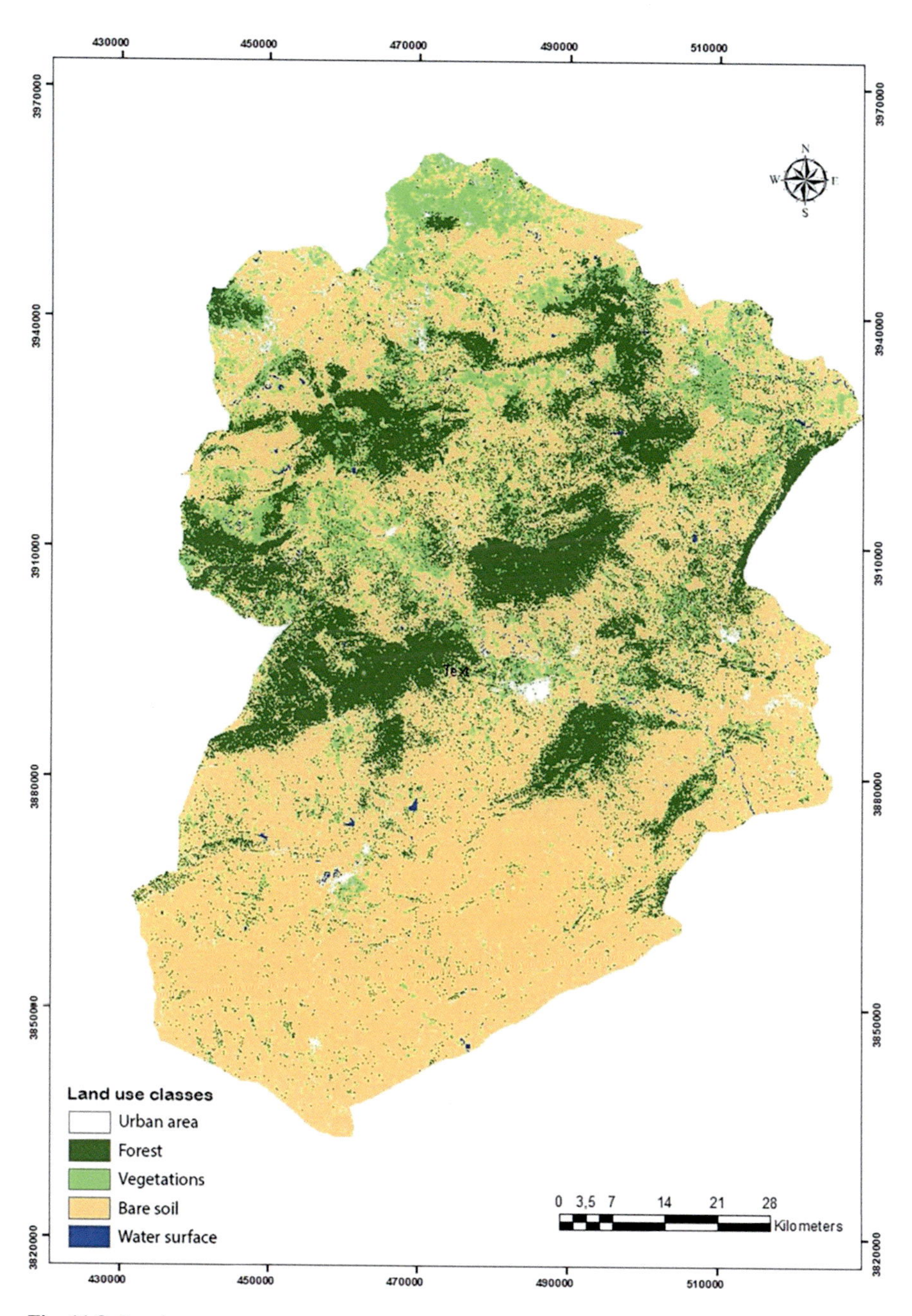

Fig. 11.8 Land use map

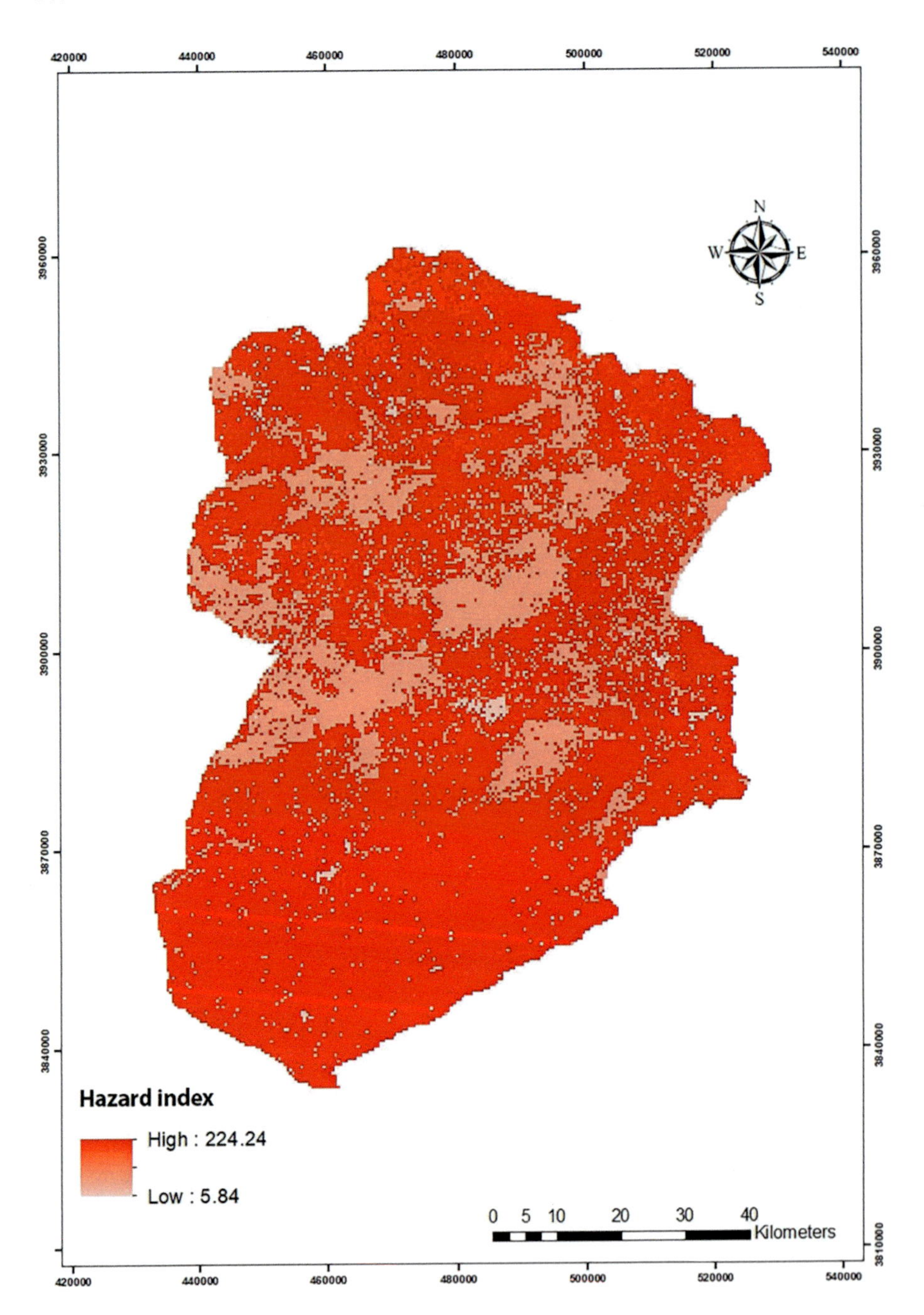

Fig. 11.9 Hazard index map

Table 11.1 Weights calculated using IDS for each parameter of vulnerability and Hazard

Parameter	Weight
Vulnerability	
Pente	0.5052
Réseau hydrographique	1.083
Perméabilité	1.7052
Hazard	
Land use	1.1
Unplanned urban areas	1.02
Rain intensity	1.86

11.4.4 Discussion

The multicriteria approach for risk assessment is very relevant and simplified since it considers the impact of factors: Density of hydrographic network, slope, permeability, rain intensity, land use and unplanned urban zones. But if considering the methodology for the elaboration of the majority of thematic maps of the different parameters, there are errors caused by interpolation. So, it is suggested to minimize errors in the spatial distribution of thematic maps of different parameters and also in the vulnerability, hazard and risk maps.

11.5 Conclusion

The floods constitute a naturel multidimensional disaster. So, the purpose of this chapter was to present a multi -attribute model for Kasserine flood risk assessment. The methodology developed here provides decision makers a spatial visualization of multi-attribute risk.

The results reveal a high risk in the majority of flat areas with high hydrographic network density and in urban areas and especially in the unplanned urban areas.

The comparison of results to the training data for validation demonstrates that not only the physical context and risk situation are relevant for flood risk assessment, but also the socio-economic and cultural context. Also, some other factors should be involved in the risk assessment such as, the quality of services, the state of infrastructure and particularly the routes. These parameters should be involved in future works in order to reflect better the reality of the area.

Also, attention must be paid to areas at risk and in particular to areas of unplanned urban areas and not covered by the sanitation network. However, this study deserves to be refined by adding flood forecasting for different return periods (after 10 years, 20 years, 50 years) in order to prepare decision makers against the impending floods.

It is also suggested.

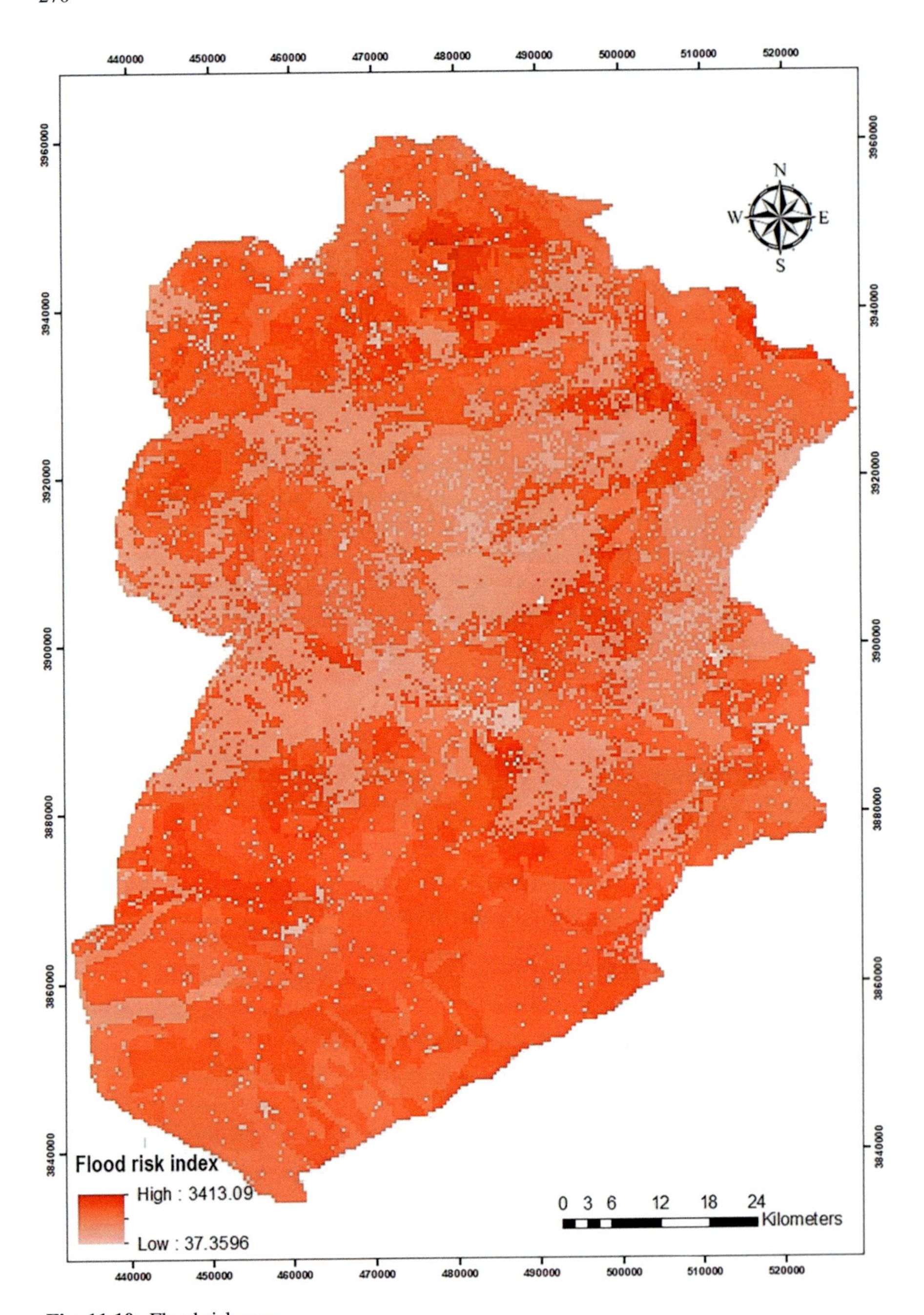

Fig. 11.10 Flood risk map

Aknowledgments This work was supported and financed by the national research project for the encouragements of young researchers PEJ "PEJC 01-19PEJC02" whose coordinator is Dr. Salwa SAIDI the first and the corresponding author. It is also supported in part by the Directorate of Dams and Hydraulic Works of the ministry of agriculture of Tunisia.

References

Ajim Ali, S. K., Parvin, F., Bao, P. Q., Vojtek, M., Vojteková, J., Costache, R., Nguyen, T., Nguyen, H. Q., Ateeque, A., Ghorbani, M. A. (2020). GIS-based comparative assessment of flood susceptibility mapping using hybrid multi-criteria decision-making approach, naïve Bayes tree, bivariate statistics and logistic regression: A case of Topľa basin, Slovakia, Ecological Indicators Volume 117, October 2020, 106620.

AUGT. (2015). L'urbanisation dans le Grand-Tunis, Tunisie, rapport inedit, p. 20.

Bates, P. D. (2004). Remote sensing and flood inundation modelling. *Hydrological Processes, 18*(13), 2593–2597. https://doi.org/10.1002/hyp.5649

Commission of the European Communities (COM). (2006). *Proposal for a Directive of the European Parliament and of the Council on the Assessment and the Management of Floods 2006/0005(COD)*. Accessed at http://eur-lex.europa.eu/legal-content/EN/TXT/?uri=CELEX:52006PC0015, March 2015.

Costache, R., Pham, Q. B., Sharifi, E., Linh, N. T. T., Abba, S. I., Vojtek, M., Vojteková, J., Nhi, P. T. T., & Khoi, D. N. (2020). Flash-flood susceptibility assessment using multi-criteria decision making and machine learning supported by remote sensing and GIS techniques. *Remote Sensing, 12*, 106.

CRDA. (2018). Etude hydrologique de la plaine de Kasserine, Unpublished report, 20p.

Das, S. (2019). Geospatial mapping of flood susceptibility and hydro-geomorphic response to the floods in Ulhas basin, India. *Journal of Remote Sensing Applications: Society Environment, 14*, 60–74.

Ghanbarpour, M., Salimi, S., & Hipel, K. (2013). A comparative evaluation of flood mitigation alternatives using GIS based river hydraulics modelling and multicriteria decision analysis. *Journal of Flood Risk Management, 6*(4), 319–331.

Haq, M., Akhtar, M., Muhammad, S., Paras, S., & Rahmatullah, J. (2012). Techniques of remote sensing and GIS for flood monitoring and damage assessment: A case study of Sindh province, Pakistan. *The Egyptian Journal of Remote Sensing Space Science, 15*(2), 135–141.

Hassen, I., Gibson, H., Hamzaoui-Azaza, F., Negro, F., Rachid, K., & Bouhlila, R. (2016). 3Dgeological modeling of the Kasserine Aquifer System, Central Tunisia: New insights into aquifer-geometry andinterconnections for a better assessment of groundwater resources. *Journal of Hydrology, 2016*. https://doi.org/10.1016/j.jhydrol.2016.05.034

HIRA. (2017). Hazard identification and risk assessment,NRV hazard mitigation plan: Update 2017.

Jaafari, A., Najafi, A., Pourghasemi, H., Rezaeian, J., & Sattarian, A. (2014). GIS-based frequency ratio and index of entropy models for landslide susceptibility assessment in the Caspian forest, northern Iran. *International Journal of Environmental Science Technology, 11*(4), 909–926.

Kim, S. M., Choi, Y., Suh, J., Oh, S., Park, H. D., Ho Yoon, S., & Go, W. R. (2012). ArcMine: A GIS extension to support mine reclamation planning. *Computational Geosciences, 46*, 84–95. https://doi.org/10.1016/j.cageo.2012.04.007

Mandavgade, N. K., Kalbande, V.N., Bilawane, R.R., Kanojiya, M.T., Padole, C.U. (2021) *AHP for ranking effect of qualitative factors in uncertainty measurement of material testing, materialstoday proceedings*. Available online 13 March 2021, In Press, Corrected Proof.

Nitin, S., Avinash, K. S., & Ankur, C., (2021). Modeling supplier selection in the era of Industry 4.0. *Benchmarking: An International Journal, 28*(5), 1809–1836. https://doi.org/10.1108/BIJ-12-2018-0441

Papaioannou, A., Santesso, N., Suzanne, N., Morin, F. S., Adachi, J. D., Crilly, R., Giangregorio, L. M., Jaglal, S., Josse, R. G., Kaasalainen, S., Katz, P., Moser, A., Pickard, L., Weiler, H., Whiting, S., Skidmore, C. J., & Cheung, A. M. (2015). Recommandations en vue de la prévention des fractures dans les établissements de soins de longue durée. *CMAJ, 187*(15), E450–E461. https://doi.org/10.1503/cmaj.151124

Patel, D. P., & Srivastava, P. K. (2013). Flood hazards mitigation analysis using remote sensing and GIS: Correspondence with town planning scheme. *Journal of Water Resources Management, 27*(7), 2353–2368.

Pradhan, B. (2009). Groundwater potential zonation for basaltic watersheds using satellite remote sensing data and GIS techniques. *Central European Journal of Geosciences, 1*, 120–129.

Rahmati, O., Zeinivand, H., Besharat, M., & Risk. (2016). Flood hazard zoning in Yasooj region, Iran, using GIS and multi-criteria decision analysis. *Journal of Geomatics, Natural Hazards, 7*(3), 1000–1017.

Saaty, T. L. (1977). A scaling method for priorities in hierarchical structures. *Journal of Mathematical Psychology, 15*, 234–281. https://doi.org/10.1016/0022-2496(77)90033

Saidi, S., Bouri, S., & Ben Dhia, H. (2011). Sensitivity analysis in groundwater vulnerability assessment based on GIS in the Mahdia – Ksour Essaf Aquifer, Tunisia: A validation study. *Hydrological Sciences Journal, HSJ, 56*(2), 1–17.

Saidi, S., Hosni, S., Manai, H., Jlassi, F., Bouri, S., & Anselme, B. (2017). GIS-based multi-criteria analysis and vulnerability method for the potential groundwater recharge delineation, case study of Manouba phreatic aquifer, NE Tunisia. *Environmental Earth Sciences, 76*(15). https://doi.org/10.1007/s12665-017-6840-1(IF:1.87)

Saidi S., Ghattassi A., Anselme B., and Bouri S. (2018) GIS based multi-criteria analysis for flood risk assessment: Case of Manouba Essijoumi Basin, NE Tunisia, H. M. El-Askary et al. (eds.), Advances in remote sensing and geo informatics applications, advances in science, Technology & Innovation, https://doi.org/10.1007/978-3-030-01440-7_64.

Sanyal, J., & Lu, X. (2004). Application of remote sensing in flood management with special reference to monsoon Asia: A review. *Journal of Natural Hazards, 33*(2), 283–301.

Sedghiyan, D. , Ashouri, A., Maftouni, N., Xiong, Q., Rezaee, E., Sadeghi, S. (2021). Prioritization of renewable energy resources in five climate zones in Iran using AHP, hybrid AHP-TOPSIS and AHP-SAW methods, Sustainable Energy Technologies and Assessments, Volume 44, April 2021, 101045.

Strobl, R., Forte, F., & Lonigro, T. (2012). Comparison of the feasibility of three flood risk extent delineation techniques using geographic information system: Case study in Tavoliere delle Puglie, Italy. *Journal of Flood Risk Management, 5*(3), 245–257.

Sutradhar, S., Mondal, P., Das, N. (2021). Delineation of groundwater potential zones using MIF and AHP models: A micro-level study on Suri Sadar sub-division, Birbhum District, West Bengal, India, Groundwater for Sustainable Development, Volume 12, February 2021, 100547.

Tehrany, M. S., Pradhan, B., & Jebur, M. N. (2014). Flood susceptibility mapping using a novelensemble weights-of-evidence and support vector machine models in GIS. *Journal of Hydrology, 512*, 332–343.

Tehrany, M. S., Pradhan, B., Mansor, S., & Ahmad, N. (2015). Flood susceptibility assessment using GIS based support vector machine model with different kernel types. *Journal of Catena, 125*, 91–101.

Ting, S. C., & Cho, D. I. (2008). An integrated approach for supplier selection and purchasing decisions. *International Journal of Supply Chain Management, 13*, 116–127.

Wanders, N., Karssenberg, D., Roo, A. d., De Jong, S., & Bierkens, M. (2014). The suitability of remotely sensed soil moisture for improving operational flood forecasting. *Journal of Hydrology Earth System Sciences, 18*(6), 2343–2357.

White, I., Kingston, R., & Barker, A. (2010). Participatory geographic information systems and public engagement within flood risk management. *Journal of Flood Risk Management, 3*(4), 337–346.

Chapter 12
Applications of Remote Sensing Techniques in Earthquake and Flood Risk Assessment in the Cyrenaica Region, Al Jabal Al Akhdar Area, NE Libya

Mohammed Afkareen and Jamal Zamot

Abstract Earthquakes and flooding are naturally occurrences events and they are known as natural hazards. These hazards considerably cause multiple harms and threaten human life. Libya is accounted among those countries that suffer from earthquakes and floods as natural hazard. Libya is located within the active plate boundaries of the African plate and the European plate, where these two continents are in a relative subducting motion. As consequence, the boundaries areas of these plates are seismically active. Furthermore, Libya is also characterized by an arid to semi-arid climate, which is punctuated by periods of thunderstorms and intensive rain in winter time on high land areas. The north-eastern part of Libya, which is called a Cyrenaica region, has experienced varying degrees severe and destructive earthquakes and floods. In this chapter, remote sensing and GIS applications have been used in linking the earthquakes and their relationship to local faults and linear features to areas most affected in the event of an earthquake in Cyrenaica. Besides that, geomorphic parameters of selected wadies have been analysed for risk assessment of flash flood on certain areas of the Al Jabal Al-Akhdar area.

12.1 Introduction

The Mediterranean region is a complex region where both of the African plate and the European plate are in a relative subducting motion. The African plate is slowly subducting beneath the European plate in the Hellenic Arc Subduction Zone. Within this area many earthquakes of great strength were recorded and caused a lot of damage in both the Greek islands and the Turkish state (Lagesse et al., 2017). The

M. Afkareen
Sirte Oil Company (SOC), Marsa El-Brega, Libya

J. Zamot (✉)
Earth Sciences department, University of Benghazi, Benghazi, Libya
e-mail: Jamal.Zamot@uob.edu.ly

M. M. Al Saud (ed.), *Applications of Space Techniques on the Natural Hazards in the MENA Region*, https://doi.org/10.1007/978-3-030-88874-9_12

Cyrenaica region is located near this seismically active area, and as consequences it is considered as an earthquake zone that is affected by large earthquake activity.

In addition, Libya is characterized by a semi-arid climate that punctuated by periods of thunderstorms and intensive rain in winter time. These kinds of heavy showers are well observed in highland areas, especially on Jabal Nafusah in the west and Jabal Al Akhdar in the east (Elfadli, 2009; El-Tantawi, 2005). In the Cyrenaica region, the averages annual rain fall are from 200 to 500 mm. The highest recording of rainfall in Libya was measured with about 850 mm on Jabal Al Akhdar area, which is an integral part of Cyrenaica region (Elfadli, 2009). Moreover, the flash flood is a well-known phenomenon that frequently occurred in Al Jabal Al Akhdar area and it causes many damages yearly, therefore, it can be considered as the main natural hazard in Al Jabal Al Akhdar area of Cyrenaica region.

However, the modern techniques of remote sensing show a successful approach in forecasting and vulnerability analysis, besides to damage assessment, caused by natural disasters. In this chapter, firstly, we focus on linking earthquakes and their relationship to local faults and linear features that extracted from digital elevation models of the areas most affected in the event of an earthquake. Next in order, many studies have used remote sensing and GIS applications for analysing the geomorphic parameters of some wadies in Al Jabal Al Akhdar area to assess the risk of rain water discharge. Here, three wadies are namely wadi Darnah, wadi Ar Ramlah and wadi Ash- Sharif; have been chosen as a representative for a risk assessment of flood in the area.

12.2 Geological Setting

The Jabal Al Akhdar area is located as an integral part of Cyrenaica region in the northeast of Libya. It is bordered by the Egyptian borders to the east, the desert region to the south and the Mediterranean Sea to the north and west directions. The study area is dominantly by two tectonic provinces, the Jabal Al Akhdar Uplift and fold belt in the north and the Cyrenaica Platform in the south. Al Jabal Al Akhdar started as basin and inverted to be uplifted mountain in two stages during Santonian and middle Eocene (Hallett, 2002). The exposed rocks in the Al Jabal Al Akhdar are made up of marine carbonate units ranging in age from the Late Cretaceous to Late Miocene (El Hawat & Shelmani, 1993). The total thickness of these units is about 2000 m, and they rest uncomfortably on each other due to their affecting by uplifted tectonic (Fig. 12.1; Hallett, 2002; El Hawat & Shelmani, 1993). The deposition environment of these carbonate rocks is mainly shallow to deep marine with Some or few evaporite rocks due to the Messinian event in the late Miocene (Hallett, 2002; El Hawat & Shelmani, 1993).

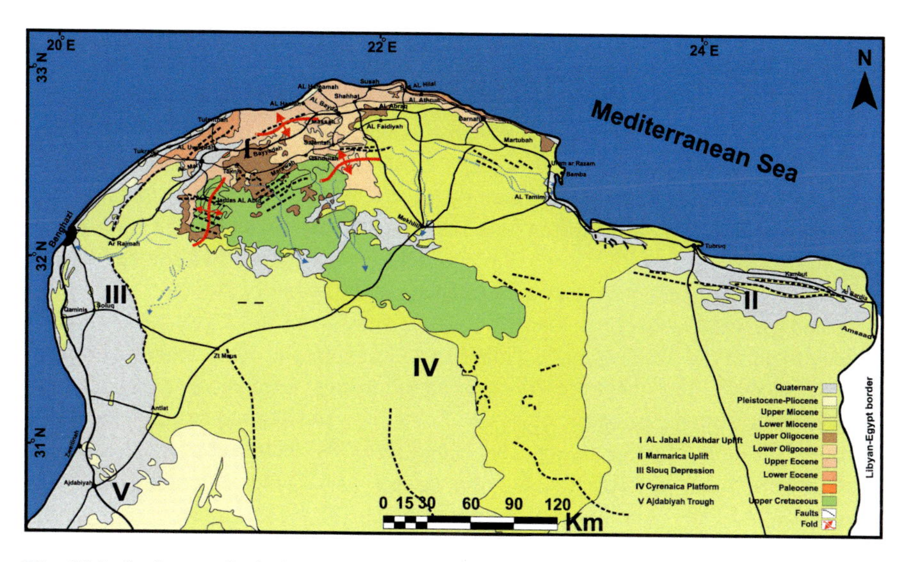

Fig. 12.1 Surface geological map with structure components of north-eastern Libya. (Shaltami et al., 2020)

12.3 Methodology

For earthquake assessment, some data have been collected from the Shuttle Radar Topographic Mission (SRTM) Multi- resolution (30 & 90 m), also Seismicity data are obtained from the United States Geological Survey (USGS) (https://earthquake.usgs.gov). In addition, the linear features extraction has done through maps of shadows and compared them with existing geological maps. For flood risk assessment, Digital Elevation Model (DEM) have been collected over the study area. This DEM for wadi Ar Ramlah and wadi Darnah was taken from ALOS PALSAR produced by (ASF) (ASF, 2020); with a spatial resolution equals to 12.5 meters, and 30 m as spatial resolution produced by (SRTM) for wadi Ash- Sharif. In order to visualize the digital data, ArcMap software were used to place location data collected by the GPS device, together with determination the maximum distance that flood reached to on a base map.

12.4 Seismicity of NE Libya

Earthquakes occur suddenly and without warning, and may cause a lot of issues in the infrastructure or deaths in the most severe earthquakes. Considering the vulnerability of Libya's infrastructure and despite a divergence of the earthquakes in time periods, the earthquakes cause many casualties and seriously damages. According to Hassen (1983) Libya has been divided into four regions in terms of earthquake magnitude, where the central and the north-eastern regions of Libya are considered the most active areas and the southern region is a region of limited activity.

The collected seismicity data from the United States Geological Survey (USGS) from (1955 to 2020) suggested that 35 earthquakes were recorded in the Cyrenaica region (Fig. 12.2a), and the most of these earthquakes are shallow (0–40 km). In addition, from the relationship between the depth and the earthquakes magnitude in Cyrenaica region (Fig. 12.2b), it is found that approximately half of the earthquakes that were recorded are with 3.3 to 5.3 magnitudes at the depth of 10 kilometres, while the largest earthquakes were recorded (5.6) are at a depth 25 kilometres.

It can be easily observed that all seismic activity in the Cyrenaica region is concentrated in the northern part, while in the southern part there are no records of seismic activity. The majorities of these earthquakes were recorded in offshore slightly apart from the coastline between Darnah city and Tubruq city. Nevertheless, although the area that extending from Benghazi City to the city of Al-Bayda is considered a fairly stable area, but the largest severe recent earthquakes were recorded in Al-Marj area; (e.g., Al-Marj earthquake in 1963).

Most earthquakes in the Al-Marj area (Fig. 12.4a, d), occurred approximately along the same line that extends from Tulmithah City to Taknis City around Al-Marj area. All these earthquakes occurred at the same depth of 10 kilometres, except the largest magnitude 5.6 that recorded at 25 kilometres.

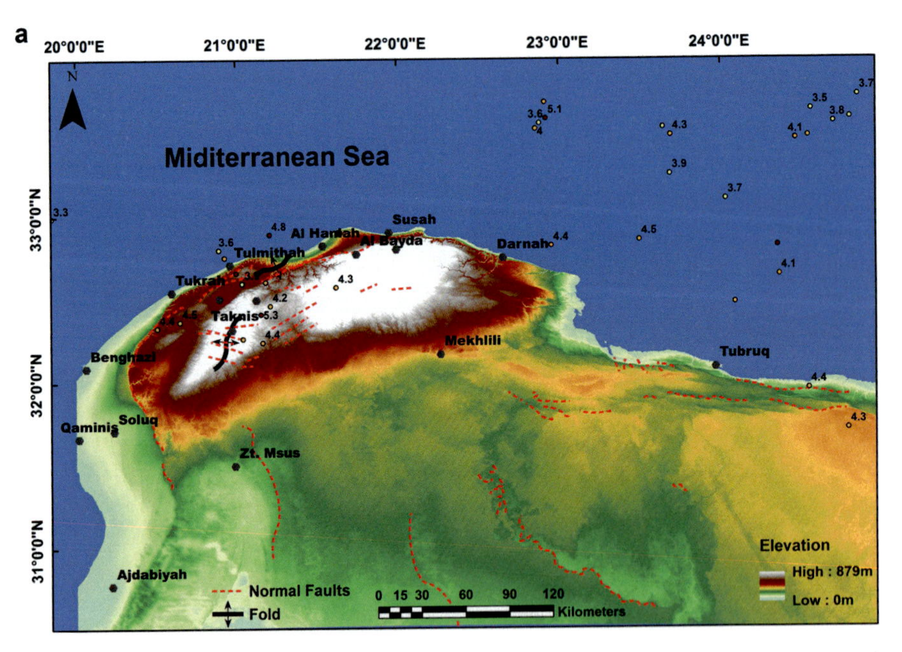

Fig. 12.2 (**a**) DEM of the north-eastern region of Libya shows the distribution of earthquakes and their magnitude; (**b**) Plot diagram illustrates depth and magnitude of earthquakes in north-eastern region of Libya

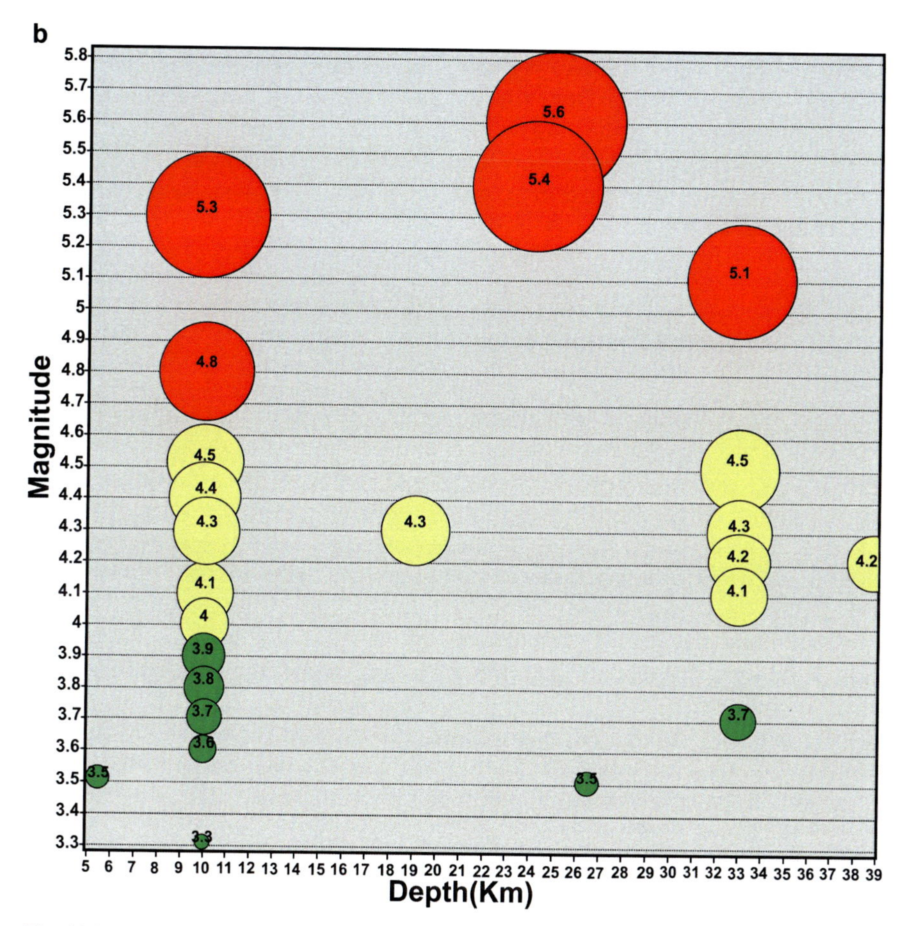

Fig. 12.2 (continued)

12.5 Results and Discussion

12.5.1 Active Deformation of NE Libya

Cyrenaica is formed by five-structural categories; Soluq Depression, Ajdabiyah Trough, Al Jabal al Akhdar Uplift, Cyrenaica Platform and Marmarica Uplift (Fig. 12.1). These structural deformations have resulted in a variety of structural elements of faulting and folding (El-Arnauti et al., 2008). The most active deformation structure is; Al Jabal al Akhdar Uplift, Marmarica Uplift and Soluq Depression.

Al Jabal al Akhdar area has been described as an anticlinorium inversion structure and the main folding took place during the Santonian and further uplifted in the middle Eocene (El Hawat & Shelmani 1993; Chrisite, 1955). Furthermore, the Marmarica Uplift in the northeast of Cyrenaica, are covered by mainly carbonates rocks of Tertiary age with a gentle slope. It represents a faulted step-down of the northern margin of the Cyrenaica Platform. The northern part of Marmarica Uplift is affected by a zone of normal faults with northwest orientation, while the south part is more stable and no faults showing on the surface of the region (El Deftar& Issawi, 1977). The Soluq Depression has very limited faults as the region was not exposed to significant tectonics after deposition, and as a result most of the sediments were preserved (Hallett, 2002).

12.5.2 Tectonic Geomorphology

Tectonic geomorphology is defined as the application of geomorphology to tectonic problems. It includes the study of landform assemblages and landscape evolution as well as development of process-response models for areas and regions affected by recent tectonic activity (Keller & Rockwell, 1984). Moreover, the terrain is formed and affected by the processes of faults and distinct shapes such as steep scarps, folds, elongate ridges offset terraces, and linear valleys, and deflected, offset and uplifted streams (Gupta, 2018).

Accordingly, the tectonic surface maps of the Al-Marj area (Fig. 12.4c, d and Fig. 12.5b) show that the Jardas Al Abid is highly dominated by faulting. The main trend of these structures is, ENE–WSW (Arsenikos et al., 2013). It can clearly observe that the offset of the liner valleys is substantially cut by an escarpment that resulted in forming triangular facets. This scarp is considered as fault.

Further, the fault scarps are prominent on the two escarpments from the northern part of Al Jabal al Akhdar, and they can be identified in the coastal direction in the east-west direction enclosing Al-Marj City (Fig. 12.5a; El Amawy et al., 2009; Campbell, 1968). Besides to the aforementioned fault scarps, two large anticlines folds are observed in the Al- Marj area between Jardas al Abid and in the east of Taknis City.

12.5.3 Lineament Extraction with Density

A lineament or Linear features can be defined simply as the shapes that appear straight line or curved feature on satellite images. It can be the result of man-made, such as roads or railroads, or the result of structural natural phenomena such as faults or fractures (Bashe, 2017). By comparing the existing faults and the linear features that extracted from digital elevation models (Fig. 12.3a, b) the interested area can be divided into three basic zonation:

Zone 1, It is the most intense in terms of severity and concentration of both faults and linear features, which is what is emphasized in this part of chapter.

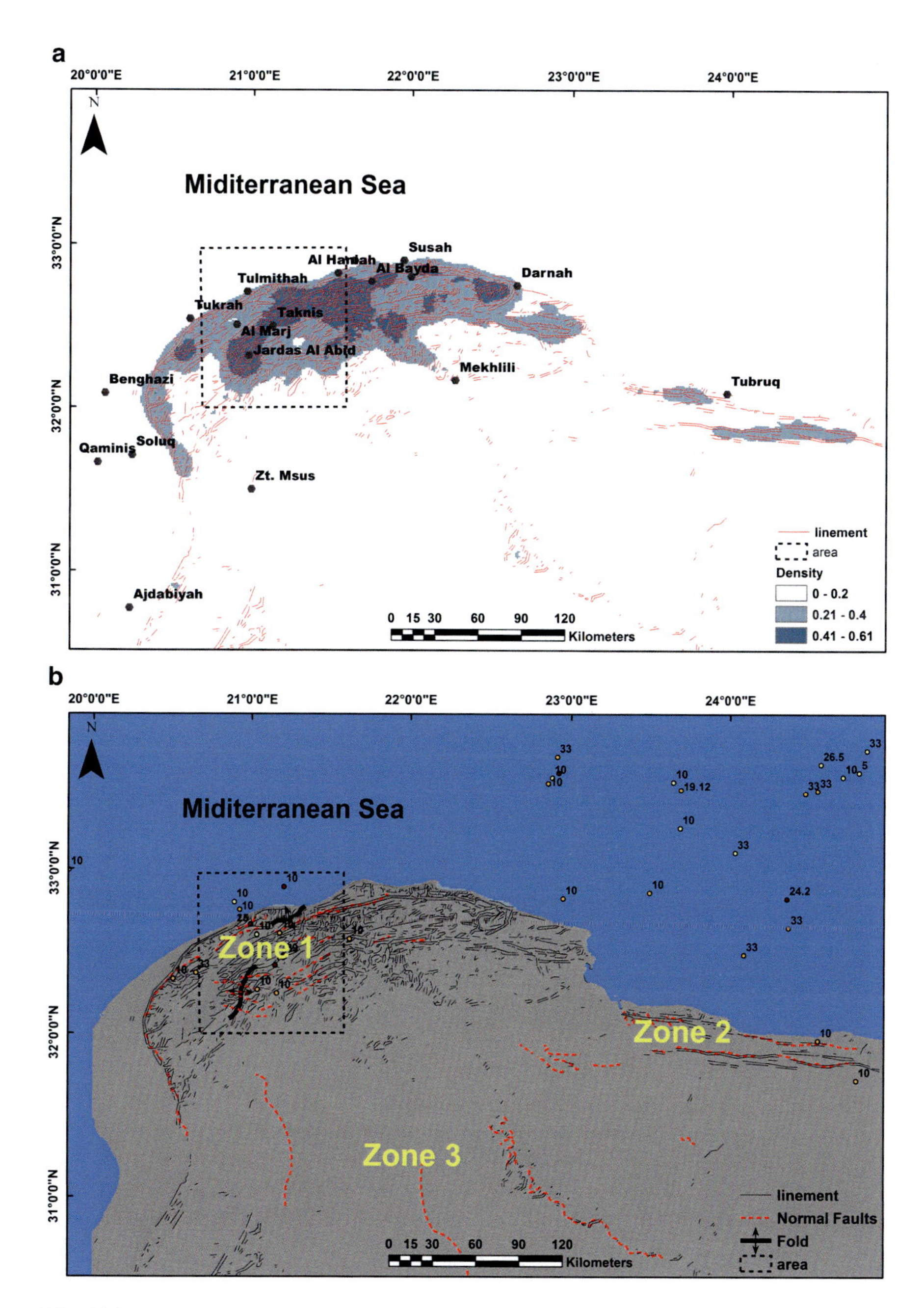

Fig. 12.3 (**a**) Lineament density map. (**b**) Lineament map with faults distribution and zonation based on severity and concentration of both faults and linear features

Zone 2, It is considered a limited concentration in terms of linear features and faults, with faults spreading clearly to the east-west direction, but the occurrence of earthquakes in a nearby area in marine areas has affected in some way the occurrence of some earthquakes.

Zone 3, This region has a very limited concentration of faults and linear features, as it is considered the most stable region in addition to its distance from the marine earthquake zone.

The linear data that was extracted from STRM and converted into the shadow models (Fig. 12.4a, b and c), found that most of the linear data are in a clear direction towards the Jardas Al Abid and eastern of the Tulmithah City and Taknis City. This is evident in the wide fold process of carbonate rocks in Jardas Al Abid, and in

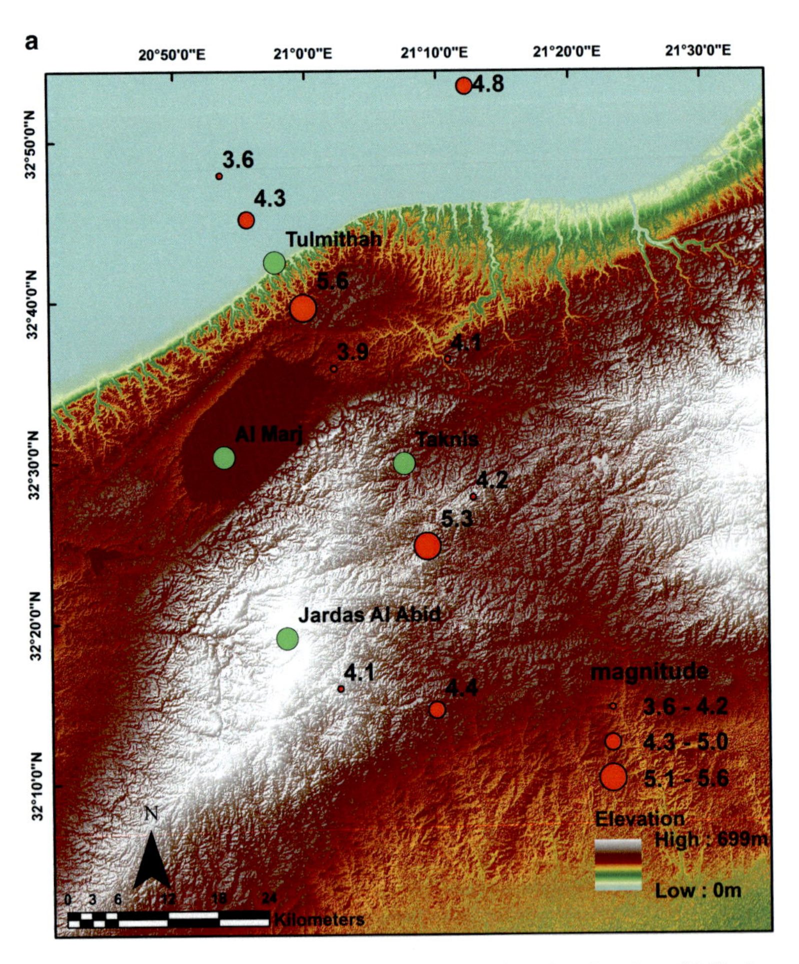

Fig. 12.4 (**a**) Displays Zone (1) with its recorded magnitudes of earthquakes. (**b**) Shadow relief map illustrates the depth of the recorded earthquake. (**c**) Lineament map shows faults distribution in the zone (1). (**d**) Lineament density map of the zone (1)

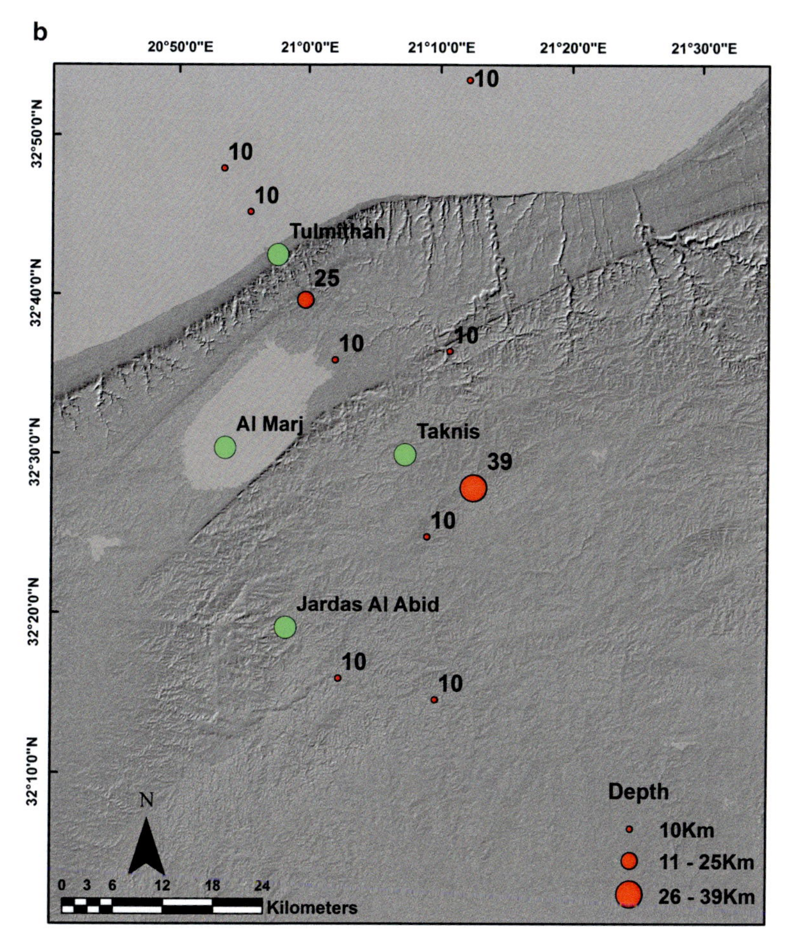

Fig. 12.4 (continued)

remarkably coincidence with most of the faults that were identified in the previous studies and in most of the existing geology maps.

The lineament density that extracted was categorized into three classes (Fig. 12.4c, d); (0–0.3 Km/Km^2, 0.4–0.5 Km/Km^2, and 0.6–0.8 km/km^2). Lineament with densities ranging from (0 to 0.3 Km/Km^2) covers about (25%) of the study area, which is part of the Al-Marj city. The low range of lineament density is due to massive accumulation of the soil and sediments that play a significant role in the absence of deductive linear features.

Lineament with densities ranging from (0.4 to 0.5 Km/Km^2) covers almost (50%) of the study area. This range can be the result of the influence of streams, valleys or linear ridges.

The third lineament densities are between 0.6 and 0.8 km/km^2, covers (25%) of the interested area. These ranges are concentrated between two locations; Jardas Al

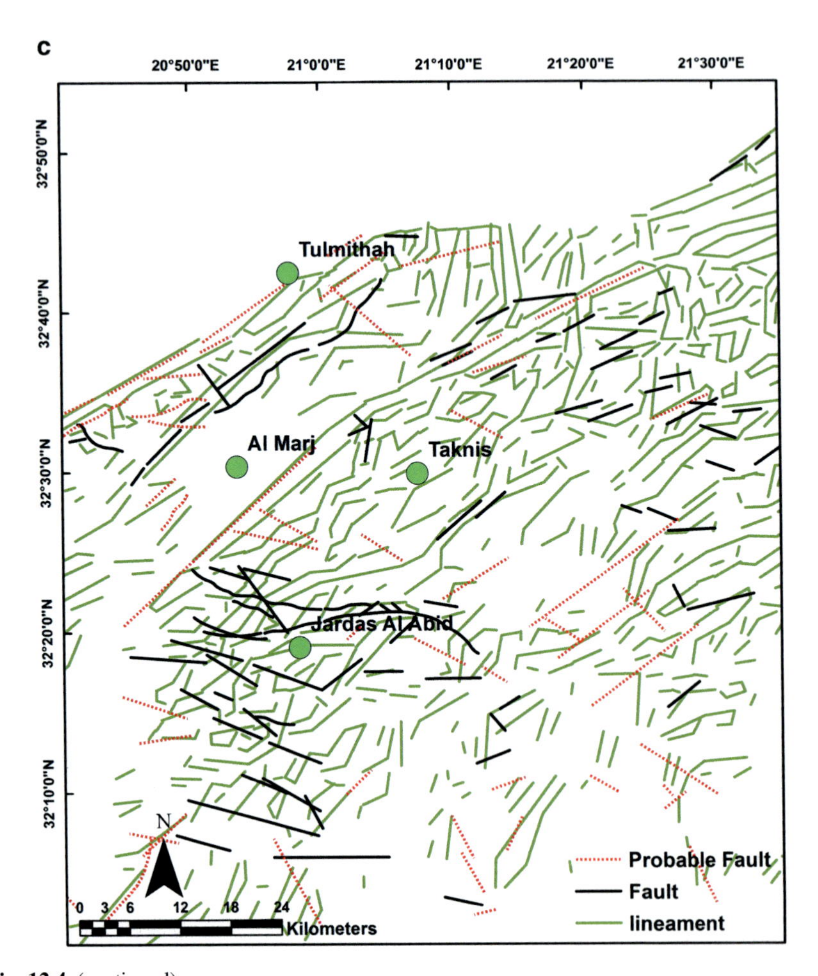

Fig. 12.4 (continued)

Abid and eastern the Tulmithah City and the Taknis City. It could be considered a fault zone of the area.

However, three topographic profiles had drawn (Fig. 12.5); the first profile is across in a direction of southwest to northeast. It crosses the Jardas Al Abaid area in the direction of the coast, where this region is affected by the intensity of faults resulting from deformation of folding. Likewise, this cross section represents two escarpments probable fault scarps to the coast region. The rest other profiles are drawn (2)(3) in the direction of northwest to southeast across the Al Marj City, Taknis City and Tulmithah City. These profiles have jointly crossed the activity line of earthquakes in the area (Fig. 12.4a, b), and reflect obviously two escarpments' lines as fault scarps.

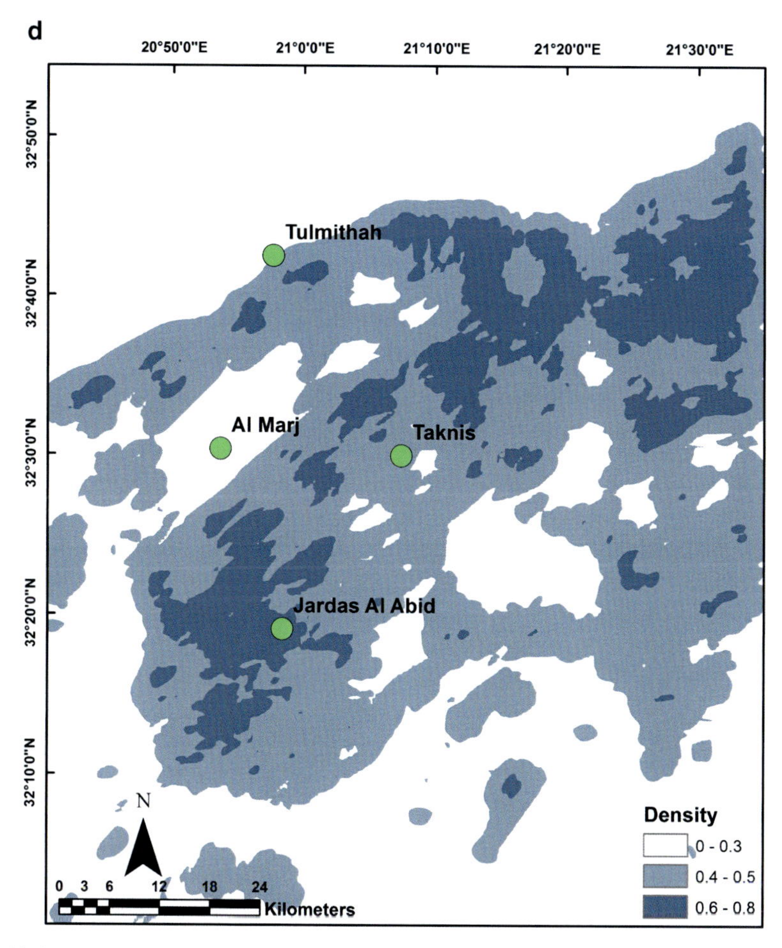

Fig. 12.4 (continued)

12.6 Flash Flood in Al Jabal Al Akhdar Area

Flash floods are defined as rising in water over a short period of time either during or within a few hours of the rainfall (Doswell, 2003). Besides that, a relatively small amount of rain or the intensity and duration of the rainfall with a low rate of infiltration can trigger flash flooding. Flooding can occur virtually anywhere, in steep rocky terrain or even within heavily urbanized regions (Youssef et al., 2011; Doswell, 2003). Many several factors have relevance to the occurrence of a flash flood ranging from topography, geomorphology, drainage, engineering structures, and climate. In fact, these are not all the factors that contributing the flood, there are other interrelated factors which influence flash floods severity, especially in desert areas, such as water loss (evaporation and infiltration) drainage networks, rainfall

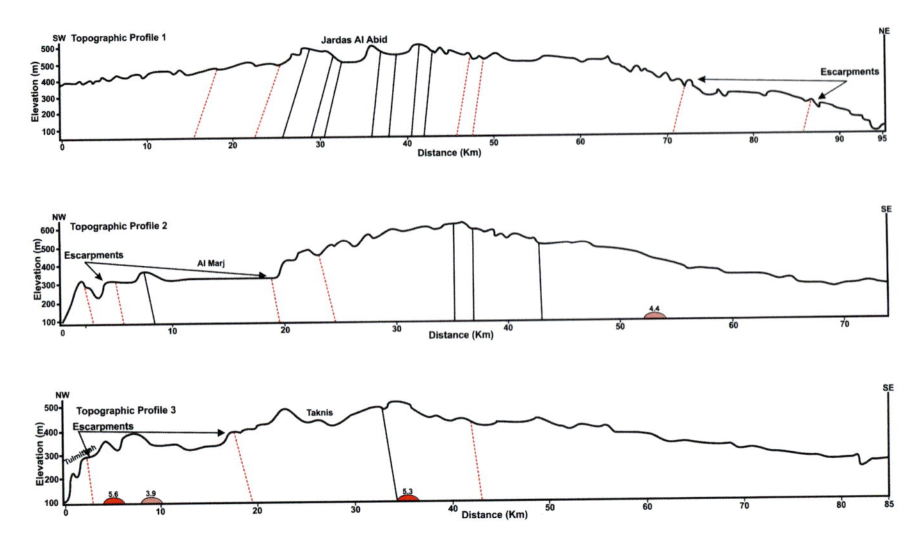

Fig. 12.5 Three topographic profiles dissect the fault scarps and uplifts along the zone (1)

characteristics, drainage orders (Youssef et al., 2011; Doswell, 2003). One of the documented floods in Al Jabal Al Akhdar area is on the 27th of September in 2018, flood swept through the village of Mikhili and caused many damages (Zamot & Afkareen, 2020). The water level reached about 1.5 m above the ground surface, where number of houses were fully and others partially devastated in addition to the hospital in the village. The water washed away some cars that were trying to pass through the flood, the hospital wall was demolished, some trees were uprooted and unfortunately two victims have died.

12.7 Location of Study Wadies

The selected wadies are namely; wadi Darnah, wadi Ar Ramlah, and wadi Ash Sharif. wadi Darnah, and wadi Ash Sharif are placed in northern part of the Al Jabal Al Akhdar area and terminated in Mediterranean Sea, while wadi Ar Ramlah is located in the southern part of the Jabal Al Akhdar and discharged into a flat desert area in the south (Fig. 12.6).

12.8 Results and Discussion

According to (Singh et al., 2019) linear aspect, relief aspect, and aerial aspect of the river basin are generally the main three categories of morphometric parameters. These include basin area, perimeter, basin length, stream order and stream length,

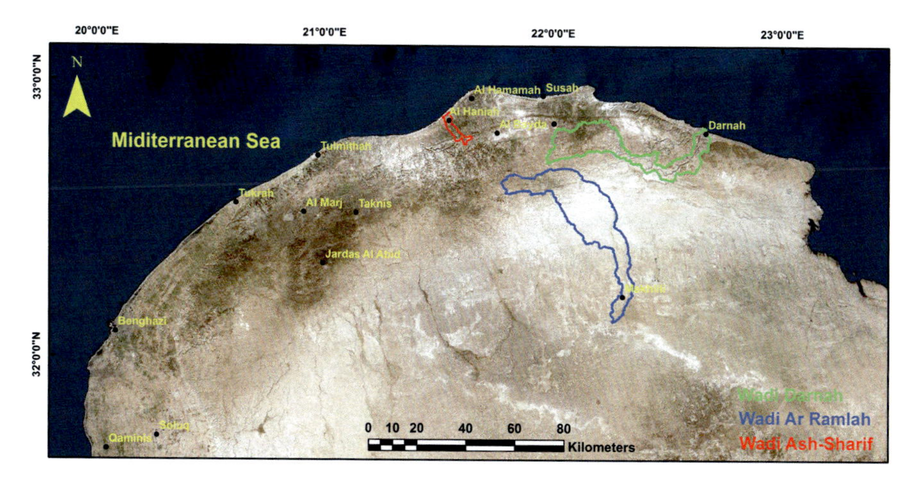

Fig. 12.6 Illustrates the location map of the three studied wadies

bifurcation ratio, basin relief, relief ratio, ruggedness number, drainage density, stream frequency, drainage texture, form factor, circulatory ratio, elongation ratio, length of overland flow, Infiltration number, dissection Index (DI), and constant channel maintenance. The main results of morphometric analysis and relief parameters of the studied wadies are listed in the Tables 12.1 and 12.2 respectively.

12.8.1 Geomorphological Aspect

Al Jabal Al Akhdar area is more or less eroded by discontinuously flooded valleys (wadies), where most of these wadies that are located in the southern part are flatten, broaden and shallowing with a few meters deep, while those in the north are deeper and narrower (Röhlich, 1974). The studied wadies passe through different types of marine carbonate rocks, besides to the quaternary deposits that cover the large part of the Al Jabal Al Akhdar area by alluvial sediments and terra-rossa soil.

Wadi Darnah is one of the longest wadies in the northern of Al Jabal Al Akhdar, it starts from Al Abraq Village at an elevation 889 m and ending at elevation 27 m in Darnah city where it is discharged into the Mediterranean Sea. Its length is about 77.773 km with water catchment area 554.89 km^2, and it has a fifth order of streams (Fig. 12.7).

Wadi Ar Ramlah has an about 100 km length, which extends from the Sidi Al Hamri in the north to about 20 km into the south of the Mikhili village to an area known as Balta Al Ramlah where the wadi is terminated. The elevation of wadi Ar Ramlah at Sidi Al Hamri is about 883 m above sea level, and ending at an elevation about 169 m above sea level. Wadi Ar Ramlah is considered as one of the longest wadies in south part of Al Jabal Al Akhdar area, and it has a sixth order of streams (Fig. 12.8).

Table 12.1 Shows results of some calculated morphometric parameters at study area

Morphometric parameters	Formula	Results of Wadi Darnah	Results of Wadi Ar-Ramlah	Results of Wadi Ash-Sharif	Refernces
Basin length (L)	GIS software	77.773 km	90.84 km	14.024 km	Schumm (1956)
Basin perimeter (P)	GIS software	279.255 km	279.039 km	48.228 km	Schumm (1956)
Basin area (A)	GIS software	554.89 km^2	832.99 km^2	41.072 km^2	Schumm (1956)
Drainage density (Dd)	Dd = Lu / A; where Lu total length of streams; A area of watershed	0.9 km/km^2	2.62 km/km^2	2.5 km/km^2	Horton (1932)
Stream frequency (sf)	Sf = nu / A; where nu total number of streams; A area of watershed	0.77 n/km^2	7.03 n/km^2	5.04 n/km^2	Horton (1932)
Texture ratio (Rt)	Rt = ΣNu/P, where nu the total number of streams, P perimeter of watershed.	1.53 n/km	20.97 n/km	4.29 n/km	Schumm (1956)
Form factor (Rf)	Rf = A/(L)2; where A area of watershed; L basin length.	0.092	0.10	0.20	Horton (1932)
Circularity ratio (Rc)	Rc = 4πA/P 2; π = 3.14, A area of watershed, P perimeter of watershed.	0.089	0.14	0.22	Miller (1953)
Elongation ratio (re)	Re = 2√(A/π)/L; where π = 3.14, A area of watershed, L length of watershed.	0.34	0.358	0.51	Schumm (1956)
Length of overland flow (Lof)	Lof = 1/2Dd; Dd drainage density	0.45	0.19	0.2	Horton (1945)
Infiltration number (in)	In = Dd * sf; sf stream frequency, Dd drainage density.	0.69	18.4	12.6	Faniran (1968)
Constant channel maintenance (Ccm)	Ccm = 1 / Dd; where Dd drainage density	1.11	0.38	0.39	Schumm (1956)

Wadi Ash- Sharif is about 14 km long with water catchment area of 41 km^2, it extends nearly from Al-Bayda city in the south at an elevation 527 m, pass through Al Haniah village and ending into Mediterranean Sea in the north with a fourth order of streams (Fig. 12.9).

Table 12.2 Shows results of some calculated relief parameters at study area

RELIEF PARAMETERS	Wadi Darnah	Wadi Ar Ramlah	Wadi Ash Sharif	Formula	References
Maximum Elevation in the area (hMax)	889 m	883 m	527 m	–	–
Minimum Elevation in the area (hMin)	27 m	169 m	0 m	–	–
Basin relief (H)	862 m	714 m	527 m	H = hMax - hMin	Schumm (1956)
Relief ratio (Rf)	11.08	7.85	37.57	Rf = H/L	Schumm (1956)
Ruggedness number (Rn).	0.78	1.87	1.31	Rn = H* Dd	Melton (1957)
Dissection Index (DI)	0.97	0.8	1	DI = H/ hMax	Schumm (1956)

12.8.2 *Slope, Aspect, and Relative Relief of the Study Wadies*

The rain water that need time to enter in the river beds for making a network of the river basin will be run-off under the effecting of slope elements, (Chow, et al., 1994). Therefore, one of the most significant factors for morphometric analysis and watershed development in geomorphological studies is slope analysis. The slope degree of the studied wadies varies from (2.6 to 62) which the low degree indicates a nearly flat area while the high degree shows the highest slope degree determined in the area, (Figs. 12.7, 12.8 and 12.9).

Furthermore, the Aspect determines the direction of the terrain to which it faces, so that affecting the pattern of precipitation, distribution of vegetation and biodiversity in the study area (Khakhlari & Nandy, 2016). The compass direction of the aspect was derived from the output raster data value. 0 is true north; a 90° aspect is to the east, an 180° is to the south. The wadi Ar Ramlah is mainly dominated by two facing slopes, which are south facing slope and west facing slope. Wadi Darnah is with northeast to southwest facing slope, while wadi Ash- Sharif is southeast to southwest facing slopes (Figs. 12.7, 12.8 and 12.9).

As a result, the direction of the facing slopes in these wadies has low moisture content and high evaporation rate.

The topographical characteristic of the study area is determined by using the relative relief of its catchment area (Singh et al., 2019). Wadi Ar Ramlah has a low relief designated in the southeast side suggests that this area of the basin is flat to gentle slope type, and form the shadow relief map, it is with low structure effect. On the other hand, the relief and shadow relief maps of wadi Darnah and wadi Ash-Sharif show high structure effect with steep slope in the north of the wadi Darnah and in the south of wadi Ash- Sharif (Figs. 12.7, 12.8 and 12.9).

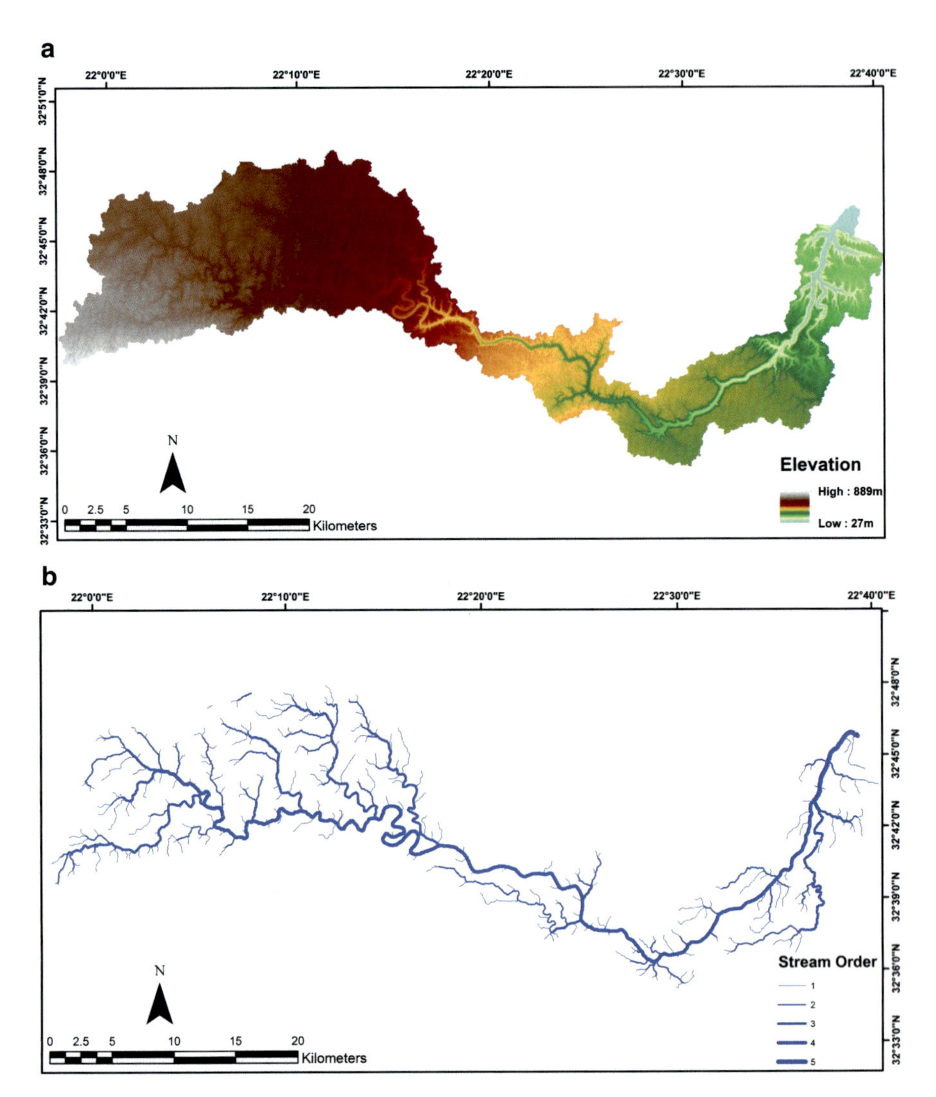

Fig. 12.7 Illustrates different types of geomorphic maps of wadi Darnah; (**a**) Elevation map; (**b**) Stream orders; (**c**) Stream frequency; (**d**) Shadow relief map; (**e**) Slope map; and (**f**) Aspect map

12.8.3 Aerial Morphometric Parameters

12.8.3.1 Drainage Density (Dd)

According to (Bhat et al., 2019) said that a very useful parameter to understand the landscape dissection, runoff potential or travel time of water in a basin, infiltration capacity of the land, relief, underlying lithology, climatic conditions, and vegetation

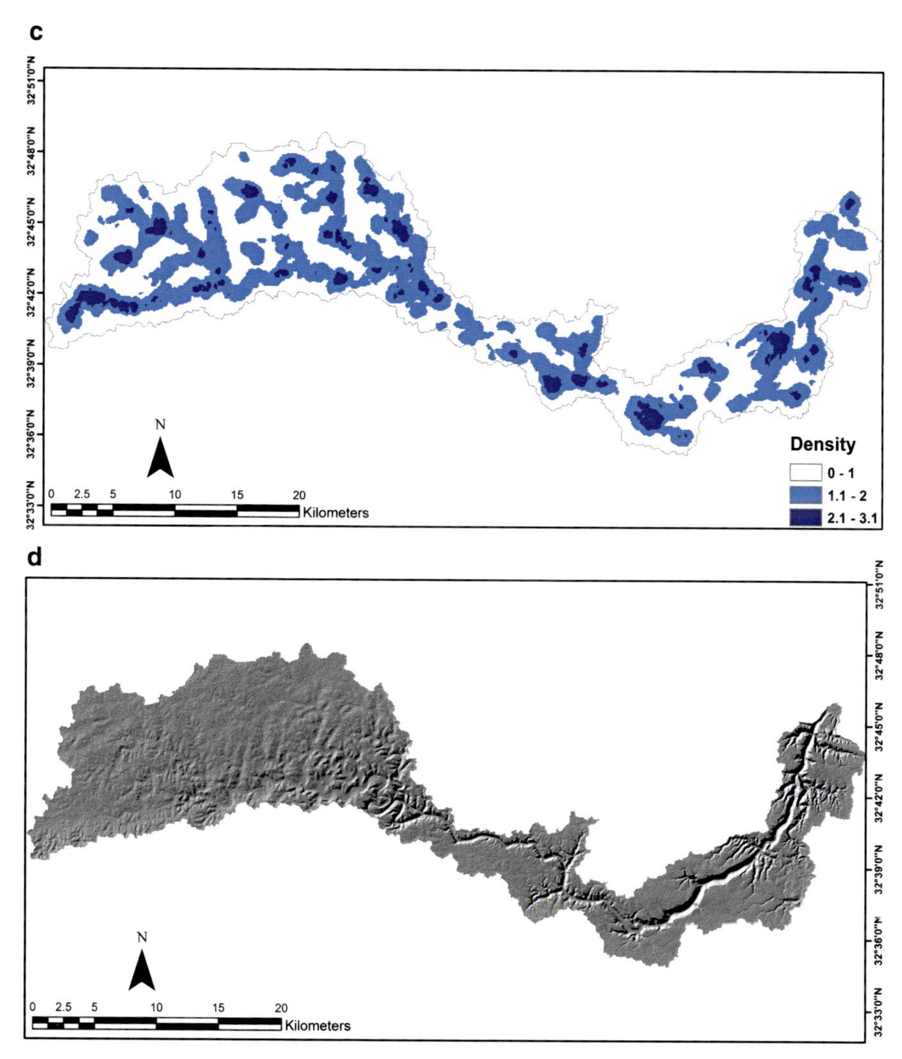

Fig. 12.7 (continued)

cover of the basin, is a Drainage density which also indicates the closeness of spacing of channels. Drainage density is calculated as the total length of streams of all orders per unit area divided by the area of drainage basin (Bhat et al., 2019). Moreover, high and low Drainage density values depend on sub-surface material either impermeable or permeable rocks, vegetation density, relief, surface runoff and infiltration capacity. Wadi Ar Ramlah and wadi Ash- Sharif have high drainage density values which reveal high runoff surface, fine drainage texture with impermeable subsurface, while Wadi Darnah has a low drainage density value with (0.9 km/km^2) indicates low runoff with denser vegetation.

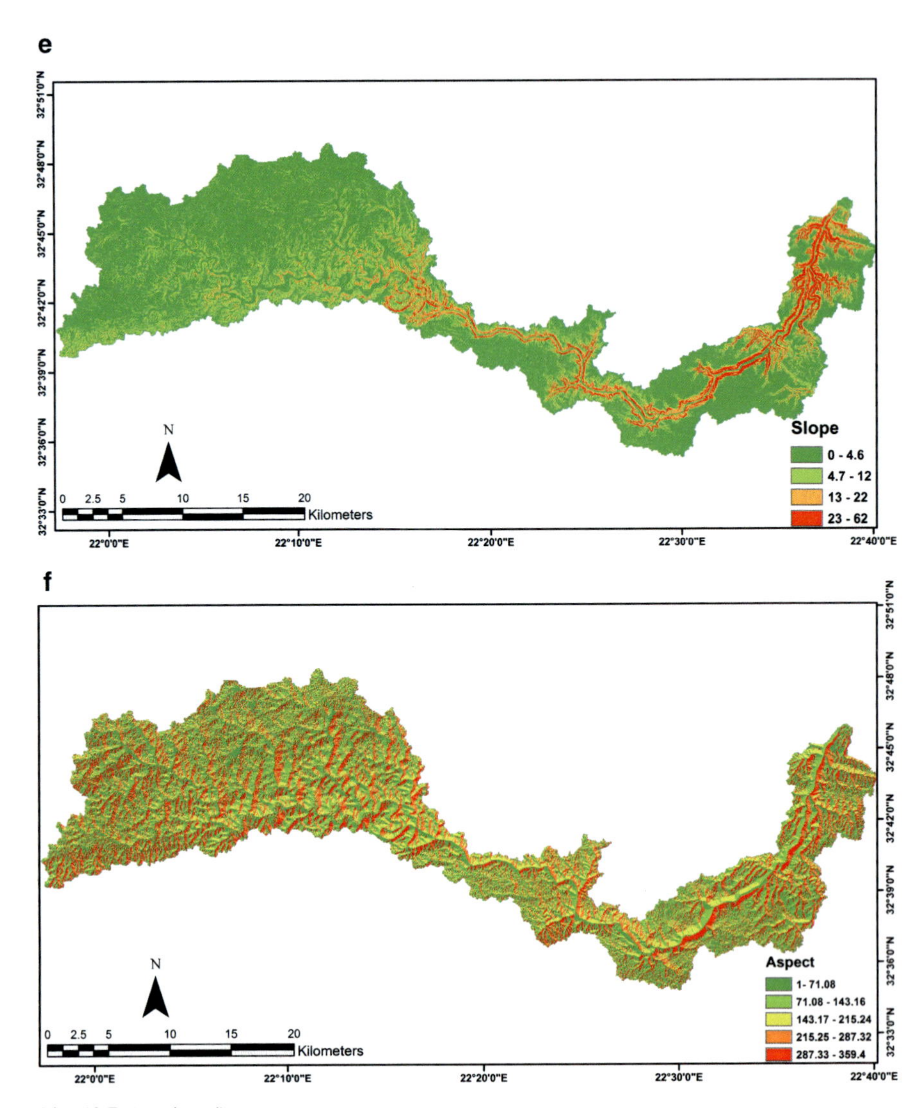

Fig. 12.7 (continued)

12.8.3.2 Texture Ratio (Rt)

Texture ratio depends on several factors such as climate, rainfall, vegetation, rock and soil type, infiltration capacity, relief, drainage density, and stage of development, and it is obtained as a ratio between total number of streams and area of a basin (Fenta et al., 2017). Drainage texture can be classified into; coarse texture (< 4 / km), intermediate (4 to 10 / km), fine (10–15 / km), and very fine texture (> 15 / km) (Fenta et al., 2017). The drainage texture values of wadi Ar Ramlah and wadi Ash-

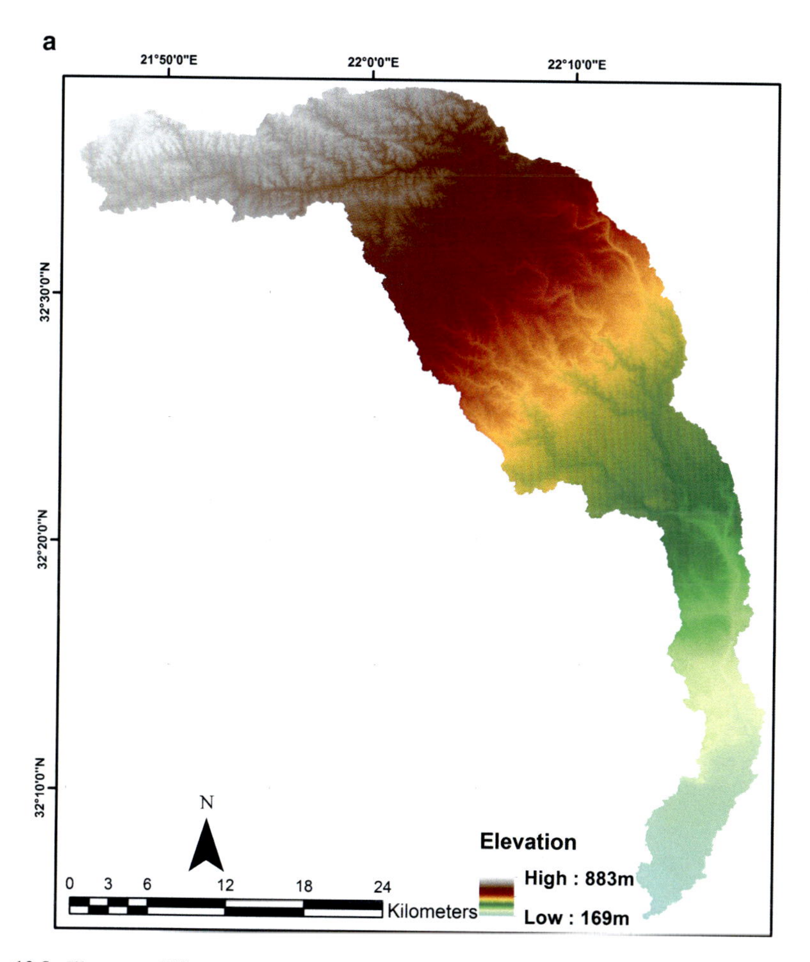

Fig. 12.8 Illustrates different types of geomorphic maps of wadi Ar Ramlah; (**a**) Elevation map; (**b**) Stream orders; (**c**) Stream frequency; (**d**) Shadow relief map; (**e**) Slope map; and (**f**) Aspect map

Sharif are higher than the drainage texture value of wadi Darnah, where higher values indicate impermeable sub-surface material, high relief conditions, and low infiltration capacity with very fine texture. Although wadi Darnah and Wadi Ash-Sharif pass through almost the same type of lithology which is Darnah Formation, but they have different values of drainage texture.

The significant low drainage texture value of wadi Darnah could be related to the diagenetic process of the lithology with highly affecting of joints structure, which display a coarse texture with medium permeable subsurface materials.

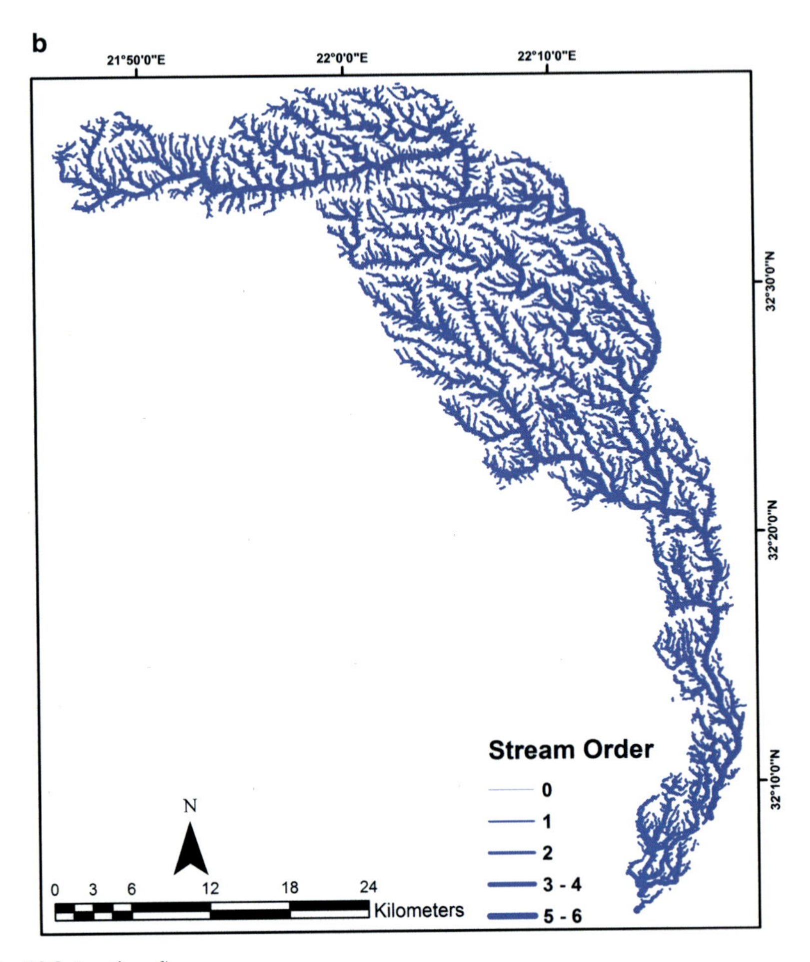

Fig. 12.8 (continued)

12.8.3.3 Stream Frequency (Sf)

The stream frequency was first introduced by Horton (1932), it is expressed as the ratio of the total number of streams in a drainage basin to the area of that basin. The stream frequency depends on the nature of rock and soil permeability of the area and it is used as an index of various stages of landscape development (Biswas, 2016). In the study area, the stream frequency values of wadi Ar Ramlah, wadi Darnah and wadi Ash- Sharif are (7.03, 0.77, 5.04) respectively. The stream frequency of wadi Ramlah and wadi Ash- Sharif are greater than 3 that shows a high run-off on medium-to-high relief of low permeability. The stream frequency value of Wadi Darnah is lower which indicates low runoff.

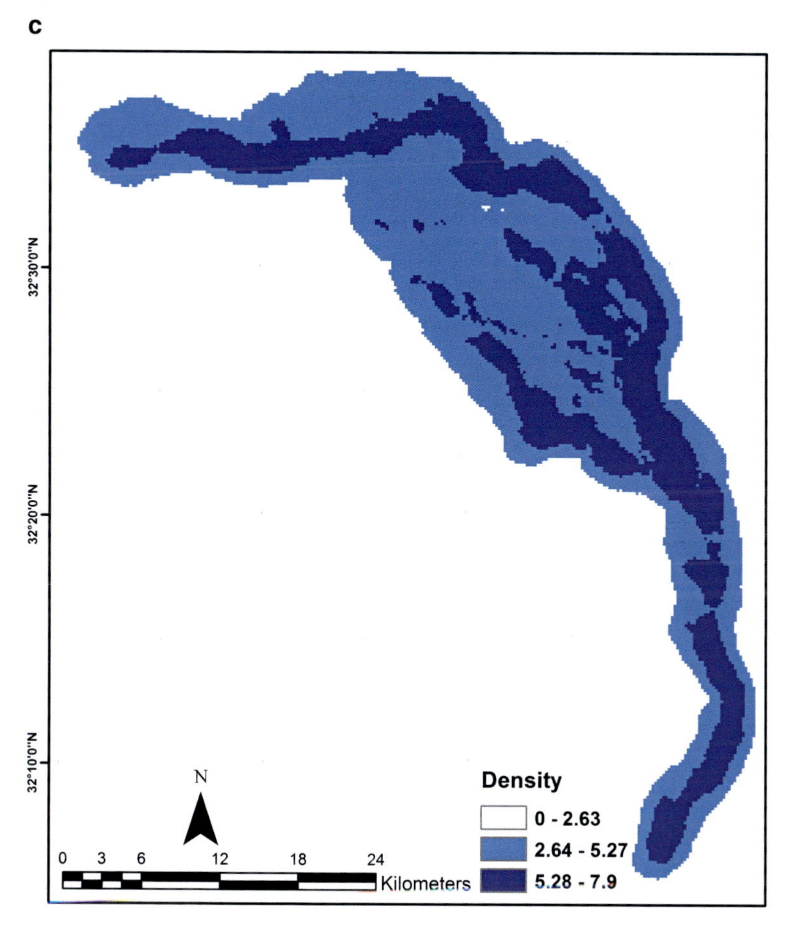

Fig. 12.8 (continued)

12.8.3.4 Infiltration Number (In)

It is a parameter of infiltration capacity of the basin. Lower infiltration numbers indicate higher infiltration and lower run-off (Prabhakaran & Raj, 2018). Measurement of the infiltration number for wadi Ar Ramlah and wadi Ash- Sharif resulted in a high value of (>12.59), indicating that the drainage basin of the both wadies are capable of producing high runoff, while wadi Darnah is with low value (0.69) that indicates on high infiltration with low run off.

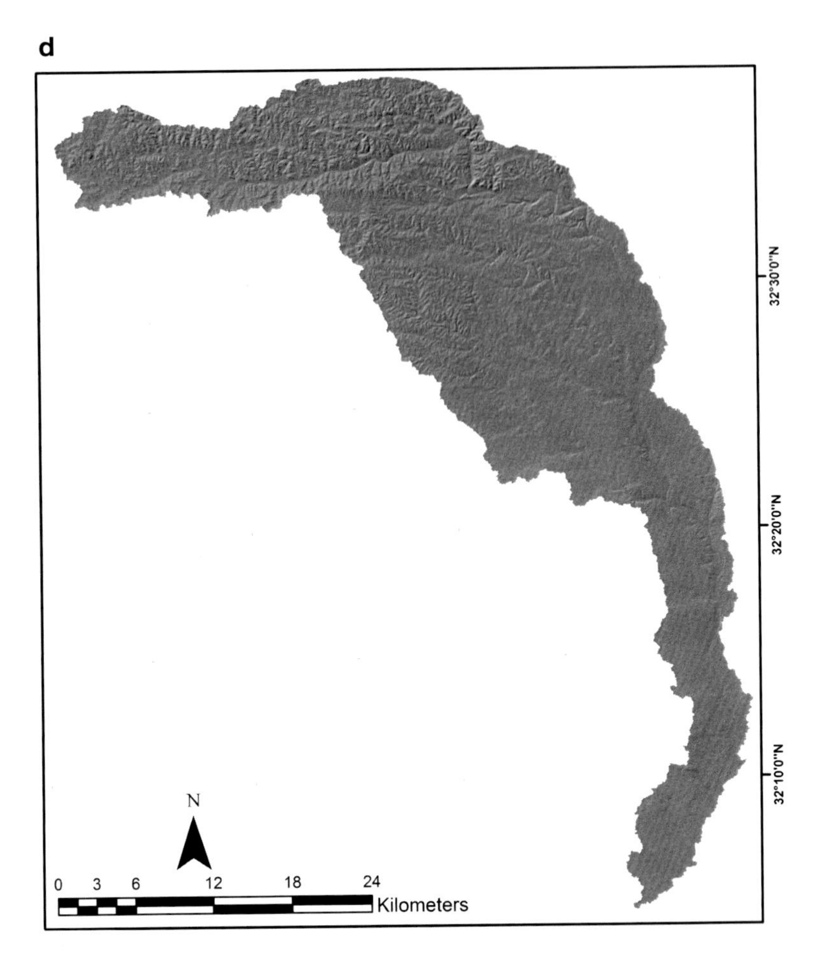

Fig. 12.8 (continued)

12.8.3.5 Form Factor (Rf)

According to (Singh et al., 2019) are believed that Form factor is a useful parameter to obtain a relationship of flow intensity of drainage basins along with their peak discharge, where high Form factor values occur in the basins having potential to produce high peak flows in short duration and low Form factor values are vice versa. The values of Form factor in wadi Ar Ramlah and wadi Ash- Sharif are generally low, but they are higher than the value of wadi Darnah. These Form factor values of wadi Ar Ramlah and wadi Ash- Sharif are indicating more elongated nature with higher peak run off of shorter duration than wadi Darnah.

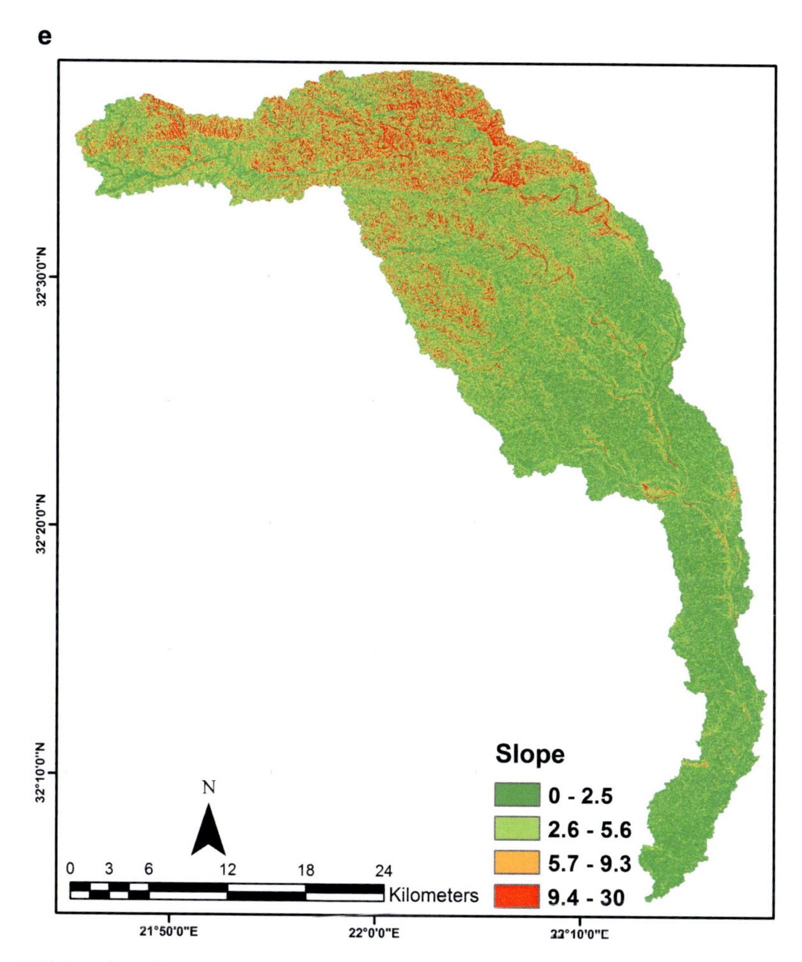

Fig. 12.8 (continued)

12.8.3.6 Elongation Ratio (Re)

The Elongation ratio is a parameter describes the shape of the basin (Schumm, 1956). The lower value of the elongation ratio indicates low infiltration capacity and high run-off conditions and vice versa (Singh et al., 2019). Whereas, the higher value elongation ratio describes a more circular shaped basin and vice-versa (Talampas & Cabahug, 2015). From the calculated elongation values of wadi Ar Ramlah and Wadi Darnah, a value of 0.35 depicts a less elongated basin shape with low relief, while wadi Ash- Sharif is slightly elongated with high relief with a value of 0.51.

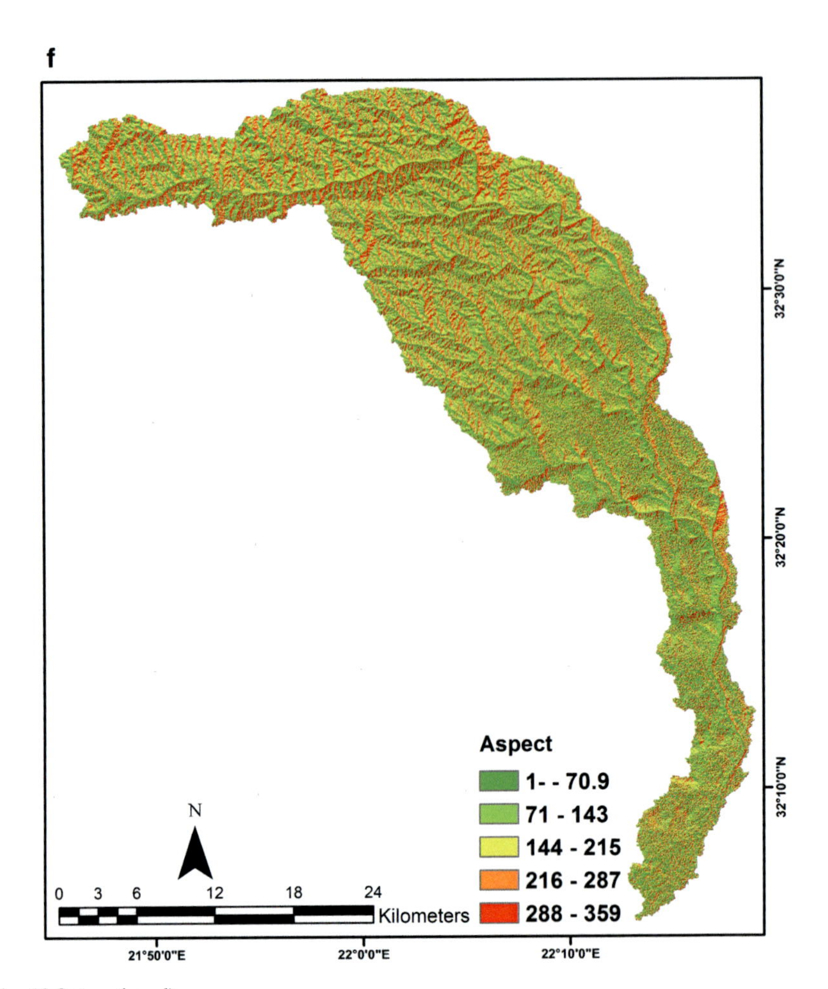

Fig. 12.8 (continued)

12.8.3.7 Circularity Ratio (Rc)

Runoff in circular shape basins gets more time to stay, therefore, circular-to-elongate basin is inversely related to their character of movement (rapid or slow) of run-off to outlet and infiltration (Singh et al., 2019). The higher value of Circularity ratio represents more circularity in the shape of the basin and vice-versa (Talampas & Cabahug, 2015). Values of Circularity ratio that ranging between 0.6 and 0.8 represent the steep ground slope and high relief, whereas values near to one correspond to low relief (Strahler, 1964; Miller, 1953). In the wadi Ar Ramlah, the circularity ratio is (0.14) and wadi Ash- Sharif is (0.22) indicating impermeable surface resulting in lower peak flow for longer duration. The Wadi Darnah show

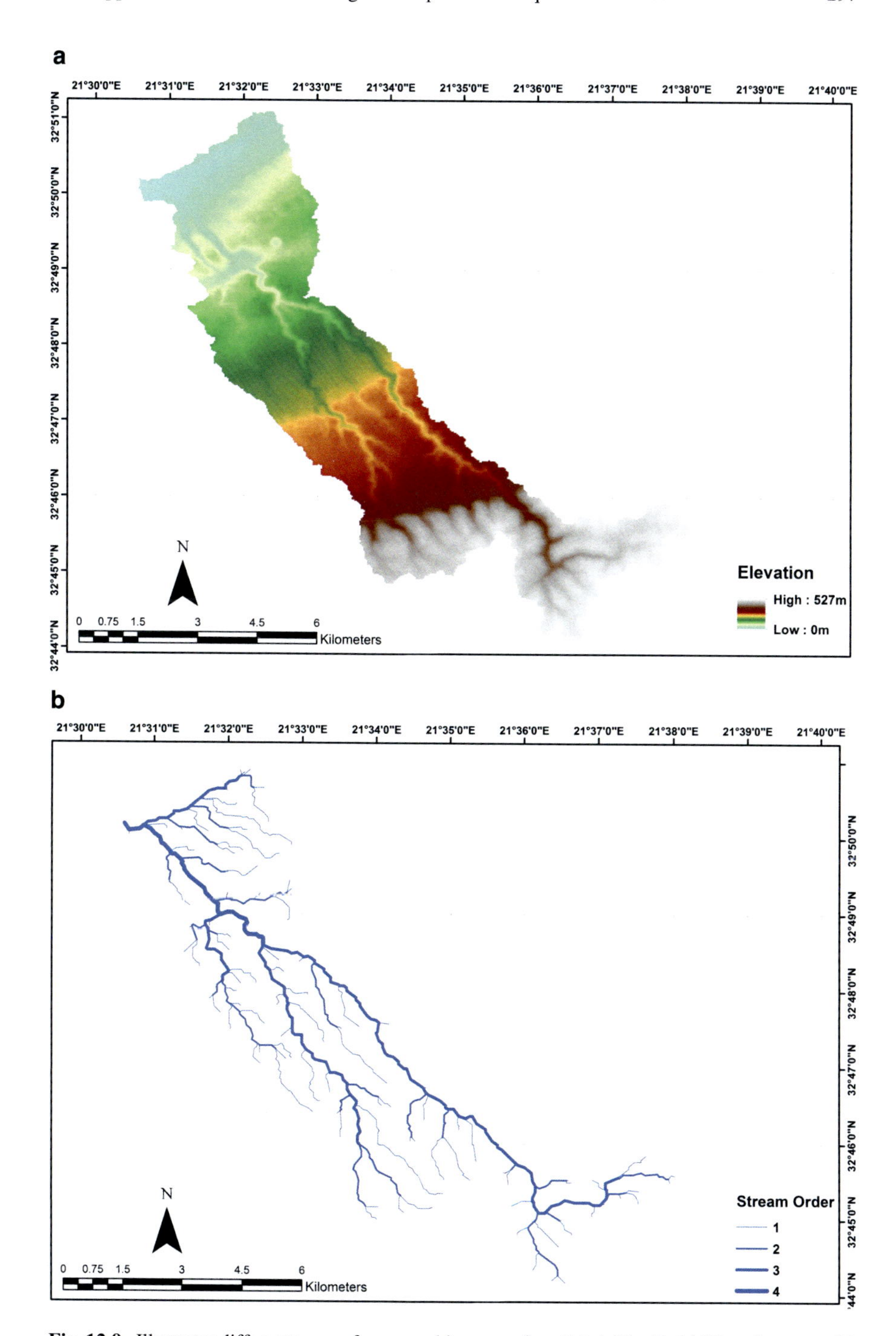

Fig. 12.9 Illustrates different types of geomorphic maps of wadi Ash Sharif; (**a**) Elevation map; (**b**) Stream orders; (**c**) Stream frequency; (**d**) Shadow relief map; (**e**) Slope map; and (**f**) Aspect map

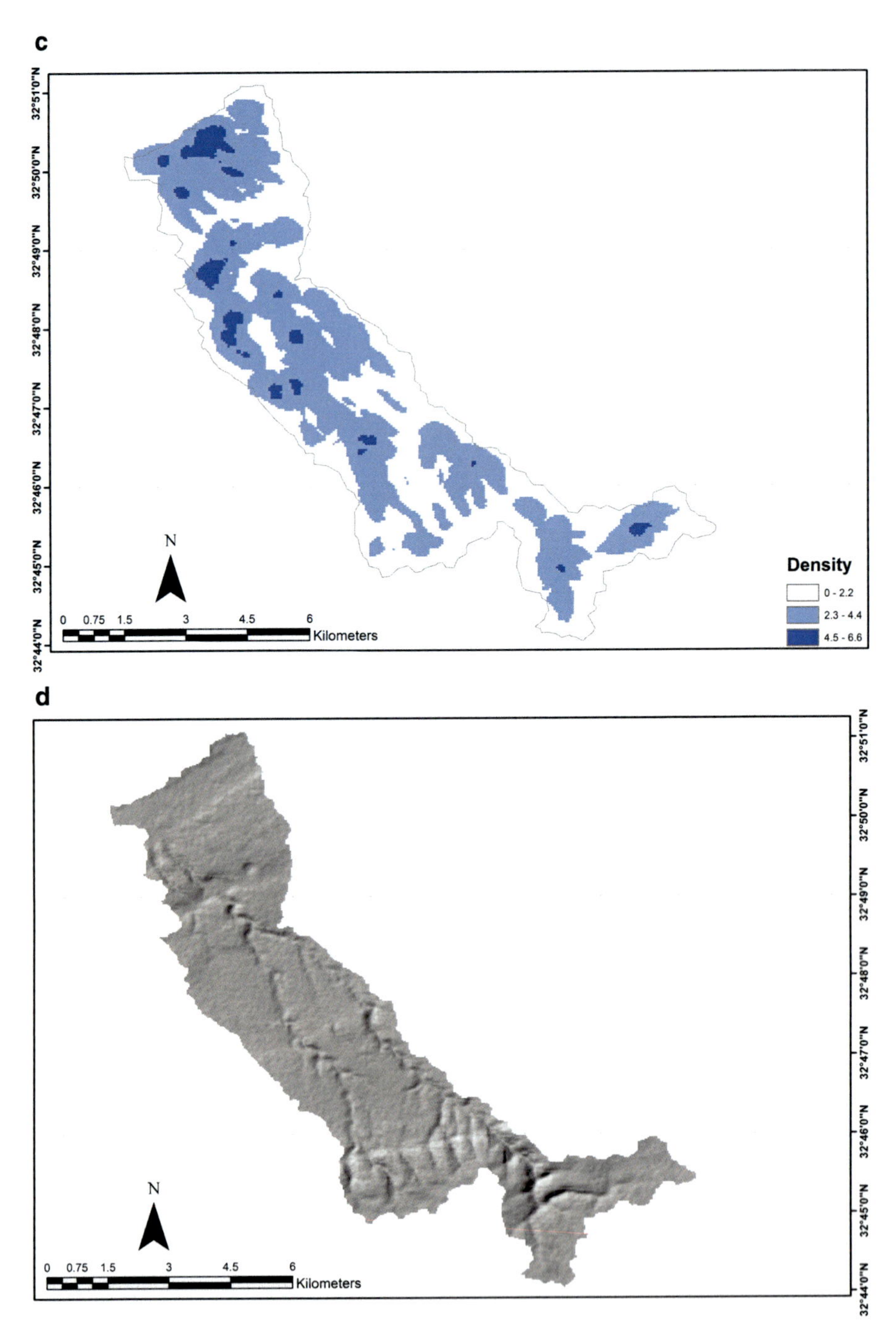

Fig. 12.9 (continued)

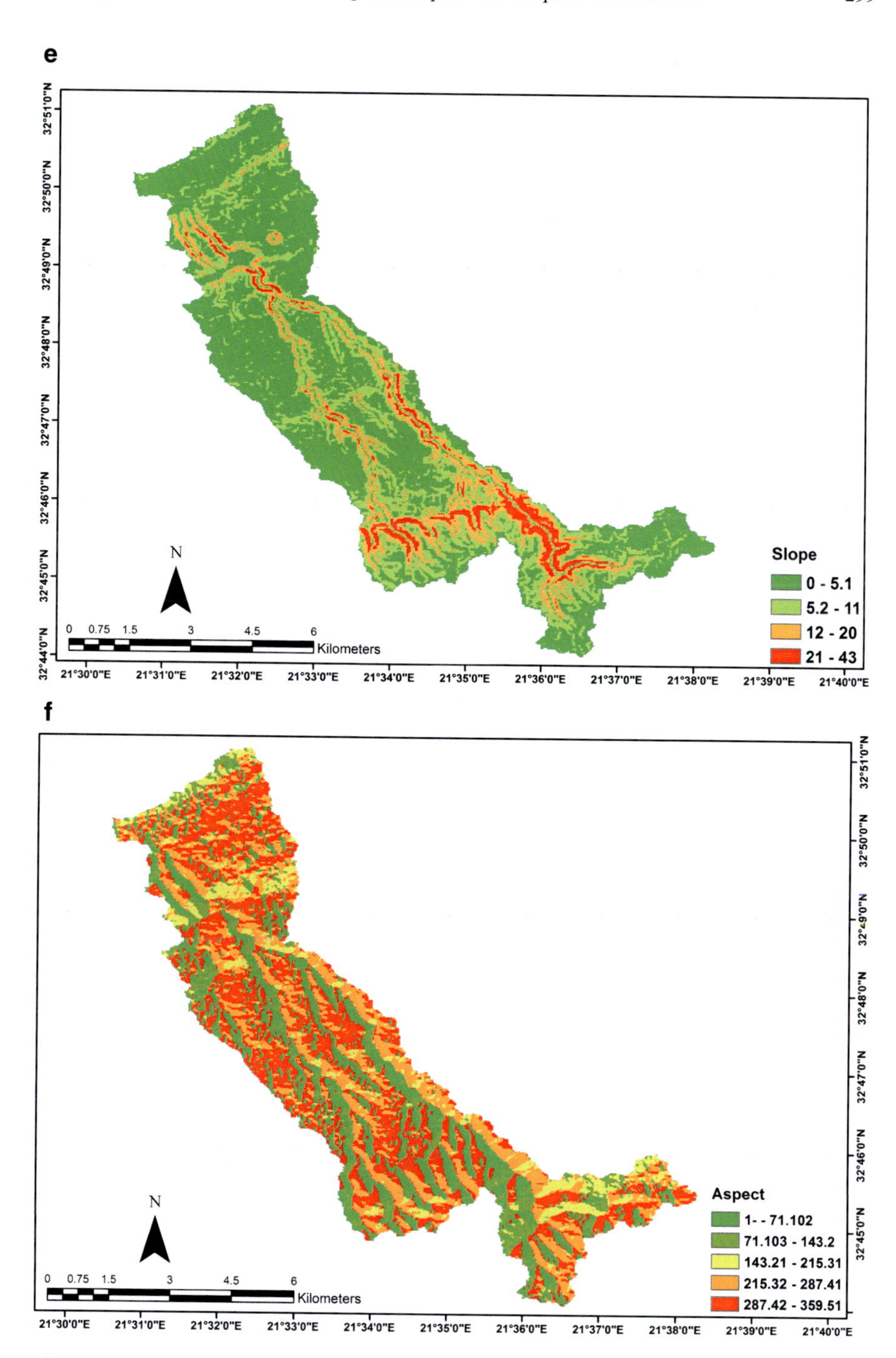

Fig. 12.9 (continued)

more elongated shape with very low value of Circularity ratio (0.089) resulting in higher peak flow for shorter duration.

12.8.3.8 Length of Overland Flow (Lof)

Length of overland flow is a length of water over the ground before it gets concentrated into certain stream channels (Sukristiyanti et al., 2018). The lower value of the length of overland flow parameter represents a well-developed drainage network.

In the study area, the Length of overland value of wadi Ar Ramlah and wadi Ash-Sharif is almost (0.2) showing relatively youthful stage of the drainage development, while wadi Darnah is measured at (0.45) an indication that the drainage basin will have a quicker surface runoff will enter the streams.

12.8.3.9 Constant of Channel Maintenance (ccm)

Constant of channel maintenance (Ccm) is the inverse of drainage density and expressed with dimension of square per unit. Drainage basins having lower values of Constant of channel maintenance will have higher value of drainage density. The most affecting factors of the constant of channel maintenance are rock type, permeability, vegetation, relief and duration of rainfall. The low values (0.38 and 0.39) of constant of channel maintenance of wadies Ar Ramlah and Ash- Sharif indicate low permeability, moderate slope, and high surface run-off. Computed value of Constant of channel maintenance of wadi Darnah is (1.11) would indicate that the drainage basin has a relatively high erodibility, medium permeability, steep slopes and high surface runoff.

12.8.3.10 Bifurcation Ratio

The Table 12.3 illustrates the overall Bifurcation ratios of the various stream orders of the studied wadies. A lower Bifurcation ratios range between 3 and 5 suggests that the structure does not exercise a dominant influence on the drainage pattern, while higher Bifurcation ratio greater than 5 indicates some sort of geological control. If the Bifurcation ratio is low, the basin produces a sharp peak in discharge and if it is high, the basin yields low, but extended peak flow (Dikpal et al., 2017). In general, the flat terrains have Bifurcation ratios 2, whereas mountainous or highly dissected terrains have values greater than 3. The results of Bifurcation ratios of the study wadies are ranging from 1.68 flat terrains with low structure effect in wadi Ar Ramlah, to greater than 3.0 in wadi Darnah and wadi Ash- Sharif that show high dissected terrains with structure control.

Table 12.3 Illustrates the overall Bifurcation ratios of the various stream orders of the study wadies

Rb = Nu/Nu +1, where Nu = Total number of stream segments of order	Bifurcation ratios (Rb) of Wadi Ar Ramlah	Bifurcation ratios (Rb) of wadi Darnah	Bifurcation ratios (Rb) of wadi Ash-Sharif
1st /2nd	2.2	2.25	2.1
2nd/ 3rd	1.8	1.52	1.1
3rd/4th	1.8	1.28	6.4
4th/5th	1.6	10.14	–
5th/6th	1.0	–	–
Mean bifurcation ratios	1.68	3.8	3.2

12.8.4 *Relief Morphometric Parameters*

The morphometric investigation of the relief parameters of the basin includes Basin Relief (H), Relief Ratio (Rf), Ruggedness Number (Rn) and Dissection Index (DI).

12.8.4.1 Relief Ratio (Rh)

Relief Ratio (Rh) is the difference in the elevation of the highest and lowest points in a watershed. Relief ratio (Rh) of water catchment of the study basins is various, (7.85) for Ar Ramlah catchment area, indicates that the basin is low to moderate relief and slope, whereas Darnah and Ash Sharif water basins are with high values (>11) illustrating that the area have high potential energy for transporting water and sediment downslope due to the steep slope of high relief.

12.8.4.2 The Ruggedness Number (Rn)

The ruggedness number indicates the extent of the instability of land surface (Strahler, 1957). The high value of the ruggedness number shows a rugged topography, highly susceptible to soil erosion and structurally complex. In the case of the study basins, the ruggedness number (Rn) values are considered moderate, which indicates the moderate soil erosion in this area with the slight complexity of structure Table 12.3.

12.8.4.3 Dissection Index (DI)

According to Singh (2000), It is an important morphometric indicator of the nature and magnitude of dissection of terrain or vertical erosion. Dissection index (DI) is expressing the ratio of the maximum relative relief to maximum absolute relief. The

Table 12.4 Summarize relation between results of morphometric parameters and risk flood

MORPHOMETRIC PARAMETERS	SURFACE RUN-OFF			PERMEABILITY			RISK		
	I	II	III	I	II	III	I	II	III
Drainage Density	Low	High	High	High	Low	Low	Low	High	High
Stream Frequency	Low	High	High	Low	Low	Low	Medium	High	High
Texture Ratio	Low	High	High	High	Low	Low	Low	High	High
Form Factor	Low	High	High	High	Low	Low	Low	High	High
Circularity Ratio	Low	High	High	High	Low	Low	Low	High	High
Elongation Ratio	Low	High	High	High	Low	Low	Low	High	High
Length of Overland Flow	High	Low	Low	Low	High	High	High	Low	Low
Infiltration Number	Low	High	High	High	Low	Low	Low	High	High
Constant Channel Maintenance	High	High	High	Medium	Low	Low	High	High	High

I Wadi Darnah
II Wadi Ar Ramlah
III Wadi Ash-Sharif

High (red)
Medium (yellow)
Low (green)

Dissection index value is between zero, which indicates on complete absence of dissection or vertical erosion, to one that reveals vertical cliff. Generally, the areas with high DI indicate high relative relief where slope of the land is steep and unstable that results in enhanced erosion. On the contrary, low DI corresponds with low relative relief, and with the subdued relief or old stage where the land is flat and more stable (Mustak, 2012). The Dissection index of Wadi Ash Sharif is (1) and wadi Darnah is (0.97) which indicate the basin is a highly dissected, whereas the Ar Ramlah is (0.8) which indicates the basin is a moderately/highly dissected.

However, the parameter values that extracted by using GIS applications have a great role in understanding the relationship between the drainage morphometric and risk of flash flood as shown in Table 12.4.

12.9 Conclusion

Remote sensing and GIS techniques had shown a great help in risk assessment of earthquake in Cyrenaica region. By comparing the linear features inferred from models of digital elevations and local faults in the study region; it is found that there is a great agreement between both of them in terms of focus and spread. It was also found that the first zone is the most concentrated zone by faults, and it is the most affected by earthquakes occurrence. However, the area with high intensity of faults is considered to be the region most affected by earthquakes, and vice versa. Furthermore, the GIS techniques have a great support in drainage characterization of runoff with systematically analysis the morphometric parameters in understanding the relationship between the drainage morphometric and risk of flash flood. The studied wadies show similarities in some results of their morphometric analysis and dissimilarity in others. Generally, it can easily say that most wadies in the Jabal Al Akhdar area have a potential to produce either high peak flows in short duration, or low peak flows in longer duration depending on different geomorphic parameters. However, the most effective geomorphic parameters that make most wadies in Jabal Al Akhdar area flooded are; the relief topography and slopes, where other parameters are considered. The most vulnerable areas that are expected to be flooded are those in low-lying lands.

References

Arsenikos, S., de Lamotte, D. F., Chamot-Rooke, N., Mohn, G., Bonneau, M. C., & Blanpied, C. (2013). Mechanism and timing of tectonic inversion in Cyrenaica (Libya): Integration in the geodynamics of the East Mediterranean. *Tectonophysics El Sevier*, 319–329.

Bashe, B. (2017). Groundwater potential mapping using remote sensing and GIS in Rift Valley Lakes Basin, Weito Sub Basin. *Ethiopia International Journal of Scientific & Engineering Research, 8*(2), 43.

Bhat, M. S., Alam, A., Ahmad, S., Farooq, H., & Ahmad, B. (2019). Flood hazard assessment of upper Jhelum basin using morphometric parameters. *Environmental Earth Science, 78*(2), 54.

Biswas, S. S. (2016). Analysis of GIS based morphometric parameters and hydrological changes in Parbati River basin, Himachal Pradesh. *India Journal of Geography Natural Disasters, 6*(175), 2167–0587.

Campbell, A. S. (1968). The Barce (Al Marj) earthquake of 1963. In F. T. Barr (Ed.), *Geology and archaeology of Northern Cyrenaica, Libya* (pp. 183–195). Petroleum Exploration Society Libya.

Chow, V. T., Maidment, D. R., Mays, L. W. & Saldarriaga, J. G., 1994. *Hidrología aplicada* (No. 551.48 C4H5).

Christie, A. M. (1955). *Geological report on Cyrenaica, Libya*. American Overseas Oil Company Reports.

Dikpal, R. L., Prasad, T. R., & Satish, K. (2017). Evaluation of morphometric parameters derived from Cartosat-1 DEM using remote sensing and GIS techniques for Budigere Amanikere watershed, Dakshina Pinakini Basin, Karnataka. *India Applied Water Science, 7*(8), 4399–4414.

Doswell III, C. A. (2003). *Flooding*. Encyclopedia of atmospheric sciences.

El Amawy, M. A., Muftah, A. M., & Abdelmalik, M. B. (2009). Karst development and structural relationship in the tertiary rocks of the western part of al Jabal al Akhdar, ne Libya: a case study in qasr Libya area. *3td International Symposium Karst Evolution in the South Mediterranean Area* (pp. 173–189).

El Deftar & Issawi (1977). Geological map of Libya, 1:250000, Sheet: Al Bardin, NH 35-1, Explanatory Booklet, Industrial Research Center (IRC), Tripoli, Libya, 97p.

El-Arnauti, A., Lawrence, S. R., Mansouri, A. L., Sengör, A. M. C., Soulsby, A., & Hassan, H. (2008). Structural styles in NE Libya. In M. J. Salem, K. M. Oun, & A. Essed (Eds.), *Geology of East Libya* (pp. 153–178). Gutenberg Press Ltd.
Elfadli, K. I. (2009). *Precipitation data of Libya*. Libyan National Meteorological Center.
El-Hawat, A. S. & Shelmani, M. (1993). *Short notes and Guidebook on the geology of Al Jabal Al Akhdar, Cyernaica, NE Libya*. 1st Symposium on the Sedimentary basins of Libya, Geology of Sirt basin. Earth Science Society of Libya (ESSL), 70.
El-Tantawi, A. M. M. (2005). *Climate change in Libya and desertification of Jifara Plain: using geographical information system and remote sensing techniques* (Doctoral dissertation, Verlag nicht ermittelbar).
Faniran, A. (1968). The index of drainage intensity: a provisional new drainage factor. *Australian Journal of Science, 31*(9), 326–330.
Fenta, A. A., Yasuda, H., Shimizu, K., Haregeweyn, N., & Woldearegay, K. (2017). Quantitative analysis and implications of drainage morphometry of the Agula watershed in the semi-arid northern Ethiopia. *Applied Water Science, 7*(7), 3825–3840.
Gupta, R. P. (2018). *Remote sensing geology* (3rd ed., p. 437). Springer.
Hallett, D. (2002). *Petroleum geology of Libya* (p. 503). Elsevier.
Hassen, H. (1983). *Seismicity of Libya and related problems*. Master thesis. Civil Engineer Department, Colorado state University, pp. 108.
Horton, R. E. (1932). Drainage-basin characteristics. *Transactions of the American Geophysical Union, 13*, 350–361.
Horton, R. E. (1945). Erosional development of streams and their drainage basins. *Geological Society of America Bulletin, 56*, 275–370.
Dataset: ASF DAAC 2008, ALOS-1 PALSAR_Radiometric_Terrain_Corrected_low_res; Includes Material © JAXA/METI 2020. https://vertex.daac.asf.alaska.edu/#. Accessed 12 Nov 2020.
Keller, E. A., & Rockwell, T. K. (1984). *Tectonic geomorphology, quaternary chronology, and paleo seismicity* (pp. 203–239). Springer.
Khakhlari, M., & Nandy, A. (2016). Morphometric analysis of Barapani river basin in Karbi Anglong District, Assam. *International Journal of Scientific and Research Publications, 6*(10), 238–249.
Lagesse, A. Free, M., & Lubkowski, Z. (2017). *Probabilistic seismic hazard assessment for Libya*. 16th World Conference on Earthquake, 16WCEE 2017, Santiago Chile, January.
Melton, M. A. (1957). *An analysis of the relations among elements of climate, surface properties, and geomorphology*. Columbia University New York.
Miller, V.C. (1953). *Quantitative geomorphic study of drainage basin characteristics in the Clinch Mountain area, Virginia and Tennessee*. Technical report (Columbia University. Department of Geology); no. 3.
Mustak, S. K. (2012). *Measurement of dissection index of Pairi River Basin using remote sensing and GIS*. The National Geographical Journal of India, BHU, Varanasi, UP.
Prabhakaran, A., & Raj, N. J. (2018). Drainage morphometric analysis for assessing form and processes of the watersheds of Pachamalai hills and its adjoinings, Central Tamil Nadu. *India Applied Water Science, 8*(1), 1–19.
Röhlich, P. (1974). Geological map of Libya, 1: 250 000. Sheet Al Bayda (NI 34–15). Explanatory Booklet. Ind. Res. Cent., Tripoli.
Schumm, S. A. (1956). Evolution of drainage systems and slopes in badlands at Perth Amboy, New Jersey. *Geological Society of America Bulletin, 67*(5), 597–646.
Shaltami, O. R., Fares, F. F., Errishi, E. L., & Oshebi, F. M. (2020). *Isotope geochronology of the exposed rocks in the Cyrenaica Basin, NE Libya* (p. 139). Springer.
Singh, S. (2000). *Geomorphology* (p. 642). Prayag Pustak Bhawan.
Singh, S., Kanhaiya, S., Singh, A., & Chaubey, K. (2019). Drainage network characteristics of the Ghaghghar River Basin (GRB), Son Valley, India. *Geology, Ecology, and Landscapes, 3*(3), 159–167.

Strahler, A. N. (1957). Quantitative analysis of watershed geomorphology. *Eos, Transactions, American Geophysical Union, 38*(6), 913–920.

Strahler, A. N. (1964). Part II. Quantitative geomorphology of drainage basins and channel networks. In *Handbook of applied hydrology* (pp. 4–39). McGraw-Hill.

Sukristiyanti, S., Maria, R., & Lestiana, H. (2018). Watershed-based morphometric analysis: A review. In *IOP conference series: earth and environmental science* (Vol. 118, p. 012028). IOP Publishing.

Talampas, W. D., & Cabahug, R. R. (2015). Catchment characterization to understand flooding in Cagayan De Oro River Basin in Northern Mindanao, Philippines. *Mindanao Journal of Science and Technology, 13*.

U.S. Geological Survey. (2020). *Earthquake lists, maps, and statistics*. Accessed 18 Jan 2020 at URL https://www.usgs.gov/natural-hazards/earthquake-hazards/lists-maps-and-statistics

Youssef, A. M., Pradhan, B., & Hassan, A. M. (2011). Flash flood risk estimation along the St. Katherine road, southern Sinai, Egypt using GIS based morphometry and satellite imagery. *Environmental Earth Sciences, 62*(3), 611–623.

Zamot, J., & Afkareen, M. (2020). *Geomorphological parameters by remote sensing and GIS techniques* (A case study of flash flood in Mikhili Village, Al Jabal Al Akhdar, NE of Libya). Paper presented at the Forth International Conference for Geospatial Technologies – Libya GeoTec 4, Tripoli, Libya, 3–5 March 2020.

Chapter 13
Applications of Remote Sensing for Flood Inundation Mapping at Urban Areas in MENA Region: Case Studies of Five Egyptian Cities

Karim I. Abdrabo, Mohamed Saber, Sameh A. Kantoush, Tamer ElGharbawi, Tetsuya Sumi, and Bahaa Elboshy

Abstarct This chapter focuses on using various remote sensing data for monitoring floods and developing risk maps. It covers a wide range of issues, reviews remote sensing data types, processing techniques, and discussing the limitations and challenges of using remote sensing images in flood monitoring, especially in MENA region. Furthermore, the chapter presents a number of previous attempts of flood monitoring in the MENA region clarifying the data they depend on and the extent of reaching reliable results. The main aim of this chapter is to highlight the role of the available remote sensing data remotely sensed, including optical data, multispectral data, and Synthetic Aperture Radar (SAR) data in supporting flood-related research and investigation such as monitoring and mapping flood events and risk where is the lacking the observational data at the arid regions.

Keywords Satellite-based data · Flood inundation mapping · MENA region · Optical · SAR · Remotely sensed data · Urban area

K. I. Abdrabo (✉)
Faculty of Urban and Regional Planning, Cairo University, Giza, Egypt

Department of Urban Management, Graduate School of Engineering, Kyoto University, Kyoto, Japan
e-mail: m.karim.ibrahim@cu.edu.eg

M. Saber · S. A. Kantoush · T. Sumi
Disaster Prevention Research Institute (DPRI), Kyoto University, Kyoto, Japan
e-mail: mohamedmd.saber.3u@kyoto-u.ac.jp; kantoush.samehahmed.2n@kyoto-u.ac.jp; sumi.tetsuya.2s@kyoto-u.ac.jp

T. ElGharbawi
Department of Civil Engineering, Suez Canal University, Ismailia, Egypt

B. Elboshy
Architectural Engineering Department, Faculty of Engineering, Tanta University, Tanta, Egypt
e-mail: bahaa.elboshi@f-eng.tanta.edu.eg

M. M. Al Saud (ed.), *Applications of Space Techniques on the Natural Hazards in the MENA Region*, https://doi.org/10.1007/978-3-030-88874-9_13

13.1 Introduction

The United Nations (UNISDR, 2015) pointed out that 43% of natural disasters that occurred globally from 1995 to 2015 were water-related disasters, affecting more than half (56%) of all people. The socioeconomic effects correlated with floods are recorded and documented in the developed and developing countries (Bisht et al., 2018; Martín-Vide & Llasat, 2018; Ozturk et al., 2018). Moreover, according to the International Disaster Database (EM-DAT; www.em-dat.be), floods are the most frequent disaster with the highest impact in terms of the number of people affected. Flash floods cause the devastating impacts; however, it is more severe in developing countries such as MENA region. During the past few decades, the frequencies of extreme events have increased in the Middle Eastern North African (MENA) Region (Zhang et al., 2005). The Arab region is characterized by increasing the frequency and intensity of extreme storm events which resulting in devastating flash floods, along with drought threats (Abdrabo et al., 2020a; Saber et al., 2020; Saber & Habib, 2016). This might be attributed to climate change or human impacts; however, the reasons are still not well understood. In spite of the multiple evidence and the growing awareness of the flood risks, the modeling capacity of flood dynamics remains poor, which is mainly related to the availability of data. Flash flood risk mitigation requires precise and accurate flood monitoring measures for helping hazard management (Arora et al., 2020). One of such measures is mapping the inundation areas, which is considered as one of the main concerns among scientists and governments around the globe (Ali et al., 2019, 2020). Flood inundation maps are used mainly for (1) Forecast scenarios; (2) Mitigation and planning – flood risk analyses (3) Timely response; (4) Damage assessment; and (5) Environmental and ecological assessments. Flood inundation mapping are generally difficult and considered more difficult in MENA region due to the difficulty of accessing the affected areas, that consequently affect the performance of the hydrological modeling which requires a detailed observational dataset for calibration and validation (Abdrabo et al., 2020b; Abushandi & Merkel, 2011; Hall et al., 2014; Kilpatrick & Cobb, 1985; Lin, 1999; Pilgrim et al., 1988; Rodier & Roche, 1978:197; Wheater et al., 2007).

In this context, Remote Sensing (RS) is an extremely useful source of observation data that could overcome the decline in field surveys and observational stations, especially in MENA region. RS plays an important role in all the phases of flood hazard management, from preparedness, emergency management, and civil protection phases and up to damage assessment for flood risk reduction. RS data provide huge advantages: low costs, data acquisition reliability, overcoming the local difficulties such as site accessibility etc. Moreover, it can play a key role in the calibration and validation of hydrological and hydraulic models in addition to providing real-time flood mapping and monitoring applications (Domeneghetti et al., 2019; Haq et al., 2012). Although the number of state-of-the-art and innovative research studies in these areas is increasing, the full potential of RS in enhancing flood mapping, modeling, and prediction has not been exploited (Saber et al., 2010;

Sanyal & Lu, 2004; University of Waterloo et al., 1993). However, as is often the case, new opportunities and applications pose new challenges: the nature of the data, the different size, and scale of the objects and the processes which can be now investigated, even the sheer quantity of available data, all require new or more powerful tools to be suitably dealt with.

RS primarily in the form of satellite and airborne imagery and altimetry such as (Resurs-P, GeoEye, and WorldView, or from drones). The different uses of RS products in the case of flood monitoring are as follows: (1) elevation data such as Shuttle Radar Topography Mission (SRTM) and digital elevation model (DEM) from WorldView-2 stereo pair imagery; (2) The land use/land cover and soil properties which can be obtained from fused ASTER multispectral and ALOS-PALSAR Synthetic Aperture Radar (SAR) data; (3) Rainfall products such as the Tropical Rainfall Measuring Mission (TRMM) (Kummerow et al., 1998) (Chen et al., 2015; Kneis et al., 2014; Ochoa et al., 2014; Prasetia et al., 2013), the Climate Prediction Centre morphing method (CMORPH) from NOAA CPC (Joyce et al., 2004), (PERSIANN) from the University of California (Hsu et al., 1997; Yoshimoto & Amarnath, 2017), Global Satellite Mapping of Precipitation (GSMaP) from Japan Science and Technology Agency (JST) and Japan Aerospace Exploration Agency (JAXA), IMERG LATE, and IMERG EARLY from NASA which are two new rainfall data products. There are many other remotely sensed meteorological products that are publicly available and occasionally updated with ground-based information, are commonly known as global datasets. Utilizing such high-resolution, multi-temporal data give the chance to enhance the performances of the forecast, alert, and post-event monitoring of inundation events (Refice et al., 2018b). The integration of remotely sensed data (such as Data Terrain Models (DTMs), flood extent, river width, land cover, water level, etc.) with flood modeling significantly enhances the prediction results (Haq et al., 2012; Refice et al., 2018a). Although the previous studies showed encouraging results using different types of RS data combined with in-situ data, many challenges face such applications from an uncertain point of view.

In the present chapter, we will start with an overview of flood monitoring systems and remote sensing approaches, followed by the potential and limitations of open satellite data for flood mapping. Finally, we present some applications of remote sensing on flash floods in MENA region at five Egyptian cities.

13.2 Overview of Remote Sensing Approaches Used in Flood Monitoring

Flood monitoring activities can be divided into three sets, according to the stage of operations with respect to the event occurrence. (1) Forecast scenarios; (2) Timely response and Emergency monitoring; (3) Damage assessment (Hutter, 2006). According to each activity, the type of RS approaches and data used in each activity is determined. Forecasts rely mainly on meteorological information, so low- to medium-resolution optical data are usually implicated. In emergency monitoring, the emphasis is clearly on fast response and relatively high resolution, so a potentially wide variety of sensors can be involved, working in both the optical and microwave spectral regions. The third type of application is the one that involves the most advanced processing techniques to defining the spatial and temporal changes of factors, which, in return, control flood generation and risk. Multi-temporal and multisensory data allow a temporal and spatial reconstruction of flood inundation, from the beginning until the end when all the inundated areas return dry.

Remote Sensing data is produced through two main components: the sensors and the platform on which the sensor is installed. There are three types of platform, surface platforms such as ladder and tall building, Arial platforms such as aircraft and balloons; and spaceborne platforms are mainly satellites and space shuttles (Liang et al., 2012).

For the sensors, there are two types of sensors, passive and active. Passive sensors detect natural radiation that is emitted or reflected by the observed object. Reflected sunlight is the most common reflect radiation and is sensed by passive sensors. Typical passive sensors include three parts; radiometer to measure the electromagnetic radiation in the visible, infrared, or microwave spectral bands, imaging radiometer (scanner) to provide a two-dimensional array of pixels from which an image may be produced; and spectroradiometer to measure the radiance in multiple spectral bands (Liang et al., 2012). The Active sensors are emitted electromagnetic radiation to illuminate the observed object. They use a pulse of energy sent by the sensor and receive reflectance of this pulse. The types of measurement tools are included in active sensors are radar which uses a transmitter operating at microwave frequencies to emit electromagnetic radiation, and a directional antenna to measure the time of arrival of reflected pulses from observed objects for determining the distance, synthetic-aperture radar (SAR) which is considered a side-looking radar imaging system that uses the relative motion between an antenna and the Earth surface (Fig. 13.1) to synthesize a very long antenna by combining signals received by the

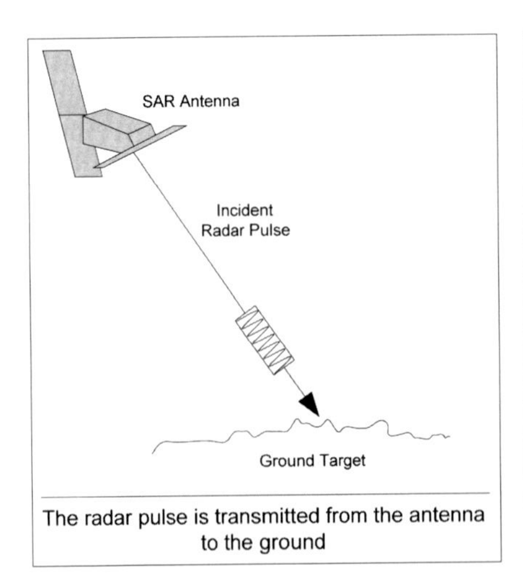

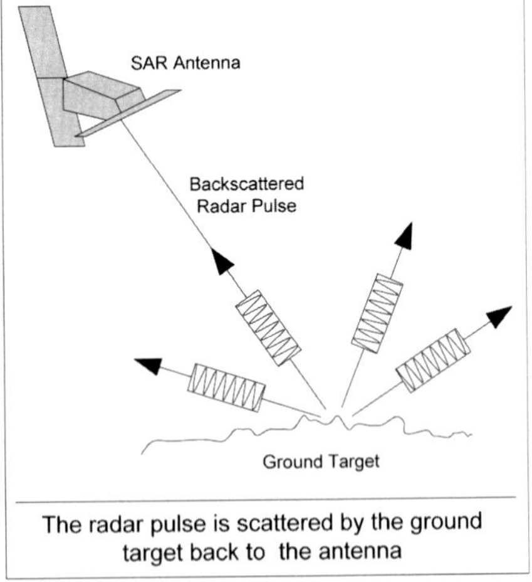

Fig. 13.1 SAR Image production process. https://crisp.nus.edu.sg/~research/tutorial/mw.htm

radar as it moves along its flight track for obtaining high spatial resolution imagery, synthetic interferometric aperture radar (InSAR) which compares two or more amplitude and phase images over the same area received during different passes of the SAR platform at different times, scatterometer which is a high-frequency microwave radar designed specifically to determine the normalized radar cross-section of the surface. LIDAR (Light Detection and Ranging) is an active optical sensor that uses a laser in the ultraviolet, visible, or near-infrared spectrum to transmit a light pulse and a receiver with sensitive detectors to measure the backscattered or reflected light, Laser Altimeter which is a laser altimeter that uses lidar to measure the height of the instrument platform above the surface (Liang et al., 2012).

The specifications of the platform and sensors determine the characteristics of the produced data, including spatial, spectral, temporal, and radiometric resolutions. Spatial resolution refers to the number of pixels representing the construction units of a digital image (Athanasiou et al., 2017). The spatial resolution could identify by the smallest object has been resolved by the sensor, which is also the area of the sensor's field of view (Liang et al., 2012). Figure 13.2 illustrates the difference between the lower and higher levels of spatial resolution.

Spectral Resolution, which represents the range of the electromagnetic spectrum, could be observed by the sensor. The spectral resolution is determined by the number and the narrowness of bands. When the number of bands lies between 3 and 10, this could describe as multispectral resolution, where the hyperspectral resolution includes the number of bands that reach the thousands (Jenice Aroma & Raimond, 2015). Figure 13.3 illustrates the different types of spectral resolution.

Temporal resolution refers to the repeat of the imaging cycle, which means the frequency of processing the same areas with the sensor. Orbit pattern and satellite sensor's design determine the frequency characteristics (Liang et al., 2012). Radiometric resolution represents the smallest energy differences that have been observed from the electromagnetic reflectance. It describes the sensor's ability to detect small changes in radiance and depends on how the continuous upwelling radiance signal is converted to discrete, digital image data. For example, The highly sensitive detector

10 m resolution, 10 m pixel size

30 m resolution, 10 m pixel size

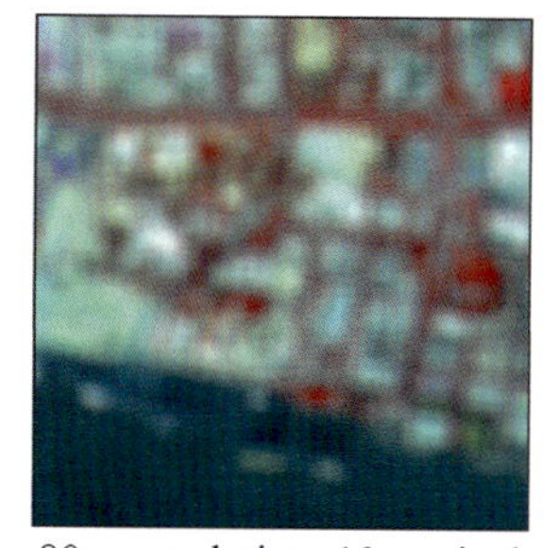

80 m resolution, 10 m pixel size

Fig. 13.2 The difference between the lower and higher levels of spatial resolution. https://crisp.nus.edu.sg/~research/tutorial/image.htm

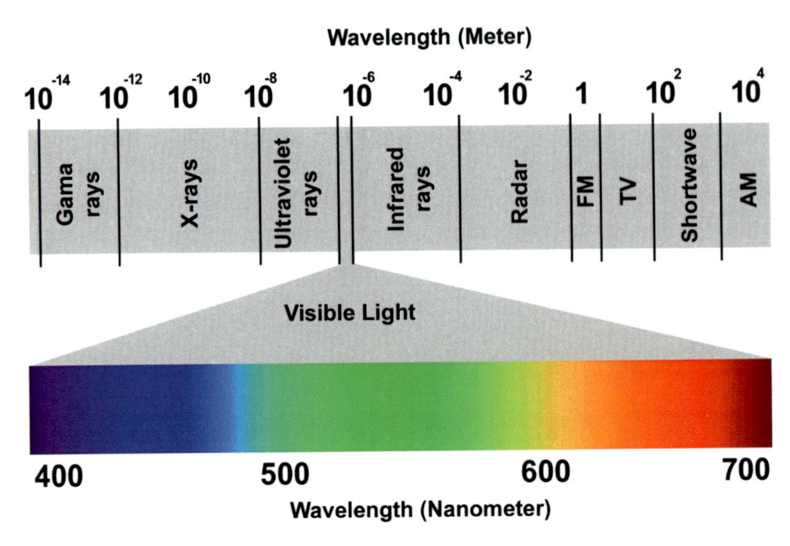

Fig. 13.3 The different types of spectral resolution. https://www.satimagingcorp.com/services/resources/characterization-of-satellite-remote-sensing-systems/

with 12-bit radiometric resolution could be more precise in investigating the water depth of the changing of a channel (Re & Capolongo, n.d.).

13.2.1 Data Types Used in Flood Monitoring

At the MENA region, and due to the lack of data, most flood monitoring research and applications depends on the free available satellite images available remote sensing data. The most widely used satellite images were reviewed from the literature are listed in Table 13.1 (Jenice Aroma & Raimond, 2015).

Depending on the used sensor and techniques, the remote sensing data could be divided into different types. These types are mainly including optical data, multi-spectral data, synthetic aperture radar (SAR) data, and vegetation and water indices. The following section identifies the production process of these types, the uses of each type, and the main differences between them.

13.2.2 Optical Data

Optical imaging depending on the visible, near-infrared, and shortwave infrared spectrums to produce the imagery types such as panchromatic, multispectral, and hyperspectral as shown in Fig. 13.4 (Zhu et al., 2018). Optical data is considered a crucial data source in the investigation of flood extension and evolution. Clean water

Table 13.1 A review for the most used remote sensing data in flood monitoring

Satellite Name	Lunched by	Purpose	Details
Landsat	National Aeronautics and Space Administration (NASA)	Land cover, forest and agricultural applications	The Landsat 7 images are of 30 m spatial resolution for multispectral and 15 m for panchromatic mode (PAN) images
Resourcesat	Indian space research organization (ISRO) offers		Available to the public users with 23.5 m and 56 m resolution, respectively
Terra	Nasa	Collecting information about both the earth surface and atmosphere	- advanced Spaceborne thermal emission and reflection Radiometer (ASTER) - clouds and Earth's radiant energy system (CERES) Multi-angle imaging Spectroradiometer (MISR), moderate resolution imaging Spectroradiometer (MODIS) - measurements of pollution in the troposphere (MOPITT).
Aqua	Nasa	Collecting information about the earth's water cycle, glaciers, and atmosphere	
Calipso	Nasa	Atmospheric, aerosol activity, and effective climate research	The Lidar instrument combined with passive infrared and visible imagers could capture the cloud movement and aerosol properties [
Earth Observing-1	Nasa		Hyperspectral imager could measure up to 200 wavelengths
Quickbird	Digital globe		Very high-resolution satellite images of 60 cm in PAN and 2.4 m in multispectral images
Formosat	National Space Organization (NSPO) of the republic of China.		High-resolution images of 2 m in PAN and 8 m in multispectral images
Spot	French organization named spot image		Offer 2.5 m to 5 m in PAN and 10 m in multispectral. Mode
Ikonos	GeoEye organization [20].		Providing high resolution multispectral and PAN images of 4 m and 1 m, respectively
Sentinel	European Space Agency (ESA)	Land cover change analysis and natural disasters monitoring applications	Provides all-weather data both day and night on radar imaging

(continued)

Table 13.1 (continued)

Satellite Name	Lunched by	Purpose	Details
Kalpana			Produces three bands of visible, thermal infrared, and water vapor infra-red images using a very high-resolution radiometer (VHRR) with a resolution of 2 × 2 km [22].
Radarsat	Canadian Space Agency (CSA)	Landcover operations on mines, icebergs, and underground water explorations	Offers SAR data which

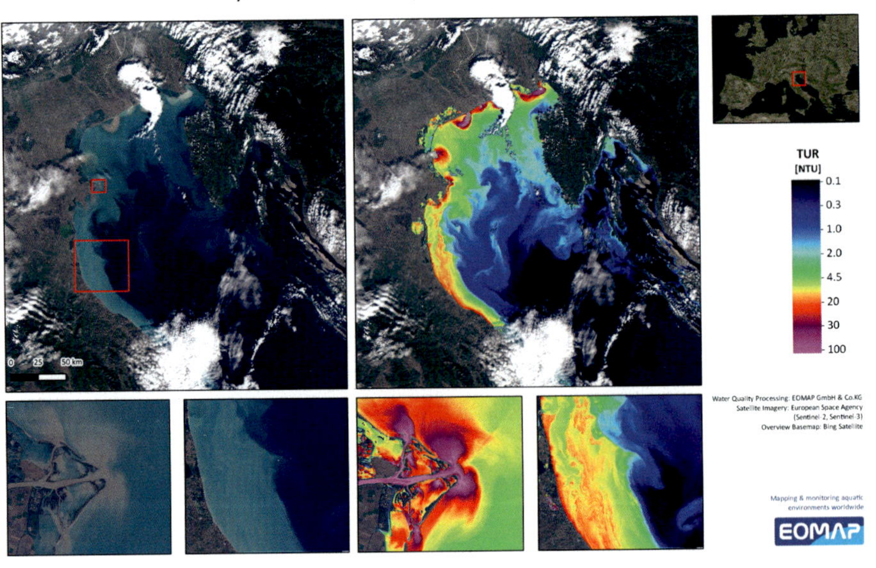

Fig. 13.4 "Left images, captured on October 31, 2018, show the large river sediment inflows into the Northern Adriatic Sea in "true" color. The right images display the turbidity levels assessed with EOMAP's EO processing system" (https://www.eomap.com/using-satellite-data-for-flood-monitoring/)

surfaces absorb most of the electromagnetic energy. Therefore, in optical images, the water area could be recognized easily where it appears as dark areas. However, in cases such less clean water and increasing reflections from water recognition is becoming more difficult, and using several spectral bands data can help (Fig. 13.5). Cloud coverage is the main obstacle that faces the use of optical data in flood monitoring, which are often associated with flood events (Re & Capolongo, n.d.).

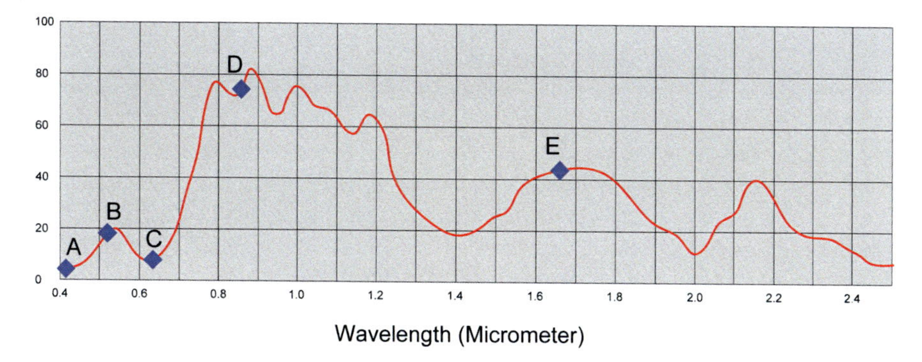

Fig. 13.5 Typical Reflectance Spectrum of Vegetation. The labeled arrows indicate the common wavelength bands used in optical remote sensing of vegetation: (**a**): blue band, (**b**): green band; (**c**): red band; (**d**): near IR band; (**e**): short-wave IR band (https://crisp.nus.edu.sg/~research/tutorial/optical.htm)

13.2.2.1 Multispectral Data

Multispectral remote sensing data is produced by a multispectral sensor. The used sensors have multichannel detectors; each channel is sensitive to radiation within a narrow wavelength band. These sensors produce multilayer image which contains both the brightness and spectral information for the objects. A hyperspectral sensor collects and processes information from 10 to 100 spectral bands. The resulting images can be used in recognizing objects, identify materials, and detect elemental components (Zhu et al., 2018). A multispectral sensor could be useful in flood monitoring with a presence of cloud coverage and within dense urban areas (Vissers, 2007).

13.2.2.2 Synthetic Aperture Radar (SAR) Data

Synthetic aperture radar (SAR) is useful in mapping the object's reflectivity with high spatial resolution through the emission and reception of electromagnetic pulses. The SAR data have various applications such as detecting objects and their geographic location, estimation of environments geophysical properties (i.e. certain dimensions, moisture content, roughness, and density) (Ditchfield, 1966).

Synthetic aperture radar data are a valuable resource for monitoring flood events. Most of the flood events have been accompanied by the widespread presence of clouds; the long-wavelength in SAR system could propagate through these clouds, which provied accurate data images.SAR sensors have achieved unprecedented resolutions and repetitivity of acquisition so that their application to flood monitoring is receiving mounting interest (Re & Capolongo, n.d.).

13.2.2.3 Vegetation and Water Indices

The spectral composition of remote sensing spectral data provides information about the physical properties of soil, water, and vegetation features in terrestrial environments. Remote sensing techniques, models, and indices are designed to convert this spectral information into a form that is easy to interpret (Bannari et al., 2017). Indices are considered a compact form of data that can effectively ensure the presence or absence of water. The indices are common identified as quantitative comparisons between the response of each ground pixel in different bands of the electromagnetic spectrum (Re & Capolongo, n.d.).

Several indices have been developed, such as the Normalized Difference Water Index (NDWI), proposed by McFeeters in 1996 to detect surface waters in wetland environments and investigate the surface water extent(Mcfeeters, 2013). Also, the Normalized Difference Vegetation Index (NDVI) is applied to estimate the level of crop's growth and detect the drought rate of vegetation (Mcfeeters, 2013). On the other hand, water emission in the infrared is generally lower than in the red part of the spectrum, so water has an inverse NDVI behavior with respect to both vegetated and unvegetated land. This makes NDVI a suitable tool to detect water surfaces rapidly (Re & Capolongo, n.d.).

13.3 Potential and Limitations of Open Satellite Data for Flood Mapping

13.3.1 Cloud Coverage Problem

In spite of the great potential that remote sensing in flood management, some limitations face its use. For example, the presence of cloud cover during the flood event has been reported as the major challenge in the use of optical remote sensing in flood management. According to Sanyal and Lu (2004), using SAR is a better option because of the higher penetration power of the radar pulse, overcoming the problem. However, its use, especially in developing countries has been constrained by its high prices as well as limited coverage (Application of Remote Sensing and Geographical Information Systems in Flood Management: A Review). Figure 13.6 provide an example of the cloud coverage effect on the remote sensing data.

Furthermore, spatial resolution has a crucial effect on flood monitoring. The spatial characteristics of the flood inundation area could constrain the use of satellite images with lower spatial resolution. When the inundated area is small, it cannot be observed by low-resolution images. Also, urban areas or dense forests complicate the process of flood detection. However, the high-resolution data and field surveys are crucial for reliable mapping (Potential and limitations of open satellite data for flood mapping).

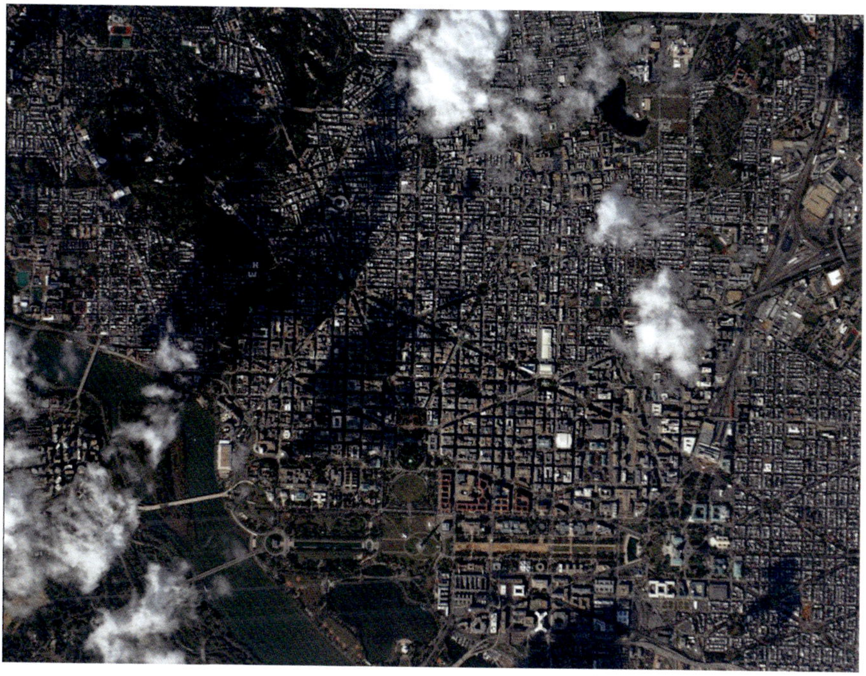

Fig. 13.6 The Sentinel-2 cloudless layer combines over 80 trillion pixels collected during differing weather conditions between May 2016 and April 2017. Image: ESA. (https://medium.com/planet-stories/cloudless-an-open-source-computer-vision-tool-for-satellite-imagery-6f4daaa4851f)

13.3.2 The Problem of Temporal Resolution in Flood Management

Temporal resolution represents a further challenge that faces the use of remote sensing in flood monitoring. The low temporal resolution causes limited availability of imageries in time-space, seasonal variations, and different technical limitations. For example, low temporal resolution may cause to not capture the peak of the flood event where most radar images are taken sometime before it. In addition, an area flooded by a small stream has a very short co-flood time interval which necessitates the higher temporal resolution. There is an essential need for a more consistent monitoring strategy in terms of frequency and timeliness of remote sensing data collection. The inadequate frequency of image collection is one of the most important limitations. It is found that the closer the time between when the image data were collected and when the event peak occurred, the more reliable the detection of maximum flood extents and depths.

13.3.3 The Problem of Detecting Flooding in Urban Areas

Floodwaters can be detected with good precision by exploiting several typical characteristics of inland water surfaces with respect to dry areas. One is the reduced reflectance of clean and calm water with respect to land areas. This behavior is common to virtually all the optical spectral ranges, as long as the acquisition is far from the specular direction. An additional way to distinguish the presence of surface water is given by the availability of reflectance information in the infrared thermal spectral bands. This is also generally low for water surfaces with respect to land areas. Both these methods rely on assumptions that are broadly fulfilled when monitoring flood events occurring over non-urban land areas, especially when using low or medium-resolution data. For instance, artificial surfaces may have very low reflectance and thus be mistaken for flooded areas. This is more likely to occur in highly complex environments like urban areas, where pavements and tarmacs may have a wide range of reflectance behaviors during a flood event. This may also include flooded surfaces exhibiting artificially high reflectance, which could be due, for instance, to a shallow water layer over a bright pavement or to specular reflections from surrounding buildings. Such cases could be successfully solved by sensors having infrared detection bands. In fact, urban areas are admittedly among the most complicated land cover types for many remote sensing applications, in virtually all regions of the electromagnetic spectrum, with flood monitoring making no exception (Re & Capolongo, n.d.).

13.4 Application of Remote Sensing on Flash Flood and Extreme Rainfall Events in MENA Region

13.4.1 Area of Interest

The surface area of the Arab region is about 13,781,751 km^2. It consists of 22 countries as shown in the map (Fig. 13.8a). The population has increased by the rate of 2% every year from 2002 to 2020. The average annual rainfall over the Arabian region varies from 0 to 1800 mm, while the average evaporation rate is more than 2000 mm/year (https://data.worldbank.org/indicator/SP.POP.GROW?view=map (2021); Saber et al., 2017b). The Arab region is suited in the northern hemisphere with semi-arid to arid climatic conditions. The total average rainfall (Fig. 13.7a) estimated from GSMaP shows spatial variability with a low precipitation rate over the region (Saber et al., 2017a). The aridity index was estimated for the region from The Consortium for Spatial Information (CGIAR-CSI) [8)] as shown in (Fig. 13.7b). In most Arabian countries, during the last 7 years, Wadi flash floods (WFF) became catastrophic and more frequent in both space and time (Fig. 13.8a & b).

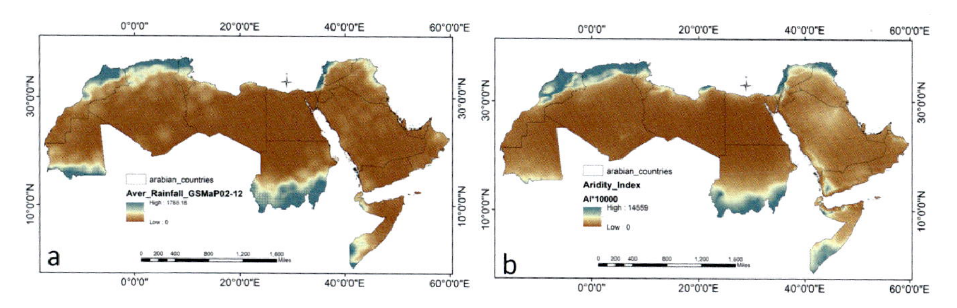

Fig. 13.7 Average rainfall estimated from Global Satellite Mapping of Precipitation (GSMaP) data (**a**) and Aridity Index (**b**) estimated from Global Aridity Index developed by the Consortium for Spatial Information (CGIAR-CSI) 2002–2012

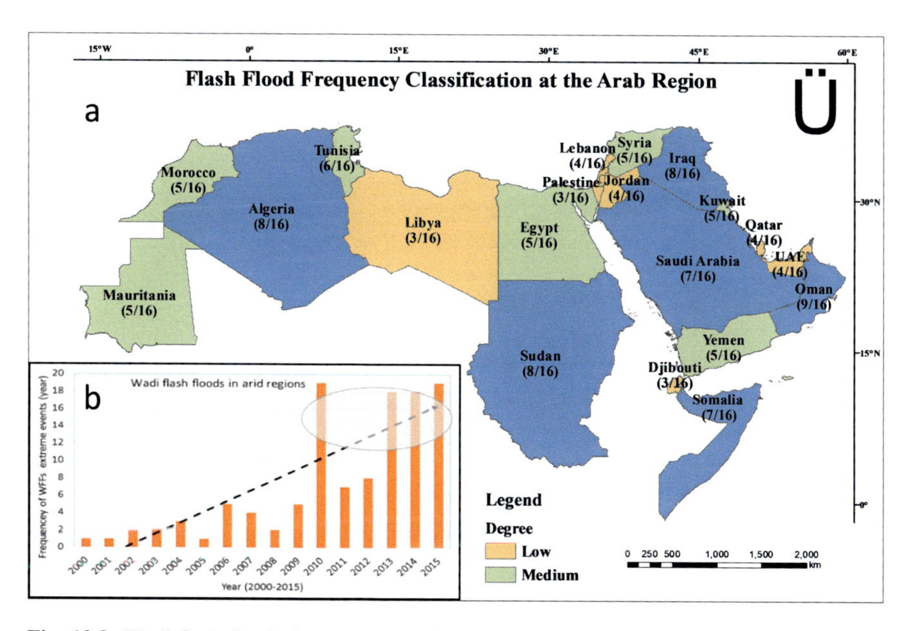

Fig. 13.8 Wadi flash floods frequency and classification spatially (**a**) and temporally (**b**) at the Arab region

13.4.2 *Case Study of Cities of (Al Arish, Ras Gharib, Al Saloum, Drunka, Hurghada)*

According to the UNDP, by 2050, two-thirds of the world's population is likely to live in cities. Urban flooding is already a major risk for cities. Increasing impervious surface area, inadequate stormwater drainage, and aging infrastructure all contribute. As a result, growing urban populations will face a greater risk of flooding from extreme weather events. Using satellite data allows individuals and organizations to develop better plans for handling floods. This can include developing better early warning techniques, better plans for rescue and relief, and more effective long-term infrastructure planning.

In Egypt, about half of the yearly precipitation falls from December to March. Precipitation is generally very low throughout the country, although it averages more than 200 mm/year along the Mediterranean coastline. Most of Egypt is a desert and is classified as arid, except for the Mediterranean coast, semi-arid. There are four climate regions in Egypt: Nile Valley (from Cairo to Assiut, from Assiut to Edfu and from Edfu to Nasser Lake), Eastern Desert (Red Sea Region), Sinai Peninsula (South Sinai and North Sinai), and Matrouh Governorate (Salloum Plateau). Our goal is to generalize the results for the whole country. Therefore, representative samples with different climatic regions in Egypt and the selection of case studies were conducted based on two factors: the history of hazardous flash floods and climatic conditions. Accordingly, the selected cities are the city of "Ras Gharib" representing the Red Sea and Eastern Desert region, the city of "Al-Arish" representing the region of the Sinai Peninsula, "Drunka village" in Assiut governorate representing the Nile Valley and Delta region, and the city of "Salloum" that is representative of the Mediterranean region (Saber et al., 2020).

Ras Gharib is the second-largest city in Red Sea Governorate and the most important Egyptian city in oil production. It is located 150 km to the north of Hurghada on the Red Sea coast (Fig. 13.4d). Ras Gharib is considered to have a desert climate. During the year, there is virtually no rainfall, with an average of about 5 mm. The average annual temperature is about 22.2 °C (Saber et al., 2020).

The Sinai Peninsula is located in the northeast of Egypt between latitudes 27°43′ to 29°55′ and longitude 32°39′ to 34°52′. Al-Arish (Fig. 13.4b) is located on the coast of the Mediterranean Sea. It has a tropical climate with a rainfall average of about 3262 mm, even during the driest months. The average annual temperature is 23.8 °C (Saber et al., 2020).

Salloum is a small Egyptian border city near the western border of Egypt with Libya. It is located on the Mediterranean coast (Fig. 13.4d). The climate in Salloum is considered a desert climate. The average annual temperature is 19.6 °C, and the average annual rainfall is 150 mm (Saber et al., 2020).

Drunka is one of the villages of Asyut Center in Asyut Governorate in Egypt (Fig. 13.4c). The climate in Asyut is called a desert climate, with a precipitation average of about 2 mm, and the average annual temperature is 22.6 °C (Saber et al., 2020).

Hurghada is an essential center for tourist and mining activities: it lies directly on the Red Sea coast, and it is bounded by latitudes 270° 10′ and 270° 30' N and longitudes 330° 30′ and 330° 52′ E. Hurghada almost every year, causing loss of life and significant damage. Accordingly, the city has become one of the most vulnerable areas to such events near the Red Sea. Satellite rainfall data show that this trend increased in Hurghada from 1983 to 2019. Additionally, the city has the highest mean annual maximum daily precipitation in Egypt. Since 2000, numerous urban flash flood events have occurred along the Egyptian Red Sea coast, which has experienced 30 medium and large events this century (Gado, 2017). There has been an increase in the exposure of the city to flood risk during winter (rainy season) from October to February due to convective rainfall (Table 13.1).

Additionally, since 1996, several urban flash flood events have been recorded in the city and its vicinity. Inhabited areas, main roads, military campuses, and tourist buildings have been severely affected. Moreover, environmental contamination due to water flooding, especially in the inhabited lowland areas, has occurred (Abdrabo et al., 2020a) (Fig. 13.9).

13.4.3 Remote Sensing Data

Many types of data sets were collected and analyzed in this study (Table 13.2 and Fig. 13.10). The lack of hydrological and meteorological data in the Egyptian cities necessitated the use of hydrological modeling to predict flood depth and the spatial extent and identify sites with high risk. The RRI model used several remote-sensing data, including a digital elevation model (DEM) with an accuracy of 12.5 m, an LC map, and historical daily rainfall records. The resolution of the rainfall data was as follows: ($0.04° \times 0.04°$)-hourly based data for the 2014 (5-year REP) and 2016 (10-year REP) events. The resolution of the rainfall data was ($0.25° \times 0.25°$)-daily based data on the 1996 (50-year REP) event. LC was mapped from Sentinel (2A) with a 30 m resolution. These data sets were used to produce the inundation maps for the 5, 10, and 50 REPs in Hurghada.

Regarding model calibration and validation, photos during the event from different local newspapers were used. One of the authors (S.A.K.) conducted fact-finding and field investigations, reconnaissance-level inventories for topographic maps, and site visits to obtain the ground truth of the interpretations from imagery. From 2014 to 2015, we visited several specific urban sites, reviewed the proposed layouts of buildings and infrastructure, and provided comments to developers regarding avoiding urban flash flood risk and other environmental impacts. We have direct knowledge of the urban flash flood history of Hurghada over the past three decades, from the early 1980s to 2019.

13.4.4 Methodology

The workflow of the hazard module for floods using the RRI model is summarized in three steps. First, the daily spatial rainfall intensities for different hazard scenarios were obtained based on PERSIANN-Climate Data Record (CDR) for the 1996 event and the PERSIANN-Cloud Classification System (CCS) for the 2014 and 2016 events with their highest resolutions, i.e., 25 km and 4 km, respectively. Second, the DEM was obtained from Advanced Land Observation System-Phased Array Synthetic Aperture Radar (ALOS-PALSAR) data available from the Alaska Satellite Facility (ASF) Distributed Active Archive Center (DAAC) with a 12.5 m resolution

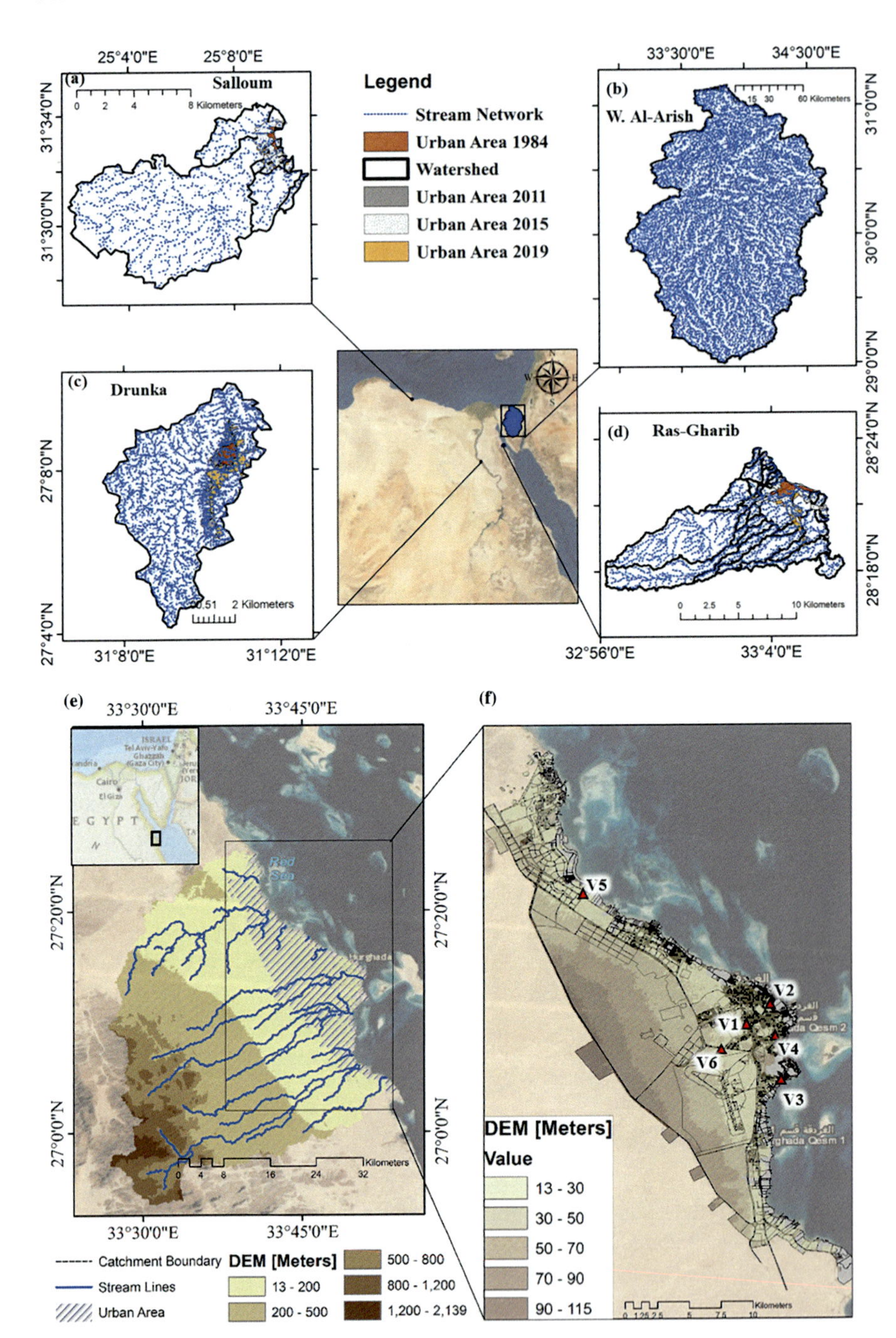

Fig. 13.9 Location maps showing the target cities and related wadi catchments and stream network developed from DEM (Sentinel-2) by GIS: (**a**) Salloum, (**b**) Wadi Al-Arish, (**c**) Drunka, (**d**) Ras Gharib, (**e**) the Hurghada catchment area, and (**f**) Hurghada city

Table 13.2 Material descriptions

Data type	Date	Format	Data source	Derived data
ASTER-ALSO-PALSAR (12.5 m spatial resolution)	2020	Geotiff	[84]	Topographic and hydrological parameters
Rainfall (scale 0.04° × 0.04°) (hourly based)	2020	PERSIANN-CCS	[85]	Rainfall distribution during the 2014 and 2016 events
Rainfall (scale 0.25° × 0.25°) (daily based)		PERSIANN-CDR	[86]	Rainfall distribution during the 1994 event
Sentinel (2A) (30 m resolution)	2019	Geotiff	[87]	Land cover types

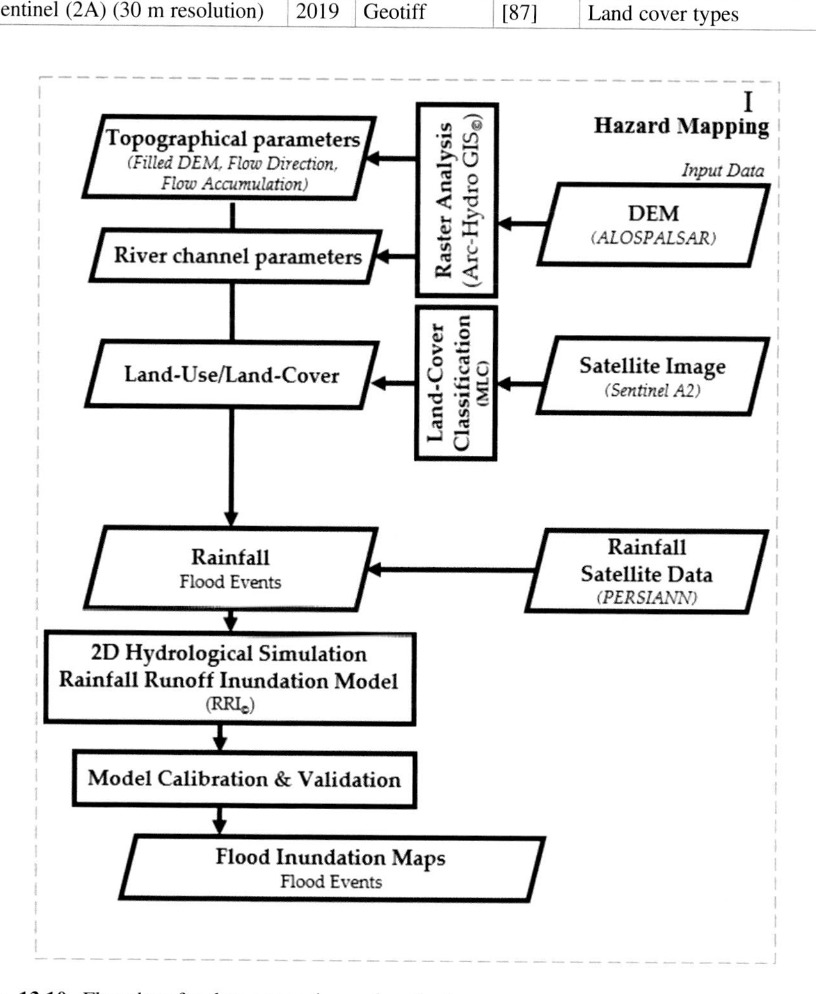

Fig. 13.10 Flowchart for data processing and methods

(ASF, 2006). The original DEM was processed using the Arc-hydro tool of ArcMap 10.6.1 to obtain the filled DEM, and the flow direction and flow accumulation were extracted. The Arc-hydro tool was used later to identify and extract the drainage features in the study areas, such as the flow direction, flow accumulation, stream networks, and watershed delineation required as inputs for the RRI model. Third, the LC map was created based on the Sentinel-2A satellite data from 2019, which were corrected based on Google Earth satellite images (General Organization for Physical Planning (GOPP), 2013, 2018). The initial parameters of the RRI model were assigned based on the validated parameters in arid regions (Abdel-Fattah et al., 2016). Finally, the RRI input raster maps for rainfall, topography, and LC were converted into ASCII files with their original resolutions, while the hazard maps (inundation depths) had the same resolution as the DEM utilized (12.5 m × 12.5 m).

- Model calibration and validation

For more comprehensive and accurate results, the RRI model was calibrated and validated. Due to the lack of observed data, the calibration and validation processes were performed based on reported images of the simulated events in each city. Regarding the land use parameters in the RRI model, the city was classified into three types of land use: desert, vegetation, and urban, in order to determine the model parameters for different cases.

13.4.5 Results and Discussions

In the case mentioned above studies, remote sensing data were used as input to distributed hydrological models to predict streamflow at the microscale scale catchments better. The results as shown in Figs. 13.6 and 13.7, which is already described in (Abdrabo et al., 2020b; Saber et al., 2020), showed encouraging results using global datasets combined with in-situ data. Moreover, it showed that model results using the remote sensing products combined with in-situ data were generally accurate (Figs. 13.11, 13.12 and 13.13).

13.5 Conclusion

In this chapter, a summary of the most important aspects of detailed flood monitoring through remote sensing has been attempted. We reviewed the basics of flood monitoring practices, focusing on the commonly accepted standards for definition, detection, and updating of flood maps. We then listed the primary sources of data that are commonly used in flood monitoring activities, spanning through optical and microwave instruments, the main sources of remotely sensed data used in this field.

The importance of using satellite-based products was discussed for application in arid regions where the data are not available. We focused mainly on the general idea of the main data processing techniques used to extract flood inundation maps and evolution information from earth observation data and to integrate remotely sensed data into hydrological inundation models, as well as to use remote sensing-derived

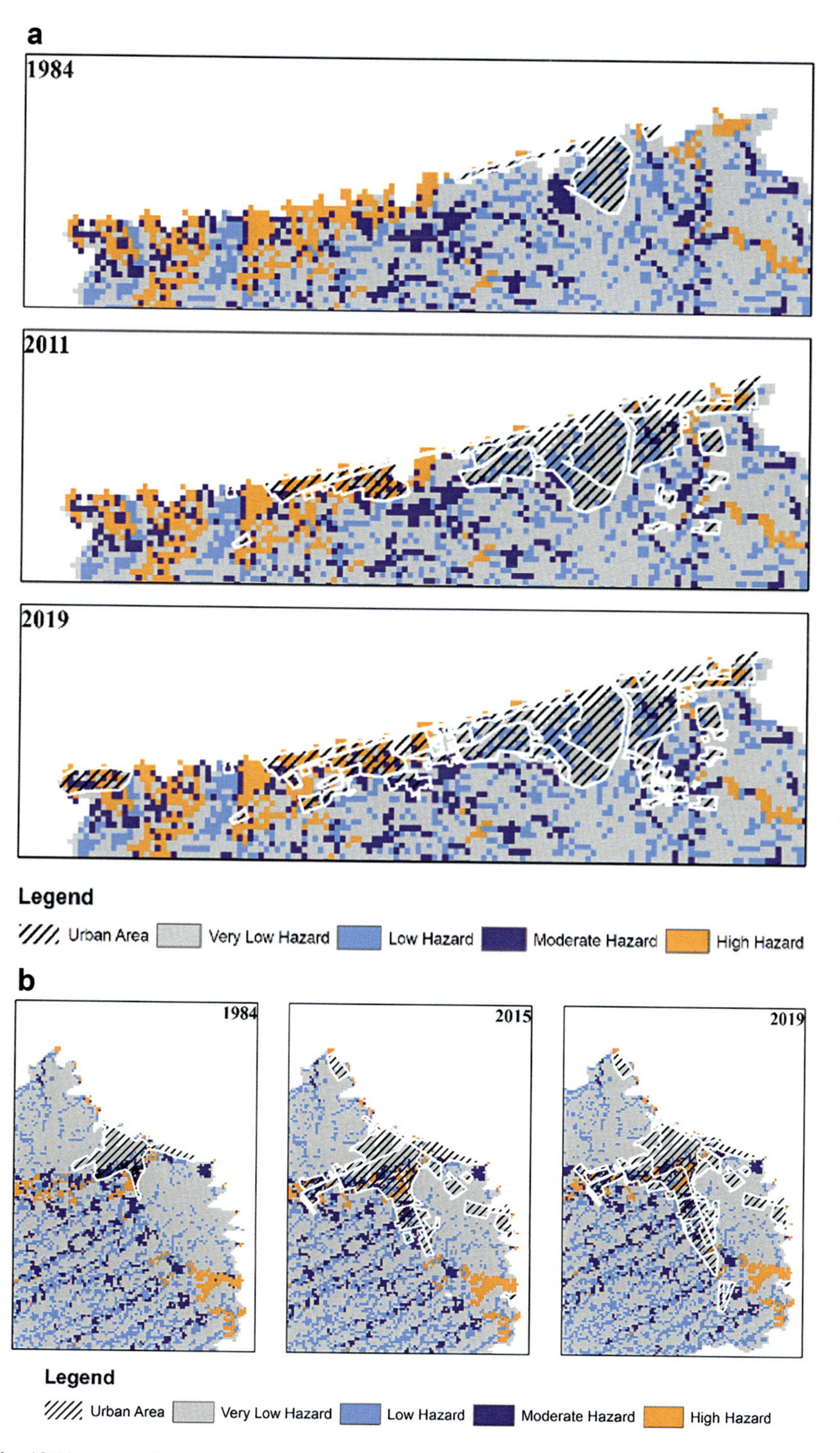

Fig. 13.11 Inundation maps showing the hazard levels affecting the urban areas (**a**) and estimates of the vulnerable areas for flood hazard categories from1984-2019 in (**a**) Al-Arish, (**b**) Ras Gharib, (**c**) Salloum, and (**d**) Drunka

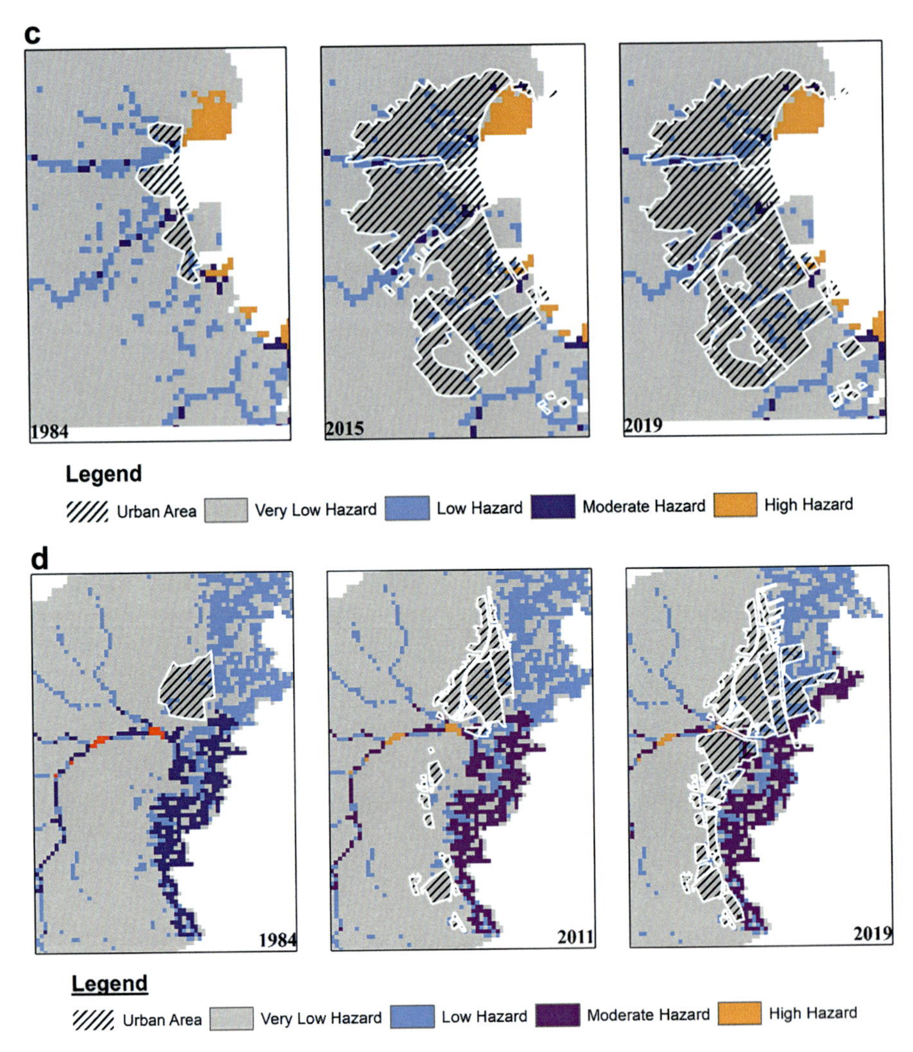

Fig. 13.11 (continued)

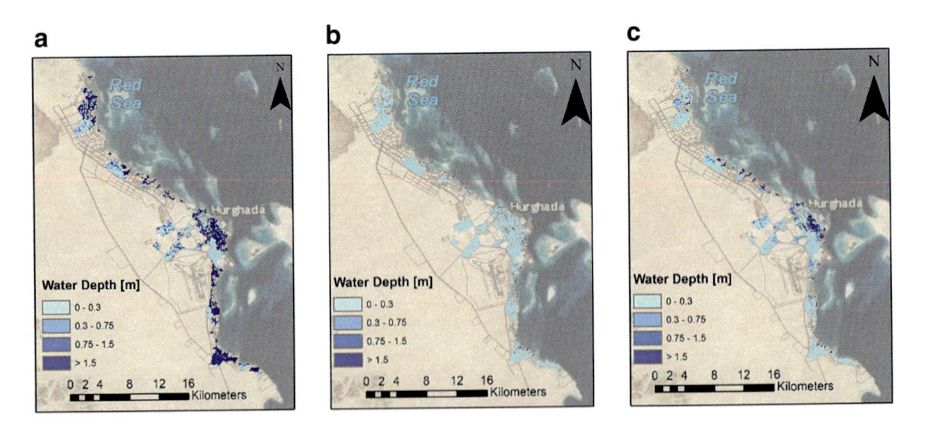

Fig. 13.12 Flood inundation maps (RRI model) for Hurghada in 1996, 2014 and 2016 (**a–c**)

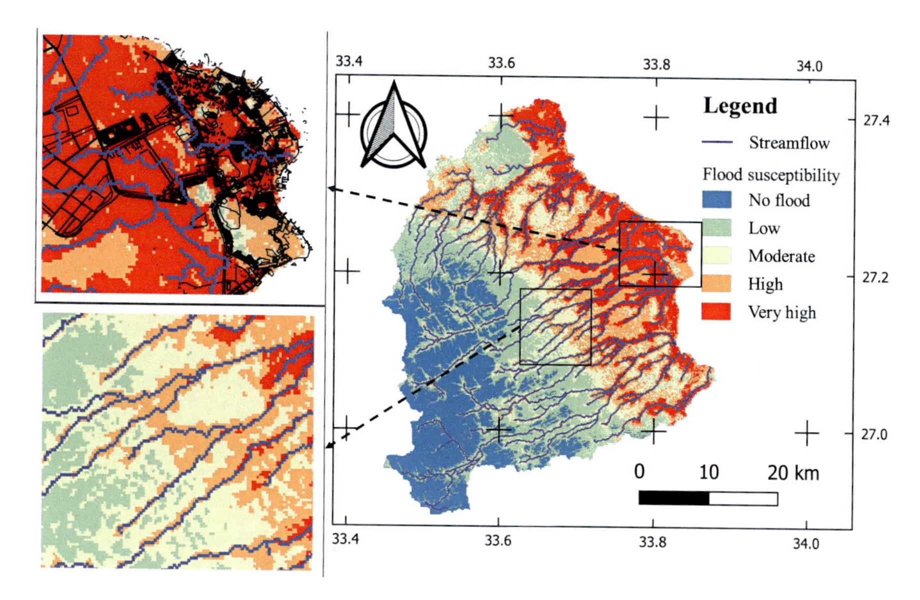

Fig. 13.13 Hurghada Flood susceptibility maps based on the Machine learning method (LightGBM) in 2021. (Source: Saber et al., 2021)

information to enrich our knowledge of flood processes and aid decision-making systems.

References

(ASF), Alaska Satellite Facility. (2006). *DEM 30 m*. Retrieved March 12, 2020 (https://search.asf.alaska.edu/#/?dataset=ALOS&zoom=7.334585016709559¢er=28.682014,29.484976&polygon=POINT(29.7851%2031.1334)&resultsLoaded=true&granule=ALPSRP141900610-KMZ).

Abdel-Fattah, M., Kantoush, S. A., Saber, M., & Sumi, T. (2016). *Hydrological modelling of flash flood in Wadi Samail, Oman*. Kyoto university.

Abdrabo, K. I., Kantoush, S. A., Saber, M., Sumi, T., Habiba, O. M., Elleithy, D., & Elboshy, B. (2020a). Integrated methodology for urban flood risk mapping at the microscale in ungauged regions: A case study of Hurghada, Egypt. *Remote Sensing, 12*(21), 3548. https://doi.org/10.3390/rs12213548

Abdrabo, K. I., Kantoush, S. A., Saber, M., Sumi, T., Habiba, O. M., Elleithy, D., & Elboshy, B. (2020b). Integrated methodology for urban flood risk mapping at the microscale in ungauged regions: A case study of Hurghada, Egypt. *Remote Sensing, 12*(21), 3548. https://doi.org/10.3390/rs12213548

Abushandi, E. H., & Merkel, B. J. (2011). Application of IHACRES rainfall-runoff model to the Wadi Dhuliel arid catchment, Jordan. *Journal of Water and Climate Change, 2*(1), 56–71.

Ali, S. A., Khatun, R., Ahmad, A., & Ahmad, S. N. (2019). Application of GIS-based analytic hierarchy process and frequency ratio model to flood vulnerable mapping and risk area estimation at Sundarban region, India. *Modeling Earth Systems and Environment, 5*(3), 1083–1102.

Ali, S. A., Parvin, F., Pham, Q. B., Vojtek, M., Vojteková, J., Costache, R., Linh, N. T. T., Nguyen, H. Q., Ahmad, A., & Ghorbani, M. A. (2020). GIS-based comparative assessment of flood susceptibility mapping using hybrid multi-criteria decision-making approach, Naïve Bayes tree, bivariate statistics and logistic regression: A case of Topľa Basin, Slovakia. *Ecological Indicators, 117*, 106620.

Arora, A., Arabameri, A., Pandey, M., Siddiqui, M. A., Shukla, U. K., Bui, D. T., Mishra, V. N., & Bhardwaj, A. (2020). Optimization of state-of-the-art fuzzy-metaheuristic ANFIS-based machine learning models for flood susceptibility prediction mapping in the middle ganga plain, India. *Science of the Total Environment, 750*, 141565. https://doi.org/10.1016/j.scitotenv.2020.141565

Athanasiou, L. S., Fotiadis, D. I., & Michalis, L. K. (2017). Propagation of segmentation and imaging system errors. *Atherosclerotic Plaque Characterization Methods Based on Coronary Imaging, 151–66*. https://doi.org/10.1016/b978-0-12-804734-7.00008-7

Bannari, A., Morin, D., Bonn, F. & Huete, A. R.. 2017. *A review of vegetation indices*. 7257(April). https://doi.org/10.1080/02757259509532298.

Bisht, S., Chaudhry, S., Sharma, S., & Soni, S. (2018). Assessment of flash flood vulnerability zonation through geospatial technique in high altitude Himalayan watershed, Himachal Pradesh India. *Remote Sensing Applications: Society and Environment, 12*, 35–47.

Chen, S., Junjun, H., Zhang, Z., Behrangi, A., Hong, Y., Gebregiorgis, A. S., Cao, J., Baoqing, H., Xue, X., & Zhang, X. (2015). Hydrologic evaluation of the TRMM multisatellite precipitation analysis over Ganjiang Basin in humid Southeastern China. *IEEE Journal of Selected Topics in Applied Earth Observations and Remote Sensing, 8*(9), 4568–4580.

Ditchfield, C. R. (1966). Microwave radiometry. *Journal of Navigation, 19*(4), 503–521. https://doi.org/10.1017/S0373463300047639

Domeneghetti, A., Schumann, G. J. P., & Tarpanelli, A. (2019). Preface: Remote sensing for flood mapping and monitoring of flood dynamics. *Remote Sensing, 11*(8), 943. https://doi.org/10.3390/rs11080943

Gado, T. A. (2017). *Statistical characteristics of extreme rainfall events in Egypt* (pp. 18–20).

General Organization for Physical Planning (GOPP). (2018). *General strategic plan of 2027 for proposed future urbanization area for the city of Tanta*.

General Organization for Physical Planning (GOPP), Ministry of Housing. (2013). *General strategic plan of 2027 for proposed future urbanization area for the city of Hurghada, Red Sea*. Cairo, Egypt.

Hall, J., Arheimer, B., Borga, M., Brázdil, R., Claps, P., Andrea Kiss, T. R., Kjeldsen, J. K., Kundzewicz, Z. W., & Lang, M. (2014). Understanding flood regime changes in Europe: A state-of-the-art assessment. *Hydrology and Earth System Sciences, 18*(7), 2735–2772.

Haq, M., Akhtar, M., Muhammad, S., Paras, S., & Rahmatullah, J. (2012). Techniques of remote sensing and GIS for flood monitoring and damage assessment: A case study of Sindh Province, Pakistan. *The Egyptian Journal of Remote Sensing and Space Science, 15*(2), 135–141. https://doi.org/10.1016/j.ejrs.2012.07.002

Hsu, K.-l., Gao, X., Sorooshian, S., & Gupta, H. V. (1997). Precipitation estimation from remotely sensed information using artificial neural networks. *Journal of Applied Meteorology, 36*(9), 1176–1190.

Hutter, G. (2006). Strategies for flood risk management–A process perspective. In *Flood risk management: Hazards, vulnerability and mitigation measures* (pp. 229–246). Springer.

Jenice Aroma, R., & Raimond, K. (2015). A review on availability of remote sensing data. In *Proceedings – 2015 IEEE international conference on technological innovations in ICT for agriculture and rural development, TIAR 2015* (pp. 150–155). TIAR. https://doi.org/10.1109/TIAR.2015.7358548

Joyce, R. J., Janowiak, J. E., Arkin, P. A., & Xie, P. (2004). CMORPH: A method that produces global precipitation estimates from passive microwave and infrared data at high spatial and temporal resolution. *Journal of Hydrometeorology, 5*(3), 487–503.

Kilpatrick, F. A., & Cobb, E. D. (1985). *Measurement of discharge using tracers*. Department of the Interior, US Geological Survey.

Kneis, D., Chatterjee, C., & Singh, R. (2014). Evaluation of TRMM rainfall estimates over a large Indian River Basin (Mahanadi). *Hydrology and Earth System Sciences, 18*(7), 2493–2502.

Kummerow, C., Barnes, W., Kozu, T., Shiue, J., & Simpson, J. (1998). The tropical rainfall measuring mission (TRMM) sensor package. *Journal of Atmospheric and Oceanic Technology, 15*(3), 809–817.

Liang, S., Li, X., & Wang, J. (2012). *A systematic view of remote sensing*.

Lin, X. (1999). Flash floods in arid and semi-arid zones. *Technical Documents in Hydrology*.

Martín-Vide, J. P., & Llasat, M. C. (2018). The 1962 flash flood in the Rubí stream (Barcelona, Spain). *Journal of Hydrology, 566*, 441–454.

Mcfeeters, S. K. (2013). *Using the normalized difference water index (NDWI) within a geographic information system to detect swimming pools for mosquito abatement: A practical approach* (pp. 3544–3561). https://doi.org/10.3390/rs5073544

Ochoa, A., Pineda, L., Crespo, P., & Willems, P. (2014). Evaluation of TRMM 3B42 precipitation estimates and WRF retrospective precipitation simulation over the Pacific–Andean region of Ecuador and Peru. *Hydrology and Earth System Sciences, 18*(8), 3179–3193.

Ozturk, U., Wendi, D., Crisologo, I., Riemer, A., Agarwal, A., Vogel, K., López-Tarazón, J. A., & Korup, O. (2018). Rare flash floods and debris flows in Southern Germany. *Science of the Total Environment, 626*, 941–952.

Pilgrim, D. H., Chapman, T. G., & Doran, D. G. (1988). Problems of rainfall-runoff modelling in arid and semiarid regions. *Hydrological Sciences Journal, 33*(4), 379–400.

Prasetia, R., As-syakur, A. R., & Osawa, T. (2013). Validation of TRMM precipitation radar satellite data over Indonesian region. *Theoretical and Applied Climatology, 112*(3), 575–587.

Re, A., & Capolongo, A. D. (n.d.) *Monitoring through remote sensing*.

Refice, A., D'Addabbo, A., & Capolongo, D. (Eds.). (2018a). *Flood monitoring through remote sensing*. Springer.

Refice, A., D'Addabbo, A., & Capolongo, D. (2018b). Methods, techniques and sensors for precision flood monitoring through remote sensing. In *Flood monitoring through remote sensing* (pp. 1–25). Springer.

Rodier, J., & Roche, M. (1978). River flow in arid regions. *Hydrometry: Principles and Practices, 453*.

Saber, M., & Habib, E. (2016). Flash floods modelling for Wadi System: Challenges and trends. In *Landscape dynamics, soils and hydrological processes in varied climates* (pp. 317–339). Springer.

Saber, M., Hamaguchi, T., Kojiri, T., & Tanaka, K. (2010). *Flash flooding simulation using hydrological modeling of Wadi Basins at Nile River Based on satellite remote sensing data* 16.

Saber, M., Kantoush, S. A., & Sumi, T. (2017a). *Assessment of water storage variability using grace and gldas data in the Arabian countries considering implications for water resources management* (pp. 13–18).

Saber, M., Alhinai, S., Al Barwani, A., Ahmed, A. L.-S., Kantoush, S. A., Habib, E., & Borrok, D. M. (2017b). Satellite-based estimates of groundwater storage changes at the Najd Aquifers in Oman. In *Water resources in arid areas: The way forward* (pp. 155–169). Springer. https://data.worldbank.org/indicator/SP.POP.GROW?view=map (2021)

Saber, M., Abdrabo, K. I., Habiba, O. M., Kantosh, S. A., & Sumi, T. (2020). Impacts of triple factors on flash flood vulnerability in Egypt: Urban growth, extreme climate, and mismanagement. *Geosciences, 10*(1), 24.

Saber, M., Boulmaiz, T., Guermoui, M., Abdrado, K. I., Kantoush, S. A., Sumi, T., Boutaghane, H., Nohara, D., & Mabrouk, E. (2021). Examining LightGBM and CatBoost models for Wadi flash flood susceptibility prediction. *Geocarto International*, 1–26.

Sanyal, J., & Lu, X. X. (2004). Application of remote sensing in flood management with special reference to Monsoon Asia: A review. *Natural Hazards, 33*(2), 283–301.

UNISDR, C. 2015. The human cost of natural disasters: A global perspective.

University of Waterloo, Nick Kouwen, Eric Soulis, and University of Waterloo. (1993). Remote sensing inputs for flash flood forecasting in urban areas. *Journal of Water Management Modeling*. https://doi.org/10.14796/JWMM.R175-08

Vissers, M. (2007). K & C Science Report – Phase 1 Tropical Forest and Wetlands Mapping , Case Study Borneo. (January).

Wheater, H., Sorooshian, S., & Sharma, K. D. (2007). *Hydrological modelling in arid and semi-arid areas*. Cambridge University Press.

Yoshimoto, S., & Amarnath, G. (2017). Applications of satellite-based rainfall estimates in flood inundation modeling—A case study in Mundeni Aru River basin, Sri Lanka. *Remote Sensing, 9*(10), 998.

Zhang, X., Aguilar, E., Sensoy, S., Melkonyan, H., Tagiyeva, U., Ahmed, N., Kutaladze, N., Rahimzadeh, F., Taghipour, A., & Hantosh, T. H. (2005). Trends in Middle East climate extreme indices from 1950 to 2003. *Journal of Geophysical Research: Atmospheres, 110*(D22).

Zhu, Lingli, Juha Suomalainen, Jingbin Liu, Juha Hyyppä, and Haggren Kaartinen. 2018. "A Review: Remote Sensing Sensors."

Chapter 14
Application of Remote Sensing and Geographic Information System Techniques to Flood and Rainwater Harvesting: Case Study of Sennar, Sudan

Mohamad M. Yagoub and Sharaf Aldeen Mahmoud

Abstract Construction of the Grand Ethiopian Renaissance Dam will have positive and negative impacts on the people living downstream. Among possible impacts is the probability of flooding of human settlements along the Blue Nile. Reduction of water level is also a possible adverse effect of this dam on the agricultural schemes along the Blue Nile; the Gezira scheme, one of the largest irrigation projects in the world, could be affected by such reduction.

The objective of this study is twofold. The first objective is to identify cities and villages near the Sennar Dam (in Sudan) that may be affected by potential fluvial (river) floods. The location of the study area near the dam exposes it to a probable dam problem and, consequently, societal, environmental, and economic damages. Therefore, identifying vulnerable locations is crucial for reducing risk to lives and properties, for preparing contingency action plans, and for making the area more resilient. The second objective is to select rainwater-harvesting (RWH) sites that can be used for water-supply backup while minimizing possible water-level reduction. The use of remote-sensing and Geographic Information System (GIS) techniques in this study has enabled access to unique sources of information to better characterize hazards and risks.

Keywords Fluvial flood · Rainwater harvesting · Sennar Dam · Sudan · Remote sensing · GIS

M. M. Yagoub (✉) · S. A. Mahmoud
Department of Geography and Urban Sustainability, College of Humanities and Social Sciences, UAE University, Abu Dhabi, UAE
e-mail: myagoub@uaeu.ac.ae

M. M. Al Saud (ed.), *Applications of Space Techniques on the Natural Hazards in the MENA Region*, https://doi.org/10.1007/978-3-030-88874-9_14

14.1 Introduction

A large number of cities and villages in Sudan were impacted by fluvial floods during August–September 2020 (UNITAR, 2021). More than 80 people died and 380,000 were affected (FloodList, 2020). Water levels in the Blue Nile are higher than those in 1906 and close to the 1988 levels. This situation has created an urgent need for identifying potential areas that could be flooded in the future.

Global efforts that integrate remote sensing and a Geographic Information System (GIS) to create flood maps as preventive tools are increasing (Prinos, 2008; Wright, 2014). For example, in the U.S., the Federal Emergency Management Agency creates interactive maps for flood hazard or risk. Such maps support community resilience by providing data, building partnerships, and supporting long-term hazard-mitigation planning (FEMA, 2020). The UK Environment Agency (2020) provides real-time maps that show fluvial-flood risks. The Global Flood Awareness System (2020) of the European Commission Copernicus Emergency Management Service has an early-warning system that can be used to reduce flood impact. Other initiatives include the Global Flood Partnership, which provides a worldwide forum for exchange of experience about flood prediction, modeling, and flood risk (Global Flood Partnership, 2021). The United Nations Office for Disaster Risk Reduction (2021) plays a greater role in coordinating and following up with governments on implementation of the Sendai Framework. Accordingly, our study falls within the context of "Think globally, act locally."

The study area includes the Sennar Dam, which is exposed to a probable flooding in case there is a problem at the Grand Ethiopian Renaissance Dam (GERD). Previous studies around the world have showed the potential risk of dam failure due to natural disasters or inadequate maintenance and management (Butt et al., 2013; Cenderelli, 2000; de Paiva et al., 2020; FEMA, 2013; Finn, 2008; Harrigan et al., 2020). Types of dam failure include dam body instability, canal gushing and soil flow, dam cracking, dam overtopping, and pilot-system failure (Escuder-Bueno et al., 2016; Kuo et al., 2007; Michailidi & Bacchi, 2017; Micovic et al., 2016; Yang et al., 2020). Discharge from a dam is governed by the maximum reservoir water level, reservoir capacity, and maximum discharge capacity. The seasonal variation in Blue Nile discharge ranges from 60 m^3/s in a very low year to >10,000 m^3/s at the peak of a high flood (Plusquellec, 1990).

Floodwater can be used to enable communities to make smart water-management decisions even during times of relative water abundance. Flooding can help resurrect wetlands and slow down climate change. Rainwater harvesting (RWH) includes gathering and storing rainwater to be used for drinking, irrigation, and livestock (Abdulrazzak, 2003; Awawdeh et al., 2012; Critchley et al., 1991; Galarza-Molina et al., 2015; Gould and Nissen-Petersen, 1999; Jha et al., 2014; Singh et al., 2016; Tiwari et al., 2018). RWH can also improve sustainability and reduce the impact of climate change (African Development Bank, 2010; UNEP, 2009). RWH has been gaining momentum worldwide. For example, Inamdar et al. (2013) used GIS to determine suitable locations for stormwater collection in Melbourne, Australia. They used runoff, open space, and accumulated catchments as criteria. They concluded that Royal Park ranked high for RWH as it had the largest water demand and the drainage outlets were close to the site.

Elhag and Bahrawi (2014) performed a study in the Kingdom of Saudi Arabia to improve RWH in terms of groundwater recharge. They created a spatiotemporal cloud map for evaluating prior-delineated watersheds that are beneath mostly cloudy skies all year. They used satellite imagery (EnviSat and Landsat), a digital elevation model (DEM), and soil-moisture and geological data. Singhai et al. (2017) used a combination of satellite images, land-use and soil data, and multi-criteria decision analysis to identify natural depression sites for RWH and found that forested areas have low-runoff potential whereas built-up areas have high-runoff potential.

Once potential RWH sites are identified, a suitable method can be employed to store water. Among the methods used are bound or terrace practices in areas with low-infiltration soil, small dams, micro-catchments, and haffirs. A haffir is an artificial excavation into which surface-water runoff is diverted during the rainy season. Its size depends on the location, hydrology, soils, and rainfall (Salih et al., 2016).

14.2 Study Area

The area around Sennar city was selected because of the availability of data, location of Sennar Dam (~260 km south of Khartoum), and the importance of the dam to the Sudanese economy (Gezira irrigation scheme).

The area includes Sennar and many other villages. Historically, the city of Sennar has been known as a major hub since the era of the Funj Chronicle (1504–1821). The city is at lat 13.539510 N and long 33.611298 E and is within UTM Zone 36 N, with an average altitude of 434.87 m (Fig. 14.1). The area includes schools, a university, hospitals, administration offices, and many banks. Most of its population (~1.5 million) depend on trade, agriculture, livestock, and services.

The average temperature in winter is 25 °C; in summer, it is 40 °C. The rainy or wet season in the study area starts in May and ends in October.

Flooding in the Blue Nile commences in May and reaches its climax during middle to late August–September. The average monthly rainfall (based on the period 2016–2019) ranges within 20–100 mm and reaches its maximum during August.

Built on the Blue Nile, the Sennar Dam was completed in 1925. Its main objective is to supply water for the Gezira irrigation scheme (Bernal, 1997). Its total storage capacity is 930 million cubic meter (Roseires Dam reservoir holds 3 billion cubic meter). The Sennar Dam is 3025-m long with a maximum height of 40 m. Two

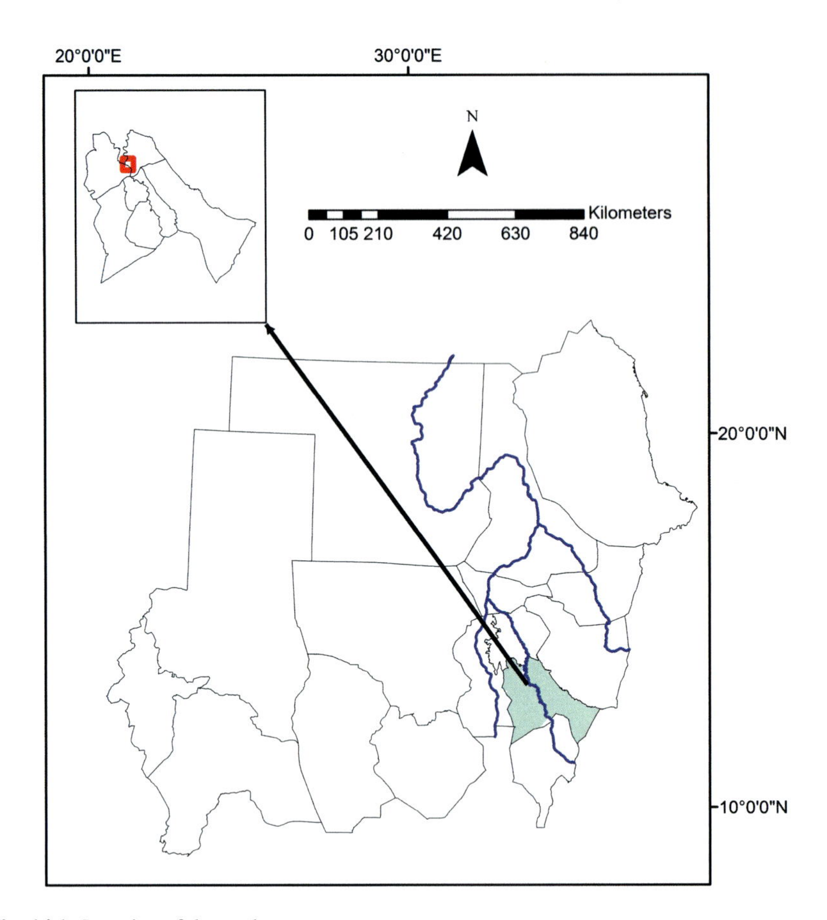

Fig. 14.1 Location of the study area

canals with numerous branches (4300 km) are used to irrigate the Gezira and Managil schemes. The minimum level in the reservoir to maintain maximum flow in the Gezira canal is 417.2 m while the top water level of the reservoir is 421.7 m (Plusquellec, 1990). A road and a rail route run along the length of the dam (The Engineer, 1924).

GERD is physically not within the study area, but its impact is the main objective of this study. The GERD has a planned full supply elevation of 640 m, 74 billion cubic meter of reservoir storage, and 6000 megawatt of installed generation capacity (Wheeler et al., 2016). The benefits of GRED could be in reduction of the flooding risk to Sudan, reduction of sediment in the river, reliability of flows, hydropower benefits, improved depth for navigation and reduced pumping costs for water users (Wheeler et al., 2016). However, on the other side GERD may reduce downstream water availability, impact on agriculture in Sudan, reduce land fertility due to the reduction of nutrient-rich sediment, and possible environmental impacts (Beyene, 2013).

14.3 Data

Data collected to identify settlements at risk and suitable RWH locations are shown in Table 14.1.

14.4 Methodology

The methodology for identifying potential areas that can be impacted by fluvial floods was based on land-use and elevation data, the topographic wetness index (TWI), and proximity to Blue Nile (1–3 km). Five experts were consulted for setting the weight for each factor. Based on their recommendations resulting from their use of the Analytical Hierarchy Process (Saaty, 1980; Teknomo, 2020). Land use, elevation, TWI, and proximity to Blue Nile were assigned the weights of 0.2, 0.2, 0.3, and 0.3, respectively. All data were saved in raster format, and ArcGIS software was used in the subsequent analysis. Values in each layer were coded to a common scale of 1 (very low risk) to 5 (very high risk), and a weighted overlay was generated for all the layers. The result was classified into high, medium, and low-risk zones, and the number of people in each zone was calculated.

Land-use map generated using the Sentinel-2 satellite image of May 22, 2020 was one of the inputs. Sentinel-2 satellite data has a better resolution (10 m for Bands 2, 3, 4, and 8) than Landsat 8 (30 m). The month of May was selected because it is summertime and hence less cloudy. Bands 2 (blue), 3 (green), and 4 (red) were merged to generate a composite image. The ISO cluster unsupervised classification method was used to create a signature file, which is used to aid supervised classification. Training sites for barren (30 samples), built-up (40 samples), farms (30 samples), gardens and orchards (30 samples), and water (30 samples) were digitized using the ArcGIS software. A higher number of training samples was taken for built-up areas in case of the mixture of buildings with trees. The maximum-likelihood

Table 14.1 Data collected to identify settlements at risk and suitable rainwater-harvesting sites

	Data	Source	Format
1.	30-m-resolution digital elevation model (DEM)	U.S. Geological Survey Earth Explorer https://earthexplorer.usgs.gov/	Raster
2.	Sentinel-2 Tile Number: T36PWA Date: May 22, 2020	Copernicus Open Access Hub https://scihub.copernicus.eu/	Raster
3.	Land use	Sentinel-2 satellite image	Vector
4.	Population	Sennar Municipality	Vector
5.	2-m-resolution DEM from aerial photos obtained in 2010	Sennar Municipality	Raster
6.	Monthly rainfall data (2009–2018)	World Weather Online https://www.worldweatheronline.com/	Text

classifier algorithm was used for supervised classification. To check the accuracy of classification, 500 points were used. A majority filter (eight neighbors) was used to smoothen class boundaries and remove small isolated regions. The Boundary Clean was used to increase the spatial coherency of the classified image and connect adjacent regions (ESRI, 2017). Small isolated regions on the classified image were removed by applying the region/group method, which is a generalization process.

The normalized difference vegetation index (NDVI) (Richards, 2013) was calculated from Sentinel-2 data using band 4 as the red band (10-m resolution) and Band 8a as the infrared band (20-m resolution) (ESA, 2020):

$$\text{NDVI} = (\text{Infrared band} - \text{Red band})/(\text{Infrared band} + \text{Red band}) \tag{14.1}$$

NDVI was used to identify water comprising the Blue Nile. NDVI values were found to be between -1.0 and 1.0. Negative values represent clouds and water, while high positive values (0.6–0.8) indicate temperate and tropical rainforests. Raster calculator was used to identify areas with NDVI values close to 0 or negative (Blue Nile water).

30-m-resolution DEM was used initially to obtain a quick overview of the topography of the area. Sinks and peaks within the DEM were removed using the fill tool. The methodology for identifying potential RWH sites was based on creating morphometric variables from the high-resolution (2-m) DEM derived from aerial photographs (Table 14.1). The variables included surface slope, flow direction, flow accumulation, and TWI. TWI provides information about flood and surface ponding at a reasonable cost compared to hydrologic and hydraulic modeling (Ballerine, 2017; Beven et al., 1979; Buchanan et al., 2014; Dilts, 2015; Qin et al., 2011; Wolock and McCabe, 1995; Yagoub et al., 2020). TWI was generated in ArcGIS software (using the D8 algorithm to model flow into a single downslope cell). TWI results were passed through a filtering, smoothing, and generalization process similar to that used for land-use classification. The large sites identified from TWI were manually digitized for calculation of areas, addition of other attributes, and cartographic output. The sites were further assessed for determining the land-use type (and whether a site is a vacant land), proximity to Blue Nile, and total area.

14.5 Results

Water (Blue Nile) and soil could not be distinguished in the supervised classification of the Sentinel-2 satellite image from May 22, 2020 as they have similar spectral characteristics. The confusion resulted from the brown color of the sediment-laden river. Generally, every year, rain starts in Ethiopia during April–May and continues in Sudan until September–October. This indicates that it is difficult to classify Blue Nile water for almost 6 months during most years. To address this problem, NDVI data were derived and water was identified as having negative NDVI values. Then, the water was used as training samples. This method achieved better results than taking training samples from water without the use of NDVI. The NDVI values for the May 22, 2020 image ranged between −0.49 and 0.79.

Table 14.2 shows the confusion-error matrix. Classification accuracy of the land-use map is 95.6% and Kappa statistic is 0.92; the classification accuracy meets the recommended accuracy of ≥85% suggested by Foody (2002). A large portion of the area (77%) is classified as farmland and fallow fields (Table 14.3); this includes agricultural schemes in Eastern Sennar and part of Gezira. Built-up areas were assigned a high flood-potential value (scale of 5) while barren areas were given low values (Table 14.3).

Elevation is an important parameter in describing floodwater. The elevation in the area ranges within 405–436 m above mean sea level (Fig. 14.2; Table 14.4). Low and very low areas comprise 10% of the area (Table 14.4). The Blue Nile was masked, and its elevation was found to range within 405–428 m (Fig. 14.2). Because this shows reservoir-water-level height downstream and upstream from the dam, the difference in height can be used for calculating the storage volume of the dam (Butt et al., 2013). The obtained value of 428 m may represent islands inside the reservoir, given that the water level of the reservoir is kept at 421.7 m.

There are 11 villages and seven districts within Sennar city (population: 32,950) that fall within the high flood-potential zone (Table 14.5; Figs. 14.3 and 14.4), which includes 37 mosques, 36 schools, eight clinics, and one church. This indicates that a large number of people and infrastructure may be impacted by potential fluvial

Table 14.2 Confusion matrix

Class	Barren	Built-up	Farms	Gardens and orchards	Water	Total	User's accuracy
Barren	58	9	1			68	0.852
Built-up		38				38	1.0
Farms	2	3	326			331	0.984
Gardens and orchards	4			28	1	33	0.848
Water	2				29	31	0.935
Totals	66	50	327	28	30	501	0.0
Producer's accuracy	0.8787	0.7600	0.9969	1	0.9666	0.0	0.956
Kappa	0.92						

Table 14.3 Land use as of May 22, 2020

Class	Scale	Area (km^2)	Percentage
Barren	2	34	10
Built-up	5	23	6
Farmland and fallow fields	3	271	77
Gardens	4	26	7
Total		354	100

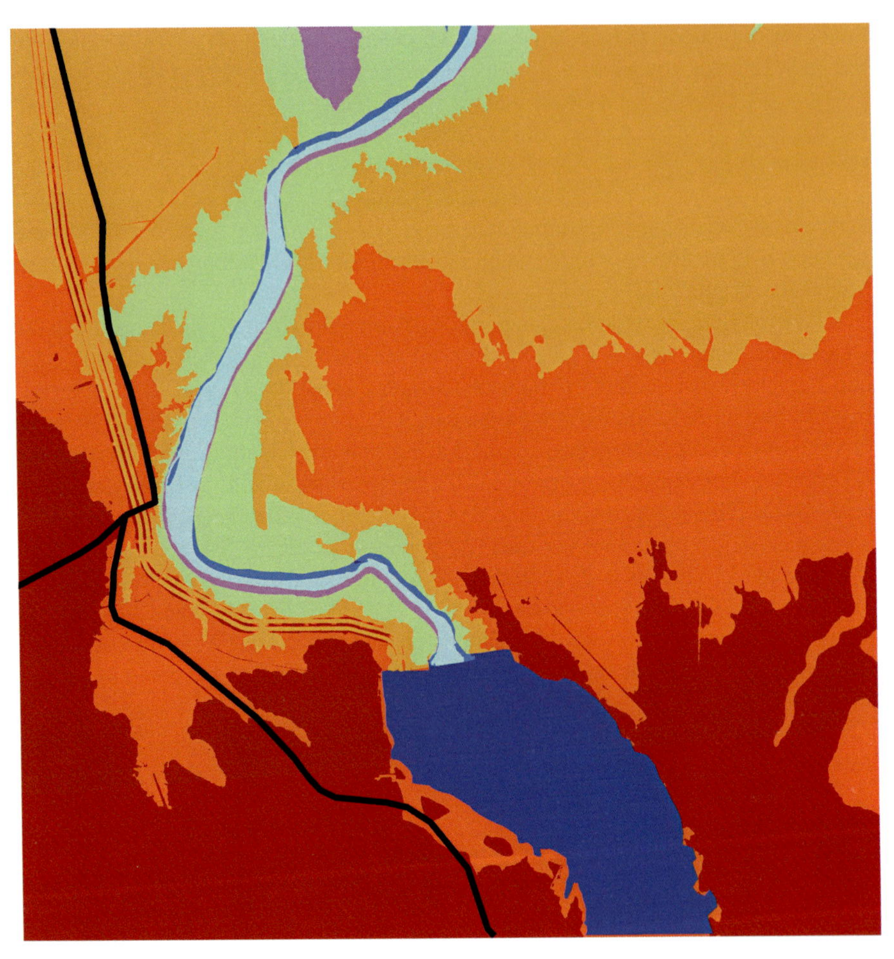

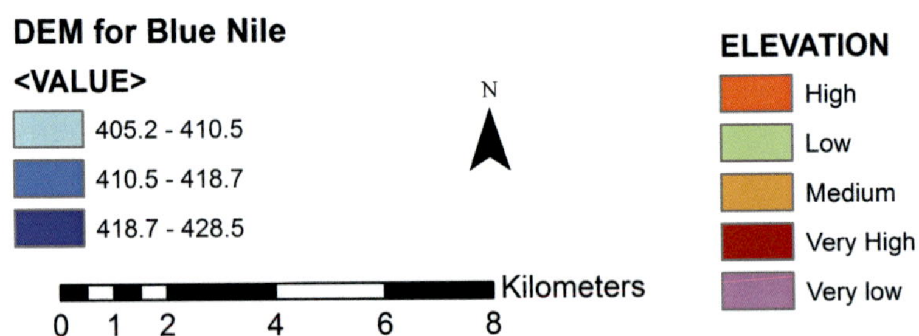

Fig. 14.2 Elevation categories

flooding. The area around Sennar Dam can get flooded if the water at the dam exceeds the flood level and the inflow is more than the discharge capacity. Furthermore, possible dam failure would affect the Gezira irrigation scheme, which supplies water to an area of over 882,000 ha where many farmers grow crops (cotton, wheat,

Table 14.4 Elevation classes

Value (m)	Scale	Area (km^2)	Percentage	Elevation class
405.1–411.3	5	4	1	Very low
411.3–418.4	4	33	9	Low
418.4–423.0	3	113	32	Medium
423.0–425.9	2	104	29	High
425.9–436.1	1	100	28	Very high
Total		354	100	

Table 14.5 Areas at potential fluvial flood zone

	Village/district name	Flood potential level	Population potentially exposed to flood
1.	Al Tikina	High[a]	682
2.	Al Islahia	High[a]	135
3.	Al Jinina	High[a]	3432
4.	Karima Abdel Rasoul	High	1483
5.	Karima Bahar	High	2168
6.	Al Arakeen	High	2318
7.	Banat	High	305
8.	Neel District (Sennar city)	High	3228
9.	Gala East District (Sennar city)	High	4418
10.	Taktok District (Sennar city)	High	1804
11.	Morba 22 and 23 district (Sennar city)	High	2749
12.	Soug District (Sennar city)	High	156
13.	Morba 21 South District (Sennar city)	High	1458
14.	Morba 21 North District (Sennar city)	High	1769
15.	Sharif Bajboj	High	1747
16.	Kasab Grabi	High	2282
17.	Kasab Jaleen	High	1545
18.	Kasab Danagla	High	1271
			32,950

[a]Areas at higher risk because they are downstream from and close to the dam

groundnuts, and sorghum) and on which traders and livestock depend (Bernal, 1997; Gaitskell, 1959). Therefore, there is an urgent need for backup plans and scenarios on how to deal with reduction of water level, flood spill from GERD, or dam failure, taking into account that the Blue Nile contributes ~57% of the yield of the Nile (Wheeler et al., 2016). Insufficient water supply in the past (1984–1985) resulted in the cessation of wheat cultivation in the area (Plusquellec, 1990).

Contributing ~35% of the Gross National Product, the Gezira scheme has been the backbone of the Sudanese economy for many years.

TWI values in the area range between 1 and 25 (Table 14.6; Fig. 14.5), whereas in a study conducted in the U.S. (Illinois), TWI indicators were found to range between −3 and 30 although they varied according to topography (Ballerine, 2017). Higher

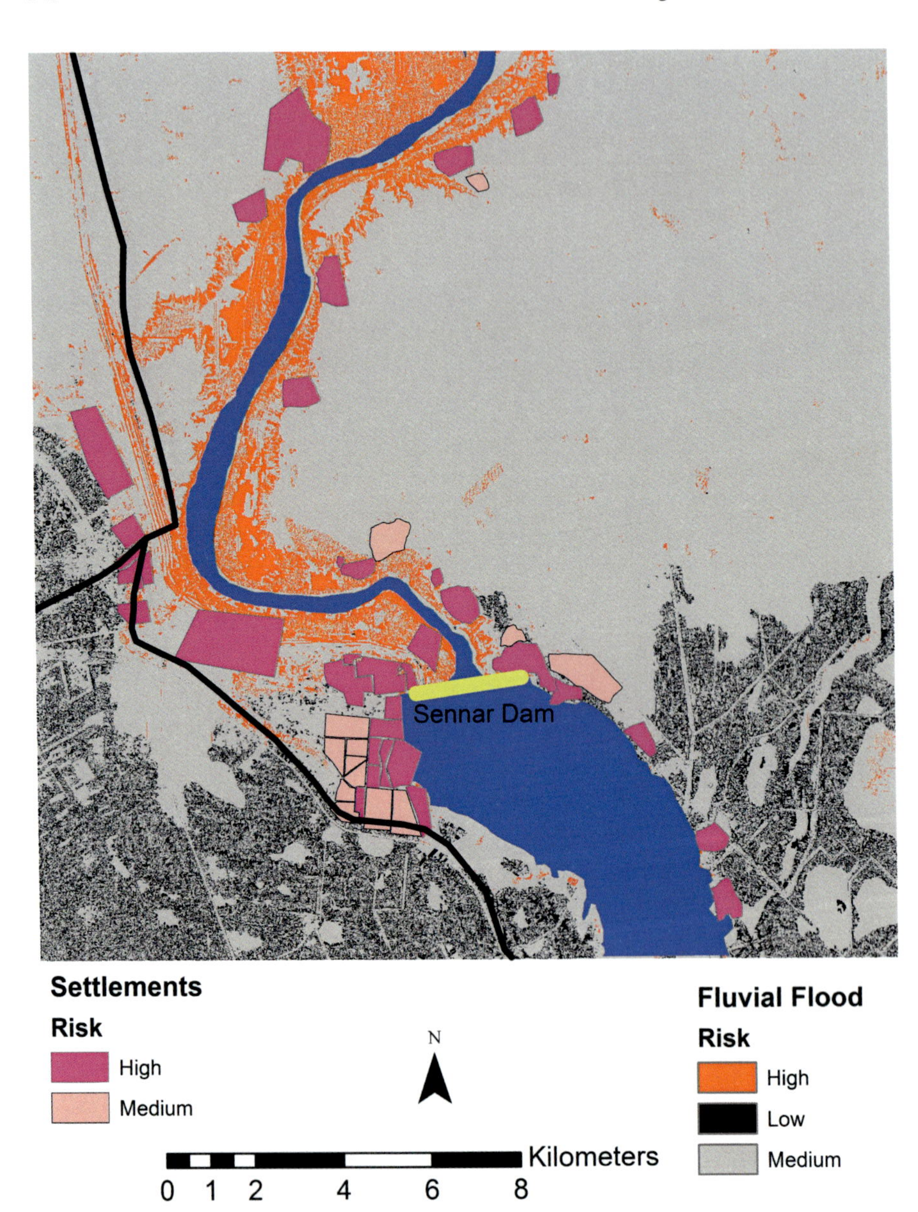

Fig. 14.3 Areas in potential fluvial-flood zone

TWI values occur where greater upslope areas drain into a gentle local slope, and lower values indicate less potential for ponding (Wolock & McCabe, 1995). Based on TWI values, the threshold value for TWI to be classified as potential RWH sites was set to 12. This threshold value is set manually with the aid of symbolization and may vary from one region to another depending on topography. Thus, TWI values of

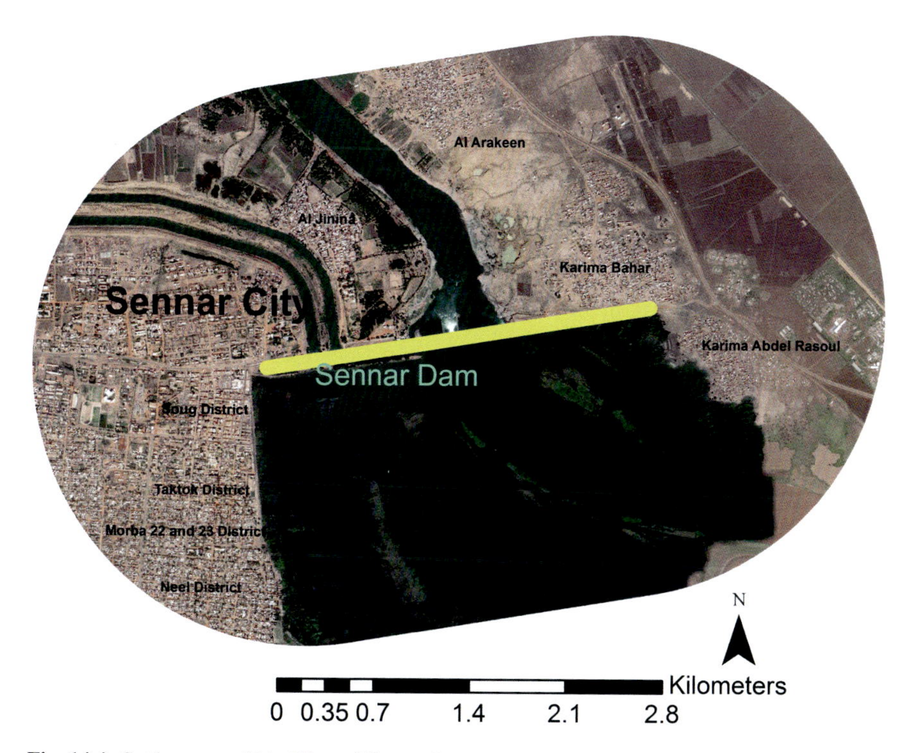

Fig. 14.4 Settlements within 2 km of Sennar Dam

Table 14.6 Topographic wetness index values

Value	Scale	Area (km^2)	Percentage
1–7	1	122	34
7–9	2	99	28
9–12	3	75	21
12–15	4	49	14
15–25	5	9	3
	Totals	354	100

>12 are assigned high scale values (Table 14.6; Fig. 14.5). Seven potential RWH sites were identified using TWI as the main criterion (Fig. 14.6). Sites 1–4 are close to the Blue Nile and can be conserved as ponds, wetland, or floodplain areas to divert water in case of fluvial flooding. Site 6 (Table 14.7; Fig. 14.6) has the largest area, but it is not suitable for RWH because it is surrounded by built-up areas, and the site can be used for future expansion of the city. Sites 5 and 7 are best for RWH and should be further assessed in terms of soil and ownership. The haffir method is proposed for potential RWH sites because it is less expensive. In addition, trees can be planted around such RWH sites. Check dams, which are small dams constructed across a valley or channel to lower the velocity of flow and to recharge groundwater, are also recommended to be constructed in valleys or khors. (Sandbags filled with pea gravel and logs can be used.).

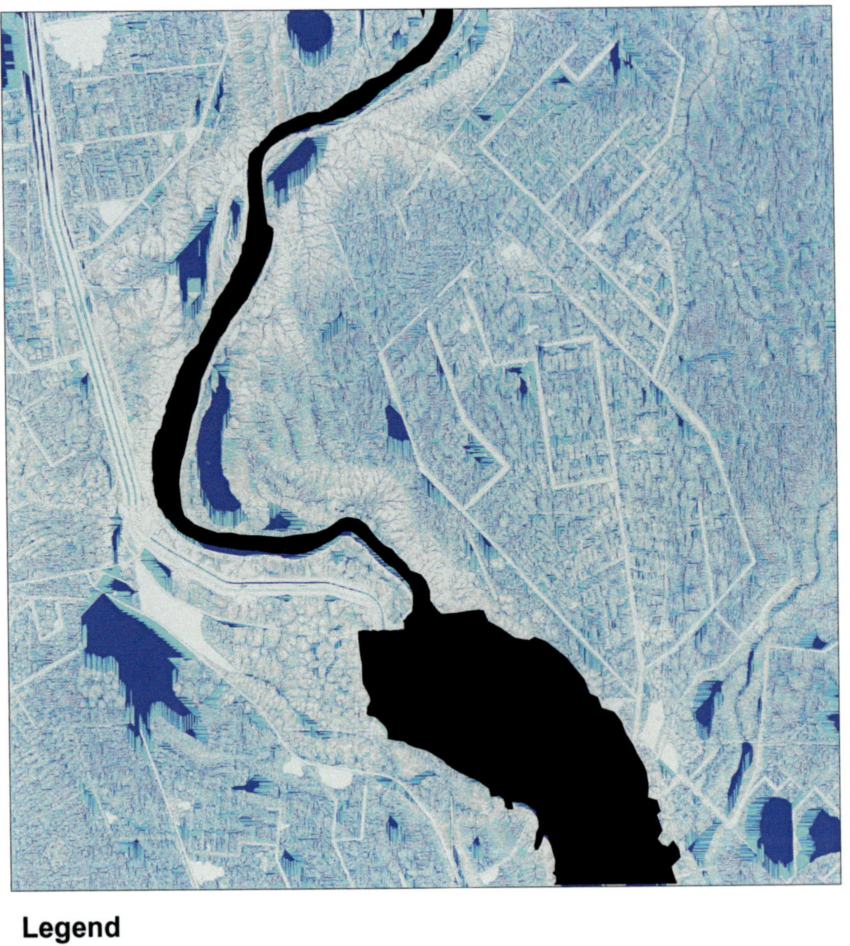

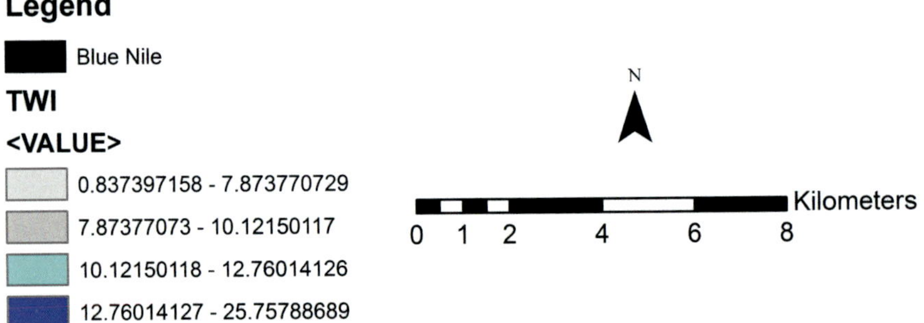

Fig. 14.5 Topographic wetness index values

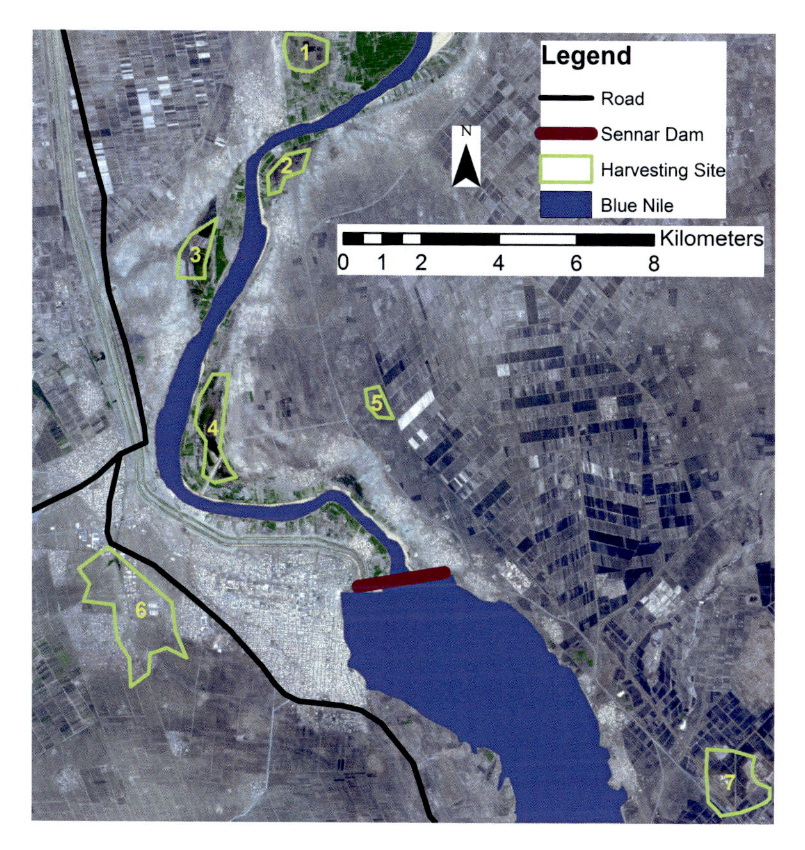

Fig. 14.6 Possible rainwater-harvesting sites

Table 14.7 Possible rainwater-harvesting sites

Site number	Area (km^2)	Percentage	Distance from Blue Nile (km)	Potential use
1	0.8646	8.39	1–2	Floodplain areas
2	0.5521	5.36	1	Floodplain areas
3	0.8676	8.42	1	Floodplain areas
4	1.6019	15.54	1	Floodplain areas
5	0.3306	3.21	2–3	Rainwater harvesting
6	4.1412	40.17	2–3	Excluded
7	1.9520	18.93	2–3	Rainwater harvesting
Total	10.3101	100		

14.6 Conclusion

In this study, settlements at a potential fluvial flood around Sennar Dam in Sudan were identified. The obtained results require further evaluation but can be used as a general guide for a sudden dam breach, insurance, impact assessment, emergency management, and developing effective flood plans. Seven possible RWH sites were also delineated. These sites will increase the coping capacity of the population with respect to shocks produced from rainfall variation and reduction of water flow due to GERD construction. The sites can be used for agriculture, livestock, and other purposes, and they will have a positive socioeconomic impact on the livelihoods of the inhabitants. The accuracy of the obtained results is affected by the population count and the accuracy of the land-use classification and DEM. TWI results depend on the DEM cell size, slope algorithm, flow-direction algorithm, and errors during either the acquisition or processing of aerial photographs.

The main causes of fluvial floods in Sudan during August–September, 2020 were high water level and building construction along the banks of the rivers. The buildings near the banks narrowed the rivers' width and altered their natural flow. Therefore, urban planning needs to take this into account, and collaboration and coordination among various stakeholders is needed (dam managers, urban planners, civil-defense officers, climate specialists, etc.). Furthermore, public awareness about flood hazards is very important. Future study could include using hydraulic and hydrological models to generate water-surface profiles for a given flood event (such as a 50-year flood), to determine the height of water across the river at a given location, to characterize the propagation of flood waves, and to map effects in the valley downstream from the dam structure. Similar studies can be conducted concerning the multipurpose Roseires Dam.

Acknowledgments The authors are grateful to UAE University Research Affair.

Office for support. The authors are also indebted to Prof. Mashael M. Al Saud (Space Research Institute, King Abdel Aziz City for Science and Technology) for her encouragement and support to participate in a book entitled "Applications of Remote Sensing on the Natural Hazards in the MENA Region". Mr. Shehab Majud is acknowledged for his following up with the editing process.

References

Abdulrazzak, M. (2003). *Water harvesting practices in selected countries of the Arabian Penisula.* Conference on water harvesting and the future development, Khartoum, Sudan, 19–20 August 2003.

African Development Bank. (2010). *Assessment of best practices and experience in water harvesting: Rainwater harvesting handbook* (75 pp.). ADB, Cote d'Ivoire, Open File Rep.

Awawdeh, M., Al-Shraideh, S., Al-Qudah, K., & Jaradat, R. A. (2012). Rainwater harvesting assessment for a small size urban area in Jordan. *International Journal of Water Resources and Environmental Engineering, 4*, 415–422.

Ballerine, C. (2017). *Topographic Wetness Index, Urban Flooding Awareness Act, action support: Will and DuPage Counties, Illinois*. Contract report 2017-02, April 2017, Illinois state. https://www.ideals.illinois.edu/handle/2142/98495. Accessed 5 Jan 2021.

Bernal, V. (1997). Colonial moral economy and the discipline of development: The Gezira Scheme and "Modern" Sudan. *Cultural Anthropology, 12*(4), 447–479. https://doi.org/10.1525/can.1997.12.4.447

Beven, K. J., Kirkby, M. J., & Seibert, J. (1979). A physically based, variable contributing area model of basin hydrology. *Hydrological Sciences Bulletin, 24*, 43–69.

Beyene, A. (2013). *Reflections on the Grand Ethiopian Renaissance Dam*. http://www.opride.com/oromsis/news/horn-of-africa/3664-reflections-on-the-grandethiopian-renaissance-dam. Accessed 18 Jan 2021.

Buchanan, B. P., Fleming, M., Schneider, R. L., Richards, B. K., Archibald, J., Qiu, Z., & Walter, M. T. (2014). Evaluating topographic wetness indices across Central New York agricultural landscapes. *Hydrology and Earth System Sciences, 18*(8), 3279–3299. http://www.hydrol-earth-syst-sci.net/18/3279/2014/hess-18-3279-2014.pdf. Accessed 8 Jan 2021.

Butt, M. J., Umar, M., & Qamar, R. (2013). Landslide dam and subsequent dam-break flood estimation using HEC-RAS model in northern Pakistan. *Natural Hazards, 65*, 241–254. https://doi.org/10.1007/s11069-012-0361-8

Cenderelli, D. A. (2000). Floods from natural and artificial dam failures. In E. E. Wohl (Ed.), *Inland flood hazards* (pp. 73–103). Cambridge University Press.

Critchley, W., Siegert, K., Chapman, C., & Finkel, M. (1991). *Water harvesting: A manual for the design and construction of water harvesting schemes for plant production* (154 pp.). Food and Agriculture Organization of the United Nations, Open File Rep.

de Paiva, C. A., da Fonseca, S. A., & do Prado, F. J. F. (2020). Content analysis of dam break studies for tailings dams with high damage potential in the Quadrilátero Ferrífero, Minas Gerais: Technical weaknesses and proposals for improvements. *Natural Hazards, 104*, 1141–1156. https://doi.org/10.1007/s11069-020-04254-8

Dilts, T. E. (2015). *Topography tools for ArcGIS 10.1*. University of Nevada Reno. https://www.arcgis.com/home/item.html?id=b13b3b40fa3c43d4a23a1a09c5fe96b9. Accessed 4 Jan 2021.

Elhag, M., & Bahrawi, J. A. (2014). Conservational use of remote sensing techniques for a novel rainwater harvesting in arid environment. *Environment and Earth Science, 72*, 4995–5005.

Environmental Systems Research Institute (ESRI). (2017). ArcGIS Desktop 10.5.1. ESRI Incorporation.

Escuder-Bueno, I., Mazza, G., Morales-Torres, A., & Castillo-Rodriguez, J. T. (2016). Computational aspects of dam risk analysis: findings and challenges. *Engineering, 2*(3), 319–324.

European Space Agency (ESA). (2020). *The spatial resolution of SENTINEL-2*. https://sentinel.esa.int/web/sentinel/user-guides/sentinel-2-msi/resolutions/spatial. Accessed 15 Jan 2021.

Federal Emergency Management Agency (FEMA). (2013). *Federal guidelines for dam safety emergency action planning for dams*. https://www.fema.gov. Accessed 7 Jan 2021.

Federal Emergency Management Agency (FEMA). (2020) *Risk mapping, assessment and planning (Risk MAP)*. https://www.fema.gov/flood-maps/tools-resources/risk-map. Accessed 7 Jan 2021.

Finn, H. (2008). *Dam failure and inundation modeling: Test case for ham dam*. Summary report, project conducted by "DHI Gulf" for UAE Ministry of Environment & Water.

FloodList. (2020). *Sudan – Over 80 dead and 380,000 affected as floods worsen*. https://floodlist.com/africa/sudan-khartoum-floods-august-2020. Accessed 2 Sept 2020.

Foody, G. M. (2002). Status of land cover classification accuracy assessment. *Remote Sensing of Environment, 80*(2002), 185–201.

Gaitskell, A. (1959). *Gezira: a story of development in the Sudan*. Faber and Faber.

Galarza-Molina, S., Torres, A., Moura, P., & Lara-Borrero, J. (2015). CRIDE: A case study in multi-criteria analysis for decision making support in rainwater harvesting. *International Journal of Information Technology & Decision Making, 14*, 43–67. https://doi.org/10.1142/S0219622014500862

Global Flood Awareness System (GloFAS). (2020). www.globalfloods.eu. Accessed 7 Jan 2021.

Global Flood Partnership. (2021). https://gfp.jrc.ec.europa.eu/. Accessed 18 Jan 2021.
Gould, J., & Nissen-Petersen, E. (1999). *Rainwater catchment systems for domestic supply: Design, construction and implementation*. Intermediate Technology Publications.
Harrigan, S., Zsoter, E., Alfieri, L., Prudhomme, C., Salamon, P., Wetterhall, F., Barnard, C., Cloke, H., & Pappenberger, F. (2020). GloFAS-ERA5 operational global river discharge reanalysis 1979–present. *Earth System Science Data Discussions*. https://doi.org/10.5194/essd-2019-232
Inamdar, P. M., Cook, S., Sharma, A. K., Corby, N., O'Connor, J., & Perera, B. J. (2013). A GIS based screening tool for locating and ranking of suitable stormwater harvesting sites in urban areas. *Journal of Environmental Management, 128*, 363–370.
Jha, M., Chowdary, V., Kulkarni, Y., & Mal, B. (2014). Rainwater harvesting planning using geospatial techniques and multicriteria decision analysis. *Resources, Conservation and Recycling, 83*, 96–111. https://doi.org/10.1016/j.resconrec.2013.12.003
Kuo, J. T., Yen, B. C., Hsu, Y. C., et al. (2007). Risk analysis for dam overtopping—Feitsui reservoir as a case study. *Journal of Hydraulic Engineering, 133*(8), 955–963.
Michailidi, E. M., & Bacchi, B. (2017). Dealing with uncertainty in the probability of overtopping of a flood mitigation dam. *Hydrology and Earth System Sciences, 21*(5), 2497–2507.
Micovic, Z., Hartford, D. N. D., Schaefer, M. G., & Barker, B. L. (2016). A non-traditional approach to the analysis of flood hazard for dams. *Stochastic Environmental Research and Risk Assessment, 30*(2), 559–581.
Plusquellec, H. (1990). *The Gezira irrigation scheme in Sudan: Objectives, design, and performance*. World Bank Technical Paper Number 120. The World Bank.
Prinos, P. (2008). *Review of flood hazard mapping*. European Community sixth framework programme for European Research and Technological Development. FLOODsite. www.floodsite.net. Accessed 14 Dec 2020.
Qin, C. Z., Zhu, A. X., Pei, T., Li, B.-L., Scholten, T., Behrens, T., & Zhou, C.-H. (2011). An approach to computing topographic wetness index based on maximum downslope gradient. *Precision Agriculture, 12*(1), 32–43. https://doi.org/10.1007/s11119-009-9152-y
Richards, J. A. (2013). *Remote sensing digital image analysis: an introduction*. Springer. https://www.springer.com/gp/book/9783642300615. isbn 978-3-642-30062-2
Saaty, T. L. (1980). *The analytic hierarchy process*. McGraw Hill.
Salih, A., Ahmed, A. S., & Yousif, T. A. (2016). *Water harvesting practices and techniques in Sudan from perspective of remote sensing and GIS*. Water harvesting symposium, University of Khartoum, Khartoum, Sudan.
Singh, L., Jha, M., & Chowdary, V. (2016). Multi-criteria analysis and GIS modelling for identifying prospective water harvesting and artificial recharge sites for sustainable water supply. *Journal of Cleaner Production, 142*, 1436–1456. https://doi.org/10.1016/j.jclepro.2016.11.163
Singhai, A., Das, S., Kadam, A. K., Shukla, J. P., Bundela, D. S., & Kalashetty, M. (2017). GIS-based multi-criteria approach for identification of rainwater harvesting zones in upper Betwa sub-basin of Madhya Pradesh, India. *Environment, Development and Sustainability, 21*, 777–797. https://doi.org/10.1007/s10668-017-0060-4
Teknomo, K. (2020). *Analytic hierarchy process (AHP) tutorial*. Available online: https://people.revoledu.com/kardi/tutorial/. Accessed 7 Jan 2021.
The Engineer. (1924, September 26). The Sennar Dam and Gezira Irrigation Scheme. *The Engineer*, pp. 349–50. https://www.gracesguide.co.uk/Special:Search?search=Sennar+dam&fulltext. Accessed 4 Jan 2021.
Tiwari, K., Goyal, R., & Sarkar, A. (2018). GIS-based methodology for identification of suitable locations for rainwater harvesting structures. *Water Resources Management, 32*, 1811–1825. https://doi.org/10.1007/s11269-018-1905-9
UK Environment Agency. (2020). *Check your long-term flood risk*. https://flood-warning-information.service.gov.uk/long-term-flood-risk/map. Accessed 4 May 2020.
UNEP. (2009). *Rainwater harvesting: A lifeline for human well-being* (80 pp.). United Nations Environment Programme. Open File Rep.

UNITAR (United Nation Institute for Training and Research). (2021). *Satellite detected water extent between 9 & 15 September 2020 in Sennar, White Nile, and Al Jazira States, Sudan*. https://www.unitar.org/maps/map/3123. Accessed 3 Jan 2021.

United Nations Office for Disaster Risk Reduction (UNDRR). (2021). *About the Sendai Framework*. https://www.unisdr.org. Accessed 7 Jan 2021.

Wheeler, K. G., Basheer, M., Mekonnen, Z. T., Eltoum, S. O., Mersha, A., Abdo, G. M., Zagona, E. A., Hall, J. W., & Dadson, S. J. (2016). Cooperative filling approaches for the Grand Ethiopian Renaissance Dam. *Water International, 41*(4), 611–634. https://doi.org/10.1080/02508060.2016.1177698

Wolock, D. M., & McCabe, G. J. (1995). Comparison of single and multiple flow direction algorithms for computing topographic parameters in TOPMODEL. *Water Resources Research, 31*(5), 1315–1324.

Wright, R. (2014). *What goes into a flood map: Infographic*. https://www.fema.gov/ Accessed 9 Sept 2020.

Yagoub, M. M., Aishah, A., Elfadil, A. M., Periyasamy, P., Reem, A., Salama, A., & Yaqein, A. (2020). Newspapers as a validation proxy for GIS modeling in Fujairah, United Arab Emirates: Identifying flood-prone areas. *Natural Hazards, 104*(1), 111–141. https://doi.org/10.1007/s11069-020-04161-y

Yang, K., Chen, F., He, C., Zhang, Z., & Long, A. (2020). Fuzzy risk analysis of dam overtopping from snowmelt floods in the nonstationarity case of the Manas River catchment, China. *Natural Hazards*. https://doi.org/10.1007/s11069-020-04143-0

Chapter 15
Drought Assessment Using GIS and Remote Sensing in Jordan

Nezar Hammouri

Abstract Drought is an anticipated consequence of climate change with diverse impacts on humans and ecosystems. Jordan with its limited natural and water resources have witnessed several drought events during the last decades with varying severity and magnitudes. Several studies recently disclosed that the country will suffers from additional severe drought events in the next decades that will threat the country's economic and social development plans. However, it is possible to reduce these impacts through assessing current drought conditions and forecasting future behavior of drought.

Recently, Geographic Information Systems (GIS) and Remote Sensing have played a great role in drought vulnerability assessment and drought mitigation plans. Therefore, this chapters will present an integrated approach for drought assessment in Jordan that based on GIS and Remote Sensing techniques. This includes the use of different types of sensors and images (For example recent Landsat OLI sensor, Advanced Very High-Resolution Radiometer (AVHRR) of National Oceanic and Atmospheric Administration (NOAA)), developing drought indices using GIS, which includes meteorological indices that based on meteorological data (SPI), and agricultural indices that based on satellite images (NDVI), develop drought vulnerability maps for Jordan using GIS modeling tools.

Furthermore, this chapter will present the use of WFP ICA context analysis approach that aims at assessing the vulnerability of drought events based on administrative scale, this will help in highlighting the most vulnerable governorate to drought risks in the country.

N. Hammouri (✉)
Faculty of Science, University of Sharjah, Sharjah, United Arab Emirates (UAE)

Prince El Hassan Bin Tala Faculty for Natural Resources and Environment,
The Hashemite University, Zerqa, Jordan
e-mail: nhammouri@sharjah.ac.ae; nezar@hu.edu.jo

M. M. Al Saud (ed.), *Applications of Space Techniques on the Natural Hazards in the MENA Region*, https://doi.org/10.1007/978-3-030-88874-9_15

15.1 Introduction

Among several natural disasters, drought was classified as the most complex and least understood phenomena that resulted in numerous agricultural, hydrological, and socioeconomical consequences (Hagman, 1984). In the last few decades, drought events become severer, with magnified impacts due to global warming and climate change. Figure 15.1 shows spatial distribution over the continents, where drought almost hitting every place on this planet.

As consequences of cumulative drought effects, about 160 million people were affected by water related disasters like drought, where about 13,500 were killed between 1996 and 2015. This great loss in human lives is accompanied by economic losses that were estimated to hits 30 billion USD yearly. With the aggregated impacts of climate change, severe drought events are expected initiates conflicts and migrations within the affected countries and communities (Ligtvoet et al., 2018).

Despite the lack of specific and global definition for the term "Drought", (IPCC, 2012) defined drought as "*A period of unusually dry weather that lasts long enough to cause a shortage in a region's water supply*". This definition implies the conditions that are critical to cause drought event which include time factor and water shortage that resulted from precipitation deficiencies.

Based on degree and duration of precipitation deficiencies and associated impacts, three types of drought can be identified, these include (Wilhite & Glantz, 1985):

1. **Meteorological Drought**: This type of drought is occurring when noticeable drop-in long-term average precipitation over a specific area for a relatively short period. Usually, this is the first stage of drought events and is the least in severity.

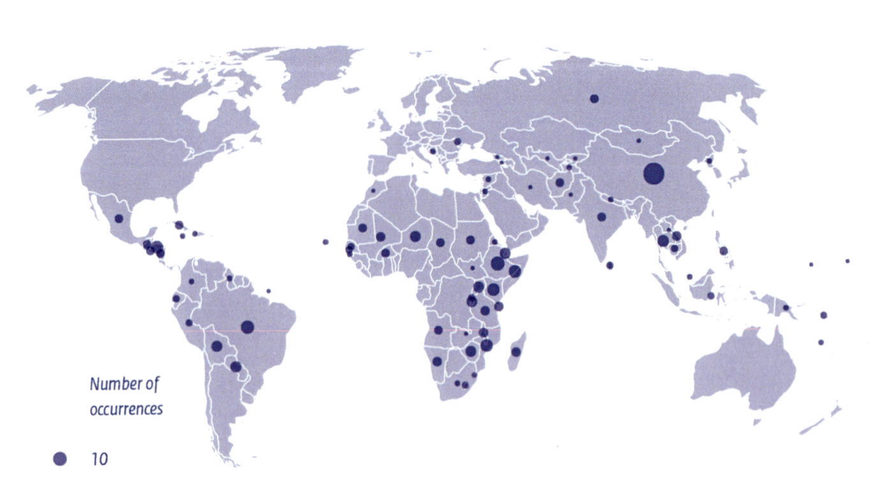

Fig. 15.1 Global spatial distribution of drought events around the globe with more frequent events in the southern hemisphere (Ligtvoet et al., 2018). (Data source: CRED, 2018)

2. **Agricultural drought**: If precipitation deficiencies last for longer time, this may lead to soil and vegetation moisture drop causing great damage for both.
3. **Hydrological drought**: This is the most severe type of drought where preparation deficiencies presets for longer time that is sufficient to cause a major drop in groundwater level, and lowering water level in lakes, rivers, reservoirs, and great deterioration of water quality.
4. **Socio-economic drought**: Drought events will lead to deficiencies in water supply and increase in water demand that may lead to widening the gap between demand and supply. This situation may lead to drop in crops productivity and livestock and other activities associated with food production which will affects food security.

This classification sort drought severity chronological, where any drought event begins meteorologically, and by time, if precipitation deficit continues form longer times, drought my developed to socio -economic drought in worst case scenarios.

A major challenge in drought assessment came from the complexity in recognizing the beginning and ending of specific drought event. In most cases drought impacts may persists even after ending of drought event. In addition, drought concept varies among regions of differing climates. The nature of drought impacts is non-structural and may spread over large areas that makes assessment and response very difficult, and mitigation actions less understandable (Dracup et al., 1980).

In the Middle East and North Africa (MENA) Region, the climate is predominantly temperate with hot and dry summers, frequent droughts, and relatively dry winters. These regions are considered as the driest and most water stressed region in the world (Damania et al., 2017).

The population of MENA countries comprises about 6.3% of the world's population, however, they get access to only about 1.4% of the world's renewable freshwater (Roudi-Fahimi et al., 2002). The recent decades had witnessed the driest periods in the past millennium (Cook et al., 2016; Touchan et al., 2011), in addition, MENA countries was exposed to severe drought events that can be attributable partially to anthropogenic impacts of climate change (Bergaoui et al., 2015). These impacts combined with abnormal demographic growth in most MENA countries added additional stresses on the already exhausted water resources causing a severe water scarcity (Damania et al., 2017).

Jordan as many countries in MENA region is tackling a severe water scarcity because of its reliance on a highly fluctuating annual precipitation (Abu-Allaban et al., 2014). The country was recently classified as the second lowest country in the world with limited water resources. The available water resources per capita are falling because of population growth, it is projected that it will fall from less than 160 m^3/capita/year at present to about 90 m^3/capita/year by 2025 (MWI & GTZ, 2005), putting Jordan in the category of an absolute water shortage (Al-Qinna et al., 2011).

15.2 Techniques for Drought Vulnerability Assessment

It is important to differentiate between the terms "**risk**", "**hazard**", "**exposure**" and "**vulnerability**" when conducting drought assessment, with notable differences between Disaster Risk Reduction (DRR) community and the Climate Change Adaptation community (CCA) (Brooks, 2003). These two scientific communities developed two different approaches to assess drought risk hazards (Tánago et al., 2016; Naumann et al., 2018), these are: **The Outcome or Impact Approach** (CCA community) which based on the relationships between stressor and response (Peduzzi et al., 2009), and **The Contextual or Factor Approach** (DRR community) which is based on intrinsic social or economic factors that define the vulnerability. The former approach relies on the use of quantitative measures of historical impacts as proxies for the vulnerability estimation (Brooks, 2003), and considers the vulnerability as endpoint of drought analysis, while the later approach considers drought vulnerability as the starting point. This allows understanding and explaining the susceptibility of exposed people, assets, and environment to the damaging effects of a drought (OECD, 2008).

Based on these two approaches, drought risk can be defined as a function of the **drought natural hazard**, the **exposed** assets, and the inherent drought **vulnerability** of the exposed social or natural system (Fig. 15.2), where vulnerability is defined as an inherent property of a system that exists independently of the external hazard (Brooks, 2003). Table 15.1 summarizes the main characteristics of these three components as well as relevant data needed to represent them (Vogt et al., 2018).

Drought vulnerability assessment depends on analyzing set of parameters and variables that are directly contributed to initiate or terminate a drought event, these

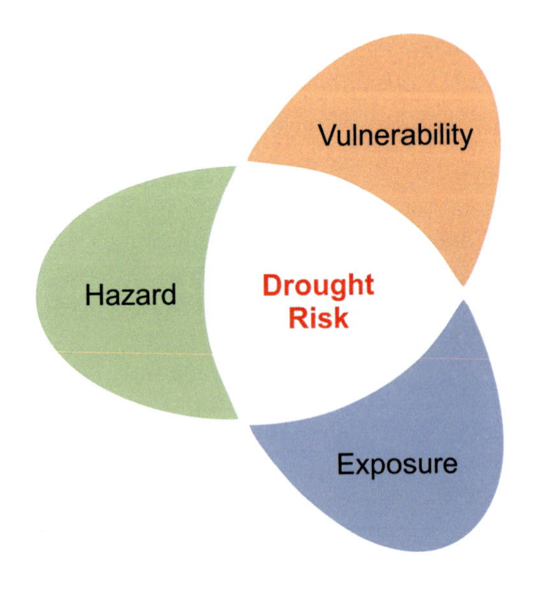

Fig. 15.2 Quantification of the drought risk by integrating the components hazard, vulnerability, and exposure (IPCC, 2014)

Table 15.1 Components of drought risk analysis

Component	Characterization	Data
Hazard	Magnitude of a hydrometeorological deficit	Meteorological, hydrological and/or biophysical indicators
Exposure	Amount of elements and assets subject to a drought hazard	Quantity and location of human population, infrastructure, economic activities and/or ecosystems
Vulnerability	Sensitivity of exposed elements to damaging effects of droughts	Composite indicators that include social, economic, environmental and/or infrastructural components
Overall risk	Potential damages and losses from droughts to a specific asset	Measured in a probabilistic scale as a combination of the drought magnitude or severity, level of exposure and vulnerability. Linked to intervention policies and management plans

Adapted from Van et al. (2017)

parameters are grouped into three categories (WMO & GWP, 2016); "**Drought indicators**", "**Drought indices**" and "**Drought triggers**."

Drought Indicators are usually physical variables and/or parameters that can be used to describe drought conditions. An example of such variables includes rainfall, temperature, runoff, groundwater and reservoir levels, and soil moisture. These parameters can be directly for drought assessment as these are always sensitive to specific drought event and can be used to estimate current and past drought events based on available quality records of these indicators.

Drought Indices are simplified mathematical models are representations that might be used to approximate drought severity. These models include mathematical relationships among different drought indicators. These models or indices quantify drought severity and magnitude over certain period, in addition, these indices describe the state of drought spatially using different scales and spatial resolutions. A common practice when developing drought indices is to use remotely sensed data as they provided the necessary data based on the desired scale or spatial resolution, and GIS which facilitate the spatial modeling of drought and spatial visualization of drought severity.

Standard precipitation Index (SPI), Normal Difference Vegetation Index (NDVI), Palmer Drought Severity Index (PDSI) and Combined Drought Index (CDI) are examples of widely used drought indices.

Drought Triggers: Looks for a certain values or threshold of drought indicator that is responsible for even drought event initiation or termination and related management responses. As for example, if the precipitation drops below the fifth percentile for two consecutive months will be equivalent to Level 4 Drought Severity. Drought triggers usually specifies the following factors (WMO & GWP, 2016): Indicator or index value, Time period, Spatial scale, Drought level and Advancing/ Receding drought conditions (Phase in or phase out).

Recently, many drought indices were used by different researchers and decision makers around the world to assess drought conditions for specific regions and periods of time. Selecting suitable indicators and indices will vary greatly from one country to another, however, the best practice in this regard is to use different thresholds with different combinations of inputs. This may require a preliminary study to determine which indicators/indices are best suited to the timing, area and type of climate and drought (WMO & GWP, 2016). In addition, the data availability compared with specific drought index data requirements will play a great role in selecting specific drought index.

To facilitate the process of selecting suitable drought indicators and indices, The Integrated Drought Management Program (IDMP) and Partner in (2016) developed a guiding table that is based on the available literature and online documents. Wide range of drought indices and indicators were classified based on several factors including Meteorology, Soil Moisture, Hydrology, Remote Sensing, and composite or modelled. Furthermore, a 'traffic-light' approach was used to classify these indices and indicators based on 'ease of us' as follows:

Green Indices: which characterized by its ease of use, simple and flexible data requirements and usually free software is available for download.

Yellow Indices: these indices require multi-variables calculations, furthermore, it needs programming skills as usually it is not available for public domain.

Red Indices: these are the most complex indices and requires code development based on the available literature, and usually these are not widely used.

Table 15.2 provide a sample of these indices, while the full list of this table is available on line at: "https://www.droughtmanagement.info/literature/GWP_Handbook_of_Drought_Indicators_and_Indices_2016.pdf".

15.3 Remote Sensing and GIS in Drought Risk Monitoring

The recent technological advances in Earth Observation (EO) Satellites, Remote Sensing, and GIS lead to widening the application of these tools in many environment and climate monitoring application, including drought assessment, and drought monitoring as well.

Earth Observation (EO) Satellites refers to the use of remote sensing technologies to monitor earth geosphere, cryosphere, hydrosphere, atmosphere, biosphere, and anthroposphere. EO satellites depend on the use of satellite-mounted payloads to gather imaging data about the Earth's systems. These systems generate images with varying spatial, temporal, and spectral resolutions that complies with the requires details and scale of the analysis. With the availability low-cost high-performance hardware and software for digital image processing of remotely sensed data collected by EO satellites, these products become an essential part of any framework for assessing and monitoring drought risks. According to Wardlow (2009), there are many advantages of using remote sensing for drought assessment and monitoring

Table 15.2 Examples of meteorological drought indices classification

Meteorology	Page	Ease of use	Input parameters	Additional information
Aridity Anomaly Index (AAI)	11	Green	P, T, PET, ET	Operationally available for India
Deciles	11	Green	P	Easy to calculate; examples from Australia are useful
Keetch–Byram Drought Index (KBDI)	12	Green	P, T	Calculations are based upon the climate of the area of interest
Aridity Index (AI)	15	Yellow	P, T	Can also be used in climate classifications
China Z Index (CZI)	16	Yellow	P	Intended to improve upon SPI data
Crop Moisture Index (CMI)	16	Yellow	P, T	Weekly values are required
Agricultural Reference Index for Drought (ARID)	23	Red	P, T, Mod	Produced in south-eastern United States of America and not tested widely outside the region
Crop-specific Drought Index (CSDI)	24	Red	P, T, Td, W, Rad, AWC, Mod, CD	Quality data of many variables needed, making it challenging to use
Reclamation Drought Index (RDI)	25	Red	P, T, S, RD, SF	Similar to the Surface Water Supply Index, but contains a temperature component

Adapted from WMO & GWP (2016)
Where, *AWC* available water content, *CD* crop data, *ET* evapotranspiration, *Mod* modelled, *P* precipitation, *PET* potential evapotranspiration, *Rad* solar radiation, *RD* reservoir, *S* snowpack, *SF* streamflow, *T* temperature, *Td* dewpoint temperature, *W* wind data

applications, these include: Spatial continuous measurements across large geographic areas, frequent revisit time for image acquisition, and historical record of conditions. For example, Landsat Programs which launched in 1972 provides free images for any spot on earth every 16 days for free. This provides an invaluable source of remotely sensed data that can be used efficiently in drought assessment and monitoring.

The most used EO platforms that used recently in drought studies include:

1. **MODIS Vegetation Index Products (NDVI and EVI)**: MODIS vegetation indices, produced on 16-day intervals and at multiple spatial resolutions, provide consistent spatial and temporal comparisons of vegetation canopy greenness, a composite property of leaf area, chlorophyll, and canopy structure. Two vegetation indices are derived from atmospherically corrected reflectance in the red, near infrared, and blue wavebands that can be used to assess Agriculture Drought, these are; Normalized Difference Vegetation Index (NDVI) and Enhanced Vegetation Index (EVI).
2. **Landsat Missions**: The Landsat program offers the longest continuous global record of the Earth's surface; it continues to deliver visually stunning and

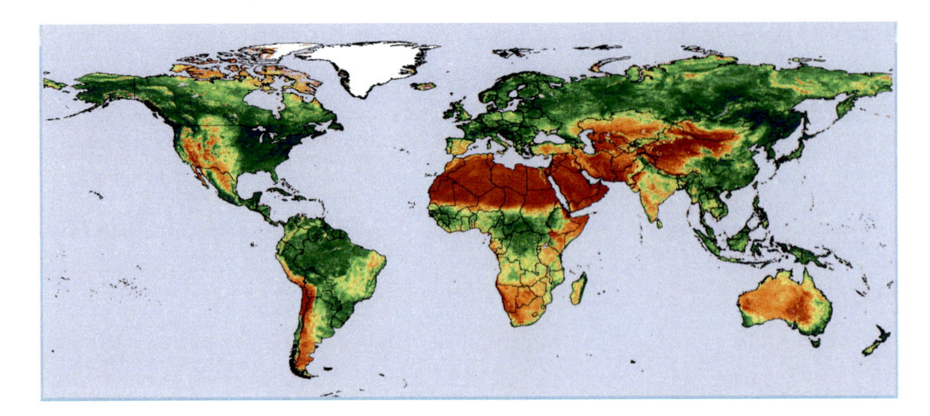

Fig. 15.3 GIMMS AVHRR Global NDVI third generation monthly and biweekly data including quality flags for 1982–2014. Downloaded from Clarks Labs website at https://clarklabs.org/additional-products/global-archive-2-avhrr/. Accessed on 15 March 2021

scientifically valuable images of our planet. Over a period of about 50 years, Landsat program has collected spectral information from Earth's surface, creating a historical archive unmatched in quality, detail, coverage, and length.

3. **Advanced Very High-Resolution Radiometer (AVHRR)**: The Advanced Very High Resolution Radiometer (AVHRR) is a broad-band, four or five channel (depending on the model) scanner, sensing in the visible, near-infrared, and thermal infrared portions of the electromagnetic spectrum. This sensor is carried on NOAA's Polar Orbiting Environmental Satellites (POES), beginning with TIROS-N in 1978. The Advanced Very High-Resolution Radiometer (AVHRR) instrument from the National Oceanic and Atmospheric Administration (NOAA) has produced a 20+ year historical record of global NDVI observations (Wardlow, 2009). Figure 15.3 shows GIMMS AVHRR Global NDVI third generation monthly and biweekly data including quality flags for 1982–2014. In this map that the data were provided by NASA GIMMS as biweekly data. Monthly NDVI images for the same period were developed using maximum value composite. This figure clearly shows the drought situation in MENA countries as indicated by brown color.
4. **Copernicus Sentinel Missions**: Copernicus is the European Union's Earth Observation Programmes, looking at our planet and its environment. It offers information services based on satellite Earth Observation and in situ (non-space) data. Seven missions of Copernicus Sentinel since 2016 where lunched where the latest one was Copernicus Sentinel-6 in 2020. Researchers of a Swiss company are testing Copernicus Sentinel-1 data to better calibrate the Enhanced SAR Vegetation Index (ESVI) for drought conditions, which could help insurance plans for farmers (ESA Sentinel News: https://sentinel.esa.int/).
5. **Rainfall Estimates from Rain Gauge and Satellite Observations (CHIRPS)**: CHIRPS was a product of collaboration between scientists at the USGS Earth Resources Observation, and Science (EROS) Center to provide a reliable data that

can be used efficiently in drought applications like drought early warning systems, drought trend analysis, and seasonal drought monitoring. CHIRPS incorporates in-house climatology, CHPclim, 0.05° resolution satellite imagery, and in-situ station data to create gridded rainfall time series for trend analysis and seasonal drought monitoring (Funk et al., 2015).

15.4 Drought Assessment in Jordan

Jordan with its limited water resources has suffered from chronic drought events that greatly affected its economic growth and social development (WMO, 2014). Drought events becomes more frequents in different time spans over the recent history of the country.

In the last decades, the severity of droughts events and its magnitude becomes more apparent especially in the last 30 years (Al-Qinna et al., 2011; Hammouri & El-Naqa, 2007). These drought events are associated with recurring summer heat waves, which could be a consequence of global warming (Abu Hajar et al., 2019).

Rajsekhar and Gorelick (2017) studied the increasing drought in Jordan due to climate change and cascading Syrian land-use impacts on reducing transboundary flow. Based on this study, results reveled that significant increase in drought incidence and severity in the future (for RCPs 4.5 and 8.4) compared to the baseline period (Fig. 15.4).

Abu Hajar et al. (2019) assessed drought conditions in Jordan using standardized precipitation index (SPI), it was found that the year from 1998 to 1999 was the driest year among the studied years with an obvious trend towards increasing drought severity in the recent years. Al-Qinna et al. (2011), analyzed drought conditions in Jordan under current and future climate scenarios using drought indices. The analysis showed that the country faced frequent non-uniform drought periods in an irregular repetitive manner during the last 35 years. In addition, drought severity, magnitudes and life span increased with time from normal to extreme levels especially at last decade reaching magnitudes of more than 4 (Fig. 15.5). Results reveled that future drought events will become more intensive at the northern and southern desserts with 15% rainfall reduction factor, followed by 10% reduction at the lowlands, and 5% at the highlands based on Global Climate Models analysis.

Al-Bakri et al. (2019) assessed the combined drought index (CDI) and the standardized precipitation index (SPI) in terms of their correlation with crop production and yield in Jordan. In addition to mapping the spatial extent of drought vulnerability at different administration levels (subdistrict, district, and governorate). Results indicated that the use of 6-months CDI (November–April) for the purpose of drought assessment would be more convenient than the use of annual CDI or CDI for periods of 3 months.

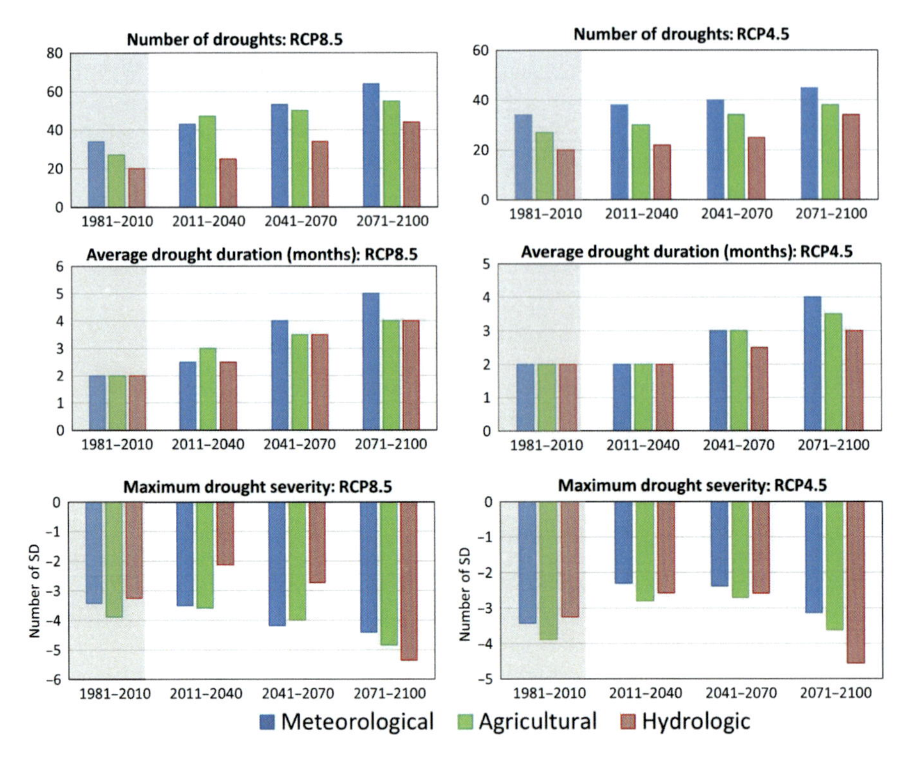

Fig. 15.4 Predicted drought characteristics for Jordan based on Representative Concentration Pathway (RCP 4.5 and 8.5) (Rajsekhar & Gorelick, 2017)

In this work, three methods will be used to assess meteorological agriculture drought for the country, these are:

1. Standard Precipitation Index (SPI).
2. Normal Difference Vegetation Index (NDVI).
3. Rainfall Estimates from Rain Gauge and Satellite Observations (CHIRPS) data.

15.4.1 Standard Precipitation Index (SPI)

Standard Precipitation Index (SPI) is one of the most widely used index to characterize meteorological drought. It was developed by McKee et al. (1993). SPI requires only precipitation records for any given location to develop a probability of precipitation that can be computed at selected time scales that may vary from 1 to 48 months. Guttman (2007) concluded that the availability of longer data records will result in more robust probability distribution as more extreme wet and extreme dry events will be embedded. SPI requires at least 20 years of precipitation record, however, its recommended to use 30 years of data (WMO & GWP, 2016).

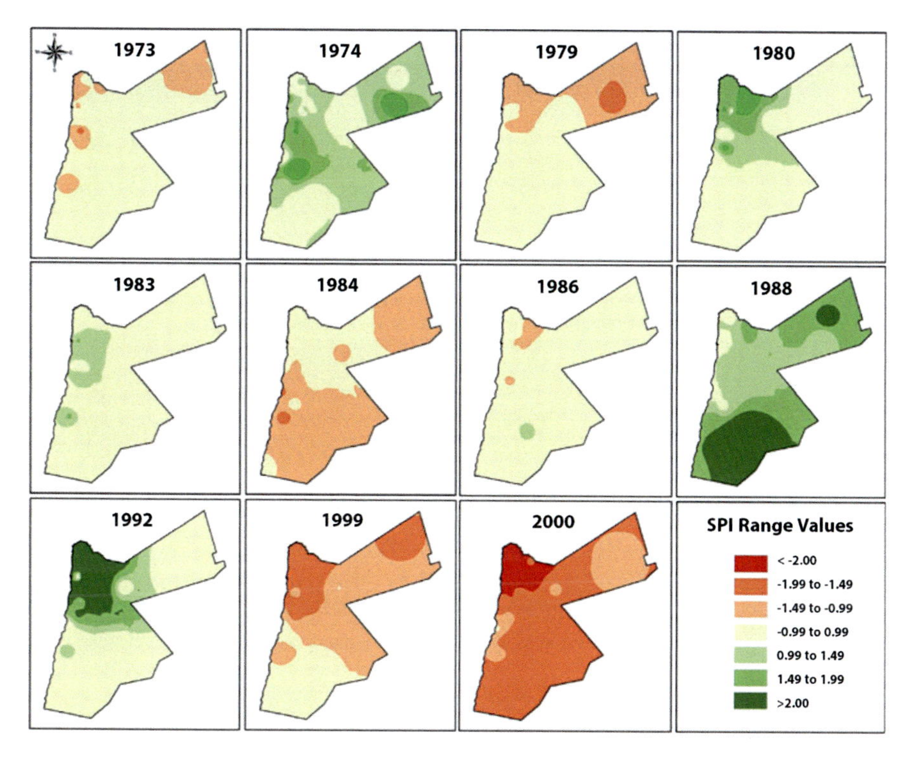

Fig. 15.5 Spatial distribution of drought events in Jordan based on SPI index (Al-Qinna et al., 2011)

SPI Index is computed based on the following equation (McKee et al., 1993):

$$\mathrm{SPI} = \frac{\left(x_{ij} - X_{im}\right)}{\sigma} \tag{15.1}$$

Where: X_{ij} is the seasonal precipitation at *i*th rain gauge station and *j*th observation, X_{im} the long-term seasonal mean and σ is its standard deviation.SPI Calculator is a free program that can be run on Windows-based personal computers and download at: National Drought Mitigation Center – SPI Program.

In this study, daily records of precipitation for 120 meteorological stations for 37 years (from 1981 to 2018) were used to develop SPI drought index for Jordan. The obtained SPI values were classified according (McKee et al., 1993) drought severity classes as shown in Table 15.3. Results of SPI analysis for Jordan are shown in Table 15.4. As indicated from this table, may stations in Jordan are facing more severe drought conditions with increasing duration, as for examples station AL0016, AL0018 and many other stations. It is very clear from this table the most of the 120 station are facing moderate to severe drought conditions, and some of these facing extreme drought conditions.

Table 15.3 SPI classification scheme (McKee et al., 1993)

SPI value	Drought category
>= 2.00	Extreme wet
1.50 to 1.99	Severe wet
1.00 to 1.49	Moderate wet
0.50 to 0.99	Mild wet
–0.49 to 0.49	Near normal
–0.99 to –0.50	Mild drought
–1.49 to –1.00	Moderate drought
–1.99 to –1.5	Severe drought
<= –2.0	Extreme drought

Table 15.4 Results of SPI analysis for selected stations in Jordan

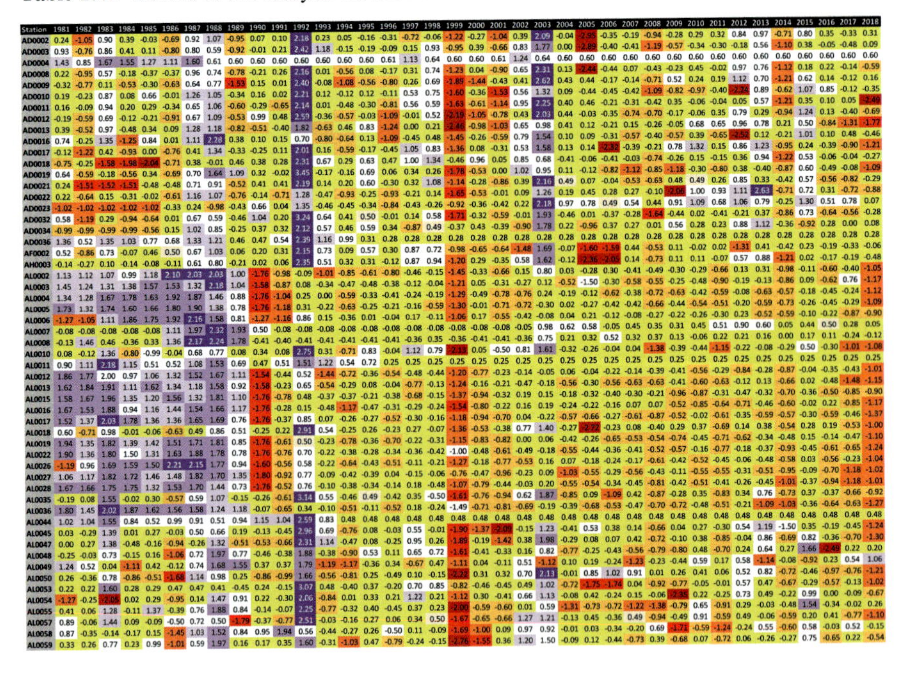

Station	1981	1982	1983	1984	1985	1986	1987	1988	1989	1990	1991	1992	1993	1994	1995	1996	1997	1998	1999	2000	2001	2002	2003	2004	2005	2006	2007	2008	2009	2010	2011	2012	2013	2014	2015	2016	2017	2018
AD0002	0.24	-1.05	0.90	0.39	-0.03	-0.69	0.92	1.07	-0.95	0.07	0.10	2.18	0.23	0.05	-0.16	-0.31	-0.72	-0.06	-1.22	-0.27	-1.04	0.39	2.09	-0.04	-2.95	-0.35	-0.19	-0.94	-0.28	0.29	0.32	0.84	0.97	-0.71	0.80	0.35	-0.33	0.31
AD0003	0.93	-0.76	0.86	0.41	0.11	-0.80	0.80	0.59	-0.92	-0.01	0.21	2.42	1.18	-0.15	-0.19	-0.09	0.15	0.93	-0.95	0.39	-0.66	0.83	1.77	0.00	-2.89	-0.40	-0.41	-1.19	-0.57	-0.34	-0.30	-0.18	0.56	-1.10	0.38	-0.05	-0.48	0.09
AD0004	1.43	0.85	1.67	1.55	1.27	1.11	1.60	0.61	0.60	0.60	0.60	0.60	0.60	0.60	0.60	0.61	1.13	0.64	0.60	0.60	0.61	1.24	0.64	0.60	0.60	0.60	0.60	0.60	0.60	0.60	0.60	0.60	0.60	0.60	0.60	0.60	0.60	0.60
AD0008	0.22	-0.95	0.57	-0.18	-0.37	-0.37	0.96	0.74	-0.78	-0.21	0.26	2.16	0.01	-0.56	0.08	-0.17	0.31	0.74	-1.23	0.04	-0.90	0.65	2.31	0.13	-2.44	-0.44	0.07	-0.43	-0.23	0.45	-0.02	0.97	0.76	-1.12	0.18	0.22	-0.14	-0.59
AD0009	-0.32	-0.77	0.11	-0.53	-0.30	-0.63	0.64	0.77	-1.53	0.15	0.01	2.40	-0.08	-1.08	-0.56	-0.80	0.26	0.69	-1.89	-1.44	-0.43	0.41	2.62	0.43	0.44	-0.17	-0.14	-0.71	0.52	0.24	0.19	1.12	0.70	-1.21	0.62	0.14	-0.12	0.16
AD0010	0.19	-0.23	0.87	0.08	0.66	-0.01	1.26	1.05	-0.34	0.16	0.02	2.21	0.12	-0.12	0.12	-0.11	0.53	0.75	-1.60	-0.36	-1.53	0.56	1.32	0.09	-0.44	-0.45	-0.42	-1.09	-0.82	-0.97	-0.40	-2.24	0.89	-0.62	1.07	0.85	-0.12	-0.35
AD0011	0.16	-0.09	0.94	0.20	0.29	-0.34	0.65	1.06	-0.60	-0.29	-0.65	2.14	0.01	-0.48	-0.30	-0.81	0.56	0.59	-1.63	-0.61	-1.14	0.95	2.25	0.40	0.46	-0.21	-0.31	-0.42	0.35	-0.06	-0.04	0.05	0.57	-1.21	0.35	0.10	0.05	-2.49
AD0012	-0.19	-0.59	0.69	-0.12	-0.21	-0.91	0.67	1.09	-0.53	0.99	0.48	2.59	-0.36	-0.57	-0.03	-1.09	-0.01	0.52	-2.19	-1.05	-0.78	0.43	2.03	0.44	-0.03	-0.35	-0.74	-0.70	-0.17	-0.06	0.35	0.79	0.29	-0.94	1.24	0.13	-0.40	-0.69
AD0013	0.39	-0.52	0.97	-0.48	0.34	0.09	1.28	1.18	-0.82	-0.51	-0.40	1.82	-0.63	0.46	0.83	-1.24	0.00	0.21	-2.46	-0.98	-1.03	0.65	0.98	0.41	0.12	-0.21	0.15	-0.26	-0.05	0.68	0.65	0.96	0.78	0.21	0.50	-0.84	-1.31	-1.77
AD0016	0.74	-0.25	1.35	-1.25	0.84	0.01	1.11	2.28	0.38	0.10	0.15	0.70	-0.80	-0.64	0.13	-1.09	-0.45	0.48	-1.45	-0.26	-0.59	0.79	1.54	0.10	0.09	-0.31	-0.57	-0.40	-0.57	0.39	-0.65	-2.52	0.12	-0.21	1.01	0.10	0.48	-0.46
AD0017	-0.12	-1.22	0.42	-0.93	0.00	-0.76	0.41	1.34	-0.33	-0.25	0.11	2.01	0.16	-0.59	-0.17	-0.45	1.05	0.83	-1.36	0.08	-0.31	0.53	1.58	0.13	0.14	-2.32	-0.39	-0.21	0.78	1.32	0.15	0.86	1.23	-0.95	0.24	-0.39	-0.90	-1.21
AD0018	-0.75	-0.25	-1.58	-1.98	-2.04	-0.71	0.38	-0.01	0.46	0.38	0.28	2.31	0.67	0.29	0.63	0.47	1.00	1.34	-0.46	0.96	0.05	0.85	0.68	-0.41	-0.06	-0.41	-0.03	-0.74	-0.26	0.15	-0.15	0.36	0.94	-1.22	0.53	-0.06	-0.04	-0.27
AD0019	0.64	-0.59	-0.18	-0.56	0.34	-0.69	0.70	1.64	1.09	0.32	-0.02	3.45	-0.17	-0.16	0.69	0.06	0.34	0.26	-1.78	-0.53	0.00	1.02	0.95	0.11	-0.12	-0.82	-1.12	-0.85	-1.18	-0.30	-0.80	0.38	-0.40	-0.87	0.60	-0.49	-0.08	-1.09
AD0021	0.24	-1.51	-1.52	-1.51	-0.48	-0.48	0.71	0.91	-0.52	0.41	0.41	2.19	0.14	0.20	0.60	-0.30	0.32	1.08	-1.14	-0.28	-0.86	0.39	2.16	0.49	0.07	-0.04	-0.53	-0.63	0.48	0.49	0.26	0.85	0.33	-0.42	0.57	-0.56	-0.82	-0.29
AD0022	0.22	-0.64	0.15	-0.31	-0.02	-0.61	1.16	1.07	-0.76	-0.14	-0.71	1.28	-0.47	-0.93	-0.25	-0.93	-0.21	0.14	-1.65	-0.53	-0.01	0.09	1.26	0.19	0.45	0.28	0.27	-0.10	-2.06	1.00	0.93	1.11	2.63	-0.71	0.72	0.31	-0.72	-0.88
AD0023	-1.02	-1.02	-1.02	-1.02	-1.02	-0.33	0.24	-0.98	-0.43	0.66	0.04	1.35	-0.46	-0.45	-0.34	-0.84	-0.43	-0.26	-0.92	-0.36	-0.42	0.22	2.18	0.97	0.78	0.49	0.54	0.44	0.91	1.09	0.68	1.06	0.79	-0.25	1.30	0.51	0.78	0.07
AD0032	0.58	-1.19	0.29	-0.94	-0.64	0.01	0.67	0.59	-0.46	1.04	0.20	3.24	0.64	0.41	0.50	-0.01	0.14	0.58	-1.71	-0.32	-0.59	-0.01	1.93	-0.46	0.01	-0.37	-0.28	-1.64	-0.44	0.02	-0.41	-0.21	0.37	-0.86	0.73	-0.64	-0.56	-0.28
AD0034	-0.99	-0.99	-0.99	-0.99	-0.56	0.15	1.02	0.85	-0.25	0.37	0.32	2.12	0.57	0.46	0.59	0.34	-0.87	0.49	-0.37	0.43	-0.39	-0.90	1.78	0.22	-0.96	0.37	0.27	0.01	0.56	0.28	0.23	0.88	1.12	-0.36	-0.92	0.28	0.00	0.08
AD0036	1.36	0.52	1.35	1.03	0.77	0.68	1.33	1.21	0.46	0.47	0.54	2.39	1.16	0.99	0.31	0.28	0.28	0.28	0.28	0.28	0.28	0.28	0.28	0.28	0.28	0.28	0.28	0.28	0.28	0.28	0.28	0.28	0.28	0.28	0.28	0.28	0.28	0.28
AF0002	0.52	-0.86	0.73	-0.07	0.46	0.50	0.67	1.03	0.06	0.20	0.31	2.15	0.73	0.09	0.57	0.30	0.87	0.72	-0.98	-0.65	-0.64	-1.48	1.69	-0.07	-1.60	-1.59	0.44	-0.53	0.11	-0.02	0.02	-1.31	0.41	-0.42	0.23	-0.19	-0.33	-0.06
AH0003	-0.14	-0.27	0.10	-0.14	-0.08	-0.11	0.61	0.80	-0.21	0.02	0.06	2.35	0.51	0.32	0.31	-0.12	0.87	0.94	-1.20	0.29	-0.35	0.58	1.62	-0.12	-2.36	-2.05	0.14	-0.73	0.11	0.11	-0.07	0.57	0.88	-1.21	0.02	-0.17	-0.19	-0.48
AL0002	1.13	1.12	1.07	0.99	1.18	2.10	2.03	2.03	1.00	-1.76	-0.98	-0.09	-1.01	-0.85	-0.61	-0.80	-0.46	-0.15	-1.45	-0.33	-0.66	0.15	0.80	0.03	-0.28	0.30	-0.41	-0.49	-0.30	-0.29	-0.66	0.13	0.31	-0.98	-0.11	-0.60	-0.40	-1.05
AL0003	1.45	1.24	1.31	1.38	1.57	1.53	1.32	2.18	1.04	-1.58	-0.87	0.08	-0.34	-0.47	-0.48	-0.38	-0.12	-0.04	-1.21	0.05	-0.31	-0.27	0.12	-0.52	-1.50	-0.30	-0.58	-0.55	-0.25	-0.48	-0.90	-0.19	-0.13	-0.86	0.09	-0.62	0.76	-1.17
AL0004	1.34	1.28	1.67	1.78	1.63	1.92	1.87	1.46	0.88	-1.76	-1.04	0.25	0.00	-0.59	-0.33	-0.41	-0.24	-0.19	-1.29	-0.49	-0.78	-0.76	0.24	-0.19	-0.12	-0.62	-0.38	-0.72	-0.63	-0.42	-0.59	-0.08	-0.63	-0.57	-0.18	-0.45	-0.24	-1.12
AL0005	1.73	1.32	1.74	1.60	1.66	1.80	1.90	1.38	0.78	-1.76	-1.18	0.31	-0.22	-0.63	-0.25	-0.21	-0.16	-0.59	-1.30	-0.01	-0.71	-0.72	-0.30	0.02	-0.27	-0.42	-0.42	-0.66	-0.44	-0.54	-0.51	-0.20	-0.59	-0.73	-0.26	-0.45	-0.29	-1.09
AL0006	-1.27	-1.05	1.11	1.86	1.75	1.92	2.16	1.58	0.81	-1.27	-1.16	0.86	0.15	-0.36	0.01	-0.04	0.17	-0.11	-1.06	0.17	-0.55	-0.42	-0.08	0.04	0.21	-0.12	-0.08	-0.27	-0.22	-0.26	-0.30	0.23	-0.52	-0.59	-0.10	-0.22	-0.87	-0.90
AL0007	-0.08	-0.08	-0.08	-0.08	-0.08	1.11	1.97	2.32	1.93	0.50	-0.08	-0.08	-0.08	-0.08	-0.08	-0.08	-0.08	-0.08	-0.08	-0.08	-0.08	-0.05	0.98	0.62	0.58	-0.05	0.45	0.35	0.31	0.45	0.51	0.90	0.60	0.05	0.44	0.50	0.28	0.05
AL0008	-0.13	1.46	0.46	-0.36	0.33	1.36	2.17	2.24	1.78	-0.41	-0.40	-0.41	-0.41	-0.41	-0.41	-0.41	-0.36	0.35	-0.36	-0.41	-0.41	-0.36	0.75	0.21	0.32	0.52	0.32	0.37	0.13	-0.06	0.22	0.21	0.16	0.00	0.17	0.11	-0.24	-0.12
AL0010	0.08	-0.12	1.36	-0.80	-0.99	-0.04	0.68	0.77	0.08	0.34	0.08	2.75	0.31	-0.71	0.83	-0.04	1.12	0.79	-2.13	0.05	-0.50	0.81	1.61	-0.32	-0.26	-0.04	0.04	-1.38	-0.39	-0.44	-1.15	-0.22	-0.08	-0.29	0.50	-0.30	-1.01	-1.08
AL0011	0.90	1.11	2.18	1.15	0.51	0.52	1.08	1.53	0.69	0.47	0.51	1.51	1.22	0.54	0.72	0.25	0.25	0.25	0.25	0.25	0.25	0.25	0.25	0.25	0.25	0.25	0.25	0.25	0.25	0.25	0.25	0.25	0.25	0.25	0.25	0.25	0.25	0.25
AL0012	1.86	1.77	2.00	0.97	1.06	1.32	1.52	1.67	1.11	-1.54	-0.44	0.52	-1.44	-0.72	-0.36	-0.54	-0.48	-0.44	-1.20	-0.77	-0.23	-0.14	-0.05	0.06	-0.04	-0.22	-0.14	-0.39	-0.41	-0.56	-0.29	-0.84	-0.28	-0.87	-0.04	-0.35	-0.43	-1.01
AL0013	1.62	1.84	1.91	1.11	1.62	1.34	1.18	1.58	0.92	-1.58	-0.23	0.65	-0.54	-0.29	0.08	-0.04	-0.77	-0.13	-1.24	-0.16	-0.21	-0.47	-0.18	-0.56	-0.30	-0.56	-0.63	-0.63	-0.41	-0.60	-0.63	-0.12	0.13	-0.66	0.02	-0.48	-1.48	-1.15
AL0015	1.58	1.67	1.96	1.35	1.20	1.56	1.32	1.81	1.10	-1.76	-0.78	0.48	-0.37	-0.37	-0.21	-0.38	-0.68	-0.15	-1.37	-0.94	-0.32	0.19	0.15	-0.18	-0.32	-0.40	-0.30	-0.21	-0.96	-0.87	-0.31	-0.47	-0.32	-0.70	-0.36	-0.50	-0.85	-0.90
AL0016	1.67	1.53	1.88	0.94	1.16	1.44	1.54	1.66	1.17	-1.76	-0.28	0.15	-0.48	-1.17	-0.47	-0.31	-0.29	-0.24	-1.54	-0.80	-0.22	0.16	0.19	-0.24	-0.22	-0.16	0.07	0.07	-0.52	-0.85	-0.64	-0.71	-0.46	-0.60	-0.02	0.22	-0.85	-1.17
AL0017	1.52	1.37	2.03	1.78	1.36	1.36	1.65	1.69	0.76	-1.76	-0.37	0.85	0.07	-0.26	-0.27	-0.52	-0.30	-0.16	-1.18	-0.94	-0.70	0.04	-0.22	-0.57	-0.66	-0.27	-0.61	-0.87	-0.52	-0.02	-0.61	-0.35	-0.59	-0.57	-0.30	-0.59	-0.46	-1.37
AL0018	0.60	-0.71	0.98	-0.01	-0.06	-0.63	0.49	0.86	0.51	-0.25	0.22	2.91	0.54	-0.25	0.26	-0.23	0.27	-0.07	-1.36	-0.53	-0.38	0.77	1.40	-0.27	-2.72	-0.23	0.08	-0.40	0.29	0.37	-0.69	0.14	0.38	-0.54	0.28	0.19	-0.53	-1.00
AL0019	1.94	1.35	1.82	1.39	1.42	1.51	1.71	1.81	0.85	-1.76	-0.61	0.50	-0.23	-0.78	-0.36	-0.70	-0.22	-0.31	-1.15	-0.83	-0.82	0.00	0.06	-0.42	-0.26	-0.65	-0.53	-0.54	-0.74	-0.45	-0.71	-0.62	-0.34	-0.48	0.15	-0.14	-0.47	-1.10
AL0022	1.90	1.36	1.80	1.50	1.31	1.63	1.88	1.78	0.78	-1.76	-0.76	0.70	-0.22	-0.38	-0.28	-0.34	-0.36	-0.42	-1.00	-0.48	-0.61	-0.49	-0.18	-0.55	-0.44	-0.36	-0.41	-0.52	-0.57	-0.16	-0.77	-0.18	-0.37	-0.93	-0.45	-0.61	-0.65	-1.24
AL0026	-1.19	0.96	1.69	1.59	1.50	2.21	2.15	1.77	0.94	-1.60	-0.56	0.58	-0.22	-0.64	-0.43	-0.51	-0.11	-0.21	-1.27	-0.18	-0.77	-0.53	0.16	0.07	-0.18	-0.24	-0.17	-0.61	-0.42	-0.52	-0.45	-0.06	-0.48	-0.58	0.03	-0.56	-0.23	-1.04
AL0027	1.06	1.17	1.82	1.72	1.46	1.48	1.82	1.70	1.35	-1.80	-0.92	0.77	-0.09	-0.42	-0.39	0.04	-0.15	-0.06	-0.76	-0.47	-0.96	-0.23	0.09	-1.03	-0.55	-0.29	-0.56	-0.43	-0.11	-0.55	-0.55	-0.31	-0.51	-0.95	-0.09	-0.70	-1.18	-1.02
AL0028	1.67	1.66	1.75	1.75	1.32	1.53	1.70	1.44	0.73	-1.76	-0.52	0.76	-0.10	-0.38	-0.34	-0.14	0.18	-0.48	-1.07	-0.79	-0.44	-0.03	0.20	-0.55	-0.54	-0.34	-0.45	-0.81	-0.42	-0.51	-0.41	-0.26	-0.45	-1.01	-0.37	-0.94	-1.18	-1.01
AL0035	-0.19	0.08	1.55	-0.02	0.30	-0.57	0.59	1.07	-0.15	-0.26	-0.61	3.14	0.55	-0.46	0.49	-0.42	0.35	-0.50	-1.61	-0.76	-0.94	0.62	1.87	-0.85	0.09	-1.09	0.42	-0.87	-0.28	0.35	-0.83	0.34	0.76	-0.73	0.37	-0.37	-0.66	-0.92
AL0036	1.80	1.45	2.02	1.87	1.62	1.56	1.58	1.24	1.18	-0.07	-0.65	0.34	-0.10	-0.51	-0.11	-0.52	0.18	-0.24	-1.49	-0.71	-0.81	-0.69	-0.19	-0.39	-0.68	-0.53	-0.47	-0.70	-0.72	-0.48	-0.51	-0.21	-1.09	-1.03	-0.36	-0.64	-0.63	-1.27
AL0044	1.02	1.04	1.55	0.84	0.52	0.99	0.91	0.51	0.94	1.15	1.04	2.59	0.83	0.48	0.48	0.48	0.48	0.48	0.48	0.48	0.48	0.48	0.48	0.48	0.48	0.48	0.48	0.48	0.48	0.48	0.48	0.48	0.48	0.48	0.48	0.48	0.48	0.48
AL0045	0.03	-0.29	1.39	0.01	0.27	-0.03	0.50	0.66	0.19	-0.13	-0.45	2.96	0.69	-0.76	0.08	-0.03	0.55	-0.01	-1.90	-1.37	-2.09	-0.15	1.23	-0.41	0.53	0.38	0.14	-0.66	0.04	0.27	-0.30	0.54	1.19	-1.50	0.35	-0.19	-0.45	-1.24
AL0047	0.00	0.27	1.38	-0.48	-0.16	-0.94	-0.26	1.32	-0.51	-0.53	-0.66	2.31	1.14	-0.47	0.08	-0.25	0.95	0.26	-1.89	-0.19	-1.42	0.38	1.98	-0.29	0.08	0.07	0.42	-0.72	-0.10	0.38	-0.85	-0.04	0.86	-0.69	0.82	-0.36	-0.70	-1.30
AL0048	-0.25	-0.03	0.73	-0.15	0.16	-1.06	0.72	1.97	0.77	-0.46	-0.38	1.88	-0.38	-0.90	0.53	0.11	0.65	0.72	-1.61	-0.41	-0.33	0.16	0.82	-0.77	-0.25	-0.43	-0.56	-0.79	-0.80	0.48	-0.70	0.24	0.64	0.27	1.66	-2.49	0.22	0.20
AL0049	1.24	0.52	0.04	-1.11	0.42	-0.12	0.74	1.68	1.55	0.37	0.37	1.79	-1.19	-1.17	-0.36	0.34	-0.67	0.47	-1.11	0.04	-0.11	0.51	-1.12	0.10	0.19	-0.24	-1.23	-0.23	-0.44	0.59	0.17	0.58	-1.14	-0.08	-0.92	0.23	0.54	1.06
AL0050	0.26	-0.36	0.78	-0.86	-0.51	-1.68	1.14	0.98	0.25	-0.86	-0.99	1.66	-0.56	-0.81	0.25	-0.49	0.10	-0.15	-2.22	0.31	0.32	0.70	2.13	-0.01	0.85	1.02	0.91	0.01	0.26	0.41	0.06	0.52	0.82	-0.72	-0.46	-0.97	-0.76	-1.21
AL0053	0.22	0.22	1.60	0.28	0.29	0.47	0.47	0.41	-0.45	0.24	-0.15	3.07	0.48	-0.40	0.37	-0.20	0.70	0.85	-0.82	-0.46	-0.45	0.49	1.02	-0.72	-1.75	-1.74	0.04	-0.92	-0.77	-0.05	-0.01	0.57	0.47	-0.67	-0.29	-0.57	-0.13	-1.02
AL0054	-1.27	-0.25	-2.05	0.02	0.29	-0.95	0.14	1.47	0.91	0.22	-0.30	2.06	-0.84	0.01	0.33	0.21	1.22	0.21	-1.12	0.30	-0.41	0.66	1.13	-0.08	0.42	-0.24	0.15	-0.06	-2.35	0.22	-0.25	0.73	0.49	-0.22	0.99	0.00	-0.09	-0.67
AL0055	0.41	0.06	1.28	-0.11	1.37	-0.39	0.76	1.88	0.84	-0.14	-0.07	2.25	-0.77	-0.32	0.40	-0.45	0.37	0.23	-2.00	-0.59	-0.60	0.01	0.59	-1.31	-0.73	-0.72	-1.22	-1.38	-0.79	0.65	-0.91	0.29	-0.03	-0.48	1.54	-0.34	-0.02	0.26
AL0057	0.89	-0.06	1.44	0.09	-0.09	-0.50	0.72	0.50	-1.79	-0.37	-0.77	2.51	-0.03	-0.16	0.27	0.06	0.34	0.50	-1.67	-0.65	-0.66	1.27	1.21	-0.13	0.45	-0.36	0.49	-0.94	-0.49	0.91	-0.59	0.49	-0.06	-0.59	0.20	0.41	-0.77	-1.10
AL0058	0.87	-0.35	-0.14	-0.17	0.15	-1.45	1.03	1.52	0.84	0.95	1.94	0.56	-0.44	-0.27	0.26	-0.50	0.11	-0.09	-1.69	-1.00	0.09	0.97	0.92	-0.01	0.03	-0.34	-0.20	0.69	-1.71	-0.59	-1.24	-0.24	0.55	-0.60	0.58	-0.03	0.52	-0.15
AL0059	0.33	0.26	0.77	0.23	0.99	-1.01	0.59	1.97	0.16	0.17	0.35	1.60	-0.31	-1.03	0.47	-0.79	-0.24	-0.15	-2.76	-1.55	0.36	1.20	1.50	-0.09	0.12	-0.44	-0.73	0.39	-0.68	0.07	-0.72	0.06	-0.26	-0.27	0.75	-0.65	0.22	-0.54

15.4.2 *Normal Difference Vegetation Index (NDVI)*

The use of NDVI in drought assessment studies was developed from work done by Tarpley et al. (1984) and Kogan (1995) with the National Oceanic and Atmospheric Administration (NOAA) in the United States. NDVI was first suggested by Tucker (1979) as an index of vegetation health and density:

$$\mathrm{NDVI} = \frac{\lambda_{\mathrm{NIR}} - \lambda_{\mathrm{R}}}{\lambda_{\mathrm{NIR}} + \lambda_{\mathrm{R}}} \quad (15.2)$$

Where: λ_{NIR} and λ_{R} are the reflectance in the near infrared (NIR) and (RED) bands, respectively.NDVI is the most used vegetation index, it reflects vegetation vigor, percent green cover, Leaf Area Index (LAI) and biomass. NDVI values are ranging between -1 and 1, values below 0 indicates water bodies, while values greater than 0 represent percent of vegetation cover in a cell within a raster.

NDVI uses only two bands and is not very sensitive to influences of soil background reflectance at low vegetation cover. Also, it has a lagged response to drought because of a lagged vegetation response to developing rainfall deficit due to residual moisture stored in the soil (Jensen, 1996).

As mentioned earlier, NDVI data can be downloaded from different sources like Advanced Very High-Resolution Radiometer (AVHRR), or it possible to generate NDVI maps using NIR and Red bands in multispectral systems like Landsat.

It is important to mention that NDVI itself **does not reflect drought or non-drought conditions**, However the severity of a drought (or the extent of wetness, on the other end of the spectrum) may be defined as NDVI deviation from its long-term mean ($\mathrm{DEV_{NDVI}}$). The deviation is calculated as the difference between the NDVI for the current time step (e.g., January 1995) and a long-term mean NDVI for that month (e.g., an 18-year long mean NDVI of all Januaries from 1981 to 2003) for each Pixel based on the following equation:

$$\mathrm{DEV_{NDVI}} = \mathrm{NDVI}_i - \mathrm{NDVI_{mean,m}} \quad (15.3)$$

Where: NDVI_i is the NDVI value for month (i) and $\mathrm{NDVI_{mean,m}}$ is the long-term mean NDVI for the same month (m).When $\mathrm{DEV_{NDVI}}$ is negative, it indicates the below-normal vegetation condition/health and, therefore, suggests a prevailing drought situation. Greater the negative departure, the greater the magnitude of a drought (Jensen, 1996).

In this study, USGS EROS Archive – Vegetation Monitoring – EROS Visible Infrared Imaging Radiometer Suite (eVIIRS) data was downloaded from Earth Resources Observation and Science (EROS) Center through USGS EarthExplorer website (https://earthexplorer.usgs.gov/). These data are available as a composite of the last 10 days interval produced every 5 days from 2012 to 2020. These data were processed using ArcGIS software to clip the full scenes that covers the whole area of

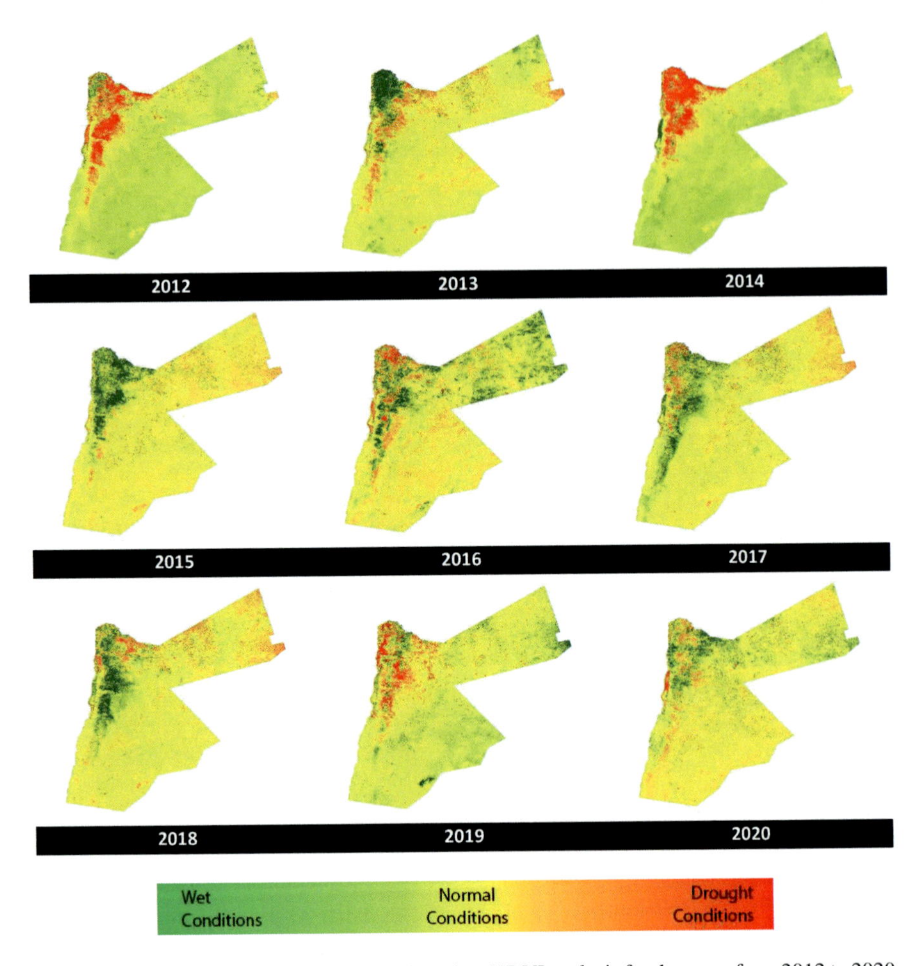

Fig. 15.6 Agriculture drought assessment based on NDVI analysis for the years from 2012 to 2020

Africa. Spatial Analyst Tools were used to calculate mean NDVI and DEV_{NDVI} as described in Eq. (15.3).

Visualization of the obtained DEV_{NDVI} results is shown in Fig. 15.6. This figure shows the varying agriculture drought conditions for February in Jordan, which is the most important rainy month within the winter season in Jordan that extends from November to April. The analysis revealed that the country faced several drought events in different years as for example in 2012, 2014 and 2019.

15.4.3 Drought Analysis Based on Rainfall Estimates from Rain Gauge and Satellite Observations (CHIRPS) Data

As discussed earlier, CHIRPS is a blended product combining different data sets including precipitation climatology, quasi-global geostationary TIR satellite observations from the Climate Hazards Group (CHG) and the National Climate Forecast System version 2, and in situ precipitation observations (Funk et al., 2015).

Based on WFP (2019), it is possible to used CHIRPS data to assess drought conditions in a region based on analyzing number of poor growing seasons (NGPS). In this study, drought data were prepared using Rainfall Estimate (RFE) from Rain Gauge and Satellite Observations which downloaded from Climate Hazards Centre InfraRed Precipitation with Station data (CHIRPS) provided by USGS FEWS NET Data Portal (https://earlywarning.usgs.gov/fews). Dekadal data (10 days precipitation) were downloaded from 2009 to 2019 rainy seasons. The downloaded raster data sets were processed based on RFE values for each season to provide a classification as poor or normal growing seasons, and the number of poor growing seasons (NPGS) summed over the time-period assessed.

Based on this analysis, values of NGPS for Jordan was ranging between 28 and 640, these ranges were reclassified based on Natural Break (Jenks, 1967) method into drought severity classes as shown in Table 15.5. Figure 15.7 shows the resultant drought vulnerability analysis for Jordan. As this figure shows, the northern and northwestern parts of the country are highly vulnerable to drought risks, knowing that these areas are located within areas that receives the highest precipitation in the country (Fig. 15.8).

To assess drought vulnerability on governorate level, NPGS were used to calculate the zonal statistics using Spatial Analyst Tools with ArcGIS environment. This enables the calculations of drought vulnerability for vector data, which is the governorates polygons, based on NPGS raster (WFP, 2019). Figure 15.9 shows the projected drought vulnerability for Jordan's governorates. This figure clearly identifies the governorates and districts that are facing highest drought vulnerability. For example, Irbid governorates and its districts are facing high and very high

Table 15.5 Drought vulnerability classification based on number of poor growing seasons (NGPS)

Drought hazard based on NGPS					
NGPS value	<50	50–122	122–272	272–490	>490
Drought hazard (Drought vulnerability)	1	2	3	4	5
Drought severity	**Very low**	**Low**	**Medium**	**High**	**Very high**

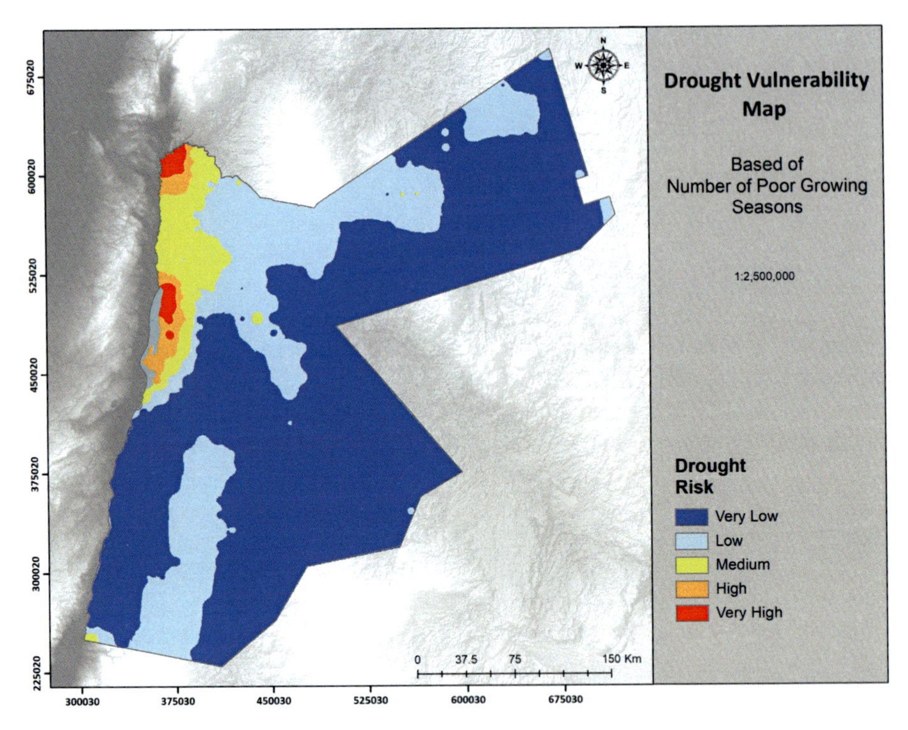

Fig. 15.7 Drought vulnerability mapping based on number of poor growing seasons (NGPS) from 1980 to 2019

drought vulnerability. Knowing that this governorate is the second highly populated area in Jordan, and lots of people their work and live on agriculture activities and livestock's. This should alert decision makers on different administrative levels to shed the lights on this situation and pay more attention towards establishing the suitable regulations and action plans that will protect farmers in such areas and propose suitable adaption measures and action plans for drought management to reduce the impacts of these drought events.

15.5 Conclusions

Drought is a natural phenomenon that threats social and economic developments plans for many countries around the world especially in countries with limited water resources and developing countries. Jordan as many other counties in the MEAN region encountered many drought events in the past and it is expected to have more severe drought events in the future with elevated impacts as a results of climate change.

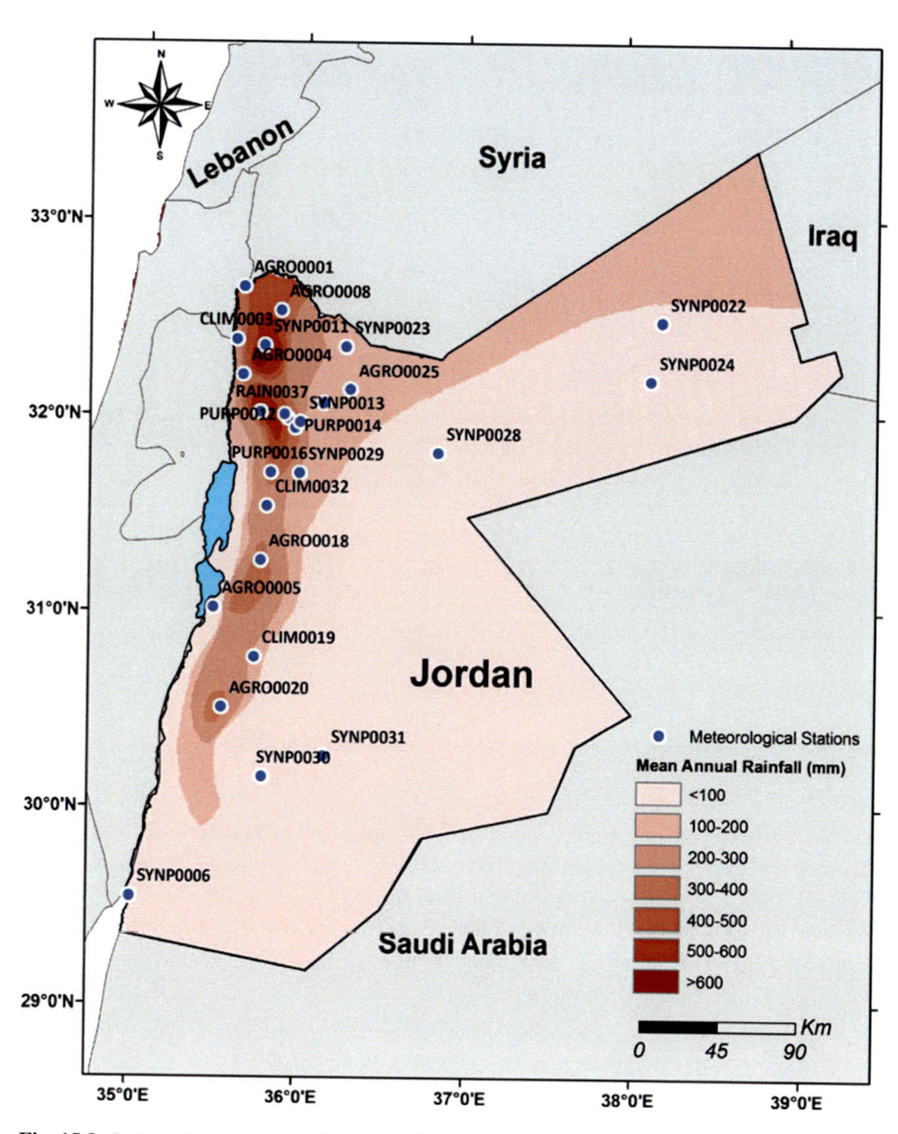

Fig. 15.8 Isohyetal map showing the spatial distribution of precipitation in Jordan (Al-Qinna et al., 2011)

This chapter reviewed the latest techniques and tools that can be used efficiently in assessing and monitoring drought events and consequent impacts. Important tools include the use of remote sensing multispectral sensors and GIS.

In addition, this chapter presented the result of drought assessment in Jordan based on three different techniques, these are: Meteorological Drought based on Standardized Precipitation Index (SPI) Method, Agriculture Drought based on

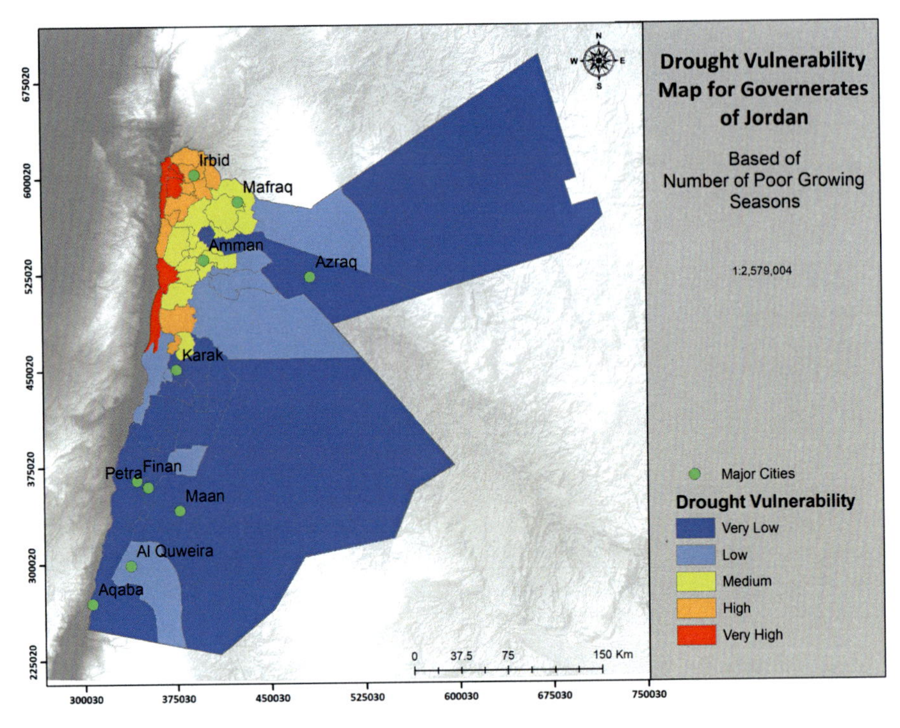

Fig. 15.9 Drought vulnerability mapping for governorates and districts of Jordan

Normal Difference Vegetation Index (NDVI), and drought assessment based on Rainfall Estimates from Rain Gauge and Satellite Observations (CHIRPS).

Daily data from 120 meteorological stations were used to develop SPI drought index from 1981 to 2018. Results of SPI modeling indicated that many stations in many locations are facing an increasing number of meteorological drought events with increasing severity and duration.

Data from USGS EROS Archive were downloaded using USGS EarthExplorer portal to assess agriculture drought conditions in the country. These data include scenes for February that covers the whole country. Using GIS, NDVI was calculated from 2012 to 2020, and $DEVN_{DVI}$ was calculated for each year. The visualized results shows that country faced several agricultural drought events especially in the years 2012, 2014 and 2019.

Another drought assessment method was done based on method described by (WFP, 2019). In this method, data from Rainfall Estimates from Rain Gauge and Satellite Observations (CHIRPS) were used to assess drought conditions in Jordan based on analyzing number of poor growing seasons (NGPS). Resultant maps showed clearly that the country facing drought conditions in 2019 especially in the northern and northwestern parts of the country, where most of the Jordan's agricultural activities are located.

These results were also projected on governorates and districts level to point out which administrative units within the country is facing drought conditions. Results indicated that many districts in the country are facing high and very drought conditions, which needs to have collaborated efforts on the country and local levels to develop suitable adaptation measures and action plans that aims at reduction the impacts of increasing drought events.

Results presented from this work and from other researchers works indicated clearly that new technological tools like EO Satellites, Multi-spectral, spatial and temporal images, GIS and spatial analysis tools are of great value in drought assessment and monitoring studies, and provides a decision-support tools that aid in planning and managing drought, and relevant actions of plans needed to reduce the impacts of drought.

References

Abu Hajar, H., et al. (2019). Drought assessment and monitoring in Jordan using the standardized precipitation index. *Arabian Journal of Geosciences, 12*, 417.

Abu-Allaban, M., El-Naqa, A., Jaber, M., & Hammouri, N. (2014). Water scarcity impact of climate change in semi-arid regions: A case study in Mujib basin, Jordan. *Arabian Journal of Geosciences, 8*(2), 951–959.

Al-Bakri, J. T., Al-Karablieh, E., & Al Naimat, M. J. (2019). Assessment of Combined Drought Index and mapping of drought vulnerability in Jordan. *Journal of Engineering Research and Application, 3*(Series I), 59–68.

Al-Qinna, M., Hammouri, N., Obeidat, M., & Ahmad, F. (2011). Drought analysis in Jordan under current and future climates. *Climatic Change, 106*, 421–440.

Bergaoui, K., et al. (2015). The contribution of human-induced climate change to the drought of 2014 in the southern Levant region. *Bulletin of the American Meteorological Society, 96*(12), 66–70.

Brooks, N. (2003). *Vulnerability, risk and adaptation: A conceptual framework*. Working Paper 38. Tyndall Centre for Climate Change Research, University of East Anglia.

Cook, B., et al. (2016). Spatiotemporal drought variability in the Mediterranean over the last 900 years. *Journal of Geophysical Research – Atmospheres, 121*(5), 2060–2074.

CRED. (2018). *Emergency events database (EM-DAT)*. Centre for Research on the Epidemiology of Disasters.

Damania, R., et al. (2017). *Uncharted waters*. World Bank Group.

Dracup, J., Lee, K., & Paulson, E. (1980). On the definition of droughts. *Water Resources Research, 16*, 297–230.

Funk, C., et al. (2015). The climate hazards infrared precipitation with stations—A new environmental record for monitoring extremes. *Scientific Data, 2*, 150066.

Guttman, N. B. (2007). Comparing the Palmer Drought Index and the Standardized Precipitation Index. *Journal of the American Water Resources Association, 4*, 1752–1688.

Hagman, G. (1984). *Prevention better than cure: Report on human and natural disasters in the third world*. Swedish Red Cross.

Hammouri, N., & El-Naqa, A. (2007). Drought assessment using GIS and remote sensing in Amman-Zarqa Basin, Jordan. *Jordan Journal of Civil Engineering, 1*(2), 142–152.

IPCC. (2012). *Managing the risks of extreme events and disasters to advance climate change adaptation. Special report of Working Groups I and II of IPCC*. Cambridge University Press.

IPCC. (2014). *Climate change 2014: Synthesis report. Contribution of Working Groups I, II and III to the fifth assessment report of the Intergovernmental Panel on Climate Change*. [Core Writing Team, R.K. Pachauri and L.A. Meyer (eds.)]. IPCC.

Jenks, G. F. (1967). The data model concept in statistical mapping. *International Yearbook of Cartography, 7*, 186–190.

Jensen, J. (1996). *Introductory digital image processing: A remote sensing perspective* (2nd ed.). Prentice Hall, Inc.

Kogan, F. (1995). Droughts of the late 1980s in the United States as derived from NOAA polar-orbiting satellite data. *Bulletin of the American Meteorology Society, 76*(5), 655–668.

Ligtvoet, W., et al. (2018). *The geography of future water challenges*. PBL Netherlands Environmental Assessment Agency.

McKee, T., Doesken, N., & Kleist, J. (1993). The relation of drought frequency and duration to time scales. In *Proceedings of the ninth conference on applied climatology* (pp. 233–236). American Meteorological Society.

MWI & GTZ. (2005). *National water master plan*. Hashemite Kingdom of Jordan, Natural Resources Authority.

Naumann, G., Carrão, H., & Barbosa, P. (2018). Indicators of social vulnerability to drought. Chapter 6. In *Drought: Science and policy, part II: Vulnerability, risk and policy* (pp. 111–125). Wiley-Blackwell.

OECD. (2008). *Handbook on constructing composite indicators, methodology and user guide*. JRC European Commission.

Peduzzi, P., Dao, H., Herold, C., & Mouton, F. (2009). Assessing global exposure and vulnerability towards natural hazards: The Disaster Risk Index. *Natural Hazards and Earth System Sciences, 9*, 1149–1159.

Rajsekhar, P., & Gorelick, P. (2017). Increasing drought in Jordan: Climate change and cascading Syrian land-use impacts on reducing transboundary flow. *Science Advances, 3*(8). https://doi.org/10.1126/sciadv.1700581

Roudi-Fahimi, F., Creel, L., & De Souza, R.-M. (2002). *Finding the balance: Population and water scarcity in the Middle East and North Africa*. Population References Bureau.

Tánago, I., et al. (2016). Learning from experience: A systematic review of assessments of vulnerability to drought. *Natural Hazards, 80*(2), 951–973.

Tarpley, J., Schneider, S., & Money, R. (1984). Global vegetation indices from the NOAA-7 meteorological satellite. *Journal of Climate and Applied Meteorology, 23*, 491–494.

Touchan, R., et al. (2011). Spatiotemporal drought variability in northwestern Africa over the last nine centuries. *Climate Dynamics, 37*, 237–252.

Tucker, C. J. (1979). Red and photographic infrared linear combinations for monitoring vegetation. *Remote Sensing of Environment, 8*(2), 127–150.

Van, L., et al. (2017). Climatological risk: Droughts. In K. Poljanšek, F. M. Marín, T. De Groeve, & I. Clark (Eds.), *Science for disaster risk management 2017: Knowing better and losing less. EUR 28034 EN*. Publications Office of the European Union. Chapter 3.9.

Vogt, J., et al. (2018). *Drought risk assessment. A conceptual framework. EUR 29464 EN*. Publications Office of the European Union.

Wardlow, B. (2009). *Remote sensing and drought monitoring, an overview, options for Mali, and future directions drought monitor workshop*. Bamako, Mali, September 14–17, 2009.

WFP. (2019). *Integrated context analysis (ICA) – GIS training manual (drought)*. WFP.

Wilhite, D., & Glantz, M. (1985). Understanding the drought phenomenon: the role of definitions. *Water International, 10*(3), 111–120.

WMO. (2014). *WMO statement on the status of the global climate in 2014*. WFP.

WMO & GWP. (2016). *Handbook of drought indicators and indices (M. Svoboda and B.A. Fuchs). Integrated drought management programme (IDMP), integrated drought management tools and guidelines*. Series 2nd ed. WMO & GWP.

Chapter 16
Using Remote Sensing Techniques for Identifying the Environmental and Quantitative Indices of Drought in Tihama, Yemen

Ali Dhaif Allah

Abstract Drought remains the most frequent and serious environmental threat in the Middle East area. In Yemen, drought has negatively been affecting both livelihood and sustainable development of the country. Since it is a natural part of the climate, it mainly depends on rainfall. This research aims at monitoring and assessing the drought risk through the changes in the vegetation cover and sand dunes deposit in the area using remote sensing and GIS. Landsat TM5 of 1985 and OLI8 of 2015 were used to evaluate the environmental and Quantitative indicators of drought using Normalized Difference Vegetation Index (NDVI) to recognize the progressive decline of vegetation. Additionally, the Standardized Precipitation Index (SPI) was used for temporal evaluation of the situation of drought. Through the index maps generated in a (GIS) environment, the results showed an increase of 26% in the area under severe drought during the period from 1985 to 2015. Similarly, the areas under moderate drought also increased to approximately 64%. Within the study period, the area also recorded 74% increase in sand dunes deposit. The analysis results by SPI-6 showed that the year 1991 was the worst years of drought experienced by the study area. Based on the fact that the study area is the most important agricultural areas in Yemen, it is recommended to further study the drought and its impact on agricultural crops in the area.

Keywords Drought · Remote sensing · GIS · Standardized precipitation index (SPI) · NDVI · Tihama plain · Yemen

16.1 Introduction

Drought is one of the greatest destructive natural dangers for human, environmental and economic terms. It leads to water and food crisis, famines, human exodus, loss of human life, land degradation and low productivity of agricultural crops (Kirono

A. D. Allah (✉)
Department of Geography, Faculty of Arts, Thamar University, Dhamar, Yemen

M. M. Al Saud (ed.), *Applications of Space Techniques on the Natural Hazards in the MENA Region*, https://doi.org/10.1007/978-3-030-88874-9_16

et al., 2011). Moreover, drought causes yearly a rate of $6–8 billion of universal losses by natural catastrophes, and drought affects the livelihood of a large number of people more than the effect of any other type of natural disasters. These include humanitarian disasters, economic losses, and stresses on natural ecosystems across the globe (Keyantash & Dracup, 2004). In addition to political and societal effects especially, in less economically developed nations that have limited adaptive capacity (Shahid & Behrawan, 2008). For instance, between 1991 and 1992, affected nearly 20 million of the population of South Africa as a result of the deficit in grain production due to drought, the deficit amounted to more than 6.7 million tons (Dinar & Keck, 2000).

In many Arab countries, droughts have become a more frequent and a serious bluster to humanitarian safety. Over the last three decades, nearly 50 million people have been affected in the Arab region by climatic disasters events, with losses estimated at $11.5 billion (Verner, 2012). This is a very hazardous problem which will increase causing serious threat on various aspects of life including humanitarian disasters, economic losses, and stresses on natural ecosystems (Kaniewski et al., 2012).

Over the last 30 years, Yemen experienced four periods of drought, 1979–1981, 1983–1984, 1990–1991, 2007–2009 (Miyan, 2015)caused a lot of damages on Yemeni economy, which largely relies on agricultural resources. According to the Ministry of Agriculture (2009), almost 73.5% of the population lives in rural areas and works in the agricultural sector, and thus significantly depends on appropriate weather conditions to their livelihoods. According to Escwa (2005), the period of drought 1990–1991 was the worst in the modern history of Yemen because this period of drought synchronized with the Gulf war in 1991, which forced nearly 800,000 Yemeni workers to return to Yemen. Hence, workers' remittances decreased, leading to economic growth retreat, and growing the inflation rate and foreign debt, because of these negative effects decreased the possibility of Yemen to cope drought and mitigation.

During that period of drought, agricultural production reduced severely, economic growth has been affected due to the low agricultural production contribution in the gross domestic product. For instance, irrigated agriculture such as vegetables registered a decline in production by 16%, as cereals yields dropped sharply, where production decreased both of millet, sorghum and barley by 33%, 34%, and 38% respectively. As for livestock have witnessed a marked decrease by 11% as a result of the lack of pasture and forage due to drought (Escwa, 2005). While that about 43% of the population living below the poverty line, according to the estimates of 2009. It is expected to increase the number of hungry people in Yemen between 80,000 and 270,000 people by 2050 (Wiebelt et al., 2011), because of the severity of frequency of droughts and changing climate.

To the best knowledge of the researcher, there is no scientific study on drought in Yemen, as well as the lack of numbers that accurately describe how the drought affects the Yemenis (Bogan, 2014; Escwa/Un-Desa, 2013; Miyan, 2015). However, there is information about the drought in Yemen found in non-local studies. These studies did not provide sufficient information on the drought in Yemen, they also did not address the drought indicators, through which intensity and frequency of droughts in Yemen can be identified. What is reported in those studies is just general information, where no details about the causes, effects, and severity of drought in Yemen could be touched. Therefore, the current study tries to bridge such a gap and as a first step to highlight the drought conditions in Yemen, as an attempt to ring the alarm bell to alert the dangers left behind by this serious phenomenon on the environment, economy and society.

This study focuses on the changes in vegetation cover (NDVI) and creeping of sand dunes to assess the drought risk in Tihama plain using remote sensing and GIS. By the fact that the study area is the most important agricultural areas in the country, it is necessary to study the changes in environmental indicators to know the drought conditions in the region. Therefore, two types of environmental indicators have been used, as they have a close and direct relationship with drought; these indicators are vegetation cover and sand dunes.

16.2 Literature Review

Drought is the most complex of all natural hazards, despite the attempts at unification, several definitions of drought continue to be employed (Wilhite, 2000). For instance, Linsley et al. (1959) defined drought as a sustained period of time without significant rainfall. While (Palmer, 1968) described drought is "a period of more than some particular number of days with precipitation less than some specified small amount" (p.2). Rossi (2003)defined drought as the occasional and recurring situation with a strong reduction compared to the normal values of water availability for a significant period of time and over a wide area. Similarly, it is defined by Wilhite (2008)as a recurrent feature of climate that is characterized by temporary water shortages relative to normal supply, over an extended period a season, year, or several years, in a wide region.

Hence, drought is commonly classified into; meteorological, agricultural, hydrological and socioeconomic droughts (Ams, 2004). Meteorological drought is a more common and natural event, whereas agricultural, hydrological and socioeconomic droughts emphasize more the human or social aspects (Eklund & Seaquist, 2015). The sequence begins with meteorological drought; persistent, dry conditions may induce agricultural, hydrological and water resources droughts (Andreadis et al., 2005; Vidal & Wade, 2009).

16.2.1 *Drought at Global Level*

Drought has a major effect on cultivation in terms of decreases in economic activity, agricultural productivity and drinking water stock in life-threatening cases that has controlled the famine (Roy & Hirway, 2007). The International Board on Climate

Change has noted that the yearly average of waterway overflow and water disposal are projected to decline by 10%–13% over certain waterless and semiarid areas in normal and low-slung opportunities, snowballing the occurrence, strength, and duration of drought, along with its related impacts (Pozzi et al., 2013). Inopportunely many nations do not have satisfactory incomes deliver an early warning but need external funding to provide the essential early threatening evidence for risk management (Cavatassi et al., 2011). Therefore, in an organized world, the essential for data on a global measure is critical for accepting the view of the failures in agricultural production, food safety, possible for civil battle and connected impacts on food prices (Forster et al., 2012).

Globally, given the expectation related to the universal climate change, might result from increasing drought problem. Kogan (1997) conducted identify drought in order to knowledge comprehensive on the drought impacts in North America, Europe, and Australia, particularly for some region with the mainland affected by drought. Allowing to the historical viewpoint by examining how the drought has varied over several regions of the world during the last millennium. Miyan (2015)his study showed that many people impacted by the natural disaster in developed countries. They lost their economy and life from a natural disaster. One of the reasons behind natural disaster is drought. As reported, that the developing countries have more effect from natural disaster compared to the developed countries.

According to Eriyagama et al. (2009), severe droughts affected vast regions of Asia, with Western India, Southern and Central Pakistan. The Asian regions have been among the continual drought-prone regions of the world. Afghanistan, India, Pakistan and Sri Lanka have described droughts at least once in each 3-year period in the past five periods. In 2012, Pakistan confirmed emergency in provinces Mirpur Khas and Tharparkar districts are in severe drought and many people had to be migrated (Tsubo et al., 2009). One of the important doubts with respect to rotation and rainfall with the present understanding of climate change in the monsoon regions remains. Therefore, in Asia has this drought problem until facing those societies. They have implemented the project. However, unable to satisfy of projects until suddenly change climate and water cycle (Hazell, 2007).

16.2.2 Drought in West Asia and North Africa (WA/NA)

According to the Epidemiology and Centre of Research, the percentage of the population affected by drought reached 51% of all other disasters combined. The drought has an increasingly common and major effect on human security located in arid areas of West Asia and North Africa (Karrou & El Mourid, 2008). This circumstance illustrates the severity of concern of drought. The WA/NA is the most affected region 83% in the world of the people in this area is unnatural by drought (El Kharraz et al., 2012).

The West Asia/North Africa region is severely vulnerable to natural disasters such as, droughts, floods, earthquakes, heat waves and sandstorms, which leads to increased significantly economic and social losses because of the events of natural disasters (World-Bank, 2014). During the past three decades, most of the Arab regions are severely vulnerable to natural disasters such as, droughts, floods, earthquakes, heat waves and sandstorms. During the past three decades, the Arab region has faced an increase in natural disasters events, which amounted to more than 276 disasters, killing 100,000, affecting 10 million and rendering nearly 1.5 million people homeless (Erian, 2011). Thus, drought is considered the major disaster among of all other disasters where it affects around 38.09 million people (Erian, 2012).

In addition, Asia was associated with arid climate in 2000, because of drought, desertification, sandstorms and water scarcity. Agricultural zones were affected due to droughts which were estimated in rural and urban areas as more than 40 million hectares (Mishra & Singh, 2010). On the other hand, through food and water shortages, drought is also thought to have caused the displacement of one million environmental refugees in Niger is 1985 (Gemenne, 2011)and 5 million in the African Sahel in 1995 (Myers, 2002). In 2000, in each of Sudan, Somalia and Kenya respectively 8, 6 and 2 million people who officially considered at hazard of famine, in addition to several million in other countries (Myers & Kent, 2001). As for losses of human life, for example, 450,000 deaths in Ethiopia and Sudan in 1984 directly attributed to drought (Guha-Sapir et al., 2004).

In many Arab countries, droughts have become a more frequent and a serious bluster to humanitarian safety. It is a very hazardous problem, and will increase causing serious threat on various aspects of life, taking into account that most of the Arab states are suffering from frailty of their ecosystems, and facing intense risks of deterioration of vegetation cover, soil, and depletion of water resources on per day continuously. This is coupled with the rushing increase in population growth accompanied by an increase pressure on natural resources in this region (Erian, 2012). As the number of population in Arab world 125 million in 1970 to exceed 280 million in 2000 and it is expected that up to more than 500 million in 2030, and hence many countries are already living under water stress conditions (United-Nations, 2003). In addition to climate change, this is the major cause for the Arab region droughts. Climate models show that over the last 30 years temperatures in the Arab region have been increasing 50% faster than global averages. Climate changes will affect more than 340 million populations in the Arab zone. More than 100 million are poor and least able to resist these changes. Over the past three decades, nearly 50 million people were affected in the Arab region by climatic disasters events, with losses estimated at $11.5 billion (Verner, 2012).

16.2.3 Drought in Yemen

According to U.S (1982), Yemen climate was affected by drought during the periods 1967/1969 and 1972/1974, which led to a threefold increase in food imports, despite the increase in agricultural production after the end of the drought in 1971. In 1983 and 1984, the Yemen's economy dropped significantly, as a result of the drought that prevailed over most of the country over that period. The drought-reduced total output of grain by more than half in 1983, but food availability remained at the normal level, due to the increased grain imports from outside the country. Generally, in 1983–1984 Yemen's agriculture dropped remarkably due to the severe drought (Usda, 1986).

What makes the matter even worse, Yemen is the poorest and least-developed country among the whole countries of the Middle East. In the recent years, Yemen has suffered from changes in rainfall patterns, higher temperatures and increased frequency of droughts. For example, the drought which took place in 1990–1991 had a significant effect on the socio-economic situation in Yemen, resulting in wide agricultural losses and a serious increase in the number of poor households in rural zones, and low revenue contribution of the agricultural production in the gross domestic product of the country (Escwa/Un-Desa, 2013).

Likewise, once again in 1990–91, the agricultural sector recorded substantial losses, because of the significant reduction in agricultural crops production and increased poverty in rural areas because of the drought (Escwa, 2005). In addition, in the same period, the drought had dangerous reflections on the food security of a large segment of the population. In general, many studies have found that there is a significant lack of understanding and awareness of the drought and its effect, as well as the capacity of mitigating it in Yemen (Escwa/Un-Desa, 2013).

Additionally, there is a number of weaknesses in the drought management system in Yemen, including the lack of technical capacity to analyze drought data and processed, if any, difficulty to get data on drought, no a regional exchange of information on drought as there is no early warning system for drought monitoring (Bogan, 2014). The drought of the period from 2007 to 2009 was the most influential which seriously damaged the socio-economic development in Yemen. It has become a real disaster sweeping the most areas of the country, leaving behind many of negative damages on the economy and society (Miyan, 2015). Noaman et al. (2013) clarified that due to the water scarcity and severity drought in Yemen, the per capita water resource does not exceed 195 cubic meters per year. It is considerably less than the water poverty line of 1000 cubic meters per year. "Due to the lack of research, there are no numbers that accurately describe how drought is affecting Yemenis" (Miyan, 2015). It has become a real disaster in Yemen. Due to the lack of studies, there are no specific numbers that accurately describe the drought in Yemen. Thus, it is hard understand how drought affects the environment, economy, and society in Yemen.

16.2.4 Drought in the Study Area

In Tihama plain, drought is one of the factors affecting soil formation and characteristics of vegetation, where the plants take thistles form and short weeds to be able to resistance drought. For example, the permanent drought which dominates the western part of the study area is not allowed to grow a good vegetarian cover with the exception of some plants which are capable of withstand the drought like Suaeda and Haloxylon plants. In addition to some annual weeds which grow during the rainy season, and because these plants are the main source for grazing in the western part of the study area. The occurrence of drought leads to deterioration of pastures and then the livestock exposure to danger (Al-Washli, 1989). Drought is also the most important causes of environmental degradation in the region (Taher, 2004).

According to Abdullah (2010), the periods of the eighties and nineties and the first 5 years of the twenty-first century are among the most drought periods impact on the Tihama Plain. The study reported that the vegetation, soil, and groundwater are the most affected factors by drought. While another study about desertification in a part of Tihama plain reported that the drought of the most important factors causing desertification and land degradation in the region (Dhaifallah, 2012).

As could be noted above, it is obvious that Yemen, including the study area, has been suffering from frequent drought, and also numerous of social, environmental and economic problems as a result of the low precipitation rates and frequent droughts. In contrast, there is no previous detailed study or sufficient information about the drought in Yemen in general and Tihama plain in particular. Thus, there is an urgent need to study the drought in Yemen, in order to fill such a gap.

16.3 Location and Climate of the Study Area

Tihama Plain is located on the west part of Yemen between latitude 12°.5–16°.5 north of the Equator and between longitude 42°.5–43°.5 east of Greenwich, and lies about 226 km of the capital city, Sana'a. The study area covers a total area of 25,314.93 km^2 (Fig. 16.1).

With regard to climate, the study area has arid and semi-arid climate, the rains in the region as low overall, where ranging between 50 and 600 mm per year, because of its location within trough of the Red Sea as well as the low level of the surface compared to the neighbouring mountain blocks. In addition, the rains are fluctuating from year to another depending on the conditions of the various pressures over the land areas and adjacent surface water bodies.

According to the yearly precipitation factor; the study area is often divided into three rainfall regions; arid region (western Part) the average rainfall between 50 and 200 mm per year, Semi-arid region (middle part) the annual rainfall 200–400 mm, and semi-wet region (eastern part) 400–600 mm annually (Abdullah, 2010). Almost the rains in the study area are falling over all the months; however, the heavy rains fall in the months (July, August and September) because these months are the summer months which represents the main rainfall season over most land in Yemen. For this reason, the biggest quantity of rains falls on the study area in the summer. As for temperatures are ranging 37 °C in summer and 24 °C in winter (Assage, 2003).

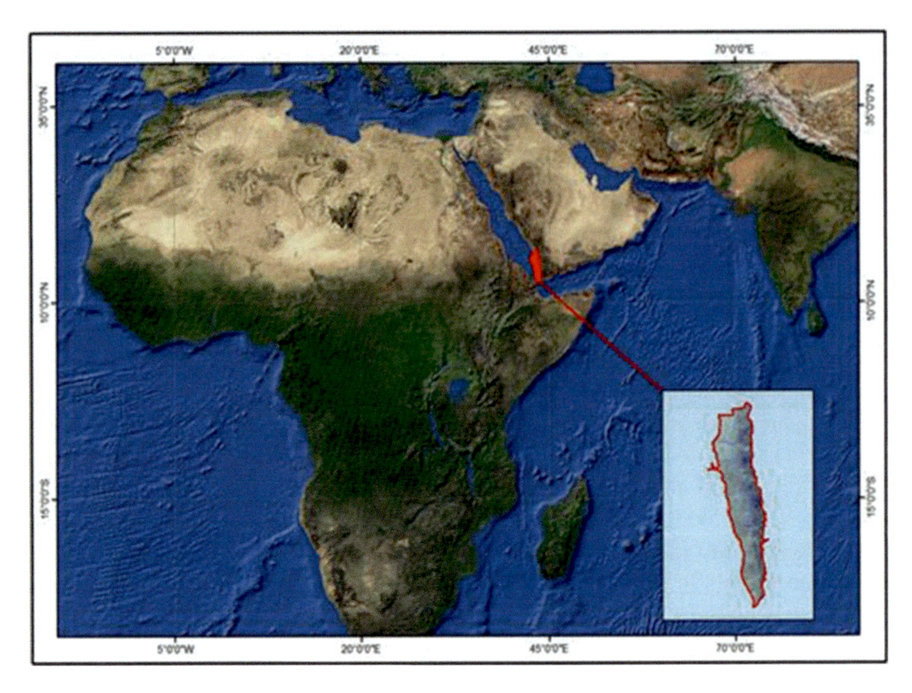

Fig. 16.1 Location of the study area

16.4 Methodology

NDVI index is one of the most important indicators available on a large scale. In addition, it is more reliable and used widely in monitoring and assessing the drought. In this study, the NDVI index was applied to know the changes on the vegetation cover which was used to assess the drought. Besides, the change in the sand dunes areas studied for the same purpose. The month of October was chosen for each of 1985 and 2015 in order to assess the changes in the vegetation cover, because this month is considered the growth month for vegetation. The same applied to assess the change in the sand dunes areas.

Moreover, the study adopted the NDVI index to determine the change in the green spaces and dunes areas on the satellite image processing using GIS software 10.2. Also, calculated the natural variability and then classification of the results to drought categories, according to the values that were obtained. Two environmental indicators of drought, vegetation and sand dunes were monitored in this study as follows.

16.4.1 Vegetation

Vegetation cover is a very strong environmental indicator of drought due to the negative correlation between rainfall amount and duration on one hand, and vegetation density on the other. To monitor and assess the vegetation of the area, multi-temporal remote sensing data was used. Landsat Thematic Mapper (TM5) of October 1985 and Landsat Operational Land Imager (OLI8) of October 2015 were downloaded from the official website of the United State Geological Survey (USGS) via the Earth Explorer. The choice of 1985 and 2015 Images was informed by the need to assess the condition of both vegetation and sand dunes between these two dates as they mark the beginning and the end of the study periods. However, due to the problem of data availability, and lack of cloud free images for 1990, 1995 and 2010, only, the 1985 and 2015 images were used. The images were then subjected to pre-processing such as band stacking, haze and noise removal and conversion of digital number (DN) to reflectance values (RV) with aid of Erdas Imagine 2014 version image processing software. Later, Normalised Difference Vegetation Index (NDNI) was computed for the 2 images to highlight the vegetation condition of the study area in 1985 and 2015.

NDVI was calculated using the formula: NDVI = NIR − RED/NIR + RED. Where:

NIR = The amount of near infrared light reflected by the vegetation and captured by the satellite sensor.

RED = The amount of red light in the visible spectrum that is reflected by the vegetation and captured by the satellite sensor. Using Landsat 5 Image, NDVI is thus calculated as:

NDVI = Band4 – Band3/Bnd4 + Band3. While using Landsat 8 OLI image it is calculated as: Band5 – Band4/Band5 + Band4. This is based on the fact that, healthy vegetation will absorb most of the visible light that falls on it, and reflects a large portion of the near-infrared light. Unhealthy or sparse vegetation reflects more visible light and less near-infrared light. Bare soils on the other hand reflect moderately in both the red and infrared portion of the electromagnetic spectrum. The NDVI algorithm subtracts the red reflectance values from the near-infrared and divides it by the sum of near-infrared and red bands. Theoretically, NDVI values are represented as a ratio ranging in value from −1 to 1 but in practice extreme negative values represent water, values around zero represent bare soil and values over 0.6 represent dense green vegetation. NDVI has been widely used to study changes in the spatial pattern of vegetation (Matson & Bart, 2013; Vogelmann et al., 2012; Zhao et al., 2015).

16.4.2 Sand Dunes

The second environmental indicator of drought investigated by this research is sand dunes. The reason beyond assessing this significant indicator is due to the positive correlation between rainfall amount and duration on one hand, and sand dune deposit on the other. The same images, Landsat Thematic Mapper (TM5) of October 1985 and Landsat Operational Land Imager (OLI8) of October 2015 were used in monitoring and assessing sand dune deposit in the study area. After the afore mentioned pre-processing, appropriate band combinations were used to discriminate and highlight sand dunes from other land cover types for visual observations. Green, near infrared and short wave infrared bands corresponding to bands 2, 4, and 7 for Landsat TM 5 and bands 3, 5, and 7 for Landsat OLI 8 were used. These bands show high reflective variability of desert surface and therefore easily highlight sand deposits. Later, colour composite images were produced and using enhancement techniques such as contrast stretching and spatial filtering spatial information contained in the two images were enhanced making it easy to identify dunes as they stand out clearer and brighter. On-screen digitizing of dune was then performed in ArcGIS 10.2 environment. A vector file was later created that contained the locations and number of digitized dunes. Subsequently, the vector file was used to create a continuous surface of dune density maps through interpolation in ArcGIS environment. This was later used to calculate relative percentage of dune deposit in each year. Figure 16.2.showa the flow chart below summarized the major steps involved in this research.

16.4.3 Method of Rainfall Data Collection

The monthly rainfall data of 5 meteorological stations for the time period 1980–2010 were acquired from the Public Authority of Agricultural Research (Table 16.1), which covered the study area as shown in Fig. 16.3.

16.4.4 Method of Data Analysis

To determine the drought years, Standard Precipitation Index (SPI) was calculated using SPI calculator software obtained from the National Drought Mitigation Centre (NDMC).

The SPI was calculated using 6 months' time scale which is appropriate for the determination of drought.

The formula for determining drought years as used in this research is:

X = Number of months in drought
Rainfall in the study area = 12 month
Heavy Rainfall = 3 month (July, August and September)
Normal dry = 12–3 = 9 = No Drought
When X = 10 or above = Drought year

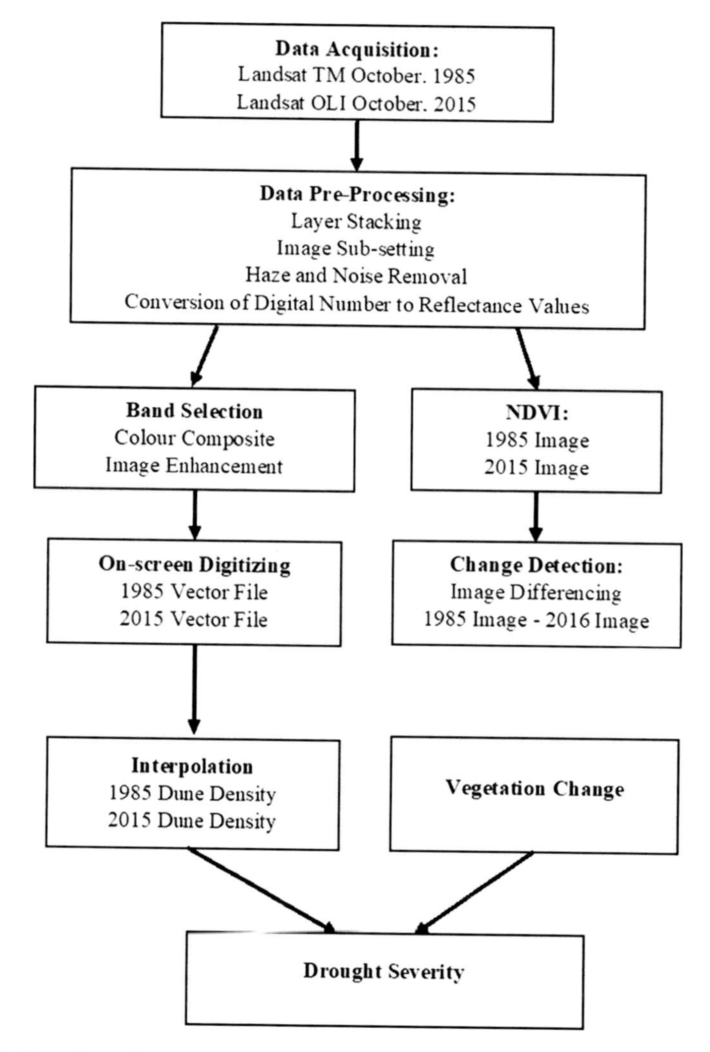

Fig. 16.2 Flow chart of the research methodology on environmental indicators

Table 16.1 Meteorological stations used in the study

Stations	Longitude	Latitude	Elevation (m)
Zuhrah	42°.02’	15°.43’	100
Bajil	43°.16’	15°.02’	120
Kalifah	43°.18’	14°.53’	140
Hodiedah	42°.59’	14°.44’	10
Jerbeh	43°.27’	14°.09’	250

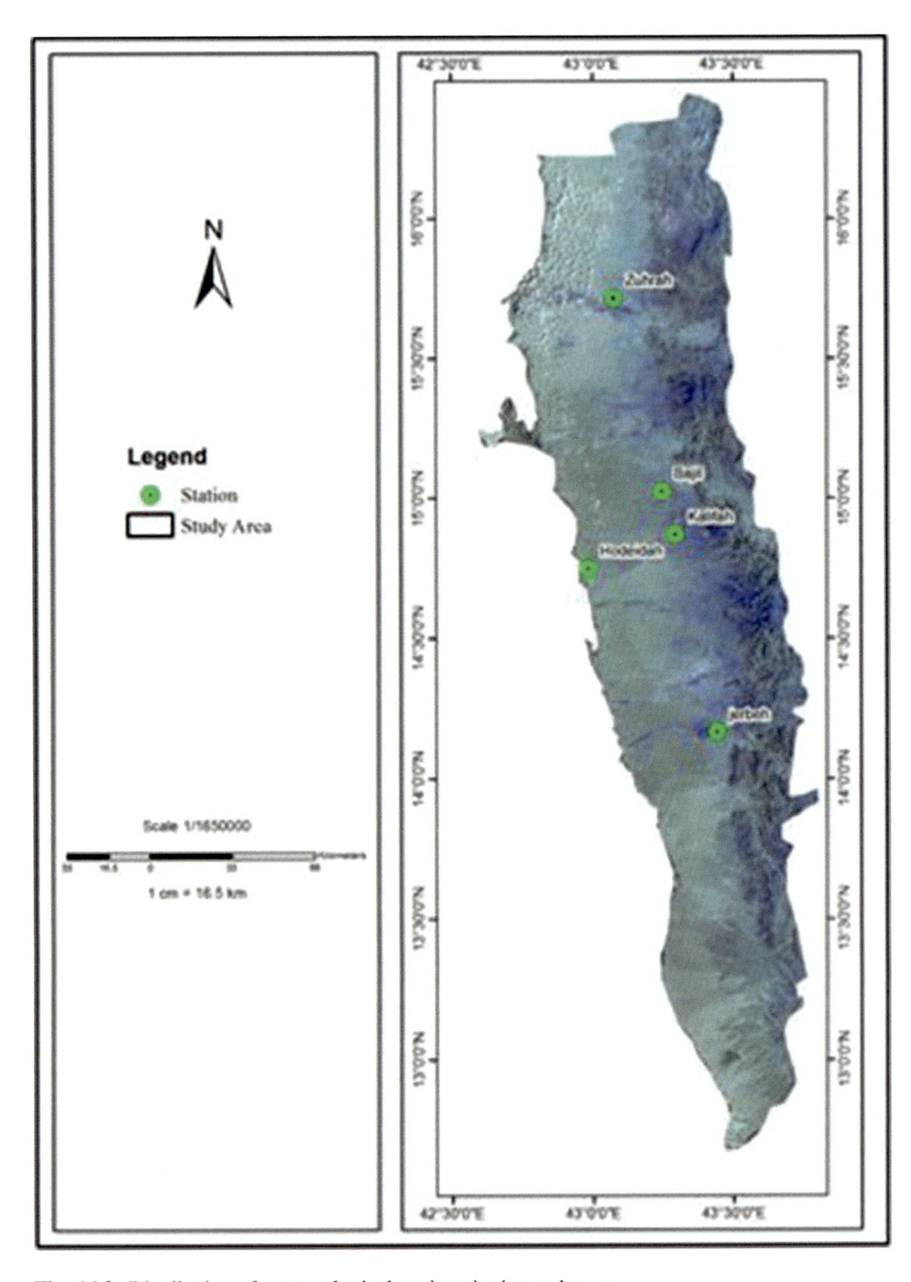

Fig. 16.3 Distribution of meteorological stations in the study area

16.4.5 Explanation

Usually, under normal circumstances Yemen Experience 3 months of heavy rainfall in July, August and September. Therefore, any year with a total of 9 dry months is considered as a normal situation. On the other hand, any year with more than 9 dry months is considered as a drought year. This means that, at least one months out of July, August or September did not receive sufficient rainfall as it used to be, thereby aggravating the drought condition already existing in the country.

16.4.6 Standard Precipitation Index (SPI)

Drought occurs when the SPI negatively on an on-going basis, and its severity up to −1.0 or less, and ending when the SPI becomes positively (McKee et al., 1993). Therefore, each phenomenon of drought has a period to be determined by its beginning, end, and severity, in every month the drought continues Table 16.2. Thus, the total positive SPI for all the months which takes by drought, it might be called the "intensity"of drought (WMO, 2012).

The SPI is calculated from the monthly precipitation record. The classification system proposed by McKee et al. (1993) is used to determine the severity of drought (mild, moderate, severe and very severe) based to SPI value, according to specific standards for any phenomenon of drought on any of time scales, as shown in Table 16.3.

Table 16.2 Severity of drought based on the SPI index

SPI value	Classification
2.00 to more	Extreme wet
1.50 to 1.99	Very wet
1.00 to 1.49	Moderate wet
0 to 0.99	Mild wet
0 to −0.99	Mild drought
−1 to −1.49	Moderate drought
−1.50 to −1.99	Severe drought
−2.00 or less	Extreme drought

Sources: WMO, 2012
Source: Ceglar et al., 2008

Table 16.3 Classification of the SPI values and drought categories the SPI values and drought category

SPI value	Drought category
0 to −0.99	Mild drought
−1 to −1.49	Moderate drought
−1.50 to −1.99	Severe drought
−2.00 or less	Extreme drought

Source: McKee et al., 1993

Through the SPI can track drought on multiple time scales, (i.e. 1, 3, 6, 9, 12, 24 and 48 months), at the same time SPI is flexible with respect to the selected period space (McKee et al., 1993).There is a general consensus among users that using SPI on short time scales (3 and 6 months) describes the events of meteorological drought, as well as describes events of the drought affecting agricultural practices (agricultural drought), while use it over a long period of time scales (12 and 24 months) is more appropriate for monitoring hydrological drought and management of water resources (Raziei et al., 2009).

16.4.7 *Software Used*

16.4.7.1 Use ArcGIS

After the drought years has been identified according to the method mentioned above, the results were transferred to Arc GIS 10.2 software, in order to mapping the spatial distribution of the severity of drought in the study area.

16.4.7.2 Use SPSS

SPSS 22 was used to compare the groups of study area. Therefore, the period was divided into 6 groups for each station. This includes (1980–1985, 1986–1990, 1991–1995, 1996–2000, 2001–2005 and 2006–2010). The Kruskal Wallis test was used to examine the differences between groups.

16.4.7.3 Use Microsoft Excel

The results were then transferred to Excel spread sheet 2010, in order to calculate the frequency of drought, identify drought years and draw some graphs to explain the severity of drought.

16.5 Result and Discussion

The environment indicators discusses under which, NDVI and sand dunes were used to show and analyzed both the spatial and temporal distributions as well as severity of drought in the study area. Section two, dealt with the quantitative indicators of drought under which, SPI 6 month was used to quantify, analyzed and discussed the spatial and temporal distribution of both agricultural and meteorological droughts respectively. Section three on the other hand discuss the general trend of drought in the study area by combining both agricultural and meteorological droughts to map the overall spatial distribution and severity of drought in the area.

16.5.1 Normalized Difference Vegetation Index (NDVI)

NDVI is one of the environmental indicators of drought due to the positive correlation between rainfall amount and duration on one hand, and the vegetation density on the other. That is, the higher the amount and duration of the rainfall in a particular area, the more dense the vegetation in that area would be. NDVI data of 1985 and 2015 shows a progressive increase in the intensity of drought in the area. Between 1985 and 2015, there is 26% increase in the area under severe drought from 3452.87 square kilometres in 1985 to 4341.14 square kilometres in 2015. Similarly, the area under moderate drought also increased from 7943.31 square kilometres in 1985 to 13,048.54 square kilometres in 2015, representing almost 64% increase. On the other hand, areas under mild drought and those receiving normal rain experienced a decrease during the period as they gradually transformed into mild and severe drought situations. In 1985, a total land area of 7534.79 square kilometres was experiencing mild drought, but this has declined to only 4352.23 square kilometres in 2015 representing over 73% decrease, which is transformed into moderate and severe drought conditions. Also, of the total of 6009.1 kilometre square of land that received relatively normal rainfall in 1985, about 38% (2872.22 kilometres square) have been transformed to a drought condition (Table 16.4 and Fig. 16.4).

16.5.2 Sand Dunes

Sand dunes deposit is yet another environmental indicator of drought. This is due to the negative correlation that exists between rainfall and vegetation on one hand and sand dunes deposit on the other. That is, the higher the amount and duration of rainfall, the more densely the vegetation cover will be, and thus, the lower the sand dunes deposit. On the other hand, areas with lower amount and shorter duration of rainfall, will in turn, have a relatively little or no vegetation cover, which also makes the soil more prone to erosion, transportation and deposition by the wind. This results in increased sand dunes deposit in the area. The result of this study shows

Table 16.4 Area (km^2) and percentage (%) of drought intensity using NDVI

NDVI/Drought	Area in 1985	Area in 2015	Difference km^2	Percentage %
Severe drought	3452.87	4341.14	888.27	26 (increase)
Moderate drought	7943.31	13048.54	5.105.23	64 (increase)
Mild drought	7534.79	4352.23	5.513.75	73 (decrease)
No drought	6009.1	3136.88	2.872.22	38 (decrease)

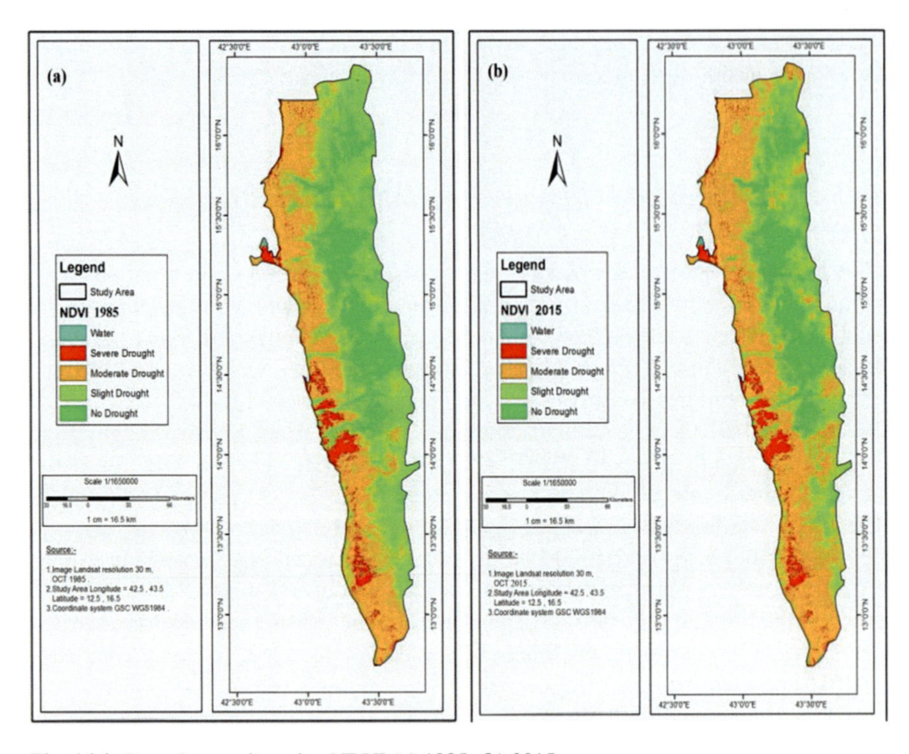

Fig. 16.4 Drought severity using NDVI (**a**) 1985, (**b**) 2015

Table 16.5 Area and percentage of sand dunes

Year	Area (km^2)	Percentage %
1985	2761.42	11
2015	4809.1	19
Difference	2047.68	74

that, there is a progressive increase in the area covered by sand dunes within the study period.

During the year 1985, of the total land area 25,314.93 square kilometres of land in the study area, the sand dunes, representing 11%, covered only about 2761.42 square kilometres. This has however, increased to 4809.1 square kilometres representing 19% of the total land area in the year 2015. This means that, within the 30 years study period, the total land area covered by the sand dunes has expanded from 2761.42 square kilometres in 1985 to 4809.1 square kilometres in 2015 representing over 74% increase. This situation suggests that there is an increasing trend in the intensity of drought in the study area over this period. (Table 16.5 and Fig. 16.5).

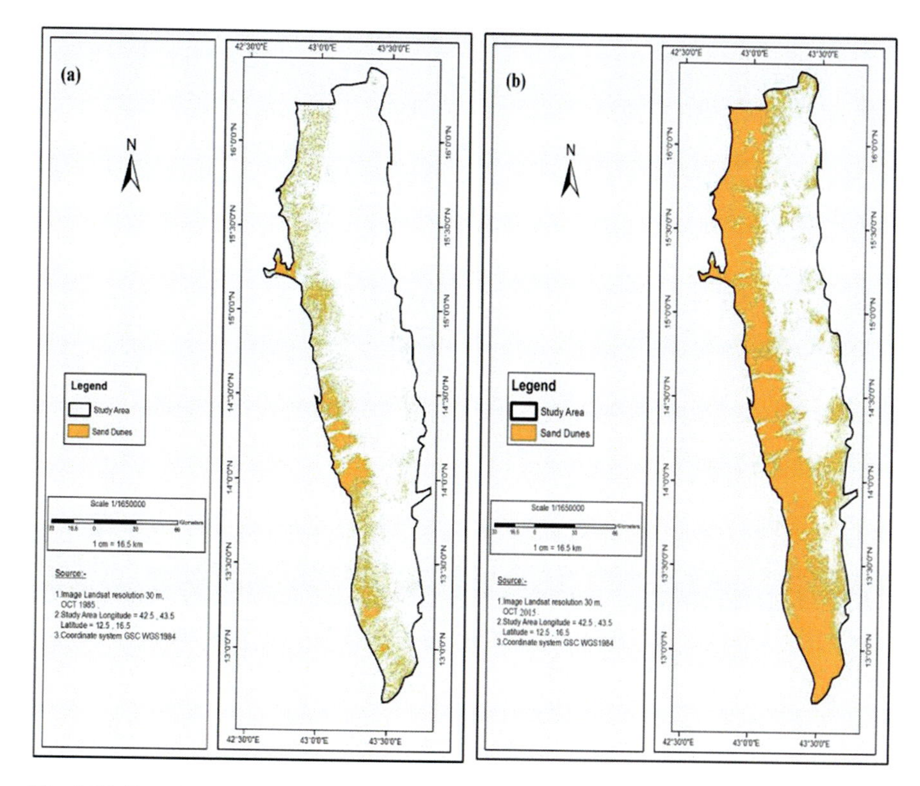

Fig. 16.5 Drought severity using sand dunes (**a**) 1985, (**b**) 2015

16.6 Temporal Drought Pattern in the Study Area

Drought occurs when the SPI negatively on an on-going basis, and its severity up to −1.0 or less, and ending when the SPI becomes positively (McKee et al., 1993). Therefore, each phenomenon of drought has a period to be determined by its beginning, end, and severity, in every month the drought continues. Thus, the total positive SPI for all the months which takes by drought, it might be called the "intensity"of drought (WMO, 2012). Through temporal analysis of drought using SPI 6, it is clear that there is variation in the severity of drought between meteorological stations under study, which the SPI values were less than (−1) during the drought years. The temporal analysis results of drought for each station in the study area as shown below:

16.6.1 Zuhrah

Form the year 1980 to 2010, a total of 8 drought years were recorded in Zuhrah station. Analysis of the SPI time series indicate that, both the intensity and frequency of the drought increased progressively. For example no drought was recorded between 1980 and 1985 and only one drought year each was recorded from 1986 to 1990 and from 1991 to 1995. However, more frequent droughts were recorded from the 2003 to 2010, during which 6 drought years were recorded in 2003, 2005, 2006, 2007, 2008 and 2009, with the year 2005 having the highest intensity of up to −3.08 SPI values (Fig. 16.6).

16.6.2 Kalifah

In kalifah, a total of six (6) drought years were recorded from 1980 to 2010. The first and the most extreme drought was recorded in the year 1991 with up to −4.34 SPI values. Another 4 years of successive droughts were recorded in 2002, 2003, 2004, 2005 and another, in 2008. The drought in 2005 was also extreme and is second in intensity after 1991, while the remaining years recorded severe droughts (Fig. 16.7).

16.6.3 Jerbeh

Jerbeh area recorded a total of 5 drought years between 1980 and 2010. The droughts occurred in 1984, 1990, 1991, 2002 and 2004. The severity of the drought

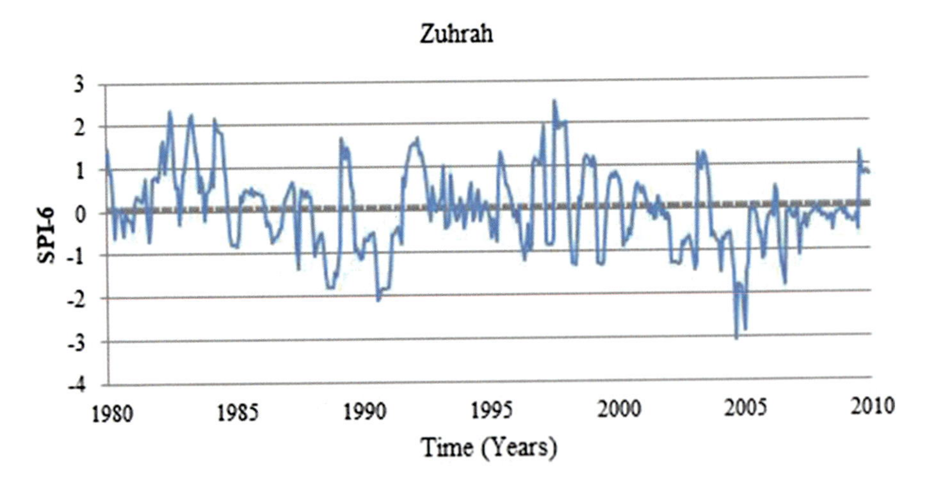

Fig. 16.6 Temporal distribution of agricultural drought in Zuhrah area

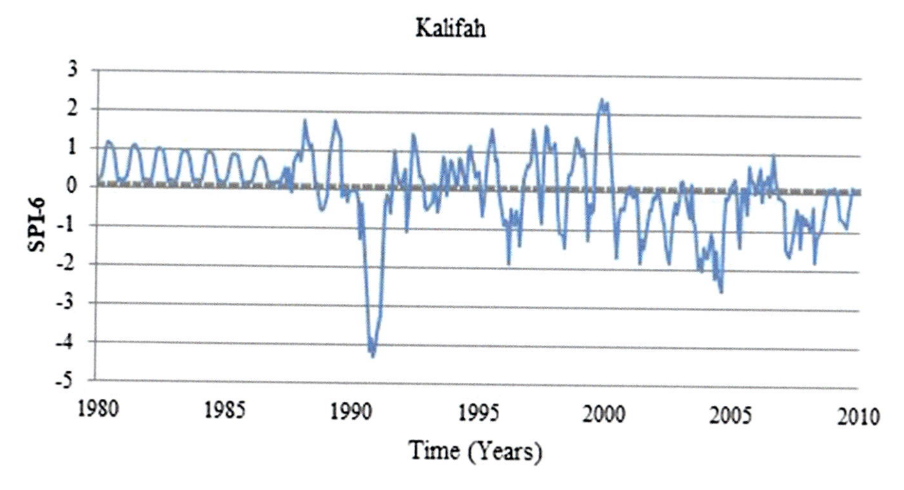

Fig. 16.7 Temporal distribution of agricultural drought in Kalifah area

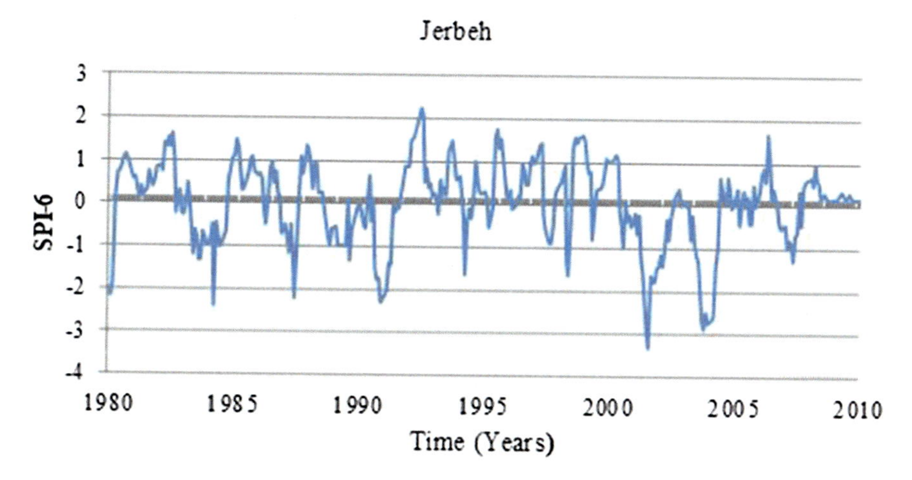

Fig. 16.8 Temporal distribution of agricultural drought in Jerbeh area

progressively increased with time from moderate droughts in 1984 and 1990 to extreme drought in 1991, 2002 and 2004. However, the 2002 drought is the most severe and extreme with the negative SPI values of −3.3, signifying an extreme drought condition. On the other hand, the 1990 drought is the mildest with the negative SPI values of – 1.33 signifying a moderate drought condition (Fig. 16.8).

16.6.4 Bajil

Bajil recorded the highest frequency of drought amongst the five station during the period under investigation. The area recorded a total of 10 drought years from 1980 to 2010. However, the drought was more frequent and more severe between 2000 and 2010, during which 6 years of persistent droughts were recorded from 2002 up to 2007. Before this droughts occurred intermittently in 1981, 1984, 1998 and 1999. The severity of the drought also increases with time from a milder drought in 1981 with SPI values of just −0.64 to a more extreme drought in 2007 with a negative SPI values of up to – 2.61, signifying an extreme drought condition (Fig. 16.9).

16.6.5 Hodeidah

In Hodeidah there appears to be two separate drought regimes. In total 7 drought years were recorded during the study period. The first regime started from 1981 to 1985 after which, a period of normal rainfall resumes from 1986 until 2005. The second drought regime was recorded from 2006 to 2010. However, the drought was not as intense as in the areas as only mild and moderate were recorded with general absence of severe drought. Moderate drought conditions with SPI values of −1.28 was recorded in1981, 1982, 2008 and 2010 while the droughts in 1984,1985 and 2006 were mild ones (Fig. 16.10).

Moreover, to make a confirmation on the results of the SPI-6 time scale, Kruskal Wallis Test to examine the differences between the 5 stations was performed. As shown in Table 16.6, the p values of all stations were < 0.05 indicating that the

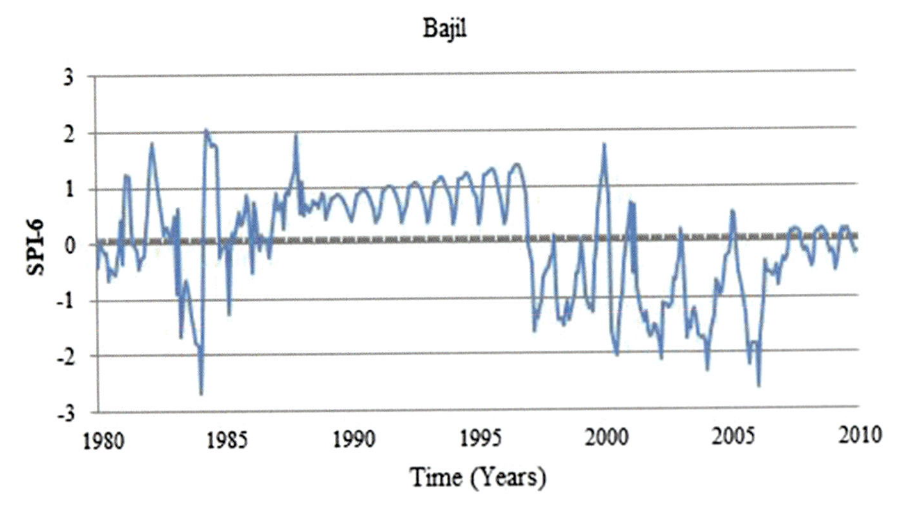

Fig. 16.9 Temporal distribution of agricultural drought in Bajil area

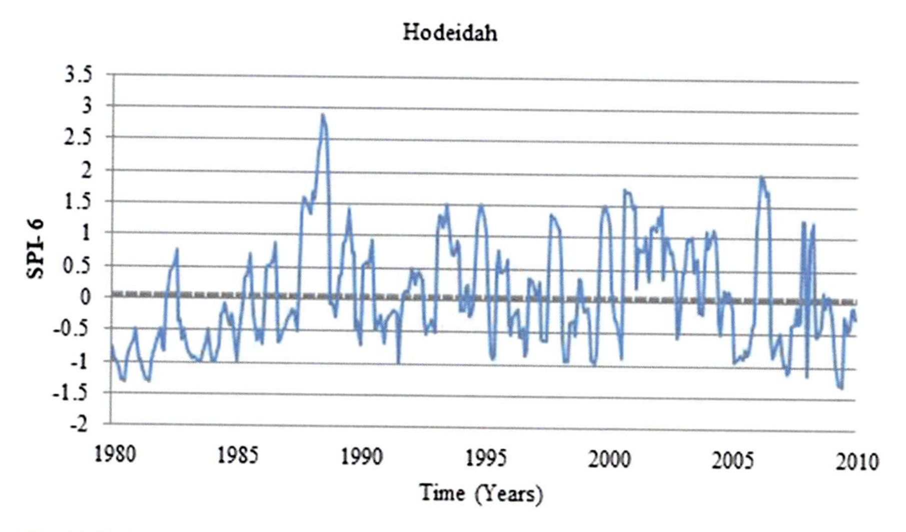

Fig. 16.10 Temporal distribution of agricultural drought in Hodeidah area

Table 16.6 Differences between stations in term of agricultural drought

	Zuhrah	Kalifah	Jerbeh	Bajil	Hodeidah
Chi-Square	47.488	97.324	46.079	141.686	119.238
df	5	5	5	5	5
Asymp. Sig.	.000	.000	.000	.000	.000

drought was different between all stations. In another words, all stations showed there are statistically significant ($P < 0.05$) variation/change in the drought intensity through the time period of the study.

The overall drought frequency in all the 5 stations indicate that on the average, near normal condition accounts for 72% in all the 5 stations put together for the 30 years period of investigation. Moderate drought account for 7%, severe drought 4% and extreme drought 2%. That is the cumulative drought period for the stations accounts for 11%. On the other hand, moderate wet for all the stations accounts for 11%, very wet, 3% and extreme wet, 1%. However, a more detailed examination of the result revealed that, out of the total of 1860 months observed in all the 5 stations, near normal condition is recorded in 1341 times, moderate drought 136 times, severe droughts 64 times and extreme drought 38 times. On the other hand, extreme wet is recorded 21 times in all the 5 stations, very wet, 58 times and 202 moderate wet (Table 16.7 and Fig. 16.11).

Table 16.7 Frequencies of drought categories based on the SPI classes

	Category						
Stations	EW	VE	MW	NN	MD	SD	ED
Zuhrah	9	14	36	264	32	13	4
Kalifah	4	9	34	277	22	15	11
Jerbeh	2	14	35	267	26	11	17
Bajil	1	7	49	242	42	25	6
Hodeidah	5	14	48	291	14	0	0
Total	21	58	202	1341	136	64	38
Percentage %	1%	3%	11%	72%	7%	4%	2%

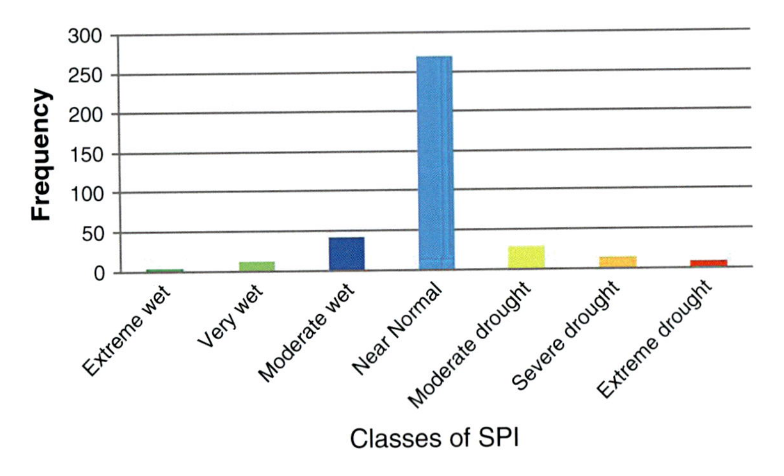

Fig. 16.11 Frequency of mean annual of SPI classes

16.7 Spatial Drought Pattern in the Study Area

The SPI 6 time scale was used for the determination of drought years, where the drought usually occurs when the SPI negatively, and its severity up to −1.0 or less, and ending when the SPI becomes positively. Accordingly, it has been shown by the results that there is spatial differences in the drought severity between meteorological stations in the study area during the drought years. It has observed the lowest SPI values during the years 1991 and 2004, so these years were considered the worst drought years. On the other hand, the year 1995 was selected as a year of wet, where that the most of SPI values were positively over the stations area.

16.7.1 1991 Drought Distribution and Intensity

The year 1991 recorded the worst drought in the last 30 years in the study area both in terms of distribution and intensity. During this year, Kalifah area recorded a

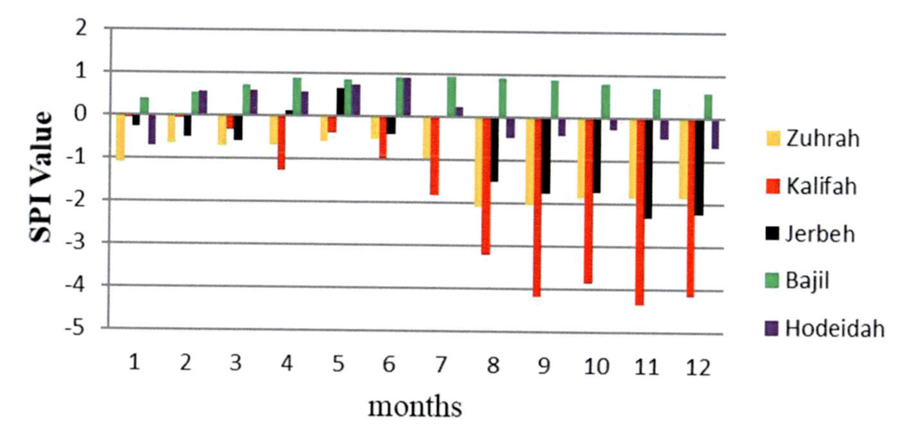

Fig. 16.12 Distribution and intensity of agricultural drought in 1991

drought intensity of up to – 4.34 in the SPI value which signifies extreme drought condition and the dryness persist for throughout the year (12 months). Zuhrah also recorded 12 month of dryness during the same year, but of lesser intensity (−2.09 SPI values). Comparably, Jerbeh recorded a more intense drought than Zuhrah (−2.32 SPI values), but of relatively shorter duration as the drought last for only 10 months of the year. The other two stations, Bajil and Hodeidah experienced a normal rainy year in 1991 (Figs. 16.12 and 16.13).

16.7.2 2003 Agricultural Drought Distribution and Intensity

In the 2003, 3 of the 5 stations namely Bajil, Kalifah and Zuhrah recorded dryness of different magnitude and duration. Bajil is driest among the 3 stations, recording 12 dry months and up to – 2.1 SPI values signifying extreme drought condition. Zuhrah also recorded 12 dry months but less intense than Bajil with – 1.28 SPI values representing moderate drought, while in Kalifah the drought last for only 10 months both more severe than Zuhrah with up to – 1.87 SPI values denoting a severe drought situation. Hodeidah and Jerbeh on the other hand, recoded relatively normal rains during the year, with Hodeidah recording up to 1.49 positive SPI values denoting moderately wet condition (Figs. 16.14 and 16.15).

16.7.3 2005 Agricultural Drought Distribution and Intensity

During the year 2005, three out of the five stations recorded extreme drought. Zuhrah, which recorded normal rainfall in 2004, turn out to be the driest in 2005 with 12 dry months and SPI values of – 3.08. it is closely followed in terms of

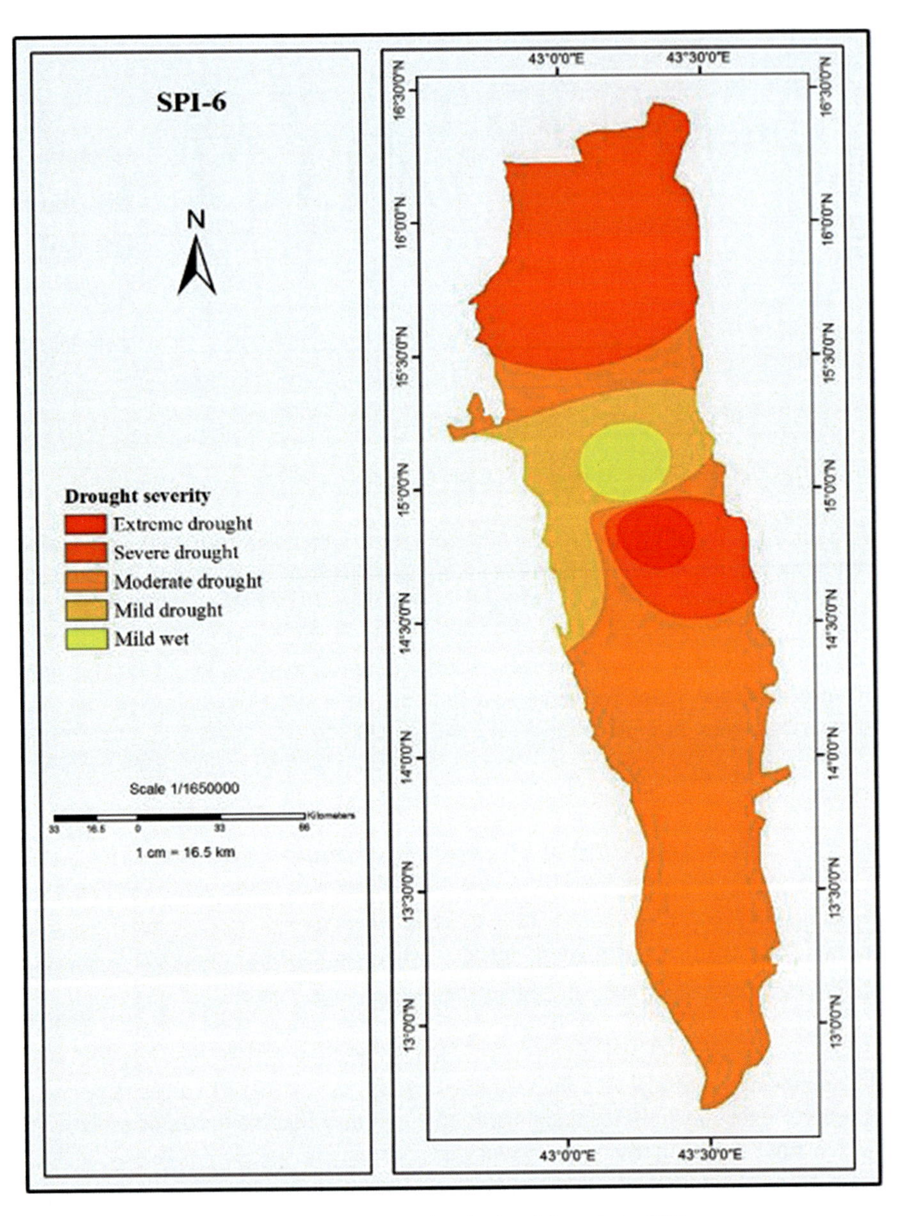

Fig. 16.13 Spatial distribution and severity of agricultural drought in 1991

drought intensity, by kalifah with 11 dry months and SPI values of – 2.56 and Bajile, with 12 dry months but – 2.33 SPI values which is lower than that of kalifah. However, all the three stations experienced extreme drought during the year. On the other hand, Jerbeh which experienced drought in 2004, recorded normal rainfall in 2005. Hodeidah also recorded normal rain in 2005 (Figs. 16.16 and 16.17).

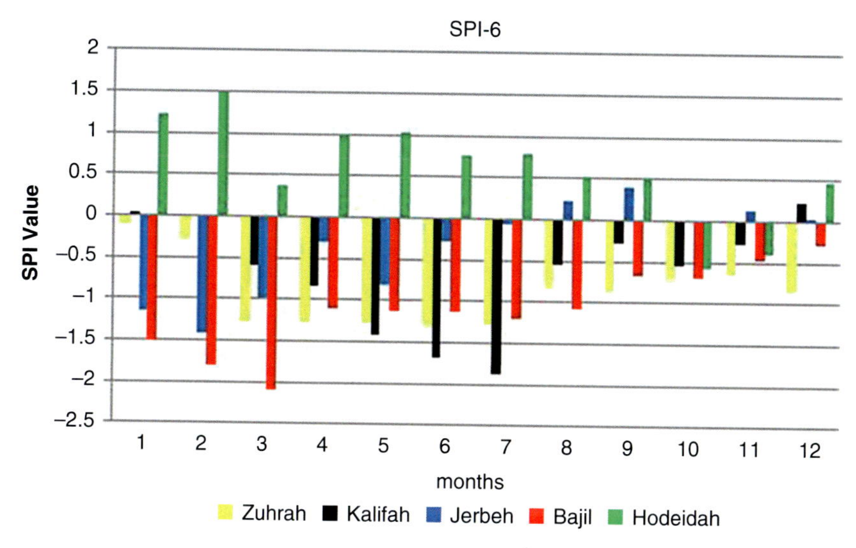

Fig. 16.14 Distribution and intensity of agricultural drought in 2003

16.8 General Agricultural Drought Frequency

The overall drought frequency in all the 5 stations indicate that on the average, near normal condition accounts for 72% in all the 5 stations put together for the 30 years period of investigation. Moderate drought account for 7%, severe drought 4% and extreme drought 2%. That is the cumulative drought period for the stations accounts for 11%. On the other hand, moderate wet for all the stations accounts for 11%, very wet, 3% and extreme wet, 1%. However, a more detailed examination of the result revealed that, out of the total of 1860 months observed in all the 5 stations, near normal condition is recorded in 1341 observations, moderate drought 136 observations, severe droughts 64 times and extreme drought 38 observations. On the other hand, extreme wet is recorded 21 observations in all the 5 stations, very wet, 58 observations and 202 moderate wet (Table 16.8 and Fig. 16.18).

16.8.1 Severity of Agricultural Drought

Dividing the study period into three phases of 10 years interval each (1980–1989, 1990–1999 and 2000–2010) revealed that the severity of drought is higher during the last phase. Generally, the area that is not affected by the drought decreases from over 60% between 1980 and 1989, to less than 38% between the years 2000 and 2010. Similarly, there is a progressive increase in the area under mild drought from close to 32% during the 1980–1989 period, to over 41% during the 2000–2010 period. This shows a transition of no drought areas to mild drought. Also, moderate drought areas

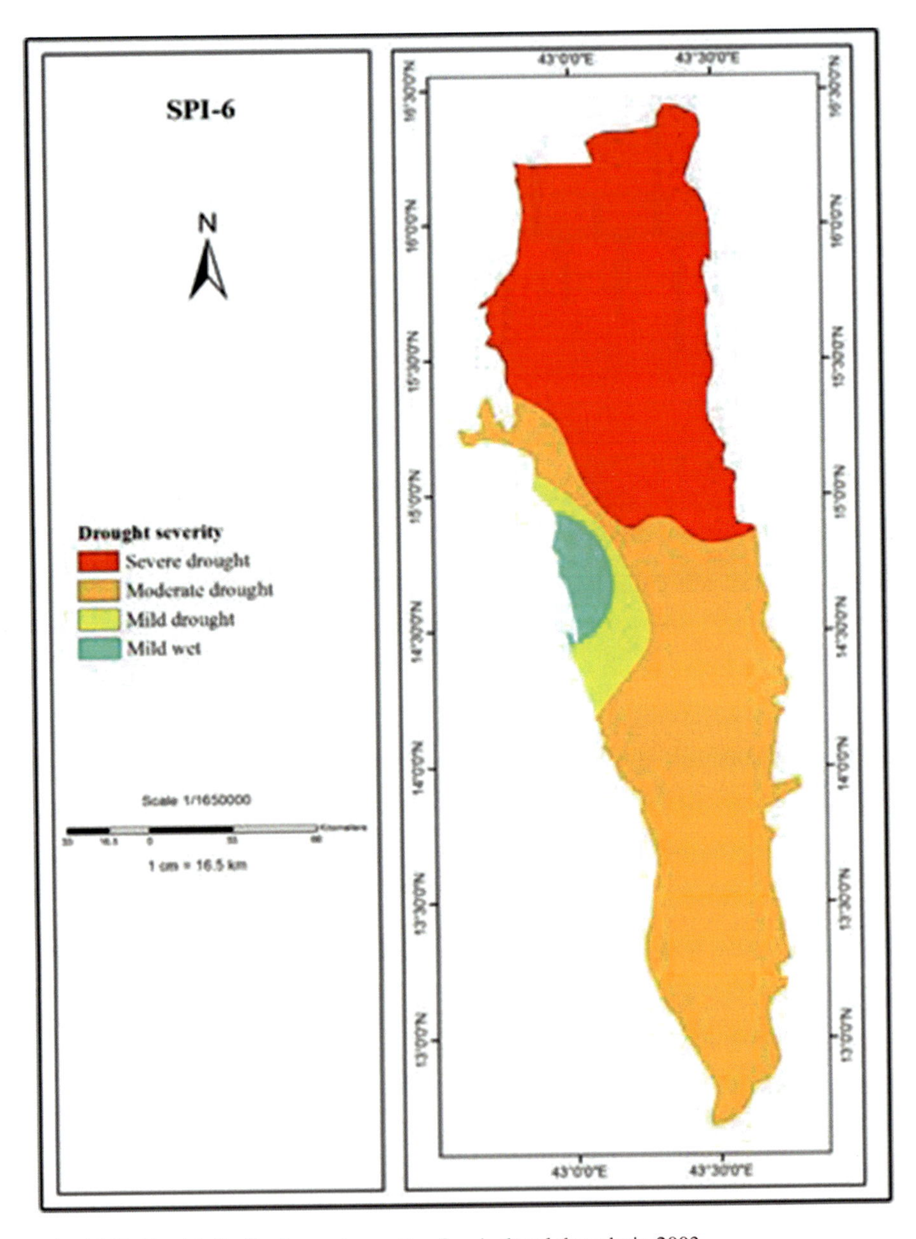

Fig. 16.15 Spatial distribution and severity of agricultural drought in 2003

progressively increased from 4.5% during the 1980–1989 period, to 4.83 in the 1990–1999 period and to over 12% during the 2000–2010 period. Severe drought also progressively increased from less than 2% from 1980 to 1989, to about 6% between 2000 and 2010. Finally, extreme drought areas also recorded a progressive

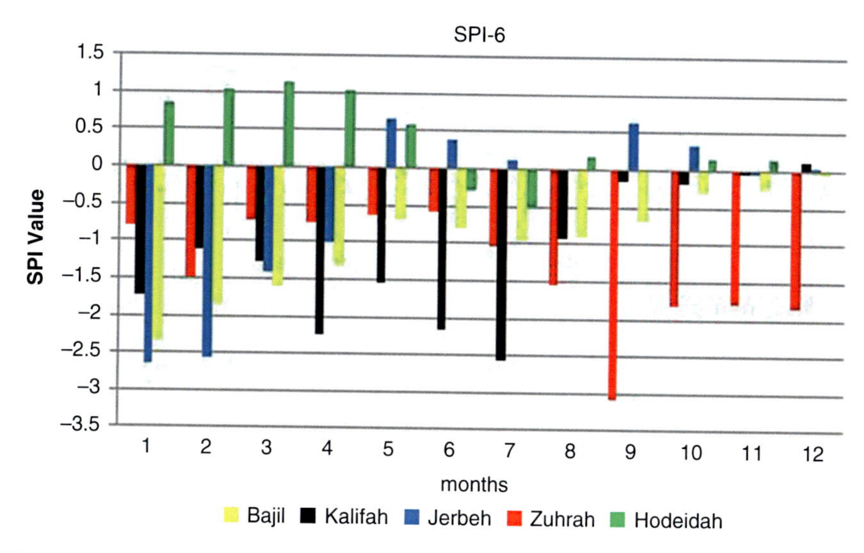

Fig. 16.16 Distribution and intensity of agricultural drought in 2005

increased from a little over 1% between 1980 and 1989, to almost 3% between 2000 and 2010 (Table 16.9 and Fig. 16.19).

16.9 Conclusion

The analysis of the above environmental indicators clearly depicts an increasing trend in the drought intensity in the study area. NDVI shows progressive decline in vegetation cover and gradual transition of areas from mild to moderate drought (64%) and from moderate to severe drought (26%). Within the study period, the area also recorded 74% increase in sand dunes deposit. These environmental indicators therefore confirm the assertion that, drought intensity is on the increase in the study area. Through temporal analysis of drought using SPI 6, it is clear that there is variation in the severity of drought between meteorological stations under study, which the SPI values were less than (−1) during the drought years. Bajil station is the most affected by the drought with a total of 10 drought years from 1980 to 2010, while Jerbeh station is the least affected with a total of 5 drought years. The analysis results by SPI 6 showed that the year1991 recorded the worst drought throughout the study period both in terms of distribution and intensity.

Moreover, unless both the individuals put concerted effort and measures in place, as well government and private organisations to arrest this ugly trend, it will continue to pose a major threat and challenges to both food security and overall development of the country. Finally, it is recommended that, the government should seriously look into the drought issue through improving resource management practices and development of drought assessment and implementation unit to help

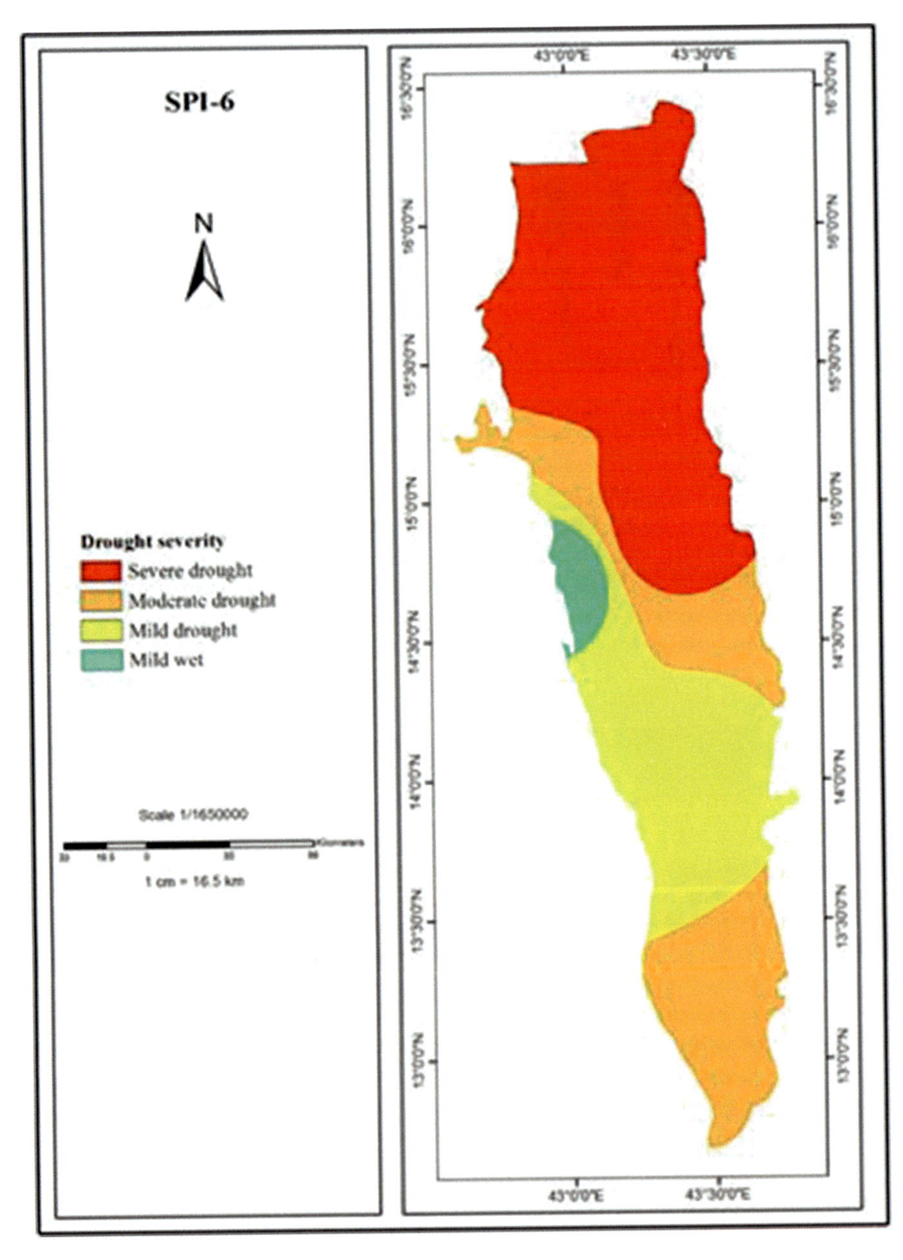

Fig. 16.17 Spatial distribution and severity of agricultural drought in 2005

minimize the adverse effects of drought. In doing this, concerned authorities should work closely with stakeholders who might be directly or indirectly affected by drought. Socio-economic data should also be taken into consideration when assessing drought risk to better understand the vulnerable groups. Based on the

Table 16.8 General frequency of SPI-6 agricultural drought

	Category						
Stations	EW	VE	MW	NN	MD	SD	ED
Zuhrah	9	14	36	264	32	13	4
Kalifah	4	9	34	277	22	15	11
Jerbeh	2	14	35	267	26	11	17
Bajil	1	7	49	242	42	25	6
Hodeidah	5	14	48	291	14	0	0
Total	21	58	202	1341	136	64	38
Percentage %	1%	3%	11%	72%	7%	4%	2%

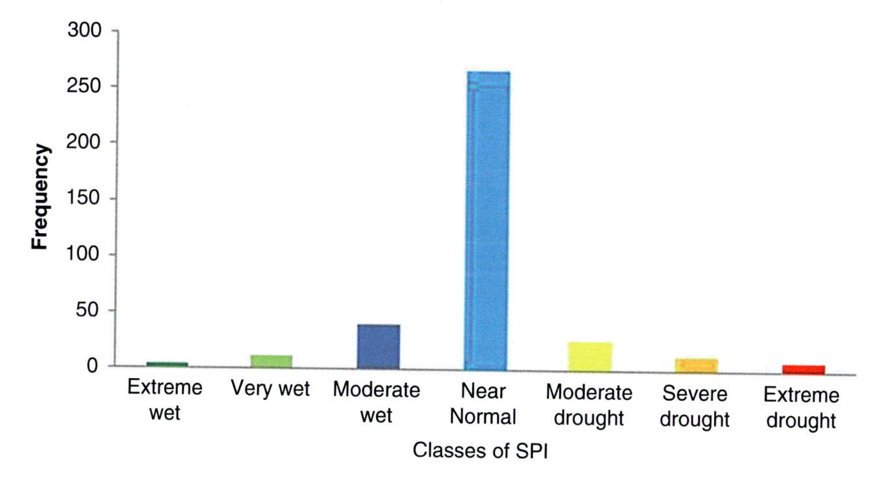

Fig. 16.18 Frequency of mean annual of SPI-6 classes

Table 16.9 General severity of SPI-6 agricultural drought

	Category				
Period	No drought %	Mild drought %	Moderate drought %	Severe drought %	Extreme drought %
1980–1989					
1990–1999	62.17	28.67	4.83	2.33	2
2000–2010	37.87	41.21	12.12	5.93	2.87

fact that the study area is the most important agricultural areas in Yemen, it is recommended to study the drought and its impact on agricultural crops in the area.

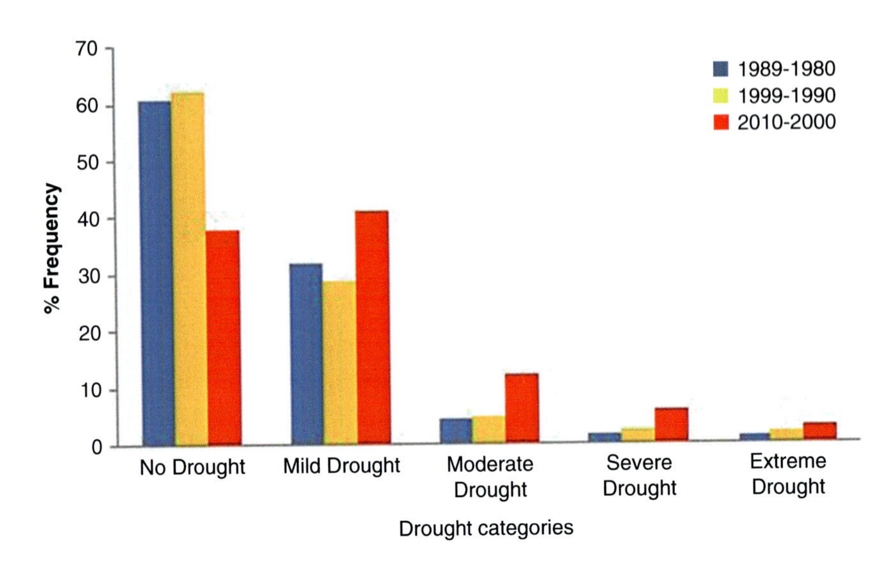

Fig. 16.19 Classification of SPI-6 agricultural drought severity for 30 years

References

Abdullah, Y. A. (2010). Environmental problems associated with climate in the middle part of Tihama plain in Yemen republic using GIS. Ph.D. Dissertation, Department of Geography, Faculty of Arts, Cairo University, Egypt.

Al-Washli, Y. A. 1989. *AL-Hodeidah Governorate – geographical study*. Master thesis, Department of Geography, Faculty of Arts, Sana'a University.

AMS. (2004). Statement on meteorological drought. *Bulletin of the American Meteorological Society, 85*, 771–773.

Andreadis, K. M., Clark, E. A., Wood, A. W., Hamlet, A. F., & Lettenmaier, D. P. (2005). Twentieth-century drought in the conterminous United States. *Journal of Hydrometeorology, 6*(6), 985–1001.

Assage, A. M. (2003). Desertification in the Tihama plain. *Journal of Yemen Geographic Society, 2*, 21.37.

Bogan, N. (2014). *Effective drought Management for Sustained Livelihoods in the Middle East. Master thesis, Nicholas School of the Environment*. Duke University.

Cavatassi, R., Lipper, L., & Narloch, U. (2011). Modern variety adoption and risk management in drought prone areas: Insights from the sorghum farmers of eastern Ethiopia. *Agricultural Economics, 42*(3), 279–292.

Ceglar, A., Zalika, C., & Lucka, K. B. (2008). Analysis of meteorological drought in Slovenia with two drought indices. *Proceedings of the BALWOIS, 2008*, 27–31.

Dhaifallah. A. A. (2012). *Spatial variation of desertification in Hodeidah Governorate*. Master thesis, Department of Geography, Faculty of Arts, Thamar University.

Dinar, A., & Keck, A. (2000). *Water supply variability and drought impact and mitigation in sub-Saharan Africa*. In D. A. Wilhite (Ed.), *Drought: A global assessment* 2 (pp. 129–148).

Eklund, L., & Seaquist, J. (2015). Meteorological, agricultural and socioeconomic drought in the Duhok Governorate, Iraqi Kurdistan. *Natural Hazards, 76*(1), 421–441.

El Kharraz, J., El-Sadek, A., Ghaffour, N., & Mino, E. (2012). Water scarcity and drought in WANA countries. *Procedia Engineering, 33*, 14–29.

Erian, W. (2011). *Drought vulnerability in the Arab region. Case study—Drought in Syria. Ten years of scarce water (2000–2010)*. Arab Center for the Studies of Arid Zones and Dry Lands and the United Nations, Secretariat of the International Strategy for Disaster Reduction.

Erian, W. (2012). *Drought vulnerability in the Arab region special case study: Syria*. (Geneva: United Nations International Strategy for Disaster Risk Reduction, 2010). Shahrzad Mohtadi,"*Climate Change and the Syrian Uprising*," Bulletin of the Atomic Scientists, 5.

Eriyagama, N., Smakhtin, V. Y. & Gamage, N. 2009. *Mapping drought patterns and impacts: A global perspective*. International Water Management Institute (IWMI) 133.

ESCWA. (2005). *Water development report 1: Vulnerability of the region to socioeconomic drought*. United Nation.

ESCWA/UN-DESA. (2013). Drought management planning in water-scarce and in transitioning-settings countries in West Asia/North Africa. https://sustainabledevelopment.un.org/content/documents/2156Drought_EGM_Working_Draft_Background_Paper_21June13.pdf

Forster, P., Jackson, L., Lorenz, S., Simelton, E., Fraser, E., & Bahadur, K. (2012). *Food security: Near future projections of the impact of drought in Asia*. The Centre for Low Carbon Futures.

Gemenne, F. (2011). Climate-induced population displacements in a 4 C+ world. *Philosophical Transactions of the Royal Society of London A: Mathematical, Physical and Engineering Sciences, 369*(1934), 182–195.

Guha-Sapir, D., Hargitt, D., & Hoyois, P. (2004). *Thirty years of natural disasters 1974–2003: The numbers*. Centre of Research on the Epidemiology od Disasters. Presses University. ISBN 2-930344-71-7

Hazell, P. (2007). *Managing drought risks in the low-rainfall areas of the Middle East and North Africa*. Cornell University.

Kaniewski, D., Van Campo, E., & Weiss, H. (2012). Drought is a recurring challenge in the Middle East. *Proceedings of the National Academy of Sciences, 109*(10), 3862–3867.

Karrou, M., & El Mourid, M. (2008). Drought management and planning strategies in semi-arid and arid agro-pastoral systems of West Asia and North Africa: A review. *Options Méditerranéennes: Série A. Séminaires Méditerranéens, 80*, 179–184.

Keyantash, J. A., & Dracup, J. A. (2004). An aggregate drought index: Assessing drought severity based on fluctuations in the hydrologic cycle and surface water storage. *Water Resources Research, 40*(9), 1–13.

Kirono, D. G. C., Kent, D. M., Hennessy, K. J., & Mpelasoka, F. (2011). Characteristics of Australian droughts under enhanced greenhouse conditions: Results from 14 global climate models. *Journal of Arid Environments, 75*(6), 566–575.

Kogan, F. N. (1997). Global drought watch from space. *Bulletin of the American Meteorological Society, 78*(4), 621–636.

Linsley, R. K., Jr., Kohler, M. A., & Spilham, J. L. H. (1959). *Applied hydrology*. McGraw Hill.

Matson, E., & Bart, D. (2013). Interactions among fire legacies, grazing and topography predict shrub encroachment in post-agricultural pa'ramo. *Landscape Ecology, 28*(9), 1829–1840.

McKee, T. B., Doesken, N. J., & Kleist, J. (1993). The relationship of drought frequency and duration to time scales. In *Eighth conference on applied climatology*, 17–22 January 1993.

Ministry of Agriculture & Irrigation. (2009). *Yearbook agricultural census, Yemen*. Retrieved from: http://www.agriculture.gov.ye/?lng=english&

Mishra, A. K., & Singh, V. P. (2010). A review of drought concepts. *Journal of Hydrology, 391*(1), 202–216.

Miyan, M. A. (2015). Droughts in Asian least developed countries: Vulnerability and sustainability. *Weather & Climate Extremes, 7*, 8–23.

Myers, N., & Kent, J. (2001). Food and hunger in sub-Saharan Africa. *Environmentalist, 21*(1), 41–69.

Myers, N. (2002). Environmental refugees: A growing phenomenon of the 21st century. *Philosophical Transactions of the Royal Society B: Biological Sciences, 357*(1420), 609–613.

Noaman, A., Petersen, G., Kiesel, J., & Wade, S. (2013). Climate change impacts on water resources in Yemen. *Journal of Earth Science and Engineering, 3*(9), 629–638.

Palmer, W. C. (1968). Keeping track of crop moisture conditions, nationwide: The new crop moisture index. *Weatherwise, 21*(4), 156–161.

Pozzi, W., Sheffield, J., Stefanski, R., Cripe, D., Pulwarty, R., Vogt, J. V., & Van Dijk, A. I. (2013). Toward global drought early warning capability: Expanding international cooperation for the development of a framework for monitoring and forecasting. *Bulletin of the American Meteorological Society, 94*(6), 776–785.

Raziei, T., Saghafian, B., Paulo, A. A., Pereira, L. S., & Bordi, I. (2009). Spatial patterns and temporal variability of drought in Western Iran. *Water Resources Management, 23*, 439–455.

Retrieved from: http://s3.amazonaws.com/zanran_storage/www.escwa.un.org/ContentPages/17951993.pdf

Retrieved from: http://pdf.usaid.gov/pdf_docs/pnaal964.pdf

Retrieved from: http://reliefweb.int/sites/reliefweb.int/files/resources/ifpridp01139.pdf

Retrieved from: http://www.escwa.un.org/information/publications/edit/upload/sdpd-05-9-e.pdf

Retrieved from: http://www.lowcarbonfutures.org/sites/default/files/Food%20Security%20%20Near%20future%20projections%20of%20the%20impact%20of%20drought%20in%20Asia.pdf

Retrieved from: https://searchworks.stanford.edu/view/3929918

Rossi, G. (2003). An integrated approach to drought mitigation in Mediterranean regions. In *Tools for drought mitigation in Mediterranean regions* (Vol. 44, pp. 3–18). Springer.

Roy, A. K., & Hirway, I. (2007). *Multiple impacts of droughts and assessment of drought policy in major drought prone states in India*. Centre for Development Alternatives.

Shahid, S., & Behrawan, H. (2008). Drought risk assessment in the western part of Bangladesh. *Natural Hazards, 46*(3), 391–413.

Taher, F. (2004). Desertification in Yemen. *Journal of Yemen Geographical Society*, (2), 67–81.

Tsubo, M., Fukai, S., Basnayake, J., & Ouk, M. (2009). Frequency of occurrence of various drought types and its impact on performance of photoperiod-sensitive and insensitive rice genotypes in rainfed lowland conditions in Cambodia. *Field Crops Research, 113*(3), 287–296.

U.S. (1982). *Agricultural sector assessment in Yemen Arab Republic*. Agency for International Development & Ministry of Agriculture of Yemen.

United Nations. (2003. *Population and development report. First issue: Water scarcity in the Arab world*. Economic and Social Commission for Western Asia. ISBN. 92-1-128268-3.

USDA. (1986). *Middle East and North Africa. Situation and outlook report* (Vol. III). United States Department of Agriculture, Economic Research Service.

Verner, D. (2012). *Adaptation to a changing climate in the Arab countries: A case for adaptation governance and leadership in building climate resilience*. World Bank Publications.

Vidal, J. P., & Wade, S. (2009). A multimodal assessment of future climatological droughts in the United Kingdom. *International Journal of Climatology, 29*(14), 2056–2071.

Vogelmann, J. E., Xia, G., Homer, C. & Tolk, B. (2012). Monitoring gradual ecosystem change using landsat time series analysis: Case studies in selected forest rangeland ecosystem. *Remote Sensing of Environment, 122*, 92–105.

Wiebelt, M., Breisinger, C., Ecker, O., Al-Riffai, P., Robertson, R., & Thiele, R. (2011). *Climate change and floods in Yemen*. International Food Policy Research institute (IFPRI), IFPRI Discussion Paper 01139 December 2011.

Wilhite, D. (2000). Drought as a natural hazard: Concepts and definitions. In D. A. Wilhite (Ed.), *Drought: A global assessment 1* (pp. 3–18). London.

Wilhite, D. A. (2008). Drought monitoring as a component of drought preparedness planning. In A. Iglesias, A. Cancelliere, D. A. Wilhite, L. Garrote, & F. Cubillo (Eds.), *Coping with drought risk in agriculture and water supply systems* (pp. 3–19). Springer.

WMO. (2012). *Standardized precipitation index user guide*. (M. Svoboda, M. Hayes, & D. Wood (Eds.)). (WMO-No. 1090), Geneva.

World Bank. (2014. *Natural disasters in the Middle East and North Africa: A regional overview*. Disaster Risk Management Unit Middle East and North Africa, Washington, DC 20433. Retrieved from: http://www.unccleam.org/sites/default/files/inventory/wb164.pdf

Zhao, X., Hu, H., Shen, H., Zhou, D., Zhou, L., Myneni, R. B., & Fang, J. (2015). Satellite-indicated long-term vegetation changes and drives in the Mongolian Plateau. *Landscape Ecology, 30*(9), 1599–1611.

Chapter 17
Assessment of Drought Impact on Agricultural Production Using Remote Sensing and Machine Learning Techniques in Kairouan Prefecture, Tunisia

Mohamed Kefi, Tien Dat Pham, Nam Thang Ha, and Kashiwagi Kenichi

Abstract Drought is a climate-induced disaster that can occur in all climatic regions with its features that varies considerably from one country to another. In addition, this natural disaster has wide-ranging impacts on water resources, ecosystems, energy, agriculture, forestry, human health and food security. Indeed, drought affects many economic activities and people. Its impacts are particularly serious in food-deficit countries with high dependence on subsistence agriculture. In Tunisia as many countries in MENA region, drought has a great effect on water problems in particular. It may conduct to increase the problem of water shortage and to affect agricultural productivities. For this reason, the Tunisian government encouraged and supported irrigated agriculture as a way to reduce the risk of drought. In this context, the improvement of water access and the irrigation for high value agricultural products were developed. Therefore, this work aims at predicting agricultural drought with the focus on its impact on olive growing farms using remote sensing data and advanced machine learning techniques. The approach developed in the current work is based on the use of multi-temporal medium resolution Landsat 8 OLI and TIRS imagery and the ensemble-based machine learning techniques for

M. Kefi (✉)
Laboratory of Desalination and Natural Water Valorisation, Water Research and Technologies Centre of Borj Cedria (CERTE), Soliman, Tunisia
e-mail: mohamed.kefi@certe.rnrt.tn

T. D. Pham
Department of Earth and Environmental Sciences, Macquarie University, Macquarie Park, NSW, Australia
e-mail: tiendat.pham@mq.edu.au

N. T. Ha
Faculty of Fisheries, University of Agriculture and Forestry, Hue University, Hue, Vietnam
e-mail: hanamthang@hueuni.edu.vn

K. Kenichi
Alliance for Research on the Mediterranean and North Africa (ARENA)
The University of Tsukuba, Tsukuba, Ibaraki, Japan
e-mail: kashiwagi.kenichi.fn@u.tsukuba.ac.jp

M. M. Al Saud (ed.), *Applications of Space Techniques on the Natural Hazards in the MENA Region*, https://doi.org/10.1007/978-3-030-88874-9_17

improving estimates of the drought indicators such as vegetation health index (VHI). The root-mean-square error (RMSE) and coefficient of determination (R2) and the cross-validation (CV) technique were employed to evaluate the performance of the proposed model. The capability of the proposed model was evaluated and compared with other machine learning algorithms, i.e., the random forests (RF), the support vector machine (SVM). The findings shows that the proposed model performed well (R^2 = 0.84, RMSE =1.38) and outperformed the remaining algorithms. It can be concluded that multi-temporal optical and thermal remote sensing combine with an advanced machine learning technique can be accurately used to estimate drought in semi-arid land areas. This approach may conduct to sustainable strategies of water resources in drought region.

Keywords Olive growing farms · Landsat 8 · Remote sensing · Drought indicators · Machine learning

17.1 Introduction

Drought is a natural disaster which has been defined as a prolonged lack or serious shortage of precipitation (Heim, 2002). Four types of drought can be identified: (1) meteorological drought, (2) hydrological drought, (3) agricultural drought, and (4) socioeconomic drought. Additionally, this natural disaster is characterized by three scopes: severity/intensity, duration and spatial distribution (Zargar et al., 2011). This phenomenon has a slow development and aggravates progressively. Its effects accumulate slowly over a long period of time (Wilhite, 1993). For this reason, its impact is serious as it has economic, environmental and social effects (Peters et al., 2002). It can also result in destitution, hunger and death (Delbiso et al., 2017). In 2019, it was estimated that about 29.2 Millions of people were affected by drought (EM-Dat, 2019). Drought is a complex and unwell understood phenomenon that affects people more than any other natural hazard (Wilhite, 1993). Additionally, the complexity of drought phenomenon may conduct to some difficulties for monitoring and quantification. In this context several drought indices were developed (Niemeyer, 2008; Zagar et al. 2011). Three groups of drought indices were considered meteorological, agricultural and hydrological (Niemeyer, 2008). However, Satellites may also contribute to assess drought. In fact, Satellite-derived information has become a necessary source of information for all phases of disaster risk management (Enenkel et al., 2020). In this context, remote sensing approaches are important and efficient tools in monitoring drought and detecting drought zone of water stress as they provide real-time spatial observations of several atmospheric and land surface variables that can be used to estimate evapotranspiration, soil moisture, and vegetation conditions (Kefi et al., 2016). Additionally, monitoring ground drought measuring parameters such as vegetation index and land surface temperature can be detected using remote sensing techniques at a large scale (Bhuiyan, 2010; Corbari et al., 2008).

In this study, we attempted to develop a novel artificial intelligence approach using advanced decision tree ensemble-based learning for predicting drought indicator for the study area. Literature review shows that advanced machine learning and optimization techniques have been widely used for environmental monitoring such as flash flooding prediction (Ngo et al., 2020, 2021), forest above-ground biomass and carbon stocks estimation (Pham et al., 2020, 2021). However, only a few studies attempted to develop a machine learning-based approach for predicting drought in Tunisia and MENA region. In Tunisia, droughts are a recurrent phenomenon and it has become more frequent. Drought occured in several years which were characterized by significant decreased rainfall (Henia, 2003). In 1999, Tunisia elaborated the first guideline of drought management. This guideline is mainly emphasized on the identification of the major drought indices, preparedness, intervention process and implementation measures (Louati et al., 2007). Additionally several institutions are involved in assessment of droughts and the implementation of appropriate measures. The most severe drought in 50 years appear for three consecutive years from 2000 to 2002. The cost of drought was estimated to about US$54 million. Cereal production which are used to maintain food security was badly affected. This problem of production conducted to augmented imports and a decline in overall economic growth (Verner et al., 2018). In addition, water resource in dams was reduced as well which have a high impact on agriculture production. For this reason, several strategies to preserve water were implemented and irrigation were developed to maintain food security and to enhance productivity in drought period. Facing of the problem of drought in Tunisia and due to its significant effects, a precise accuracy of drought prediction still remains a challenge. Thus, this work aims to fill the gaps in the current literature by integrating the multi-temporal multispectral and thermal Landsat remotely sensed data and the historical data in the proposed machine learning techniques for improving the accuracy of drought prediction and monitoring.

This research can be useful to assess temporal and spatial aspects of vegetation growing conditions and drought on one hand and to develop appropriate response against drought on the other hand.

Indeed, this research is organized as follows. Section 17.2 introduces the study area and describes the approach and data applied for the analysis in details. The main findings related to drought assessment are presented and discussed in Sects. 17.3 and 17.4. Finally, the conclusion section highlights the main outcomes of the work.

17.2 Material and Methods

17.2.1 Description of Study Areas

In order to assess the drought intensity and to determine its effect on agriculture, some olive growing farms of Kairouan was selected as study area. Kairouan is considered as one the hottest prefecture in Tunisia and it is also famous for olive

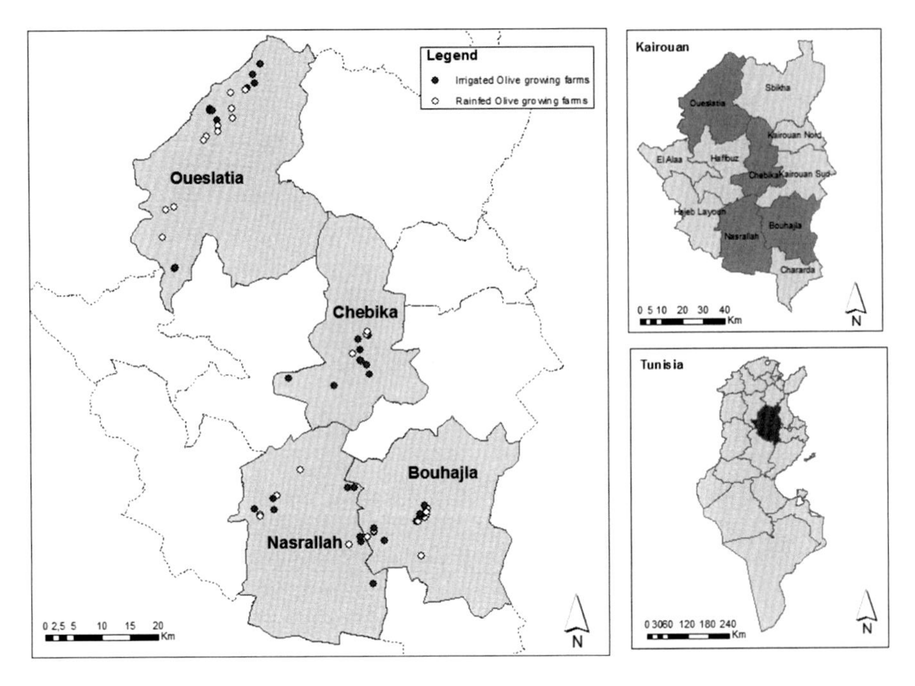

Fig. 17.1 Study areas location

production. In fact, this product is considered as the most important one in Tunisia. It has not only economic implication but also social and tradition impacts. This study was conducted in olive growing farms of Kairouan prefecture. Regarding to several factors such as cultivars, production and access to water, 4 regions (Oueslatia, Chebika, Nasrallah and Bouhajla) are selected as study areas (Fig. 17.1). About 65 rainfed and irrigated farms were visited during the field survey.

17.2.2 Materials

17.2.2.1 Data Acquisition

Multi-temporal Landsat 8 OLI optical imagery at 30 m spatial medium resolution data were used to calculate temporal changes of NDVI (Normalized Difference Vegetation Index) and RVI (Ratio Vegetation Index) and multi-temporal Lansat 8 TIRS thermal bands at 100 m spatial resolution were employed to generate the LST (Land Surface Temperature) for 2015. The five available cloud free scenes selected from June, July, August and September 2015 were included in this research. The remotely sensed images were acquired by two different satellite sensors of LANDSAT 8 including LANDSAT OLI and LANSAT 8 TIRS (Table 17.1).

Table 17.1 Acquired satellite remote sensing data

Satellite sensor	Date of acquisition	Pixel size	Path/row	Spectral resolution	Band used
Landsat 8 OLI	15-June	30 m	191/35	Multi-spectral (8 bands)	4, 5
	1-July				
	17-July				
	18-Aug				
	3-Sep				
Landsat 8 TIRS	15-June	100 m	191/35	Thermal (2 bands)	10, 11
	1-July				
	17-July				
	18-Aug				
	3-Sep				

LANDSAT 8 OLI and LANDSAT 8 TIRS images were acquired from the USGS, NASA from glovis (http://glovis.usgs.gov/).

17.2.2.2 Field Survey Data

The field data was collected during several field trips carried out throughout the year 2014/2015. The main purpose of these field trips was the collection of information on olive farms (irrigated and rainfed farms). Four regions (Ouesseltia/Chebika/Nasrallah/Bouhajla) of Kairouan prefecture were visited and we recorded ground-truth points (GTPs) using GPS (Global Positioning Systems) to identify locations of different olive farms types. In fact, a total of 65 locations were collected and recorded during the field survey.

17.2.3 Methodology

17.2.3.1 RVI and NDVI Calculation

The DN (Digital Number) value of the spectral NIR and Red bands were converted into surface reflectance value (range from 0.0 ~ 1.0) at top of atmosphere using the calibrating tool in ENVI 5.2, and then, the NDVI value and RVI were obtained by the following formulas (Eq. 17.1) and (Eq. 17.2) (USGS, 2015).

$$\mathrm{NDVI} = \frac{\rho 4 - \rho 3}{\rho 4 + \rho 3} \tag{17.1}$$

and

$$RVI = \frac{\rho 4}{\rho 3} \quad (17.2)$$

Where: $\rho 4$ and $\rho 3$ are the TOA reflectance for the near infrared band (NIR, band 5) and red band (band 4) of Landsat 8 OLI, respectively. The NDVI values range between -1 and 1.

17.2.3.2 BT and LST

We analyzed temporal changes of temperature in the olive farms using two thermal bands (band 10 and band 11) of the Landsat 8 TIRS sensor for calculating BT (Brightness Temperature) and LST (Land Surface Temperature) for five periods between June and September in 2015. A number of steps were computed using ArcGIS 10.5 to transfer the DN (Digital Number) to BT and LST.

Firstly, DN was converted from spectral radiance to BT using formula (Eq. 17.3) as suggested by the USGS (2015).

$$BT = K2/\ln(K1/L\lambda + 1) \quad (17.3)$$

Where:

BT: Top of Atmosphere Brightness Temperature in Kelvin
L_λ: Spectral radiance (Watts/(m^2*sr*μm)
K1: Thermal conversion constant for the band (K1 = 774.89 and 480.89 for band 10 and band 11, respectively from the metadata in the header file)
K2: Thermal conversion constant for the band (K2 = 1321.08 and 12001.14 for band 10 and band 11, respectively from metadata in the header file)

Secondly, LST in Kelvin retrieval from Landsat 8 TIRS was calculated using formula (Eq. 17.4)

$$LST_K = BT/[1 + (\lambda * BT/\rho)\ \ln \varepsilon] \quad (17.4)$$

Where:

LST_K: Land Surface Temperature in Kelvin
λ: Wavelength of emitted radiance (λ = 10.8 and 12 for band 10 and band 11, respectively)
$\rho = h*c/\sigma$ (1.438×10^{-2} m.K) = 14,380
ε: emissivity which is given by $\varepsilon = 1.009 + 0.047 \ln(NDVI)$

Lastly, LST in Kelvin is converted to Celsius by using formula (Eq. 17.5)

$$LST = LST_K - 273.15 \quad (17.5)$$

17.2.3.3 VCI, TCI and VHI

Three important indexes for drought indicators such as the Vegetation Condition Index (VCI), the Temperature Condition Index (TCI), and the Vegetation Health Index (VHI) were calculated using the following equations (Eqs. 17.6, 17.7 and 17.8) (Kogan et al., 2004; Unganai & Kogan, 1998)

$$VCI = 100*(NDVI - NDVI_{min})/(NDVI_{max} - NDVI_{min}) \quad (17.6)$$

$$TCI = 100*(LST_{max} - LST)/(LST_{max} - LST_{min}) \quad (17.7)$$

$$VHI = 0.5*VCI + 0.5*TCI \quad (17.8)$$

The VHI index of the study sites have been classified in Table 17.2 (Ghaleb et al., 2015). The final VHI index was reclassified between 0 and 100 using ArcGIS10.4. Then, we estimated the drought indicator based on the predicted value of VHI for the entire Kairouan prefecture using advanced machine learning techniques in Sect. 17.2.3.4.

The VCI depicts vegetation dynamics and represents the moisture condition changes. The higher values indicate the better vegetation conditions (Kogan, 1995). In contrast, the lower values corresponds the vegetation stress due to high temperature and drought condition (Bhuiyan, 2008). The index plays a crucial role in identifying the location of olive farms and distinguish irrigated areas.

In order to eliminate cloud, water, snow or ice, urban areas and bare soil, we used a threshold of a NDVI value as follows: If NDVI value $<0.1 \rightarrow$ Value = NODATA.

If NDVI value $>0.1 \rightarrow$ same value applies

Where 0.1 is the threshold value for vegetated area including olive farms. By checking the Google Earth and pixels extracted from GTP (Ground Truth Points) collected from field survey, we determined this threshold value. This threshold is similar the threshold provided by the NOAA which indicated that the NDVI at a very low values of NDVI (0.1 and below) correspond to barren areas of rock, sand, or snow. Moderate values represent shrub and grassland (0.2 to 0.3), while high values indicate temperate and tropical rainforests (0.6 to 0.8) (http://www.ospo.noaa.gov/Products/land/gvi/NDVI.html).

In the current work, agricultural drought indicators are predicted using the LST, the BT, and vegetation health indices (VHI) generated from Landsat 8 TIRS data and vegetation indices such as NDVI and RVI generated from the multi-temporal

Table 17.2 Drought classification of VHI values

Drought class	Values
Severe drought	<20
Moderate drought	<30
Mild drought	<40
No drought	≥40

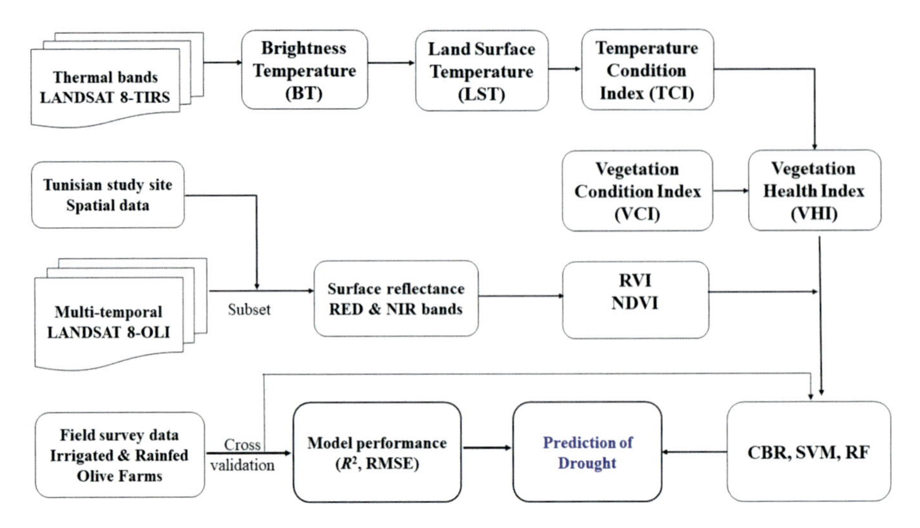

Fig. 17.2 Flow chart used to predict agricultural drought using multi-temporal optical and thermal remote sensing data

Landsat 8 OLI multispectral data (Deo & Şahin, 2015; Kefi et al., 2016; Tran et al., 2017; Valipour, 2016).

The flowchart to predict agricultural drought using remote sensing and machine learning approach is shown in Fig. 17.2.

17.2.3.4 Machine Learning Models

Catboost Regression (CBR)

The CBR is an advanced gradient boosting algorithm, which has recently been developed by Dorogush et al. (2018). This technique can handle data with categorical features and minimize the over-fitting problem by selecting the best tree structure for calculating leaf values (Dorogush et al., 2018; Prokhorenkova et al., 2018). The algorithm is a powerful ML technique that is recently implemented and released as an open-source library. This algorithm achieves superior results in both classification and regression problems (Pham et al., 2021) by the implementation of ordered boosting, which is a modification of standard boosted decision tree algorithms (Dorogush et al., 2018). In the CBR model, the random permutations of the training set and the gradients utilized for selecting an optimal tree structure are generated to enhance the robustness and prevent overfitting phenomenon of the model.

Support Vector Machine (SVM)

The SVM algorithm is a supervised learning technique, which was proposed by Vapnik (2013) based on a statistical learning theory and the kernel-based approach. This method is able to handle non-linear and in solving classification and regression tasks. The SVR converts a non-linear data space into a higher dimensional feature space through a kernel function. The key benefit of SVR is that it is able to produce high prediction accuracy with small numbers of training samples (Mountrakis et al., 2011).

Random Forests (RF)

The RF (Breiman, 2001) is currently the most well-known bagging algorithm, which employs ensemble decision trees and works effectively for regression and classification tasks. This algorithm begins with a range of bootstrap samples from the original dataset, and then each regression tree chooses the best split among the selected features. In the RF model, two-thirds of the training samples are assigned as in-bag data, and the remaining ones are used as out-of-bag (OOB) data. The OOB data can be used to generate the feature importance. Two parameters such as the number of trees and the number of features are needed to be determined in the RFR method.

Generating Machine Learning Model

The configuration and the construction of the three ML models were implemented using fivefold cross-validation (CV) technique in Python 3.7 environment. A total of 65 samples were randomly partitioned into $k = 5$ equal sized subsamples, of which one single subsample was used as the validation dataset for testing the model, and the remaining $k - 1$ subsamples are used as training data. The cross-validation process is then repeated 5 times, with each of the k subsamples used exactly once as the validation data. The five times results can then be averaged to produce a single prediction. The advantage of this technique over repeated random sub-sampling is that all observations are employed for both the training and validation phases, and each observation is used for validation exactly once, thus providing the reliable result for assessment. In the current work, we employed the CV technique to evaluate the model performance using the standard metrics in the Scikit-learn library (Pedregosa et al., 2011).

Model Evaluation

In this study, we used two standard criteria to measure statistical errors in regression models such as R^2 and RMSE to evaluate and compare the model performance of the three ML models. These statistical measures have been widely applied in modeling environmental parameter retrievals (Navarro et al., 2019; Wang et al., 2020). These standard criteria are often used to assess the differences between the actual and the predicted drought indicator (Pham et al., 2021; Vafaei et al., 2018). As the number of samples are small, we employed the fivefold CV technique in the Scikit-learn library in Python 3.7 to evaluate the ML models' performance

$$R2(y,\widehat{y}) = 1 - \frac{\sum_{i=1}^{n}(y_i - \widehat{y}_i)^2}{\sum_{i=1}^{n}(y_i - \bar{y})^2} \tag{17.9}$$

$$\text{RMSE}\,(y,\widehat{y}) = \sqrt{\frac{1}{n_{\text{samples}}}\sum_{i=0}^{n_{\text{samples}}-1}(y_i - \widehat{y}_i)^2} \tag{17.10}$$

17.3 Results

17.3.1 Multi-Temporal Vegetation Indices Analysis

We analyzed the average NDVI and RVI values using multi-temporal vegetation indexes to compare the difference between irrigated and rainfed olive farms. Our results show that NDVI values for irrigated growing olive farms during the year 2015 are higher than those of rainfed areas. Figure 17.3 shows spatial distribution map of the average NDVI of olive farms in the study sites and Fig. 17.4a, b represented the significant differences of NDVI and RVI values between two types of olive growing farms.

17.3.2 Temporal Land Surface Temperature Analysis

Analyzing the multi-temporal land surface temperature during the year 2015, we found that the average temperature of irrigated olive farms areas were lower than that of rainfed farms (Fig. 17.5a, b). Additionally, we generated the spatial distribution of the LST for irrigated and rainfed olive farms (Fig. 17.6).

In the study areas, July seems to be the hottest month and the LST on July 17 appeared the highest date in 2015 ranging from 27.28 to 46.22 °C (Fig. 17.6).

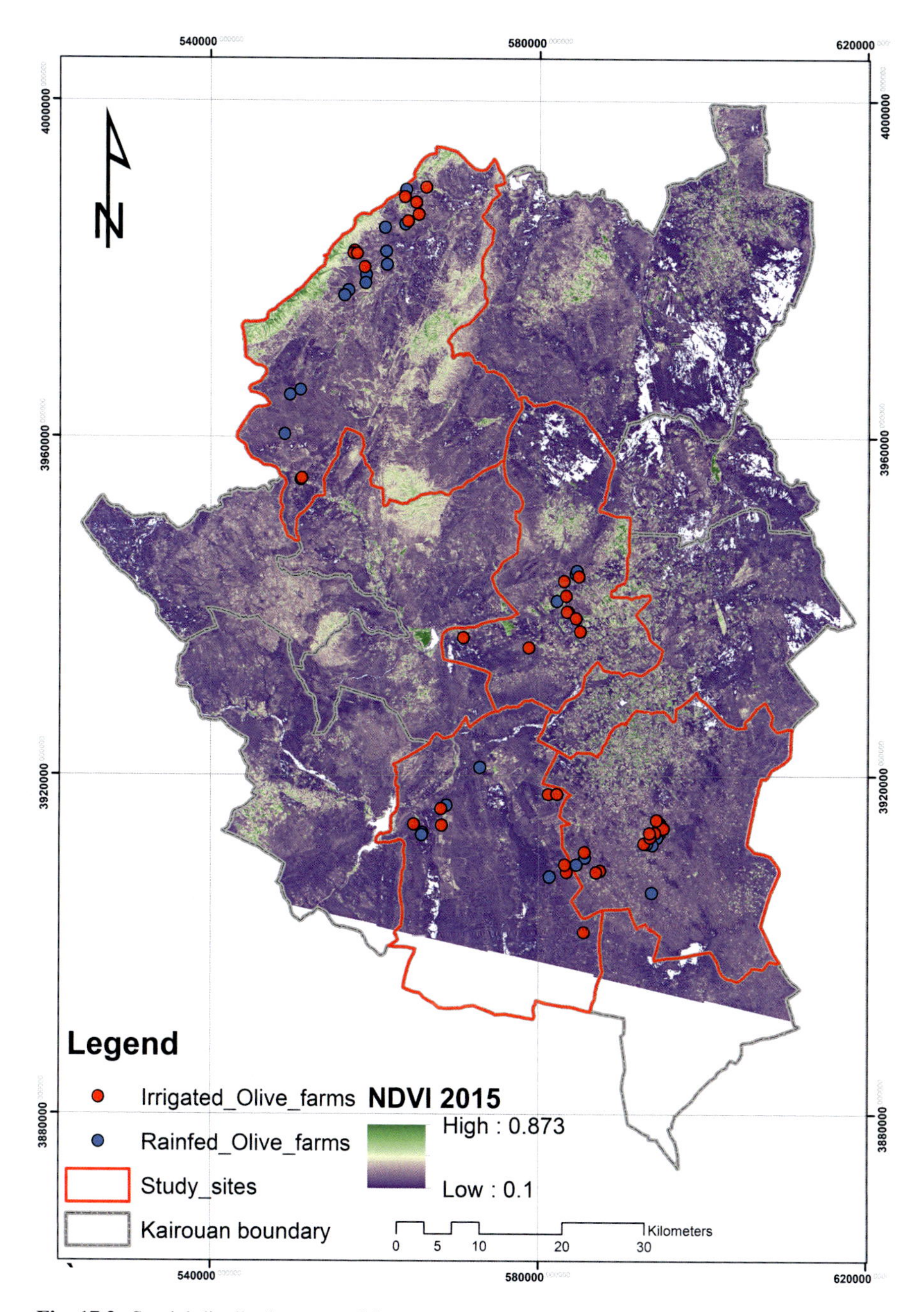

Fig. 17.3 Spatial distribution map of the mean NDVI in 2015

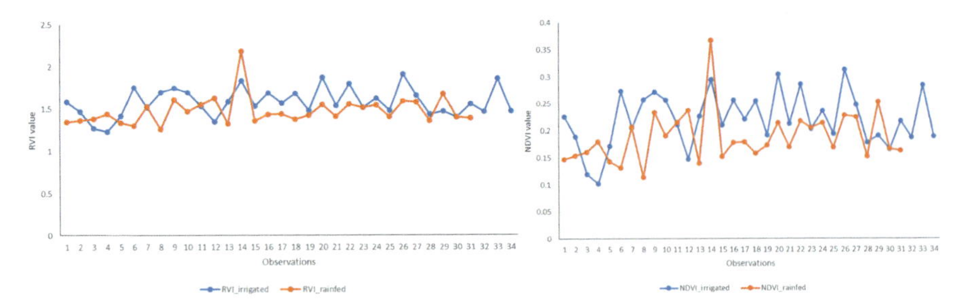

Fig. 17.4 Comparison of NDVI and RVI values between two olive farms types (**a**) NDVI Value; (**b**) RVI Value

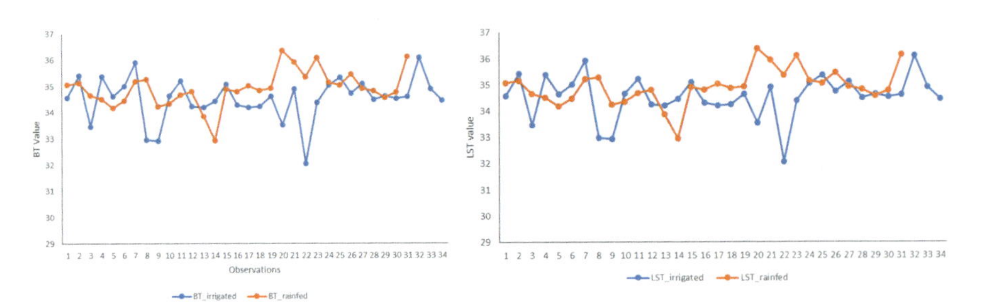

Fig. 17.5 Comparison of BT and LST between two kinds of olive farms (**a**) BT Value; (**b**) LST Value

17.3.3 Modelling Results and Comparison

Table 17.3 shows the results for modelling drought severity index. Overall, three ML algorithms show satisfactory results for predicting the drought indices in terms of R^2 and RMSE using cross-validation technique. Remarkably, the proposed model using the CBR technique yielded the highest predictive performance ($R^2 = 0.84$, RMSE $= 1.38$), followed by the SVM algorithm ($R^2 = 0.76$, RMSE $= 1.91$). The proposed model outperformed the two well-known ML algorithms for estimating the drought indicator in the study area. Nevertheless, the RF algorithm produced the lowest results ($R^2 = 0.71$, RMSE $= 2.61$) (Fig. 17.7).

We estimated the VHI index using the surface reflectance, spectral indices (NDVI, RVI) and land surface temperature (BT, LST) and generated the spatial patterns of drought index in the study area. The results show that over large areas in Kairouan prefecture were severe and moderate drought with a precise accuracy ($R^2 = 0.84$) using fivefold cross validation.

As the CBR model produced the highest predictive performance, we employed the proposed model to generate the spatial distribution of drought locations in the study sites (Fig. 17.8).

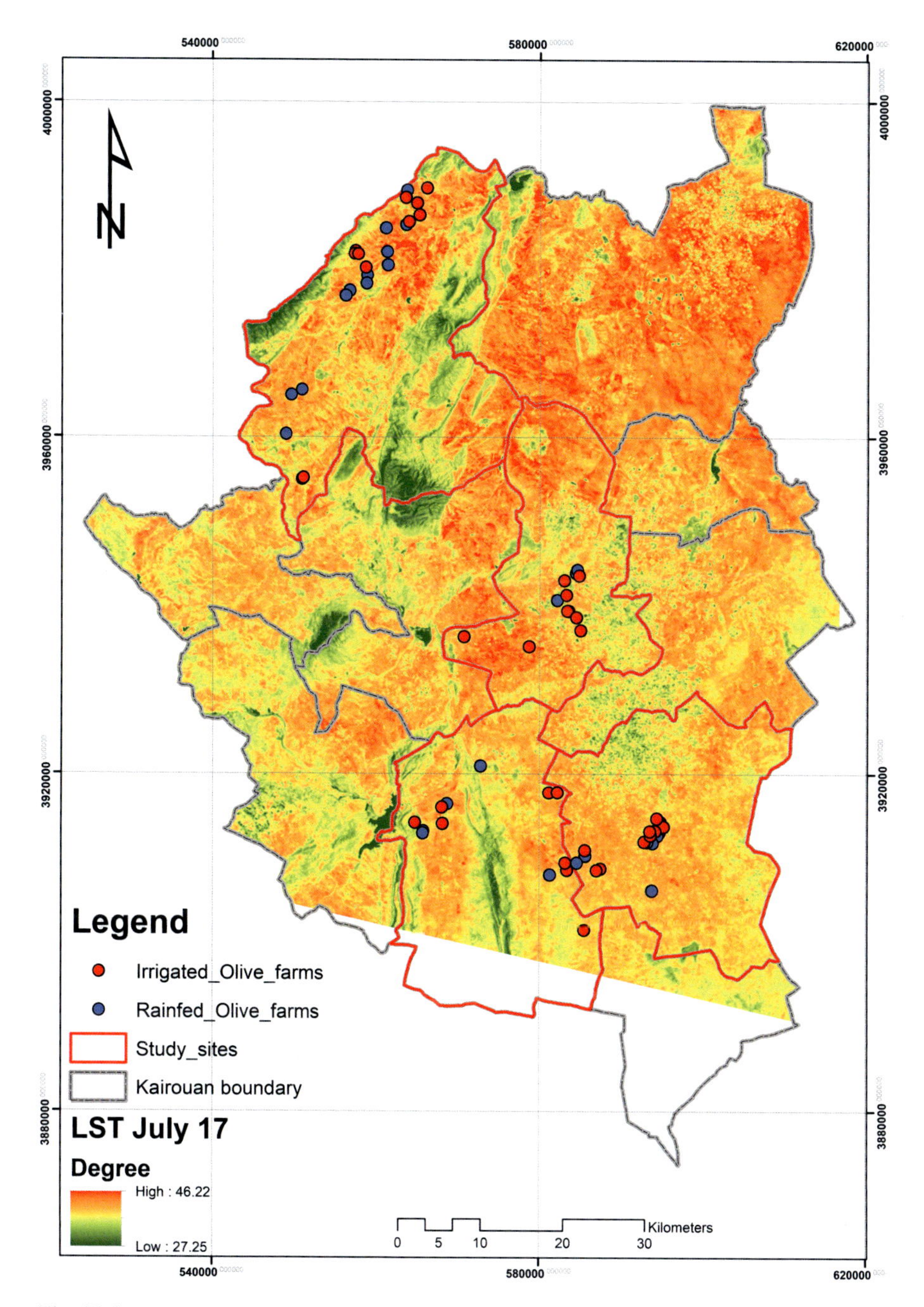

Fig. 17.6 Spatial distribution map of LST in the study area on July 17, 2015

Table 17.3 Machine learning models' performance for predicting drought indicators

No	Machine learning model	R2 testing	RMSE
1	CatBoost regression (CBR)	0.84	1.38
2	Random Forests (RF)	0.71	2.61
3	Support Vector Machine (SVM)	0.76	1.91

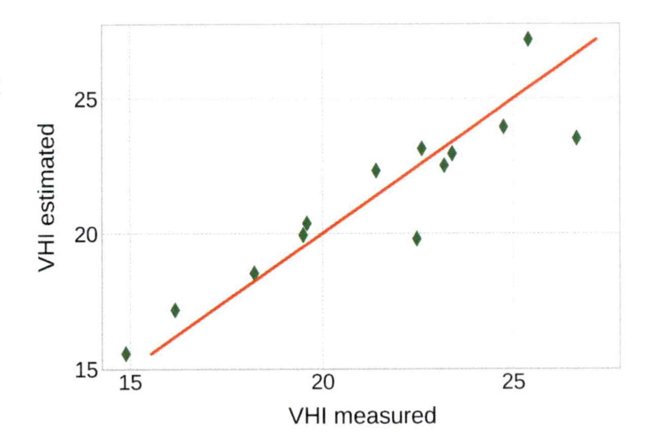

Fig. 17.7 Scatter-plot shows the predictive performance of the proposed CBR model in the testing phase

17.4 Discussion

Our results show that the mean NDVI values of irrigated growing farms are greater than those of rainfed farms (Fig. 17.4), reflecting a potential indicator in discriminating olive growing farms. The evaluation of multi-temporal BT and LST indexes conducted to assess the land surface temperature of olive growing farms, which is extremely high in dry season (July). The two indicators are also useful to assess drought conditions, which are the similar results reported by Tran et al. (2017). In addition, the results indicated that the surface temperature of irrigated farms are cooler than those in rainfed farms (Fig. 17.5), which is consistent with other studies (Ferreira & Duarte, 2019; Sandholt et al., 2002).

The assessment of drought indexes play an important role in monitoring drought severity in Kairouan prefecture, which is one of the hottest place in Tunisia. Our proposed machine learning model shows that the prefecture was exposed of moderate to severe drought with a precise accuracy ($R^2 = 0.84$) using fivefold cross validation. Our results provide a useful tools and new insights for decision makers to monitor olive growing farms production and to assess the performance of irrigation in drought periods. In addition, Masmoudi et al. (2010) emphasized that olive trees are regularly suffering of drought period which affect the production and a supplementary irrigation is required to stabilize and to maintain the production.

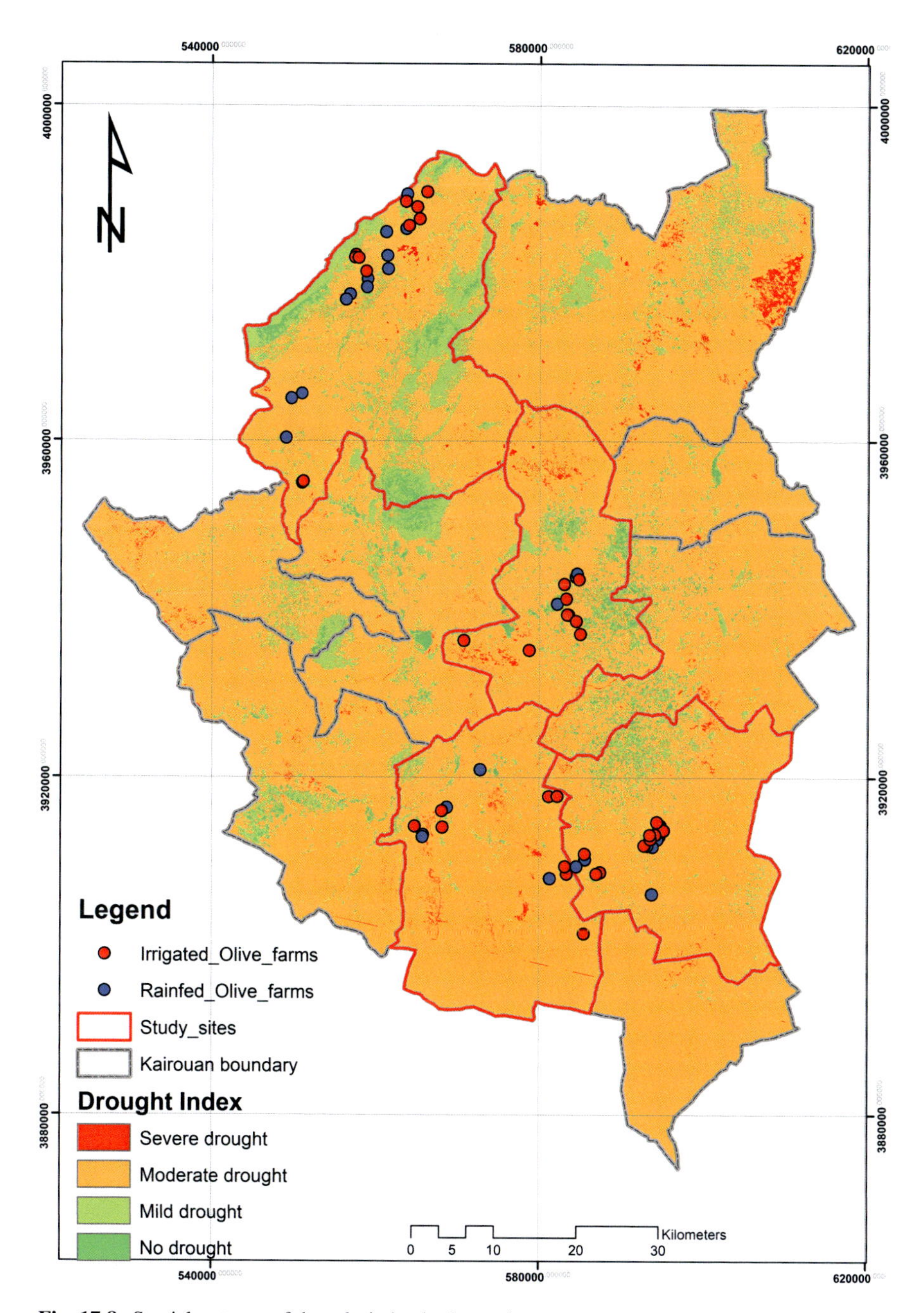

Fig. 17.8 Spatial patterns of drought index in the study area

17.5 Conclusion

Drought as a climate related disaster is difficult to avoid and to quantify. Indeed, the implementation of an efficient drought management system to reduce the impact of drought and to develop sustainable adaptation strategies is a challenge for local government. Therefore, an efficient assessment will conduct to implement suitable strategies and measures. In this study, we employed multi-temporal optical and thermal remote sensing data combined with the Catboost regression and cross-validation techniques to predict drought indicators in olive growing farms in Kairouan prefecture, Tunisia. Our findings show that the CBR model outperforms the SVM and RF techniques in the prediction of drought. This study also demonstrates that a combination of band reflectance and vegetation indices derived from Landsat 8 OLI multispectral sensor and thermal bands of Landsat 8 TIRS combined with the proposed model is able to robust drought indicator in the arid and semi-arid areas with precise results ($R^2 = 0.84$, RMSE = 1.38). As the Landsat time-series data and the programming code in Python language developed in this study can be freely accessed, the novel framework developed in the current study can be easily applied in other arid and semi-arid lands in MENA region. In Tunisia, the period of drought is increasing and its effects on water resource and agriculture productivity are serious. Therefore, the approach applied based on remote sensing and machine learning can be used as an early warning system which can be useful for the establishment of sustainable strategies to combat drought.

References

Bhuiyan, C. (2008). Desert vegetation during droughts: Response and sensitivity. *The International Archives of the Photogrammetry, Remote Sensing and Spatial Information Sciences, 37*(B8), 907–912.

Bhuiyan, C. (2010). Desert vegetation during droughts: Response and sensitivity. In *International archives of the photogrammetry, remote sensing and spatial information science*, (Vol. XXXVIII, Part 8).

Breiman, L. (2001). Random forests. *Machine Learning, 45*(1), 5–32. https://doi.org/10.1023/a:1010933404324

Corbari, C., Horeschi, D., Ravazzani, G., & Mancini, M. (2008). Land surface temperature from remote sensing and from an energy water balance model for irrigation management. In A. Santini, N. Lamaddalena, G. Severino, & M. Palladino (Eds.), *Irrigation in Mediterranean agriculture: Challenges and innovation for the next decades* (Options Méditerranéennes: Série A. Séminaires Méditerranéens; n. 84) (pp. 223–233). CIHEAM.

Delbiso, T. D., Altare, C., Rodriguez-Llanes, J. M., Doocy, S., & Guha-Sapir, D. (2017). Drought and child mortality: A meta-analysis of small-scale surveys from Ethiopia. *Scientific Reports, 7*, 2212. https://doi.org/10.1038/s41598-017-02271-5

Deo, R. C., & Şahin, M. (2015). Application of the extreme learning machine algorithm for the prediction of monthly effective drought index in eastern Australia. *Atmospheric Research, 153*, 512–525. https://doi.org/10.1016/j.atmosres.2014.10.016

Dorogush, A. V., Ershov, V., & Gulin, A. (2018). *CatBoost: gradient boosting with categorical features support*. arXiv preprint arXiv:181011363.

EM-DAT. (2019). *Natural Disasters 2019*. UCLouvain, Centre for Research on Epidemiology of Disasters CRED, USAID.

Enenkel, M., Brown, M. E., Vogt, J. V., McCarty, J. L., Reid Bell, A., Guha-Sapir, D., Dorigo, W., Vasilaky, K., Svoboda, M., Bonifacio, R., Anderson, M., Funk, C., Osgood, D., Hain, C., & Vinck, P. (2020). Why predict climate hazards if we need to understand impacts? Putting humans back into the drought equation. *Climatic Change, 162*, 1161–1176. https://doi.org/10.1007/s10584-020-02878-0

Ferreira, L. S., & Duarte, D. H. S. (2019). Exploring the relationship between urban form, land surface temperature and vegetation indices in a subtropical megacity. *Urban Climate, 27*, 105–123. https://doi.org/10.1016/j.uclim.2018.11.002

Ghaleb, F., Mario, M., & Sandra, A. (2015). Regional landsat-based drought monitoring from 1982 to 2014. *Climate, 3*(3), 563.

Heim, R. R., Jr. (2002). A review of twentieth century drought indices used in the United State. *Bulletin of the American Meteorological Society, 83*(8), 1149–1166. https://doi.org/10.1175/1520-0477-83.8.1149

Henia, L. (2003). Les grandes sécheresses en Tunisie au cours de la dernière période séculaire. In P. Arnould, & M. Hotyat (Eds.), *Eau, Environnement, Tunisie et milieux méditerranéens* (ENS Editions, pp. 25–36).

Kefi, M., Pham, T. D., Kashiwagi, K., & Yoshino, K. (2016). Identification of irrigated olive growing farms using remote sensing techniques. *Euro-Mediterranean Journal for Environmental Integration, 1*(1), 3. https://doi.org/10.1007/s41207-016-0004-7

Kogan, F. N. (1995). Application of vegetation index and brightness temperature for drought detection. *Advances in Space Research, 15*(11), 91–100. https://doi.org/10.1016/0273-1177(95)00079-T

Kogan, F., Stark, R., Gitelson, A., Jargalsaikhan, L., Dugrajav, C., & Tsooj, S. (2004). Derivation of pasture biomass in Mongolia from AVHRR-based vegetation health indices. *International Journal of Remote Sensing, 25*(14), 2889–2896. https://doi.org/10.1080/01431160410001697619

Louati, M. H., Bergaoui, M., Lebdi, F., Methlouthi, M., El Euchi, L., & Mellouli, H. J. (2007). Application of the drought management guidelines in Tunisia [Part 2. Examples of application]. In A. Iglesias, M. Moneo, & A. López-Francos (Eds.), *Drought management guidelines technical annex* (Options Méditerranéennes: Série B. Etudes et Recherches; n. 58) (pp. 417–467). CIHEAM/EC MEDA Water.

Masmoudi, C. S., Ayachi, M. M., Gouia, M., Laabidi, F., Ben Reguaya, S., Oueled Amor, A., & Bousnina, M. (2010). Water relations of olive trees cultivated under deficit regimes. *Scientia Horticulturae, 125*(4), 573–578. https://doi.org/10.1016/j.scienta.2010.04.042

Mountrakis, G., Im, J., & Ogole, C. (2011). Support vector machines in remote sensing: A review. *ISPRS Journal of Photogrammetry and Remote Sensing, 66*(3), 247–259. https://doi.org/10.1016/j.isprsjprs.2010.11.001

Navarro, J. A., Algeet, N., Fernández-Landa, A., Esteban, J., Rodríguez-Noriega, P., & Guillén-Climent, M. L. (2019). Integration of UAV, Sentinel-1, and Sentinel-2 data for mangrove plantation aboveground biomass monitoring in Senegal. *Remote Sensing, 11*(1). https://doi.org/10.3390/rs11010077

Ngo, P.-T. T., Pham, T. D., Nhu, V.-H., Le, T. T., Tran, D. A., Phan, D. C., Hoa, P. V., Amaro-Mellado, J. L., & Bui, D. T. (2020). A novel hybrid quantum-PSO and credal decision tree ensemble for tropical cyclone induced flash flood susceptibility mapping with geospatial data. *Journal of Hydrology*, 125682. https://doi.org/10.1016/j.jhydrol.2020.125682

Ngo, P.-T. T., Pham, T. D., Hoang, N.-D., Tran, D. A., Amiri, M., Le, T. T., Hoa, P. V., Bui, P. V., Nhu, V.-H., & Bui, D. T. (2021). A new hybrid equilibrium optimized SysFor based geospatial data mining for tropical storm-induced flash flood susceptible mapping. *Journal of Environmental Management, 280*, 111858. https://doi.org/10.1016/j.jenvman.2020.111858

Niemeyer, S. (2008). New drought indices. In A. López-Francos (Ed.), *Drought management: Scientific and technological innovations* (Options Méditerranéennes: Série A. Séminaires Méditerranéens; n. 80) (pp. 267–274). CIHEAM.
Pedregosa, F., Varoquaux, G., Gramfort, A., Michel, V., Thirion, B., Grisel, O., Blondel, M., Prettenhofer, P., Weiss, R., & Dubourg, V. (2011). Scikit-learn: Machine learning in python. *Journal of Machine Learning Research, 12*(Oct), 2825–2830.
Peters, A. J., Walter-Shea, E. A., Ji, L., Vina, A., Hayes, M., & Svoboda, M. D. (2002). Drought monitoring with NDVI-based standardized vegetation index. *Photogrammetric Engineering and Remote Sensing, 68*(1), 71–75.
Pham, T. D., Le, N. N., Ha, N. T., Nguyen, L. V., Xia, J., Yokoya, N., To TT, Trinh, H. X., Kieu, L. Q., & Takeuchi, W. (2020). Estimating mangrove above-ground biomass using extreme gradient boosting decision trees algorithm with fused Sentinel-2 and ALOS-2 PALSAR-2 data in can gio biosphere reserve, Vietnam. *Remote Sensing, 12*(5), 777.
Pham, T. D., Yokoya, N., Nguyen, T. T. T., Le, N. N., Ha, N. T., Xia, J., Takeuchi, W., & Pham, T. D. (2021). Improvement of mangrove soil carbon stocks estimation in North Vietnam using Sentinel-2 data and machine learning approach. *GIScience & Remote Sensing, 58*(1), 68–87. https://doi.org/10.1080/15481603.2020.1857623
Prokhorenkova, L., Gusev, G., Vorobev, A., Dorogush, A. V., & Gulin, A. (2018). CatBoost: unbiased boosting with categorical features. In *Advances in neural information processing systems* (pp. 6638–6648).
Sandholt, I., Rasmussen, K., & Andersen, J. (2002). A simple interpretation of the surface temperature/vegetation index space for assessment of surface moisture status. *Remote Sensing of Environment, 79*(2), 213–224. https://doi.org/10.1016/S0034-4257(01)00274-7
Tran, H. T., Campbell, J. B., Tran, T. D., & Tran, H. T. (2017). Monitoring drought vulnerability using multispectral indices observed from sequential remote sensing (Case study: Tuy Phong, Binh Thuan, Vietnam). *GIScience & Remote Sensing, 54*(2), 167–184. https://doi.org/10.1080/15481603.2017.1287838
Unganai, L. S., & Kogan, F. N. (1998). Drought Monitoring and Corn Yield Estimation in Southern Africa from AVHRR Data. *Remote Sensing of Environment, 63*(3), 219–232. https://doi.org/10.1016/S0034-4257(97)00132-6
USGS (2015) *Landsat 8 (L8) data users handbook*. USGS.
Vafaei, S., Soosani, J., Adeli, K., Fadaei, H., Naghavi, H., Pham, T. D., & Tien Bui, D. (2018). Improving accuracy estimation of forest aboveground biomass based on incorporation of ALOS-2 PALSAR-2 and sentinel-2A imagery and machine learning: A case study of the Hyrcanian Forest area (Iran). *Remote Sensing, 10*(2), 172.
Valipour, M. (2016). Optimization of neural networks for precipitation analysis in a humid region to detect drought and wet year alarms. *Meteorological Applications, 23*(1), 91–100. https://doi.org/10.1002/met.1533
Vapnik, V. (2013). *The nature of statistical learning theory*. Springer.
Verner, D., Treguer, D., Redwood, J., Christensen, J., McDonnell, R., Elbert, C., & Konishi, Y. (2018). *Climate variability, drought, and drought Management in Tunisia's agricultural sector*. World Bank.
Wang, D., Wan, B., Liu, J., Su, Y., Guo, Q., Qiu, P., & Wu, X. (2020). Estimating aboveground biomass of the mangrove forests on northeast Hainan Island in China using an upscaling method from field plots, UAV-LiDAR data and Sentinel-2 imagery. *International Journal of Applied Earth Observation and Geoinformation, 85*, 101986. https://doi.org/10.1016/j.jag.2019.101986
Wilhite, D. A. (1993). *The enigma of drought. Drought assessment, management, and planning: Theory and case studies* (pp. 3–15). Springer/Kluwer Academic Publishers.
Zargar, A., Sadiq, R., Naser, B., & Khan, F. I. (2011). A review of drought indices. *Environmental Reviews, 19*, 333–349. https://doi.org/10.1139/a11-013

Chapter 18
Quantitative Assessment of Land Sensitivity to Desertification in Central Sudan: An Application of Remote Sensing-Based MEDALUS Model

Abdelrahim Salih and Abdalhaleem A. Hassaballa

Abstract As desertification risk is one of the main environmental and socioeconomic issues in central Sudan, mapping and assessing the environmental sensitivity of the land to desertification is an issue of concern. The present study highlighted the Bara and Tayba localities of North of Kordofan State (central Sudan) in order to map and assess the environmental sensitivity of the land to desertification based on different indicators. The sensitivity was estimated with a modification of the widely used Environmentally Sensitive Area Index (ESAI), also known as MEDALUS, through employing 17 quantitative parameters divided into 4 main quality indices, namely, climate quality index (CQI), vegetation quality index (VQI), soils quality index (SQI), and management or socio-economic quality index (MQI). The establishment of these qualities was achieved using remote sensing supported with specialized tools in the geographical information system (GIS) software. The Area Under the Curve (AUC) approach was used to validate the model output. Four degrees of sensitivity have been produced, namely: extremely sensitive; highly sensitive; moderately sensitive; and low or non-sensitive. The results have also revealed that, approximately 21% of the study site have been classified as highly critical sensitive areas to desertification. Highly fragile areas were estimated to present 37% of the study site, whereas, moderate and potentially affected sensitive areas have covered around 31%. Eventually, the applied model has proven its applicability as a tool used for assessing as well as categorizing sensitivity of lands to desertification, especially in arid and semi-arid environments.

A. Salih (✉)
Department of Geography, Imam Mohammad Ibn Saud Islamic University (IMSIU), Al-Ahsa, Kingdom of Saudi Arabia
e-mail: aSalih@imamu.edu.sa

A. A. Hassaballa
Department of Environment & Agricultural Natural Resources, College of Agricultural and Food Sciences, King Faisal University, Al-Ahsa, Kingdom of Saudi Arabia

M. M. Al Saud (ed.), *Applications of Space Techniques on the Natural Hazards in the MENA Region*, https://doi.org/10.1007/978-3-030-88874-9_18

Keywords Desertification risk · Environmental sensitivity area index · Sudan · MEDALUS · North Kordofan

18.1 Introduction

Desertification is a serious environmental problem in arid and semi-arid regions (Verón et al., 2006). UNEP and Thomas (1992) defined desertification as "land degradation in arid, semi-arid, and dry sub-humid areas resulting from various factors, including climatic variation and human activities". Dregne (1986) in El-Baz and Hassan (El-Baz & Hassan, 1986) described it as land degradation with major challenges in the arid environment of the world. Desertification has been described by Hellden in the proceedings of the China-EU Workshop on Integrated Approach to Combat Desertification that held in China as the expansion of desert conditions (Hellden Helldén, 2003). He also concluded that the driving forces and the impacts of this phenomenon is a site specific.

Climatic variations and human activities are among the main factors that causes desertification (Verón et al., 2006). The accumulation of sand dunes/sheets on agricultural lands and urban areas is another illustration of desertification in the arid regions (Nwilo et al., 2020). Vegetation removal, soil salinity, decline in soil fertility, decline in rainfall amount and duration, overgrazing, and overcropping are factors that can lead to desertification (El-Baz & Hassan, 1986; Nwilo et al., 2020). Countries of the Sahel are more susceptible to desertification as they heavily depend on their land for food security (Nwilo et al., 2020).

According to these descriptions and definitions of desertification and their effects, it is also an environmental issue in Sudan, especially in the northern and central parts of the country. In the North Kordofan State that located in the central part of Sudan, it is where the most area that is vulnerable to desertification (Adam, 2011). In this state and according to Dawelbait (2010) and Khiry (2007), desertification is a major threat that affects environmental sustainability. According to Dawelbait (2010); Khiry (2007) and Salih (2006), the main driving forces of desertification in north Kordofan state are overgrazing, overcropping, variability of droughts, reduction in vegetation cover, sand encroachment (especially in the northeastern part of the state), soil salinity and poor land management. For example, Adam (2011) found that *Acacia Senegal* trees lost about 25% of their area from 1972 to 2007. A significant decrease in vegetation cover and increase in sand dunes were observed during the period from 1976 to 2003 (Khiry, 2007). Salih (2006) stated that reduction in livestock in terms of quantity and quality is an indicator of the decrease in biological productivity in the northern part of the North Kordofan State and this was specifically observed in Bara locality, which is located in the northern part, and accordingly this is a sing of desertification in this area. Furthermore, the decline in productivity of the land due to desertification was observed during the period from 1961 to 1973 (Olsson, 1983), and this decline can be supported by the data shown in Fig. 18.2 during the period from 1991 to 2000. Assessment of desertification, its causes and effects have been studied by several scholars at different locations in North Kordofan

State (e.g. Adam, 2011; Dawelbait & Morari, 2012; Dawelbait, 2010;Helldén, 2003; Khiry, 2007; Salih, 2006). However, a little or no information has been quantitatively provided about the lands that are susceptible or sensitive to desertification risk. Therefore, it is extremely important to assess the lands sensitivity to desertification in this area in order to implement effective reclamation programs that minimize or prevent future development of this phenomenon.

Desertification poses a serious risk to human habitats. Moreover, the risk of desertification is a major barrier to agricultural and urban development practices. As a result, analyzing environmental sensitivity and assessing the likelihood of desertification has become a major step towards sustainable development. In response to this need, different contributions have been proposed to develop efficient and economic-effective methods for desertification risk assessment. Among these approaches, the integration of remote sensing and a geographical information system (GIS) has proven particularly successful. The remote sensing images, have proven to be an excellent tool in mapping, monitoring and assessing desertification. Different models have been applied, which are based on physical and biological parameters derived from remote sensing, to map, monitor, assess and quantify desertification including its degree and severity (Dawelbait, 2010; Gao & Liu, 2010; Lamqadem et al., 2018a; Salih et al., 2021). However, one of the serious implications of the methods used by those studies is the inability to provide information about the susceptibility or sensitivity of the lands to desertification as they are established based only on one or two desertification indicators, such as albedo and soil moisture to assess desertification risk.

Alternatively, the environmental sensitivity area index (ESAI), also known as MEDALUS proposed by Kosmas et al. (1999), has been used and evaluated by several studies, and proven to be a valuable approach for monitoring desertification risk, or in identifying areas that are sensitive to desertification, and they obtained satisfactory results (Bedoui, 2020; Budak et al., 2018; Jiang et al., 2019; Lamqadem et al., 2018b; Prăvălie et al., 2017; Salvati & Bajocco, 2011). For example, (Jiang et al., 2019) have used this index for monitoring land sensitivity to desertification risk in central Asia using fourteen indicators including vegetation, climate, soil and land management quality. They concluded that ESAI provides reliable results in relation to desertification potential. Recently, Bedoui (2020) conducted a study to assess sensitivity to desertification in a region of central Tunisia by integrating different parameters derived from remote sensing and weighted values in a GIS with MEDALUS model. They found that incorporating soil, vegetation, climate and management information quality into a MEDALUS model allowed them to gain great insight into the spatial distribution of desertification risk.

Although a significant progress was made for mapping, assessing and monitoring desertification risk using the proposed model and other remote sensing-based approaches in arid and semi-arid areas, to date, no research has monitored and assessed the sensitivity to desertification in the North Kordofan State based on ESAI model. Therefore, the purpose of this study was to map and assess land susceptibility to desertification in this area, on the bases of the aforementioned indicators and ESAI model. This study also aims to provide a more detailed view of the sensitivity or susceptibility to desertification for Bara and Tayba localities. The information obtained from the ESAI map and assessment can help tremendously in combating desertification in the study site. The work has also assumed to be a step further in establishing an early warning system for the study site.

The establishment of the proposed index was based on various environmental qualities including, for example, soil, vegetation cover, climate and management quality (Kosmas et al., 1999). In this study, about seventeen desertification indicators (Table 18.1) were selected, according to the data availability, using the ESAI methodology. The methodology adopted was based on the integration between remote sensing and GIS for the indicator‘s extraction, slicing, scoring and quality mapping (Fig. 18.5).

Table 18.1 The data source and its characteristics for the study area

IDs	Data type	Parameters extracted	Source of data	Data scale and resolution
a)	Shuttle Radar Topographic Mission (SRTM) digital elevation model (DEM)	Slope, and aspect	Earth Explorer-©USGS.gov https://earthexplorer.usgs.gov/	30 m resolution (1Arc-second)
b)	Landsat-8(OLI) image, date: 01 Jan 2019	LULC, NDVI, and land surface Albedo (α)	Earth Explorer-©USGS.gov https://earthexplorer.usgs.gov/	30*30 meter
c)	Global soil data	Texture	Provided by Fischer et al. (2008) and FAO (2007)	(scale, 1: 5,000,000)
d)	Metrological data	Rainfall (mm)	WorldClim http://www.worldClim.org	1 km^2
e)	Land cover map	Agricultural intensity data	Provided by FAO (2012)	(scale, 1: 100,000)
f)	Geological map	Parent material	Provided by Ministry of Energy and Mines (1981)	(scale, 1: 2,000,000)
g)	Socio-economic data	Population density, and livestock data	AMESD (2010)	1 km
h)	Climatic data: Global aridity Index and potential Evapotranspiration (ET0) climate database v2	Aridity and evapotranspiration	(Trabucco & Robert, 2018)	1 km
i)	Google earth imagery	Digitizing the locations of pre-decertified villages (points)	Google Earth Pro software	–

18.2 An Overview of the Study Area

North Kordofan State is mostly categorized by gentle corrugation of plains having a regular altitude between 350 to 500 m MSL. This plain is mainly covered with sand dunes shared by separated hills.

The study area is located within North Kordofan State (in the middle of Sudan) between longitudes of 28′ 30 "00 – 30' 30" 00 E and latitudes of 13′ 30 "00 – 15' 30" 00 N; and about 350 km to the south-west direction from the capital city of Sudan (Khartoum). The study site covers approximately 38,837.8 km^2 represented in two parts, namely, Bara and Tayba localities (Fig. 18.1). The study area is situated primarily in just a sand belt forming a thin strip along with incursion of clay soils within some separated locations. Hills seem to be more like chains stretching over the north, west and south borders as well as in the middle of the study area. Sand dunes generally cover the majority of the northern and eastern portions of the area. It can be noticed also that; the study area is placed within a drainage system that belongs to the Nile river basin. The majority of the water courses (Wadis) within the study area tend to be ephemeral streams which in turn flow throughout a short time following the rainy period and minimal runoff actually reaches the Nile since they vanish in the desert prior to joining the Nile stream (Khiry, 2007; Adam, 2011).

The arid and/or semi-arid -desert natures are the most describing climate of the area joined with minimal periodic rain varies from below 100 mm at the north to around 350 mm at the south. The area witnesses four seasons, namely: the Autumn (the rainy season), which extends from May to October having maximum rainfall in August; from October to early December, comes a transitional period with low humidity and night low temperatures where, most of the rain-fed cultivated crops are harvested in this period; the Winter (the cold season), extends from December to the mid of February having modest temperatures and cozy humidity; and finally, the Summer (the hot dry season) which comes with common north eastern winds and lasts from March to the mid of May. Rainfall occurs in short intense quantities for 6 months from May through October, with concentration of 80 to 90% from July and up to September. The area rainfall exhibits a high variation spatially and temporally, especially for the period from 1960 to 2005 (Fig. 18.2) (Hulme, 2001), in which, the large extent on the latitude influences the length of the rainy season (Olsson, 1985). The relative humidity in its mean value ranges from 20% in the winter months to 75% throughout August, during rains. The winds within the study area blow from the north east during winter time and from south west throughout summer time. Mostly winds possess moderate speed of lower than 3 m/s, however they are very effective at shifting sands from dunes if soils are subjected to exposure.

Gentle strips of depressions covered by clay soils are dispersedly observed between the dunes. Additionally, there are several rocky outcrops, mostly at the northern part of the study area. Despite the fact that sandy soils are usually lacking nitrogen, phosphorus, organic matter, as well as other components, they preserve a lot more cropping stress. This is due to the fact that the sandy soils are quite simple to cultivate also it fits producing numerous crops for instance sorghum, millet, sesame

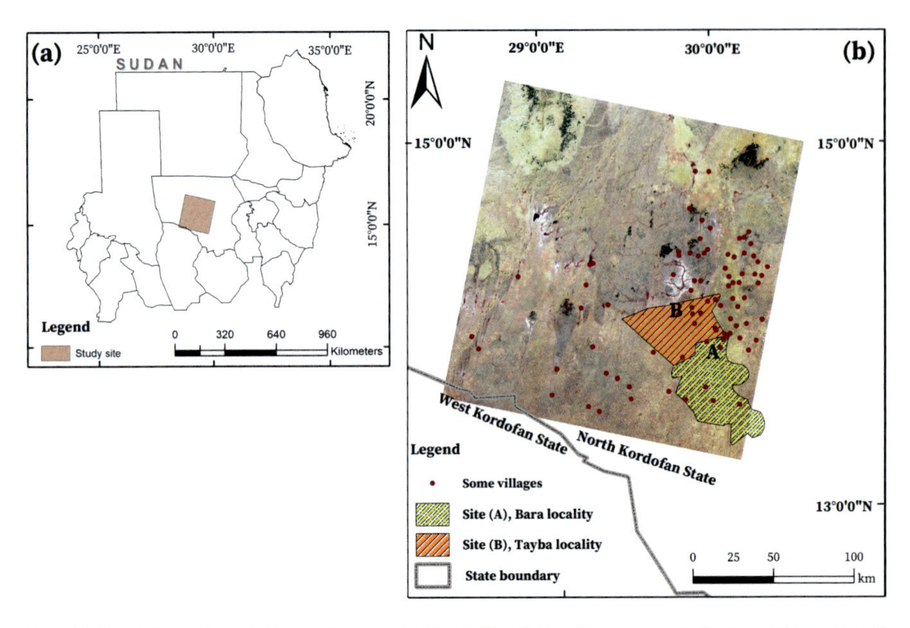

Fig. 18.1 (**a**) Location of the study area in North Kordofan State, central Sudan. (**b**) Landsat-8 (OLI)scene shown on false color composite (RGB 543) 2020 represents the actual study site in the North Kordofan State overlain by the locations of the most affected localities by desertification, namely, Bara and Tayba. The red dots show some of the villages distributed along the study site

and groundnuts. However, the challenge of the sandy soils is because they eliminate their fertility within extremely short period of time and became simply eroded and desertified.

The area is covered with vegetation in a very scattered trend, which is a consequence of the reduced quantity of rainfall. The vegetation is subjected to extreme climatic situations and ought to withstand drought that may extend for many years with minimum rain in the least (Schmidt & Karnieli, 2000). Within the eastern portion of the study area, vegetation is rare and dominated by *Maerua crassifolia, Leptadena pyrotechnica and Acacia tortilis*.

In the western portion, the vegetation cover is actually greater and denser. Trees and shrubs tend to be disturbed with available grassland, where most of the inhabitants in the area are either residents or nomads. Figure 18.3(a and b) highlights the fluctuation in the total cultivated area throughout 10 years period extended from 1989 to 2000, observed at Bara and Tayba localities, respectively. Their principal jobs are generally livestock raising and conventional farming. Rainfall quantities as well as its distribution usually are the main factors of practicing farming. However, at the north portion of the area, the majority of the tribes (i.e. Kababish, Hawaweer and Kawahle) raise animals and also grow crops in limited portions of the lands, because precipitation is extremely inconsistent and the probability of agricultural

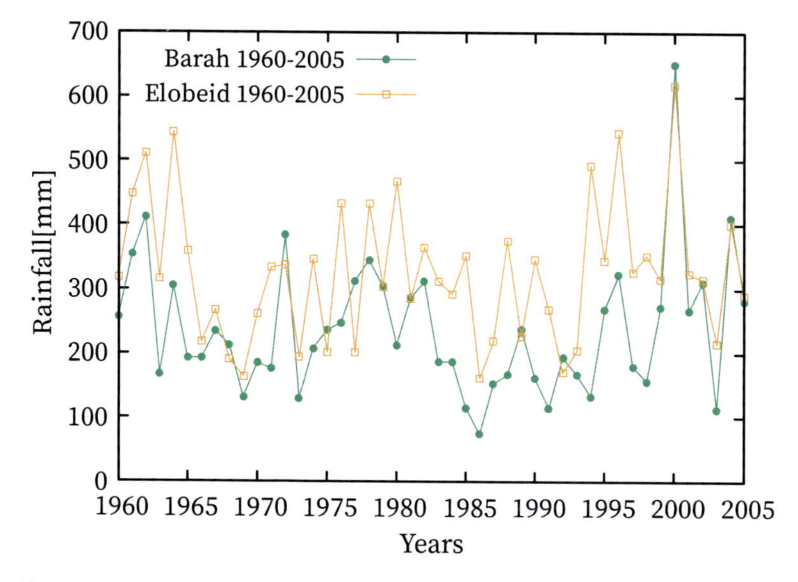

Fig. 18.2 Rainfall fluctuation in the study area between 1960 and 2005

failure could be high. Nevertheless, in the southern part the rainfed agriculture is conducted on the clay soils.

18.3 Research Approaches and Model Adopted

18.3.1 Data

The used dataset for the desertification sensitivity analysis and assessment are provided in Table 18.1 with their characteristics. Figure 18.4 shows some examples of the datasets used for the study area.

18.3.2 Parameters for Desertification Sensitivity Assessment

Several factors such as human activities, physical, and social might lead to desertification occurrence in the study area. Aiming to the assessment of lands sensitivity to desertification for different parts of the study area, geological, climatological, socio-economical, and geographical factors have been considered. To assess and produce the desertification sensitivity map, the factors that were used in the present study are the aridity index and potential Evapotranspiration, slope and aspect, rainfall (mm), population density and livestock, soil texture, surface albedo,

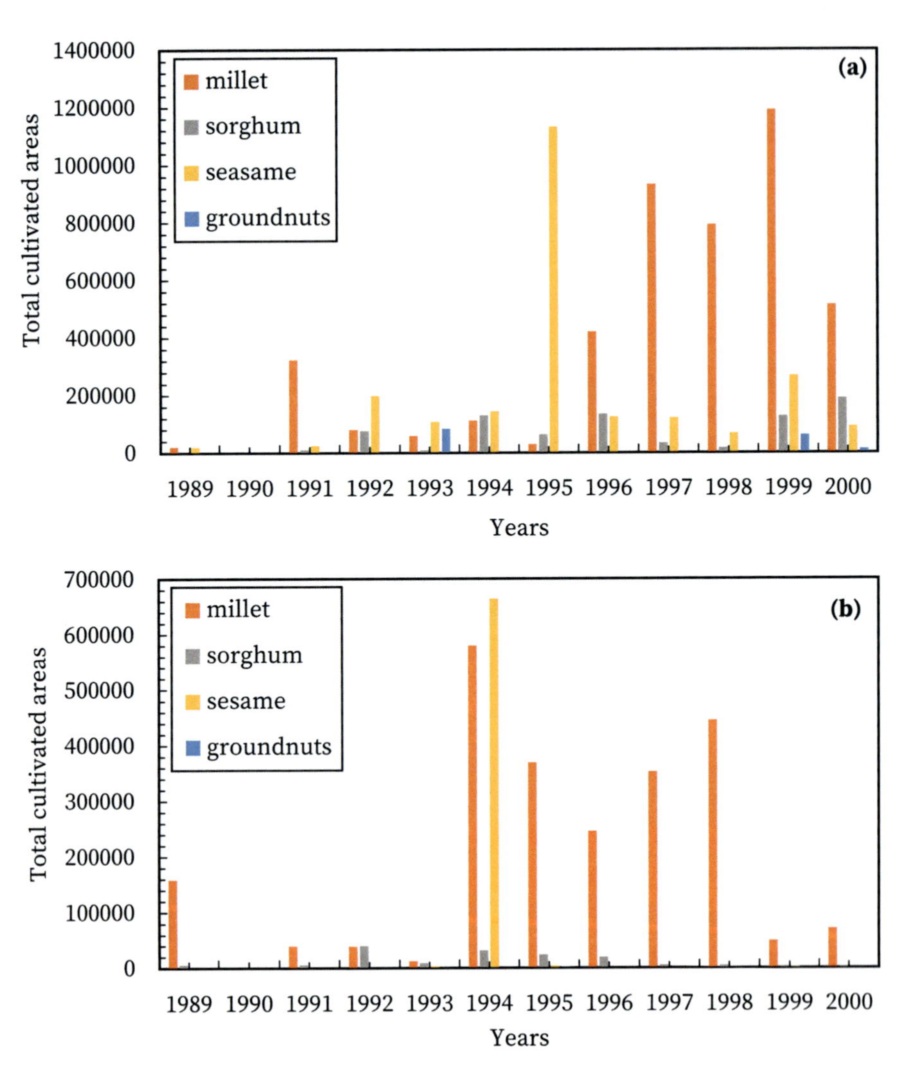

Fig. 18.3 The observed total cultivated area throughout 10 years period extended from 1989 to 2000, for (**a**) Bara and (**b**) Tayba

fractional vegetation cover (FVC), and land cover (Table 18.1). These parameters were selected along with their rates on the base of the previous literatures (e.g. Dawelbait, 2010; Khiry, 2007; Kosmas et al., 1999; Lamqadem et al., 2018b; Mohamed, 2016).

A spatial database has been created using QGIS 3.16 (the open-source software) for the preprocessing of the geological map and the Landsat-8 (OLI) image, where, SCP plugin was used for image calibration. ArcGIS 10.2 software in turn, was used for the preparation of the desertification indicators and the thematic layers. Table 18.2 shows the main parameters and classes for the ESAI and the related

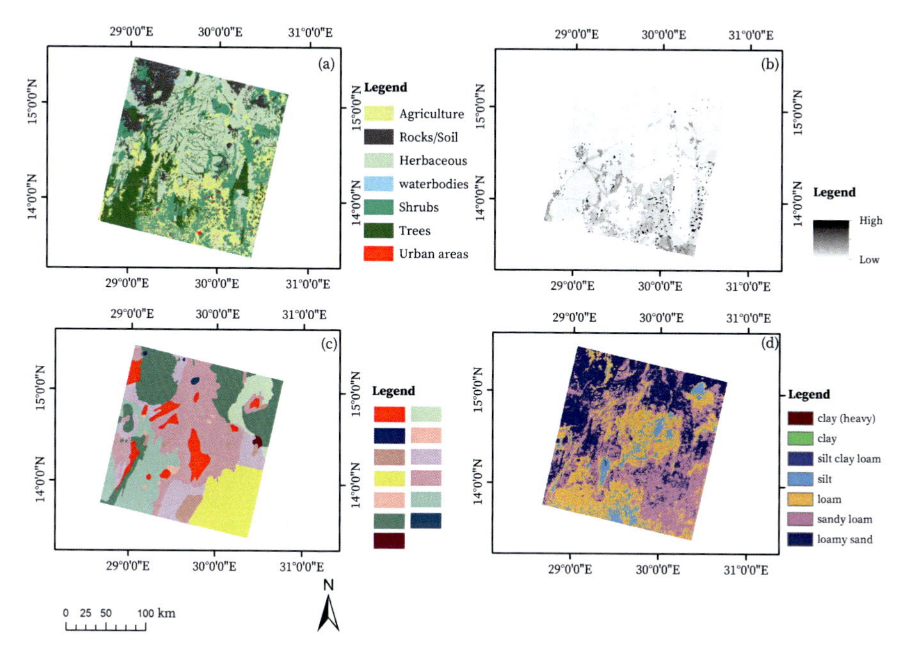

Fig. 18.4 Desertification/land degradation indicators: (**a**) land use/cover; (**b**) population density; (**c**) Geology of the area in which "the red" indicates acidic and intermediate volcanics, mainly rhyolites and trachytes, "the dark blue" indicates basic volcanic, mainly basalts, "the rose dust" indicates batholic granites, grey granites and pegmatites, "the yellow" indicates Gezira formation (unconsolidated clay, silt, sands and gravels, "the rose quartz" indicates granodiorites, "the dark green" indicates Nubian sandstones formation, "the light green" indicates Nubian sandstone formation (non outcrop), "the deep green" indicates syenites (outcrop), "the lilac" indicates undifferentiated basement complex, "the tudor rose" undifferentiated genesis group, "the turquoise" indicates undifferentiated schist group, "the steel blue" indicates undifferentiated sediment (conglomerates, sandstones, siltstones and shales), while "the purple" indicates younger granites; (**d**) representing soil texture of the study area (FAO, 2007)

indicators used for the desertification sensitivity assessment. In this study, desertification sensitivity classes, parameters and scores have entailed vegetation quality index (VQI), soil quality index (SQI), climate quality index (CQI), and management quality index (MQI) (Kosmas et al., 1999). Figure 18.5 presents the flowchart of the research methodology and model adopted for the study area.

Each single indicator of these four was established based on the parameters presented in Table 18.2. Scores from 1 (low sensitivity to desertification) to 2 (high sensitivity to desertification) were assigned to each indicator's classes by adopting the approach presented in the literature (e.g. Jiang et al., 2019; Kosmas et al., 1999; Salih, 2006). Details for these indicators and their establishment methods are given in the following paragraphs:

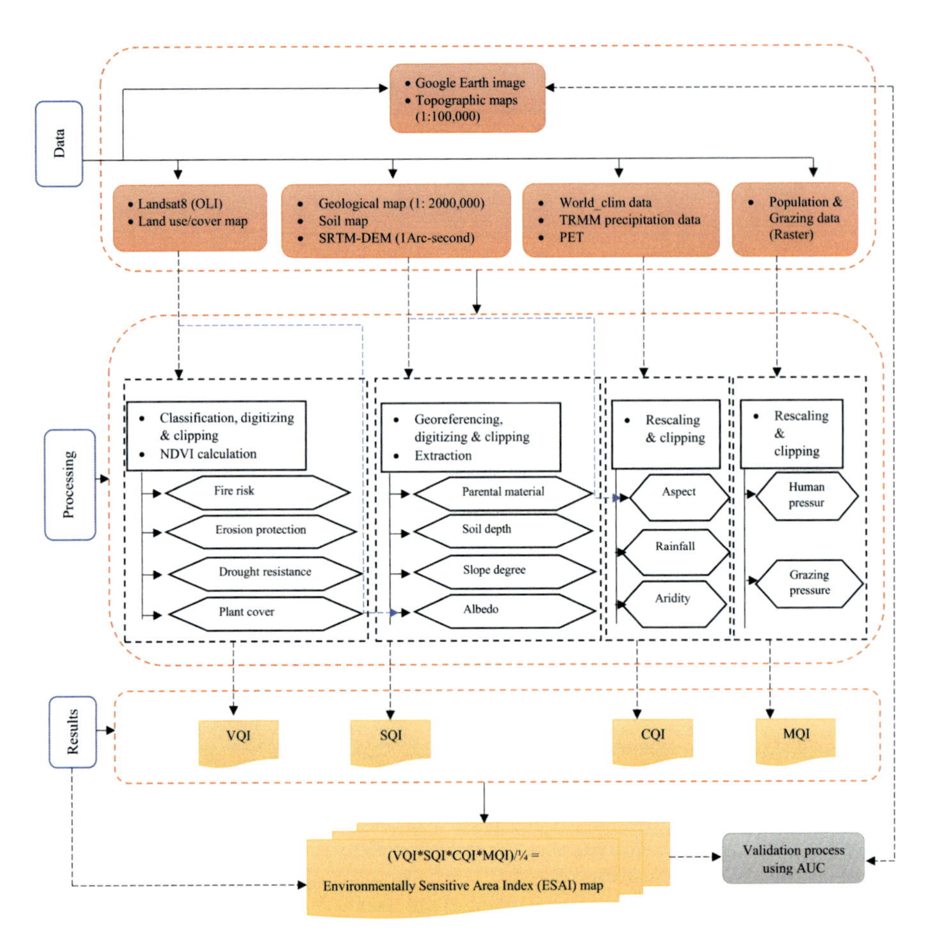

Fig. 18.5 The follow chart of the research methods adopted for the study area

18.3.2.1 Vegetation Quality Index

Vegetation cover plays a crucial role in the estimation of the desertification occurrence in arid and semi-arid environment as it associates with the reduction of soil erosion and reducing sand movement. Accordingly, VQI was calculated based on four parameters: fire risk (fire is one of the key factors that threating vegetation removal in the study site), drought resistant, erosion protection, and plant cover—fractional vegetation cover (FVC) (Table 18.2). The first three parameters are provided from the LULC map, that was obtained from classifying Landsat-8(OLI) image using the supervised classification approach with the maximum likelihood

Table 18.2 The indicators used to calculate the CQI, SQI, VQI, and the MQI and their parameters and scores

Indicator	Parameters	Class	Description of the parameters	Score	Method of categorization
CQI	Rainfall (mm)	1 2 3	High >217 Moderate 175–217 Low <175	1 1.5 2	
	Aridity(P/PE0)	1 2 3	Low aridity <0.072 Moderate aridity 0.056–0.072 High aridity <0.056	1 1.5 2	
	Aspect	1 2	N, NE, NW, flat areas S, SE, SW, E	1 2	
	Evapotranspiration (mm)	1 2 3	Low <3040 Moderate 3040-3118 High >3118	1 1.5 2	
SQI	Texture	1 2 3 4	Clay, Clay (heavy), Silt clay loam Silt Loam Loamy sand, Sandy loam	1 1.2 1.6 2	
	Soil salinity	1 2 3	Non-saline soil Moderate saline soil High saline soil	1 1.7 2	
	Slope (%)	1 2 3 4	<6 6–18 18–35 >35	1 1.2 1.5 2	
	Parent material	1 2 3	Coherent Moderate Fragile	1 1.5 2	Classification and scores were based on the method presented in (Lamqadem, Pradhan, et al., 2018b).
	Albedo (α)	1 2 3	< 0.2 0.2–0.25 > 0.25	1 1.7 2	
VQI	Fire risk	1 2 3	Rocks, Sands Bare soil Vegetation cover	1 1.5 2	
	Drought resistance	1 2 3 4	Vegetation Bare soil Rocks Sands	1 1.2 1.5 2	
	Erosion protection	1 2 3 4	Rocks Vegetation cover Bare soil Sands	1 1.2 1.8 2	

(continued)

Table 18.2 (continued)

Indicator	Parameters	Class	Description of the parameters	Score	Method of categorization
	Plant cover (%)	1 2 3 4 5	<16 16–27 27–35 35–42 >42	1 1.3 1.6 1.8 2	
MQI	Population density (people per square kilometer)	1 2 3 4 5 6	<25 25–50 50–100 100–200 200–400 >400	1 1.2 1.4 1.6 1.8 2	
	Overgrazing	1 2 3 4	< 20 heads per square km 20–60 heads per square km 60–100 heads per square km >100 heads per square km	1 1.2 1.6 2	Classification and scores were based on the method presented in (Bedoui, 2020).
	Agricultural intensity	1 2 3	Bare rocks and soil, seasonal/perennial and natural/artificial waterbodies, tress closed to sparse, urban. Herbaceous closed to sparse, Shrubs closed to sparse. Cropland (in terrestrial and aquatic land).	1 1.5 2	Classification and scores were based on the method presented in (Jiang et al., 2019).
	Policy enforcement	1 2 3	Seasonal/perennial and natural/artificial waterbodies, tress closed to sparse. Herbaceous closed to sparse, Shrubs closed to sparse. Cropland (in terrestrial and aquatic land), bare rocks and soil, urban.	1 1.5 2	

classifiers (Canada, 2008). While the fourth parameter "FVC" was calculated using Eq. (18.1) (Zhang et al., 2019), upon calculating the normalized difference vegetation index (NDVI) from Landsat-8(OLI) image.

$$FVC = \frac{NDVI - NDVI_{min}}{NDVI_{max} - NDVI_{min}} \tag{18.1}$$

Where, *min* and *max* values have been retrieved from minimum and maximum NDVI values in which, $NDVI_{min}$ represents the area covered by bare soil, while $NDVI_{max}$ represents the area that has dense vegetation cover.

The VQI was calculated based on Eq. (18.2), using the geometric mean of the score values of all the four parameters (Kosmas et al., 1999).

$$VQI = (fire\ risk \times erosion\ protection \times drought\ resistant \times plant\ cover\ (\%))^{1/4} \tag{18.2}$$

18.3.2.2 Soil Quality Index

SQI is a parameter that reflects the statement of the soil by means of its ability to store water and resists erosion (Salvati & Bajocco, 2011), especially in arid and semi-arid lands where the ability of soil to store water is very low. The SQI is calculated based on the following parameters (soil texture, soil salinity, slope (%), parent material—prepared and classified based on the method presented on Lamqadem et al. (2018b), and surface albedo) (Table 18.2). The first four parameters were proposed for the ESAI approach. SQI considered as one of the most important parameters to quantify and assess desertification as soil deterioration can lead ultimately to decline in food production. The soil texture data were obtained at a 30 arc-second spatial resolution, from Harmonized World Soil Database V 1.2, provided by FAO Soil Portal (FAO, 2007). Figure 18.2d shows the soil texture map for the study area.

The slope has a crucial role in the desertification occurrence, because the steep-to-very-steep slops can increase the potential to soil erosion by water. The slope has been computed using SRTM-DEM; also, a geological map (Fig. 18.4c) has been used for preparing the parent material classes based on the standard MEDALUS model (Kosmas et al., 1999).

In this research, soil salinity and soil brightness (Albedo) were incorporated as indicators for desertification occurrence as they influence soil quality. Surface albedo is commonly used as indicator for assessing desertification and land degradation in arid and semi-arid lands, (e.g. Budak et al., 2018; Lamqadem et al., 2018a; Salih et al., 2021; Sepehr et al., 2007). Liang albedo equation (Liang, 2001) was used to compute soil surface albedo for the study area. On the other hand, soil salinity is considered as one of the most severe environmental problems worldwide amongst the other forms of soil degradation (Allbed et al., 2014). These two parameters were both derived from Landsat-8 (OLI) image after necessary reformation. For the soil salinity derivation, the following Equation (18.3) was used (Elhag, 2016):

$$Soil\ salinity = \frac{(\beta - \mathcal{R})}{(\beta + \mathcal{R})} \tag{18.3}$$

Where, β stands for the blue band, while $\mathcal{R}$ stands for the red band.

Finally, the SQI indicator was obtained based on the geometric mean of these five parameters (Table 18.2), with Eq. (18.4) as follows:

$$SQI = (Soil\ texture \times Soil\ salinity \times slope\ (\%) \times Parent\ materials \times Albedo)^{1/5} \tag{18.4}$$

18.3.2.3 Climate Quality Index

Four variables were used to prepare the CQI, i.e., rainfall (mm), aridity, aspect and potential evapotranspiration (Table 18.2). The evapotranspiration was not considered as a MEDALUS model input factor, however, it has been added here because of the high variability of the rainfall on the study area, as rainfall is one of the high critical factors that assists in obtaining the CQI (Prăvălie et al., 2017), according to the WorldClim data (Table 18.2).

The potential evapotranspiration (PET) was incorporated into the CQI as it reflects the content of the moisture of the soil and the other conditions for the study area, therefore, it is a crucial environmental factor in desertification occurrence. On the other hand, the aridity index (AI), was obtained from the Global Aridity Index database (version 2) Lamqadem et al., (2018b). It represents one of the primary factors that lead to desertification assessment in the study area, which influences with other parameters the water availability to vegetation cover (e.g. Lamqadem et al., 2018b; Sepehr et al., 2007; Trabucco & Robert, 2018). Both Aridity and potential evapotranspiration data were available in a high resolution of 30 arc-seconds, at the periods from 1970–2000.

The rainfall pattern over the study area is characterized by a fairly regular degrease towards the north. The rainfall data were obtained in a grid format at a $0.5°*0.5°$ resolution from WorldClim global climate and weather data (https://worldclim.org/data/index.html), as described in (Fick & Hijmans, 2017), provided in mean monthly basis from 1970 to 2000. The aspect was also considered because it represents important factor in desertification occurrence (Budak et al., 2018). Slope with aspect oriented to the southeast is exposed to large amount of sunlight and heat, compared to those with northeast orientation (Lamqadem et al., 2018b) and this is where low moisture content, low vegetation cover, and high temperature with high evaporation were depicted. The other orientations of aspect had no effect on the desertification occurrence as they had a high moisture content. The aspect has been calculated based on the SRTM-digital elevation model (DEM), after a necessary reformation. The CQI was then calculated based on the Eq. (18.5) (Kosmas et al., 1999).

$$CQI = (rainfall \times aridity \times aspect \times evapotranspiration)^{1/4} \quad (18.5)$$

18.3.2.4 Management Quality Index

Different human activities can affect environmental ecosystem with different degrees. Considering the data availability, four parameters related to anthropogenic activities in the study area were taken into consideration for assessing and quantifying the management quality index (MQI): (1) population density per square.km, (2) overgrazing, (3) agricultural intensity, and (4) policy enforcement (Table 18.2). Population distribution in the study site is governed by water availability, abundance of natural resources, and social services. The southern and eastern parts of the study site are characterized by having high population density, while the other parts have low population density. To quantify and classify the second parameter, a method presented by Bedoui (2020) was followed, while for the third and the fourth parameters, a method produced by Jiang et al. (2019) was adopted. The agricultural intensity and policy enforcement were extracted from the land cover map (Fig. 18.4a), provided by National Centre for Research, Remote Sensing and Seismology Authority (RSSA) of Sudan (2011). The management quality index (MQI) was then calculated based on Eq. (18.6) using the score values (Kosmas et al., 1999) provided in Table 18.2.

$$MQI = (agricultural\ intensity \text{ x } policy\ enforcement)^{1/2} \quad (18.6)$$

18.3.2.5 ESAI Map

The thematic layers (Indicators) with their corresponding scores were eventually combined together based on the geometric mean of the indicators' values as shown in Eq. (18.7) by the means of ArcGIS software's, where, the raster calculator tool was used in order to calculate the overall score indices for the final environmental sensitivity to desertification map "ESIA" (Kosmas et al., 1999) for the study area.

$$ESAI = (CQI \times SQI \times VQI \times MQI)^{1/5} \quad (18.7)$$

The final ESAI map was classified into two class groups. The first group included the following categories: low sensitivity, moderate sensitivity, high sensitivity, and very high sensitivity areas to desertification. While the second group was classified based on the original ESAI mode (Kosmas et al., 1999), which included the following four classes: potentially-affected areas; moderately-affected areas; highly-fragile-affected areas; and highly-critical-affected areas with desertification.

18.3.2.6 Accuracy Assessment

Finally, the resultant ESAI's map and their indicators have been validated using a receiver operating characteristic curve (ROC), which is a measure the goodness-of-fit from the area under the curve (AUC), which takes values from 0 to 1 where value of 1 indicates best performance of the model used (Mandrekar, 2010).

Validation usually means comparing the maps generated by models to independent dataset. This comparison could be qualitative such as visual interpretation through a basic overlay or quantitative, carried out making use of functions such as cumulative curve (Remondo et al., 2004). The present research managed to analyze a spatial correlation between different desertification drivers, in order to depict possible kinds of spatial correspondences among them. To achieve this, success rate and prediction rate was used for examining the model compatibility and prediction capability.

The Area-Under-the Curve Validator (AUC)

Desertification vulnerability can be verified through comparing sensitivity map with training data which were employed for developing the models with the validation data that were not utilized throughout the model creation stage. The success rate curve is dependent on the concept of comparison between the prediction map along with the desertification utilized in the modeling (Chung & Fabbri, 2003), and the success rate technique can assist in determining how effectively the resulting desertification susceptibility maps have classified the areas of present desertification (Tien et al., 2012). Through the technique, the rate curves can be produced, while the area under the curve (AUC) presents the quality of the models to efficiently categorize the occurrence of existing desertification (training dataset), while the area under the predicted rate curve explains the capacity to which the proposed desertification model can predict. The AUC value mostly varies from 0.5 (50%) to 1.0 (100%), and the best model produces an AUC value in close proximity to 1.0 (perfect prediction), whilst a random prediction model produces an AUC value near to 0.5 (Swets, 1988).

Throughout the analysis, training and testing sample locations were delineated at certain scattered villages over the study region as they have been pending to desertification occurrence throughout the identified study period. To assess the relative ranks for every prediction pattern at the desertification sensitivity map, the determined index values of all desertification cells were categorized in descending order. Then the ordered cell values were split into 100 classes, with accumulated 1% intervals and then plotted against the testing points (locations of previous desertification occurrences) that associated with these cells. Eventually, the AUC was determined as the percentage of the cumulative cell areas of common association that take trapezoidal shape. The percentage of AUC was then regarded as an indicator of the model exactness for the susceptibility map creation. The research additionally intended to examine the association reliability of every unique parameter (driving factors i.e. soil quality index (SQI); climate quality index (CQI); vegetation quality index (VQI); and socio-economic or management quality index (MQI) as being controlling factors in desertification occurrence. The developed subclasses of the independent factor maps have been reclassified and plotted against the testing points utilizing AUC as well.

18.4 Desertification Sensitivity Analysis and Assessment

18.4.1 Assessment of Environmental Quality Indicators

Generally speaking, observing the environmental sensitivity to desertification map and their indicators, it becomes evident that the spatial distribution of the very high sensitivity areas is the largest in the north-western, and eastern part of the study area (Fig. 18.6).

The assessment of the VQI map (Fig. 18.6a) has proven that, approximately 74% of the study site has been located within the limits of the low quality, specifically in the northern and central parts of the study site, 18% within the moderate, and only

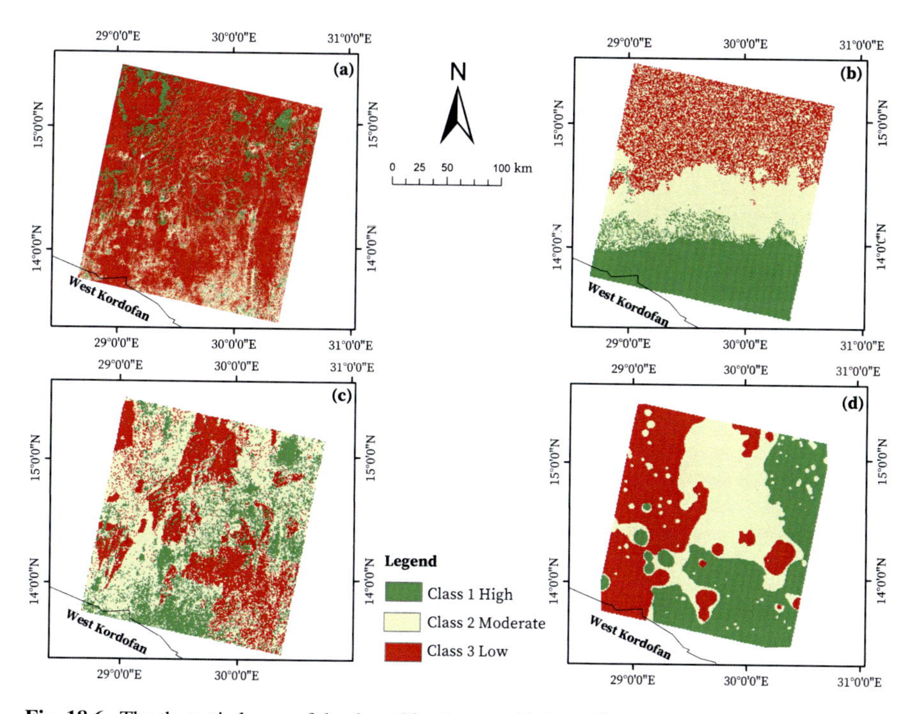

Fig. 18.6 The thematic layers of the desertification sensitivity indicators (i.e., VQI, SQI, CQI, and MQI), respectively, involved in the analysis in the study area provided for the whole test site

8% within the high-quality zone (Table 18.3). This distribution could be justified by two factors, firstly, due to the climate change— reduction in rainfall and due to land mismanagement in terms of burning, overgrazing and overcropping. Vegetation cover was degraded or depleted specially in the northern and central parts. Secondly, the local community residents in the north-eastern part of the study site practices their cultivation (in shifting cultivation modes to maintain soil fertility) mainly on the high-lands "Quz" that is already fragile and more vulnerable to wind erosion by aeolian process, which are active in the study site a long time ago.

The CQI assessment has shown that the low-quality climate with around 26%, was concentrated in the north-northwestern part of the study site (Fig. 18.6b), while the rest of the areas exhibited about 43% as a moderate quality and 29% as a high quality. It has been justified by that due to the high decrease in rainfall amounts toward the northern part of the country in general. Furthermore, these areas are already characterized by high aridity values.

The analysis of SQI (Fig. 18.6c) revealed that the high precarious conditions (30%) have been allocated in the northern, central, and eastern parts of the study site. It was noticeable that the vast majority of the areas with moderate quality class corresponding to 47% (Table 18.3) were situated within the western part. These areas are mainly covered by Aeolian deposits (Quz) with low organic matter and low water holding capacity.

In terms of MQI (Fig. 18.6d), most of the area (approximately 81%), were classified as high quality with good management and protection. While the low-quality class was found only within 2% of the total area (Table 18.3), which represented the unprotected rangelands. The high percent of quality management can be explained due to three reasons: Firstly, the agricultural production using the traditional farming systems in the study site is the main source of income for the local community. This has made them an agricultural dependent community, therefore, they cultivate their lands with variable types of crop that need high management for sustainability. Secondly, to keep their agricultural lands with high

Table 18.3 Indicators of desertification/land degradation along with its sensitivity classes, area per square kilometer and percentage

Desertification/land degradation indicators	Sensitivity class	Area per square Kilometer	Percent (%)
CQI	High	11020.49	29.60
	Moderate	16319.06	43.90
	Low	9856.72	26.50
SQI	High	8163.99	22.00
	Moderate	17696.07	47.50
	Low	11365.92	30.50
MQI	High	29622.18	81.00
	Moderate	6205.68	17.00
	Low	764.25	02.00
VQI	High	2897.97	07.78
	Moderate	6720.39	18.04
	Low	27642.94	74.18

protection, as they tend to implement a policy enforcement with high level to make their lands under control from any aggression, such as overgrazing by the nomadics. Finally, the decrease in vegetation coverage, specially during the period between 1976 and 1988 (Khiry, 2007), led the local community to adopt their own strategies to enforce more powers toward their lands. Obviously, these levels of agricultural management would have imposed an impact on the environment as mapped in Fig. 18.6d.

To provide more details about the environmental sensitivity to desertification in the study site by adopted visual interpretation techniques, two localities were selected, i.e. Bara and Tayba (Fig. 18.1), which have been the most affected areas by the land degradation because of sands encroachment and drought (Khiry, 2007). Figure 18.7 shows the thematic layers of the extracted indicators of these two localities. Considering the outcomes of the two maps in terms of VQI, (Fig. 18.7a1 and a2), it reveals that the spatial distribution of the low-quality zones is the same in both localities, with their largest distribution in the northern and north-eastern parts. This category was found to be strongly influenced by drought forces because these two localities are very poor in terms of vegetation cover, as documented earlier by Khiry (2007), in addition to suffering a severe degradation towards the north. Moreover, the fluctuation in rainfall (as shown in Fig. 18.2) had a major role on the spatial variability of vegetation along the study site, in addition to the extensive grazing.

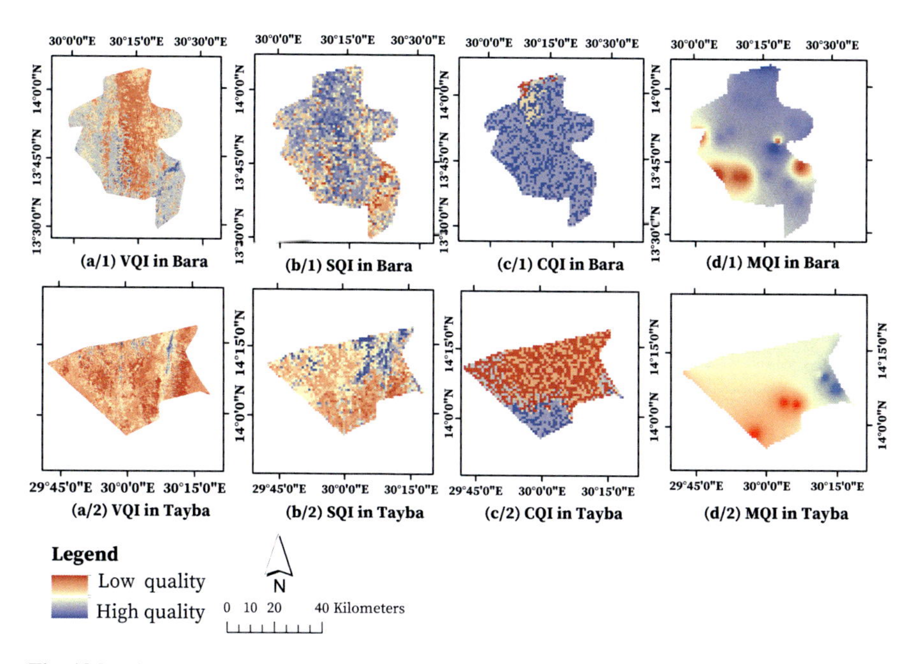

Fig. 18.7 The thematic layers of the desertification sensitivity indicators (i.e., VQI, SQI, CQI, and MQI), respectively, involved in the analysis in the study area provided for the two areas of interest, i.e., Bara and Tayba

Further, the SQI assessment map of Fig. 18.7b1 and b2, has revealed that the very-high soil quality zone has a spatial extension in the northern, central, and southern parts of Bara. However, the most critical zone was found in the eastern and central part of Bara where the El-Basheri and El-Taweel Oases are located. This result is in line with the findings of Khiry (2007) who stated that the desert formation, i.e., sand dunes was increasing with different levels around El-Bashiri oasis. On the other hand, in Tayba locality, the high-quality soil was concentrated only on the northeastern part, while the rest of the area marked by low quality zone. This situation can be explained by the means of the Aeolian process that has been active in the study site a long time ago, consequently, the soils of the two localities have been highly vulnerable to wind erosion and desert encroachment (Khiry, 2007).

Furthermore, the CQI assessment, (Fig. 18.7c1 and c2), has also shown that approximately 90% of Bara locality area was occupied by high-quality zone, except the northern part. This situation can be justified due to the high amount of rainfall southward. Conversely, approximately 85% of the lands at Tayba locality were covered by low-quality zone. Also, this can be justified due the decrease in the amount of precipitation toward the north.

Most of the areas in Bara locality are marked by high-quality management zone (Fig. 18.7d1). This zone is mainly distributed in the northern and the central parts of the areas that were pending to high-protected cultivated lands, while the southwestern part is marked by low-quality management, which corresponds to the areas covered by rangelands. On the other hand, at the northern part of Tayba locality, the area was highlighted by moderate-quality class, while the central and eastern parts were highlighted by low and high classes, respectively. These differences in quality management could be justified by deducing that the local communities are highly agricultural dependent and this dependence can yield intensive cultivation practice, and intensive overgrazing in one side of the area, where the population density was very high compared to the other side of the area.

18.4.2 Assessment of the ESAI Map for the Whole Study Site

Considering the low and high values of the final ESAI map, the study was able to categorize the area into four degrees of sensitivity to desertification using the reclassification by ASCII tool in ArcGIS 10.2 software. These have included: the potentially affected areas with low sensitivity; the moderately fragile areas with moderate sensitivity; the highly fragile areas with high sensitivity; and the very-high critical areas with very high sensitivity to desertification (Fig. 18.8). By observing Fig. 18.8, it is evident that the spatial distribution of the high and very high sensitivity zones is the largest portion over the north, north-western and central parts of the study site (Fig. 18.8). It is well noticed that the vast majority of VQI and CQI in the study site, were centralized at the very-high and the high sensitivity zones.

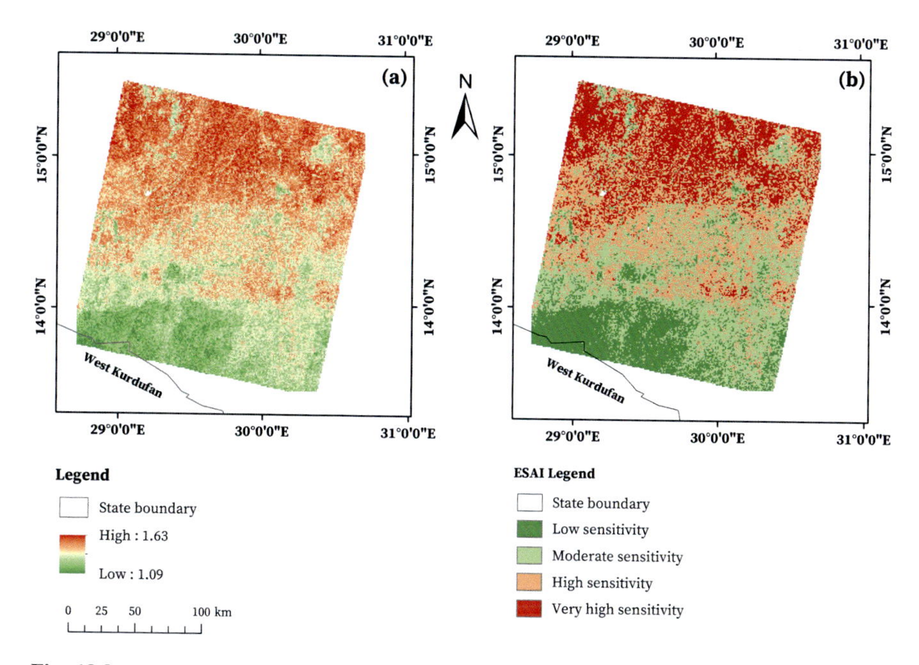

Fig. 18.8 The environmental sensitivity area to desertification: (**a**) ESAI unclassified map; (**b**) ESAI classified map. The white areas denote masked pixels

Table 18.4 The resultant areas (per square kilometer) and the percentages of the produced categories

Model used	Desertification/land degradation class	Area in square Kilometer	Percent (%)
ESAI	Potentially affected areas	4738.33	12.97
	Moderately fragile areas	10760.58	29.46
	Highly fragile areas	13359.021	36.57
	Highly critical areas	7673.03	21.00

Table 18.4 presents the areas (per square kilometer) and the percentages of the produced categories. According to Table 18.4, the percent of the high and very-high sensitivity areas "high fragile and high critical zones" occupy approximately 57.5% of the total area of the study site, i.e., more than 140,000 square kilometers.

18.4.3 Assessment of ESAI Map at Locality Level

Referring to the visual interpretation technique, it has been revealed that the high and very-high sensitivity areas were the most spreading, and noticeably observed in the northern and central parts in both localities (Fig. 18.9c and d). The very-high and high sensitivity zones constituted about 52% in Bara and about 56% in Tayba

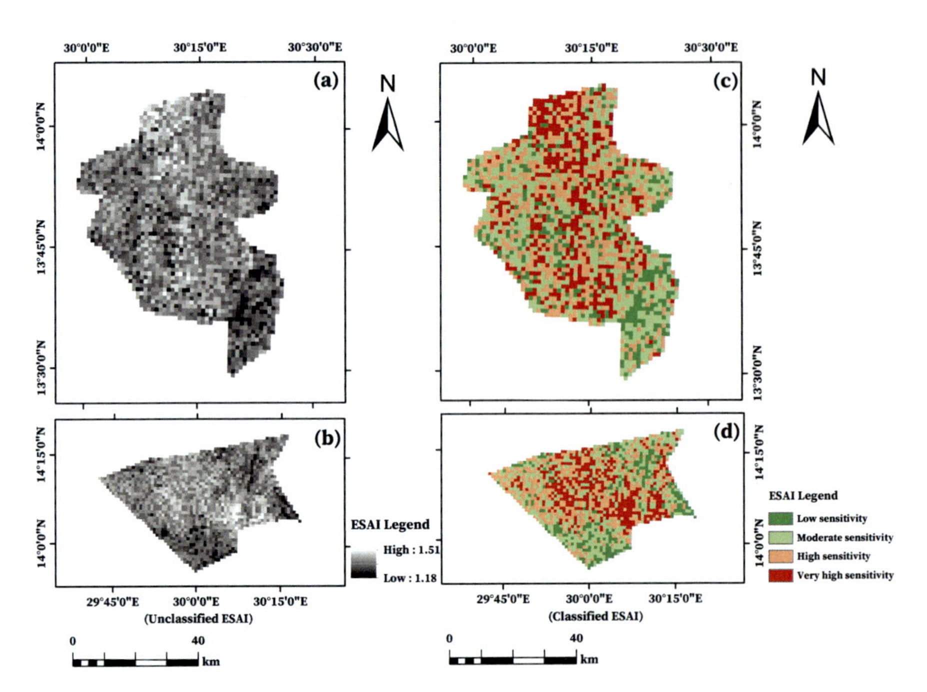

Fig. 18.9 The ESAI maps for Bara (upper) and Tayba (lower) localities, where, (**a** and **b**) are the unclassified maps, while (**c** and **d**) are classified maps

localities. As these results were subjected to the association of different factors such as soil salinity and its effect, and sand drift (Khiry, 2007) and mismanagement, and overgrazing and overuse of the land (Dawelbait, 2010), it is worth to worth to assure that sand encroachment and dunes drift are considered as the main processes that caused land degradation, especially in arid and semi-arid environment (Dawelbait & Morari, 2012; Lamqadem et al., 2018b; Salih et al., 2021). The growth of the villages, especially around water bodies, is another contributor to this result.

With regard to the spatial distribution of the low-to-moderate sensitivity zones to desertification for the two localities, both zones of sensitivity were observed mainly at the south part of localities. This result indicates that the soil salinity, sand encroachment and drought had less effects southward due to the abundance in rainfall amount and duration at these locations, in addition to the better management of natural resources and the efficient use of crop lands. According to the literatures, better understanding of the environment and the good agricultural practices on the land with the good planning can lead ultimately to a proper management of the natural resources and improves the quality of life (Dale & Mclaughlin, 1988; Balla, 1995; Pieri, 1997).

Eventually, it should be mentioned that some of the used indicators were generated from relatively low spatial resolution data (such as livestock and population data), which were found inconsistent with other datasets, those were available in moderate to high spatial resolution. This inconsistency may has affected the final ESAI map. In addition, by using supervised classification as a conventional method that is used to acquire information about the land use/cover map, it was difficult to classify the urban areas as accurately as possible, since it was mixed with some natural features at the study area. In addition, human as well as natural resources information about the study site, as well as the two localities were unavailable in terms of quantity and quality. Therefore, the study had some limitations; hence, further research is recommended (considering the advanced image classification methods, such as, spectral mixture analysis or object-based approaches) for the enhancement of the environmental sensitivity area assessment of desertification.

18.4.4 Model Validation

The AUC technique was adopted in order to verify and measure the reliability of the generated desertification sensitivity map, and also to guarantee the capability of each driving factor as being a causative component for desertification occurrence. For accuracy analysis, the testing points (truly desertified locations) were utilized and correlated spatially versus the sensitivity map. The acquired accuracy assessment has demonstrated a robust prediction of the desertification given by the resultant desertification sensitivity map, producing an AUC of 60 % (Fig. 18.10a). On the other hand, when the independent variables (desertification driving parameters) were correlated independently, a variable AUC values were obtained. This might be due to the inconsistency in surface cover distribution (especially for soil cover), as the area soils were subjected to the disturbance of different factors such as soil salinity and its effect, and sand drift (Khiry, 2007) and mismanagement. Further, land overgrazing and overuse were additionally sharing factors (Dawelbait, 2010). However, the determined prediction (AUC) rates of the separate components produced AUC of 65.2 %, 60.8 %, 54.5 %, and 52.8 % for CQI, VQI, SQI, and MQI, respectively (Fig. 18.10b). This indicates that the climatic and meteorological variables (represented in rainfall and temperature in particular) have probably played exceptional role in forming the surface cover at the area, proving to be the most predictive factor for quantifying desertification, followed by vegetation cover distribution. This can be justified by the fact that the local communities are highly agricultural dependent, which made them intensify cultivation practices.

Despite the fact that AUC varied between 50 and 65 which may be viewed as "weak to moderate model", the inability of the individual parameters has been weakened by the inconsistency of their sub-variables. For instance, the climate quality index CQI which has proven to have a moderate AUC, was hindered by a rainfall index (R) that was generated from a 250 m spatially-based data resolution and correlated with a 30 m-based aspect direction (AD) and a 250 m-based aridity index (AI) in order to form the CQI. However, the shrubby nature of the green surface cover has imposed a signature formed from a holistic surface objects, leading to the moderately acquired VQI-AUC accuracy.

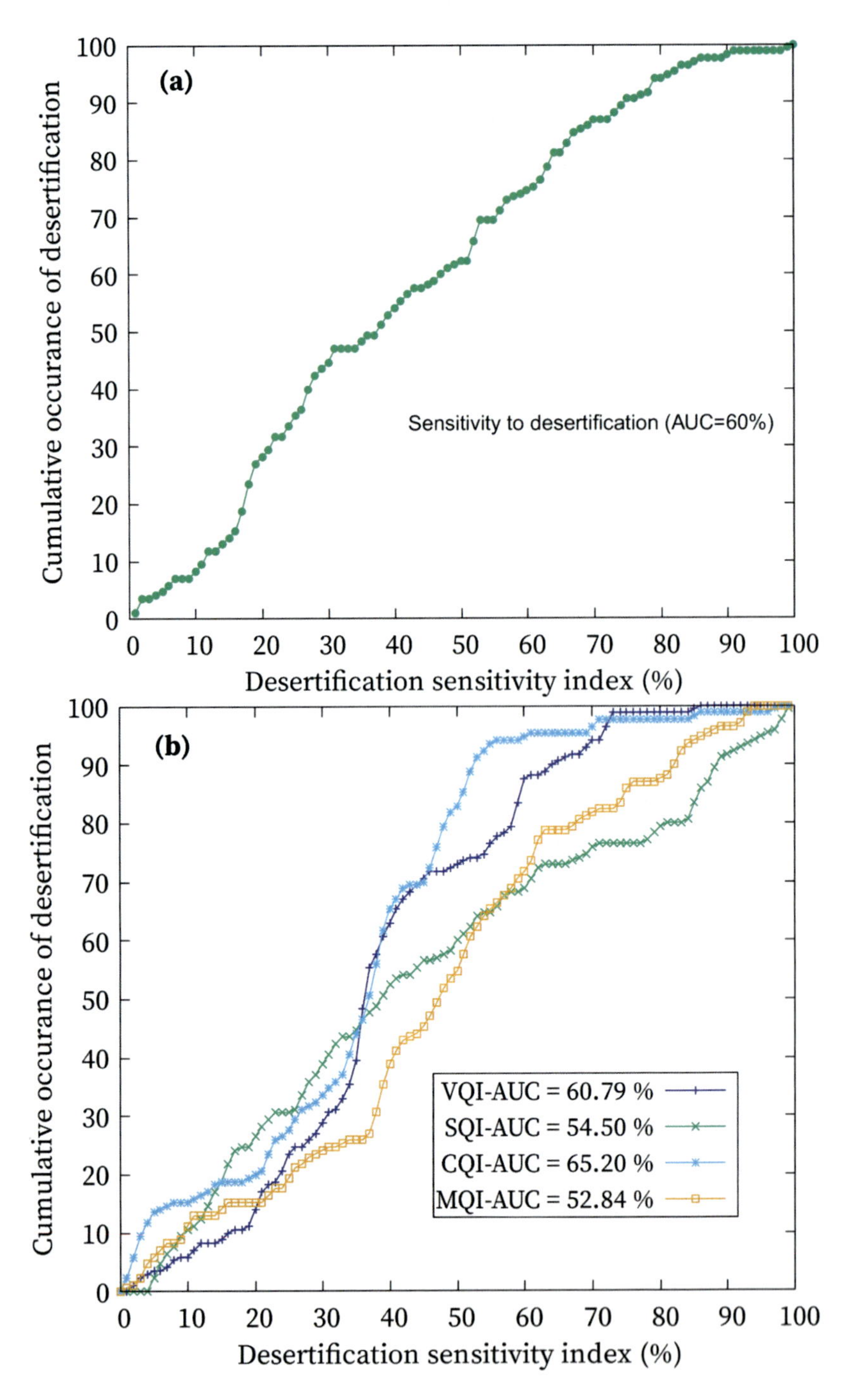

Fig. 18.10 "Parameters Occurrence Index", the spatial validation (AUC) of predicted (**a**) desertification map and (**b**) the independent parameters responsible for desertification quantification and distribution

18.5 Conclusion and Recommendations

In this study, an ESAI map was produced using the MEDALUS model and remote sensing-based indicators, which included; vegetation quality indicator (VQI), soil quality indicator (SQI), climate quality indicator (CQI) and management quality indicator (MQI). The developed model was implemented in the north Kordofan state where Bara and Tayba localities were assigned as pilot study area.

The proposed methodology has shown that 57% of the total area of the study site belongs to "high" and "very-high" sensitivity classes, which were susceptible to desertification. The northern-northwestern and central parts of the study site presented the highest sensitivity and susceptibility to desertification. They were characterized by low rainfall, high soil salinity, overgrazing, overcopping, potentially fire risk, low drought resistant, accumulation of sand dunes/sheets (especially in the area where the Bara and Tayba localities located), and low management of crop lands. The final ESAI map and based-indicators were validated using a receiver operating characteristics (ROC) curves that measure the area under the ROC curve (AUC).

The produced ESAI map presented satisfactory results. The validation of the maps indicated reliable results and acceptable accuracy of the produced map. Therefore, the proposed methodology and the ESAI map should be considered as tools by the local authorities and local communities, in addition to the national and international organizations, beside their policy makers and strategic activists to combat desertification or land degradation and enable sustainable development strategy for agriculture, grazing and other human activities. Ultimately, this will halt and reverse land degradation and halt biodiversity loss. Therefore, the study recommends the following:

1. Information related to the environment and its change can be obtained, improved and monitored through the use of remote sensing data and techniques.
2. The local community, authorities, and national and international organizations should adopt the sustainable development goals (SDGs) to combat desertification, restore degraded lands and soil in the study site through continuous monitoring and assessment of these phenomena using remote sensing satellite imagery.
3. Human activities such as shifting cultivation that is practiced in the study site should be encouraged, and efforts should be devoted to combat cutting down trees for construction or fuel purposes, because this will lead to bare soil that hasten both water and wind erosion and increase degradation. Therefore, these activities should be prevented by adopting and developing new strategies, by the local community, working on improving the environment and preservation and to improve living conditions in arid and semi-arid regions.
4. The conditions of the vegetation should be monitored and assessed regularly through remote sensing satellite imagery and site-based data to prevent depletion of vegetation cover.

5. The local communities, to keep their lands out of degradation and maintain soil productivity, should adopt sustainable land management.
6. Local communities should be involved in any developmental strategies or plans.

Acknowledgement We would like to thank all data providers (for example, FAO, AMESD, the Ministry of Energy and Mines, Geological & Mineral Resources Department, and USGS) for providing the necessary data to conduct this study.

References

Adam, H. E. (2011). *Integration of remote sensing and GIS in studying vegetation trends and conditions in the Gum Arabic Belt in North Kordofan, Sudan*. PhD Thesis, Technology University of Dresden.

Allbed, A., Kumar, L., & Sinha, P. (2014). Mapping and modelling spatial variation in soil salinity in the Al Hassa oasis based on remote sensing indicators and regression techniques. *Remote Sensing, 6*, 1137–1157. https://doi.org/10.3390/rs6021137

AMESD. (2010). *Natural habitat conservation assessment. African monitoring of the environment for sustainable development in the IGAD region ahead*. http://archives.au.int/bitstream/handle/123456789/5948/Africa.html. Accessed 29 Jan 2021.

Balla, A. M. (1995). *Management of problem soils in arid ecosystems* (1st ed., p. 272). CRC Press. https://doi.org/10.1201/9780203748411

Bedoui, C. (2020). Study of desertification sensitivity in Talh region (Central Tunisia) using remote sensing, G.I.S. and the M.E.D.A.L.U.S. approach. *Geoenvironmental Disasters, 7*, 16. https://doi.org/10.1186/s40677-020-00148-w

Budak, M., Gunal, H., Celik, I., YILDIZ, H., Acir, N., & Acar, M. (2018). Environmental sensitivity to desertification in northern Mesopotamia; application of modified MEDALUS by using analytical hierarchy process. *Arabian Journal of Geosciences, 11*, 488. https://doi.org/10.1007/s12517-018-3813-y

Canada, N. R. (2008). *Fundamentals of remote sensing – Introduction*. https://www.nrcan.gc.ca/maps-tools-publications/satellite-imagery-air-photos/remote-sensing-tutorials/fundamentals-remote-sensingintroduction/9363.html. Accessed 13 Feb 2021.

Chung, C. J. F., & Fabbri, A. G. (2003). Validation of spatial prediction models for landslide hazard mapping. *Natural Hazards, 30*(3), 451–472.

Dale, P. F., & Mclaughlin, J. D. (1988). *Land information management: An introduction with special reference to cadastral problems in Third World countries*. Clarendon.

Dawelbait, M. A. A. (2010). *Monitoring desertification in Sudan using remote sensing and GIS*. PhD Thesis, Technology University of Dresden.

Dawelbait, M., & Morari, F. (2012). Monitoring desertification in a Savannah region in Sudan using Landsat images and spectral mixture analysis. *Journal of Arid Environments, 80*, 45–55. https://doi.org/10.1016/j.jaridenv.2011.12.011

Dregne, H. E. (1986). Desertification of arid lands. In *Physics of desertification* (pp. 4–34). Springer.

El-Baz, F., & Hassan, M. H. A. (Eds.). (1986). *Physics of desertification*. Springer. https://doi.org/10.1007/978-94-009-4388-9

Elhag, M. (2016). Evaluation of different soil salinity mapping using remote sensing techniques in arid ecosystems, Saudi Arabia. *Journal of Sensors*. https://doi.org/10.1155/2016/7596175

FAO. (2007). *Digital soil map of the world*. http://www.fao.org/geonetwork/srv.html. Accessed 14 Feb 2021.

FAO. (2012). *Geospatial information for sustainable food systems, Land Cover Atlas of Sudan (2012)*. http://www.FAO.org/geospatial/en.html. Accessed 1 Feb 2021.

Fick, S. E., & Hijmans, R. J. (2017). WorldClim 2: New 1-km spatial resolution climate surfaces for global land areas. *International Journal of Climatology, 37*, 4302–4315. https://doi.org/10.1002/joc.5086

Fischer, G., Nachtergaele, F. S., Prieler, H. T., Van Velthuizen, L., & Verelst, D. W. (2008). *Global agroecological zones assessment for agriculture (GAEZ 2008)*. IIASA/FAO.

Gao, J., & Liu, Y. (2010). Determination of land degradation causes in Tongyu County, Northeast China via land cover change detection. *International Journal of Applied Earth Observation and Geoinformation, 12*, 9–16. https://doi.org/10.1016/j.jag.2009.08.003

Helldén, U. (2003). Desertification and theories of desertification control: A discussion of Chinese and European concepts. In S. Guangchang (Ed.), *Proceedings of the China-EU workshop on integrated approach to combat desertification* (pp. 94–104). Ministry of Science and Technology of China, Chuuina Association for International Sciences 3 and Technology Cooperation.

Hulme, M. (2001). Climatic perspectives on Sahelian desiccation: 1973–1998. *Global Environmental Change, 11*(1), 19–29.

Jiang, L., Bao, A., Jiapaer, G., Guo, H., Zheng, G., Gafforov, K., Kurban, A., & De Maeyer, P. (2019). Monitoring land sensitivity to desertification in Central Asia: Convergence or divergence? *Science of the Total Environment, 658*, 669–683. https://doi.org/10.1016/j.scitotenv.2018.12.152

Khiry, M. A. (2007). *Spectral mixture analysis for monitoring and mapping desertification processes in semi-arid areas in North Kordofan State, Sudan*. PhD Thesis, Technology University of Dresden. https://doi.org/10.23689/fidgeo-307.

Kosmas, C., Ferrara, A., Briassouli, H., & Imeson, A. (1999). *Methodology for mapping environmentally sensitive areas (ESAs) to desertification* (pp. 31–47).

Lamqadem, A. A., Saber, H., & Pradhan, B. (2018a). Quantitative assessment of desertification in an arid oasis using remote sensing data and spectral index techniques. *Remote Sensing, 10*, 1862. https://doi.org/10.3390/rs10121862

Lamqadem, A. A., Pradhan, B., Saber, H., & Rahimi, A. (2018b). Desertification sensitivity analysis using MEDALUS model and GIS: A case study of the oases of middle Draa Valley, Morocco. *Sensors (Basel), 18*(7), 2230. https://doi.org/10.3390/s18072230

Liang, S. (2001). Narrowband to broadband conversions of land surface albedo I: Algorithms. *Remote Sensing of Environment, 76*, 213–238.

Mandrekar, J. N. (2010). Receiver operating characteristic curve in diagnostic test assessment. *Journal of Thoracic Oncology, 5*, 1315–1316. https://doi.org/10.1097/JTO.0b013e3181ec173d

Ministry of Energy and Mines, Geological & Mineral Resources Department. (1981). *Geological map of Sudan, Khartoum, Sudan*. http://www.esdac.jrc.ec.europa.eu/ESDB_Archive/EuDASM/Africa/lists/y4_csd.html. Accessed 12 Feb 2012.

Mohamed, T. A. (2016). *Monitoring and analyzing of desertification trend in north Sudan using MODIS images FROM 2000 to 2014*. MSc thesis, Southern Illinois University at Carbondale.

National Centre for Research, Remote Sensing and Seismology Authority (RSSA) of Sudan. (2011). *Land cover/land use digital map*. http://www.ncr.gov.sd.html. Accessed 22 Feb 2021.

Nwilo, P. C., Olayinka, D. N., Okolie, C. J., Emmanuel, E. I., Orji, M. J., & Daramola, O. E. (2020). Impacts of land cover changes on desertification in northern Nigeria and implications on the Lake Chad Basin. *Journal of Arid Environments, 181*, 104190. https://doi.org/10.1016/j.jaridenv.2020.104190

Olsson, L. (1983). *Desertification or climate? Investigation regarding the relationship between land degradation and climate in the Central Sudan*. Gleerup.

Olsson, K. (1985). *Remote sensing for fuel wood resources and land degradation studies in Kordofan, The Sudan*. PhD Thesis, Lund University.

Pieri, C. (1997). Planning sustainable land management: The hierarchy of user needs. *ITC Journal, 3*(4), 223–228.

Prăvălie, R., Săvulescu, I., Patriche, C., Dumitraşcu, M., & Bandoc, G. (2017). Spatial assessment of land degradation sensitive areas in southwestern Romania using modified MEDALUS method. *CATENA, 153*, 114–130. https://doi.org/10.1016/j.catena.2017.02.011

Remondo, J., Bonachea, J., & Cendrero, A. (2004). *Probabilistic landslide hazard and risk mapping on the basis of occurrence and damages in the recent past. Landslides: Evaluation and stabilization* (pp. 125–130). Balkema, Taylor & Francis Group.
Salih, D. M. (2006). *Mapping and assessment of land use, land cover using remote sensing and GIS in North Kordofan State, Sudan*. Technology University of Dresden. https://doi.org/10.23689/fidgeo-338.
Salih, A., Hassaballa, A. A., & Ganawa, E. (2021). Mapping desertification degree and assessing its severity in Al-Ahsa Oasis, Saudi Arabia, using remote sensing-based indicators. *Arabian Journal of Geosciences, 14*, 192. https://doi.org/10.1007/s12517-021-06523-7
Salvati, L., & Bajocco, S. (2011). Land sensitivity to desertification across Italy: Past, present, and future. *Applied Geography, Hazards, 31*, 223–231. https://doi.org/10.1016/j.apgeog.2010.04.006
Schmidt, H., & Karnieli, A. (2000). Remote sensing of the seasonal variability of vegetation in a semi-arid environment. *Journal of Arid Environments, 45*, 43–59.
Sepehr, A., Hassanli, A., Ekhtesasi, M., & Bodagh, J. J. (2007). Quantitative assessment of desertification in south of Iran using MEDALUS method. *Environmental Monitoring and Assessment, 134*, 243–254. https://doi.org/10.1007/s10661-007-9613-6
Swets, J. A. (1988). Measuring the accuracy of diagnostic systems. *Science, 240*(4857), 1285–1293. https://doi.org/10.1126/science.3287615
Tien, B. D., Pradhan, B., Lofman, O., & Revhaug, I. (2012). Landslide susceptibility assessment in Vietnam using support vector machines, decision tree, and Naive Bayes models. *Mathematical Problems in Engineering, 26*. https://doi.org/10.1155/2012/974638
Trabucco, A., & Robert, Z. (2018). *Global aridity index and potential evapotranspiration (ET0) climate database v2*. https://doi.org/10.6084/m9.figshare.7504448.v3.
UNEP NM, & Thomas, D. (1992). *World atlas of desertification* (pp. 15–45). Edward Arnold.
Verón, S. R., Paruelo, J. M., & Oesterheld, M. (2006). Assessing desertification. *Journal of Arid Environments, 66*, 751–763. https://doi.org/10.1016/j.jaridenv.2006.01.021
Zhang, S., Chen, H., Fu, Y., Niu, H., Yang, Y., & Zhang, B. (2019). Fractional vegetation cover estimation of different vegetation types in the Qaidam Basin. *Sustainability, 11*, 864. https://doi.org/10.3390/su11030864

Chapter 19
A Selected Benchmark for Landslides Susceptibility Assessments in Northern Morocco

Meryem Elmoulat and Lahcen Ait Brahim

Abstract During the past two decades (2000–2020) many studies have been conducted to build strategies as preventing landslides. Most of these approaches are focusing on different methods (statistical, mathematical, and machine learning. . .etc.) for the same study area. In the present research, we are benchmarking the results of our approach at two different geographical areas by using the Frequency Ratio (FR). First, we are comparing robustness between two susceptibility assessments for smaller areas such as Ceuta (257 Km^2) and Tétouan-Ras-Mazari (590 Km^2) to find out their similarities and their differences. Second, for both areas, we are utilizing the same controlling factors e.g., lithology, landuse, fault density, drainage density, slope degrees, slope aspects, and elevation. Third, we calculated the Prediction Rates (PR) of all factors. Fourth, our results show that three factors (Landuse, Elevation, Lithology for Area-1; in addition to, landuse, elevation, and fault density for Area-2 are sufficient for establishing acceptable LSAs for each area. Fifth, the evaluation process for the models is assessed by Areas Under the Curve (AUC). The accuracy of Models-I is 78.55% and Model- II is 79.25%. Last, our strategy could be an asset for our successors interested in Landslide Susceptibility Assessments (LSAs) in mountains MENA regions.

Keywords Frequency ratio (FR) · Landslides susceptibility assessments (LSAs) · Smaller areas · And northern Morocco

19.1 Introduction

Landslides, both natural and man-made are often the cause of extensive material and human damages. These phenomena represent one of the most serious problems affecting several countries in North Africa (Morocco, Algeria, and Tunisia). Morocco, by its geographical location, is regularly exposed to climate change, the

M. Elmoulat (✉) · L. Ait Brahim
Research Unit GEORISK, LG2E Laboratory, Faculty of Sciences, University Mohammed V Rabat, Rabat, Morocco

M. M. Al Saud (ed.), *Applications of Space Techniques on the Natural Hazards in the MENA Region*, https://doi.org/10.1007/978-3-030-88874-9_19

recurrence of drought, the diversity of geological, hydrogeological, and geomorphological nature, which play an instrumental role in the dynamic of triggering Mass Movements (MM). The serious catastrophic events whose impact on the natural environment as well as on human life is nowadays concerned scientists, engineers, and geologists; because of the destructive effect on the economy and human life, which prevents the exploitation of the potential socio-economic development of nations. According to a report published by the Algerian Ministry of agriculture and rural development (Mate, 2004), there are about seven natural hazards have been identified. These risks are usually landslides that influence the development of several Algerian cities. These Mass Movements (MM) led to the destruction of several buildings and the relocation of many families. The volumes of the land involved are highly diverse, depending on the type, depth, activity, and speed of the landslides. The deeper the MM, the more the disaster is damaging. Tunisia is known for landslides that affect its mudstones and marlstone slopes (Maurer, 1976, 1979). According to a definition given by (Marthelot, 1957, 1959), the different types and causes of landslides in Tunisia, are phenomena that can cause substantial damage and significantly harming the well-being of populations by threatening human life (Kassab, 1976, 1979). In Morocco and as mentioned by (Pateau, 2014) many the regions affected by these natural disasters. Rif chain in the northern side of the Kingdom is very active from a geodynamic, geomorphological (ravinement, landslides, etc.), climatic (longer periods of drought in late summer and an intensification of torrential rains in winter) points of view. These traits have led to an increase in risks inherently related to extreme climate change recorded since the 2000s. Moreover, other ingredients are also responsible for MM like economic and socio-demographic growth. Tangier-Tétouan region of the Western Rif is increasingly populated (3,157,075 inhabitants (Haut-commissariat au Plan, 2014)), as it is crossed by newly established communication channels (Mediterranean bypass) and attracts more socio-economic activities.

In this chapter, we will use two study sites in the Tangier-Tetouan region: Ceuta city and Tétouan-Ras-Mazari. More information about these two study areas can be found in our previously published works (Elmoulat & Ait Brahim, 2018, 2020). The present manuscript is crucial to carry out our scientific approach to identifying and assessing landslides susceptibility; in order to minimize the destructive damage of these phenomena. Our choice is justified by the fact that these two sites are integrated into the province of Tetouan, which comes first in the areas severely affected by landslides in Northern Morocco. In explanation, several researchers around the world have conducted benchmarking between many statistical, mathematical, and/or computer learning techniques; in our current study, we focus on leading a Benchmark between two landslides susceptibility assessments (LSAs) of two different geographical areas by using the same statistical approach and the same type of independent factors. This present chapter is organized in four consecutive sections arranged as follows: In the first section (Sect. 19.1), we introduce the research subject that is the mapping of landslides in two different sites using the same

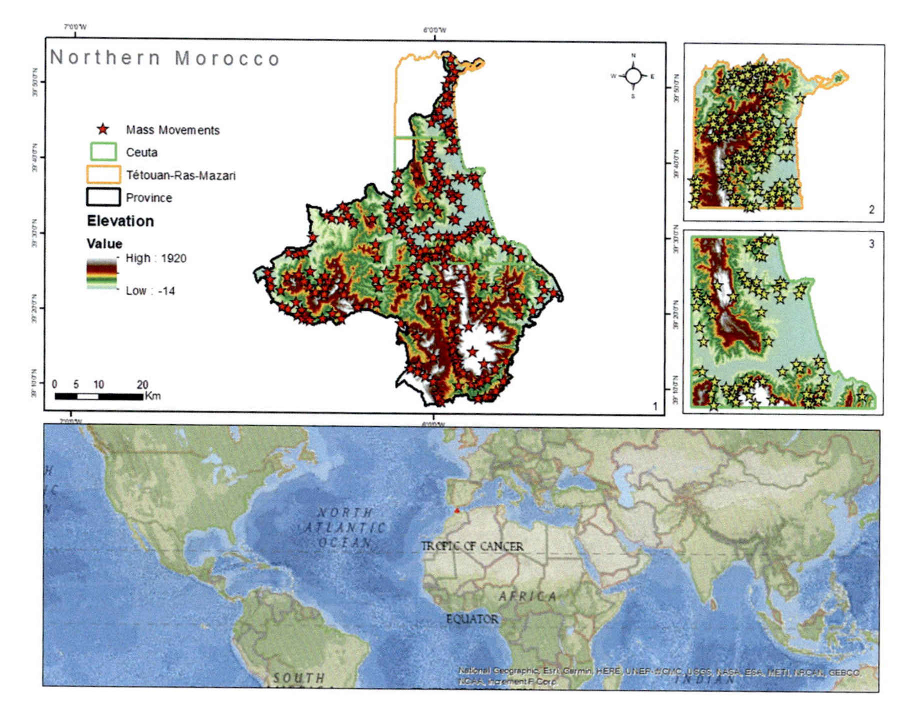

Fig. 19.1 Study areas (Area-1 "Ceuta", Area-2 "Tétouan-Ras-Mazari")

statistical approach (FR). In the second section (Sect. 19.2), in which we will explain the procedures and the techniques followed to obtain our desired results. In the third section (Sect. 19.3), we will present our findings. In the fourth section (Sect. 19.4), we provide interpretation and meaning to our previously presented data in the results section. In the fifth section (Sect. 19.5), we conclude and share an overview of our future works (Fig. 19.1).

19.2 Methods and Procedures

In this section, we are going to explain in detail how our questions are answered. This section includes several sub-sections about the implementation of our methodology. First, we provide an overview of our datasets, materials, and sources of our inputs. Second, we explain how we gathered our geodatabase and the preparation process. Third, we talk about technical details concerning the projections and provide useful information related to our data. Fourth, we cite our previous works and summarize the main information regarding the most important inputs of our modeling, which are the dependents (landslides) and independents factors (seven controlling factors). Fifth, we cite certain beneficial references about the statistical approach employed in this research. Sixth, we calculate the Prediction Rate (PR) of all our controlling factors and in relationship with landslides. Figure 19.2 depicts the detailed flowchart of our methodology to conduct this research.

19.2.1 Datasets Sources and Materials

Remote sensing imagery becomes more important in landslide studies. Comparison of images collected before and after the event gives relevant support to landslide recognition. Landslide inventory data collected from landslide mapping results and historical records can be employed as a database for susceptibility mapping using GIS and statistical approaches. According to our previously published works (Elmoulat & Ait Brahim, 2018, 2020), a summary of the sources of the datasets is provided in the table below (Table 19.1):

Based upon our previous research works (Ait Brahim et al., 2002; Elmoulat et al., 2015) and the ones conducted by our predecessors (Mastere et al., 2015) in the Rif Region seven parameters (lithology, fault density, landuse, drainage density, slope degree, slope aspect, and elevation) are mapped to run the two models (Model-I for Area- 1 and Model-II for Area-2). All these parameters are geo-processed in ArcMap environment except for lithology and landuse for the enhancement and the classification are assisted by using Erdas Imagine.

19.2.2 Production of Dependents and Independents Factors

On one hand, approximately 70% (114 centroids) out of 130 total landslides polygons for Ceuta and 70% (123 centroids) out of 165 for Tétouan-Ras-Mazari were used to train each model. On the other hand, 16 and 42 centroids were used to test model-I and Model-II, respectively. Moreover, seven parameters were used in mapping LSAs for both Models, i.e. lithology, landuse, fault density, drainage density, slope degree, slope aspects, and elevation. All thematic maps have been converted to Grid format. The pixel size of each grid is 100 m.

19.2.2.1 Lithology

Landsat ETM + 7 images are a beneficial toolset to discriminate the different spectral properties for lithological units (Ciampalini et al., 2014). They represent a great

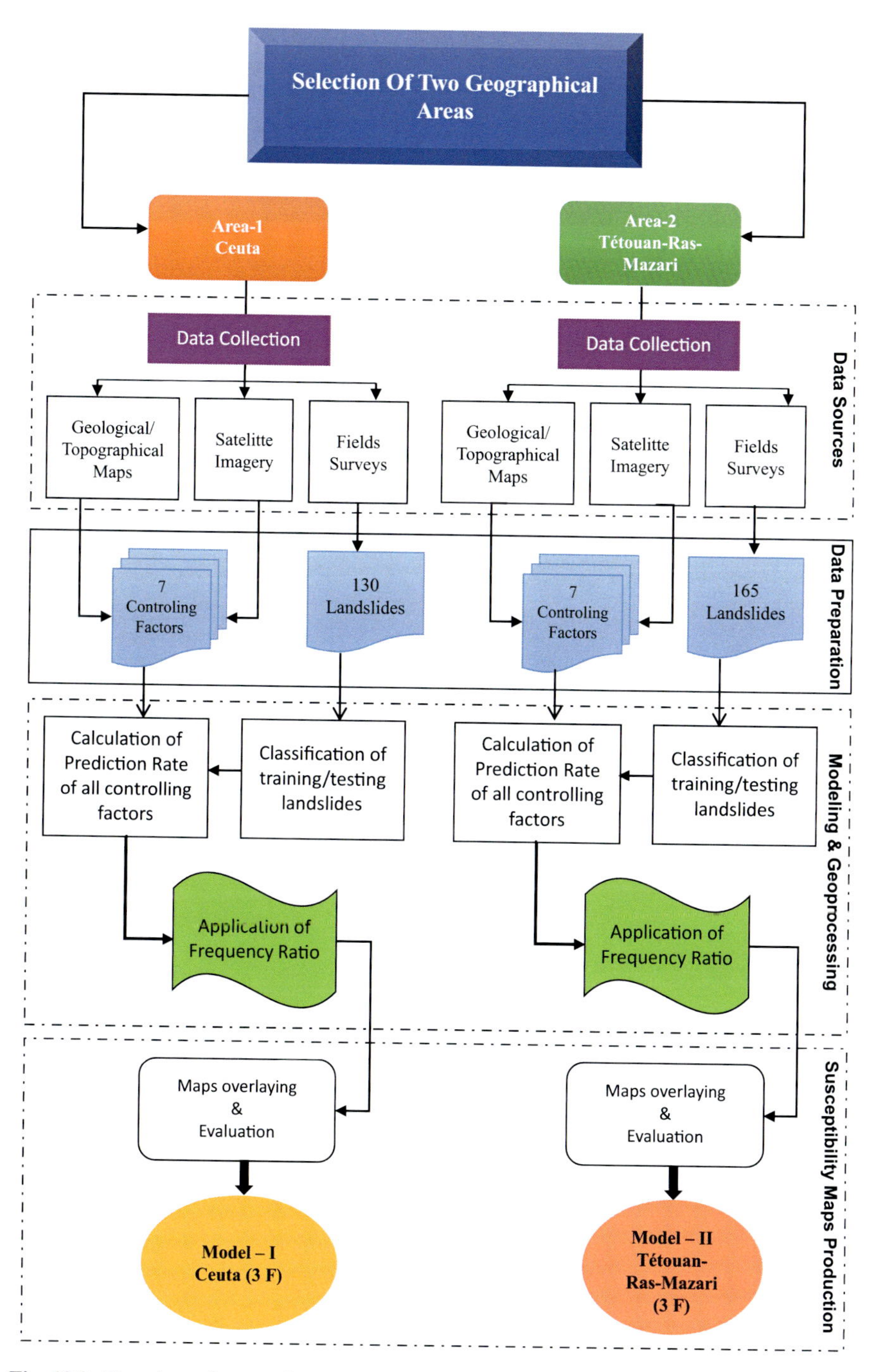

Fig. 19.2 Flowchart of our methodology

Table 19.1 Datasets sources for the two study areas

Data type	Data	Sources	Details
Dependent factor	Landslides	Field surveys	130 landslides for Ceuta and 126 for Tétouan-Ras-Mazari
Independent	Lithology	Satellite imagery: http://earthexplorer.usgs.gov Geological maps: scale of 1/50,000 (Ceuta & Tétouan-Ras-Mazari).	1- LANDSAT ETM+ images consist of seven spectral false-color composites bands with a spatial resolution of 30 m for Bands 1–5, and 7. The resolution for Band 8 (panchromatic) is 15 m. 2- Landsat TM image compose of 6 spectral bands and 1 thermal band (Band 6) with 30 m of spatial resolution.
	Fault density	Satellite imagery: http://earthexplorer.usgs.gov Geological maps: scale of 1/50,000 (Ceuta & Tétouan-Ras-Mazari).	1- LANDSAT ETM+ images consist of seven spectral false-color composites bands with a spatial resolution of 30 m for Bands 1–5, and 7. The resolution for Band 8 (panchromatic) is 15 m. 2- Landsat TM image compose of 6 spectral bands and 1 thermal band (Band 6) with 30 m of spatial resolution.
	Landuse	Satellite imagery: http://earthexplorer.usgs.gov	LANDSAT multi-sensors (TM5, ETM + 7 et ETM + 8), multi-date (1999, 2009, and 2018)
	Drainage density	Digital Elevation Model (DEM) form Shuttle Radar Topography Mission (SRTM): To generate high spatial resolution of 90 meters digital topographic model of the study area. Topographical map: Equidistance contours of 10 meters and scale of 1/50,000.	Sensor: C-band and X-Band Scene size: 1-degree latitude × 1 degree longitude Capture resolution: 1 arc second Pixel resolution: 30 m
	Slope degree	DEM from SRTM	
	Slope aspect	DEM from SRTM	
	Elevation	DEM from SRTM	

source for lithological mapping where access is difficult either for political or geomorphological reasons. For both study sites (Ceuta City and Tétouan-Ras-Mazari), we mapped lithology and classified it into five classes (Figs. 19.3a and 19.4a).

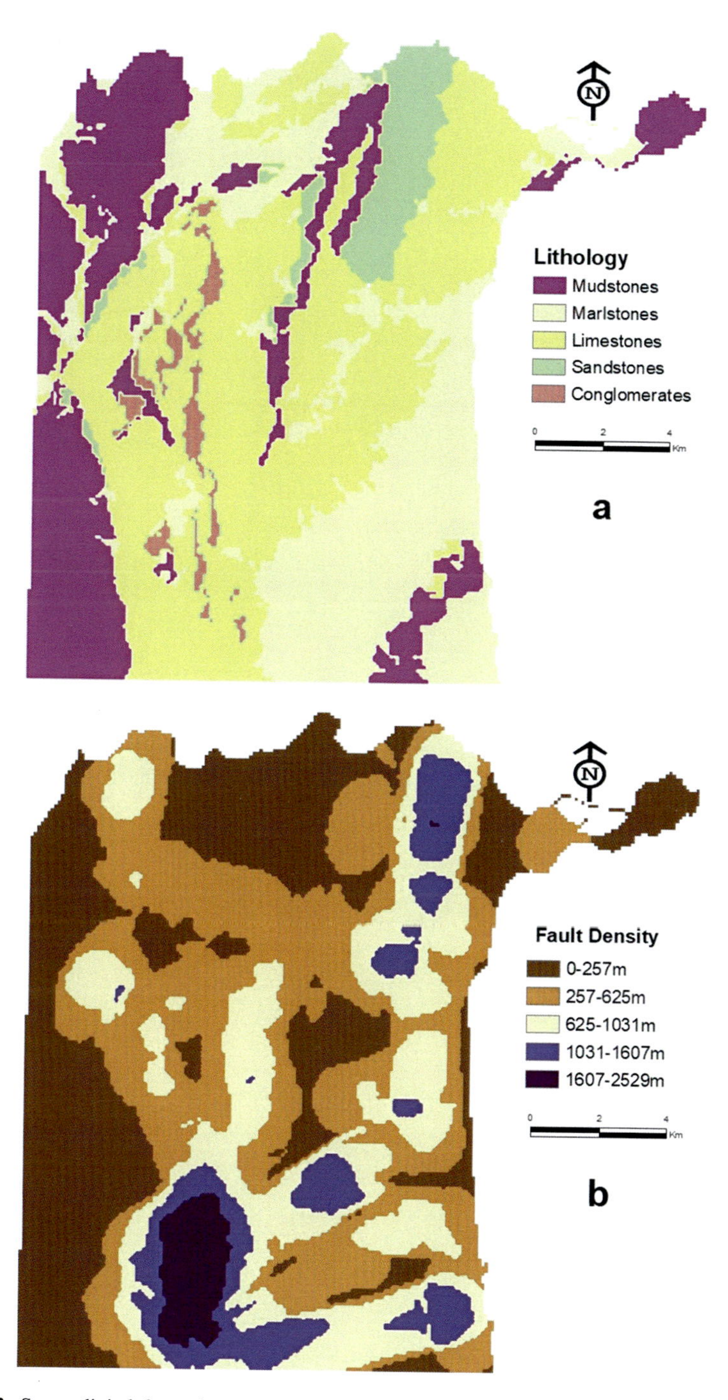

Fig. 19.3 Seven digital thematic maps of the controlling factors of Ceuta City

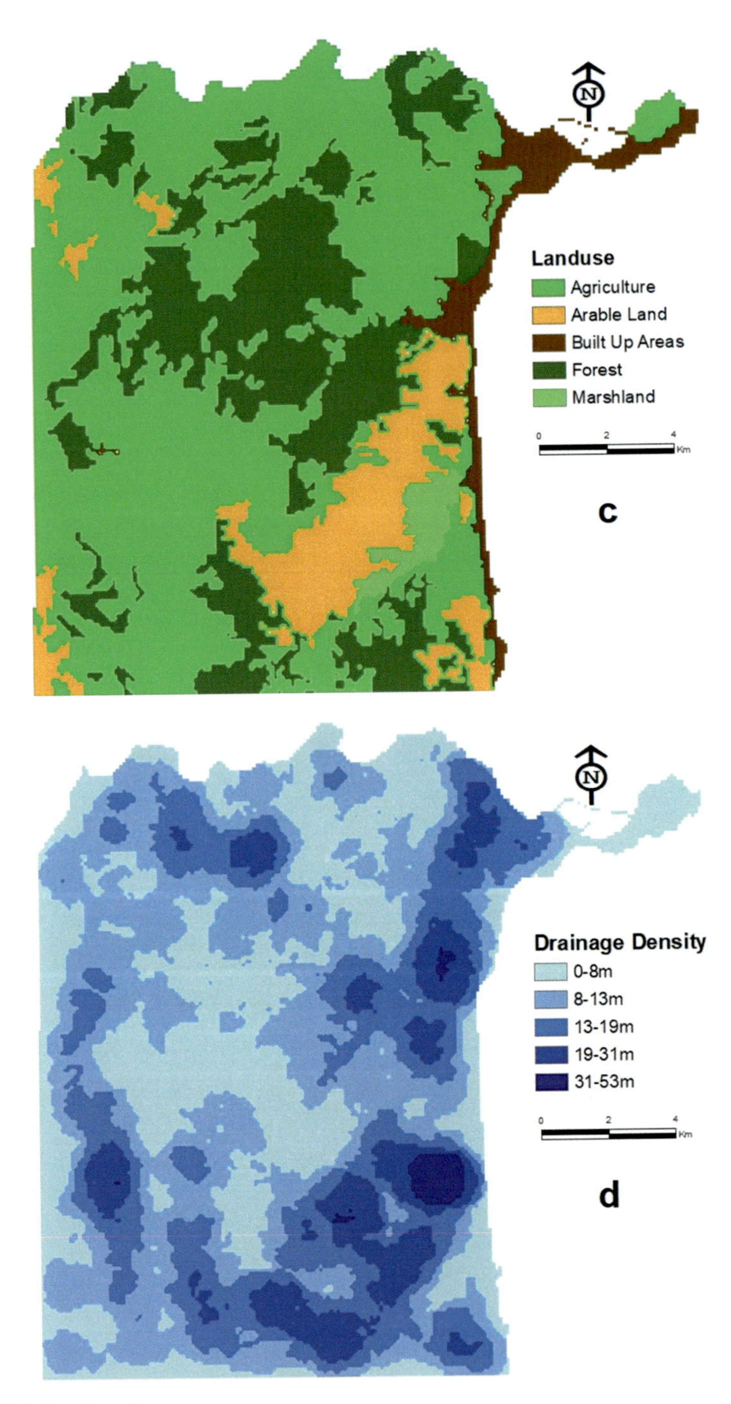

Fig. 19.3 (continued)

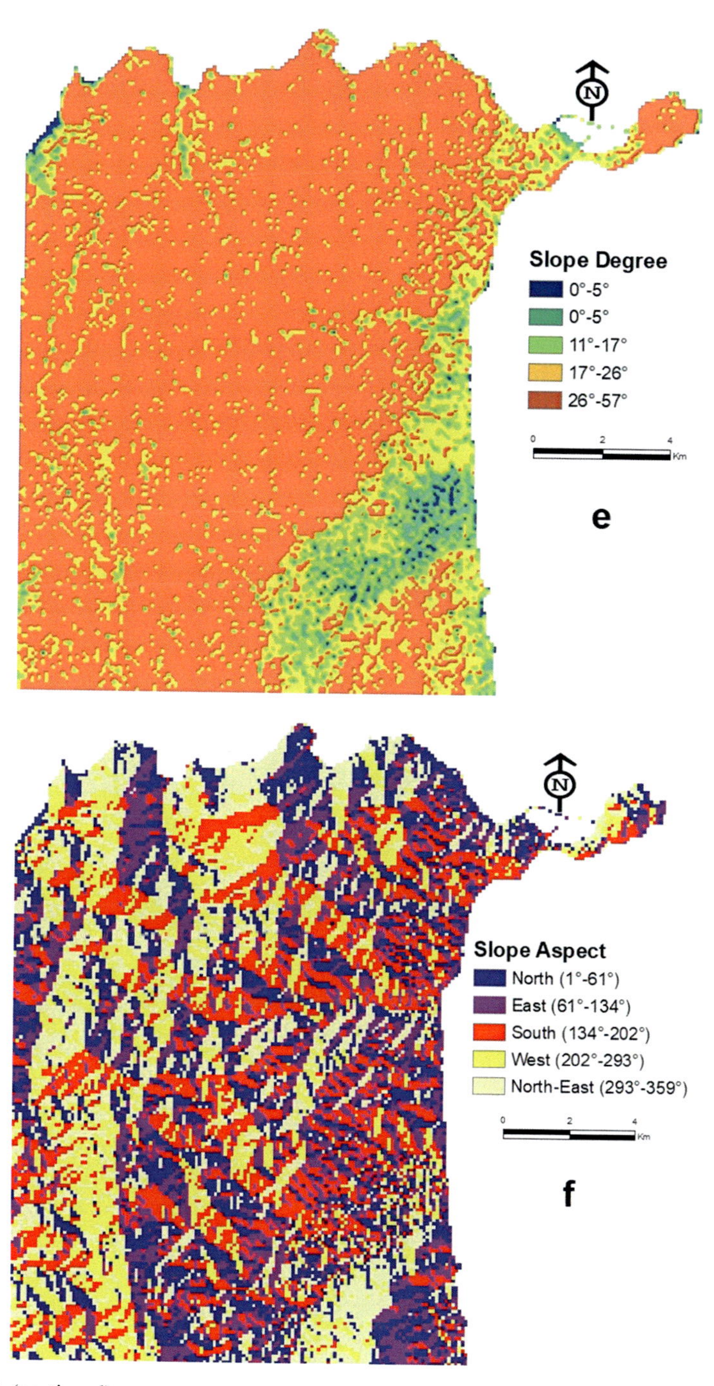

Fig. 19.3 (continued)

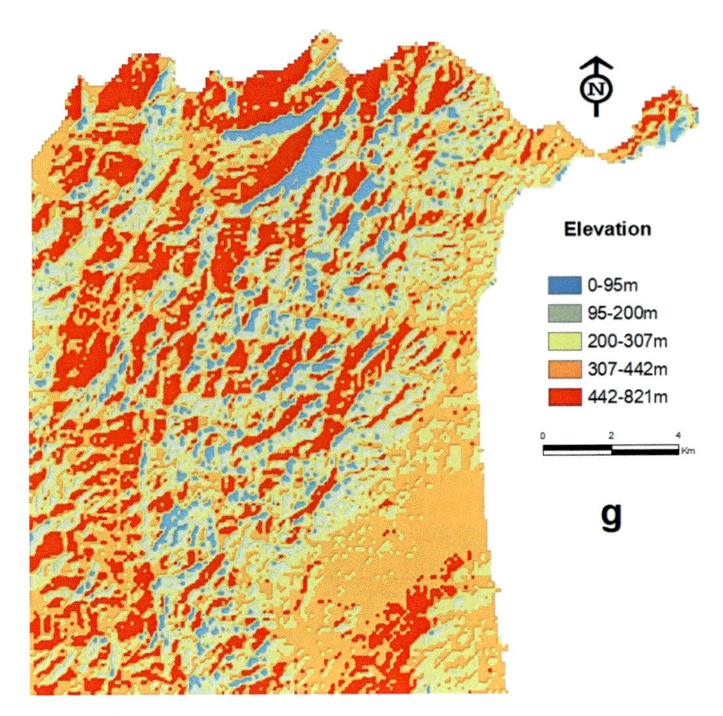

Fig. 19.3 (continued)

19.2.2.2 Fault Density

Experts (Mansour & Ait Brahim, 2005; Shi et al., 2016) in Western Rif have paved the way and allowed us to create faults map for two study areas. Faults maps of Ceuta (Fig. 19.3b) and of Tétouan-Ras-Mazari (Fig. 19.4b) are established by using geological maps, landsat+7, and validated by ground truth studies.

19.2.2.3 Landuse

Vegetation is one of the most prone areas to landslides. The landuse maps have been created based on the supervised classification of ETM+ and validated with field surveys. Figures 19.3c and 19.4c depicts the different classes of landuse factor for Area-1 and Area-2, respectively.

19.2.2.4 Drainage Density

Drainage density was automatically extracted from DEM using ArcHydro tools. The final maps of drainage density were validated using the topographic map of Ceuta and Tétouan-Ras- Mazari (Figs. 19.3d and 19.4d).

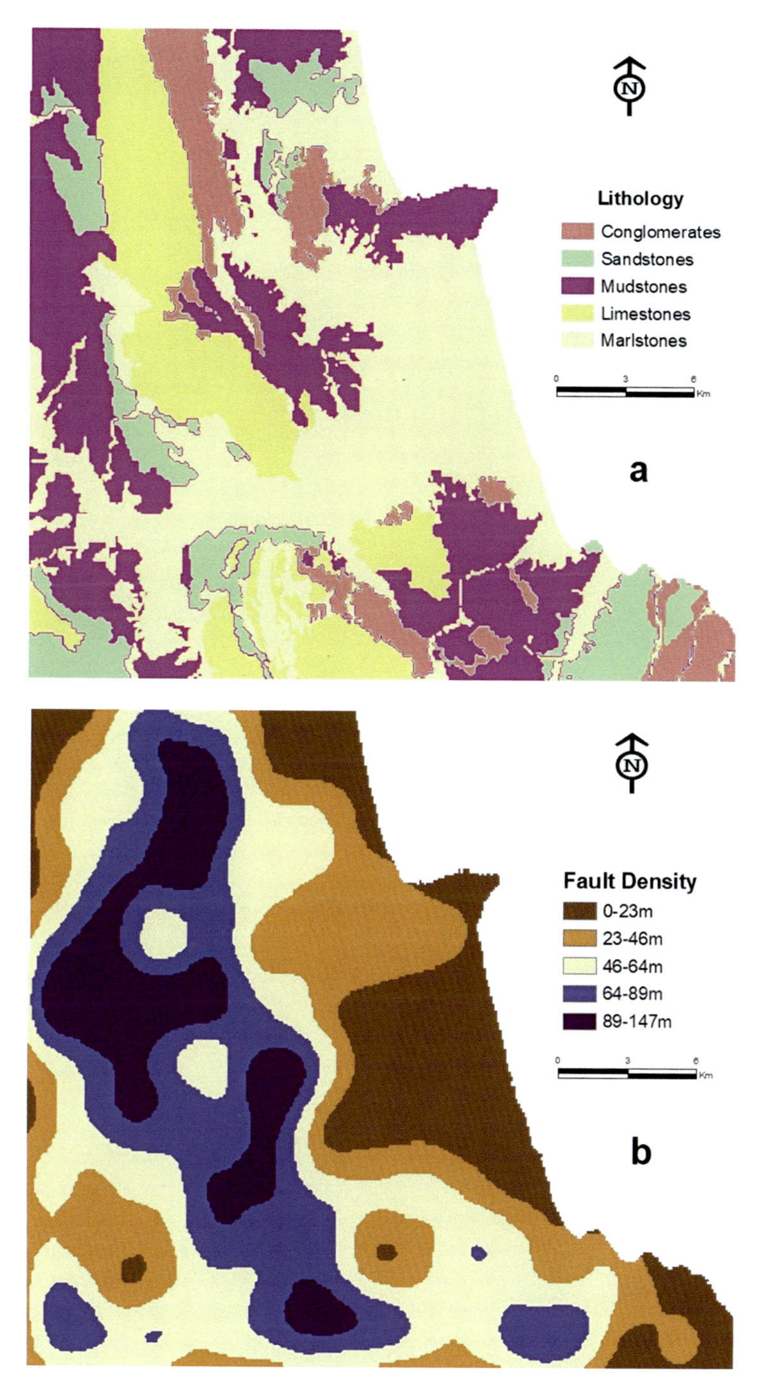

Fig. 19.4 Seven digital thematic maps of the controlling factors of Tétouan-Ras-Mazari

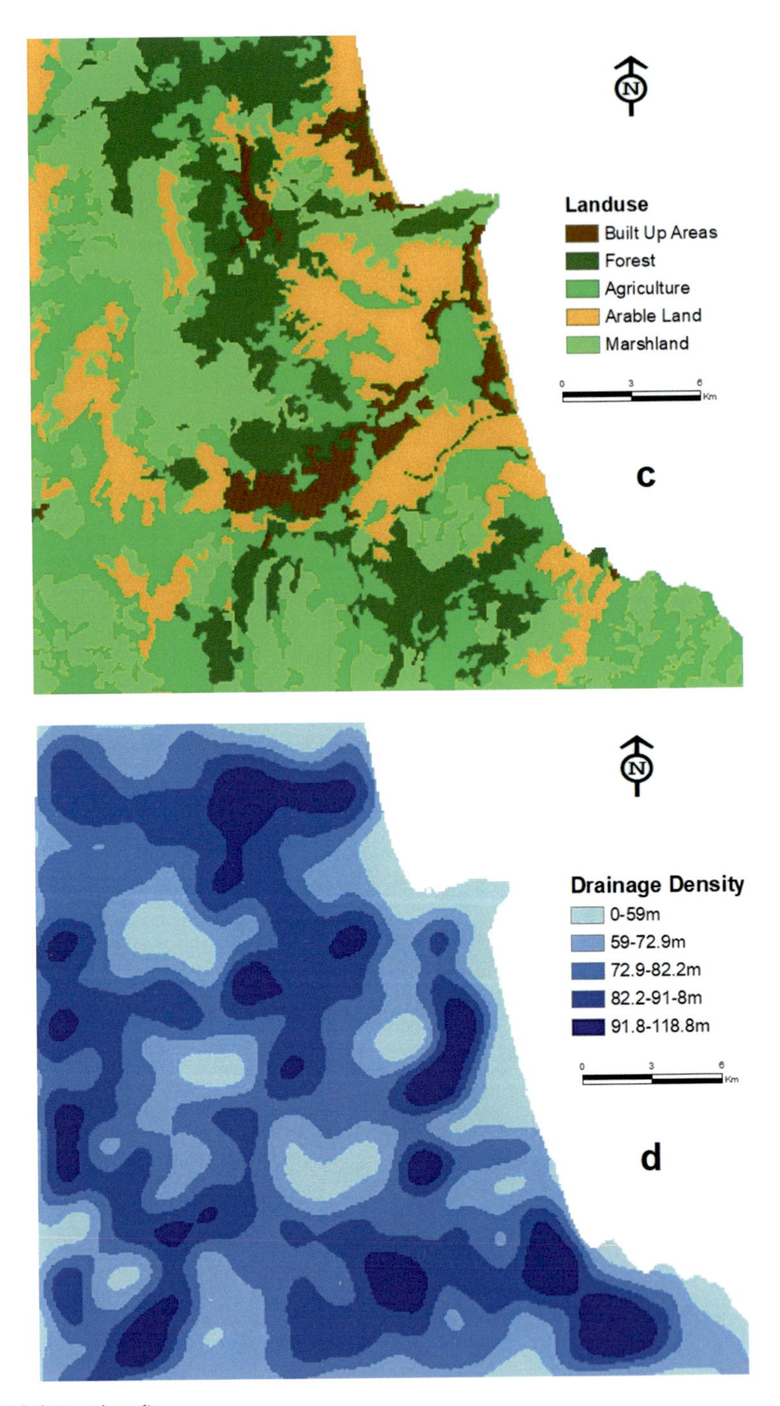

Fig. 19.4 (continued)

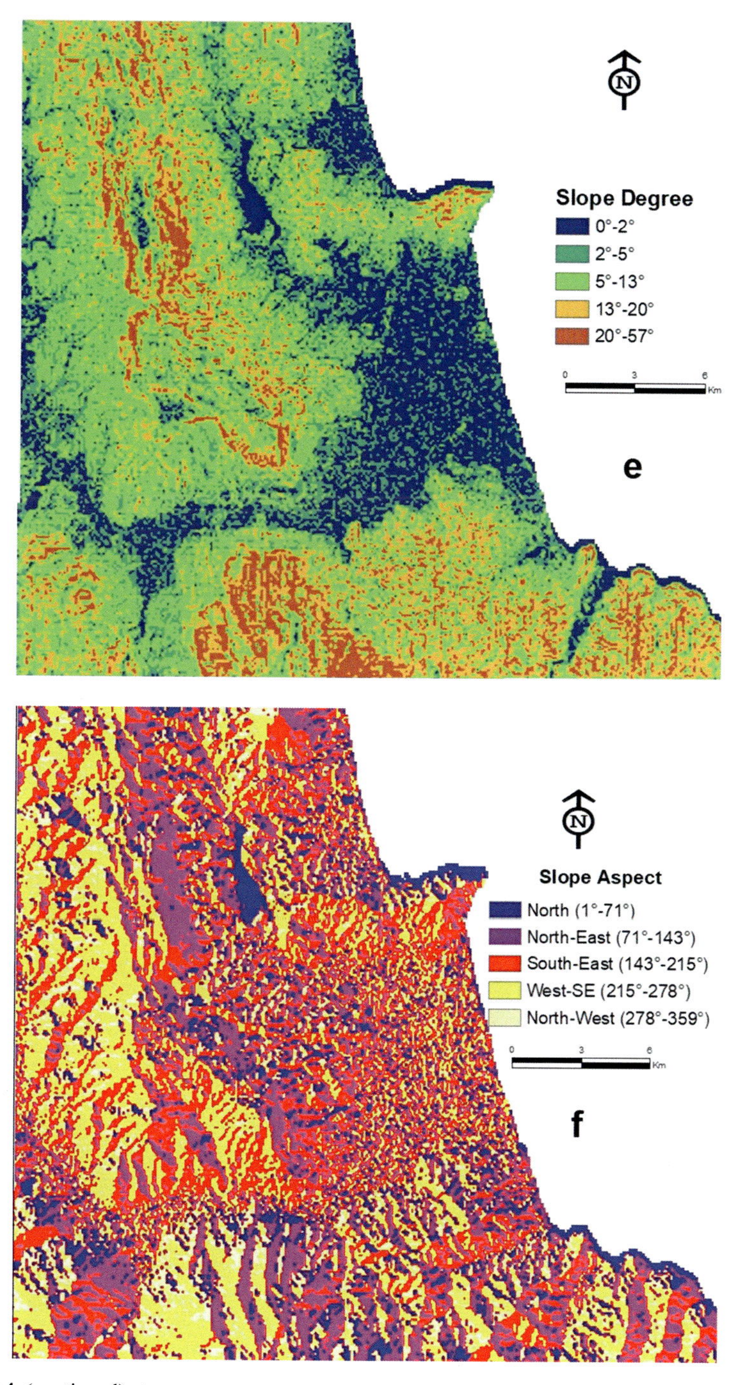

Fig. 19.4 (continued)

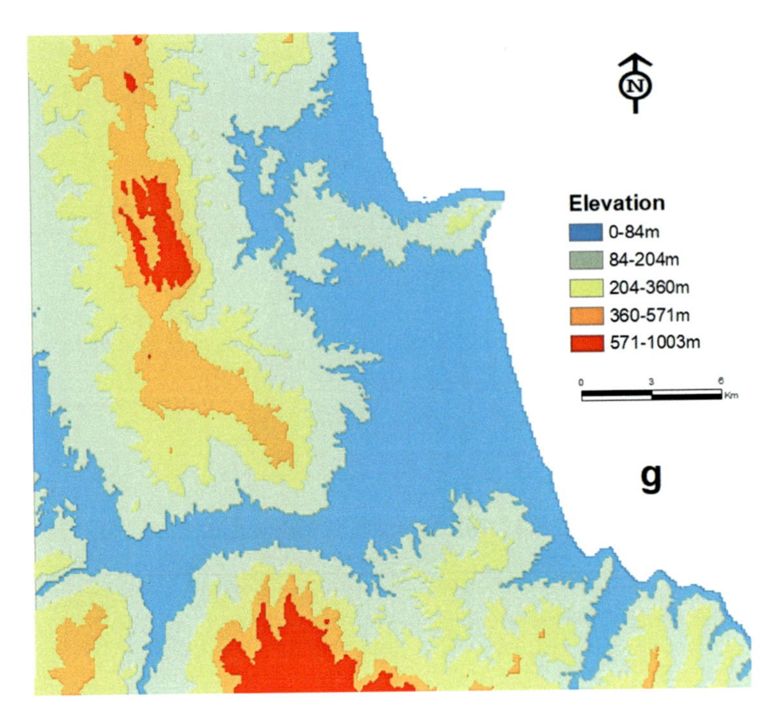

Fig. 19.4 (continued)

19.2.2.5 Slope Degrees

Several authors (Lee, 2019) suggested that slope degrees are mandatory for LSAs. Digital Elevation Model (DEM) extracted from Shuttle Radar Thermometer Mission (SRTM) of 30 m spatial resolution covering our study areas was a useful source to create slope degree maps (Figs. 19.3e and 19.4e).

19.2.2.6 Slope Aspect

The parameter indicates the orientation of the surface that faces at a specific location. Slope aspects maps are considered as a topographical factor that was assessed by using the DEM of the study areas. Figures show slope aspect maps for Ceuta and Tétouan-Ras- Mazari, respectively. of.

19.2.2.7 Elevation

Elevation represents used for points above the surface in each altitude. It used to measure the relationship between altitude above sea level and areas prone to landslides. Elevation maps for Area-1 and Area-2 are created by using the DEM and they are given by Figs.

19.2.3 Relationship Between Dependents and Independents Factors

To better understand the relationship of all the independent factors with landslide occurrences, we create Pie charts for Area-1 (Fig. 19.5) and for Area-2 (Fig. 19.6). We provide an explanation and interpretation of all the scientific reasoning behind these figures in the discussion section.

19.2.4 Method of Landslides Susceptibility Assessment

According to a review article written by (Nicu, 2017) who conducted a survey based on 776 articles during the last two decades (1999–2018), 24.7% of total articles used Frequency Ratio (FR) to produce GIS-based landslide Susceptibility Mapping. In addition to that, many authors (Kim et al., 2018; Chen et al., 2019; Zhou et al., 2016), proved that FR predicts great results in terms of landslide assessment. In other terms, this statistical approach depicts the relationships between landslides' occurrence in the past and landslides controlling factors. This being said, the greater the ratio is, the stronger the relationship between the landslides and the controlling factors can be (Kirat, 1993). For these reasons and more, we are implementing the same method as our previous research conducted over Ceuta city (Elmoulat & Ait Brahim, 2020) to map landslides susceptibility for two different areas at local scales (Area-1) and (Area-2). All the calculated values mainly ratio, FR, the absolute difference between the maximum and minimum SA values and PR for all factors and for both areas are given in Tables 19.2 and 19.3 below. More information about the equations and how we conducted our calculation can be found in (Elmoulat & Ait Brahim, 2020).

Figure 19.7 depicts the PR of each factor for both models from low to the highly predicted.

19.3 Results

This section is dedicated to only the presentation of our results. Any reading/ interpretation is provided in the discussion section.

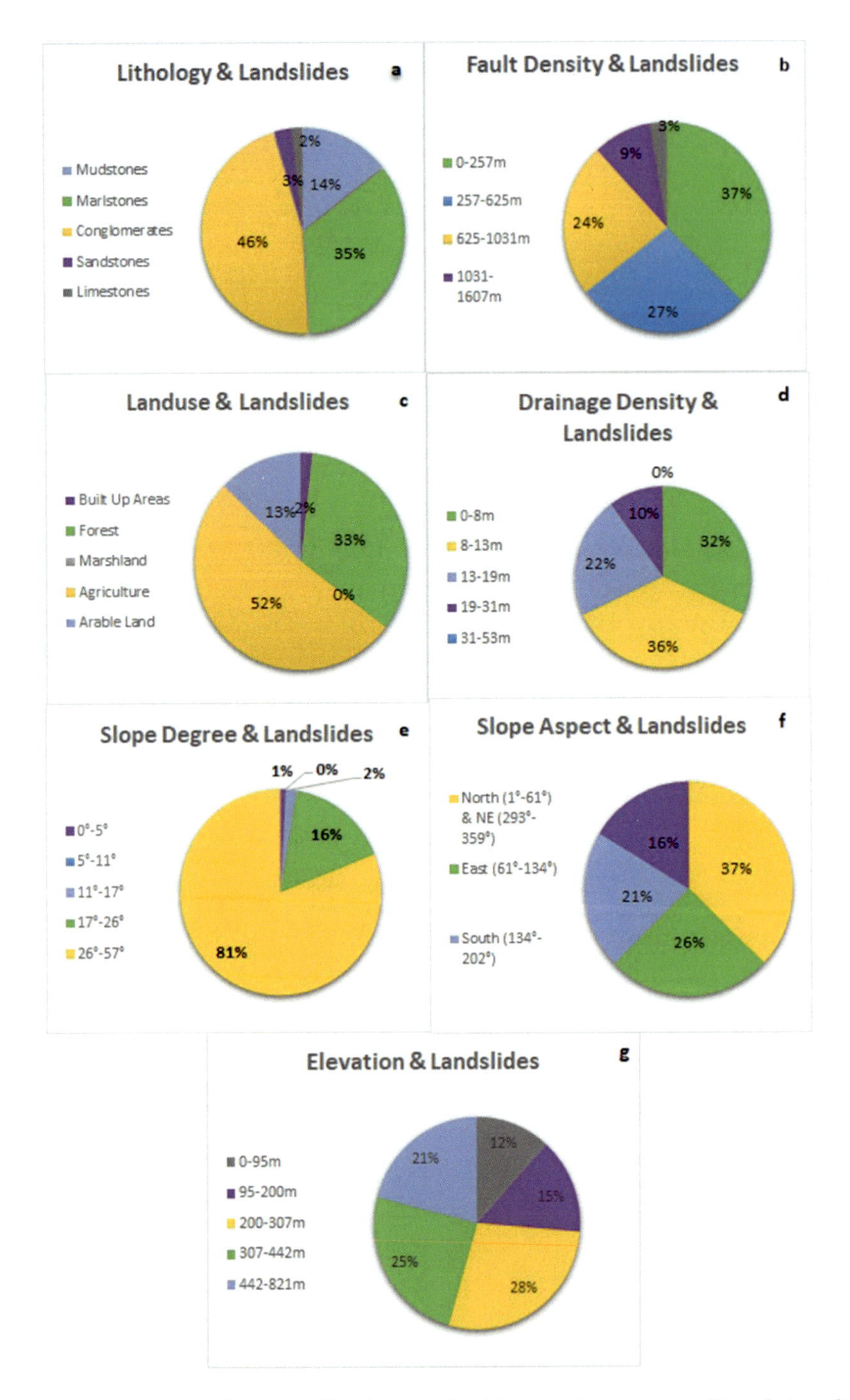

Fig. 19.5 Percentage of relationships between landslides and seven controlling factors for Ceuta City

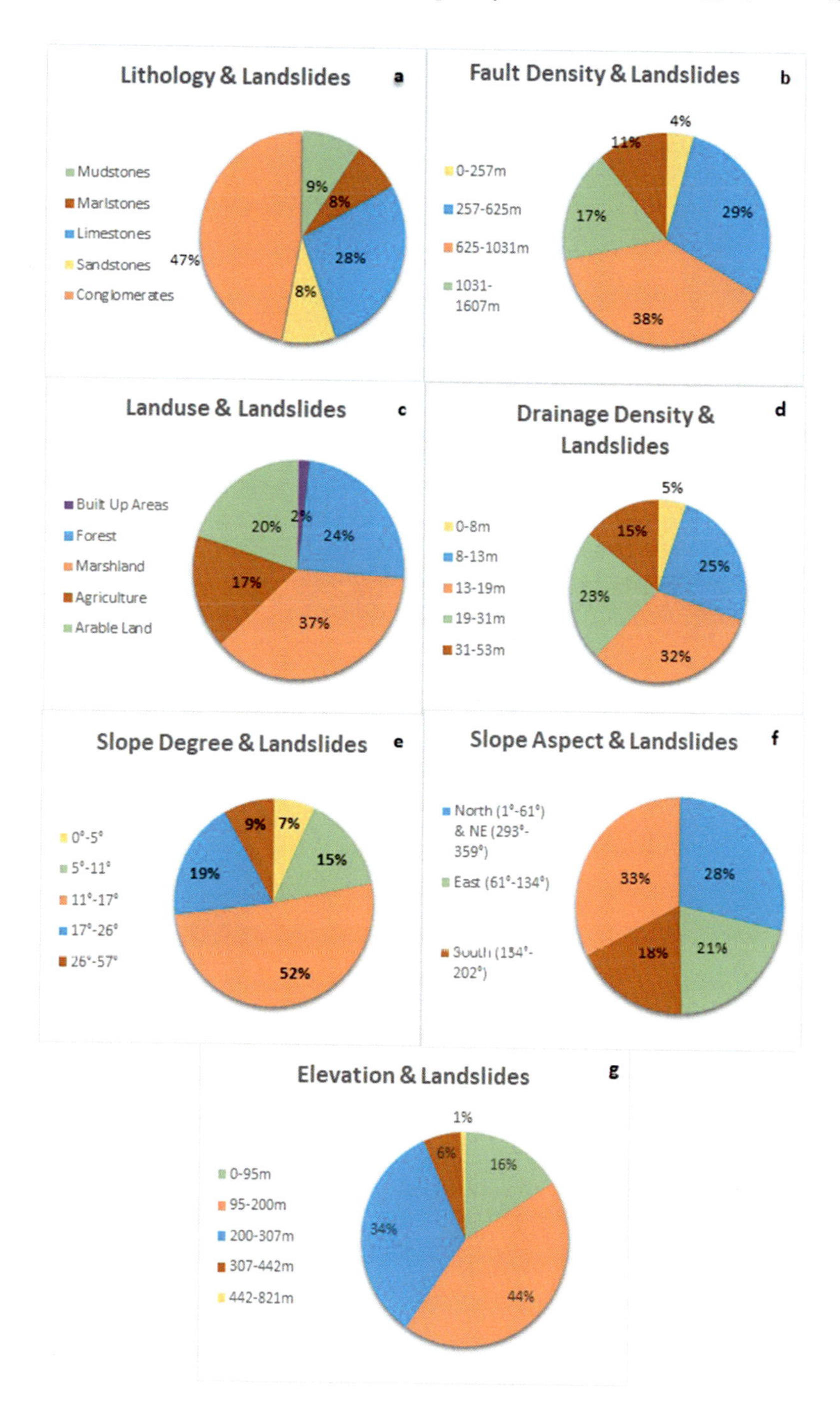

Fig. 19.6 Percentage of relationships between landslides and seven controlling factors for Tétouan-Ras-Mazari

Table 19.2 Ratios and Prediction rates for all the controlling factors of Model-I

Factors and Calsses	Ratio	FR	Min RF	Max RF	[Max- Min]	PR
Lithology			0.11	0.3	0.19	3.8
Mudstones	0.62	0.15				
Marlstones	1.28	0.30				
Limestones	1.13	0.27				
Sandstones	0.47	0.11				
Conglomerates	0.72	0.17				
Fault Density			0.18	0.23	0.05	1
0–257 m	1.15	23.02				
257–625 m	0.80	15.98				
625–1031 m	1.11	22.32				
1031–1607 m	1.03	20.61				
1607–2529 m	0.90	18.07				
Landuse			0	0.33	0.33	6.6
Built up areas	0.37	0.10				
Forest	1.17	0.33				
Marshland	0.00	0.00				
Agriculture	0.98	0.27				
Arable land	1.09	0.30				
Drainage Density			0	0.25	0.25	5
0–8 m	1.01	0.25				
8–13 m	1.01	0.25				
13–19 m	1.01	0.25				
19–31 m	1.00	0.25				
31–53 m	0.00	0.00				
Slope Degree			0	0.33	0.33	6.6
0°–5°	1.19	0.33				
5°–11°	0.00	0.00				
11°–17°	0.30	0.08				
17°–26°	0.98	0.27				
26°–57°	1.10	0.31				
Slope Aspect			0.17	0.25	0.08	1.6
North (1°–61°)	0.84	0.17				
East (61°–134°)	1.26	0.25				
South (134°–202°)	0.99	0.20				
West (202°–293°)	1.02	0.20				
North-East (293°–359°)	0.92	0.18				
Elevation			0.12	0.36	0.24	4.8
0–95 m	2.17	0.36				
95–200 m	0.95	0.16				
200–307 m	1.29	0.22				
307–442 m	0.75	0.12				
442–821 m	0.84	0.14				

Table 19.3 Ratios and prediction rates for all the controlling factors of Model-II

Factors and calsses	Ratio	FR	Min RF	Max RF	[Max–Min]	PR
Lithology			0.11	0.29	0.18	1.72
Conglomerates	1.03	0.23				
Sandstones	0.69	0.15				
Mudstones	1.02	0.22				
Limestones	0.48	0.11				
Marlstones	1.33	0.29				
Fault Density			0.05	0.30	0.25	2.40
0–23 m	0.24	0.05				
23–46 m	1.28	0.27				
46–64 m	1.42	0.30				
64–89 m	0.83	0.18				
89–147 m	0.91	0.19				
Landuse			0.07	0.29	0.33	3.16
Built up areas	0.33	0.07				
Forest	1.23	0.28				
Agriculture	1.29	0.29				
Arable land	0.82	0.19				
Marshland	0.77	0.17				
Drainage Density			0.10	0.30	0.25	2.39
0–59 m	0.50	0.10				
59–72.9 m	1.10	0.22				
72.9–82.2 m	1.06	0.21				
82.2–91.8 m	0.86	0.17				
91.8–118.8 m	1.49	0.30				
Slope Degree			0.09	0.30	0.30	2.89
0°–2°	0.41	0.09				
2° 5°	0.85	0.19				
5°–13°	1.36	0.30				
13°–20°	0.97	0.21				
20°–57°	0.93	0.21				
Slope Aspect			0.15	0.25	0.10	1
North (1°–71°)	1.22	0.24				
North-East (71°–143°)	0.76	0.15				
South-East (143°–215°)	0.80	0.16				
West-SE (215°–278°)	1.30	0.25				
North-West (278°–359°)	1.04	0.20				
Elevation			0.05	0.36	0.31	2.93
0–84 m	0.53	0.13				
84–204 m	1.28	0.31				
204–360 m	1.50	0.36				
360–571 m	0.63	0.15				
571–1003 m	0.23	0.05				

Fig. 19.7 Prediction rates curve of all the controlling factors for Model I (Ceuta) and for Model II (Tétouan-Ras-Mazari)

19.3.1 Model-I (Area-1 Ceuta)

Th results to extract from the abovementioned data in the methods section allow us to create landslides Susceptibility Assessment for Ceuta City (Fig. 19.8). Among multiple tests to create the most predictable model we used the highly predicted factors and conditionally independent, which are landuse, elevation, and lithology for Ceuta City. Afterward, we classified the final map into five classes as not susceptible, low, moderate, high, and very high by using Jenks Natural breaks classification method of ArcMap.

19.3.2 Model-II (Area-2 Tétouan-Ras-Mazari)

The susceptibility map of Area-2 (Tétouan-Ras-Mazari) applied the same rule of the selection of controlling factors as stated in previous part 3.1, which are landuse, elevation, and fault density. Figure 19.9 depict the final LSA of Tétouan-Ras-Mazari.

19.3.3 Models Validation

To understand the area of each subclass of both LSAs, we create the graphes in Fig. 19.10. More reading of thse graphes are provided in the discussion section.

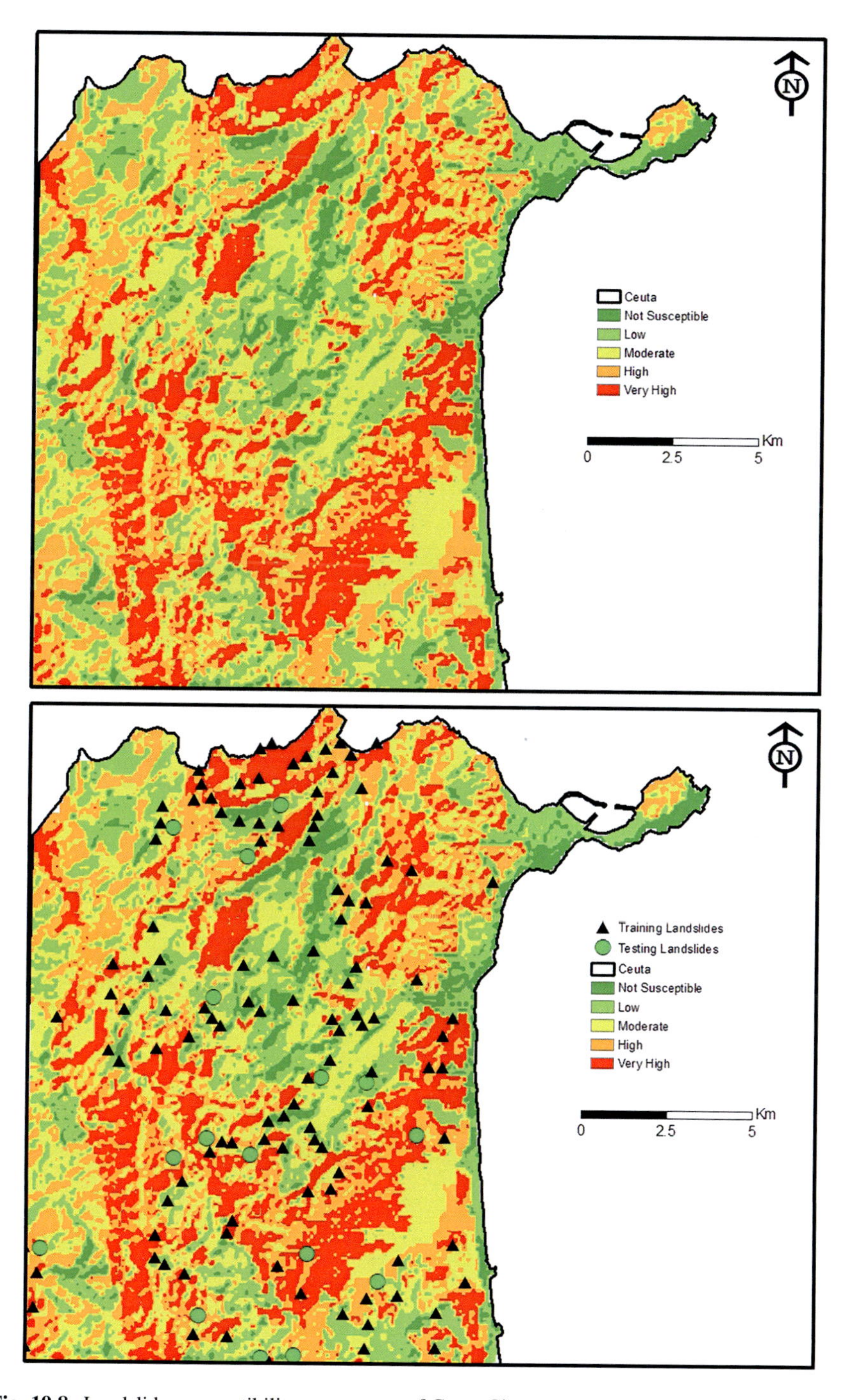

Fig. 19.8 Landslides susceptibility assessment of Ceuta City

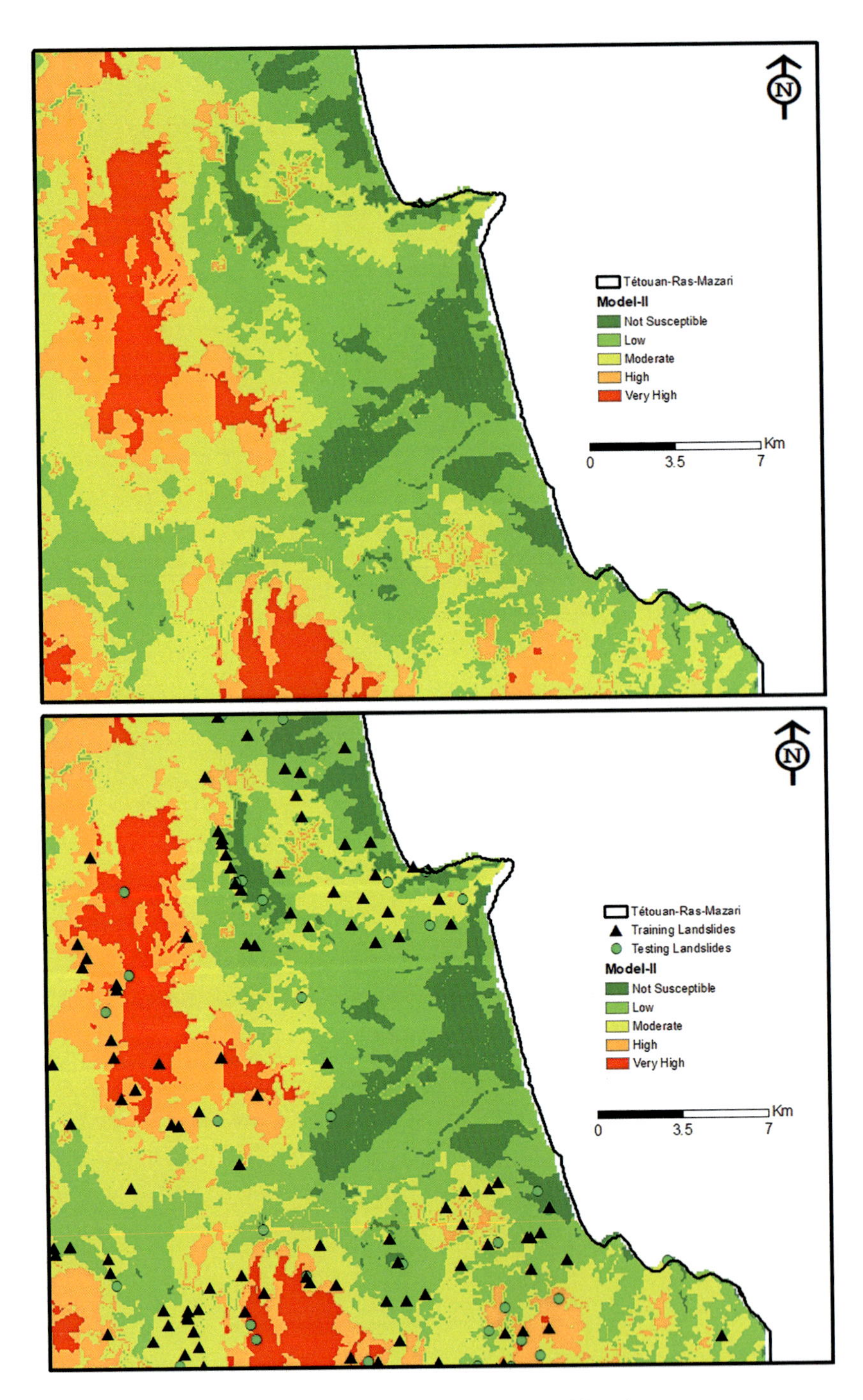

Fig. 19.9 Landslides susceptibility assessment of Tétouan-Ras-Mazari

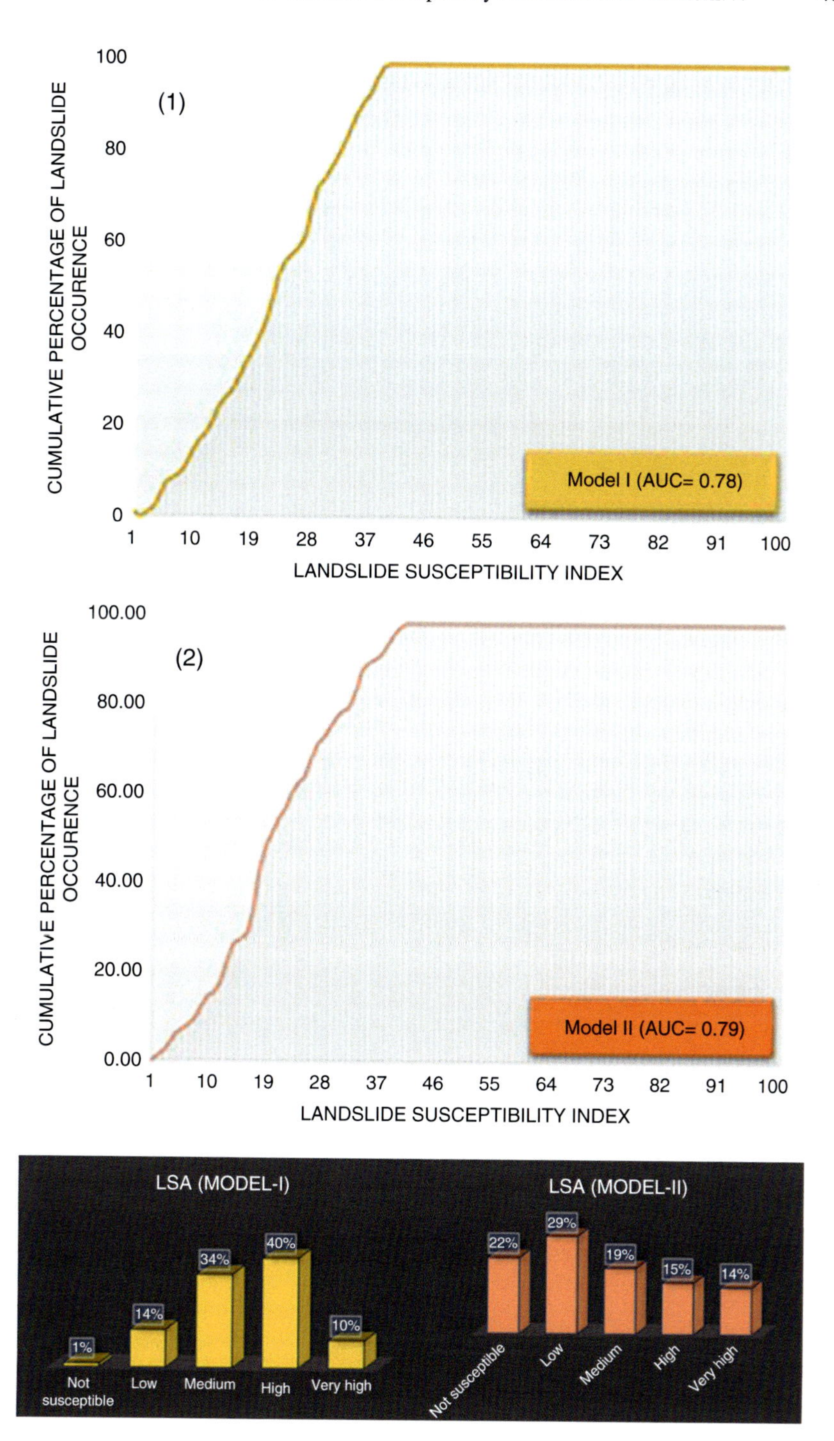

Fig. 19.10 Areas percentanges for LSAs for Area-1 and Area-2

Receiver operating characteristic (ROC) or Area Under Curve (AUC) is well know for its capability to predict the accuracy of a model based on presence and absence of a series of data. AUC of our Models are given by Fig. 19.11.

The percentages of cross-validation between the results of LSAs and the testing data are show in the Fig. 19.12.

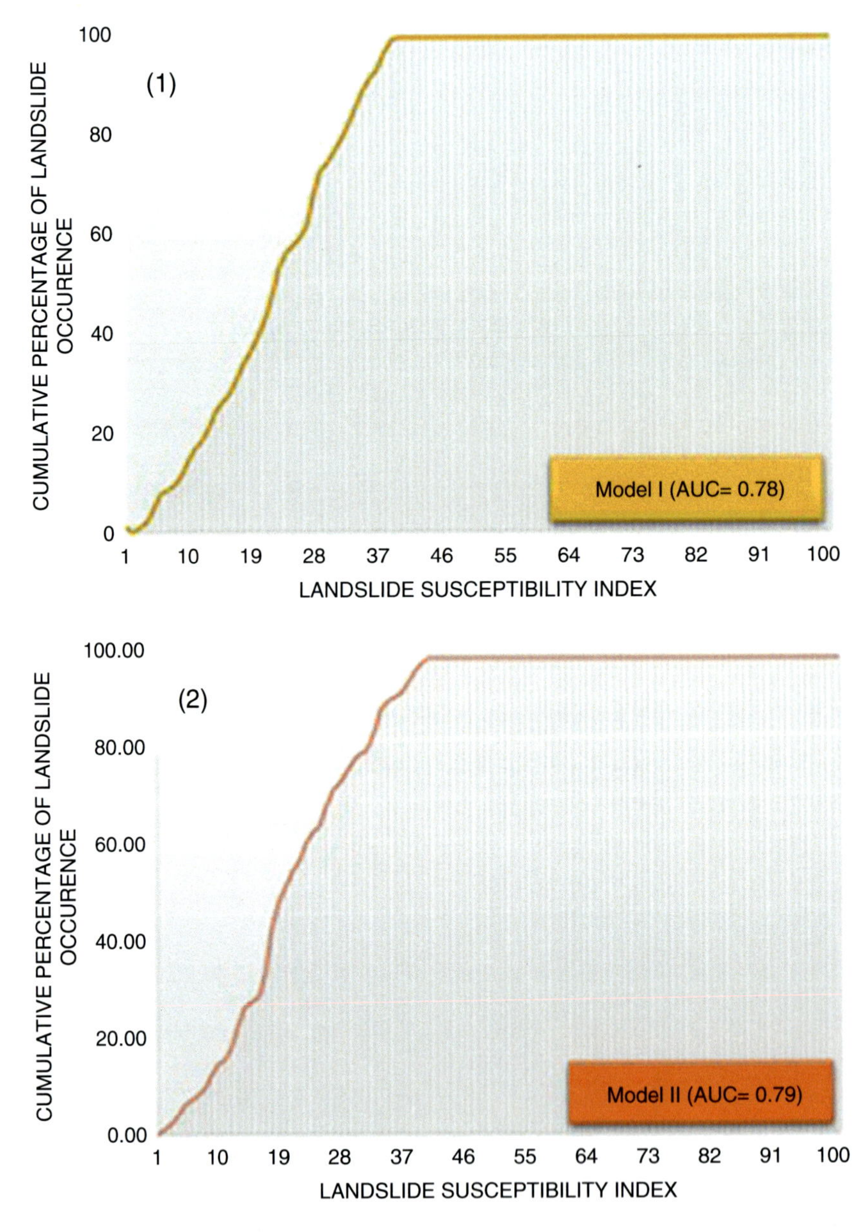

Fig. 19.11 AUC for model-I (1) and for Model-II (2)

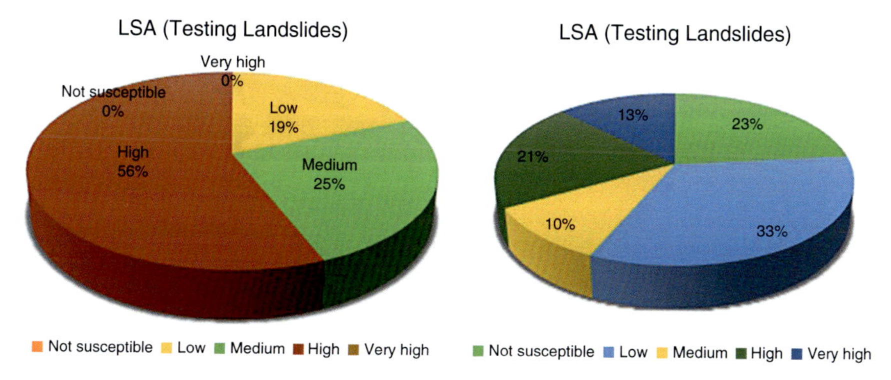

Fig. 19.12 Results of cross-validation between the results of LSAs and the testing data

19.4 Discussions

19.4.1 Relationship Between Landslides and Controlling Factors

According to the results of our analysis shown in the Figs. 19.5 and 19.6, the most striking findings to emerge from the relationship between landslides' occurrence and the explanatory factors used to predict the LSA are discussed as follow:

19.4.1.1 Landuse

The relationship between landuse and mass movement, specifically rock falls in Ceuta city and landslides in Tétouan-Ras-Mazari are shown in Figs. 19.5c and 19.6c. This type of mass movements are very abundant in agricultural land with 53%, and in the forests subclass 34% for Ceuta City. Concerning Tétouan-Ras-Mazari, the highest percentage exhibits 37% of landslides in marshland and 24% in forest. There are several possible explanations for this result. First, we are concerned about the quality of the agricultural land in Ceuta area and marshland in Tétouan-Ras-Mazari. If these two subclasses of landuse are prone to rock falls/landslides; so probably the top layer of soil is removed due to soil erosion; which is a natural process caused by the weather that harms the agricultural land. Second, we observe that the subclass of forest is also conducive to the outbreak of rock falls in this portion of the study area. Unexpectedly, forest stabilizes the equilibrium of land and therefore increases the shearing strength. However, this is not particularly surprising given the fact that a clear impact of rockfalls on forest is observed. As matter of fact, dendrochronology can show a past effect, with missing of trees' competent. To better understand the relationship between rock falls and the forest, a further macroscopic section can be used for dating rockfalls events (16).

19.4.1.2 Elevation

By looking at the Figs. 19.5g and 19.6g, we notice the relationship between MM occurrence and altitude. They show that in Ceuta moderate and high altitudes represent 28%(200–307 m) and 25% (307–442 m) respectively. This result reflects the direct relationship between MM occurrences and high altitudes. Regarding Tétouan-Ras-Mazari the two highest percentage are 44% (for low altitude (95–200 m) and 34% for medium ones (200–307 m). Thus, it can be conceivably hypothesized due to the road cutting and human activities in unstable slopes.

19.4.1.3 Fault Density & Lithology

The third factors used in our models are conditionally dependent for two different geographical areas and therefore we will be discussed them separately:

For Area-1, lithology is considered as the third highly predicted factor. Figure 19.5a reveals that conglomerates exhibits 46% of MM and marlstones 35% of MM. These values are consistent with previous results (Ait Brahim et al., 2002; Ait Brahim & Soussey, 2003; Kirat, 1993). Consequently, rock falls are intimately related to the conglomerates and marlstones of Flysch Nappes. Concerning Area-2, Faut density is the third predicted factor. Figure 19.6b provides the results of this class. We notice that highest percentage is attributed to middle class 38% (625 m–1031 m) and next comes the low class (257–625 m) that represent 28% of total MM. This expectation lends to support the general rule that the probability of landslides' occurrence increases when the fault density decreases.

19.4.2 Investigation of Landslides Susceptibility Assessment (LSA)

19.4.2.1 Model-I

Investigation of our findings throughout Figs. 19.8 and 19.10 (left) highlights that our model predicts 40% of high susceptible areas and 10% of very high ones of the total study area by using training data and three conditionally independent factors (Landuse, fault density, elevation and). In addition to that, we conduct another analysis to strengthen our confidence about the accuracy of our model and cross-validate the result of susceptibility assessment by using 30% of testing data and the statistic results shown in Fig. 19.12 (left). Our model predicts 56% of the total study area as highly susceptible. To better evaluate our model, we calculate the ROC or AUC. This test forecasts 78%.55 of landslides' occurrence. This is in a good alignment with previous findings in literature. As mentioned by (Chen et al., 2019): "AUC values between 0.7 and 0.9 indicated reasonable agreement". These findings allowed us to recommend that at smaller areas such as Ceuta city three conditionally independent factors are barely enough to predict risky areas in Northern Morocco.

19.4.2.2 Model-II

Inspection of susceptibly assessment map of Tétouan-Ras-Mazari (Fig. 19.8), statistical analysis, and prediction rate (Fig. 19.10 (left)) of this model highlights the following results:

On one hand, the susceptibility areas of the total study area are 22%, 29%, 19%, 14%, and 15% for the five classes that are not susceptible, low, medium, high, and very high; respectively. Unexpectedly, this can be explained by the high number of active landslides used to run the model (123). On the other hand, the results gathered from the analysis by using testing data. Figure 19.12 (right) highlights that 13% of very high and 21% of high areas. Contrary to expectations, this test overestimated low class and predicted 33% of the total areas. It is plausible that a number of limitations may influence the results obtained. This limitation may be related to the exact ratio of active landslides used for training/testing data. In our previous research (Elmoulat & Ait Brahim, 2018), we used a ratio of 50% of training and 50% of testing and the current model a ratio of 70% of data to run the model and 30 for validation purpose. In terms of the success rate curve, the success rate of the model is 79.25%, which is the highest value found in our analysis among four different tests using only the highly predicted factors (Figs. 19.11 and 19.2). This result shares a number of similarities with (Corona et al., 2017). Hence, we can conclude that three controlling factors which area land use, elevation, and fault density exert more power in triggering landslides in Area-2.

19.4.3 Comparison Between Model-I vs Model-II

In terms of results obtained from the AUC (Fig. 19.11) and the statical analysis conduct by using training and testing data (Figs. 19.10 and 19.12), we conclude that the two areas have two similar predicant factor that are landuse and elevation. The geographical area inserted in the same Rif region and therefore shares that same type of land covers; in addition to that in term of elevation they are slightly different. For area-1 the high rate is observed in the highest class (307–442) because of the geographic location up to the North of the African plate. Concerning area- 2 the highest class is slow class (84–204). This can be explained by the geographical location of area-2 compared to the one of area-1. Thirdly, Model-I used lithology as the third factor to predict susceptible areas, which is crucial because marlstones and congregates are dominant formation in area-1. From petrographic and structural point of view, these elements trigger landslides in the Rif region (Shi et al., 2016, Corona et al., 2017). Furthermore, Model-II called out fault density. In our previous works (Elmoulat and Ait Brahim Elmoulat & Ait Brahim, 2018) and field observations density around 0 m and 60 m are areas prone to mass movements. Our current observation is in complete agreement with our earlier findings. We would like to mention that faults zones between the Flysches Nappes and Tangier Unit highlight an important number of mass movements.

19.5 Conclusion

The present research composes of two different models to create Landslides Susceptibility Assessment (LSA) of two distinct geographical areas (Ceuta City and Tetouan-Ras-Mazari). Both LSAs are implemented by using the Frequency Ratio approach. The factors used in these models based upon the prediction rates of each controlling factor of landslides' occurrences. We selected only conditionally independent parameters whose Prediction Rates (PR) are highly classified. Indeed, conditional dependence decreases the overall accuracy of any model. Our findings reveal that landuse and elevation are major controlling factors for both models. Lithology exerts more power over Ceuta City and fault density takes the deal for Tetouan-Ras-Mazari. The overall accuracy of the two models are evaluated by AUC and cross-validation by using the testing data, which are in reasonable agreement with our predecessors.

The hypothesis of the current research is primally focused on benchmarking two susceptibility assessments for two smaller areas to determine their main similarities and differences and then gain new knowledge based upon our findings. We are aware that our results cannot be generalized, given the fact that both areas are geographically located in the same meridians, despite the distinct position of their parallels.

To further our research, we plan to extend our studies in several different regions in the world; in totally different locations (countries and why not continents). For instance, we would like to conduct another benchmarking, as soon as we collect the necessary data to do so, between susceptibility maps of areas located, for example, in Asia and/or South American; in addition to the ones already established in North Africa. The knowledge we may acquire from this research, will improve our understanding of global landslides' susceptibility mapping.

In the end, we believe in a famous saying by a Philanthropist, Petra Nemcova: "*We cannot stop natural disasters, but we can arm ourselves with knowledge: so many lives wouldn't have to be lost if there was enough disaster preparedness.*" (Nemcova, 2021).

Acknowledgments The authors are extremely grateful and thankful to:

– Dr. Olivier DEBAUCHE, Senior Researcher at Department of Computer Science, Faculty of Engineering, University of Mons, Belgium; for his unconditional support on the computing and calculation processes of this research.

- Dr. Arnaud J TEMME, Associate Professor at Department of Geography and Geospatial Sciences, Kansas State University, USA; for his guidance and direction on the reasoning process, to tackle the current work, and find out an equilibrium between what is desired to be published and the availability of our data.
- Dr. Omar F. ALTHUWAYNEE, Post-doctoral Fellow at Department of Energy and Mineral Resources Engineering, Sejong University, South Korea; for granting free access to his online course to test our results.

References

Ait Brahim, L., & Soussey, A. F. (2003). Utilisation de la télédétection pour l'analyse de la fracturation du domaine interne Rifain (Maroc): Relation avec la répartition des sources. *Télédétection, 1*, 33–47.

Ait Brahim, L., Chotin, P., Hinaj, S., Abdelouafi, A., El Adraoui, A., Nakcha, C., Dhont, D., Charroud, M., Alaoui, F. S., Amrhar, M., & Bouaza, A. (2002). Paleostress evolution in the Moroccan African margin from Triassic to present. *Tectonophysics, 357*(1–4), 187–205.

Chen, W., Yan, X., Zhao, Z., Hong, H., Bui, D. T., & Pradhan, B. (2019). Spatial prediction of landslide susceptibility using data mining-based kernel logistic regression, naive Bayes and RBF network models for the Long County area (China). *Bulletin of Engineering Geology and the Environment, 78*(1), 247–266.

Ciampalini, A., Bardi, F., Bianchini, S., Frodella, W., Del Ventisette, C., Moretti, S., & Casagli, N. (2014). Analysis of building deformation in landslide area using multisensor PSInSAR™ technique. *International Journal of Applied Earth Observation and Geoinformation., 1*(33), 166–180.

Corona, C., Lopez-Saez, J., Favillier, A., Mainieri, R., Eckert, N., Trappmann, D., Stoffel, M., Bourrier, F., & Berger, F. (2017). Modeling rockfall frequency and bounce height from three-dimensional simulation process models and growth disturbances in submontane broadleaved trees. *Geomorphology, 15*(281), 66–77.

Elmoulat, M., & Ait Brahim, L. (2018). Landslides susceptibility mapping using GIS and weights of evidence model in Tetouan-Ras-Mazari area (Northern Morocco). *Geomatics, Natural Hazards and Risk, 9*(1), 1306–1325.

Elmoulat, M., & Ait Brahim, L. (2020). Landslides susceptibility assessment using frequency ratio method, remote sensing datasets, and GIS-based techniques. In D. S. Krogh (Ed.), *Landslides: Monitoring, susceptibility and management* (pp. 69–114). Nova Science Publishers, Inc.

Elmoulat, M., Brahim, L. A., Mastere, M., & Jemmah, A. I. (2015). Mapping of mass movements susceptibility in the Zoumi region using satellite image and GIS technology (Moroccan Rif). *International Journal of Scientific & Engineering Research., 6*(2), 210–217.

Haut-commissariat au Plan. (2014). *Note sur les premiers résultats du Recensement Général de la Population et de l'Habitat*. https://www.hcp.ma/. Accessed 30 Mar 2021.

Kassab, A. (1976). Agriculture et ressources en eau en Kroumirie. IV Coll. *Géogr. maghrébine, Tunis. Cah du C.E.R.E.S*, 207–220.

Kassab, A. (1979). L'homme et le milieu naturel dans les régions de Sejnane et de Tabarca. *Méditerranée, 3e sér, 35*, 39–46.

Kim, J. C., Lee, S., Jung, H. S., & Lee, S. (2018). Landslide susceptibility mapping using random forest and boosted tree models in Pyeong-Chang, Korea. *Geocarto international., 33*(9), 1000–1015.

Kirat, M. (1993). *Essai de cartographie géomorphologique et étude des mouvements de terrain dans la vallee de l'Oued El Kbir (Province de Tetouan, Rif occidental: Maroc Septentrional)* [Test of geomorphological mapping and movement study of ground in the valley of the Wadi El Kbir (Province of Western Tetouan)] [dissertation]. Lille: University of Sciences and Technology of Lille I.

Lee, S. (2019). Current and future status of GIS-based landslide susceptibility mapping: A literature review. *Korean Journal of Remote Sensing, 1*(264), 109–117. https://doi.org/10.7780/kjrs.2019.35.1.12

Mansour, M., & Ait Brahim, L. (2005). Apport de la télédétection Radar et du MNT à l'analyse de la fracturation et la dynamique des versants dans la région de Bab Taza, Rif. *Maroc. Télédétection, 5*, 95–103.

Marthelot, P. (1957). L'érosion dans la montagne kroumir. *Revue de géographie alpine*, 273–287.

Marthelot, P. (1959). Note sur un décollement de versant dans la vallée des Atatfa (Kroumirie). *Actes du 84e Congrès national des sociétés savantes, Dijon Section de Géographie*, 61–65.

Mastere, M., Lanoë, B. V., Brahim, L. A., & El Moulat, M. (2015). A linear indexing approach to mass movements susceptibility mapping-the case of the Chefchaouen province (Morocco). *Revue internationale de géomatique., 25*(2), 245–265.

Mate. (2004). *Rapport national de l'Algérie sur la mise en œuvre de la Convention de Lutte Contre la Désertification*. Algérie: Ministère de l'agriculture et du développement rural. www.unccd-prais.com. Accessed 30 Mar 2021.

Maurer, G. (1976). Les mouvements de masse dans l'évolution des versants des régions telliennes et Rifaines d'Afrique du Nord. Actes du Symposium sur les versants en pays méditerranées 1976. *Provence.C.E.G.R.M*, Suppl, 133–137.

Maurer, G. (1979). Les milieux naturels et leur aménagement dans les montagnes humides du domaine Rifain et tellien d'Afrique du Nord. *Meditarranée*, 47–56.

Nemcova, P. (2021). https://www.quoteswave.com/. Accessed 29 Mar 2021.

Nicu, I. C. (2017). Frequency ratio and GIS-based evaluation of landslide susceptibility applied to cultural heritage assessment. *Journal of Cultural Heritage, 28*, 172–176.

Pateau, M. (2014). De l'aléa au risque naturel: cas de la région Tanger-Tétouan (Rif, Maroc). *Geo-Eco-Trop, 1*, 23–32.

Shi, J. S., Wu, L. Z., Wu, S. R., Li, B., Wang, T., & Xin, P. (2016). Analysis of the causes of large-scale loess landslides in Baoji. *China. Geomorphology., 264*, 109–117.

Zhou, S., Chen, G., Fang, L., & Nie, Y. (2016). GIS-based integration of subjective and objective weighting methods for regional landslides susceptibility mapping. *Sustainability, 8*(4), 334.

Chapter 20
A Geophysical and Remote Sensing-Based Approach for Monitoring Land Subsidence in Saudi Arabia

Abdullah Othman and Karem Abdelmohsen

Abstract Groundwater represents the main source of accessible freshwater in Saudi Arabia. Overexploitation of these resources can adversely lead to serious natural hazards in these regions. Understanding the mechanisms and factors causing these hazards whether natural or anthropogenic have become essential to safe people life and sustain the development in such arid regions. Here we applied an integrated approach (field, Interferometric Synthetic Aperture Radar [InSAR], hydrogeology, Geoinformatics) over the northern and central parts of Saudi Arabia to identify the nature, intensity, and spatial distribution of deformational features. The Lower Mega Aquifer System (LMAS) in central and northern Saudi Arabia were used as a test sites. Findings suggest that the natural recharge reduction and excessive groundwater extraction in the LMAS are the major cause of land subsidence and deformation features in the study site. Spatial and temporal correlation of radar and Gravity Recovery and Climate Experiment (GRACE) solutions indicated that sustainable extraction could be attained by reducing the current extraction rates. This approach provides replicable and cost-effective approach for ideal utilization of fossil aquifers in arid lands and for reducing deformation hazards associated with their use. Also suggesting potential areas for sustainable agricultural development that will lead to concrete actions and have an immediate humanitarian/community/environmental benefits.

Keywords GRACE: Radar interferometry · InSAR · Fossil aquifers · Land subsidence · Saudi Arabia

---

A. Othman (✉)
Department of Environmental Engineering, Umm Al-Qura University, Makkah, Saudi Arabia
e-mail: agothman@uqu.edu.sa

K. Abdelmohsen
Department of Geological and Environmental Sciences, Western Michigan University, Kalamazoo, MI, USA

M. M. Al Saud (ed.), *Applications of Space Techniques on the Natural Hazards in the MENA Region*, https://doi.org/10.1007/978-3-030-88874-9_20

20.1 Introduction

Understanding the factors that causing natural disasters and hazards whether natural or anthropogenic factors have become essential to safe people life and assessing the development in arid and semi-arid regions. Land subsidence can substantially jeopardize human lives and infrastructures. This phenomenon represents a response to the differential subsiding of the ground surface due to the extent reduction of belowground materials by compaction or collapse of subsurface cavities (Ren et al., 1989). Land subsidence a result of volume reduction, either by natural or anthropogenic activities, is typically a slow process happening over nearly large areas. The rates of subsidence have been traditionally measured using ground-based surveying methods (e.g., geodetic leveling and global positioning systems) (Galloway & Burbey, 2011). Yet, the advancement of the Interferometric Synthetic Aperture Radar (InSAR) technologies enabled cost-effective and timely estimations of land subsidence over large-scales (Gabriel et al., 1989). Examples of these applications include; some areas in California and Las Vegas such as Antelope, Santa Clara, and Las Vegas valleys (Galloway et al., 1998; Galloway & Hoffmann, 2007), The Nile Delta in Egypt (Becker & Sultan, 2009; Rateb & Abotalib, 2020) and central Arabia (Othman & Abotalib, 2019). Moreover, (InSAR) technologies have been recently used to detect and quantify sluggish, large property displacements all over the globe (Hooper, 2006; Kampes, 2006), such as technologies enabled cost-effective San Francisco Bay Area (Bürgmann et al., 2006) and central Mexico (Chaussard et al., 2014). The relationship between radar-detected subsidence and related spatial - temporal measurements sheds light on the factors that cause the reported subsidence. (Higgins et al., 2014). These factors usually include groundwater or oil/gas extraction rates, rapid urbanization (Al-Jammaz et al., 2021), decline in groundwater levels, and the distribution of different lithologies. For example, significant subsidence of over 40 mm over 35 days in the oil fields over Lost Hills and Belridge in San Joaquin Valley, California was attributed to immoderate oil production from depths of about 700 m under the ground (Fielding et al., 1998). Subsidence and related deformations such as fissures, sinkholes, and earthquakes, have been reported in many arid and semi-arid regions, where excessive groundwater extraction is a common practice. In Las Vegas Valley, between 1992 to 1997, and by using geodetic and satellites applications such as GPS and InSAR data, deformational features, and land subsidence of up to six centimeters by year were discovered due to intensive pumping of groundwater (93 km^3/year) (Bell et al., 2002). Significant land deformation and subsidence were also reported from Mashhad city northern Iran, where InSAR measurements indicated subsidence of$\sim$28–30 cm/y between 2003 and 2005 along the axis of the Mashhad valley. These rates were set on as main factors caused by extensive extraction of the aquifer system as evident by analysis of piezometric records (i.e. 65 m of water table decline since 1960s) (Motagh et al., 2007). Moreover, Earth fissures were reported from the Sarir South agricultural project over the Nubian Sandstone Aquifer System (NSAS) in Libya, where water levels dropped by more than 6 meters leading to compaction of fine-grained sediments within the aquifer materials and development of earth fissures (El Baruni, 1994). In the same context, using GPS data from two stations (period: 2006 to 2009), subsidence rates of up to 10 cm per a year were reported from the Quetta Valley in Pakistan, where subsidence was also attributed to excessive groundwater abstraction to support progressive population growth (1975: 260,000; 2010: 1.2 million) (Khan et al., 2013). Groundwater-extraction-induced deformation was reported for both confined and unconfined aquifers (Galloway et al., 1998; Othman & Abotalib, 2019). The reported deformation was either uniform and extend over large areas, or differential, causing earth fissures, damage in buildings and infrastructures (Burbey, 2002) and might involve horizontal displacement in addition to the vertical component (Burbey, 2001; Galloway & Burbey, 2011).

The largest number of subsidence cases were documented in arid lands, where precipitation was minimal, limited or no surface water resources were available, intensive irrigation programs were implemented mainly on groundwater as a source for freshwater (Maliva & Missimer, 2012) such as the Saharan-Arabian desert belt. Aquifers in the Saharan-Arabian deserts are fossil aquifers that were recharged mainly during the Quaternary wet climatic periods (Abotalib et al., 2016, 2019a). Currently, they receive only limited local recharge, which is way lower than the annual abstraction (Sultan et al., 2014). Examples of these aquifers include the Mega Aquifer System (MAS) in Saudi Arabia (Hoetzl et al., 1978; Sultan et al., 2014; Abotalib et al., 2019b), the Nubian Aquifer System in northeastern Africa (Abdelmohsen et al., 2020; Mohamed et al., 2017), the North-Western Sahara Aquifer System in northwestern Africa, and the Great Artesian Basin in Australia (Taylor et al., 2013).

Overstimulated exploitation of non-renewable groundwater in such regions, and for certain fossil aquifers, will be expected to cause artesian head downturn, high extensional stresses, and other problems.

In this study, we are investigating on the LMAS of Saudi Arabia, one of the world's largest fossil aquifer systems, examine comparable cases elsewhere, and note the similarity with our results from the LMAS to those published from similar areas around the world. The Mega Aquifer System (MAS) (surface area: ~2 million km^2) spans the Arabian Peninsula's Arabian Shelf in Saudi Arabia, United Arab Emirates (UAE), Qatar, Yemen, Oman, Jordan and Kuwait. The aquifer currently is the major water resource for mega-agricultural projects and within these reclaimed areas, several ground deformations and subsidence features were reported. Examples of these areas are the Al-Yutamah valley, Tabah village, and Najran valley (Roobol et al., 1985; Vincent, 2008; Youssef et al., 2014).

The government of Saudi Arabia started in the mid-eighties an ambitious agricultural development program in central Arabia (e.g. Al-Qassim and Ha'il regions central Saudi Arabia), which led to a raise in groundwater extraction (1.9×109 m^3/year in 1984 to 4.4×109 m^3/year in 2004) and cultivated lands (from 213×10^3 hectares in 1984 to 316×10^3 hectares in 2004). Likewise, widespread agricultural projects were employed in northern Saudi Arabia (e.g. As-Sirhan valley northern Saudi Arabia) leading to a raise in groundwater extraction from 0.095 x 10^9 m^3/year

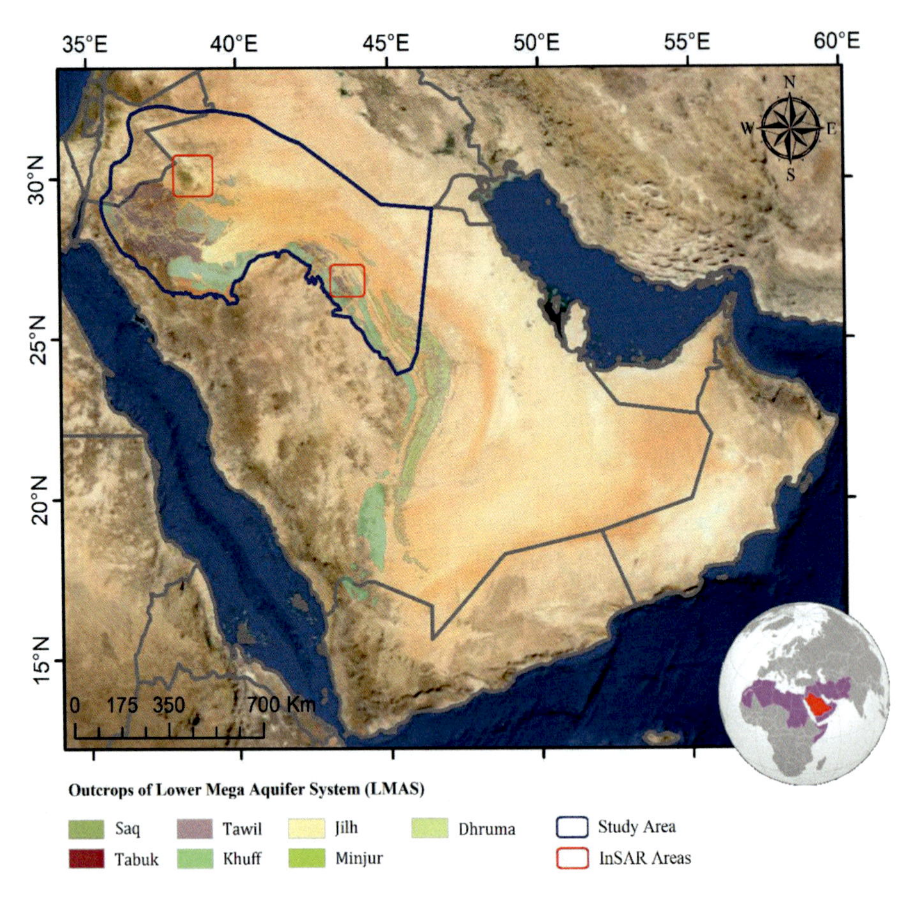

Fig. 20.1 Location map of the regional study area and the investigated InSAR areas over Wadi As-Sirhan Basin and Al-Qassim Region. The boundaries and outcrops of the Lower Miga Aquifer System

to 2×10^9 m^3/year, the cultivated areas have grown from 14×10^3 hectare in 1984 to 164×10^3 hectare in 2004 (MOAW, 1984, 2004; Abunayyan and BRGM, 2008).

Our integrated study includes a regional investigation over LMAS in northern and central parts of the KinSaudi Arabia and a local investigation over the Wadi As-Sirhan Basin (WASB) and Al-Qassim Region (Fig. 20.1). The regional investigation implemented field observations to examine deformational land features and applying spatial correlation for features with other relevant datasets in term of causal effects.

20.2 Geology and Hydrogeology of the Study Area

The Arabian Shield's basement complex crops out along Red Sea margins, creating the Red Sea mountain range (Red Sea Hills) (Fig. 20.2). Successions of Phanerozoic sedimentary overlie the basement complex unconformably, with east-dipping and approaching thicknesses of ten kilometers near the Arabian Gulf (Powers et al., 1966; Lloyd & Pim, 1990; Konert et al., 2001; Margat, 2007). The Red Sea mountains outcrops are younger to the eastward of the Gulf (Quaternary age) and older to the westward (Cambrian age) (Fig. 20.2). The Red Sea rifting exposed the basement complex and the overlaying Phanerozoic formation, which facilitates aquifer recharge within the surface exposures of these formations. The Red Sea opening rejuvenated many of the Proterozoic NW-trending Najd fault systems that affect the crystalline rocks of the Arabian Shield (Fairer, 1983; Morsy & Othman, 2020). The fault rejuvenation was dip-slip (e.g. normal faults and grabens) in response to the extensional events associated with Red Sea rifting during the Oligo-Miocene time (Kellogg & Reynolds, 1983; Giannerini et al., 1988). These faults are represented by the Kahf fault system in northern and central Saudi Arabia

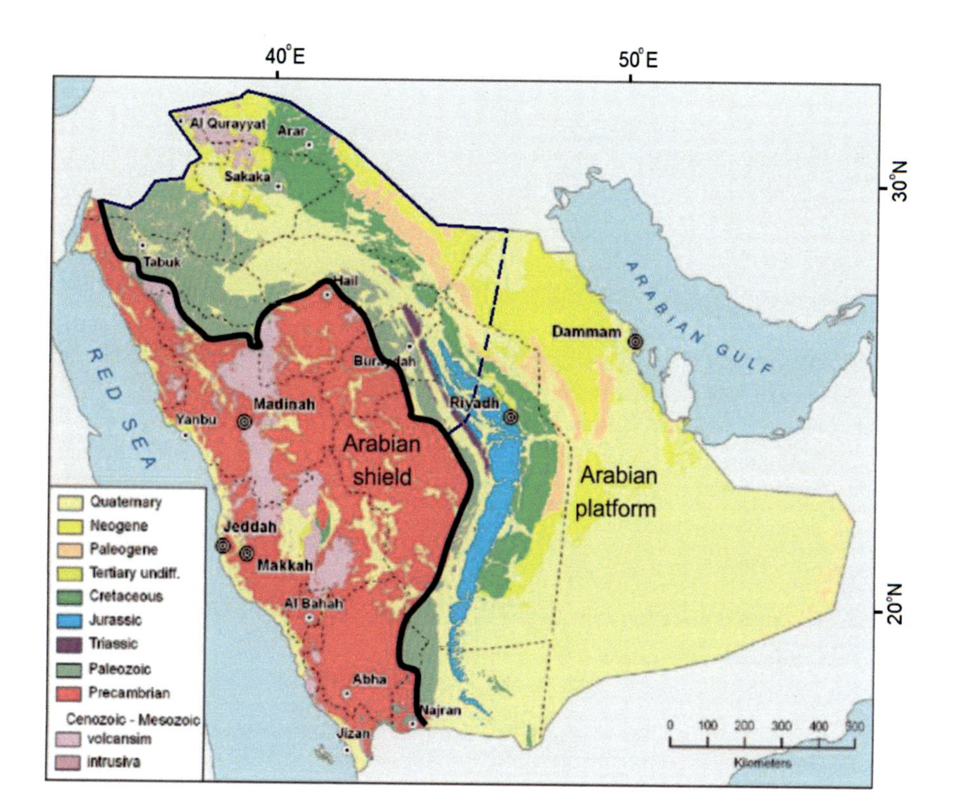

Fig. 20.2 Regional geological map of Saudi Arabia. (After Powers et al., 1966)

(Othman et al., 2018). Similar observations of dip-slip displacement on Najd faults were also reported in the Egyptian side of the shield (Sultan et al., 2011).

Phanerozoic sections contain one of the largest fossil aquifers in the world which formed from sandstone and limestone sequences (Figs. 20.1 and 20.2). The LMAS was primarily recharged during previous wet climatic conditions, but it is still receiving a modern recharge at the moment, especially in southwest Saudi Arabia, where AAP rates of up to 800 mm/y have been recorded (Alharbi et al., 2014). Groundwater Cl-36 ages attained close to 1 million years from some aquifers in the southeastern part of Saudi Arabia (Sultan et al., 2019). Those previous wet climatic conditions are completely different from current arid conditions with an average annual precipitation of 100 mm/y and minimal surface-water resources over Saudi Arabia (Barthélemy et al., 2007).

The main aquifer units within the MAS include formations of Paleozoic sandstone, Mesozoic-Cenozoic carbonates, Paleocene, and Neogene sedimentary formations (Wagner, 2011). The MAS geological have been split into two major groups separated by the anhydrite layer (Hith formation) (Othman et al., 2018). The topmost is the Upper Mega Aquifer System (UMAS), including Dammam, Rus, Umm Er Radhuma, Aruma, Wasia, Biyadh formations. Then Lower Mega Aquifer System (LMAS), which contains Dhruma, Minjur, Tawil, Tabuk, Wajid, and Saq formations. The area of MAS is ~236.3×10^4 km^2, respectively, respectively.

20.3 Methodology

We briefly describe the technical approaches applied in this study to monitor the land deformation over Saudi Arabia using (1) Field observations and land deformation database, (2) TWS and GWS extracted from GRACE temporal monthly solutions, (3) Radar Interferometric data and (4) precipitation extracted from Tropical Rainfall Measuring Mission (TRMM) data. The spatial and temporal integration of these data sets with other relevant data sets (geologic, geochemical, remote sensing) provides more evidences of the factors controlling the reported deformational features (e.g. sinkholes, fissures, and subsidence) announced in the arid environmental lands over the fossil aquifers.

20.3.1 Gravity Recovery and Climate Experiment (GRACE) Data

GRACE is a satellite mission launched in 2002 by the German Aerospace Center (DLR) and National Aeronautics and Space Administration (NASA) for the temporal global variations measurements in the Earth's gravitational field (Tapley et al., 2004) that represents the temporal variation in the TWS (Wouters et al., 2014;

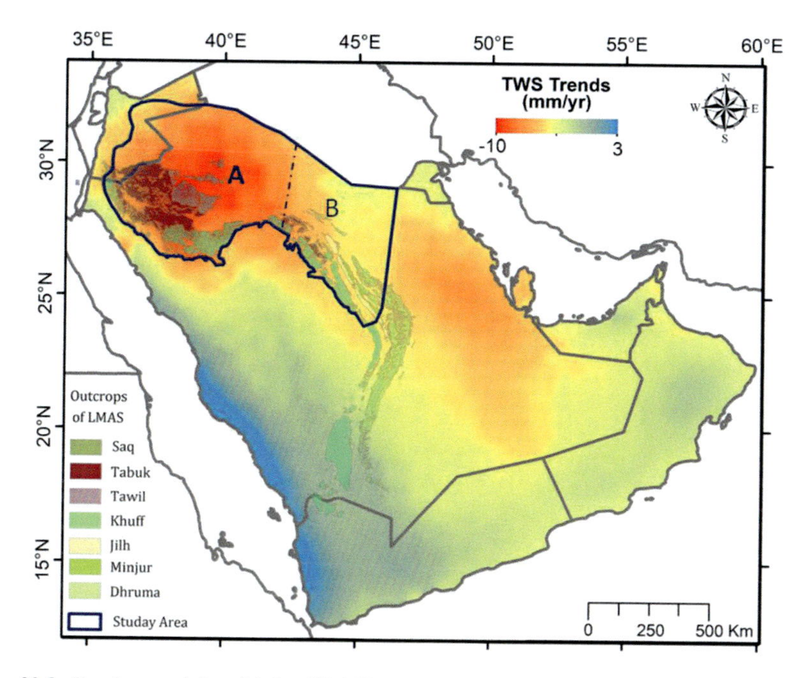

Fig. 20.3 Secular trend (mm/y) for GRACE-derived TWS for the Arabian Peninsula extracted from temporal (04/2002–03/2016) CSR 1° × 1° Mascon solutions showing high TWS depletion rates over the study areas

Ahmed et al., 2016; Mohamed et al., 2017; Ahmed & Abdelmohsen, 2018; Sultan et al., 2019; Abdelmohsen et al., 2019, 2020). TWS refers to the vertically integrated water storage measurement that includes groundwater, soil moisture storage, surface water storage, snow water storage, and canopy water storage (Strassberg et al., 2007; Abdelmalik & Abdelmohsen, 2019).

Release (05; RL05), with a spatial resolution of 0.5° × 0.5°, temporal monthly constrain solutions (mascon; Save et al., 2016) covering the investigated time period (4/2002 to 6/2016) were utilized to estimate the long-term components of TWS changes over the lower MAS and the WASB. The mascon solutions were provided by Center for Space Research (CSR) at the University of Texas (Save et al., 2016). The mascon solution doesn't require any post processing, analysis, filtering, or scaling factors (Save et al., 2016).

The long term trend TWS component was extracted by continuously fitting a linear regression function (Fig. 20.3) and the trend uncertinity was calculated using an approach described in Scanlon et al. (2016). The change in groundwater storage was calculated using a common equation of water mass:

$$\Delta GWS = \Delta TWS - \Delta SMS \tag{20.1}$$

Where ΔGWS and ΔSMS represent the variation in groundwater storage and soil moisture storage, respectively. Global Land Data Assimilation System (GLDAS) model was utilized to derive SMS. land surface modeling system developed by NASA as a simulation model for hydrologic and climatic variables (Rodell et al., 2004). Four GLDAS models to estimate the soil moisture:, Community Land Model [CLM], Noah, and MOSAIC, Variable Infiltration Capacity [VIC] were used to represent the soil moisture over the study area (Rodell et al., 2004; Dai et al., 2003). The uncertainty (σ_SM) were calculated from the standard deviation (STD) of the trends from the GLDAS four simulations. The trend uncertainty in GWS (σ_{GWS}) was calculated using standard error propagation equations:

$$\sigma_{GWS} = \sqrt{(\sigma_{TWS})^2 + (\sigma_{SM})^2} \tag{20.2}$$

20.3.2 Radar Interferometric Data (Persistent Scatterer Interferometry)

Twenty-nine descending scenes from the European Remote Sensing ENVISAT satellite were implemented. Twenty-two of those scenes covered the Wadi As-Sirhan Basin (Tabarjal and Busayta area, tracks: 221, 351 and 493) and the remaining eight scenes were acquired over Al-Qassim area (Track: 49 and 278). The investigated period for the first set was from 2003 to 2012, while the second set covered the period between 2003 and 2005. The spatial baseline ranged from 15 to 855 m concerning the master scenes. The maximum temporal baseline was 1351 days for both investigated areas, and the ranges of permanent scatterers' density from 10 to 251,351 days for both investigated areas, and the ranges of permanent scatterers' density from 10 to 25 persistent scatterers/km^2 with a coherence threshold value of 0.5 and an amplitude dispersion value of 0.4.

Persistent Scatterer Interferometry (PSI) techniques has been applied (Hooper et al., 2004, 2007) to examine the spatial variations in subsidence rates across the two sites on the Northern and southern parts of the study area (LMAS) (Fig. 20.1). The PSI technique restrict the unwrapping phase and analysis to coherent pixels, containing individual scatterers and remaining stable over the investigated period. Buildings, rock formations, rocks, houses, utility poles, and well platforms were classified as persistent scatterers in the study region using the Stanford Approach for Permanent Scatterers (StaMPS) algorithm (Hooper et al., 2007, 2012). The interferometric was processed using the StaMPS process (Hooper et al., 2004). The accurate orbit information was provided from the Delft Institute for Earth-Oriented Space Research (Scharroo & Visser, 1998).

20.3.3 *Landsat Images*

Multi-temporal Landsat (5, 7, and 8) images (path 172 and row 39) were been acquired for the following dates (February 1987, 1991, and 2000, March 2003, January 2014, and February 2017) then were analyzed to monitor the development of cultivated lands throughout the past three decades (1987 to 2017). Temporal changes in the agricultural developments areas were estimated using the NDVI (Normalized Difference Vegetation Index). Using the reflectance values from the near-infrared and red bands, an NDVI value was estimated using standard model and process (Rouse et al., 1974):

$$\mathrm{NDVI} = (\text{near infrared} - \text{red})/(\text{near infrared} + \text{red}). \tag{20.3}$$

It is a positive relationship between the values of NDVI and density of vegetation. The cultivated land areas were extracted from the Landsat imagery using a threshold NDVI value of 0.3; visual comparisons with Google Earth images validated the threshold value.

20.4 Discussion and Findings

Our analysis of the land satellite images and GRACE solutions over the largest aquifer systems in Saudi Arabia revealed that land subsidence features can be targeted excessive extraction of groundwater. In the next parts, we going to describe and explain our findings over the selected areas in relation and relevant deformation factors.

20.4.1 *The Lower Mega Aquifer System*

The excessive groundwater abstraction of the LMAS and unsustainable utilization of the aquifers was reflected in TWS from GRACE solutions. Depletions in TWS and GWS derived from GRACE over LMAS (areas: A and B) were observed. The spatial and temporal changes in TWS anomaly over Saudi Arabia and entire the Arabian platform is shown in Fig. 20.3. Positive TWS trends indicate an increase in water mass with time and vise versa in the areas of the negative TWS trends. Figure 20.3 depicts that the study area over LMAS and WASB (Fig. 20.3 area A) are experiencing a significant decrease in TWS trend (red and orange), compared to the southern areas (Fig. 20.3 area B (Al-Qassim Region)) that extend from lower depletion (yellow and orange) to near-steady statuses (blue) over the Red Sea seaboard plain. Figure 20.3 shows the temporal differences in GRACE-derived TWS, and the secular trend over the WASB in LMAS (area A) (-8.7 ± 1.1 mm/year;

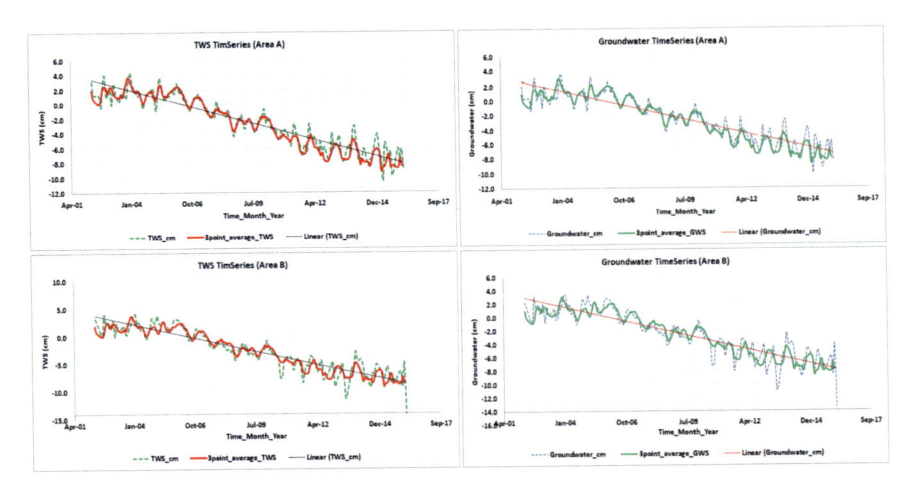

Fig. 20.4 Time series and secular trends for TWS and GWS over the area A and area B in the regional study area (LMAS)

$-3.1 \pm 0.\ 3$ km^3/year) and Al-Qassim Region (area B) (-9.5 ± 1.4 mm/year: -1.7 ± 0.2 km^3/year). Using TWS derived from GRACE solutions combined with soil moisture derived from GLDAS applied to Eq. 20.1, GWS time series for both WASB and LMAS were derived (Fig. 20.4).

The observed TWS and GWS seasonal periodical cycle were controlled mainly by groundwater abstraction. Temperatures in summers (June–August) are high (range: 28° to 37 °C), compared to winter season temperatures which range between (6° to 19 °C), and the rates of evaporation is between 25 to 28 mm/day in the summers season and as low as 0.4 mm/day in the winter seasons (Alsharhan et al., 2001). As a result, agricultural activities and groundwater abstraction got lower during the winter season (December–February) and decline in the summer seasons. Excessive abstraction in the winter season caused water-level declines, and surface land subsidence and the replenishment (gaining mass) during the summer season by infiltration and groundwater flow. Since there is a losing mass rate of -4.2 ± 0.7 km^3/year in GWS over the study basin, the extracted groundwater does not fully compensated.

The fissures near the boundaries of the subsidence bowls surrounding Busayta area in WASB and in the southeastern part of Al-Qassim area (Fig. 20.5A and B) were most probably caused by curving beam transforming around the area of deformation, the regions have experienced high lateral extension (Bell et al., 2002; Othman et al., 2018). Fissures also can result from the lateral (horizontal) forces linked with sediment compaction or horizontal pressures caused by extraction; all of these mechanisms produce lateral (horizontal) strain extended along the weakness planes (Helm, 1994). This mechanism could have caused Earth fissures that are located proximal to faults. The fissures close to the Kahf fault system in Area A and B in Figs. 20.5 and 20.6 are good examples of this deformation mechanism.

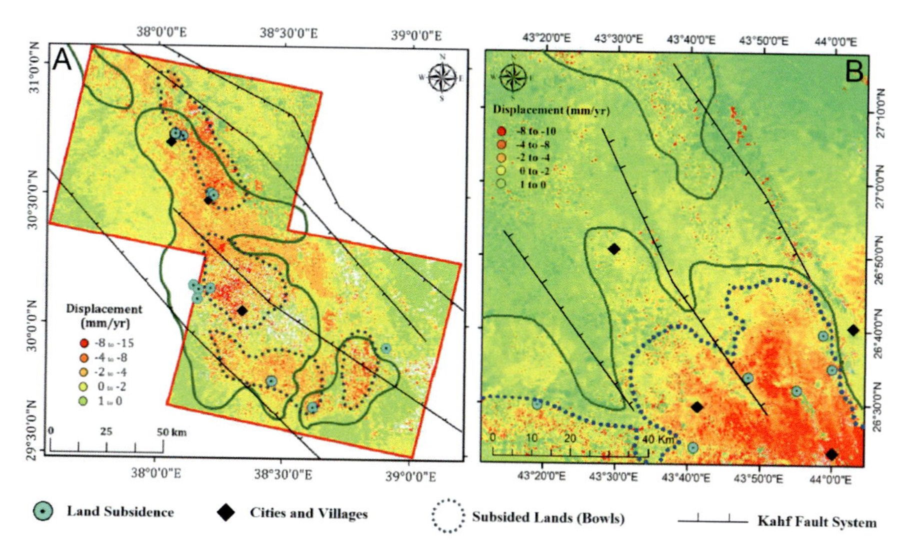

Fig. 20.5 "A" PSI results over Wadi As-Sirhan (Busayta and Tabarjal areas) "B" and the Al-Qassim region. The distribution of the major faults are also depicted with black lines. (Modified after Wallace et al., 2000)

Fig. 20.6 Field photos showing land fissures and sinkholes proximal to the land subsidence in the study areas

Our research findings are consistent with land subsidence phonomena associated with aquifers depletion which reported from numerous places around the world (e.g., Poland, 1960; Galloway et al., 1998; Hoffmann et al., 2001; Burbey, 2002; Bell et al., 2002; Christiansen et al., 2007; Assumpção et al., 2010; Galloway & Burbey, 2011; González et al., 2012; Avouac, 2012; Martínez et al., 2013; Amos et al., 2014). The detected subsided lands in these areas were attributed to increased adequate pressure that causes layer compaction and land deformation and subsidence (Donaldson et al., 1995). The effective stress balanced the downward stress effect due to the overload the overlying rock, water effective stress and pore pressure balanced the downward stress from the overlying rock and water. The changes in groundwater extraction-induced stresses trigger earthquakes and land deformations that would have occurred later when natural stress is accumulated (Avouac, 2012).

The illustrated land subsidence in the northern central part of the Kingdom of Saudi Arabia has a similar patter to other global aquifer systems in terms of theresponse to overexploitation of groundwater and the significant associated draw-down of water table such as in the Mojave Desert, California (Galloway et al., 1998), central and northeast Iran (Motagh et al., 2008) and in North China Plain (Guo et al., 2015). Such aquifers are expecting to show near-steady GWS secular trends, and stable velocity (0 radar averaged velocity) (Chew & Small, 2014). The case from LMAS is different because the fossil aquifer was primarily recharged in previous wet climatic periods yet still receives a small amount of recharge (Sultan et al., 2008). LMAS is not in near-stable state condition according to two observations. Which groundwater abstraction from the LMAS was not compensated for by replenish-ment,: (1) the depletion in TWS (-4.8 ± 0.7 km^3/year) and GWS (-4.2 ± 0.7 km^3/year) over the investigated period (Figs. 20.3 and 20.4) the radar interferometric technique studies over WASB and Al-Qassim Regions produce deformation rates of -4 to -15 mm/year (Fig. 20.5A and B). The regional deformation in the study basin is attributed mainly to a rise in effective stress that accelerate compaction and land subsidence (Fig. 20.5). The balance between the downward stress from the overload of the upper layers and the upward buoyancy force by the groundwater is challenged by the overexploitation from the Saq aquifer, where the rate of annual groundwater withdrawal exceeds 100 times the rate of annual recharge (et Miniéres, B.D.R.G., 2006). This overexploitation reduce pore pressure in the aquifer layers causing expand effective stress, which causes compaction and land subsidence in this location (Donaldson et al., 1995).

The MAS in the upper part Saudi Arabia shows the vertical hydrological con-nection between the aquifer layers due to the structural control and preferred pathways for groundwater flows (e.g. Kahf fault system) and the geological incom-patibility associated with differential deposition of aquifer layers (Othman et al., 2018) (McGillivray & Husseini, 1992), (et Miniéres, B.D.R.G., 2006). The vertical connection between the deep gaseous-rich Saq aquifer (e.g. H2S and CO2; Kalin et al., 2001) and the upper Paleozoic and Triassic hyrologic layers leads to enhanced evaporation and leakaes from Saq aquifer and dissolution of limestones formations and forming the karst topography during the wet climatic periods and associated elevation of the groundwater level (Abotalib et al., 2016). These karst remain mechanically stable under different climatic conditions untilwater table drawdown, which causes instability over the pre-existing subsurface cavitis. This kind of mechanism was reported as the primary mechanism of formation of falling features in the Arabian Peninsula to the eastern region (Abotalib et al., 2019b). It can illustrate the the formation of local land fall features in the region of study. More evidence of support the latest mentioned mechanism is supported by the field observation/description of sinkholes northern of Buriydah City in Al-Qassim Region (Youssef et al., 2016). The abovementioned mechanisms, which induced by groundwater dynamics, however controlling the study basin doesn't rule out the iffluence of the geotechnical properties of the lithological units on the surface deformation in the study basin. The illustrated features of cracks and fissures in Fig. 20.6 might be affected by the existence of near surface approximately thick layer of up to 15 m with high plastic clay beds, causes upward rupture structure when sub-ground layers damping from close agricultural activities extends these beds (Al Fouzan & Dafalla, 2013). Consequently, the mixture of dissolution and subsequent sagging and landfall of the upper layers due to the drop down of groundwater level and the existence of high plastic clay layers with the hydrological and geological factors and the availability of suitable geological and hydrological conditions (e.g. loss of buoyancy) can cause the reported regional and local landfall features in the study basin.

20.5 Summary and Implications

The LMAS, one of the highly hydrologic conductive aquifers in the Middle East and North Africa, represents an significant source of groundwater abstraction for irrigation in the upper part of Saudi Arabia. Mainly, this sandstone fossil aquifer of the Paleozoic age has been inordinately mined throughout the past 50 years, causing regional groundwater table dropdown of up to 150 m. The subsidence locations inferred from the InSAR processing maps strongly correlate with GRACE-derived TWS results suggesting that the subsidence can result from excessive groundwater extraction resources. The rapid increase in the cultivated lands between 1987 and 2012 supports this phenomenon (Fig. 20.7) that continues to our present time. Findings from the present study support that excessive fossil groundwater abstraction is critical for forming land subsidence, collapse features, and earth fissures in the study areas. However, the formation of such features might be associated with other factors. The studied approach could provide a simplified and cost-effective method to assess the surface land deformations and landscape associated with anthropogenic and/or natural induced variations in the groundwater tables under arid/semi-arid to hyper-arid conditions.

Acknowledgments The authors are grateful for Umm Al-Qura University, the University of Texas at Austin (Space Research Center), European Space Agency, Ministry of Environment, Water and Agriculture, and Saudi Geological Survey for their data, field, and logistical support.

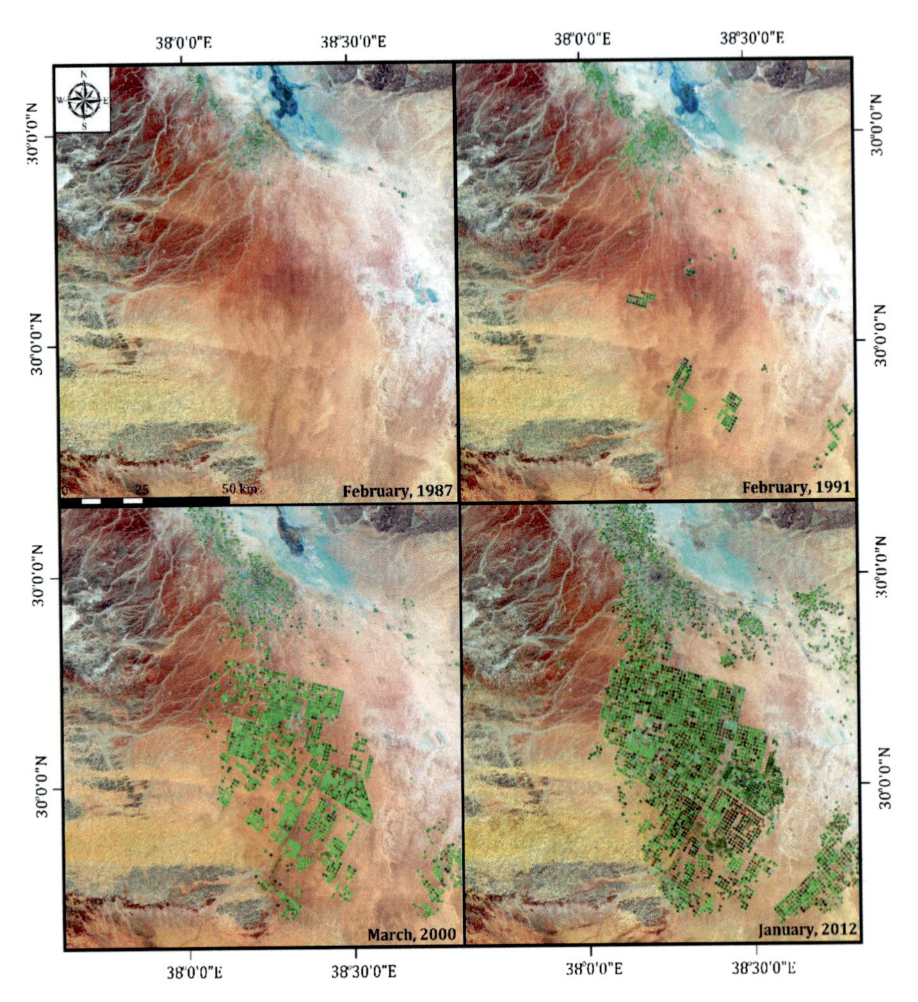

Fig. 20.7 Time series for Landsat 5, 7 and 8 scenes acquired (February 1987, February 1991, February 2000, January 2012) over Busayta area in Wadi As-Sirhan Basin (WASB). (Source: Earth Resources Observation and Science (EROS) Center)

References

Abdelmalik, K. W., & Abdelmohsen, K. (2019). GRACE and TRMM mission: The role of remote sensing techniques for monitoring spatio-temporal change in total water mass, Nile basin. *Journal of African Earth Sciences, 160*. https://doi.org/10.1016/j.jafrearsci.2019.103596

Abdelmohsen, K., Sultan, M., Ahmed, M., Save, H., Elkaliouby, B., Emil, M., Yan, E., Abotalib, A. Z., Krishnamurthy, R. V., & Abdelmalik, K. (2019). Response of deep aquifers to climate variability. *Science of the Total Environment, 677*, 530–544. https://doi.org/10.1016/j.scitotenv.2019.04.316

Abdelmohsen, K., Sultan, M., Save, H., & Abotalib, Z. A. (2020). What can the GRACE seasonal cycle tell us about lake-aquifer interactions? *Earth-Science Reviews*. https://doi.org/10.1016/j.earscirev.2020.103392

Abotalib, A. Z., Sultan, M., & Elkadiri, R. (2016). Groundwater processes in Saharan Africa: Implications for landscape evolution in arid environments. *Earth-Science Reviews, 156*, 108–136.

Abotalib, A. Z., Sultan, M., Jimenez, G., Crossey, L., Karlstrom, K., Forman, S., ... Polyak, V. (2019a). Complexity of Saharan paleoclimate reconstruction and implications for modern human migration. *Earth and Planetary Science Letters, 508*, 74–84.

Abotalib, A. Z., Heggy, E., Scabbia, G., & Mazzoni, A. (2019b). Groundwater dynamics in fossil fractured carbonate aquifers in Eastern Arabian Peninsula: A preliminary investigation. *Journal of Hydrology, 571*, 460–470.

Abunayyan BRGM. (2008). *Investigations for updating the groundwater mathematical model(s) of the Saq overlying aquifers*. Ministry of Water and Electricity.

Ahmed, M., & Abdelmohsen, K. (2018). Quantifying modern recharge and depletion rates of the Nubian Aquifer in Egypt. *Surveys in Geophysics, 39*(4), 729–751. https://doi.org/10.1007/s10712-018-9465-3

Ahmed, M., Sultan, M., Yan, E., & Wahr, J. (2016). Assessing and improving land surface model outputs over Africa using GRACE, field, and remote sensing data. *Surveys in Geophysics, 37*, 529–556.

Al Fouzan, F., & Dafalla, M. A. (2013). Study of cracks and fissures phenomenon in Central Saudi Arabia by applying geotechnical techniques. *Arabian Journal of Geosciences, 7*, 1157–1164.

Alharbi, T., Sultan, M., Sefry, S., et al. (2014). An assessment of landslide susceptibility in the Faifa area, Saudi Arabia, using remote sensing and GIS techniques. *Natural Hazards and Earth System Sciences, 14*, 1553–1564.

Al-Jammaz, A., Sultan, M., Izadi, M., Abotalib, A. Z., Elhebiry, M. S., Emil, M. K., Abdelmohsen, K., Saleh, M., & Becker, R. (2021). Land subsidence induced by rapid urbanization in arid environments: A remote sensing-based investigation. *Remote Sensing, 13*, 1109. https://doi.org/10.3390/rs13061109

Alsharhan, A., Rizk, Z., Nairn, A. E., Bakhit, D., et al. (2001). *Physical geography of the Arabian peninsula, hydrogeology of an arid region: The Arabian gulf and adjoining areas* (pp. 7–42). Elsevier.

Amos, C. B., Audet, P., Hammond, W. C., et al. (2014). Uplift and seismicity driven by groundwater depletion in Central California. *Nature, 509*(7501), 483–486.

Assumpção, M., Yamabe, T. H., Barbosa, J. R., et al. (2010). Seismic activity triggered by water wells in the Paraná Basin, Brazil. *Water Resources Research, 46*(7), W07527.

Avouac, J. P. (2012). Earthquakes: Human-induced shaking. *Nature Geoscience, 5*(11), 763–764.

Barthélemy, Y., Béon, O., Le Nindre, Y., et al. (2007). *Modelling of the Saq aquifer system (Saudi Arabia). Aquifer systems management: Darcy's legacy in a world of impending water shortage* (pp. 175–190). Taylor & Francis.

Becker, R. H., & Sultan, M. (2009). Land subsidence in the Nile delta: Inferences from radar interferometry. *The Holocene, 19*, 949–954.

Bell, J. W., Amelung, F., Ramelli, A. R., & Blewitt, G. (2002). Land subsidence in Las Vegas, Nevada, 1935–2000: New geodetic data show evolution, revised spatial patterns, and reduced rates. *Environmental & Engineering Geoscience, 8*(3), 155–174.

Burbey, T. J. (2001). Storage coefficient revisited: Is purely vertical strain a good assumption? *Groundwater, 39*, 458–464.

Burbey, T. J. (2002). The influence of faults in basin-fill deposits on land subsidence, Las Vegas valley, Nevada, USA. *Hydrogeology Journal, 10*, 525–538.

Bürgmann, R., Hilley, G., Ferretti, A., & Novali, F. (2006). Resolving vertical tectonics in the San Francisco Bay area from permanent scatterer InSAR and GPS analysis. *Geology, 34*, 221–224.

Chaussard, E., Wdowinski, S., Cabral-Cano, E., & Amelung, F. (2014). Land subsidence in Central Mexico detected by ALOS InSAR time-series. *Remote Sensing of Environment, 140*, 94–106.

Chew, C. C., & Small, E. E. (2014). Terrestrial water storage response to the 2012 drought estimated from GPS vertical position anomalies. *Geophysical Research Letters, 41*, 6145–6151.
Christiansen, L. B., Hurwitz, S., & Ingebritsen, S. E. (2007). Annual modulation of seismicity along the San Andreas Fault near Parkfield, CA. *Geophysical Research Letters, 34*(4), L04306.
Dai, Y., Zeng, X., Dickinson, R. E., Baker, I., et al. (2003). The common land model. *Bulletin of the American Meteorological Society, 84*, 1013–1023.
Donaldson, E. C., Chilingarian, G. V., & Yen, T. F. (1995). Stresses in sediments (Chapter 3). In *Subsidence due to fluid Withdrawal* (pp. 165–190). Elsevier.
El Baruni, S. (1994). Earth fissures caused by groundwater withdrawal in Sarir South agricultural project area, Libya. *Applied Hydrogeology, 2*, 45–52.
Fairer, G. M. (1983). Reconnaissance geology of the Ishash Quadrangle, Sheet 26/39C, Kingdom of Saudi Arabia, U. S. Geological Survey, No. 83–821, Reston, Virginia, Saudi Arabia.
Fielding, E. J., Blom, R. G., & Goldstein, R. M. (1998). Rapid subsidence over oil fields measured by SAR interferometry. *Geophysical Research Letters, 25*, 3215–3218.
Gabriel, A. K., Goldstein, R. M., & Zebker, H. A. (1989). Mapping small elevation changes over large areas: Differential radar interferometry. *Journal of Geophysical Research: Solid Earth, 94*, 9183–9191.
Galloway, D. L., & Burbey, T. J. (2011). Review: Regional land subsidence accompanying groundwater extraction. *Hydrogeology Journal, 19*, 1459–1486.
Galloway, D. L., & Hoffmann, J. (2007). The application of satellite differential SAR interferometry-derived ground displacements in hydrogeology. *Hydrogeology Journal, 15*, 133–154.
Galloway, D. L., Hudnut, K. W., Ingebritsen, S., et al. (1998). Detection of aquifer system compaction and land subsidence using interferometric synthetic aperture radar, Antelope Valley, Mojave Desert, California. *Water Resources Research, 34*, 2573–2585.
Giannerini, G., don Campre, R., Feraud, G., & Zakhem, B. A. (1988). Deformations intraplaques et volcanisme associe; exemple de la bordure NW de la plaque Arabique au Cenozoique. *Bulletin De La Société Géologique De France, 4*(6), 937–947.
González, P. J., Tiampo, K. F., & Palano, M. (2012). The 2011 Lorca earthquake slip distribution controlled by groundwater crustal unloading. *Nature Geoscience, 5*(11), 821–825.
Guo, H. P., Zhang, Z. C., Cheng, G. M., Li, W. P., Li, T. F., & Jiao, J. J. (2015). Groundwater-derived land subsidence in the North China plain. *Environment and Earth Science, 74*(2), 1415–1427.
Helm, D. C. (1994). Hydraulic forces that play a role in generating fissures at depth. *Bulletin of the Association of Engineering Geologists, 31*(3), 293–304.
Higgins, S. A., Overeem, I., Steckler, M. S., Syvitski, J. P., Seeber, L., & Akhter, S. H. (2014). InSAR measurements of compaction and subsidence in the Ganges-Brahmaputra delta, Bangladesh. *Journal of Geophysical Research: Earth Surface, 119*, 1768–1781.
Hoetzl, H., Felber, H., & Zoetl, J. (1978). *The quaternary development of the upper part of Wadi Ar-Rimah (Saudi Arabia). Quaternary period in Saudi Arabia* (pp. 173–182). Springer.
Hoffmann, J., Zebker, H. A., Galloway, D. L., & Amelung, F. (2001). Seasonal subsidence and rebound in Las Vegas Valley, Nevada, observed by synthetic aperture radar interferometry. *Water Resources Research, 37*(6), 1551–1566.
Hooper A J (2006) Persistent scatter radar interferometry for crustal deformation studies and modeling of volcanic deformation. , Stanford University.
Hooper, A., Zebker, H., Segall, P., & Kampes, B. (2004). A new method for measuring deformation on volcanoes and other natural terrains using InSAR persistent scatterers. *Geophysical Research Letters, 31*, L23611.
Hooper, A., Segall, P., & Zebker, H. (2007). Persistent scatterer interferometric synthetic aperture radar for crustal deformation analysis, with application to Volcán Alcedo, Galápagos. *Journal of Geophysical Research: Solid Earth, 112*, B07407.
Hooper, A., Bekaert, D., Spaans, K., & Arıkan, M. (2012). Recent advances in SAR interferometry time series analysis for measuring crustal deformation. *Tectonophysics, 514*, 1–13.

Kalin, R. M., Elliot, T., Suba, A., Katbeh, H., Rimawi, O., IA, A. L.-H., & Aleissa, K. (2001). *Sustainability of groundwater resources in water-scarce regions: From micro-to macro-scales* (Reports submitted to Natural Resources & Environment Research Institute) (p. 56p). King Abdulaziz City for Science and Technology.

Kampes, B. M. (2006). *Radar interferometry: Persistent Scatterer technique, remote sensing and digital image processing 12*. Springer.

Kellogg, K. S., & Reynolds, R. L. (1983). Opening of the Red Sea: Constraints from a palaeomagnetic study of the as Sarat volcanic field, South-Western Saudi Arabia. *Geophysical Journal International, 74*, 649–665.

Khan, A., Khan, S., & Kakar, D. (2013). Land subsidence and declining water resources in Quetta Valley, Pakistan. *Environmental Earth Sciences, 70*, 2719–2727.

Konert, G., Afifi, A. M., Al-Hajri, S., et al. (2001). Paleozoic stratigraphy and hydrocarbon habitat of the Arabian plate. *GeoArabia, 6*(3), 407–442.

Lloyd, J., & Pim, R. (1990). The hydrogeology and groundwater resources development of the Cambro-Ordovician sandstone aquifer in Saudi Arabia and Jordan. *Journal of Hydrology, 121*, 1–20.

Maliva, R., & Missimer, T. (2012). *Compaction and land subsidence. Arid lands water evaluation and Management* (pp. 343–363). Springer.

Margat, J. (2007). *Great aquifer systems of the world. Aquifer systems management: Darcy's legacy in a world of impending water shortage* (pp. 105–116).

Martínez, P. J., Marín, M., Burbey, T. J., et al. (2013). Land subsidence and ground failure associated to groundwater exploitation in the Aguascalientes Valley, Mexico. *Engineering Geology, 164*, 172–186.

McGillivray, J. G., & Husseini, M. I. (1992). The Paleozoic petroleum geology of Central Arabia. *American Association of Petroleum Geologists Bulletin, 76*, 1473–1490.

MOAW (Ministry of Agriculture and Water). (1984). *Agriculture statistical yearbook*. Department of Economic Studies and Statistics.

MOAW (Ministry of Agriculture and Water). (2004). *Agriculture statistical yearbook*. Department of Economic Studies and Statistics.

Mohamed, A., Sultan, M., Ahmed, M., et al. (2017). Aquifer recharge, depletion, and connectivity: Inferences from GRACE, land surface models, and geochemical and geophysical data. *Geological Society of America Bulletin, 129*, 534–546.

Morsy, E. A., & Othman, A. (2020). Assessing the impact of groundwater mixing and sea water intrusion on oil production in coastal oil fields using resistivity sounding methods. *Arabian Journal of Geosciences, 13*, 434. https://doi.org/10.1007/s12517-020-05469-6

Motagh, M., Djamour, Y., Walter, T., Wetzel, H., Zschau, J., & Arabi, S. (2007). Land subsidence in Mashhad Valley, Northeast Iran: Results from InSAR, levelling and GPS. *Geophysical Journal International, 168*, 518–526.

Motagh, M., Walter, T. R., Sharifi, M. A., Fielding, E., Schenk, A., Anderssohn, J., & Zschau, J. (2008). Land subsidence in Iran caused by widespread water reservoir overexploitation. *Geophysical Research Letters, 35*(16), L16403.

Othman, A., & Abotalib, A. Z. (2019). Land subsidence triggered by groundwater withdrawal under hyper-arid conditions: Case study from Central Saudi Arabia. *Environmental Earth Sciences, 78*(7), 243.

Othman, A., Sultan, M., Becker, R., et al. (2018). Use of geophysical and remote sensing data for assessment of aquifer depletion and related land deformation. *Surveys in Geophysics, 39*, 543–566. https://doi.org/10.1007/s10712-017-9458-7

Poland, J. (1960). Land subsidence in the San Joaquin Valley, California, and its effect on estimates of ground-water resources. *International Association of Hydrological Sciences, 52*, 324–335.

Powers, R., Ramirez, L., Redmond, C., & Elberg, E. (1966). *Geology of the Arabian peninsula: Sedimentary geology of Saudi Arabia. U. S. Geological Survey*. Professional Paper: a review of the sedimentary geology of Saudi Arabia, Alexandria, Virginia 560:1–147.

Rateb, A., & Abotalib, A. Z. (2020). Inferencing the land subsidence in the Nile Delta using Sentinel-1 satellites and GPS between 2015 and 2019. *Science of the Total Environment, 729*, 138868.

Ren, G., Whittaker, B., & Reddish, D. (1989). Mining subsidence and displacement prediction using influence function methods for steep seams. *International Journal of Mining Science and Technology, 8*, 235–251.

Rodell, M., Houser, P., Jambor, U., et al. (2004). The global land data assimilation system. *Bulletin of the American Meteorological Society, 85*, 381–394.

Roobol, M., Shouman, S., & Al-Solami, D. (1985). *Earth tremors, ground fractures, and damage to buildings at Tabah*. Saudi Arabian Deputy Ministry for Mineral Resources Technical Record, DGMR–TR-05–4.

Rouse, Jr J, Haas, R, Schell, J, & Deering, D. (1974). *Monitoring vegetation systems in the Great Plains with ERTS, NASA*. Goddard Space Flight Center 3d ERTS-1 Symp 1, Sect A, pp. 309–317.

Save, H., Bettadpur, S., & Tapley, B. D. (2016). High-resolution CSR GRACE RL05 mascons. *Journal of Geophysical Research: Solid Earth, 121*, 7547–7569.

Scanlon, B. R., Zhang, Z., Save, H., et al. (2016). Global evaluation of new GRACE mascon products for hydrologic applications. *Water Resources Research, 52*, 9412–9429.

Scharroo, R., & Visser, P. (1998). Precise orbit determination and gravity field improvement for the ERS satellites J Geophys res. *Oceans, 103*, 8113–8127.

Strassberg, G., Scanlon, B. R., & Rodell, M. (2007). Comparison of seasonal terrestrial water storage variations from GRACE with groundwater-level measurements from the high plains aquifer (USA). *Geophysical Research Letters, 34*(14), L14402.

Sultan, M., Sturchio, N., Al Sefry, S., et al. (2008). Geochemical, isotopic, and remote sensing constraints on the origin and evolution of the Rub Al Khali aquifer system, Arabian Peninsula. *Journal of Hydrology, 356*, 70–83.

Sultan, M., Yousef, A., Metwally, S., et al. (2011). Red Sea rifting controls on aquifer distribution: Constraints from geochemical, geophysical, and remote sensing data. *Geological Society of America Bulletin, 123*, 911–924.

Sultan, M., Ahmed, M., Wahr, J., Yan, E., & Emil, M. K. (2014). *Monitoring aquifer depletion from space: Case studies from the Saharan and Arabian aquifers. Remote sensing of the terrestrial water cycle* (pp. 347–366). Wiley.

Sultan, M., Sturchio, N. C., Alsefry, S., Emil, M. K., Ahmed, M., Abdelmohsen, K., AbuAbdullah, M. M., Yan, E., Save, H., Alharbi, T., Othman, A., & Chouinard, K. (2019). Assessment of age, origin, and sustainability of fossil aquifers: A geochemical and remote sensing–based approach. *Journal of Hydrology*. https://doi.org/10.1016/j.jhydrol.2019.06.017

Tapley, B. D., Bettadpur, S., Watkins, M., & Reigber, C. (2004). The gravity recovery and climate experiment: Mission overview and early results. *Geophysical Research Letters, 31*(9).

Taylor, R. G., Scanlon, B., Döll, P., et al. (2013). Ground water and climate change. *Nature Climate Change, 3*, 322–329.

Vincent, P. (2008). *Environmental impacts and hazards, Saudi Arabia: An environmental overview* (pp. 216–245). Taylor & Francis.

Wagner, W. (2011). The Arabian plate: Geology and Hydrogeologic characteristics (Chapter 1). In *Groundwater in the Arab Middle East* (pp. 1–54). Springer.

Wallace, C., Dinin, S., & Al-Farasani, A. (2000). *Geologic map of the Wadi as-Sirhan quadrangle, sheet 30C*. Kingdom of Saudi Arabia, Saudi Geological Survey.

Wouters, B., Bonin, J., Chambers, D., et al. (2014). GRACE, time-varying gravity, earth system dynamics and climate change. *Reports on Progress in Physics, 77*, 116801.

Youssef, A. M., Sabtan, A. A., Maerz, N. H., & Zabramawi, Y. A. (2014). Earth fissures in Wadi Najran, Kingdom of Saudi Arabia. *Natural Hazards, 71*, 2013–2027.

Youssef, A. M., Al-Harbi, H. M., Gutiérrez, F., et al. (2016). Natural and human-induced sinkhole hazards in Saudi Arabia: Distribution, investigation, causes and impacts. *Hydrogeology Journal, 24*, 625–644.

Chapter 21
Landslide Susceptibility Map Production of Aden Peninsula – South West of Yemen

Khaled Khanbari, Adnan Barahim, Ziad Almadhaji, and Sami Moheb Al-Deen

Abstract The slope stability assessment of Aden peninsula which is located at south-western part of Yemen, was carried out. The study area consist of volcanic rocks which affected by number of faults, joints and dykes. All important factors affecting slope stability in the area such as slope angle, slope height, discontinuities measurements, weathering, vegetation cover, rainfall and previous landslide were evaluated. The study was constructed based on integration of field investigation and satellite images in combination with Geographic Information System (GIS). Land use/cover and Landslide susceptibility maps were produced by using satellites images and the Landslide Possibility Index System, the correlation values were computed between the factors measured and Landslide Possibility Index values. Landslide Susceptibility classes derived into six categories while four these categories are recorded in Aden Peninsula: very high, high, moderate and low zone. The results show that rockfall is the main mode of failure in the study area. Some treatments may achieve where it is needed such as removal of rock overhang and unstable blocks, support the toe of the slope and the overhanging parts by retaining wall, erecting well seal drainage conduit and grouting of discontinuities. The outcomes of this research can be helpful for decision makers and future development planners to avoid the zones with high susceptibility during land use planning and urban extension.

K. Khanbari (✉)
Yemen Remote Sensing and GIS Center, Sana'a, Yemen

Earth Science Department, Faculty of Petroleum and Natural Resources, Sana'a University, Sana'a, Yemen

A. Barahim
Earth Science Department, Faculty of Petroleum and Natural Resources, Sana'a University, Sana'a, Yemen

Z. Almadhaji
Geological Survey and Mineral Resources Board (GSMRB), Sana'a, Yemen

S. M. Al-Deen
Yemen Remote Sensing and GIS Center, Sana'a, Yemen

M. M. Al Saud (ed.), *Applications of Space Techniques on the Natural Hazards in the MENA Region*, https://doi.org/10.1007/978-3-030-88874-9_21

Keywords Landslide · LPI · Susceptibility map · Aden Peninsula · Yemen

21.1 Introduction

Landslides resulting from large-scale rock slope failures are the major hazard in mountainous regions. In the twentieth century, disasters caused by massive rock slope failures have killed more than 50,000 people on a global basis (Evans et al., 2006). Landslide Susceptibility is one of the most reliable methods to identify landslide- prone areas (Hansen, 1984), i.e. the probability of landslide occurrence.

Landslide susceptibility mapping is essential technique to detract and prevent the landslide detriment. According to Van Westen, 1997, a landslide susceptibility analysis is an analysis that involves the determination of the spatial distribution of landslides, involves four processes: (a) the construction of a landslide inventory map, (b) the assessment of parameters that influence landslides, (c) the implementation of appropriate methods for determining the weights of each parameter and (d) the compilation of the landslide susceptibility map within a Geographic Information System (GIS) environment.

Mostly the susceptibility zonation maps are prepared on a 1:50,000 or 1:25,000 scale (Sarkar et al., 2013). Therefore, assessment of probabilistic slope failure hazard is a one part of decision analytical approach to landslides risk assessment and management.

Many methods have been developed as tools to assess landslide susceptibility, Heuristic approach is one of them, it is based on opinion of geomorphologic experts (Francipane et al., 2014). Generally this approach is divided into two phases: a direct mapping analysis, in which the geomorphologists (or geologist) determine the susceptibility in the field directly on the base of their experience, and a qualitative map combination, in which the experts use their knowledge to determine the weighting value for each class parameter in each parameter (Bartolomei et al., 2006; Puglisi et al., 2007).

Remote sensing and Geographic Information System (GIS) techniques are useful for landslide hazard zonation mapping and can help identify the area best suited for developmental activities (e.g. Saha, 2005; Van Westen et al., 2008; Gupta et al., 2008).

The study area is a volcano mountainous region, a little of landslide was recorded at the region, and it is predictable they are frequently increased due to human activities expansion. The study area is located in the eastern part of Aden governorate which includes the districts of Sirah (or Crater, non-official name), At Tawahi, Al Mu'alla, Khur Maksar, Dar Sa'd, Ash Shaykh Othman and Al Mansurah. These districts represent Aden Peninsula area (Fig. 21.1a).

Aden is an economic capital of Yemen which is located at the south-western part of Yemen, 450 km to the south of Sana'a city; capital of Yemen (Fig. 21.1a). It is characterized by mountainous to coastal area with wet hot climate. The annual average temperature ranges between 22.5 °c in January and 34.2 °c in June, the

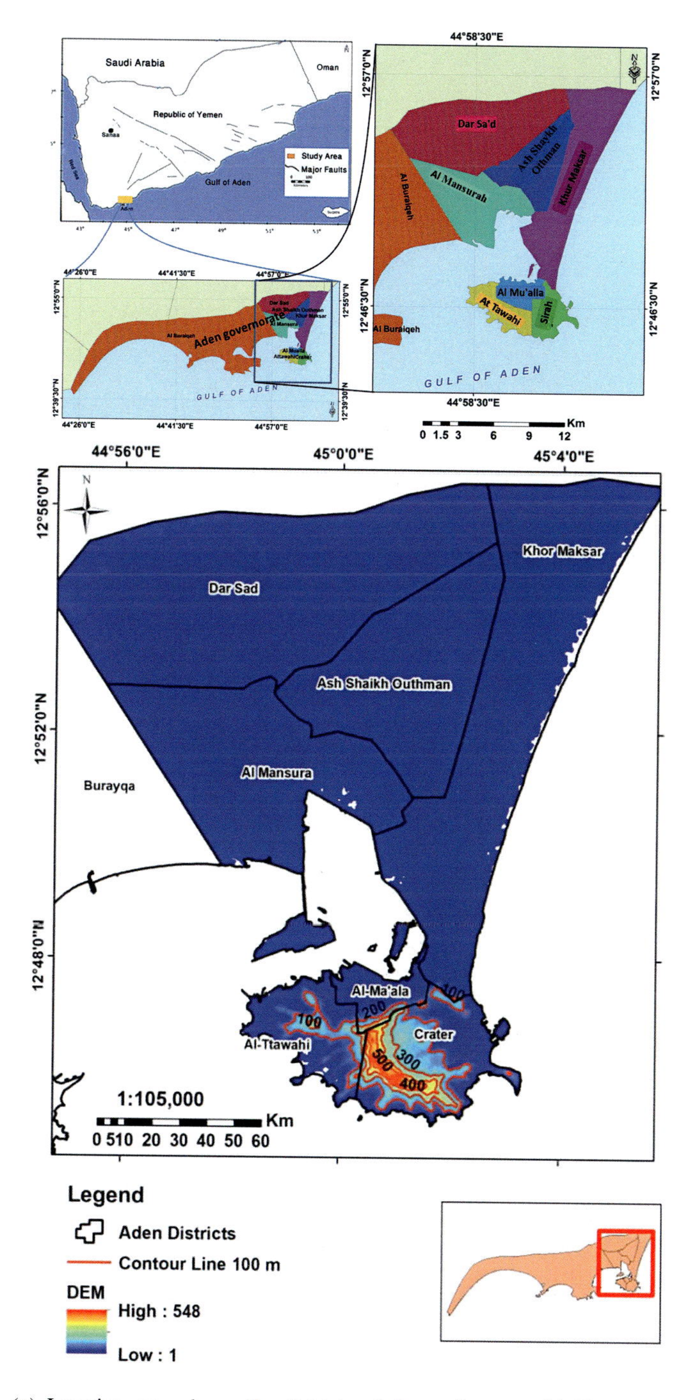

Fig. 21.1 (**a**) Location map shows the districts of the study area, (**b**) Elevation map of the study area

Fig. 21.2 Photo of the historical and archaeological Sirah Fort

annual average precipitation ranges between 56.4 mm/year in June and 304.7 mm/year in August and the annual average humidity ranges from 61.4 in June to 75.2 in August.

This research aims to produce the Landslide Susceptibility Map of Aden Peninsula area, which contain the most important monumentally tourist in Aden represented by the historical and archaeological Sirah Fort (Fig. 21.2) and Aden Tanks (Fig. 21.3), and the most important economic construction represented by the Aden Harbour.

The mountainous area is located at Sirah (or Crater), At Tawahi and Al Mu'alla districts. The elevation map which is produced from digital elevation model (DEM) (Fig. 21.1b) shows that the highest elevation (548 m) is located at the volcanic crater boundary of Sirah district. The other districts are lay's at coastal zone. The geology of the area in general characterized mainly by Cenozoic Volcanics.

21.2 Geologic Setting

21.2.1 Geology/Lithology

The geology of Aden governorate is an integral part of the composition of Yemen and the Arabian Peninsula. Aden governorate is mainly covered by Quaternary-Recent sediments, Sabkhas and Cenozoic Volcanics (Fig. 21.4). The volcanic rocks are represented by three volcanoes: Aden, Little Aden, and Ras Imran, which are belonging to the Aden Volcanic Series. These extinct volcanoes were erupted

Fig. 21.3 Photo of the Aden Tanks

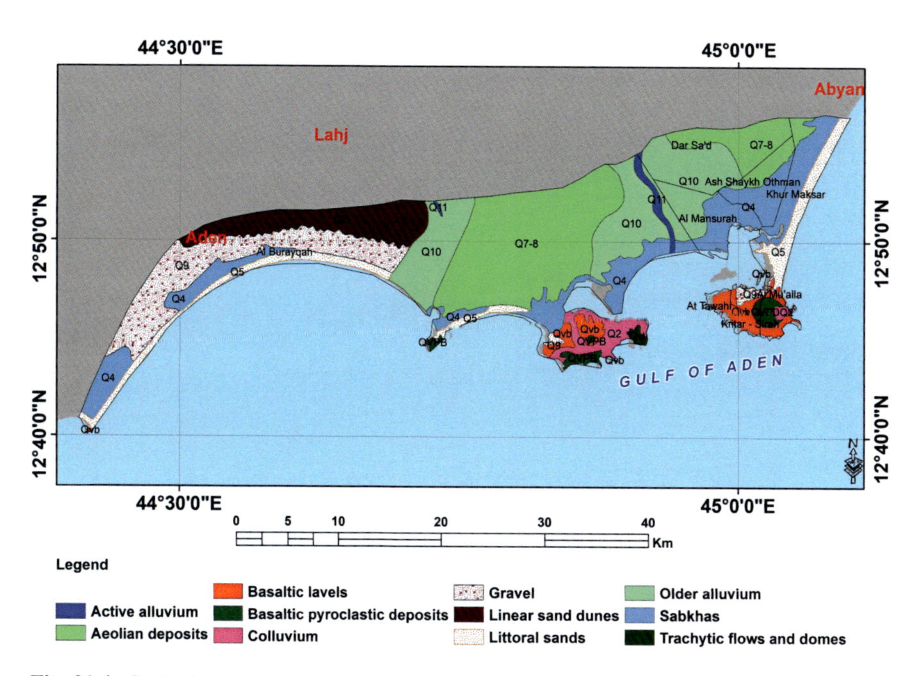

Fig. 21.4 Geological map of Aden Governorate (modified after GSMRB, 1990)

through central-vent, strato-volcanoes about 5 m.y. ago (Dickinson et al., 1969; Cox et al., 1970). Each volcano evolved through a complex cone-building stage during which the predominant rock types were trachybasalt, trachyte and peralkaline rhyolite. This was followed by periods of caldera formation during which trachyandesites, trachytes, and peralkaline rhyolites were erupted (Cox et al., 1968). The major element chemistry of the Aden Volcanic Series is intermediate between the alkaline and tholeiitic associations (Cox et al., 1970). The most acceptable 'parental' magma is mildly alkaline olivine basalt which, on fractionation, produced a series ranging from trachybasalts through trachyandesites and trachytes to rhyolites (Cox et al., 1970). A final stage of parasitic activity included small amounts of olivine-tholeiite amongst the eruptive products (Cox et al., 1968).

21.2.2 Tectonic Setting

Yemen is bounded by two active, young oceanic rifts which are resulting from the movement of the Arabian plate away from Africa in a NE direction since 40 Ma (Beydoun, 1970). The Red Sea is located between African and Arabian plates, and the Gulf of Aden between Somalian and Arabian plates. These two oceanic basins are connected within area of Afar hotspot (Fig. 21.5). The western part of Yemen which is characterized by Tertiary volcanics (Yemen Trap Series) represents a volcanic margin along the Red Sea coast and the western part of the coast of the Gulf of Aden (Khanbari & Huchon, 2010). Tard et al., 1991 show that the southern part of the volcanic margin (Aden margin) presents offshore the characteristics features of a volcanic passive margin. The volcanic activities in Yemen are related

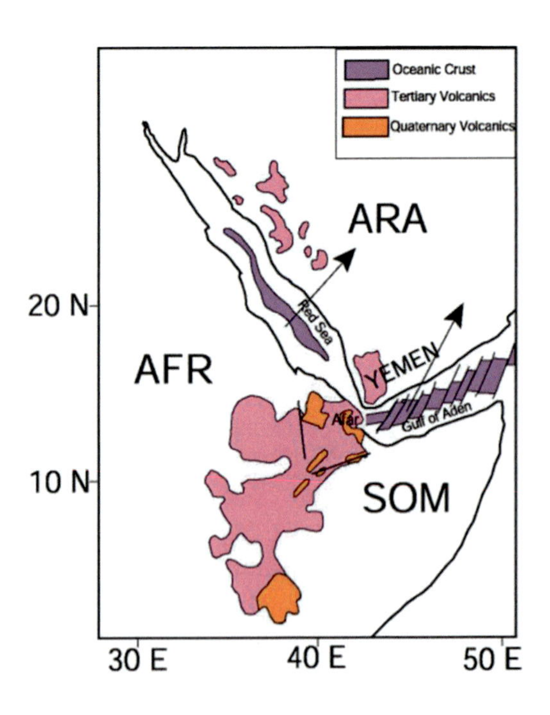

Fig. 21.5 Simplified Tectonic Map of Yemen and the surrounding areas. Arrows show the directions of relative plate motions (modified after Khanbari & Huchon, 2010)

to the Afar mantle plume that impacted the Arabia–Africa area during the Oligocene (Baker et al., 1997).

The volcanoes of Aden Volcanic Series (study area) lie along an east-west lineament parallel to the north margin of the Gulf of Aden which is a young oceanic basin. Its general trend is WSW-ENE (N70°E). The Gulf of Aden is thus oblique to the movement of Arabia toward the north-east (Dauteuil et al., 2001) (Fig. 21.5). The oceanic ridge becomes younger toward the west which reflects its propagation toward Afar hotspot (Fournier et al., 2010). The western part of Gulf of Aden which is located near the study area is characterized by high seismic activities. Although the study area is located near active ridge of the Gulf of Aden, the bibliography does not describe any great earthquakes occurred in the past, but the study area remains under the effect of the very active ridge of the Gulf of Aden (Khanbari, 2020).

In space and time the Aden Volcanic Series represents a relatively minor episode intermediate between the alkaline volcanism of the Lower Tertiary Yemen Trap Series and equally abundant tholeiitic activity in the centre of the Gulf of Aden (Cox et al., 1970). The upper part of the Yemen Trap Series is petrologically similar to the Aden Volcanic Series. Cox et al., 1970 suggest that the magma originated in the mantle.

The volcanoes of Aden Volcanic Series were active between 5 and 10 m.y. ago and remained active over periods of 1–1.5 m.y (Cox et al., 1970). It is at least suggestive that the period of magmatic quiescence between the Yemen Trap Series and the Aden Volcanic Series is coeval with the hiatus in ocean floor spreading in the Gulf of Aden (Laughton et al., 1970) and that the volcanic activity along the Aden volcanic series corresponds to the renewal of ocean floor spreading.

21.3 Methodology

This study presents landslide hazards analysis at Aden Peninsula area by using Remote Sensing, Geographic Information System (GIS) and Field data. The Remote Sensing data include Landsat-8 (OLI) image, high resolution Quick-Bird image with 60 cm resolution dated 2020 (source: Google Earth) and Digital Elevation Model (DEM) with 30 m resolution downloaded from ASTER global DEM. These data were provided by the Yemen Remote Sensing and GIS Center (YRSC). Landslide hazard map was prepared in the study area by processing and interpretation of remote sensing and field data in combination to build GIS database to evaluate the distribution of landslides.

The Landsat-8 image was processed and interpreted to construct a geological map of the study area. Both the Landsat and high resolution images were used to produce a land use/cover map of the study area. Additionally, DEM was used to calculate the slope angle and to create a slope map.

The field data was collected over a period of 20 days with the Geological Survey and Mineral Resources Board (GSMRB) staff. The main features affecting slope stability were measured and estimated in the field at 177 stations (locations) within area of 712.23 km^2 for the eastern part of Aden Governorate which is represented by Aden Peninsula. These included slope height (the vertical distance between the top and the toe of the slope at the station), slope angle, grade of fracture, grade of weathering, gradient of discontinuities, spacing of discontinuities, orientation of discontinuities, vegetation cover, water infiltration and previous landslide. Then the parameters were integrated using a heuristic method (Gemitzi et al., 2011; Francipane et al., 2014; Barahim et al., 2018) as a rating system according to the Landslide Possibility Index chart (LPI) (Bejerman, 1998 and Barahim, 2004) (Fig. 21.6); this is similar to an Analytical Hierarchy Process (AHP) approach, which is a multi-objective, multi-criteria decision-making approach that enables the user to arrive at a scale of preferences drawn from a set of alternatives (Saaty & Vargas, 1991; Saaty & Vargas, 2001; Pour & Hashim, 2017). The correlation coefficient values between each factor and LPI value were calculated by Microsoft excel software in order to understand which factors significantly affect slope instability in the study area.

All interpretations of remote sensing data and field data were implemented in ArcGIS 10.2 in order to evaluate slope instability, and to produce the landslide susceptibility map. The engineering characteristics of the rock mass were described according to the Geological Society of London Engineering Group Working Party (Anon, 1972) and the failure types were recognized.

21.4 Land Use/Cover

Land use/cover data was classified by using unsupervised classification of Landsat image and supporting with visual interpretation of high resolution image. The land use/cover map (Fig. 21.7) has been classified into seven classes such as lithological outcrops, built areas, barren lands, agricultural zones, infrastructures, wet lands and roads. Land use/cover map which represents the distribution of the various surface activities in the study area, shows that lithological outcrops cover 8.3% of the area, and that the built areas cover 14.5%. The barren lands are dominated in the study area and they represent the greatest area with covering of 36.6%. The agricultural zones are covering 15.4% of the area. The infrastructures such as the air-port and sea-ports cover 12.2%. Asphaltic and paved roads are represented in the map by lines. These roads cover an area of 6.6%. Aden Governorate is characterized by wetlands which cover 6.3% of the area. Table 21.1 shows the area and the percentage of each activity of Land use/cover map for each district.

This study will focus on the districts which are characterized by great area of lithological outcrops. They are Sirah, Al Mu'alla and At Tawahi districts. They are located close to the slopes and they subjected to slope stability hazard. These districts are bounded by the sea and the mountains. During the last 30 years, the

Landslide Possibility Index (LPI)							
No.	**Factor**	**Scale**	**Estimation**	**No.**	**Factor**	**Scale**	**Estimation**
1	SLOPE HEIGHT (M)	1 – 8 m	1	**2**	SLOPE ANGLE (°)	< 15°	0
		9 - 15 m	2			15° - 30°	1
		16 – 25 m	3			30° - 45°	2
		26 – 35 m	4			45° - 60°	3
		> 35 m	5			> 60°	4
3	GRADE OF FRACTURE (Number of Fracture)			**4**	GRADE OF WEATHERING (Alteration and compressive strength)	Fresh	0
		Sound	0			Slightly Weathered	1
		Moderately Fractured	1			Moderately Weathered	2
		Highly Fractured	2			Highly Weathered	3
		Completely Fractured	3			Completely Weathered	4
						Res. Soil	5
5	GRADIENT OF THE DISCONTI-NUITIES (°)	<° 15	0	**6**	SPACING OF THE DISCONTI-NUITIES (M)	3 <	0
		15° - 30°	1			3 - 1	1
		30° - 45°	2			1 - 0.3	2
		45° - 60°	3			0.3 - 0.05	3
		> 60°	4			0.05 >	4
7	ORIENTATION OF THE DISCONTI-NUITIES	Favourable 	0	**8**	VEGETATION COVER (%)	Void (< 20%)	0
		Unfavourable	4			Scarce (20% -60%)	1
		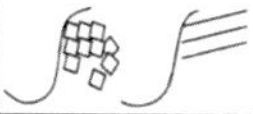				Abundant (> 60%)	2
9	WATER INFILTRATION (mm/year)	Inexistent	0	**10**	PREVIOUS LANDSLIDES (m^3/year)	Not registered	0
		Scarce	1			Registered Small volume (<3m^3/year)	1
		Abundant (>500mm/year) **Permanent**	2			Registered High volume (>3m^3/year)	2
		Abundant (>500mm/year) **Seasonal**	3				

1 + 2 + 3 + 4 ± 5 + 6 + 7 + 8 + 9 + 10 = LPI value

The LPI value is obtained by adding the estimations of attributes 1 to 10. If the orientation of the discontinuities is favourable, subtract the estimation of the gradient.

I (Nil)	(0-5)	III (Moderately low)	(11-15)	V (High)	(21-25)
II (Low)	(6-10)	IV (Moderately high)	(16-20)	VI (Very high)	(>25)

Fig. 21.6 LPI chart used in the study area (Bejerman, 1998; Barahim, 2004)

construction of building increases toward the mountainous area which increases the effect of slop stability hazard and threat the life of population.

Table 21.1 shows that the lithological outcrops cover 75.9% of the area in Sirah district, and the built areas cover 13.9% in Al Mu'alla district, lithological outcrops cover 16.5% of the area, and the built areas cover 30.9% while in At Tawahi district lithological outcrops cover 51.1% of the area, and that the built areas cover 25.3%.

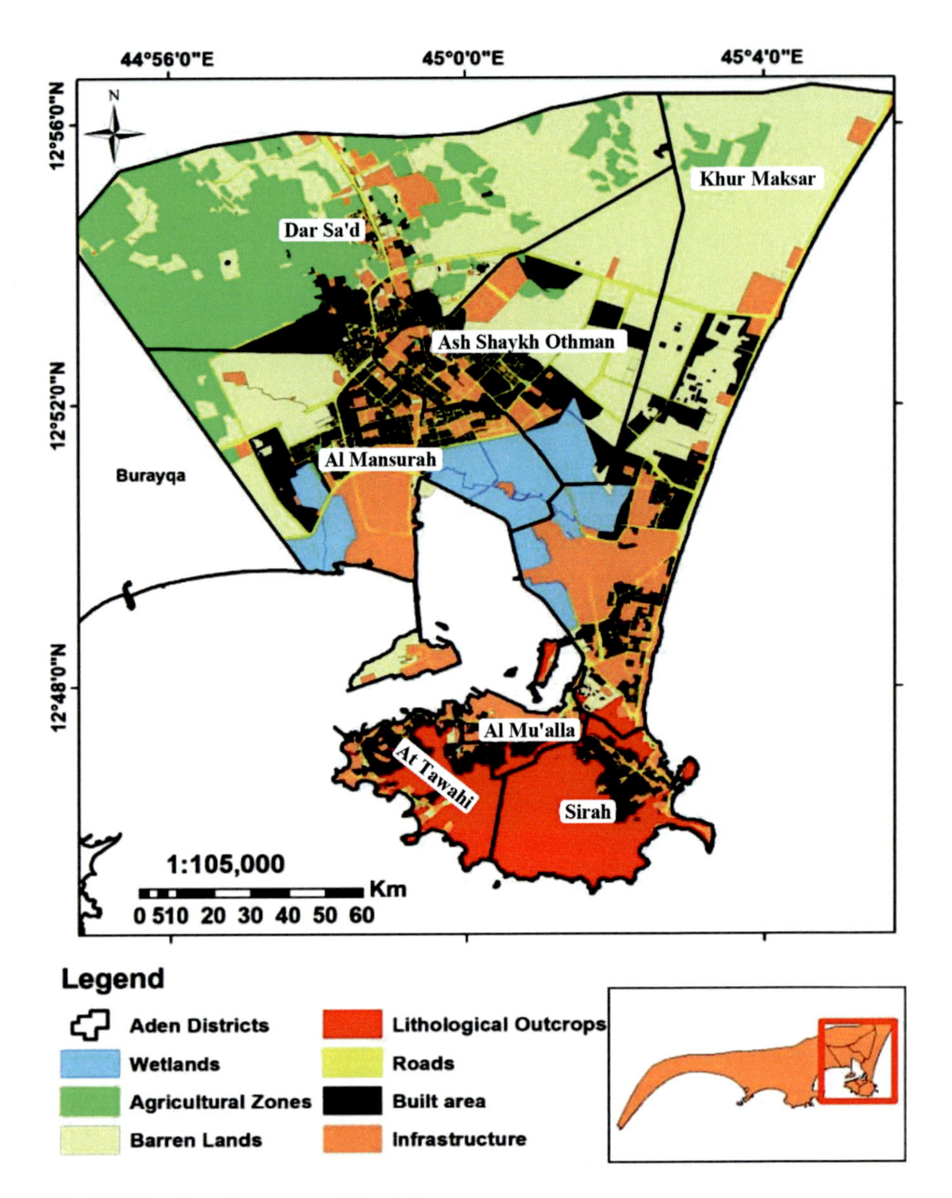

Fig. 21.7 Land use/cover map of the study area

21.5 Landslide Data Analysis, Results and Discussion

21.5.1 LPI Assessment

The total area under investigation is 210.06 km^2 and the rock formations cover an area of 18.78 km^2 which equal to 8.9% of the area. Along the study area 196 stations

Table 21.1 Land use/cover activities in the study area

District		Built area	Lithological Outcrops	Barren Lands	Agricultural Zones	Infrastructures	Wetlands	Roads	Total
Sirah	Area (m2)	2,030,449.2	11,044,224.9	147,253.3	0	925,810.7	0	409,717.4	14,557,455.5
	%	13.9	75.9	1.0	0.0	6.4	0.0	2.8	100.0
Al Mu'alla	Area (m2)	1,269,408.8	680,365.7	167,831.1	7056.9	1,610,130.8	0	376,255.8	4,111,049.1
	%	30.9	16.5	4.1	0.2	39.2	0.0	9.2	100.0
At Tawahi	Area (m2)	2,385,000.3	4,810,634.5	322,803.8	0	1,488,583.3	0	406,604.2	9,413,626.1
	%	25.3	51.1	3.4	0.0	15.8	0.0	4.3	100.0
Khur Maksar	Area (m2)	6,674,127.8	799,008.9	27,703,486.2	1,268,948.7	10,655,915	3,390,380.7	4,217,193.8	54,709,061.1
	%		1.5	50.6	2.3	19.5	6.2	7.7	100.0
Ash Shaykh Othman	Area (m2)	7,134,136.3	0	15,073,095	310,881.9	0	2,117,071.7	2,594,500.4	27,229,685.3
	%	26.2	0.0	55.4	1.1	0.0	7.8	9.5	100.0
Al Mansura	Area (m2)	5,909,449.7	0	9,246,563.5	2,278,731.8	7,939,533.3	7,705,165.7	3,394,080.3	36,473,524.3
	%	16.2	0.0	25.4	6.2	21.8	21.1	9.3	100.0
Dar Sa'd	Area (m2)	4,753,849.8	0	2,3764,944.5	28,320,266.3	2,865,985.4	0	2,374,732.4	62,079,778.4
	%	7.7	0.0	38.3	45.6	4.6	0.0	3.8	100.0
Total	Area (m2)	30,156,421.9	17,334,234	76,425,977.4	32,185,885.6	25,485,958.5	13,212,618.1	13,773,084.3	20,857,4179.8
	%	14.5	8.3	36.6	15.4	12.2	6.3	6.6	100.0

were identified, 117 of them on rock formations exposed at districts of Sirah, At Tawahi, Al Mu'alla and southern part of Khur Maksar. The other 79 stations at sand and Sabkha areas covered the northern part of Khur Maksar and the coastal regions of the Sirah, At Tawahi and Al Mu'alla districts. The distance between each station is 200 and 500 m to the north-south and east-west directions, with exception of some military establishment areas. Dar Sa'd, Ash Shaykh Othman and Al Mansurah districts have a semi horizontal surface covered by sand and Sabkha, and the landslide hazard is nil (Fig. 21.7).

According to the Geological Society Engineering Group Working Party of London (Anon, 1972) the rock masses (rock units) in the study area are identified as black colour and fine grain size, thick bedded to bulky, moderately to widely spaced joints, slightly weathered, basalt, trachybasalt, trachyte and rhyolite, which is moderately permeable and has very strong compressive strength.

Based on the LPI assessment system several plots were created using Microsoft Excel (Fig. 21.8) to show the relationship between the susceptibility category (LPI value) and the different affected features recognized, showing the correlation coefficient (R^2) for each relation. The LPI shows best correlation coefficient value with the orientation of the discontinuities ($R^2 = 0.60$), and with less significant correlation coefficient values with slope angle ($R^2 = 0.28$) and slope height ($R^2 = 0.25$) at linear equations (Fig. 21.8a, b and c), that refer to the direct link of the structure orientation and the slope.

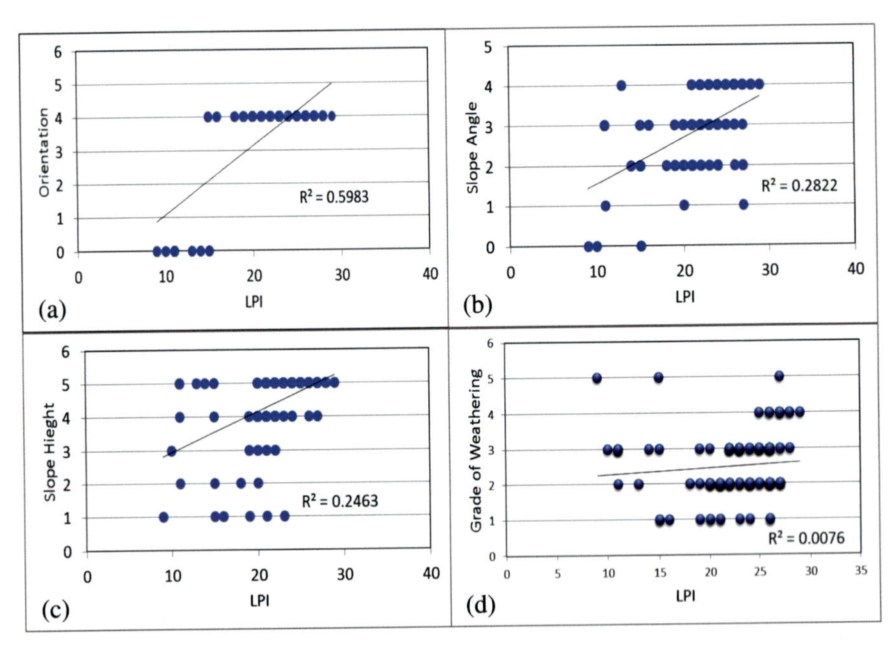

Fig. 21.8 Graphic Relationships between LPI and, (**a**) orientation of discontinuities, (**b**) slop angle, (**c**) slope height and (**d**) grade of weathering in the study area

On the other hand, poor correlations were found with the other factors such as grade of weathering ($R^2 = 0.01$) (Fig. 21.8d). This might reflect the homogeneity of the mineral component of the exposed outcrops within the study area.

The above relationships prove that the LPI in the Aden Peninsula area is mainly controlled by the factors of orientation of the discontinuities, slope angle and slope height. Digital maps were created using ArcGIS 10.2 to show the similarity of these factors (Fig. 21.9a, b and c), and the variation of the other factors i.e. grade of weathering (Fig. 21.9d) with the final LPI value (Fig. 21.9e).

21.5.2 Landslide Susceptibility Map

The LPI rating system was divided to six classes, and the landslide susceptibility classified to six categories based on the LPI values. The number and percentage of each susceptibility category for the Aden Peninsula area were calculated (Table 21.2). High and very high susceptibility zones cover 78% of the study area. Moderate susceptibility zones cover 20%, while low susceptibility zones cover 2% of it (Fig. 21.10). The discontinuities work as an importance factor for failures (Evans, 1981; Hoek & Bray, 1984; Wyllie & Mah, 2004; Sarkar et al., 2013) so the topography represented by slope angle and height.

Landslide susceptibility map in the study area was created (Fig. 21.11) based on the evaluation of landslide susceptibility categories. The susceptibility map and land use map show that the built areas cover 35.6% of the mountainous region. Most of these built areas under moderate and high susceptibility zones, which represent 15% of the total urban zone in study area.

21.5.3 Discussion

The main causes of landslide susceptibility at Aden Peninsula area are discontinuities orientation then slope angle and height; these factors may reflect the effect of topographical caldera shape and the human activities represented by slopes cut. In this study the affected LPI factors have similar correlation coefficient values compared with those computed by the same LPI rating system applied in Wadi Dhahr area which is located in the western part of Yemen and is characterized mainly by Cenozoic Volcanics and Cretaceous sandstone (Barahim et al., 2018), on the other hand, the study of little Aden (Barahim et al., in preparation) have little different correlation coefficient values for the affected LPI factors.

The variation in correlation coefficient values obtained by these three areas (Aden Peninsula, Wadi Dhahr and Little Aden) might be due to changes in the topography, rock type and (or) human activities. However, the results of these studies show that the LPI is acceptable enough to assess landslide susceptibility and furthermore it is have a good pointer for main factor affected the landslide itself.

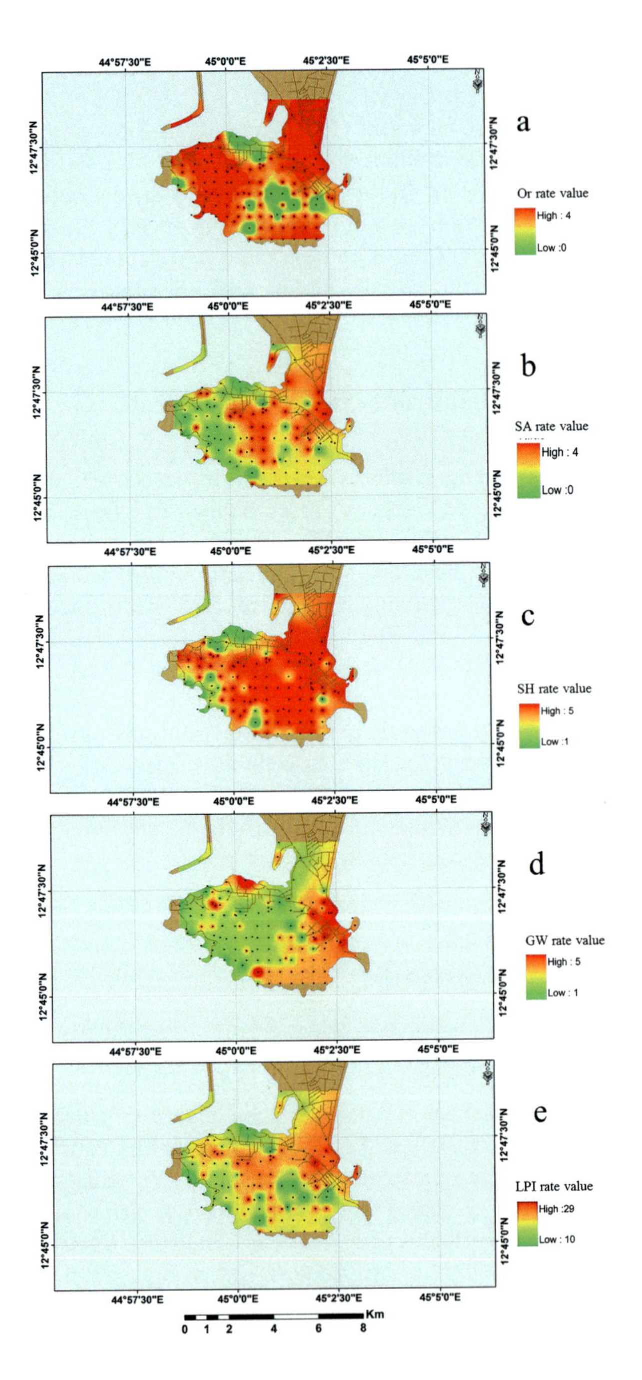

Fig. 21.9 LPI digital maps of the study area showing the distribution of: (a) orientation of the discontinuities, (b) slope angle, (c) slope height, (d) grade of weathering and (e) the final LPI value

The communal failure type in the study area is Rockfall, which is recorded at 60 stations then rolling failure which is recorded at ten stations. Sliding and toppling recorded at local areas with small scale.

Table 21.2 Number and percentage of stations for LPI classes and the susceptibility categories for the rock formation at Aden Peninsula area

LPI value	LPI class	susceptibility category	LPI for rock formation at Aden Peninsula	
			Number of stations	Percentage of susceptibility category
0–5	I	Very low	–	–
6–10	II	Low	2	2%
11–15	III	Moderately low	11	9%
16–20	IV	Moderately high	13	11%
21–25	V	High	56	48%
> 25	VI	Very high	35	30%
Total			117	100%

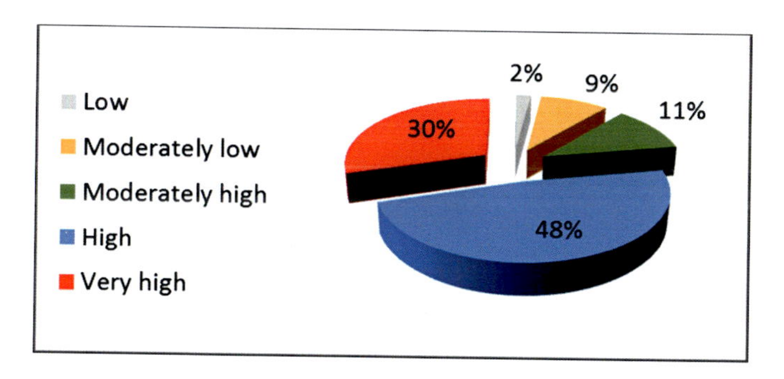

Fig. 21.10 Percentages of the susceptibility categories for rock stations at Aden Peninsula area

The last event of landslide in Sirah district was followed a heavy rainfall on February 2020. Therefore it is necessary to take some construction measures to treat the instability problem of the slopes, most appropriate for the topographical, geological and operational conditions at the site (Wyllie & Mah, 2004). The following treatments may achieve where it is needed:

1. Removal of rock overhang and unstable blocks by trim blasting.
2. Support the toe of the slope and the overhanging parts by retaining wall.
3. Erecting well seal drainage conduit.
4. Grouting of discontinuities by convenient filling material like cement in order to increase the cohesion of the different parts of the rock mass and to prevent water infiltration to the lower part of the slope.

21.6 Conclusion

The results of this research show that the LPI in Aden peninsula is mainly controlled by orientation of the discontinuities, slope height and slope angle factors. The landslide susceptibility map classifies the area into four classes of landslide

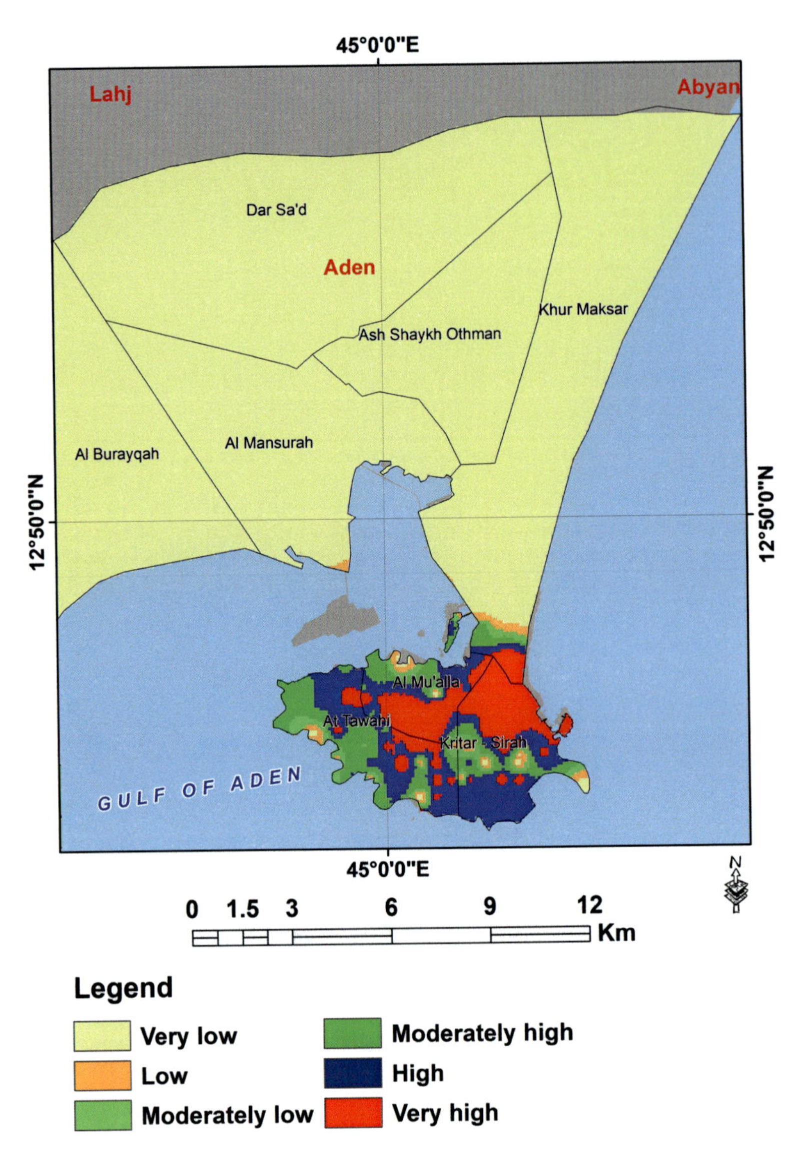

Fig. 21.11 Landslide susceptibility map of Aden Peninsula

susceptibility zones such as; very high and high hazard zones which cover 78% of the study area, the moderate hazard zones which cover 20% and 2% covered by low and nil hazard zones. Rockfall are the main modes of failure occur in the study area. Removing of unstable blocks, retaining wall and grouting are recommended as a treatment for risky unstable slopes. The outcomes of this research can be helpful for decision makers and future development planners to avoid the zones with high susceptibility during land use planning in the study area.

Acknowledgement The field work of this study was partially funded by the Geological Survey and Mineral Resources Board of Yemen (GSMRB). The satellite images were provided by Yemen Remote Sensing and GIS Centre (YRSC).

References

Anon. (1972). The preparation of maps and plans in terms of engineering geology. *Quarterly Journal of Engineering Geology, 5*(4), 293–382.

Baker, J., Menzies, M., Thirlwall, M., & Macpherson, C. (1997). Petrogenesis of quaternary intraplate volcanism, Sana'a, Yemen: Implications for plume–lithosphere interaction and polybaric melt hybridization. *Journal of Petrology, 38*, 1359–1390.

Barahim, A. A. (2004). *Slope stability study of Hajja-Amran road in Yemen and derivation of toppling equations for blocks having triangular cross-section*. Unpublished PhD theses, Baghdad University, Iraq, p. 158

Barahim A. A., Khanbari K. M., Algodami, A. F., Almadhaji, Z. A. and Adris, A. M. (2018) Slope stability assessment and landslide susceptibility map production of Wadi Dhahr area, northwest of Sana'a, Yemen. Sultan Qaboos University Journal for Science, 23(2), pp120–136

Barahim, et al. (in preparation). *Landslide susceptibility map of little Aden – South West of Yemen.*

Bartolomei, A., Brugioni, M., Canuti, P., Casagli, N., Catani, F., Ermini, L., Kukavicic, M., Menduni, G., & Tofani, V. (2006). Analisi della suscettibilità da frana a scala di bacino, Bacino del fiume Arno, Toscana-Umbria. *Italia, Giornale di Geologia applicata, 3*, 189–195.

Bejerman, N. J. (1998). Evaluation of landslide susceptibility along State Road 5, Cordoba, Argentina. In *Proceedings, 8th international congress of IAEG Vancouver*, Canada, Balkema, Rotterdam (Vol. 2, pp. 1175–1178).

Beydoun, Z. R. (1970). Southern Arabia and northern Somalia: Comparative geology. *Philosophical Transactions of the Royal Society of London. Series A, Mathematical and Physical Sciences, 267 A*, 267–292.

Cox, K. G., Gass, I. G., & Mallick, D. I. J. (1968). The evolution of the volcanoes of Aden and Little Aden, South Arabia. *Quarterly Journal of the Geological Society, 124*, 283–308.

Cox, K. G., Gass, I. G., & Mallick, D. I. J. (1970). The peralkaline volcanic suites of Aden and Little Aden, South Arabia. *Journal of Petrology, 11*, 433–461.

Dauteuil, O., Huchon, P., Quemeneur, F., & Souriot, T. (2001). Propagation of an oblique spreading center: The western Gulf of Aden. *Tectonophysics, 332*, 423–442.

Dickinson, D. R., Dodson, M. H., Gass, I. G., & Rex, D. C. (1969). Correlations of initial 87Sr/8'Sr with Rb/Sr in some late tertiary volcanic rocks of South Arabia. *Earth and Planetary Science Letters, 6*, 84–90.

Evans, R. S. (1981). An analysis of secondary toppling rock failures-the stress redistrbution method. *Quarterly Journal of Engineering Geology, 14*, 77–86.

Evans, S. G., Mugnozza, G. S., Strom, A., & Hermanns, R. L. (2006). Landslides from massive rock slope failure. IV Earth and Environmental Science Springer, Printed in the Netherlands (Vol. 49, p. 662).

Fournier M., Chamot-Rooke N., Petit C., Huchon P., Al-Kathiri A., Audin L., Beslier M.O., D'Acremont E., Fabbri O., Fleury JM., Khanbari K., Lepvrier C., Leroy S., Maillot B. and Merkouriev S. (2010) Arabia-Somalia plate kinematics, evolution of the Aden –Owen-Carlsberg triple junction, and opening of the Gulf of Aden. Journal of Geophysical Research. 115, B04102

Francipane, A., Arnone, E., Lo Conti, F., Puglisi, C., & Noto, L. V. (2014). *A comparison between heuristic, statistical, and data-driven methods in landslide susceptibility assessment: An application to the Briga and Giampilieri catchments. International conference on Hydroinformatics* (p. 8p). City University of New York.

Gemitzi, A., Falalakis, G., Eskioglou, P., & Petalas, C. (2011). Evaluating landslide susceptibility using environmental factors, fuzzy membership functions and GIS. *Global Network of Environmental Science and Technology Journal, Greece, 13*(1), 28–40.
GSMRB (Geological Survey and Mineral Resources Board). (1990). *Geological map of Aden, scale 1:250,000.*
Gupta, R. P., Kanungo, D. P., Arora, M. K., & Sarkar, S. (2008). Approaches for comparative evaluation of raster GIS-based landslide susceptibility zonation maps. *International Journal of Applied Earth Observation and Geoinformation, 10*(3), 330–341.
Hansen, A. (1984). Landslide hazard analysis. In D. Brunsen & D. B. Prior (Eds.), *Slope instability* (pp. 523–602). Wiley.
Hoek, E., & Bray, J. W. (1984). *Rock slope engineering* (3rd ed., p. 358). Institution of Mining and Metallurgy.
Khanbari, K. (2020). Seismotectonic provinces of Yemen. *Journal of Science and Space Technologies, CRTEAN, 6*, 8–22.
Khanbari, K., & Huchon, P. (2010). Paleostress analysis of the volcanic margins of Yemen. *Arabian Journal of Geosciences, 3*, 529–538.
Laughton, A. S., Whitmarsh, R. B., & Jones, M. T. (1970). The evolution of the Gulf of Aden. *Philosophical Transactions of the Royal Society A – Journals, 267*, 227–266.
Pour, A. B., & Hashim, M. (2017). Application of Landsat-8 and ALOS-2 data for structural and landslide hazard mapping in Kelantan, Malaysia. *Natural Hazards and Earth System Sciences, Copernicus Publications on behalf of the European Geosciences Union, 17*, 1285–1303.
Puglisi, C., Falconi, L., Leoni, L., Pino, P., Rasà, R., & Tripodo, A. (2007). Analisi della Suscettibilità da frana in Sicilia (1:250.000): Relazioni con scenari climatici futuri. In *Workshop. Cambiamenti Climatici e Dissesto Idrogeologico: Scenari Futuri per un Programma Nazionale di Adattamento, Napoli, 9–10, Luglio.*
Saaty, T. L., & Vargas, G. L. (1991). *Prediction, projection and forecasting* (p. 251). Kluwer Academic Publishers.
Saaty, T. L., & Vargas, G. L. (2001). *Models, methods, concepts, and applications of the analytic hierarchy process*. Kluwer Academic Publisher.
Saha, A. K. (2005). An approach for GIS-based statistical landslide susceptibility zonation with a case study in the Himalayas. *Landslides Journal, 2*, 61–69.
Sarkar, S., Kanungo, D. P., & Sharma, S. (2013). Landslide hazard assessment in the upper Alaknanda valley of Indian Himalayas. *Geomatics, Natural Hazards and Risk Journal, Taylor & Francis, 6*(4), 308–325.
Tard, F., Masse, P., Walgenwitz, F., & Gruneisen, P. (1991). The volcanic passive margin in the vicinity of Aden. *Bulletin des Centres de Recherches Exploration-Production Elf Aquitaine, 15*(1), 1–9.
Van Westen, C. (1997). *Statistical landslide hazard analysis. ILWIS 2.1 for windows application guide* (Vol. 15, pp. 73–84). ITC Publication.
Van Westen, C. J., Castellanos, E., & Kuriakose, S. L. (2008). Spatial data for landslide susceptibility, hazard, and vulnerability assessment: An overview. *Engineering Geology Journal, 102*(3–4), 112–131.
Wyllie, D. C., & Mah, C. W. (2004). *Rock slope engineering civil and mining* (4th edn.). Taylor & Francis e-Library, p. 431

Chapter 22
The Use of Earth Observation Data in Wildfire Risk Management: A Case Study from Lebanon

George H. Mitri

Abstract Wildfires in Lebanon pose an increasing threat not only to the natural environment but also to urban settings and local communities near forests. In response to this increasing threat, the Government of Lebanon endorsed a National Strategy (i.e., the Strategy) for forest fire management (Decision No. 52/2009). The Strategy acknowledged that decisions about wildfires are best made within a risk management framework including five different components, known as the 5Rs, namely (R1) Research, information and analysis; (R2) Risk modification, including fire vulnerability reduction and prevention of harmful fires; (R3) Readiness; (R4) Response, including all means of intervention for fire suppression; and (R5) Recovery, including the rehabilitation and ecological restoration of healthy forest conditions. Various tools and instruments were essentially needed to support the implementation of each component of the Strategy. These included the use of Earth Observation (EO) data and the employment of different remote sensing techniques among others. More specifically, satellite remote sensing proved to be useful for fire risk management as EO offered precise and frequent data especially that Lebanon lacked reliable national data on fire risk. In this context, the aim of this work was to review how EO contributed to fire risk management in Lebanon before (i.e., R1, R2 and R3) and after fire occurrence (i.e., R5) while supporting forecast and early detection of fires (i.e., R4). This study started by presenting EO cases which served as background information on the potential use of satellite remote sensing throughout the 5 Rs of the Strategy. The second part presented a case study from Lebanon on the actual use of EO in fire risk management.

Keywords Earth observation · Satellite imagery · Remote sensing · Wildfire · Hazard · Vulnerability · Risk management

---

G. H. Mitri (✉)
Land and Natural Resources Program, Institute of the Environment, University of Balamand, Tripoli, Lebanon

Faculty of Arts and Sciences, University of Balamand, Tripoli, Lebanon
e-mail: george.mitri@balamand.edu.lb

M. M. Al Saud (ed.), *Applications of Space Techniques on the Natural Hazards in the MENA Region*, https://doi.org/10.1007/978-3-030-88874-9_22

22.1 Introduction

Forests in Lebanon represent a unique feature in the arid environment of the Eastern Mediterranean. These forests include remnants of valuable broad-leaved trees, conifer forests and evergreen trees that cover the Lebanese mountains in patches (AFDC, 2019).

Wildfires have been increasingly damaging Lebanon's forests (Mitri et al., 2017a). Lebanon experienced in the past two decades increasing fire frequency, intensity and severity resulting in various ecological, social and economic consequences. While the increase in number of wildfires and extent of burned areas was mostly related to human activities, climate change and weather extremes worsened the situation (Salloum & Mitri, 2014). In 2020, Lebanon suffered its worst fire season. Increasing drought conditions and weather extremes facilitated the spread of fires over large areas while pushing the flames to ignite at record altitude above 2000 m (i.e., approximately 441 ha of burned high mountain lands). The unprecedented fires burned over a total area of 7132 ha while the yearly average of burned areas was 1000 ha. In addition to the high ecological damages, the recent fires had large socio-economic damages especially that privately owned large forest and agricultural areas were severely burned and agricultural crops were lost in time of a national financial and monetary crisis.

There is no doubt that drought and recurrent weather extremes contributed to the spread of fires over large areas and in some cases along high mountainous lands. In fact, a previously published study projected a significant increase in drought conditions in high mountainous areas especially in the Akkar region (Mitri et al., 2015b). The landscape was so dry and combustible from multiple years of gradual change leading to an increase in the frequency and intensity of fires. As wildfires will continue to sweep across the country threatening more lives, destroying private and public properties and infrastructure, different research and studies (i.e., involving socio-economic, environmental and climatic data) showed a dire message: this could be Lebanon's new normal (Salloum & Mitri, 2014).

In response to a deteriorating situation of wildfire occurrence and spread, a capable management of wildfire risk was essentially needed. Accordingly, a National strategy for forest fire management was developed and endorsed by the Lebanese Council of Ministers (Decision No. 52/2009). The aim of the strategy was to reduce the risk of intense and frequent forest fires whilst allowing for fire regimes that are socially, economically and ecologically sustainable. The strategy acknowledged that decisions about fire management are best made within a risk-management framework, known as the 5Rs, namely, Research, information and analysis (i.e., R1);, Risk modification, including fire vulnerability reduction and prevention of harmful fires (i.e., R2), Readiness, covering all provisions intended to improve interventions and safety in the event of fire (i.e., R3), Response, including all means of intervention for fire suppression (i.e., R4), and Recovery, including the rehabilitation and ecological restoration of healthy forest conditions (i.e., R5), and the support to individuals and communities in the short- and medium term aftermath of the fire (Mitri & Gitas, 2011). These can be grouped under three phases of fire management namely pre-fire management (i.e., R1, R2 and R3), forecast and early detection (i.e., R4) and post-fire assessment and monitoring (i.e., R5).

While Lebanon lacked the necessary technological measures and management capacities to address a number of measures related to fire management including monitoring, prediction (early warning), preparedness, prevention, suppression and restoration, Earth Observation (EO) presented various opportunities for improved wildfire risk management (Mitri & Gitas, 2011). More specifically, satellite remote sensing data and techniques were reported as effective tools for conducting different studies related to forest fire management (Chuvieco, 2009). Yet, the development of new remote sensing instruments provided a number of opportunities to advance studies and researches on forest fire management in Lebanon.

The aim of this work was to review how EO data contributed to fire risk management in Lebanon throughout three phases namely before (i.e., first phase comprising R1, R2 and R3) and after fire occurrence (i.e., third phase comprising R5) while supporting in between forecast and early detection of fires (i.e., second phase comprising R4). First, EO cases which served as background information on the potential use of satellite remote sensing throughout the 5 Rs of the Strategy were presented. Second, uses of EO along the three phases of fire risk management in Lebanon were demonstrated.

22.2 Use of Satellite Remote Sensing in Fire Risk Management: An Overview

In general, developing countries lack the necessary infrastructure to gather and process data needed to predict when and where natural disasters might occur and what the impact would be (Zorn, 2018). EO, however, responds to the need of impactful and cost-effective approaches to prepare, respond and recover from disasters, especially in developing countries. In this context, the Sendai framework made specific recommendations about the use of EO solutions to address disaster resilience (Lorenzo-Alonso et al., 2019). These included the need to promote real time access to reliable data, making use of space and in situ information and promote and enhance related services to geospatial and space-based technologies. This section provided an overview on the use of EO and remote sensing techniques within a risk management framework, known as the 5Rs.

22.2.1 Research, Information and Analysis (R1)

The strategic objective of this component was to support and promote the improvement, know-how sharing, monitoring and dissemination of knowledge on fire ecology, fire management and post-fire vegetation dynamics among all relevant actors (science/research, policy makers, land managers, grassroots' groups), bridging science and traditional knowledge. In general, satellite remote sensing provided valuable data to support research in fire risk management. This included (1) the development of effective fire monitoring systems (Jang et al., 2019), (2) the development of daily danger indices based on vegetation types, and thus, developing a comprehensive danger-rating system (Camia et al., 2006; Chuvieco & Salas, 1996), and 3) the development of annual comprehensive databases on forest fires for analytical use (Camia et al., 2010; San-Miguel-Ayanz et al., 2009; Barbosa et al., 2006). The European Forest Fire Information System (EFFIS) provided standardized European forest fire danger forecast and burned area maps for the EU Mediterranean region (San-Miguel-Ayanz et al., 2009). The system RISICO (RISchio Incendi & COordinamento) which involved remote sensing data has been used since 2003 by the Italian National Civil Protection for daily dynamic forest fire risk assessment (D'Andrea et al., 2008). In Greece, a daily forest fire risk map involving up-to-date satellite data of vegetation greenness was published by the civil protection (Gitas et al., 2004).

22.2.2 Risk Modification (R2)

The strategic objective of this component was to develop effective measures intending to reduce fire vulnerability, increase ecological and social resilience to fire, and prevent the occurrence of harmful fires and unsustainable fire regimes. Satellite remote sensing was considered a useful tool for sustaining prevention activities (Chuvieco, 2009). EO data proved to provide valuable information on type (e.g. distribution and amount of fuels) and status of vegetation in a consistent way and at different spatial and temporal scales (Lasaponara & Lanorte, 2007; Arroyo et al., 2006; Lasaponara et al., 2006). Also, satellite remote sensing proved to assist in the detailed analysis of forests status and the improvement of pre-fire management plans (Hernandez-Leal et al., 2006).

22.2.3 Readiness or Pre-Suppression (R3)

The strategic objective of this component was to undertake all possible provisions by individuals, communities and fire and land management agencies to be prepared before a fire event occurs, and improve interventions and safety in monitoring the probability of fire and detecting the event of fire. Satellite remote sensing data contributed to the conduction of a proper distribution at the landscape level of fire control infrastructures such as fire lookout towers (Catry et al., 2007; Nogueira et al., 2002), water reservoirs, forest strips with low tree density and low shrub cover, fire break areas of first and second level, forest tracks with fire break lines along them, and protection perimeters in urbanized areas (Jaiswal et al., 2002; Chuvieco & Salas, 1996).

22.2.4 *Response (R4)*

The strategic objective of this component was to quickly suppress and limit the extension of fires through the development of methods and techniques coupled with appropriate material and very well trained personnel. Satellite remote sensing helped in the development of fire behaviour models and/or combustibility models (Johnston et al., 2019) to allow fire-fighting brigades better predict the fires and manage them, thus avoiding the uncontrollable expansion of fires (Stergiadou et al., 2007; Dimitrakopoulos, 2002). Also, spatial information such as, vegetation cover density and location and defensible space of buildings, contributed to the improvement of forest fire suppression planning (Szpakowski & Jensen, 2019). In this regard, satellite remote sensing are tools that can be implemented to extract, store and process relevant information (Roy et al., 2017). In Spain, the EMERCARTO mapping viewer involving satellite remote sensing data proved to be a powerful GIS tool for the optimization and control of resources during forest fire suppression (Briones et al., 2007). Also, it is possible for countries around the Globe to submit official requests to international charters such as the International Space Charter which can provide available data from a series of satellites such as RADARSAT, ERS, EMVISAT, SPOT, LANDSAT, and DMC to help in the dissemination of timely information on a disaster occurrence.

22.2.5 *Recovery (R5)*

The strategic objective of this component was to provide support for individuals and communities in the immediate aftermath of the fire as well as in the medium and longer term efforts of community and economic renewal, and restore healthy ecological conditions of burned forest lands to facilitate the natural recovery of vegetation and increase forest resilience against future fires. Remote sensing satellite data previously assisted in 1) mapping fire affected areas and assessing the impact of fires on different vegetation types (Chuvieco et al., 2019; Mitri & Gitas, 2004), and 2) mapping fire type (Mitri & Gitas, 2006) and fire severity (Gitas et al., 2009; Mitri & Gitas, 2008). Also, satellite remote sensing data helped in implementing activities aiming at the reduction of soil erosion, and mapping forest regeneration and vegetation recovery (Fiorucci et al., 2013; Mitri & Gitas, 2010, 2013; Gitas et al., 2011; Veraverbeke et al., 2011; Gouveia et al., 2010; Vila & Barbosa, 2010; Hernandez-Clemente et al., 2009; Twele & Barbosa, 2005; Diaz-Delgado et al., 2003) for the development of post-fire active restoration/rehabilitation activities (e.g., forest landscape restoration). Eventually, EO contributed to the development of national reporting systems based on fire statistics, expanding national databases on forest fires, their occurrence, and the ecosystems where they occur (Röder et al., 2008; Viedma et al., 1997). For instance, Portugal employed an operational system for mapping burned areas from satellite remote sensing imagery (San-Miguel-Ayanz et al., 2009; Pereira & Santos, 2003).

22.3 The Role of Satellite Remote Sensing in Fire Risk Management: A Case Study from Lebanon

As Lebanon acknowledged that decisions about wildfires are best made within a risk management framework including five different components, the use of satellite remote sensing proved to be useful for fire risk management in Lebanon as EO offered precise and frequent data over relatively large areas. As shown in Fig. 22.1, EO contributed to fire risk management in Lebanon before (i.e., R1, R2 and R3) and after fire occurrence (i.e., R5). In between, EO data contributed to forecast and early detection of fires (i.e., R4).

22.3.1 Study Area Description

Located on the eastern part of the Mediterranean (Fig. 22.2), the study area encompasses the entire Lebanese territory (total area of 10,452 km^2). The country is divided into four distinct physiographic regions: the coastal plain, the Lebanon mountain range, the Beqaa Valley, and the Anti-Lebanon mountain range. The Lebanon mountain range rises steeply from the coast to mountains reaching 3088 meters above sea level (masl) and supports most of Lebanon's forests. In Lebanon, forests and other wooded land cover 24.5% of the National territory (Mitri

Pre-fire (R1, R2, and R3)	Forecast and early detection (R4)	Post-fire (R5)
Socio-economic/biophysical mapping	Fire detection	Burned area mapping, monitoring and emissions
Fuel type mapping	Fire danger forecast	Fire type and severity mapping
Fire risk mapping/ assessment	Wildfire potential in present and future	Vegetation recovery

Fig. 22.1 Contribution of satellite remote sensing to fire risk management in Lebanon

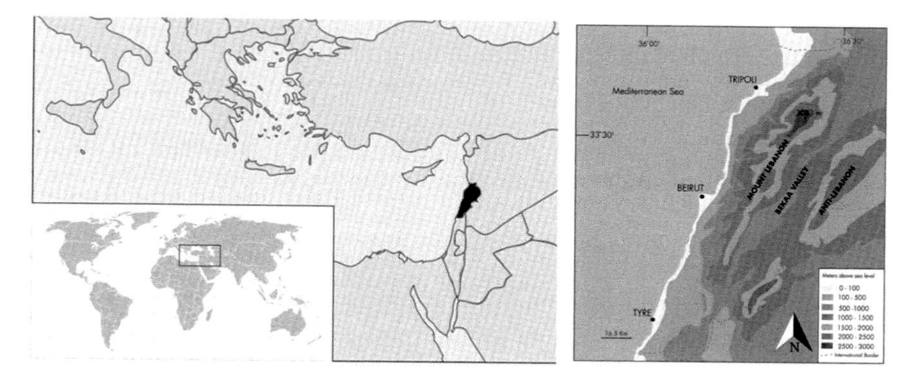

Fig. 22.2 Location of the study area

et al., 2017a). Other categories of land cover/land use include cropland, grassland, bare land, land with little vegetation and settlement. Lebanon is characterized by dry summers extending from June to November (Salloum & Mitri, 2014) with average daytime temperatures above 30 °C, and little rain (i.e., around 90% of the total annual precipitation falling between November and March).

The major forest species are *Quercus calliprinos, Quercus infectoria, Pinus pinea, Pinus brutia, Pinus halepensis, Cedrus libani, Juniperus excelsa, Abies silicica* and *Cupressus sempervirens,* among others. Like other Mediterranean countries, the largest number of fires occurs during the dry season mostly due to human activities. Around 33% of the national territory is of moderate to very high risk of fire (Mitri et al., 2017a). The average length of the fire season is approximately 147 days and the average peak month is September, yet, Lebanon is expected to face in the future an increasing number of fire occurrence and an extended length of fire seasons (Salloum & Mitri, 2014).

22.3.2 *Pre-Fire Risk Management*

EO facilitated the assessment of public perceptions towards socio-economic impact of wildfires at the level of a fireshed (Mitri et al., 2018). The methodology of work comprised an assessment of fire risk and subsequently, the identification of priority high risk villages/towns to be targeted by focused socio-economic surveys. Fire risk was assessed as a product of fire hazard and fire vulnerability (Mitri et al., 2015a). Fire hazard assessment involved the use of data mainly related to the distribution and density of forest fuel using Landsat satellite imagery. The combined use of both EO data for mapping fire risk area and socio-economic survey data allowed a more comprehensive and structured assessment of how the public perceive socio-economic impact of wildfires at the level of a fireshed.

In addition, the influence of both socio-economic and biophysical variables on fire occurrence in Lebanon was modelled (Mitri et al., 2016). Forward stepwise binary logistic regression analysis of 24 socio-economic and biophysical variables was used to predict wildfire occurrence. Many of these variables were derived from satellite remote sensing data. These included vegetation fuel type, conversion of lands and burned areas among others. The accuracy of the fire occurrence model was approximately 85% when tested on the validation data set. The probabilistic spatial output of the fire threat model was considered satisfactory given the challenges of using multi-source and multi-resolution data. More specifically, the results suggested that increasing the resolution of socio-economic data when combined with satellite data would improve modelling accuracy of fire occurrence in Lebanon.

A classification approach to generate fuel type maps using Landsat TM and ASTER imagery was presented (Mitri et al., 2011a). Fuel type classification of the ASTER image resulted in an overall accuracy of 70% with a total of six fuel type classes. Comparison of the fuel type mapping results with those from the Landsat TM imagery showed that the use of spectral and spatial information of ASTER imagery in a Geographic Object-Based Image Analysis (GEOBIA) approach resulted in the recognition of the main fuel type classes of the PROMETHEUS classification model, and the production of satisfactory results in extremely heterogeneous vegetated areas. In addition, the use of very high spatial resolution SPOT imagery (i.e., 2.5 m) for the characterization of fire risk in the Mediterranean area of North Lebanon was investigated (Salloum et al., 2013). The use of SPOT imagery together with other ancillary data such as terrain aspect and land cover/land use of the area allowed the mapping of fuel types with an overall accuracy of 88%, a very acceptable delineation of recently burned areas, and eventually fire risk mapping of unburned areas.

In addition, a forest fire risk assessment and mapping model under actual and future climatic conditions (i.e., in function of hazard, vulnerability and exposure) was developed (Zeidan et al., 2019). The methodology of work involved a stepwise approach (i.e., impact chain) to evaluate the following main components: exposure, sensitivity, hazard and adaptive capacity. Multi-sources geospatial data including satellite were collected and integrated to create composite indices.

The spatial distribution of wildfire risk in Lebanon was assessed (Fig. 22.3). More specifically, the work aimed at identifying and mapping (a) wildfire hazard, (b) wildfire vulnerability, and (c) wildfire risk (Mitri et al., 2015a). A model using multi-source satellite imagery and climatic data was developed with the use of GEOBIA. The development of the wildfire hazard map included classification of forest fuel type, combustibility, and fire spread whereas the vulnerability map included classification of demographic vulnerability (i.e., boundary, occupation and scatter indicators) and forest vulnerability (i.e., environmental and replacement values).

In 2014, a web-based decision framework was developed to improve fire risk management in Lebanon (Mitri et al., 2014a). The primary objective of the application (i.e., so-called FireLab) was to provide an online user-friendly interface for displaying data that are critical for making informed fire management decisions. Data included 257 variables related to fire activity, risk, and hazard and are generated at the municipality level for all of Lebanon using EO derived information. FireLab is

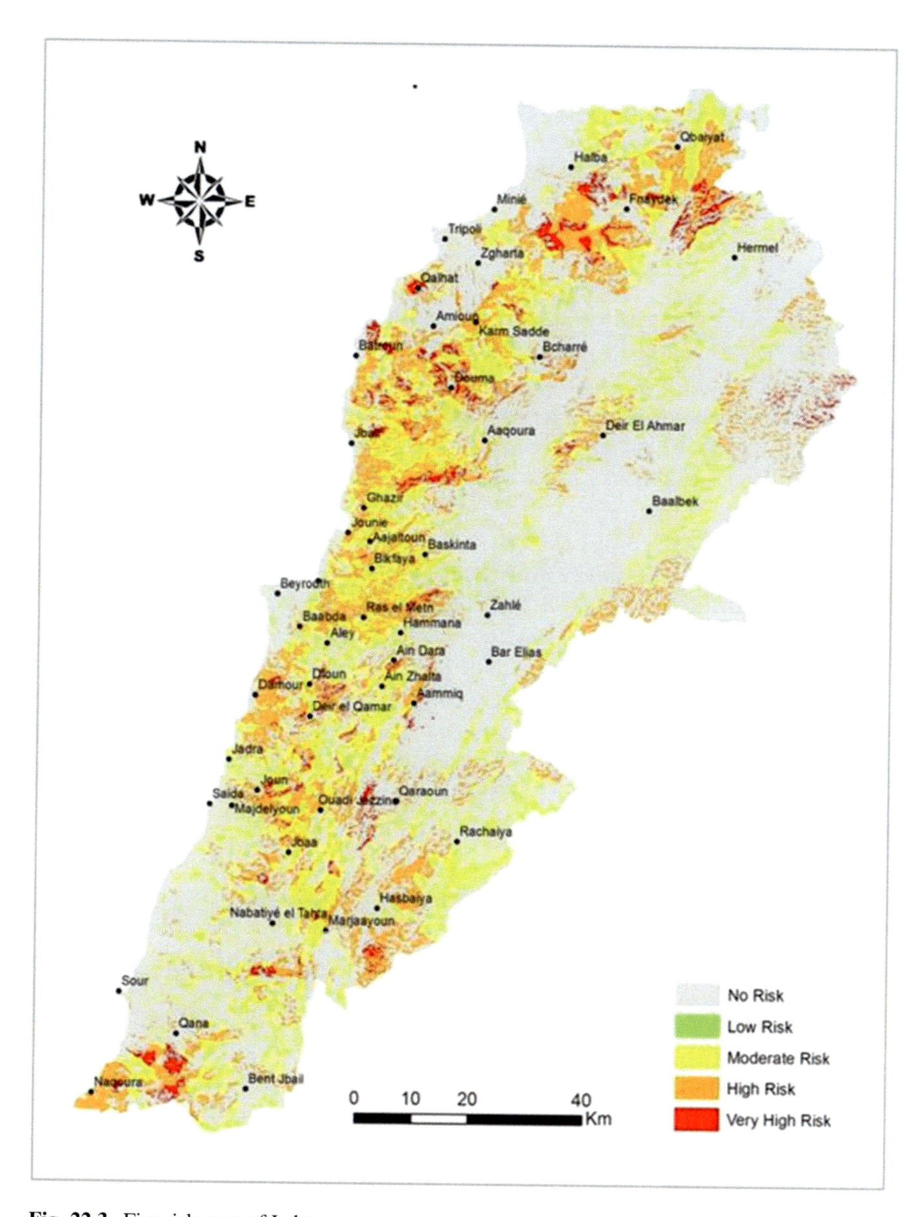

Fig. 22.3 Fire risk map of Lebanon

delivered through a web browser, making it widely accessible to the public in a format that allows users to easily display wildfire conditions and to describe and share modeled wildfire potential scenarios of current and future conditions.

Finally, fire risk associated with repetitive armed conflicts on the coastal zone in North Lebanon was assessed (Mitri et al., 2011b). The methodology of work involved the use of multi-temporal Landsat (MSS and TM) imageries acquired between 1975 and 2010. The GEOBIA approach was employed in this work. The results allowed the identification of fire risk which reflected many of the direct and indirect effects of repetitive armed conflicts on vegetated areas.

22.3.3 Forecast and Early Detection

Systems used for forest fire detection can be divided into three groups based on where the sensors are deployed: aerial, terrestrial, or combination. In Lebanon, a wireless sensor node to support early forest fire detection was developed and tested (Fig. 22.4) in an attempt to complement aerial fire detection (Sakr et al., 2014).

A new national fire danger forecast system for use as an operational fire management tool in Lebanon was also developed (Mitri, 2015; Mitri et al., 2017a). Accordingly, the customization of an index for fire danger forecasting involved the use of both Lebanon's fire risk map and Fire Weather Index (FWI) forecasts. Lebanon's initial fire risk map (Mitri et al., 2015a) was previously developed by employing geospatial biophysical and climatic data in addition to the land cover/land-use map. FWI data were produced by the European Forest Fire Information System (EFFIS) by adopting the Canadian FWI and by using weather forecasts in addition to observed synoptic weather data (San-Miguel-Ayanz et al., 2012). A comparison of FWI forecasts and fire occurrence (i.e., observed fire events) during the fire season of 2015 resulted in sensitivity (76.2%) and specificity (18.6%) estimates. The automated daily fire danger forecasts (Fig. 22.4) are expected to serve as essential tools for both scientific and operational purposes.

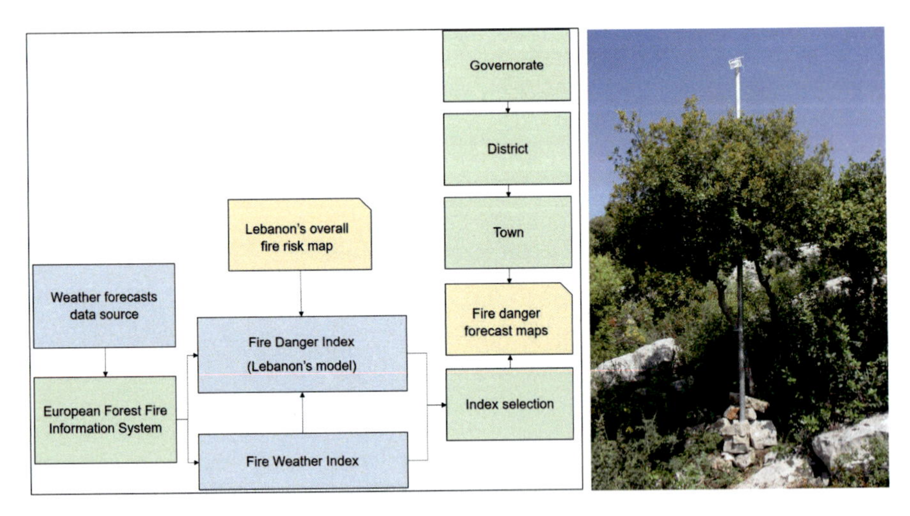

Fig. 22.4 Flowchart of the national fire danger forecast system interface (left) which can be supplemented with ground-based deployed sensor nodes for early detection and prediction of fire (right)

Furthermore, the spatial distribution of wildfire potential was estimated with the use of geo-spatial climatic data. Accordingly, the temporal and spatial variabilities of wildfire potential were investigated (Mitri et al., 2014b). This involved the use of monthly Keetch-Byram Drought Index (KBDI) maps for all Lebanon. The KBDI maps were derived from monthly rainfall and temperature data downscaled from global climatic data, and employed in an object-based approach. This modelling effort provided a better understanding of potential wildfire temporal and its spatial variability, information critical for developing effective fire prevention policies and adaptation strategies. A comparison between the average monthly values (2006–2010) of the Fire Weather Index (FWI) for Eastern Mediterranean and the monthly KBDI average values for Lebanon showed a strong correlation, thus indicating an increase in wildfire potential during the fire season (Mitri et al., 2015b). The comparison between the relative areas of current and future wildfire potentials showed a significant increase in KBDI values affecting a total area of 1143 km^2 (441.3 sq. mi) in the country.

In addition, forest fire occurrence was predicted by using a reduced number of monitored features, and eliminating the need for weather prediction mechanisms (Sakr et al., 2011). The conducted work compared two artificial intelligence based methods, namely artificial neural networks (ANN) and support vector machines (SVM), utilizing a reduced set of weather parameters. The results demonstrated the ability to predict forest fire danger with a limited number of inputs, however, the work concluded about the need to predict forest fire occurrence in more fire seasons and in different study areas. This necessitated the use of extensive fire occurrence datasets acquired from various EO sources. The use of Artificial Intelligence (Sakr et al., 2010 in combination with satellite data represents another opportunity for advancing forest fire prediction.

22.3.4 Post-Fire Risk Management

Most recently, charters and emergency services were activated in response to disastrous fires in Lebanon. Most of these counted on the use of space solutions for improved disaster resilience. They established the necessary connections between data providers, information developers, and end users to ensure that decision-makers can benefit from satellite technology in the disaster resilience community. In this context, The International Charter for Space and Major Disasters is a global mechanism for countries to access satellite imagery in support of their disaster response activities. The charter can provide rapid access to free data from a virtual constellation of satellites owned by space agencies and satellite operators. In addition, the Copernicus Emergency Management Service (Copernicus EMS) provides rapid mapping and information for emergency response in relation to different types of disasters. These include meteorological hazards, geophysical hazards, deliberate and accidental man-made disasters and other humanitarian disasters as well as prevention, preparedness, response and recovery activities. In 2019, the International

Charter Space and Major Disasters estimated fire hotspots density in Lebanon and Syria as detected by NASA's Visible Infrared Imaging Radiometer Suite (VIIRS). Produced maps illustrated satellite-detected fire hotspots based on the analysis of VIIRS accessed via NASA FIRMS, between October 15 and 16, 2019. A total of 30 hotspots were detected in Lebanon and 121 hotspots were detected in three analyzed Governorates (i.e., Homs, Lattakia and Tartous) in Syria (UNITAR-UNOSAT, 2019). Also, the European Commission's Copernicus Emergency Management Service (EMSR396) was activated on 15 October 2019 to provide satellite maps of disastrous fires occurring across the country. Fire danger forecast maps were timely produced and communicated to Lebanon for use in further actions.

In a different type of application, time-series Landsat imagery were employed for mapping burned areas, and consequently well documented methodological approaches were applied for estimating greenhouse gas (GHG) emissions from fires at the national level (Mitri et al., 2017b). The methodology of work comprised the use of a semi-automated model of GEOBIA for mapping yearly burned areas and, subsequently, estimating GHG emissions (Mitri & Karam, 2016). The overall classification accuracy of the model was found to be 86% (MoE/UNDP/GEF, 2015).

The yearly temporal pattern of fire activity and its relationship to weather in Lebanon during the past decade were investigated (Salloum & Mitri, 2014). Fire seasonality was determined using fire occurrence data from 2001 to 2011. More specifically, data showing the location (latitude and longitude) and date of fire hotspot events between 2001 and 2011 were also acquired through the Global Fire Information Management System (GFIMS), which integrates remote sensing and GIS technologies to deliver MODIS hotspot/fire locations (Davies et al., 2011). Although the GFIMS fire records are not the most comprehensive and accurate data on fire occurrence, such data are best suited for this study because of the lack of accurate and officially published fire occurrence data in the country.

Potential post-fire land degradation risk in Lebanon was modeled in order to plan and prioritize post-fire management actions (Mitri et al., 2020). The study included (1) mapping "burn severity" of the year 2019 using Landsat 8 and Sentinel 2-A imagery and (2) modeling post-fire land degradation risk using GEOBIA approach. Mapping burned areas of the 2019 fire season involved the use of a differenced Normalized Burn Ratio (dNBR) of Sentinel 2-A imagery. Another dNBR image was produced with the use of Landsat 8 imagery. Mapping overall burn severity was conducted based on cross-mapping between the two dNBR images which involved EO sources of different spatial and spectral characteristics. Potential post-fire degradation risk was modelled in GEOBIA with the combined use of burn severity and topographic data (i.e., slope gradients). In addition, various field data were collected to produce a Composite Burn Index (CBI) of fire-affected sites. The combined use of multi-temporal satellite remote sensing images (i.e., for deriving seasonal post-fire degradation risk in winter and spring following a fire event) and their corresponding ancillary data (i.e., CBI and soil texture) allowed the prioritization of fire affected sites for implementing post-fire restoration measures (Fig. 22.5).

In a another study, Sentinel-2A imagery were employed in combination with other ancillary data for mapping (1) plant composition and abundance, and

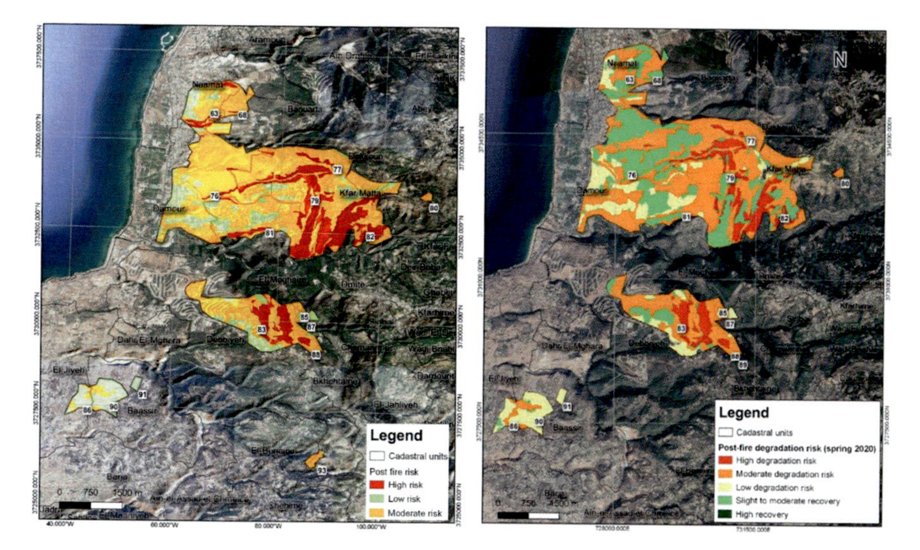

Fig. 22.5 Post-fire degradation risk in winter (left) and spring (right) following the fire event

(2) phyto-sociability (i.e., aggregation) of plant species (Mitri et al., 2017c). This contributed to assessing the effect of fire regime on vegetation cover in the Mediterranean forest landscape of Lebanon's western mountain range. The accuracy assessment results showed an overall average accuracy of approximately 60% for mapping plant abundance and phyto-sociability. A careful investigation of the results would provide further insights towards improving the performance of the GEOBIA model using the Sentinel-2A imagery.

Another model for operational monitoring of vegetation cover dynamics in fire-prone Mediterranean ecosystem with the use of multi-temporal Landsat images was developed (Mitri & Fiorucci, 2012). The work involved (1) mapping the extent of burned areas, and (2) monitoring post-fire vegetation recovery within fire affected areas. Multi-temporal Landsat images were employed. Image pixels located within objects previously classified as burned areas were monitored 16 years and 20 years after the fire events. The developed approach was flexible enough to be employed for a continuous monitoring of post-fire vegetation dynamics with the combined use of field data (EL Halabi et al., 2014).

22.4 Conclusions

The review of previously conducted researches and studies proved the usefulness of EO in supporting fire managers and planners who are involved in the implementation of Lebanon's National strategy for forest fire management. It is expected that the use of EO data and remote sensing techniques in each of the Strategy's components will result in strengthened capacities of public authorities and units involved in forest fire management.

Clearly, there is a great potential for including the wide range of current and newly emerging EO products in an operational monitoring of fire risk in Lebanon. Given, a relatively small area of the national territory, very few satellite scenes are needed to cover the entire national territory. In addition, EO products have the potential of improving and complementing conventional field and statistical data collection (e.g., the annual fire reports of Lebanon), as well as provide new types of other information of relevance (e.g., environmental).

Overall, EO provides (1) synoptic view of the national territory's surface, (2) regular and repeatable observations, and (3) cost-effectiveness for monitoring remote and inaccessible areas. Most importantly, countries with limited financial resources like Lebanon can extremely benefit from EO data which are increasingly made free using open data policies. In this context, data from the Sentinel satellite programme were made free for all by the Copernicus programme of the European Commission (EU). Also, the United States Geological Survey (USGS) made the large archive of Landsat data free for all to use. However, Lebanon has not benefited yet from the widespread use of data architectures and servers on the internet, therefore enabling operational cloud computing of its national datasets in the forest and forest fire sectors.

22.5 Key Recommendations

Below are some key recommendations, which outline how the full potential of using EO for fire risk management should proceed:

1. Communicate and demonstrate the potential of EO so that it can be fully exploited in fire risk management at all levels. This potential has not been clearly communicated to date, in particular to statisticians who need the data the most. It is worth mentioning that the time scales of EO data availability have significantly increased. Forest fire indicators requiring historical baselines are now continual, comprising data records over several decades (e.g., Landsat time series spanning from the late 1970s to the present). In addition, the precision of EO data was improved at the point of image acquisition, therefore the accuracy of EO derived products such as burned areas, fire severity, and vegetation recovery is potentially higher. Currently, the main limitation is where EO data can be stored and easily accessed and in a format ready to be used.
2. Improve information flows to inform decision-making related to fire risk management. For instance, Lebanon was able to derive precise daily fire danger forecast maps but the flow of generated information from EO to geospatial datasets and into the hands of decision-makers was not efficiently streamlined. Accordingly, potential bottlenecks (e.g., where information is exchanged between governmental agencies and local authorities and agencies) should be

addressed and accordingly, design appropriate strategies to improve data flow (e.g., one example would be the use of spatial data infrastructures).
3. Establish partnerships between the central administration of statistics (CAS) and geospatial experts working on fire risk assessment. This is expected to enable the potential of EO to be fully realized within the national statistical systems. Especially if EO data were to be merged with traditional sources of data such as field survey data. Establishing partnerships amongst the different national agencies and academic/research institutions is a first step towards building an effective collaboration.
4. Incentivize commercial EO data providers to collaborate with governmental and non-governmental organizations on the provision of very high resolution earth observations. Such products of very high resolution satellite imagery are essentially needed for certain applications in monitoring the impact of fires on terrestrial ecosystems (e.g., fire severity, forest regeneration and vegetation recovery).
5. Overall, improve the cooperation between governmental authorities and national/local agencies within the country, and increase their awareness of EO products and their role in improved fire risk management.

References

AFDC. (2019). *State of Lebanon's forests 2018* (Mitri, G. Ed.). Beirut: Association for Forests, Development and Conservation/Ministry of Agriculture/Ministry of Environment/Food and Agriculture Organization/United Nations Development Programme/International Union for Conservation of Nature/Lebanon Reforestation initiative.

Arroyo, L., Healy, S., Cohen, W., & Manzanera, J. (2006). Using object-oriented classification and high-resolution imagery to map fuel types in a Mediterranean region. *Journal of Geophysical Research, 11*, 1–10.

Barbosa, P., Kucera, J., Strobl, P., Vogt, P., Camia, A., & San-Miguel-Ayanz, J. (2006). European forest fire information system (EFFIS) - rapid damage assessment: Appraisal of burnt area maps in southern Europe using modis data (2003 to 2005). *Forest Ecology and Management, 234*, S218. https://doi.org/10.1016/j.foreco.2006.08.245

Briones, F. A., Fernandez, J., Dominguez, A., Molino, R., Quesada, C., & Alferez, C. (2007). *Herramientas para la gestión de la extinción de incendios forestales*. Visor de cartografía para Emergencias: EMERCARTO. Seville: International Wildland Fire Conference, pp. 13–17.

Camia, A., Barbosa, P., Amatulli, G., & San-Miguel-Ayanz, J. (2006). Fire danger rating in the European forest fire information system (EFFIS): Current developments. *Forest Ecology and Management., 234*, S20. https://doi.org/10.1016/j.foreco.2006.08.036

Camia, A., Durrant, T., & San-Miguel-Ayanz, J. (2010). The European fire database: Development, structure and implementation. In D. X. Viegas (Ed.), *VI International conference on forest fire research*.

Catry, F. X., Rego, F. C., Santos, T., Almeida, J., & Relvas, P. (2007). *Forest fires prevention in Portugal-using GIS to help improving early fire detection effectiveness, wild-fire 2007, Seville, Spain*.

Chuvieco, E. (2009). *Earth observation of wildland fires in mediterranean ecosystems*. Heidelberg/Dordrecht/London/New York: Springer.

Chuvieco, E., & Salas, J. (1996). Mapping the spatial distribution of forest fire danger using GIS. *International Journal of Remote Sensing, 10*(3), 333–345.

Chuvieco, E., Mouillot, F., van der Werf, G., San Miguel, J., Tanase, M., Koutsias, N., García, M., Yebra, M., Padilla, M., Gitas, I., Heil, A., Hawbaker, T., & Giglio, L. (2019). Historical background and current developments for mapping burned area from satellite earth observation. *Remote Sensing of Environment, 225*, 45–64. https://doi.org/10.1016/j.rse.2019.02.013

D'Andrea, M., Dal Pra, S., Angelini, F., Fiorucci, P., Gaetani, F. Mazzetti, P., and Verlato, M. 2008. *RISICO: A grid architecture for high resolution nation-wide forest fire risk assessment. Geophysical Research Abstracts* (Vol. 10), EGU2008-A-07307, 2008.

Davies, D., Ilavajhala, S., Wong, M., Molinario, G., Justice, C., Latham, J., & Martucci, A. (2011). The global fire information management system: Transformation to an operational system at UN FAO. In *Working paper FM/27/E, Sun City, South Africa: FAO*, p. 79.

Diaz-Delgado, R., Lloret, F., & Pons, X. (2003). Influence of fire severity on plant re-generation through remote sensing imagery. *International Journal of Remote Sensing, 24*(8), 1751–1763.

Dimitrakopoulos, G. (2002). Mediterranean fuel models and potential fire behaviour in Greece. *International Journal of Wildland Fire, 11*, 127–130.

EL Halabi, A., Mitri, G., & Jazi, M. (2014). Monitoring post-fire regeneration of Pinus brutia in Lebanon. In D. X. Viegas (Ed.), *Advances in forest fire research* (pp. 564–568). Imprensa da Universidade de Coimbra.

Fiorucci, P., Mitri, G. Franciosi, C., Lima, M., Chiara, F., Biondi, G., & D'andrea, M. (2013). Modelling vegetation succession in fire affected areas in the Mediterranean. In *9th international workshop of the EARSeL Special Interest Group (SIG) on Forest Fires*, 15–17 October 2013 (pp. 153–156). UK: University of Leicester.

Gitas, I., Mitri, G., Avyikou, I., & Diamanti, E. (2004). *Vegetation greenness mapping of Greece using MODIS imagery. MODIS workshop*, June 24–25. University of Valladolid, Spain.

Gitas, I., De Santis, A., & Mitri, G. (2009). Remote sensing of burn severity. In E. Chuvieco (Ed.), *Earth observation of wildland fires in mediterranean ecosystems* (pp. 129–148). Springer.

Gitas, I., Mitri, G., Veraverbeke, S., & Polychronaki, A. (2011). Advances in remote sensing of post-fire monitoring (a review). Remote sensing of biomass: principles and applications/book 1 (T. E. Fatoyinbo, Ed.). INTECH publisher: In press, ISBN 978-953-307-315-6.

Gouveia, C., DaCamara, C., & Trigo, R. (2010). Post-fire vegetation recovery in Portugal based on spot/vegetation data. *Natural Hazards and Earth System Sciences, 10*, 673–684.

Hernandez-Clemente, R., Navarro, R., Hernandez-Bermejo, J., Escuin, S., & Kasimis, N. (2009). Analysis of postfire vegetation dynamics of mediterranean shrub species based on terrestrial and NDVI data. *Environmental Management, 43*, 876–887.

Hernandez-Leal, A., Arbelo, M., Gonzalez-Calvo, A. (2006). Fire risk assessment using satellite data, advances in space research, *Advances in Space Research, 37*(4), 741–746.

Jang, E., Yoojin, K., Jungho, I., Dong-Won, L., Jongmin, Y., & Sang-Kyun, K. (2019). Detection and monitoring of forest fires using Himawari-8 Geostationary Satellite Data in South Korea. *Remote Sensing, 11*(3), 271. https://doi.org/10.3390/rs11030271

Jaiswal, R. K., Mukherjee, S., Raju, K. D., & Saxena, R. (2002). Forest fire risk zone map-ping from satellite imagery and GIS. *International Journal of Applied Earth Observation and Geoinformation, 4*, 1–10.

Johnston, J., Paugam, R., Whitman, E., Schiks, T., & Cantin, A. (2019). Remote sensing of fire behavior. In S. L. Manzello (Ed.), *Encyclopedia of wildfires and wildland-urban Interface (WUI) fires*. Springer. https://doi.org/10.1007/978-3-319-51727-8_174-1

Lasaponara, R., & Lanorte, A. (2007). Remotely sensed characterization of forest fuel types by using satellite ASTER data. *International Journal of Applied Earth Observations and Geoinformation, 9*, 225–234.

Lasaponara, R., Lanorte, A., & Pignatti, S. (2006). Characterization and mapping of fuel types for the mediterranean ecosystems of Pollino National Park in southern Italy by using hyperspectral MIVIS data. *Earth Interact, 10*(13), 1–11.

Lorenzo-Alonso, A., Utanda, A., Aulló-Maestro, M., & Palacios, M. (2019). Earth observation actionable information supporting disaster risk reduction efforts in a sustainable development framework. *Remote Sens, 11*(1), 49.

Mitri, G. (2015). Advancing Lebanon's fire danger forecast. In K. Themistocleous, D. Hadjimitsis, I. Gitas, & L. Boschetti (Eds.), *10th EARSeL forest fire special interest workshop sensors, multi-sensor integration, large volumes: New opportunities and challenges in forest fire research*, 2–5 November 2015. Limassol, p. 33.

Mitri, G., & Fiorucci, P. (2012). Towards monitoring post-fire vegetation cover dynamics in the Mediterranean with the use of object-based image analysis of Landsat images. In Y. Ban (Ed.), *1st workshop on temporal analysis of satellite images*, 24–25 May 2012 Mykonos Island, Greece.

Mitri, G., & Gitas, I. (2004). A semi-automated object-oriented model for burned area map-ping in the Mediterranean region using Landsat-TM imagery. *International Journal of Wildland Fire, 13*(3), 367–376.

Mitri, G., & Gitas, I. (2006). Fire type mapping using object-based classification of IKO-NOS imagery. *International journal of wildland Fire, 15*(4), 457–462.

Mitri, G., & Gitas, I. (2008). Mapping the severity of fire using object-based classification of Ikonos imagery. *International Journal of Wildland Fire, 17*(3), 431–442.

Mitri, G., & Gitas, I. (2010). Mapping post-fire vegetation regeneration using EO-1 Hyperion. *IEEE Transactions on Geoscience and Remote Sensing, 48*(3), 1613–1618.

Mitri, G., & Gitas, I. (2011). The role of remote sensing in the implementation of Lebanon's National strategy for forest fire management. In J. San-Miguel Ayanz, I. Gitas, A. Camia, & S. Oliveira (Eds.), *8th international workshop on advances in remote sensing and GIS applications in forest fire management remote sensing of forest fires, from local to global assessments*, 20–21 October 2011, Stresa, Italy, pp. 209.

Mitri, G., & Gitas, I. (2013). Mapping post-fire forest regeneration and vegetation recovery using a combination of very high spatial resolution and hyperspectral satellite imagery. *International Journal of Applied Earth Observation, 20*, 60–66.

Mitri, G. H., & Karam, J. (2016). Mapping greenhouse gas emissions and removals from the land use, land use change, and forestry sector at the local level. In *GEOBIA 2016: Solutions and synergies*, 14 September 2016–16 September 2016. University of Twente Faculty of Geo-Information and Earth Observation (ITC).

Mitri, G., Nader, M., & Salloum, L. (2011a). Fuel type mapping in the mediterranean region of North Lebanon using object-based image analysis of ASTER imagery. In J. San-Miguel Ayanz, I. Gitas, A. Camia, S. Oliveira (Eds.), *8th international workshop on advances in remote sensing and GIS applications in forest fire management remote sensing of forest fires, from local to global assessments,* 20–21 October 2011, Stresa, Italy, p. 39.

Mitri, G., Nader, M., Van der Molen, I., & Lovett, J. (2011b). The use of satellite imagery for the assessment of fire risk associated with repetitive armed conflicts in North Lebanon. In P. Duce, & D. Spano (Eds.), *International conference on fire behaviour and risk modelling*. Alghero, Italy, 4–6 October 2011, p. 159.

Mitri, G., Jazi, M., Antoun, E., McWethy, D., Kahaleh, R., & Nader, M. (2014a). The development of a web-based application for improved wildfire risk management in Lebanon. In D. X. Viegas (Ed.), *In. Advances in forest fire research* (pp. 1276–1280). Imprensa da Universidade de Coimbra.

Mitri, G., Jazi, M., & McWethy, D. (2014b). Investigating temporal and spatial variability of wildfire potential with the use of object-based image analysis of downscaled global climate models. *South-Eastern European Journal of Earth Observation and Geomatics, 3*(2S), 251–254.

Mitri, G., Jazi, M., McWethy, D. (2015a). Assessment of wildfire risk in Lebanon using geographic object-based image analysis. Photogrammetric Engineering & Remote Sensing Vol. 81, No. 6, June 2015, pp. 499–506.

Mitri, G., Jazi, M., & McWethy, D. (2015b). Assessing Lebanon's wildfire potential in association with current and future climatic conditions. In: R. E. Keane, M. Jolly, R. Parsons, Russel; K. Riley (Eds.), 2015 *Proceedings of the large wildland fires conference; May 19–23, 2014;*

Missoula, MT. Proc. RMRS-P-73. Fort Collins: U.S. Department of Agriculture (USDA), Forest Service, Rocky Mountain Research Station, pp. 318–321.
Mitri, G., Antoun, E., Saba, S., & McWethy, D. (2016). Modelling forest fire occurrence in Lebanon using socio-economic and biophysical variables in object-based image analysis. In *GEOBIA 2016: Solutions and synergies*, 14 September 2016-16 September 2016, University of Twente Faculty of Geo-Information and Earth Observation (ITC).
Mitri, G., Saba, S., Nader, M., & McWethy, D. (2017a). Developing Lebanon's fire danger forecast. *International Journal of Disaster Risk Reduction, 24*, 332–339.
Mitri, G., Nader, M., & Lea, K. (2017b). Mapping burned areas from time series of Landsat imagery for estimating greenhouse gas emissions at the National level. In *11th EArsel forest fires SIG workshop: New trends in forest fire research incorporating bid data and climate change modeling* 25–27 September 2017, Chania, Greece, p. 43.
Mitri, G., Beshara, J., Nehme, M., & Nassar, A. (2017c). Mapping plant abundance and Phyto-sociability using Sentinel-2 satellite imagery. In *11th EArsel forest fires SIG workshop: New trends in forest fire research incorporating big data and climate change modeling*, 25–-27 September 2017, Chania, Greece, p. 9.
Mitri, G., Beshara, J., & Nehme, M. (2018). In D. X. Viegas (Ed.), *Risk assessment and reflections on socio-economic perception of wildfires at the fireshed level. Advances in forest fire research 2018* (pp. 1160–1164). Imprensa da Universidade de Coimbra. https://doi.org/10.14195/978-989-26-16-506
Mitri, G., Nasrallah, G., Gebrael, K., Beshara, J., & Nehme, M. (2020). Assessment of post-fire land degradation risk in Lebanon's 2019 fire affected areas using remote sensing and GIS. In *Proceedings of SPIE 11524, eighth international conference on remote sensing and geoinformation of the environment (RSCy2020), 115240T* (26 August 2020). https://doi.org/10.1117/12.2571119
MoE/UNDP/GEF. (2015). *National greenhouse gas inventory report and mitigation analysis for the land use*. Land-Use Change and Forestry Sector in Lebanon.
Nogueira, G.S., Ribeiro, G.A., Ribeiro, C.A.A.S, Silva, E.P. 2002. Installation of fire detection towers using the GIS system, R. Árvore, Viçosa-MG, v.26, n.3, p.363–369.
Pereira, J., & Santos, M. (2003). *Fire risk mapping and burned area mapping in Portugal* (p. 64). Lisboa.
Röder, A., Hill, J., Duguy, B., Alloza, J. A., & Vallejo, R. (2008). Using long time series of Landsat data to monitor fire events and post-fire dynamics and identify driving factors. A case study in the Ayora region (Eastern Spain). *Remote Sensing of Environment, 112*(1), 259–273. https://doi.org/10.1016/j.rse.2007.05.001
Roy, P. S., Behera, M. D., & Srivastav, S. K. (2017). Satellite remote sensing: Sensors, applications and techniques. *Proceedings of the National Academy of Sciences/A. Section A, Physical Sciences, 87*, 465–472. https://doi.org/10.1007/s40010-017-0428-8
Sakr, G., Elhajj, I., & Mitri, G. (2010). Artificial intelligence for forest fire prediction: A comparative study. In D. Viegas (Ed.), *VI International conference on forest fire research*, 15–-18 November 2010. Coimbra, Portugal (ADAI), p. 37.
Sakr, G., Elhajj, I., & Mitri, G. (2011). Efficient forest fire occurrence prediction for developing countries using two weather parameters. *International Scientific Journal Engineering Applications of Artificial Intelligence, 24*(5), 888–894.
Sakr, G., Ajour, R., Khaddaj, A., Saab, B., Salman, A., Helala, O., Elhajj, I., & Mitri, G. (2014). Forest fire detection sensor node. In D. X. Viegas (Ed.), *Advances in forest fire research* (pp. 1395–1406). Imprensa da Universidade de Coimbra.
Salloum, L., & Mitri, G. (2014). Assessing the temporal pattern of fire activity and weather variability in Lebanon. *International Journal of Wildland Fire., 23*(4), 503–509. https://doi.org/10.1071/WF12101
Salloum, L., Mitri, G., & Gitas, I. (2013). The use of very high spatial resolution SPOT imagery for fire risk characterization. In *9th international workshop of the EARSeL Special Interest Group (SIG) on forest fires, 15–17 October 2013* (pp. 112–116). UK: University of Leicester.

San-Miguel-Ayanz, J., Pereira, J., Boca, R., Strobl, P., Kucera, J., & Pekkarinen, A. (2009). Forest fires in the European mediterranean region: Mapping and analysis of burned areas. In C. Emilio (Ed.), *Earth Observation of wildland fires in mediterranean ecosystems* (pp. 189–204). Springer.

San-Miguel-Ayanz, J., Schulte, E., Schmuck, G., Camia, A., Strobl, P., Liberta, G., et al. (2012). Comprehensive monitoring of wildfires in Europe: the European Forest fire information system (EFFIS). *In Approaches to Managing Disaster – Assessing Hazards, Emergencies and Disaster Impacts*. https://doi.org/10.5772/28441

Stergiadou, A., Valese, E., & Lubello, D. (2007). Detailed cartography system of fuel types for preventing forest fires. In *Proceedings of the 6th workshop on forest fires, EARSeL* (pp. 120–124).

Szpakowski, D. M., & Jensen, J. L. R. (2019). A review of the applications of remote sensing in fire ecology. *Remote Sens, 11*, 2638. https://doi.org/10.3390/rs11222638

Twele, A., & Barbosa, P. (2005). Monitoring vegetation regeneration after forest fires using satellite imagery. In Oluiþ (Ed.), *New strategies for European remote sensing*, (pp. 41–49). Rotterdam.

UNITAR-UNOSAT. (2019). *Fire Hotspots Based on the Analysis of VIIRS*. https://reliefweb.int/sites/reliefweb.int/files/resources/UNOSAT_A3_Natural_Portrait_FR20191016LBN_Hotspots_20191017.pdf. Accessed on 8 Apr 2021.

Veraverbeke, S., Lhermitte, S., Verstraeten, W., & Goossens, R. (2011). A time-integrated MODIS burn severity assessment using the multi-temporal differenced normalized burn ratio (DNBRMT). *International Journal of Applied Earth Observation and Geoinformation, 13*(1), 52–58.

Viedma, O., Meliá, J., Segarra, D., & Garcia-Haro, J. (1997). Modeling rates of ecosystem recovery after fires by using Landsat TM data. *Remote Sensing of Environment, 61*(3), 383–398.

Vila, G., & Barbosa, P. (2010). Post-fire vegetation regrowth detection in the Deiva Marina region (Liguria-Italy) using Landsat TM and ETM+ data. *Ecological Modelling, 221*, 75–84.

Zeidan, S., Bacciu, V., Nassif, N., Mitri, G., Jezzini, N., & Spano, D. (2019). *Mapping wildfire risk in Lebanon: Challenging a stepwise approach for effective purposes. ClimRisk19 – Climate risk: Implications for ecosystem services and society, challenges, solutions*, October 23rd–25th, 2019 in Trento, Italy.

Zorn, M. (2018). Natural disasters and less developed countries. In S. Pelc & M. Koderman (Eds.), *Nature, tourism and ethnicity as drivers of (de)marginalization: insights to marginality from perspective of sustainability and development, perspectives on geographical marginality* (pp. 59–78). Springer. https://doi.org/10.1007/978-3-319-59002-8_4. (March 4, 2021).

Chapter 23
Avalanche Hazards with Mitigation in Turkey and Qualitative Risk Assessment for Snow Avalanches in Ayder (Rize, NE Turkey) Using Combination of GIS, Remote Sensing Techniques and Field Studies

Tayfun Kurt

Abstract According to records of the Disaster and Emergency Management Presidency, avalanches have killed 30 people per year in Turkey over the last 30 years. For example, an avalanche occurred in *Görmeç,* in the province of Siirt, on February 01, 1992, which killed 97 people. This paper provides information about avalanche fatalities and avalanche mitigation works in Turkey as well as qualitative risk assessment method for snow avalanches in a defined region. Figure of avalanche hazard situation is presented to construct a picture of the potential threats. In the continuation, a regional-scale qualitative risk assessment is applied in a data-sparse region exposed to snow avalanches in NE Turkey. Input data included information on previous avalanche events, avalanche release and run-out areas derived from desktop review and field mapping, and an inventory of elements at risk. It is shown how categorization of elements at risk may be useful to obtain reliable results using a scoring approach. The method may foster risk management strategies in remote areas of less-developed countries, may support the choice of appropriate mitigation and may also improve risk communication and awareness. The obtained map provides are reliable and easy to understand information where avalanches constitute risky situation in regional scale as well as where new avalanche paths may develop under favourable conditions in defined region.

Keywords Qualitative risk assessment · Snow avalanches · Prioritization · Avalanche mitigation in Turkey

---

T. Kurt (✉)
Independent researcher, Istanbul, Turkey

M. M. Al Saud (ed.), *Applications of Space Techniques on the Natural Hazards in the MENA Region*, https://doi.org/10.1007/978-3-030-88874-9_23

23.1 Introduction

The threat of snow avalanches to the human environment is obvious, and has been reported from mountain regions throughout Europe over decades (Pfister, 1998; Fraefel et al., 2004; Fuchs et al., 2015). Apart from the ongoing discussion on leisure accidents affecting the ski industry, considerable parts of settlements and infrastructure are at risk. In order to reduce losses for the latter, in Central European mountain regions the concept of risk has been introduced. The concept of risk is based on a functional relation between hazard, elements at risk as well as vulnerability, and is embedded through the risk management approach in the framework of mitigation and adaptation planning (Kienholz et al., 2004). Also on the international level, starting with the 1990s IDNDR Decade (UN, 1989), the emphasis was shifted from hazard to risk assessment, focusing on both the physical and socio-economic dimension and thus a wider understanding of threats originating from natural hazards, recently pursued by the Hyogo Framework (UNISDR, 2007) and the Sendai Framework (UN/ISDR, 2015; Klein et al., 2019).

Snow avalanches compromise citizens inhabiting mountain areas as well as buildings and infrastructure. Nevertheless, while incidents are increasingly well documented from mountain regions of Central Europe (Fuchs et al., 2015; Bründl et al., 2010; Laternser & Schneebeli, 2002; Höller, 2007; Haeberli et al., 2015), data remains fragmentary in remoter mountain regions such as those in Russia (Shnyparkov et al., 2012; Fuchs et al., 2017) and Central Asia, to name the most prominent (Aydın et al., 2014). This issue has also repeatedly been claimed for Turkish mountain regions (Borhan & Kadioğlu, 1998; Gürer, 1998) because, mainly in the eastern part of Anatolia and in the eastern Black Sea region, an average of 24 fatalities is recorded per year, most of them inside buildings buried by snow avalanches (Aydın et al., 2014). This is a high number, compared to other European mountain regions were only a low number of fatalities is chronicled Fuchs et al., 2013). According to available data, avalanche events have increased in Turkey since the 1990s. For example, during the winter of 1992, 158 avalanche events were recorded, 453 people were killed, and 108 people suffered injuries. Gürer (2002). The Turkish government has responded and became particularly aware of the destructions caused by snow avalanches in the late twentieth century, and has subsequently defined multiple institutional responsibilities for mitigating snow avalanches. As in many other European countries, these are split with respect to different competencies regarding (a) the disaster management cycle in prevention and emergency response, and (b) according to different categories of elements at risk (e.g., settlement areas vs. transport infrastructure).

Moreover, these extraordinary avalanche events gave impetus to avalanche control works, which represent the traditional engineering way of hazard management, but also to land-use planning and local structural protection (Karsli et al., 2009). Authorities have also become particularly aware of strategies aiming at a decrease in future risk resulting from natural hazards (Bicer, 2016) including regional planning, land improvement, information and education (Yuksel et al., 2011).

While risk assessment is feasible and target-oriented if considerable data on hazard, vulnerability, and risk is available, it still remains challenging in data-sparse regions (Keiler & Fuchs, 2016). Because the risk concept is pillared by a quantifying function of the probability of occurrence of a hazard with a given magnitude and frequency (the scenario p_{Si}) and the related consequences on elements at risk *j*, regularly specified by their individual monetary value (A_{Oj}), the related (physical) vulnerability in dependence on scenario *i* ($v_{Oj,\ Si}$) and the probability of exposure ($p_{Oj,\ Si}$) of elements at risk *j* exposed to scenario *i* (Eq. 23.1, see (Fuchs et al., 2015).

$$R_{i,j} = f\left(p_{Si}, A_{Oj}, v_{Oj,Si}, p_{Oj,Si}\right) \tag{23.1}$$

Rooted in both technical and economic risk analyses, this quantitative definition of risk provides a framework for probabilistic risk assessment. The risk equation in terms of the combination of hazard and consequences only differs in terms of spatial resolution (scale), and can be assessed quantitatively or qualitatively. While specific risk would result in one individual value of potential losses for a given hazard probability (Varnes, 1984; Fell et al., 2008), the total (cumulative) risk represents the overall losses, such as the expected number of lives lost and persons injured, or the damage to property and disruption of economic activities for a given area and over a specific period in time (Bründl et al., 2010). Even if such quantitative information on natural hazard risk is an important milestone for the development of tailored management strategies Holub & Fuchs, 2009; Kurt, 2014), the concept poses several challenges because of its scale-dependent approach and the underlying data requirements.

Apart from temporal and spatial dynamics of avalanche risk (Fuchs et al., 2015), the main challenge is rooted in the system boundaries of risk assessment (Powell, et al., 2016; Andretta, 2014), which turns into a challenge of scale if snow avalanche risk is assessed regionally (Bründl et al., 2010). Nevertheless, also on a regional scale, risk includes the probability of multiple losses caused by snow avalanche impact over a definite period of time within a certain area (Shnyparkov et al., 2012). However, due to a lack in avalanche inventories, a purely quantitative risk assessment remains fragmentary (Keiler & Fuchs, 2016), and therefore more qualitative approaches are often used (Dahl et al., 2010). Qualitative hazard identification and risk assessment can effectively assist risk managers in priority-setting and politicians in decision-making, in particular in remoter regions where available data is sparse or even unavailable (Shnyparkov et al., 2012; Rogelis et al., 2016; Ding et al., 2016; Gruber & Mergili, 2013).

In order to close the knowledge gap on avalanche hazards in Turkey, snow avalanche risk, and to provide necessary background information for practitioners to manage this threat, a qualitative approach was undertaken to assess snow avalanche risk in this region. Both, hazard and elements at risk were quantified on a scale of 1:25,000 by using modelling approaches and a GIS environment, and the results were visualized and validated by available incident information and field surveys.

23.2 Avalanche Triggering Factors

Topographic factors, vegetation factors and weather conditions such as terrain, meteorological, and snowpack factors may be very important in the occurrence of avalanches (Höller, 2007; Hendrix et al., 2005; Hageli & McClung, 2007; Schweizer et al., 2003). For example, avalanches generally occur on slopes between inclination 28° and 55° (McClung & Schaerer, 2006; Miklau & Sauermoser, 2011). Avalanches normally are not triggered on slopes outside this range because snow masses tend not to accumulate on such slopes (Sullivan et al., 2001).

Slope aspect can be particularly important factor due to snowmelt, solar radiation and wind loading on the snowpack (Grimsdottir & McClung, 2006; Cooperstein et al., 2004; McClung & Schaerer, 2006) . For example, south-facing slopes in the northernhemisphere can be especially dangerous in the spring for occurrence of wet snowavalanche when heated by the sun (Dubayah, 1994). North-facing slopes may be slowerto stabilize than slopes facing in other directions (Daffern, 2009). Another example, leeward slopes, slopes facing away from the wind, are dangerous because this is where the snow collects and may form an unstable slab (Meloysund et al., 2007). On the other hand, windward slopes that face the wind generally have less snow and are usually more stable (McClung & Schaerer, 2006; Dubayah, 1994). Moreover, because of solar radiation and wind-drifted snow, the strength and thickness of the snow cover and distribution of weak layers can vary with the aspect (Grimsdottir & McClung, 2006). Statistical work indicated that; most avalanches fell in northeastern an eastern aspects, which are the lee and shady aspects (Grimsdottir & McClung, 2006).

Snowpack forms from layers of snow that accumulates in geographic regions and high altitudes where the climate includes cold weather for extended periods during the year (Broulidakis, 2013).

Snowpack factors are snowpack depth and structure, such as hardness, layering, crystal forms and free water content (McClung & Schaerer, 2006). According to Gaume et al. (2013), the spatial variability of snowpack properties has an important impact on snow slope stability and thus on avalanche formation. For example, as the snow falls it settles in layers of varying strength and weakness. Because numerous layers constitute a snowpack, it is important to understand the properties of each layer of the weak layer such as overlying load, densities, temperature gradient and crystal types, because each one forms under varying weather conditions and will bond to approaching layers differently (Jamieson & Johnston, 2001). Weak layers deep in the snowpack can cause avalanches even if the surface layers are strong or well-bonded (Jamieson et al., 2003).

Vegetation cover is among the factors affecting avalanches by increasing or decreasing friction on the surface (Butler, 1972; Simonson et al., 2010). Smooth slopes with pasture can accelerate formation of avalanches due to lack of resistance (Tunçel, 1990). If there is vegetation cover in the form of shrubs in the avalanche release zones, the vegetation cover can hold the snow mass and delay the initiation of avalanche. According to Brang (2001), dense forests are also effective to reduce avalanches by preventing the avalanche release in initiation zones. On the other hand, vegetation analysis can be used to survey past avalanches and to estimate the frequency and intensity of snow-slide events for specific avalanche path locations and time periods of interest (Simonson et al., 2010; Burrows & Burrows, 1976; Carrara, 1979; Mears, 1992; Casteller et al., 2007; Bebi et al., 2009).

Wind loading may occur without precipitation, by scouring of snow on exposed windward slopes and subsequent deposition of this scoured snow on lee slopes (Schweizer et al., 2003; McClung & Schaerer, 2006). Variations in wind speed and snow drift can be important that they form layers of different density or harness creating stress concentrations within the snowpack (McClung & Schaerer, 2006). For example, it has been assumed that snow drift peaks at a wind speed of about 20–25 m s^{-1} and decreases with even higher wind speeds (Schweizer et al., 2003).

Temperature is a decisive factor contributing to avalanche formation, particularly in situations without loading (Schweizer et al., 2003). The temperature of the weather and of the snowpack has an increasing effect on the risk of avalanche. According to McClung and Schaerer (2006), the mechanical properties of snow are highly temperature dependent. In general, there are two important groups of competing effects: metamorphism (depending on temperatures) and mechanical properties (excluding metamorphism effects) including snow hardness, fracture propagation potential and strength (Schweizer et al., 2003). The probability of powder snow avalanche is high in the presence of low weather temperature and wind McClung and Schaerer (2006). On the other hand, temperature rises in the spring can cause wet snow avalanches (McClung & Schaerer, 2006; Tunçel, 1990). Another example regarding temperature is solar radiation because it can prepare conditions for avalanches to initiate (Grimsdottir & McClung, 2006). Solar radiation can present greater risk situations in alpine regions than lower down, due to the open terrain (Grimsdottir & McClung, 2006; Cooperstein et al., 2004). Surface hoar forms when relatively moist air over a cold snow surface becomes oversaturated with respect to the snow surface, causing a flux of water vapor, which condenses on the surface (McClung & Schaerer, 2006).

23.3 Avalanche History of Turkey

Turkey consists of 7 regions (Fig. 23.1). The mean altitude of Turkey is 1141 m a.s. l., more than three times avalanches higher that of Europe (300 m a.s.l.), with a mean slope angle of 10º. Altitudes higher than 1500 m with slopes greater than 27° cover 5.1% of the total area (Aydın et al., 2014; Elibüyük & Yılmaz, 2010). Northeast, east and southeast parts of Turkey are the most endangered areas based on recorded avalanche events between 1950 and 2019 (AFAD, 2020, see Fig. 23.2). Considering

Fig. 23.1 Turkey consists of 7 regions

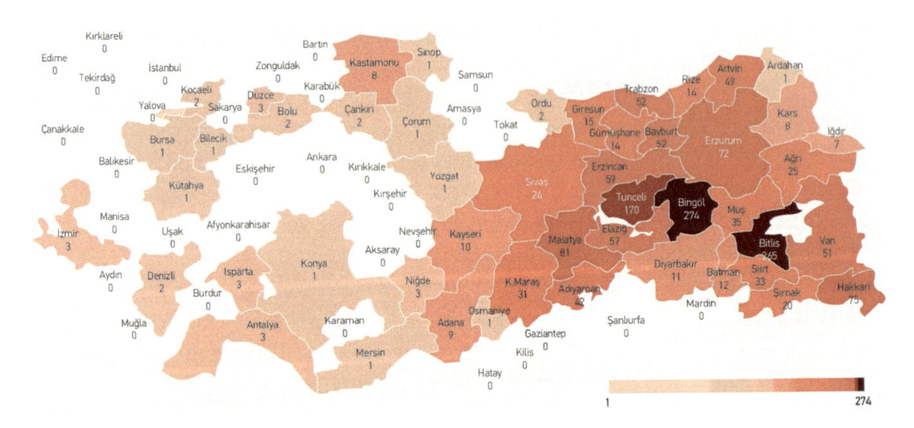

Fig. 23.2 Cities affected by avalanches in Turkey between 1950 and 2019 with number of cases (AFAD, 2020)

a lot of snowfalls in the mountainous areas, generated elevation and inclination map of Turkey based on SRTM (Shuttle Radar Topography Mission) database and maximum snow height map also verify that higher inclination and altitude including heavy snowfall trigger more avalanches in Turkey (Figs. 23.3 and 23.4). The earliest recorded avalanche fall in Turkey occurred in 1890 (Varol & Yavaş, 2006). Based on avalanche records from AFAD that consist of all types of documentation (reports, photos etc.) from 1890 to 2014, 1997 avalanche events occurred, and more than 1446 people were killed by avalanches. Based on AFAD records, on average, avalanche events have caused 30 deaths in Turkey every year as well as damage to villages, settlements, infrastructure and forests over the last 30 years (Fig. 23.5).

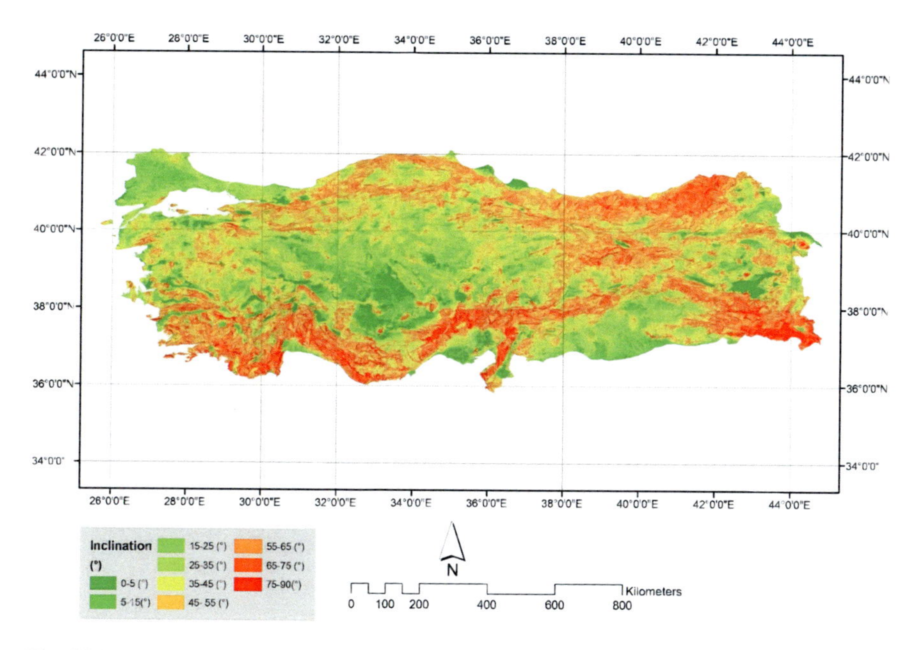

Fig. 23.3 Inclination map of Turkey Generated based on SRTM database

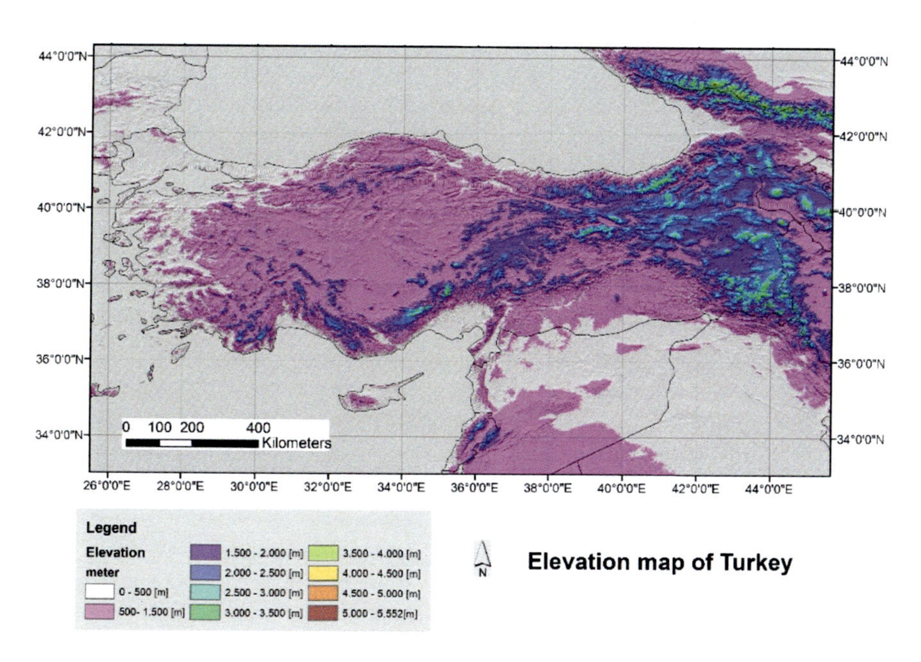

Fig. 23.4 Elevation map of Turkey Generated based on SRTM database

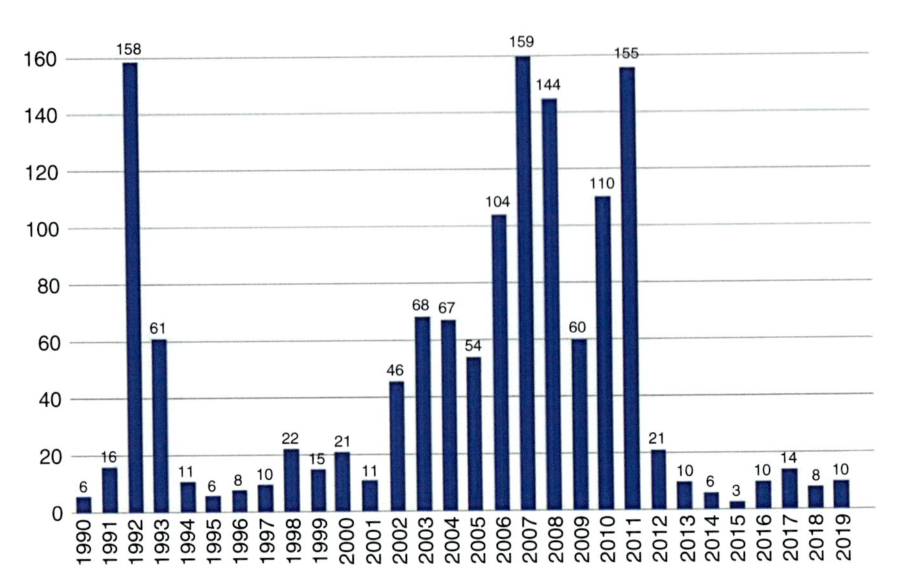

Fig. 23.5 Avalanche fatalities between 1950 and 2019 (AFAD, 2020)

Avalanche observations of past and present avalanche activity are of the utmost importance for any avalanche forecasting operation and avalanche control concepts (Laternser & Schneebeli, 2002). According to Gürer (2002), prior to the 1950s avalanche events were not recorded in Turkey by authorities unless they caused deaths or injuries. In order to create a database for all disasters including avalanches, TABB (Turkish National Disaster Archive) was established with support from AFAD (Disaster and Emergency Management Presidency) in 2004.

In order to create a database on Turkey as a whole, including previous disasters, Turkish National Disaster Archive (TABB) was established in 2004. Written documents, reports, photos, and other types of information have been collected from the following governmental bodies: AFAD, General Directorates of Food Control, Gendarmerie General Commands, Police, Turkish Atomic Energy Authority and Media. Then, some criteria imposed for records by TABB: the exact date if known, in the absence of a certain date, the number of fatalities, in the absence of fatalities, the number of injured. Specifications are searched within these data parameters. However, so far all the collected data has not been entered into TABB databank and the monitoring process is behind schedule.

During the winter of 1992, a total of 158 avalanche events were recorded, 443 people were killed, and 108 people suffered injuries in Turkey (Gürer, 2002). The distribution of several major avalanche events of 1992 and 1993 are shown in Table 23.1. One reason for avalanche fatalities in Turkey was the heavy snowfall. For example, the heavy snowfall (Fig. 23.6) in January and February caused many

Table 23.1 Major avalanche events of 1992 and 1993 in Turkey

Date	Deaths	Details
02.01.1992	20	Due to heavy snowfall, 20 inhabitants were killed in *Karabeya* village, *Yüksekova*, province of Hakkari.
21.01.1992	10	10 inhabitants were killed in *Kesmetaş* Village, *Şirvan*, province of Siirt.
01.02.1992	97	71 soldiers in Turkish army and 26 inhabitants in the *Görmeç* village were killed due to avalanche fall in province of Siirt.
07.02.1992	55	31 person in *Boğazören* village, *Beytüşşebap*, province of Şırnak; 13 people in different villages in province of Batman, 5 people in *Erimli*, province of Elazığ, 6 people were died in province of Bingöl and Diyarbakır.
08.02.1992	21	15 inhabitants in *Çığlıca* village, province of Şırnak, 6 inhabitants were killed in *Tatlıca*, province of Batman
21.02.1992	32	32 soldiers were killed in both *Eruh* and *Uludere*, province of Siirt
25.02.1992	26	Due to heavy snowfall, 26 inhabitants in *Anaköy* Village, *Gevaş*, province of Van.
18.01.1993	59	59 inhabitants were killed and 21 were injured due to avalanche event in *Üzengili* village, province of Bayburt
25.02.1993	26	26 inhabitants were killed in Anaköy, province of Van
27.02.1993	6	6 passengers were killed on the Hakkari-Van main road due to avalanche event

avalanche events in eastern Turkey (Gürer, 2002). Many main roads were closed and many villages were affected by avalanches.

One of the most deadly avalanches in Turkey occurred in the winter of 1992. A brigade of 71 Turkish soldiers and 26 inhabitants were killed by an avalanche in the village of *Görmeç*, province of Siirt on 01.02.1992. According to Borhan and Kadıoğlu (Borhan & Kadioğlu, 1998), the day the Görmeç avalanche happened, the weather was rainy and snowy. Another detail of this avalanche that contributed to the number of deaths was that there was no avalanche control structure in the region.

Another deadly avalanche occurred in the village of *Üzengili,* Bayburt Province, on January 18th, 1993, at 07:45 am. This avalanche killed 59 people and destroyed 62 buildings. According to Taştekin (2003), on January 16th, the weather temperature was – 5.0 °C, the next day air the temperature dropped by 10 degrees and became −15 °C (Fig. 23.7). As a result, the snow surface cooled. During the daytime of January 17th, the air temperature increased. Snowfall during the night of January 17th, (between 21:00 and 07:00) (Figs. 23.7 and 23.8). New snow precipitation caused an additional load and the previous snow layer could not carry new snow, and in the morning of the 18th, the Üzengili avalanche occurred.

The Palandoken ski resort is located on the northern slopes of the Palandoken range in Erzurum province; skiing is possible for 150 days in a year, skiing altitude is 2200–3176 m. So far, 9 skiiers lost their lives between 1996–2006 and three more skiers in total (Fig. 23.9).

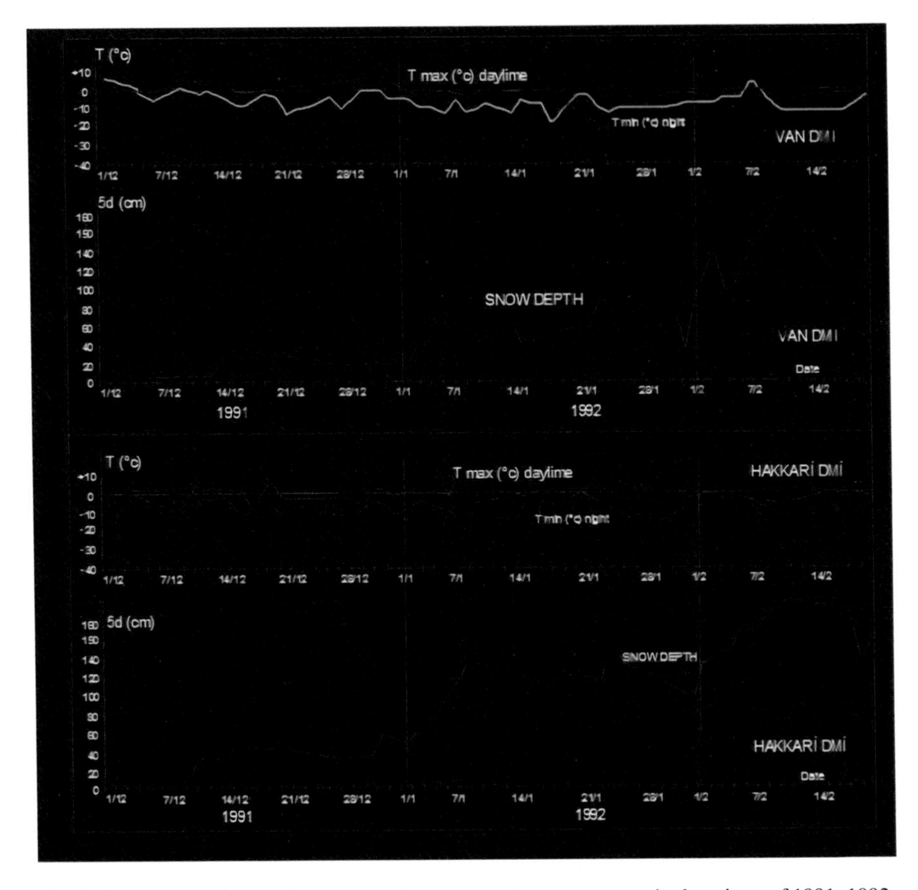

Fig. 23.6 Snow depths, maximum and minimum weather temperature in the winter of 1991–1992, in Hakkari and Van Province (Gürer, 2002)

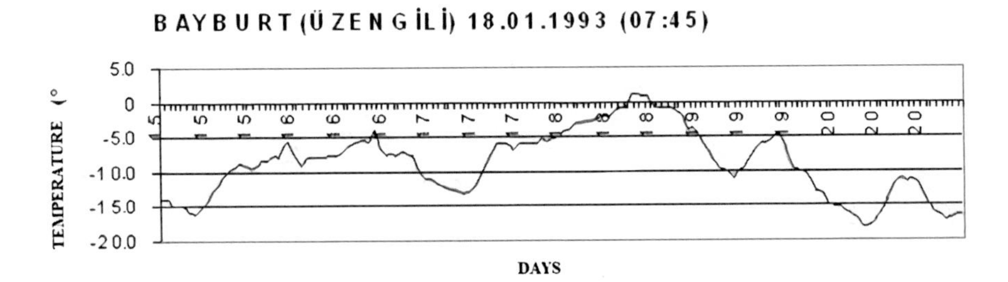

Fig. 23.7 Air temperature of Bayburt Üzengili in 18.01.1993 (Taştekin, 2003)

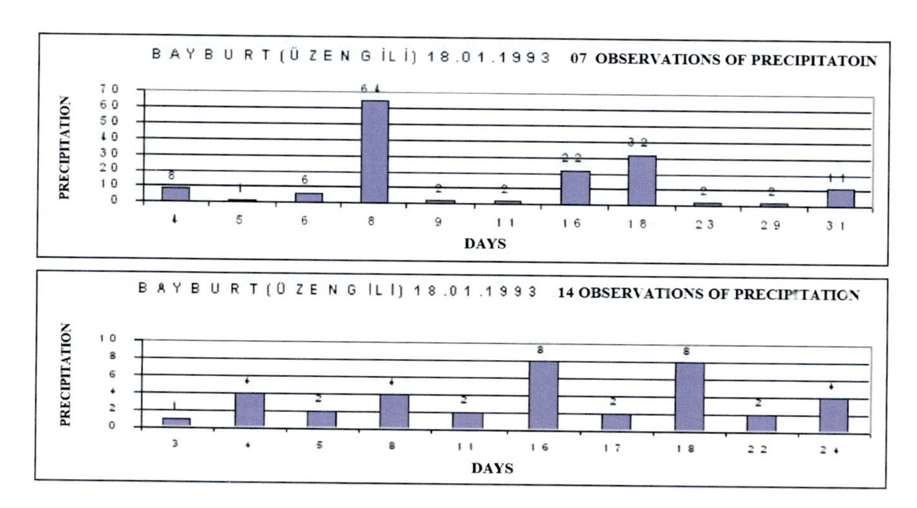

Fig. 23.8 According to 7 and 14 of records, precipitation in Bayburt, Üzengili (Taştekin, 2003)

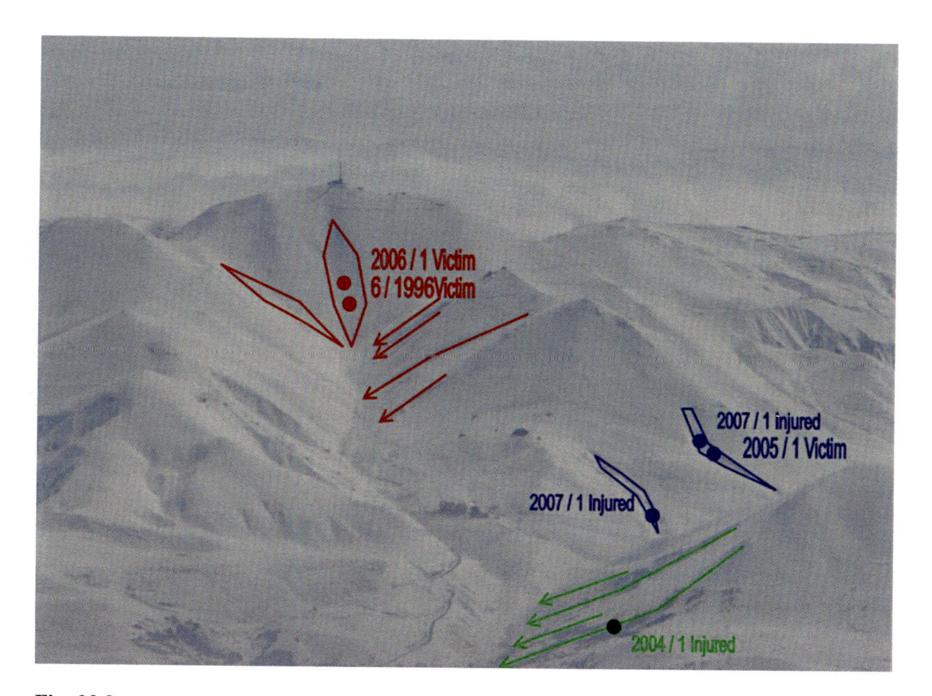

Fig. 23.9 Avalanche locations in Palandöken skiing centre in Erzurum province (Kurt, 2017)

23.4 Avalanche Mitigation in Turkey

Until today, different governmental bodieshave been dealt with avalanches to keep people and property in safe. Today governmental bodies have been dealing with avalanche control issue, especiallyover the past few decades, in order to control avalanches and keep people and their property safe in Turkey such as the Ministry of Forestry and Water Affairs, *General Directorate of Forestry (OGM); General Directorate of Combating Desertification and Erosion (ÇEM);* Ministry of Interior, *Disaster and Emergency Management Presidency* Mostly passive avalanche control methods are used instead of permanent control methods due to absence of organized avalanche control service (Table 23.2).

23.4.1 Avalanche Mitigation Projects in Settlement Areas in Turkey

If an avalanche occurs in an area of no settlements, no property, or no traffic, it does not constitute risk. Hence, avalanche protection may not be necessary in these uninhabited areas. On the other hand, if an avalanche presents hazard, a decision has to be made quickly to ensure maximum safety of endangered objects in the hazardous zone. So, avalanche protection works reduce the hazard avalanches pose to human lives and properties. Today, numerous endangered settlements still have no protection in Turkey.

Table 23.2 Organizations responsible for snow avalanche management in the Republic of Turkey (Kurt, 2017)

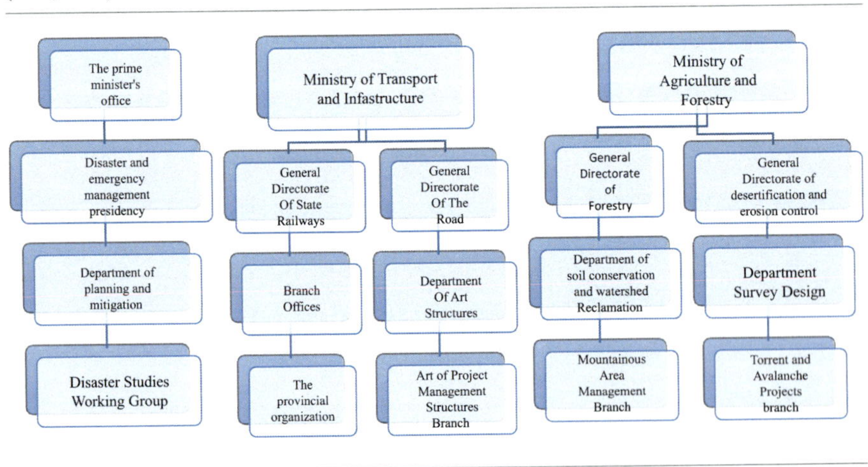

Avalanche protection may be divided into temporary and permanent measures (McClung & Schaerer, 2006). Temporary measures are applied for short periods when avalanches are expected to occur. On the other hand, permanent measures usually require expense for engineering works but perform without the need for a daily hazard evaluation (McClung & Schaerer, 2006). There are very few snow pack supporting structures or plans to protect settlements in Turkey beyond the current method of reforestation.

In the starting zones, avalanche control methods are implemented to prevent the start of avalanches or limit the snowpack motion that can be triggered by snow movement due to steep slopes, with the help of supporting structures, snow fences or other type of apparatus (McClung & Schaerer, 2006; Höller, 2007; Margreth et al., 2007). TheUzungöl project was the first recorded avalanche-control project implemented in 2004 by OGM (General Directorate of Forest). This project comprised 6680 meters of steel snow fences, 3340 meters of snow breakers, and mini-piles constructed in the avalanche release zones (Figs. 23.10 and 23.11).

Avalanche protection may be divided into temporary and permanent measures (McClung & Schaerer, 2006). Temporary measures are applied for short periods when avalanches are expected to occur. On the other hand, permanent measures usually require expense for engineering works but perform without the need for a daily hazard evaluation (McClung & Schaerer, 2006). There are very few snow pack supporting structures or plans to protect settlements in Turkey beyond the current method of reforestation.

In the starting zones, avalanche control methods are implemented to prevent the start of avalanches or limit the snowpack motion that can be triggered by snow movement due to steep slopes, with the help of supporting structures, snow fences or other type of apparatus (McClung & Schaerer, 2006; Margreth et al., 2007; Höller, 2007). The Uzungöl project was the first recorded avalanche-control project implemented in 2004 by OGM (General Directorate of Forest). This project comprised 6680 meters of steel snow fences, 3340 m of snow breakers, and mini-piles constructed in the avalanche release zones (Figs. 23.10 and 23.11).

Fig. 23.10 Snow nets used as supporting structures in release zone, Uzungöl, province of Trabzon

Fig. 23.11 Snow fences and mini-piles in Uzungöl, Trabzon

Fig. 23.12 Steel snow bridges in Trabzon (Photo: Anonym)

In recently, steel snow bridges have been constructed also in Çaykara, Karaçam, in Trabzon Province in 2016 (Fig. 23.12).

23.4.2 Avalanche Hazard Zoning and Mapping in Turkey

Avalanche hazard maps can give an idea of the safety level of a certain area in regard to the risk of natural disaster (avalanches, rock falls, or torrents) (Holub & Fuchs,

Table 23.3 Types of avalanche levels used in Turkey

Avalanche zone	Details
Red	High risk level. No construction is allowed and unsuitable are for residence.
Blue	Middle avalanche risk level, permanent use for settlement and infrastructure is possible but with additional safety measures
White	No avalanche risk

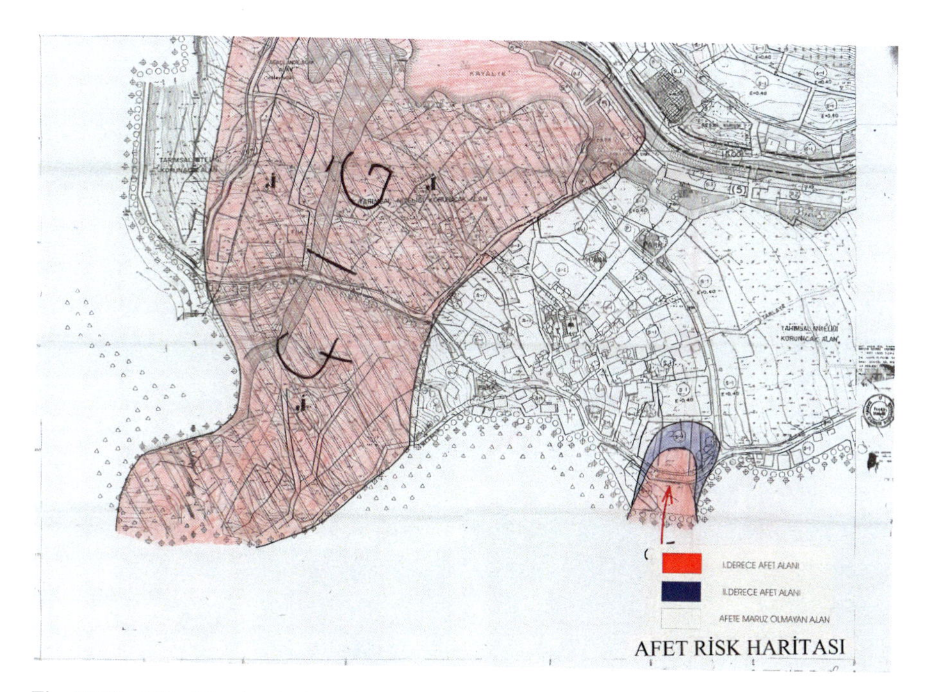

Fig. 23.13 Offical Avalanche hazard zoning in Turkey

2009). Avalanche hazard zoning is used in Turkey to prevent buildings being constructed in areas endangered by avalanches and to indicate avalanche prone areas.

Three different risk levels are used for avalanche hazard maps in Turkey (Varol & Şahin, 2006) (Table 23.3).

On the other hand, avalanche hazard zonings are marked on maps for all avalanche prone areas that are designated as avalanche paths (avalanche areas), active avalanche paths, potential avalanche flow tracks, and possible avalanche flow tracks (Fig. 23.13). These maps are drawn based on avalanche chronology, topographic properties (aspect, slope etc.), and vegetation cover.

23.4.3 Avalanche Control in Winter Resorts

Most common avalanche control methods for winter resorts in the regions affected are avalanche forecasting, control programs, closure of ski paths and warning signs at defined locations. Also, in the event of heavy snowfall, methods in use include artificial avalanche release by using the Gazex system (Fig. 23.14) under controlled conditions in order to trigger smaller, less-destructive avalanches, closure of avalanche paths. For example, the Gazex system is being used in the Erzurum Palandöken ski resort.

23.4.4 Avalanche Mitigation Near Main Roads

Avalanche hazards can be reduced by defense structures such as avalanche galleries (Fig. 23.15), snow sheds or snow breakers (Figs. 23.16 and 23.17), and closure of main roads. KGM is responsible for construction and safety of roads in Turkey. In this context, KGM takes measures against avalanches by constructing avalanche galleries and snow sheds (Fig. 23.17).

Fig. 23.14 Artificial avalanche release by using Gazex, Palandöken, Erzurum Province (Kurt, 2017)

Fig. 23.15 Avalanche gallery (28 meter long) in Elazığ Province (Kurt, 2017)

Fig. 23.16 Snow shed (206 meter long) in Province of Hakkari (Kurt, 2017)

Fig. 23.17 Effectiveness of snow sheds built by KGM in Turkey (Kurt, 2017)

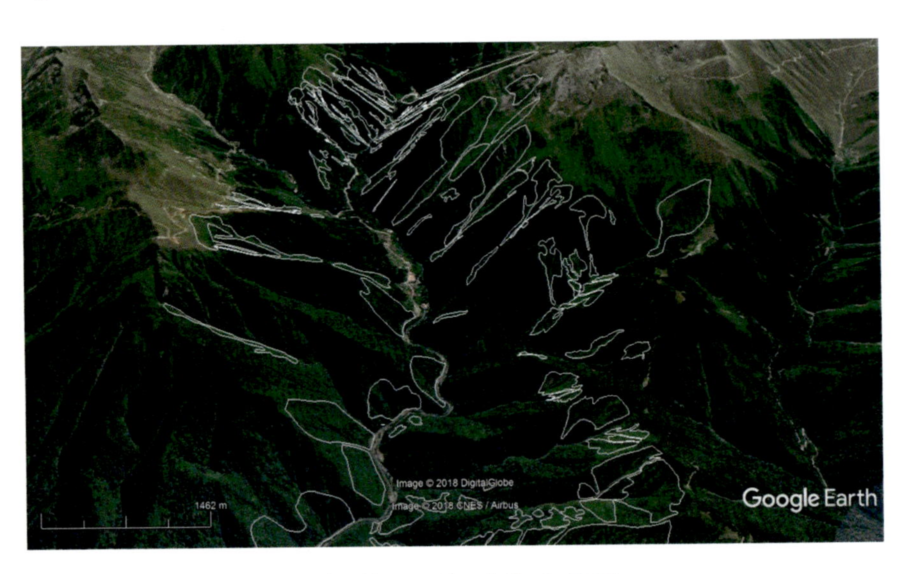

Fig. 23.18 Overwiev of the test site. (Source: Google Earth, 2018)

23.5 Qualitative Risk Assesment: A Case Study

23.5.1 Study Area

The study area is located in Ayder, a recreational area in the Province of Rize, NE Turkey, and a hot spot of winter tourism (Figs. 23.18 and 23.19). Such data-limited region is located in NE Turkey, where approximately 4 million people inhabit the

Fig. 23.19 Overview of Ayder, showing considerable construction activities (Kurt, 2017)

mountain region of the Pontian Alps, stretching from the Black Sea coast over a length of around 300–400 km NE and around 200 km inland. The region is characterized by rural settlements, and mountain topography; the highest elevation is mount Ağrı Dağı (3932 m asl) at the Georgian border. The properties (such as buildings, infrastructure, etc.) are continuously exposed to snow avalanches during the winter season. The area covers a total of 7800 hectares and is located between 1300 and 2450 m asl with steep gradients and considerable snowfall during winter.

The Ayder area was officially declared as tourism region by the Turkish government in 1987. As a consequence, heavy land development took place and a considerable number of elements at risk were located adjacent to potential avalanche run-out areas, which clearly shows the dynamic component of risk in the area (see Fig. 23.19).

Land cover in the area is dominated by forest. Snow avalanches occur due to natural factors such as inclination and aspect supplemented by vegetation; while herbaceous plants dominate most of the potential avalanche release areas and parts of the transit tracks, the overall forest cover is dominated by spruce (Picea orientalis L.), beech (Fagus orientalis L.), chestnut (Castanea sativa Mill.), and alder (Alnus glutinosa L.). Silent witnesses can be observed throughout the study area; they represent information on the spatial occurrence of snow avalanches. Hence, taking characteristics of stand structure by age classification, tree heights, soil regime, and canopy cover, available information on avalanche activities was gathered (Figs. 23.19, 23.20 and 23.21).

Fig. 23.20 "J" shaped trunks due to snow gliding (Çelik & Kurt, 2012)

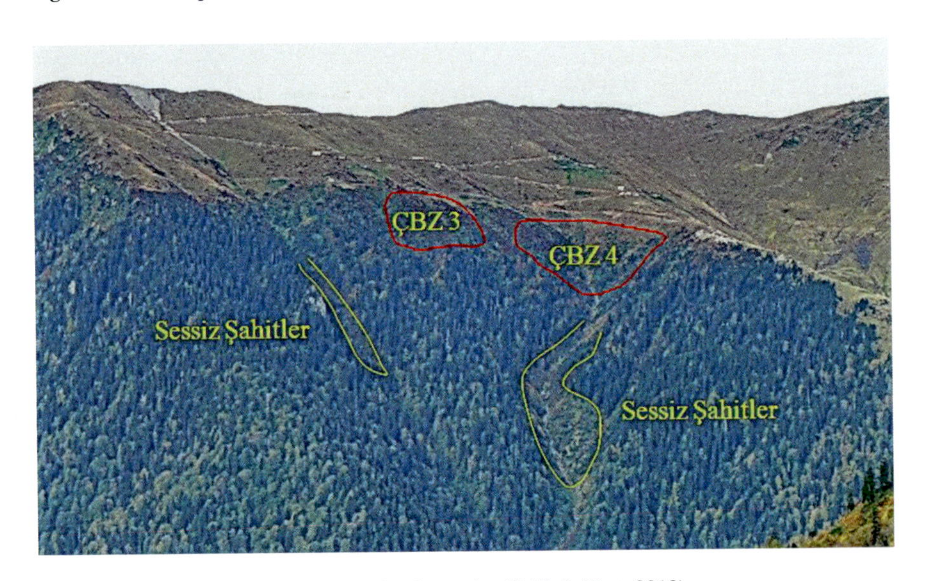

Fig. 23.21 Silent witnesses on the avalanche tracks (Çelik & Kurt, 2012)

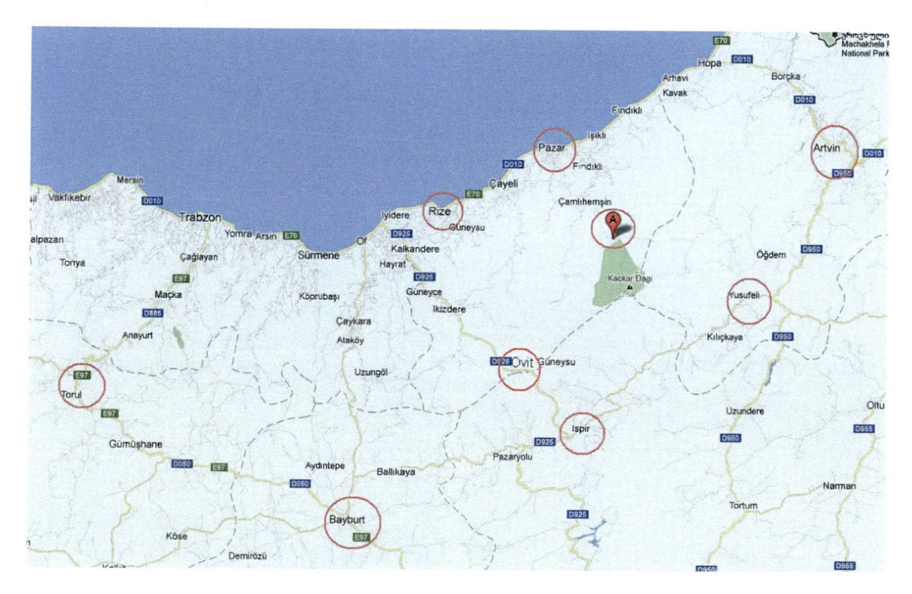

Fig. 23.22 Locations of Çamlıhemşin, Ovit, Pazar, Rize, Artvin, İspir, Bayburt and Torul meteorological stations (Çelik & Kurt, 2012)

23.5.2 Meteorological Characteristics

To get information about maximum snow heights meteorological data needed for hazard assessment was taken from the Çamlıhemşin weather station. This station, though not being the closest to the test site, was taken due to higher representativeness than closer stations such as Rize and Pazar, the latter being located at the shoreline (Fig. 23.22).

According to the data, maximum snow height amounts to 2.1 m for the period between 1984 and 1992. Because of the relatively high difference in altitude between the Çamlıhemşin weather station and the test site, however, this value was height adjusted and supplemented by local knowledge. As a consequence, maximum snow heights of 5.0 m were taken as being representative for Ayder, as verified by inhabitants whose houses were repeatedly buried by snow during the winter.

23.5.3 Avalanche History

Documented information about avalanche events provides the most reliable information about avalanche activity. According to reports by the local population, and supplemented by an assessment of terrain characteristics during field works, major deposition areas have been identified (Fig. 23.23).

Fig. 23.23 Run out zones of avalanches according to inhabitants

The largest avalanche events in Ayder occurred in 1947 and 1992. Run out zones of these avalanches are indicated in Fig. 23.24. While the archives did not report any loss for the 1947 avalanche, presumably due to a lack of elements at risk located in the run-out areas during that time, the powder snow avalanche of 1992 destroyed one hotel and damaged three wooden houses (Çelik & Kurt, 2012).

To fill the knowledge gap in avalanche documentation, desktop reviews and fieldwork was undertaken and 67 avalanche release areas were detected, most of them located in an elevation of around 2100 m asl. Subsequently, a susceptibility map was deduced linking potential release areas to possible run-out areas (Fig. 23.25).

23.5.4 Method

The applied method is composed of three main steps including risk identification, risk analysis, and risk evaluation. As risk identification is the initial step for risk management, data related to the avalanche susceptibility has to be made available so that hazard and consequence analysis can be executed (Erener & Düzgün, 2013). Determining the avalanche-prone areas provide useful information for identifying future avalanche occurrences. Consequently, as a first step, avalanche susceptibility

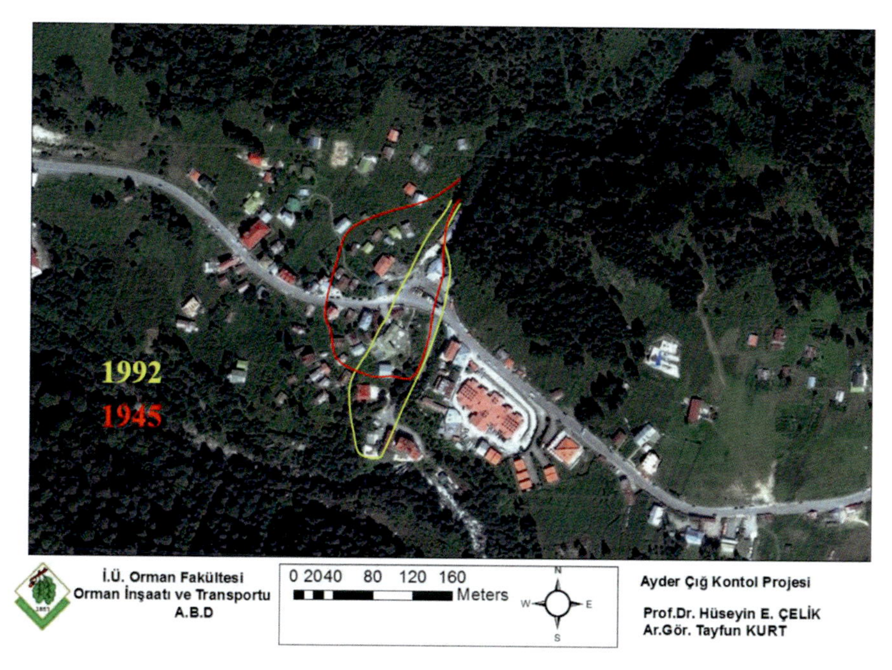

Fig. 23.24 Run-out zones of the 1947 and 1992 avalanches gliding (Çelik & Kurt, 2012)

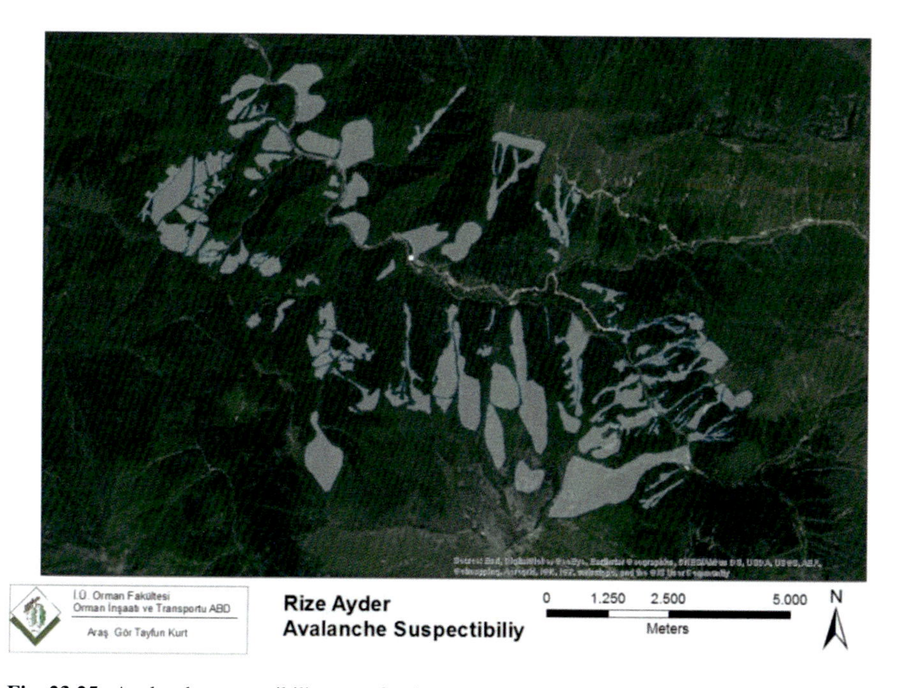

Fig. 23.25 Avalanche susceptibility map for the study area

Table 23.4 Calculation of risk scores

R=Occurrence of avalanche x Effect level			Effect level of avalanche				
			Highest	High	Medium	Low	Lowest
			5	4	3	2	1
Occurrence of avalanche	Highest	5	25	20	15	10	5
	High	4	20	16	12	8	4
	Medium	3	15	12	9	6	3
	Low	2	10	8	6	4	2
	Lowest	1	5	4	3	2	1
Low							
Medium							
Hight							

Table 23.5 Evaluation of avalanche occurrence in the study area

Definition of avalanche occurrence	Frequency (avalanche events) [%]
Highest	≥ 1.0 (every year)
High	0.1> to <1,0 (every 5 years)
Medium	0.02> to ≤0.1 (every 10 years)
Low	0.02> to ≤0.01 (every 50 years)
Lowest	≤ 0.01 (every 100 years)

maps were transferred into hazard maps by integrating avalanche trigger probabilities with using satellites images (such as Google Earth), aerial photos. In a second step, qualitative risk maps (based on a scoring) were obtained by an evaluation of hazard, vulnerability, and consequences for each element at risk, following the procedure outlined in Remondo et al. (2005) see Table 23.4).

In order to deduce risk maps, a two-step approach was followed. In the first step, the occurrence of avalanches was evaluated for all avalanche tracks. One of the important phenomena is determining the occurrence probability of a specific triggering event (such as periods of intense snowfall), which will trigger the avalanche for a given time period (Bellaire et al., 2013; Laternser & Schneebeli, 2002). In contrast, the frequency of an avalanche is defined as the period of time in which an avalanche of defined magnitude reaches a specific run-out area or threshold (Haeberli et al., 2015). In this sense, documented information on previous avalanche events provided the most reliable data with regard to avalanche history. However, because only one avalanche control project from 2012 was available for the study area (Çelik & Kurt, 2012), the frequency of avalanches was evaluated empirically based on information received from the local population and analysing irregularities in the vegetation cover as once every year; every 5 years; every 10 years; every 50 years; and every 100 years (Table 23.5).

In the second step, the possible effects of snow avalanches were defined based on fieldwork and information obtained during interviews from inhabitants. As presented in Table 23.6, buildings, roads, other infrastructure and citizens inhabiting exposed buildings were considered during the analysis. Therefore, elements at risk were classified into three categories representing information on health, economy and the environment. Each category was evaluated using five different levels of severeness (lowest, low, medium, high, and highest). The category health included the number of people at risk; the category economy included the number of buildings exposed and the length of major roads, and the category of environment showed the affected vegetation (i.e. forest cover) in hectares (Table 23.6). After evaluating the effect level for individual avalanche paths and run-out areas in each category, the most severe avalanche path showing the highest risk was determined based on an overlay in GIS.

23.5.5 Results

The results of the qualitative risk assessment in the study area are presented in Table 23.7. Based on the avalanche susceptibility mapping and the compilation of hazard maps, qualitative risk information was obtained using the scoring and classification approach outlined above. The results were verified by comparing them to existing cadastral maps, where possible (see Figs. 23.26 and 23.27). More than 3 km of road are endangered by multiple avalanche tracks, and a total of 37 buildings is endangered. Based on a desktop review, around 100 individuals are at risk, and almost 1100 hectares of forest are threatened. As such, the results indicate those areas with higher risk levels, and information can be used by land-use planners and for hazard and risk management so that investments in technical protection will become more efficient.

The avalanche path with the highest risk is CBZ no. 63, endangering two restaurants, an internet cafe, a gift shop, a small grocery store, five hotels that can host 350 tourists overnight, a highway bridge, and a car park. The endangered part of the main road is 200 m long and 6 m wide (Fig. 23.27).

Overall the result of the qualitative risk assessment using a 1:25,000 scale is a ranking of risk for the study area. In comparing the all release zones, the development of the risks of the all studied avalanche tracks differ considerably (Table 23.8). As a result, qualitative risk scores were calculated on a continuous scale where numerical values indicate the risk priority of avalanche tracks respectively.

According to the calculation of risk scores,

- for the traffic infrastructure the avalanche track no (CBZ):11 has shown to have the highest risk score;
- for exposed buildings as well as with respect to citizens at risk, the avalanche track no 17 has shown to have the highest risk score;
- and in terms of environment (protected area) the highest risk score was also avalanche track no11.

Table 23.6 Classification and characterisation of risk indicators in the study area

Classification	Effect level of avalanche				
	Lowest	Low	Medium	High	Highest
Health (Number of persons who may be affected, number or leisure centres)	**Number of persons $\leq$ 1** and/or **Leisure Centre** (N) $\leq$ **1**	**2 $\geq$ number of persons $\leq$ 4** and/or **1 $\geq$ leisure Centre** (N) $\leq$ **5**	**5 $\geq$ number of persons $\leq$ 9** and/or **6 $\geq$ leisure Centre** (N) $\leq$ **10**	**10 $\geq$ number of persons** $\leq$ and/or **11 $\geq$ leisure Centre** (N) $\leq$ **19**	**Number of persons $\geq$ 21** and/or **Leisure Centre** (N) $\geq$ **20**
Environment (Protected areas that may be affected)	**Protected areas** (ha) $\leq$ **20**	**21 $\geq$ protected areas** (ha) $\leq$ **50**	**51 $\geq$ protected areas** (ha) $\leq$ **75**	**76 $\geq$ protected areas** (ha) $\leq$ **150**	**Protected areas** (ha) $\geq$ **151**
Economy (Length of road that may be affected, number of personal property)	**Property** (N) $\leq$ **1** and/or **Length of road** (m) $\leq$ **100**	**2 $\geq$ property** (N) $\leq$ **5** and/or **101 $\geq$ length of road** (m) $\leq$ **200**	**6 $\geq$ property** (N) $\leq$ **9** and/or **201 $\geq$ length of road** (m) $\leq$ **300**	**10 $\geq$ property** (N) $\leq$ **19** and/or **301 $\geq$ length of road** (m) $\leq$ **400**	**Property** (N) $\geq$ **20** and/or **Length of road** (m) $\geq$ **401**

Table 23.7 Assessment of exposed elements at risk in the study area

Number of avalanche path	CBZ (Avalanche path) No	Length of road (Economy)	Number of property (Economy)	Number of persons (Health)	Natural Park, Forested area (Protected area)	Avalanche occurrence
1	1	290 m	0	0	37 hectares	3
2	2,3	0	0	0	29 hectares	4
3	4, 5, 6, 7, 8	0	0	0	52 hectares	4
4	9, 10, 11, 12, 13	315 m	0	0	63 hectares	3
5	14	0	0	0	47 hectares	3
6	15, 16	0	0	0	35 hectares	3
7	17, 18, 19, 20, 21	0	0	0	78 hectares	3
8	22, 23, 24, 25, 26, 27, 28, 29, 30	0	1	0	188 hectares	3
9	31	0	0	0	34 hectares	4
10	38, 39, 40, 41, 42, 43	500 m	0	0	58 hectares	5
11	32, 33, 34, 35, 36, 37, 44, 45	1200 m	25		130 hectares	5
12	46, 47, 48	0	0	0	56 hectares	2
13	49, 50, 51, 52, 53	180 m	0	0	30 hectares	2
14	54	175 m		0	9 hectares	2
15	55, 56	0	0	0	20 hectares	2
16	57, 58, 59	200 m	5	0	56 hectares	1
17	**60, 61, 62, 63**	**200 m**	**6**	**100**	**75 hectares**	**3**
18	64	110 m	0	0	27 hectares	2
19	65, 66, 67	150 m	0	0	73 hectares	2

23.6 Conclusion and Discussion

In avalanche control, the first required data is the snowpack-related records. It has been shown that risk mapping requires accuracy, validity, reliability, and data accuracy. These issues, depend on the overall availability of data as well as on the institutional framework, such as the prevailing legal situations and economic resources. Moreover, a multi-disciplinary approach also requires knowledge on different topics such as for example a statistical background to assess past avalanche events, an understanding of geology and silviculture, and a basic understanding of economics so that the value of elements at risk can be reliably assessed.

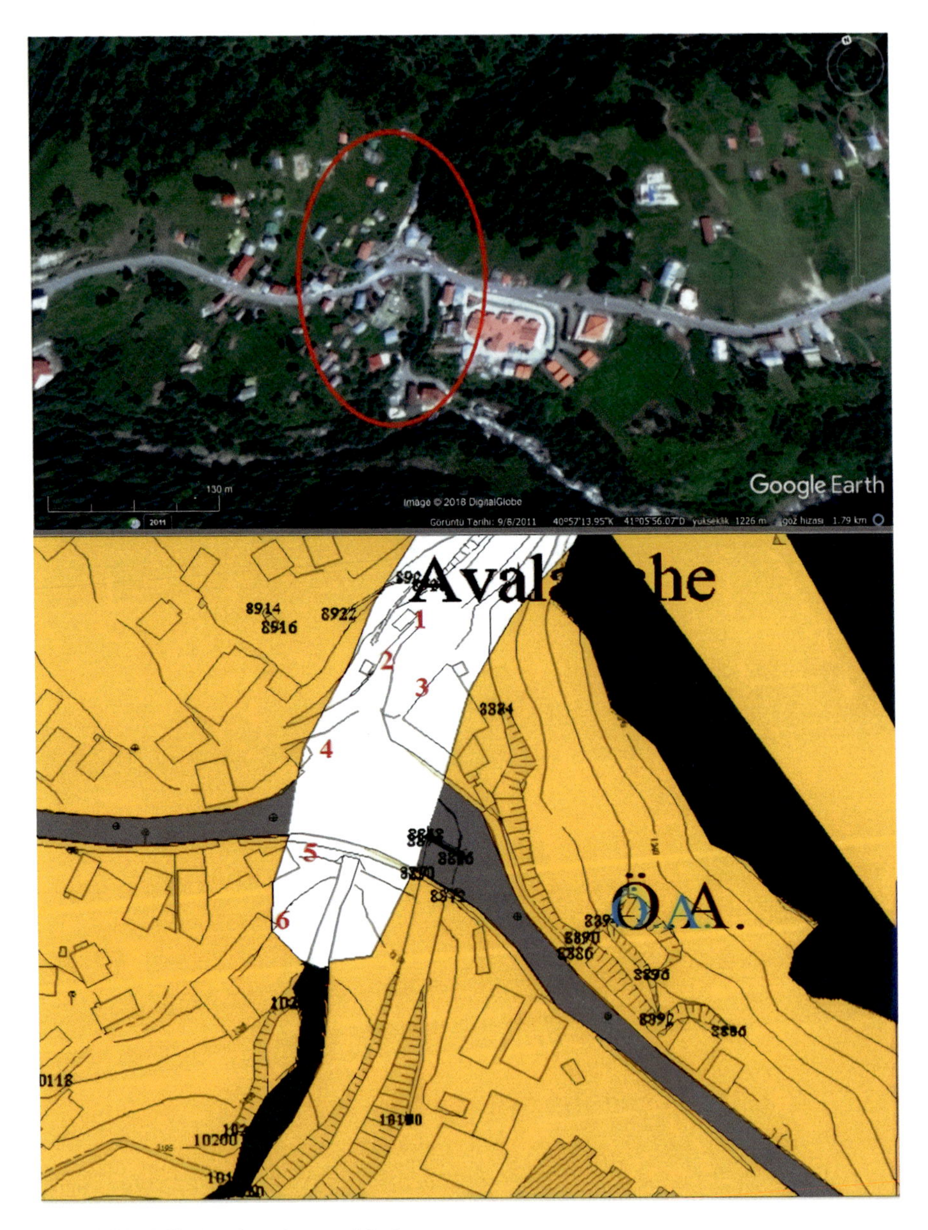

Fig. 23.26 Offical cadastral map of Ayder

The study particularly revealed the lack of snowpack records in possible avalanche starting zones, and associated snowpack characteristics such as density, snow depths etc. Because Turkish meteorological stations were generally established before in or around the cities. Therefore, there is lack of records regarding starting

Fig. 23.27 Elements at risk according to (Çelik & Kurt, 2012)

Table 23.8 Overview on the avalanche tracks with the highest risk scores in the study region

Avalanche Track no	CBZ No (Avalanche track)	Length of road (Economy)	Avalanche occurrence	Being effected level	**Risk Score**
11	32, 33, 34, 35, 36, 37, 44, 45	1200 m	5	5	**25**
Avalanche Track no	CBZ no	Number of building (economy)	Avalanche occurrence	Being effected level	**Risk score**
17	60, 61, 62, 63	6	3	2	**8**
Avalanche Track no	CBZ no	Number of human (health)	Avalanche occurrence	Being effected level	**Risk score**
17	60, 61, 62, 63	100	3	5	**15**
Avalanche Track no	CBZ no	Natural park (environment)	Avalanche occurrence	Being effected level	**Risk score**
11	32, 33, 34, 35, 36, 37, 44, 45	130 hectares	5	4	**20**

zones with respect to snowpack properties such as density, snow depths etc. In recent years, the Turkish government has started to close this gap by establishing a denser monitoring network regarding meteorological parameters, such as snow density, depth and temperature, wind velocity and direction, precipitation. Moreover, there is a strong need to further extend an avalanche incident system containing all types of avalanche events occurring in Turkey. Such a system should be standardized and should include necessary information such as date, location, damage height and loss type, and other relevant details (Kurt, 2017). For example, according to the TABB records, the earliest recorded snow avalanche in Turkey dates on 01.01.1968 in Elazığ, while Varol and Yavaş (2006) reported an earlier snow avalanche incident already in 1890. This clearly shows the overall data inconsistency with respect to historic events.

The second step in avalanche control is preparing the risk maps There are risk maps prepared and a few in the process of being prepared but if one takes into account the number of potential avalanche areas, much remains to be done. In order to achive this aim, Turkey should create its own avalanche guidelines to determine technical standards guide and reference that establish the details for creating and upholding an effective avalanche control. Avalanche hazard mapping may be improved by using various levels to indicate risk levels in terms of impact pressure of the snow mass. On the other hand, According to (Li, 1998), new generation high-resolution satellite images will provide strong geometric capabilities. So, required departments (ÇEM, OGM, KGM, AFAD) should specialize and produce precise maps (large-scale maps) such as those created using airborne laser scanners or terrestrial laser scanners in order to increase accuracy of topography analysis (i.e. slope, aspect analysis, avalanche modelling).

By using qualitative risk mapping approaches in data-sparse mountain regions, preliminary information for decision-makers can be performed. This information can be further refined during future campaigns by e.g. using more quantitative approaches once data gaps are closed. Nevertheless, as shown by this case study from NE Turkey, even a qualitative ranking allows for priority-setting in risk mitigation, and such information can also be useful with respect to necessary cost-benefit analyses of technical mitigation required.

Turkish Government has tried to close these gaps but the speed should be increased. Authorities are now keen to avoid future avalanche damage further to large avalanches over the last two decades. For instance After the Üzengili avalanche that occurred in 1992 in the Bayburt province, authoritiesdecided to evacuated these villages: Üzengili, Yaylapınar, Kavlatan, Harmanözü and Dumlu. These villages (889 persons) were relocated to safe areas (Report, 2011). Training for avalanche control has gained importance and some forest engineers have been sent abroad by faculties of forestry and the Forest Ministry. Some avalanche control projets are being implemented by governmental bodies and the private sector. The forest ministry has been implementing snow nets, and micro piles since the early 2000s in Trabzon, KGM have been constructing snow galleries and snow fences in eastern Turkey and artificial avalanche release is being used in ski resorts in the private sector. In addition to these precautions, some hotels have forbidden skiing in areas at risk. The areas at risk are marked by warning signs. This circumtances shows that there are a number of governmental bodies responsible for avalanche protection (i.e., AFAD, OGM, ÇEM, and KGM), and this multi-department situation could cause uncertainties. There should be a single organization responsible for avalanches to prevent complexity in making risk maps, preparing avalanche control projects, and implementing projects similar to the Austrian model. Aıustria has the Austrian Service for Torrent and Avalanche Control (Die Wildbach und Lawinenverbauung), which is part of the forest department. This office only deals with avalanches and torrents, and arranging. Morever, it has been shown that even if originally relying on a quantitative conceptualisation, risk can be successfully computed in data-scarce regions. In particular in mountain areas of less-developed countries, qualitative information on hazard and risk can be used based on scoring in order to assess snow avalanche risk. As also reported in Dahl et al. (2010), these approaches are useful for a preliminary assessment, and in a further step the establishment of an observation networks and the collection of reliable data on snow heights, terrain characteristics and avalanches conditions can be targeted. This will enhance the efficiency of forecasting and warning services adapted to local or regional conditions, and the education of experts in risk mitigation and management.

Acknowledgements This work received funding from the Scientific Research Projects Unit of Istanbul University Project number: 47628. The present work partly contains some results presented in the doctoral dissertation completed by Tayfun Kurt at the, Istanbul University. The practical study, which lasted a few months, was performed at the Ayder, Rize Province. Special thanks to Priv.-Doz. Dr. Sven Fuchs (BOKU, Vienna, Austria). Finally I might not know where the life's road will take me, but walking with You, Allah, through this journey has given me strength.

References

AFAD. (2020). *Afet kapsamında 2019 yılına bakış ve doğa kaynaklı olay istatistikleri* (A view of 2019 within the scope of disasters and nature-borne event statistics). Disaster and Emergency Management Presidency (In Turkish).

Andretta, M. (2014). Some considerations on the definition of risk based on concepts of systems theory and probability. *Risk Analysis, 34*(7), 1184–1195. https://doi.org/10.1111/risa.12092

Aydın, A., Bühler, Y., Christen, M., & Gürer, I. (2014). Avalanche situation in Turkey and back calculation of selected events. *Natural Hazards and Earth System Sciences, 14*(5), 1145–1154. https://doi.org/10.5194/nhess-14-1145-2014

Bebi, P., Kulakowski, D., & Rixen, C. (2009). Snow avalanche disturbances in forest ecosystems – State of research and implications for management. *Forest Ecology and Management, 257*, 1883–1892. https://doi.org/10.1016/j.foreco.2009.01.050

Bellaire, S., Jamieson, B., & Stetham, G. (2013). *Relating avalanche activity to climate change and coupled ocean-atmospheric phenomena. Davos atmosphere and cryosphere assembly DACA-13, Davos, Switzerland.*

Bicer, C. T. (2016). Mass movement hazard map guidelines for land use at Turkey. In *37th Asian Conference on Remote Sensing, ACRS 2016. Proceedings of a meeting held 17–21 October 2016, Colombo, Sri Lanka* (Vol. 1. Asian Association on Remote Sensing, Pathumthani, pp. 723–731).

Borhan, Y., & Kadioğlu, M. (1998). Synoptic analysis of eastern and southeastern Anatolia avalanches. *Turkish Journal of Engineering and Environmental Sciences, 22*(4), 345–352. E-ISSN: 1303-6157.

Brang, P. (2001). Resistance and elasticity: promising concepts for the management of protection forests in the European Alps. *Forest Ecology and Management, 145*, 107–119. https://doi.org/10.1016/S0378-1127(00)00578-8

Broulidakis, N. (2013). *Thermal analysis of the de Cordova snow house exhibit*. Dissertation, Worcester Polytechnic Institute.

Bründl, M., Bartelt, P., Schweizer, J., Keiler, M., & Glade, T. (2010). Review and future challenges in snow avalanche risk analysis. In I. Alcántara-Ayala & A. Goudie (Eds.), *Geomorphological hazards and disaster prevention* (pp. 49–61). Cambridge University Press. https://doi.org/10.1017/CBO9780511807527.005

Burrows, C. J., & Burrows, V. L. (1976). *Procedures for the study of snow avalanche chronology using growth layers of woody plants*. Boulder: University of Colorado Institute of Arctic and Alpine Research Occasional Paper 23. https://instaar.colorado.edu/uploads/occasional-papers/OP23-Burrows-and-Burrows-1976.pdf

Butler, D. R. (1972). Snow avalanche path terrain and vegetation, glacier National Park. *Montana Arctic and Alpine Research*, 17–32. https://doi.org/10.1080/00040851.1979.12004114

Carrara, P. E. (1979). The determination of snow avalanche frequency through tree-ring analysis and historical records at Ophir, Colorado. *Geological Society of America Bulletin, 90*, 773–780. https://doi.org/10.1130/0016-7606(1979)90<773:TDOSAF>2.0.CO;2

Casteller, A., Stockli, V., Villalba, R., & Mayer, A. C. (2007). An evaluation of dendroecological indicators of snow avalanches in the Swiss Alps. *Arctic, Antarctic, and Alpine Research, 39*, 218–228. https://doi.org/10.1657/1523-0430(2007)39[218:AEODIO]2.0.CO;2

Çelik, H. E., & Kurt, T. (2012). *Rize Çamlıhemşin Ayder Çığ Kontrolu Uygulama Projesi, ÇEM 13 03 53 01*. Ankara: Çölleşme ve Erzonyonca Mücadele Genel Müdürlüğü.

Cooperstein, M. S., Birkeland, K. W., & Hansen, K. J. (2004). The effects of slope aspect on the formation of surface hoar and diurnally recrystalized near-surface faceted crystals: Implications for avalanche forecasting. In *Proceedings of the 2004 international snow science workshop* (pp. 83–93). https://arc.lib.montana.edu/snow-science/objects/issw-2004-083-093.pdf. Accessed on 04 Jan 2021.

Daffern, T. (2009). *Backcountry avalanche safety*. Rocky Mountain Books. ISBN 978- 1-897522-54-7.

Dahl, M.-P., Mortensen, L., Veihe, A., & Jensen, N. (2010). A simple qualitative approach for mapping regional landslide susceptibility in the Faroe Islands. *Natural Hazards and Earth System Sciences, 10*(2), 159–170.

Ding, M., Heiser, M., Hübl, J., & Fuchs, S. (2016). Regional vulnerability assessment for debris flows in China – A CWS approach. *Landslides, 13*(3), 537–550. https://doi.org/10.1007/s10346-015-0578-1

Dubayah, R. C. (1994). Modeling a solar radiation topoclimatology for the Rio Grande River Basin. *Journal of Vegetation Science, 5*(5), 627–640. https://doi.org/10.2307/3235879

Elibüyük, M., & Yılmaz, E. (2010). Altitude steps and slope groups of Turkey in comparison with geographical regions and subregions. *Journal of Geographical Sciences, 8*, 27–55. https://doi.org/10.1501/Cogbil_0000000104

Erener, A., & Düzgün, H. S. (2013). A regional scale quantitative risk assessment for landslides: Case of Kumluca watershed in Bartin. *Turkey Landslides, 10*(1), 55–73. https://doi.org/10.1007/s10346-012-0317-9

Fell, R., Corominas, J., Bonnard, C., Cascini, L., Leroi, E., & Savage, W. (2008). Guidelines for landslide susceptibility, hazard and risk zoning for land-use planning. *Engineering Geology, 102*(3–4), 85–98. https://doi.org/10.1016/j.enggeo.2008.03.022

Fraefel, M., Schmid, F., Frick, E., & Hegg, C. (2004). 31 Jahre Unwettererfassung in der Schweiz. In M. Mikoš & D. Gutknecht, (Eds.), *Internationales symposion interpraevent, Riva del Garda, May 24–27, 2004* (pp. I/45–56). https://www.dora.lib4ri.ch/wsl/islandora/object/wsl:17874. Accessed 23 Jan 2021.

Fuchs, S., Keiler, M., Sokratov, S. A., & Shnyparkov, A. (2013). Spatiotemporal dynamics: The need for an innovative approach in mountain hazard risk management. *Natural Hazards, 68*(3), 1217–1241. https://doi.org/10.1007/s11069-012-0508-7

Fuchs, S., Keiler, M., & Sokratov, S. (2015). Snow and avalanches. In C. Huggel, M. Carey, J. J. Clague, & A. Kääb (Eds.), *The high-mountain cryosphere: Environmental changes and human risks* (pp. 50–70). Cambridge University Press. https://doi.org/10.1017/CBO9781107588653.004

Fuchs, S., Shnyparkov, A., Jomelli, V., Kazakov, N., & Sokratov, S. (2017). Editorial to the special issue on natural hazards and risk research in Russia. *Natural Hazards: Journal of the International Society for the Prevention and Mitigation of Natural Hazards, 1*, 1–16. https://doi.org/10.1007/s11069-017-2976-2

Gaume, J., Schweizer, J., Herwijnen, A. V., Chambon, G., Eckert, N., & Naaim, M. (2013). The effect of spatial variations of snowpack properties on snow slope stability: A mechanically-based statistical approach. In *International Snow Science Workshop Grenoble – Chamonix Mont-Blanc* (pp. 57–60). https://hal.archives-ouvertes.fr/hal-00951325

Grimsdottir, H., & McClung, D. (2006). Avalanche risk during backcountry skiing-an analysis of risk factors. *Natural Hazards, 39*, 127–153. https://doi.org/10.1007/s11069-005-5227-x

Gruber, F. E., & Mergili, M. (2013). Regional-scale analysis of high-mountain multi-hazard and risk indicators in the Pamir (Tajikistan) with GRASS GIS. *Natural Hazards and Earth System Sciences, 13*(11), 2779–2796. https://doi.org/10.5194/nhess-13-2779-2013

Gürer, İ. (1998). International cooperation for solving the avalanche problem in Turkey. *Natural Hazards, 18*(1), 77–85. https://doi.org/10.1023/A:1008013710228

Gürer, İ.(2002). *Türkiye'de Yerleşim Yerlerine Yönelik Kar ve Çığ Problemleri. THM Türkiye Mühendislik Haberleri Sayı 420-421-422/2002/4-5-6.* http://www.imo.org.tr/yayinlar/dergi_goster.php?kodu=173&dergi=13. Accesed 05 Jan 2021.

Haeberli, W., Whiteman, C., & Shroder, F. J. (2015). *Snow and ice-related hazards, risks, and disasters. A general framework, snow and ice-related hazards, risks and disasters* (pp. 1–34). Elsevier, , ISBN: 978-0-12-394849-6 (Chaper 9).

Hageli, P., & McClung, D. M. (2007). Expanding the snow-climate classification with avalanche-relevant information: Initial description of avalanche winter regimes for southwestern Canada. *Journal of Glaciology., 53*(181). https://doi.org/10.3189/172756507782202801

Hendrix, J., Owens, I., Carran, W., & Carran, A. (2005). Avalanche activity in an extreme maritime climate: The application of classification trees for forecasting. *Cold Regions Science and Technology, 43*(1–2), 104–116. https://doi.org/10.1016/j.coldregions.2005.05.006

Höller, P. (2007). Avalanche hazards and mitigation in Austria: A review. *Natural Hazards, 43*(1), 81–101. https://doi.org/10.1007/s11069-007-9109-2

Holub, M., & Fuchs, S. (2009). Mitigating mountain hazards in Austria – Legislation, risk transfer, and awareness building. *Natural Hazards and Earth System Sciences, 9*(2), 523–537. https://nhess.copernicus.org/articles/9/523/2009/.

Jamieson, B., & Johnston, C. D. (2001). Evaluation of the shear frame test for weak snowpack layers. *Annals of Glaciology, 32*(1), 59–69. https://doi.org/10.3189/172756401781819472

Jamieson, B., Geldsetzer, T., & Stethem, C. (2003). Forecasting for deep slab avalanches. *Cold Regions Science and Technology, 33*(2), 275–290. https://doi.org/10.1016/S0165-232X(01)00056-8

Karsli, F., Atasoy, M., Reis, S., Demir, O., & Gökçeoğlu, C. (2009). Effects of land-use changes on landslides in a landslide-prone area (Ardesen, Rize, NE Turkey). *Environmental Monitoring and Assessment, 156*(1–4), 241. https://doi.org/10.1007/s10661-008-0481-5

Keiler, M., & Fuchs, S. (2016). Vulnerability and exposure to geomorphic hazards – Some insights from mountain regions. In M. Meadows & J.-C. Lin (Eds.), *Geomorphology and society* (pp. 165–180). Springer.

Kienholz, H., Krummenacher, B., Kipfer, A., & Perret, S. (2004). Aspects of integral risk management in practice – Considerations with respect to mountain hazards in Switzerland. *Österreichische Wasser- und Abfallwirtschaft, 56*(3–4), 43–50.

Klein, J. A., Tucker, C. M., Steger, C. E., Nolin, A., Reid, R., Hopping, K. A., Yeh, E. T., Pradhan, M. S., Taber, A., Molden, D., Ghate, R., Choudhury, D., Alcántara-Ayala, I., Lavorel, S., Müller, B., Grêt-Regamey, A., Boone, R. B., Bourgeron, P., Castellanos, E., ... Yager, K. (2019). An integrated community and ecosystem-based approach to disaster risk reduction in mountain systems. *Environmental Science & Policy, 94*, 143–152. https://doi.org/10.1016/j.envsci.2018.12.034

Kurt, T. (2014). Assessment of the Perchertal avalanche in Tyrol. *Austria Turkish Journal of Earth Sciences, 23*(3), 339–349. https://doi.org/10.3906/sag-1303-128

Kurt, T. (2017). *Imprrovement of technical capacity in avalanche risk management in Turkey.* Dissertation, İstanbul University, Institute of Graduate Studies in Science and Engineering, Departmant of Forest Engineering.

Laternser, M., & Schneebeli, M. (2002). Temporal trend and spatial distribution of avalanche activity during the last 50 years in Switzerland. *Natural Hazards, 27*(3), 201–230. https://doi.org/10.1023/A:1020327312719

Li, R. (1998). Potential of high-resolution satellite imagery formational mapping products. *Photogrammetric Engineering and Remote Sensing, 64*, 1165–1170.

Margreth, S., Leuenberger, F., Lundström, T., Auer, M., & Meister, R. (2007). *Defence structures in avalanche starting zones, technical guideline as an aid to enforcement. Environment in Practice no. 0704, Federal Office for the Environment, Bern; WSL Swiss Federal Institute for Snow and Avalanche Research SLF, Davos*. 134 pp. https://www.wsl.ch/fileadmin/user_upload/SLF/Permafrost/Bauen_im_Permafrost/Lawinenverbau_im_Anbruchgebiet_E.pdf

McClung, D. M., & Schaerer, P. A. (2006). *The avalanche handbook* (3rd ed.). The Mountaineers Books. ISBN-10: 9780898868098.

Mears, A. I. (1992). Snow avalanche hazard analysis for land-use planning and engineering. *Colorado Geological Survey Bulletin, 49*, 55.

Meloysund, V., Lisa, K. R., Hygen, H. O., Hoiseth, K. V., & Leira, B. (2007). Effects of wind exposure on roof snow loads. *Building and environment, 42*(10), 3726–3736. https://doi.org/10.1155/2018/7018325

Miklau, F. R., & Sauermoser, S. (2011). *Handbuch Technischer Lawinen shutz, Ernst und Sohn Gmbh, Berlin, 2011*. ISBN: 978-3-433-02947-3.

Pfister, C. (1998). *Raum-zeitliche Rekonstruktion von Witterungsanomalien und Naturkatastrophen 1496–1995*. Schlußbericht NFP31 edn. vdf Hochschulverlag an der ETH, Zürich.

Powell, J. H., Mustafee, N., Chen, A. S., & Hammond, M. (2016). System-focused risk identification and assessment for disaster preparedness: Dynamic threat analysis. *European Journal of Operational Research, 254*(2), 550–564. https://doi.org/10.1016/j.ejor.2016.04.037

Remondo, J., Bonachea, J., & Cendrero, A. (2005). A statistical approach to landslide risk modeling at basin scale; from landslide susceptibility to quantitative risk assessment. *Landslides, 2*, 321–328. https://doi.org/10.1007/s10346-005-0016-x

Report (2011) *İl ve çevre durum raporu, Bayburt Valiliği Bölge ve Şehircilik İl Müdürlüğü* (City and the environment report, Bayburt governor's office, environment and urban planning directorate of cities) Environment and urban planning ministry, 237 pp, 2011. (In Turkish).

Rogelis, M. C., Werner, M., Obregón, N., & Wright, N. (2016). Regional prioritization of flood risk in mountainous areas. *Natural Hazards and Earth System Sciences, 16*(3), 833–853. https://doi.org/10.5194/nhess-16-833-2016

Schweizer, J., Jamieson, J. B., & Schneebeli, M. (2003). Snow avlanche formation. *Reviews of Geophysics, 41*, 4/1016. https://doi.org/10.1029/2002RG000123

Shnyparkov, A. L., Fuchs, S., Sokratov, S. A., Koltermann, K. P., Seliverstov, Y. G., & Vikulina, M. A. (2012). Theory and practice of individual snow avalanche risk assessment in the Russian arctic. *Geography, Environment, Sustainability, 5*(3), 64–68. https://doi.org/10.24057/2071-9388-2012-5-3-64-81

Simonson, S. E., Greene, E. M., Fassnacht, S. R., Stohlgren, T. J., & Landry, C. C. (2010). Practical methods for using vegetation patterns to estimate avalanche frequency and magnitude. In *International Snow Science Workshop Proceedings, Lake Tahoe, CA, USA* (pp. 548–555). https://arc.lib.montana.edu/snow-science/objects/ISSW_P-045.pdf

Sullivan, R., Thomas, P., Veverka, J., Malin, M., & Edgett, K. S. (2001). Mass movement slope streaks imaged by the Mars Orbiter Camera. *Journal of Geophysical Research: Planets (1991–2012), 106*(E10), 23607–23633. https://doi.org/10.1029/2000JE001296

Taştekin, A. T. (2003). *Meteoroloji ve çığ*. Ankara: DMİ Genel Müdürlüğü. https://www.mgm.gov.tr/FILES/genel/makale/meteorolojivecig.pdf. Accessed 23 Jan 2021.

Tunçel, H. (1990). Avalanches as natural hazard and avalanches in Turkey. *Atatürk Kültür, Dil Ve Tarih Yüksek Kurumu, Coğrafya Bilimleri ve Uygulama Kolu Coğrafya Araştırmaları Dergisi, 1*(2), 71–98. http://tucaum.ankara.edu.tr/wp-content/uploads/sites/280/2015/08/cadata2_3.pdf

UN. (1989). *International decade for natural disaster reduction: Resolution/adopted by General Assembly*. A-RES-44-236. https://digitallibrary.un.org/record/82536. Accessed 23 Jan 2021.

UN/ISDR. (2015). *Sendai framework for disaster risk reduction 2015–2030*. United Nations Office for Disaster Risk Reduction. https://www.unisdr.org/files/43291_sendaiframeworkfordrren.pdf. Accessed 23 Jan 2021.

UNISDR. (2007). *Hyogo framework for action 2005–2015: Building the resilience of nations and communities to disasters (A/CONF.206/6)*. United Nations Office for disaster risk reduction. https://www.unisdr.org/files/1037_hyogoframeworkforactionenglish.pdf. Accessed 23 Jan 2021.

Varnes, D. (1984). *Landslide hazard zonation: A review of principles and practice* (Vol. 3, Natural Hazards). Paris: UNESCO. ISBN: 92-3-101895-7.

Varol, N. Ö., & Şahin, D. (2006). *Türkiye'de Çığ Afeti Zararlarını Azaltma Çalışmaları TMMOB Afet Sempozyumup* (pp. 395–404). https://www.imo.org.tr/resimler/ekutuphane/pdf/3921.pdf

Varol, N. Ö., & Yavaş, Ö. M. (2006). *Türkiye'de Çığ Olaylarının Değerlendirilmesi. 59. Türkiye Jeoloji Kurultayı* p. 359). https://www.jmo.org.tr/resimler/ekler/3f61f3a8034cbfb_ek.pdf

Yuksel, I., Arman, H., Goktepe, F., & Ceribasi, G. (2011). Flood management to prevent flooding damages in western Black Sea region in Turkey. *International Journal of Physical Sciences, 6*(29), 6759–6766. https://doi.org/10.5897/IJPS11.1183

Chapter 24
Space Techniques to Recognize Seismological and Geomorphological Features in Libya

Somaia Suwihli

Abstract Seismogenic geomorphology features are the surface evidence of tectonic deformation and earthquake zones, and they can be used to identify and refine the location and characteristics of seismic sources. This is, in particular, useful in the early stages of earthquake hazard studies in tectonic environments where multidisciplinary data may not be available. With the appearance of operational remote sensing, it becomes possible to efficiently and accurately map seismic activities induced ground changes and reactivated landslides. Also, it is significant to notice that remote sensing provides unbiased recording of the events. In this study, remotely-sensing imagery provider called DigitalGlobe were used to identify seismogenic geomorphologic features in different locations in Libya. Two different scaled grids were used on a Libya map to help classify and compilation of imagery and sites at closer inspection and lower elevation in tandem with latitude-longitude locations. Remotely sensed data from DigitalGlobe, one of the largest private satellite data imagery providers in the world, has the potentiality in identification of localized seismogenic geomorphologic features, which include graben, faultline scarps, displacement of offset, en-echelon faulting, strike-slip fault, offset stream channels and wadis. When seismic activity is often sporadic and/or long-term like in Libya, the identification of surface features caused by or related to seismic activities represents a vital surrogate for the recognition and documentation of earthquake magnitude and frequency.

Keywords Remotely-sensed images · Earthquakes · Geomorphologic features · Libya

S. Suwihli (✉)
Department of Geography, University of Benghazi, Benghazi, Libya

Department of Geosciences, University of Arkansas, Fayetteville, AR, USA

M. M. Al Saud (ed.), *Applications of Space Techniques on the Natural Hazards in the MENA Region*, https://doi.org/10.1007/978-3-030-88874-9_24

24.1 Introduction

Libya is located in the Northeast African Tectonic Zone where it is bordered to the East by the Gulf of Aqaba and Red Sea. The plate tectonic models suggest that the African plate is moving northward relative to the Eurasian plate at a rate of order of 6 mm per year (Salamon et al., 2003; Reilinger et al., 2006). Corti et al. (2006) suggested that the Red Sea Rift is transferring compressing to western Egypt and Libya, generating significant earthquakes along the southeastern Mediterranean where earthquakes have effects on nature, human life, and man-made structures (Ambraseys et al., 1995; Mourabit et al., 2014).

Since ancient times, Libya has experienced earthquakes in varying degrees. However, big seismic events that took place and caused comprehensive destruction are relatively few and limited in Libya. Historical and archaeological studies showed that the Cyrenaica region in north-eastern Libya, experienced destructive earthquakes in 262 A.D. and 365 A.D., and they caused the complete destruction of the city of Shahhat killing most of its residents. (Goodchild, 1968). Sabratha, located about 70 kilometers west of Tripoli on the Mediterranean coast, was hit by tremendous, destructive earthquakes in 306 and 310 A.D., and again in 365 A.D. (di Vita-Evrard, 1999). Several major earthquakes occurred before the 1900s in Libya where many towns and villages were destroyed. They were in Sabha Territory, the adjacent Fezzan Province (704), Tripoli (1183, 1656, 1685) (Ambraseys, 1984), near the Libyan-Egyptian border (1811) (Kebeasy, 1980), and in Murzuq (1853, 1860) (Ambraseys et al., 1995). But the 1935 Hun Graben earthquake 7.1 (MI) was described as the strongest earthquake in Libya, and it was one of the largest earthquakes in Africa's history (Johnston, 1989; Suleiman & Doser, 1995). The city of Al-Marj in the region of Al-Jabal Al-Akhdar on the coast of northeastern Libya was razed by the 1963 earthquake of moderate magnitude (5.6 Ml) (Gordon & Engdahl, 1963; Minami, 1965). The coastal zones of northwestern and northeastern Libya continue to be seismically active where several tremors of 4 to 5.7 magnitude were recorded between 1990 and 2020 (USGS, 2020), see Fig. 24.1.

24.2 Remote Sensing of Natural Disasters

In different research disciplines, a lot of quality and valuable information about biophysical characteristics and human activities has been used with the development in remote sensing science. In the event of a natural disaster, more advanced remote sensing methodologies energetically and quickly were developed. This has linked remote sensing with spatial measurements and accessory data for more accurate mapping, faster analysis, and more effective forecasting. Moreover, remote sensing represents unique data and valuable information to observe changing conditions on Earth at various spatial and temporal scales (Chen et al., 2012). By offering frequent, accurate, and almost instantaneous data, remote sensing provides needed

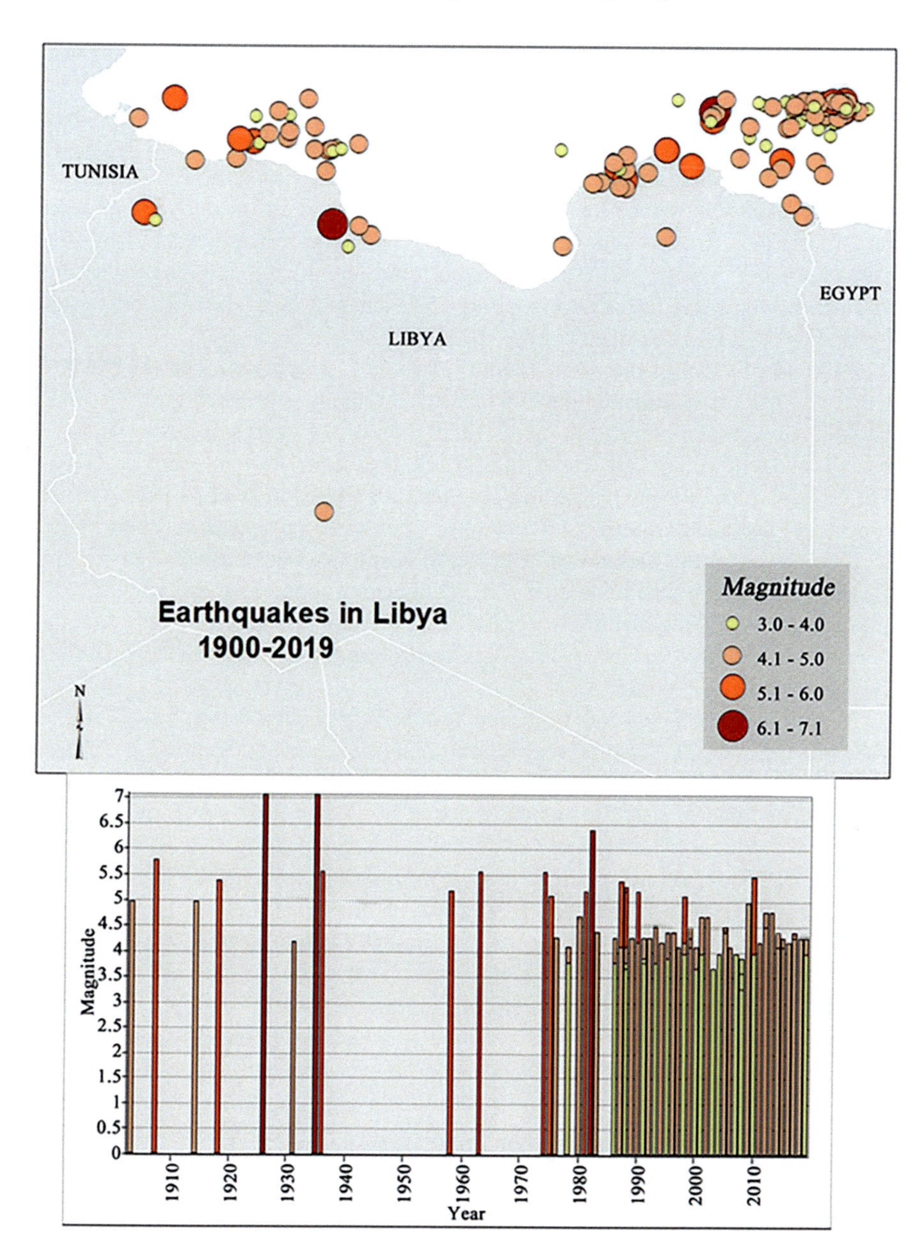

Fig. 24.1 Earthquakes affecting Libya from 1900 to 2019 (M ≥ 4.0). This map and graph illustrate the distribution of the earthquakes that have occurred in or near Libya (20°–34°N, 10°–25°E) with color-coded events identified chronologically and spatially. Note the recent swarms (low magnitude 3–4) between 1990 and 2019 which are more frequent than previous events (Suwihli & Paradise, 2020)

reconnaissance as well as quantitative and sustained measurements. Under challenging disasters, such as the large area extent of an impacted area, the potential shortage of on-the-ground access because of damaged infrastructure and remoteness, remote sensing might be the only way to see what is happening on the ground. While it is not possible to prevent natural disasters from happening, it is possible to manage a potential disaster to reduce losses through a four-part process of mitigation, preparation, response, and recovery (Lewis, 2009). With the further technical evolution of remote sensing (Chen et al., 2012), its applications are increasingly used in natural disaster management, from risk modelling and vulnerability analysis to early warning and to damage assessment (Lewis, 2009).

From a landslide-inventory-based and GIS-based framework, Chau et al. (2004) employed historical landslide data of the area, coupling with geological, geomorphological, climatic, population, and rainfall data to analyzed systematic landslide hazard in Hong Kong. By using high-resolution digital terrain models (DTM), Tarolli et al. (2012) extracted geomorphic features related to shallow mass-wasting processes (landslide crowns, bank erosion). Their study area was in a part of Rio Cordon Basin, in the Dolomites, a mountainous area in the Eastern Italian Alps. Based on Digital Elevation Model (DEM) matching algorithm, Rui et al. (2017) presented a landslide detection method in variable mountainous regions. The researchers resulted that large-scale landslides can be detected based on robust DEM matching.

Remotely sensed data has been also extensively applied to volcanoes‘studies (Hooper et al., 2012). Various observations included sill and dike intrusion (Jónsson et al., 1999; Fukushima et al., 2005; Pedersen & Sigmundsson, 2006), inflation and deflation of inferred magma chambers (Lu et al., 2000; Fialko & Simons, 2001; Pritchard & Simons, 2002; Wicks et al., 2006), eruption (Froger et al., 2004; Yun et al., 2007), and faulting (Amelung et al., 2007). Solano-Rojas et al. (2017) used the new generation of X-band SAR satellites – Interferometric SAR (InSAR) – to monitor deformation in the Popocatepetl Volcano in Central Mexico. They presented deformation results from 2012 to 2016. Effusion rate and estimates of lava volume are critical for assessing volcanic hazard, and the choice of appropriate methods depends on the accessibility to the volcanic area to make field measurements, and how often they can be repeated. Because Volcán El Reventador (Ecuador) is inaccessible except few locations, in addition to planimetric field Naranjo et al. (2016) used topographic satellite radar-based measurements of lava flow thicknesses and volumes for activity between 2002 and 2009 at El Reventador. These measurements of lava thickness by a small set of Synthetic Aperture Radar (SAR) interferograms allowed the retrieval of the shape of the compound lava flow field and show that it was subsiding by up to 6 cm/year in 2009 (Naranjo et al., 2016).

24.3 Remote Sensing in Seismology Research

Remote sensing data has been applies in seismology research very early. The first use was related to structural geology and geomorphology. Clark (1978) used aerial photographs (high-altitude, oblique, taken by the U.S. Air Force at 60,000 feet) to find active faults. Clark studied the Garlock Fault in the Mojave Desert of South California. In some parts of south the USSR, Trifonov (1984) examined three aspects of aerial and space images application in neotectonics: survey, study, and mapping, study of the deep-seated structures, in addition to seismic risk studies of neotectonically active zones. Gupta et al. (1994) analyzed a pre- and post-earthquake LISS-II sensor data set of Uttarkashi area to detect the surface effects of the Uttarkashi 20 October 1991 Earthquake, and he observed that there were areas on the post-earthquake image were higher while others were lower than on the per-earthquake image. Also, he found that there was an evident change in discharge of the Ganga River after the earthquake. However, because of the spatial resolution of data was not high (36.25 m) at that time, landslides caused by earthquake were not clearly identified. Whereas, later, the IRS-1C-PAN sensor with a high resolution (5.8 m) provided a special opportunity to map surface changes and landslides induced by earthquake. Saraf (1998) used IRS-PAN pre- and post-earthquake remote sensing data sets of two different dates to survey the Jabalpur earthquake (22 May 1997) induced changes. Pre-earthquake IRS-LISS-III digital data then was used to study the structural framework of the study area. Saraf (2000) studied the change deformations after the Chamoli 29 March 1999 Earthquake ($M = 6.3$) hit Himalaya, India. He mapped landslides induced by this earthquake and compared the surface deformations by using IRS-1C-PAN pre- and post-earthquake images. A pseudo color transform (PCT) approach for the assessment of earthquake damage was successfully evaluated on IRS-PAN per- and post- earthquake data set to accurately identify the landslides and also to assess quantitative damage.

In addition, the interference patterns which are formed by differencing two synthetic aperture radar (SAR) images enable the researchers to record different geophysical phenomena. The fringes made by the topographic relief can be removed by using a digital elevation model (DEM). By comparing couples of images for different periods of time, it has been possible to identify the geophysical signal and interferometric artifact. Massonnet et al. (Massonnet et al., 1993, 1994; Massonnet & Feigl, 1995) applied pair-wise logic around the 1992 Landers, California earthquake region using SAR images acquired by the ERS-1 satellite, and that by comparing pairs of images for different periods of time. The map of the coseismic displacement field which was generated by this technique brought new insights into the nature of deformation caused by earthquakes (Zebker et al., 1994; Peltzer et al., 1994). Peltzer et al. (1994) analyze the coseismic surface displacement field in the vicinity of the fault trace. He used the interferometric map created by Massonnet et al. (1993). Price and Sandwell (1998) studied the same earthquake area, but the used different method which brought out short-wavelen features to reveal previously unrecognized strain patterns.

A new method was used to automatically determine the earthquake source process and near-source strong ground motions. Dreger and Kaverina (2000) used regional and local distance data to study the October 16, 1999 Hector Mine earthquake (MW7.1) in California. Broadband displacement data were inverted to resolve the fault plane ambiguity independent of aftershock location and surface faulting information and for the seismic moment tensor.

Satellite remote sensing systems provide spatially continuous information to observe the tectonic field, and that can assist the understanding of specific fault systems. Combining ground network data with remote sensing data provides a better understanding of displacements and slip models validation that is used in a regional position of tectonic strain. Cakir et al. (2003) used combined tectonic landscape observations and Synthetic Aperture Radar (SAR) data to determine an improved model of the displacement associated with the 1999 Izmit Turkey Earthquake, which ruptured the North Anatolian Fault at the eastern end of the Marmara Sea. Talebian et al. (2004) used ENVISAT radar data to map surface displacements due to the 2003 Bam, Iran Earthquake (Mw 6.5). They detected that >2 m of displacement occurred at depth on a blind strike-slip fault which had not previously been identified.

Fu et al. (2004) uncovered the geomorphology and geometry of an active fault associated with 2003 Bam Earthquake in SE Iran near the towns of Bam and Baravat. They mapped the active fault that ruptured during the earthquake by using three-dimensional (3D) pre- and post-event images created from Advanced Spaceborne Thermal Emission and Reflection Radiometer (ASTER) visible and near infrared (VNIR) data.

By satellite remote sensing observations, it has been possible to have insights into how much energy is released by earthquakes and other deformation modes and how stress is transferred between fault systems from depth to the surface. Argus et al. (2005) used synthetic aperture radar (SAR) interferometry and global positioning system (GPS) geodesy to study the relationship between interseismic strain accumulation and horizontal anthropogenic motions in metropolitan Los Angeles.

Earthquake prediction has been an intense scientific research field. The scientists consider a number of geophysical and geochemical phenomena as an earthquake precursor. The earthquake space research supposes that such disastrous natural phenomenon relates to: surface temperature increase, gas and aerosol exhalation, electromagnetic disorder in the ionosphere, and Earth's deformation (Tronin, 2006).

The employment of satellite thermal Infrared measures has been used as an indicator of seismic activity. Ouzounov and Freund (2004) studied solid Earth–atmosphere interactions and possibly solid Earth–seafloor interactions prior to the occurring of strong earthquakes. They analyzed land surface temperature (LST) and sea surface temperature (SST) data, and infrared (IR) emissivity. An evidence for correlations between solid Earth processes and atmosphere/ocean dynamics prior to major earthquakes was found, particularly for a thermal anomaly pattern which is apparently related to pre-seismic activity. Many attempts to determine the thermal anomalies and detect the pre-earthquake anomalous thermal infrared signal were made (Tronin et al., 2002, 2004; Tramutoli et al., 2005, 2015; Bellaoui et al., 2017).

In addition, Gorny et al. (2020) analyzed satellite thermal images of the Earth's surface within the spectral range of 10.5–11.3 mkm over some linear structures of the Middle-Asian seismically active region. In certain individual zones of some major tectonic dislocations there, A retrospective analysis of a continuous series of observations of the outgoing IR radiation flux has shown that appear from time to time positive anomalies of IR radiation, for example at the point of intersection of the Talasso-Ferghana and Tamdy-Tokrauss faults. The time of the appearance of these positive anomalies coincides with the activation of faults over which there has been detected an increase of the outgoing IR radiation flux. In the Tien Shan, in 1984 the majority of crustal earthquakes were accompanied by the appearance of a positive anomaly of the IR radiation at the point of the intersection of the faults.

24.4 Tectonic/Seismic Geomorphology of Libya

The major regional structures of Libya are represented by basins (the Ghadamis, Sirte, Murzuq (Murzuk), and Al Kufrah Basins) and uplifts (Nafusah, Cyrenaican, and Haruj Uplifts). Several large basalt flows and flood basalts of Cenozoic age exist, and they represent some of the most visible North African landform features from space (Figs. 24.2a and 24.2b). Almost one third of the country is covered by sand, gravel, and rubble which are moving and blanketing the surface from eolian, fluvial, and downslope movement. Over time, different parts of the country have been faulted, tilted, or warped into arches or basins, or covered by outpourings of lava. Two groups of faults dominate the middle section of Libya. The fault group in the north influences the shape of the Gulf of Sirte. Near the two fault groups' intersection, the extensive flood basalt fields are located. These two fault trends are roughly parallel with the great rift system in the Gulf of Suez and East African zones (Conant & Goudarzi, 1967). In Libya, recent fault and deformation regions are a part of the active faults of North Africa which are concentrated generally on the northwest side of Africa, and that belongs to the Alpine–Himalayan belt, and mainly in the Atlas Orogenic System (Skobelev et al., 2004).

Campbell (1968) explained that there is a close correlation between the seismic activities and the existing structure and tectonic in Libya. The Hun Graben, a prominent Rift Valley, extends north from Hun in central of the country to Qaddahia in the northwest; the faults trend NNW-SSE to NW-SE. The faults' margins of this graben are marked by scarps and substantial fault displacement, giving them visible extents, and they could be recognizable at the surface (Abdunaser, 2015). Throughout the history, the Hun Graben has been one of the seismically most active areas in the country (Suleiman & Doser, 1995); in 1935, a series of strike-slip earthquakes occurred along the eastern margin graben fault (Abdunaser & McCaffrey, 2015). The first tremor (and aftershocks) of the 1963 Earthquake of Al-Marj, strongly suggested that the earthquake was a result of fault movement (Campbell, 1968).

Fig. 24.2a Satellite image of Haruj as seen from space (Google Earth, 2015b)

Fig. 24.2b Volcanoes in the central of Haruj. Photo from Norbert Brügge (2015)

24.5 Recognizing Seismological and Geomorphological Features in Libya

In the last few years, the development and fast expansion of International Computer Networks have provided an impetus to disseminate and manage information and data, and with this integration, remote sensing has become real time and interactive tool. With the appearance of operational remote sensing, it becomes possible to map earthquake induced ground changes and reactivated landslides efficiently and accurately (Saraf, 2000).

In this chapter, satellite images from Google Earth were used to identify seismogenic geomorphology features in Libya. The satellite data were produced from the imagery provider called DigitalGlobe and mosaiced in Google Earth. DigitalGlobe is the global leader in geospatial information, satellite imagery, and location-based intelligence. The company has a big collection of extensive very high resolution (VHR) satellite sensors in its orbiting instrumentation which includes (a) GeoEye-1 has color 1.64 m multi-spectral, and 41 cm black and white, (b) WorldView-1 with 50 cm black and white, (c) WorldView-2 with color 1.84 m multi-spectral, and 46 cm black and white, (d) WorldView 4–3 with color 1.24 m multi-spectral spatial resolution, 31 cm black and white, and 0.31 m spatial resolution, (e) QuickBird which has 0.65 m spatial resolution and (f) CNES' Pleiades-1A data which has 0.5 m spatial resolution (Sivakumar, 2017).

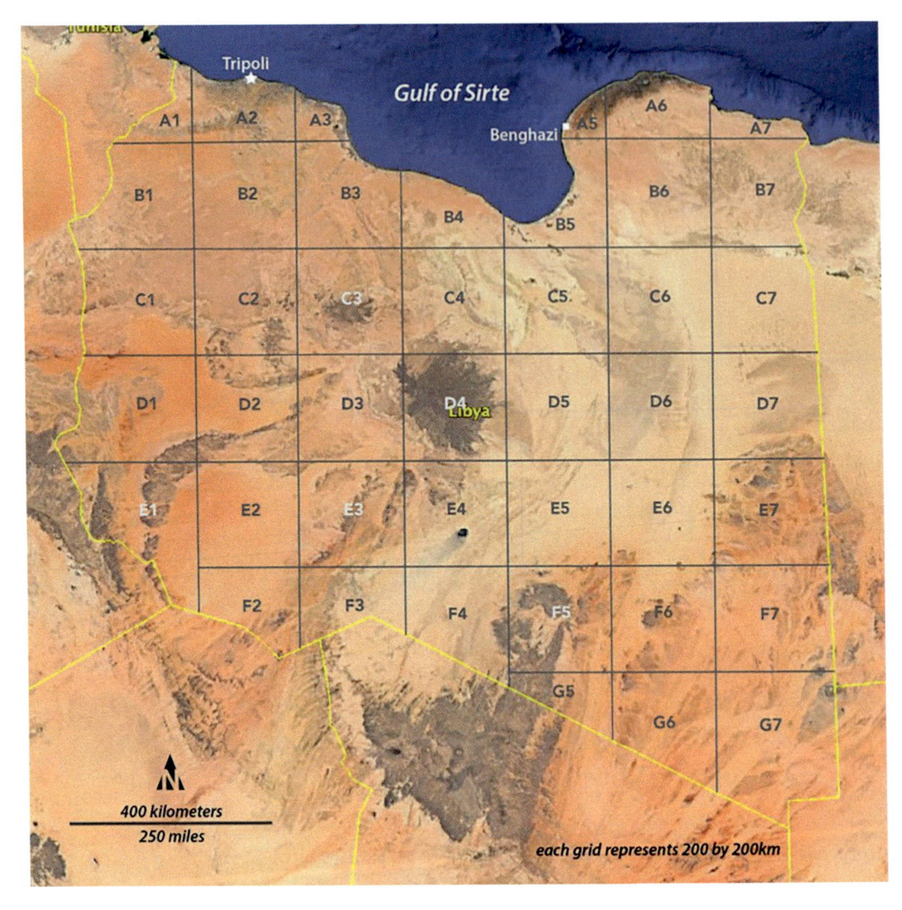

Fig. 24.3 An overlain grid on a satellite image where each grid represents 200 by 200 km (small scale). The grid is used to help classify and archive image classification and compilation in tandem with latitude-longitude locations (Google Earth, 2015b)

The Google Earth data provide accurate and mission-critical information about changes that occur on earth features. Such data support a wide variety of uses which include mapping and analysis, mission-planning and management, environmental monitoring, and near real time natural hazard monitoring. Hence, by using this vast very high resolution we can understand our changing planet in order to save lives, resources, and time. In this chapter, Google Earth represents an ideal platform for the location and analysis of surface geomorphology features related to seismic activities.

Two grids of different scales were used on a Libya map. First one was a broad grid where each grid represents 200 by 200 km (small scale) (Fig. 24.3), and that to help classify and archive image classification and compilation in tandem with latitude-longitude locations. Second grid was tighter where each grid represents 20 by 20 km (larger scale) (Fig. 24.4) to make the classification and compilation of imagery and sites at closer inspection and lower elevation (nadir) more facilitate.

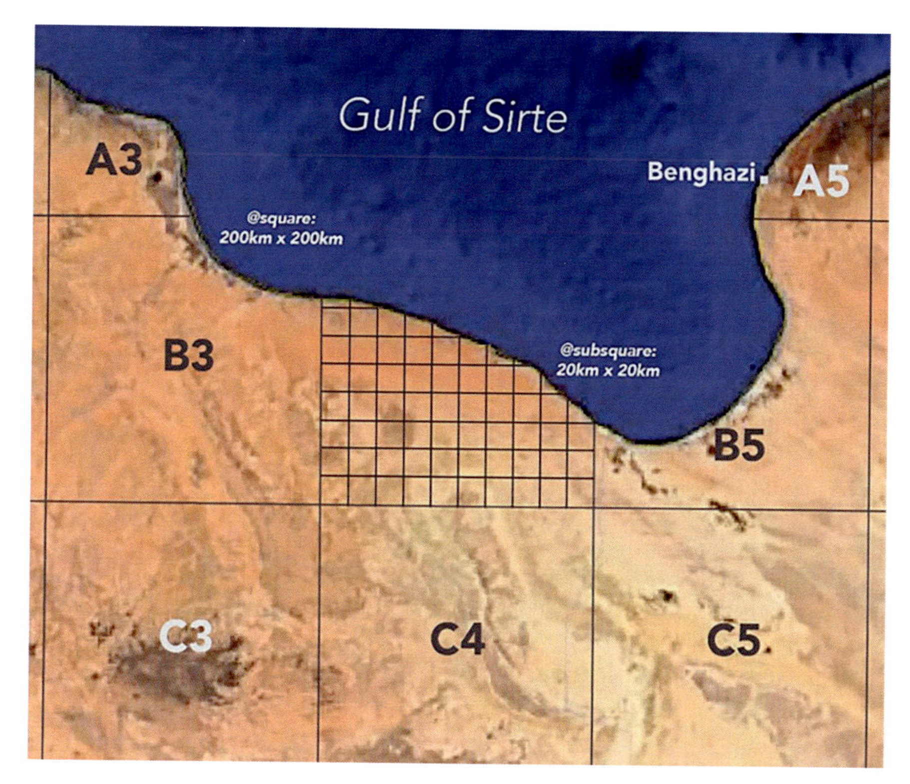

Fig. 24.4 Tighter grid overlain on a satellite image representing 20 by 20 km (big scale) to facilitate the classification and compilation of imagery and sites at closer inspection and lower elevation (Google Earth, 2015b)

In different sites in the country, through DigitalGlobe data a number of seismogenic geomorphologic features were identified. In northeast Al Jabal Al Gharbi (western Libya), the identification showed that several stream channels were cut by a number of parallel faultline, where the displacement created offset wadi channels. Also, en-echelon faulting was located (Fig. 24.5). In Haruj Volcanic field, several geomorphic features were identified. In the southwest of the field, small flanking en-echelon (possible transform faulting), a large strike slip fault, and faultline scarps were identified in 26°26'N, 16°51′E (Fig. 24.6). North of the field, distinctive en-echelon faulting, faultline scarps graben complex, and offset wadi channels developed in 28°58'N, 17°48′E and 28°56'N, 17°49′E (Figs. 24.7 and 24.8). In eastern and northeastern Sirte, en-echelon faulting, faultline scarps, offset wadi channels indicative of an active fault displacement, and large complex graben feature trending SE-NW were located in 30°13'N, 18°28′E, and faultline scarps and displacement of en-echelon faulting in 29°44'N,18°04′E (Figs. 24.9 and 24.10). In Al Jabal Al Akhdar (northeast Libya), offset stream channels and wadis were located in Al-Shuqluf, Al-Khirbet, and Amira Valleys between Al Hamamah and Susah in 32°54'N, 21°45′E (Fig. 24.11). These features indicate the surface presence of

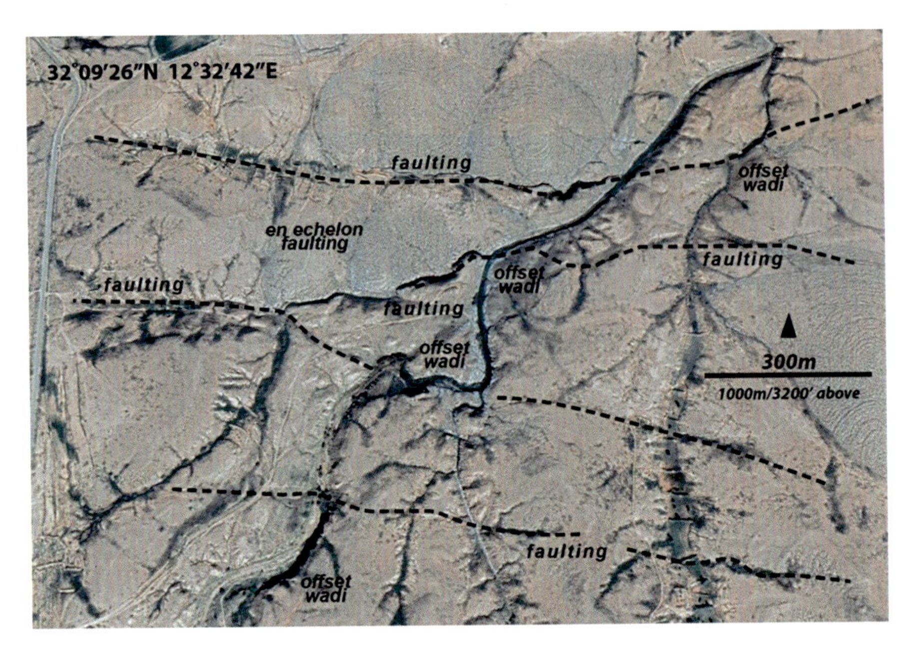

Fig. 24.5 Matched image displaying (**a**) en-echelon faulting, (**b**) faultings, (**c**) and offset wadi channels located in 32°09' N 12°32′ E northeast Al Jabal Al Gharbi western Libya (Google Earth, 2019a)

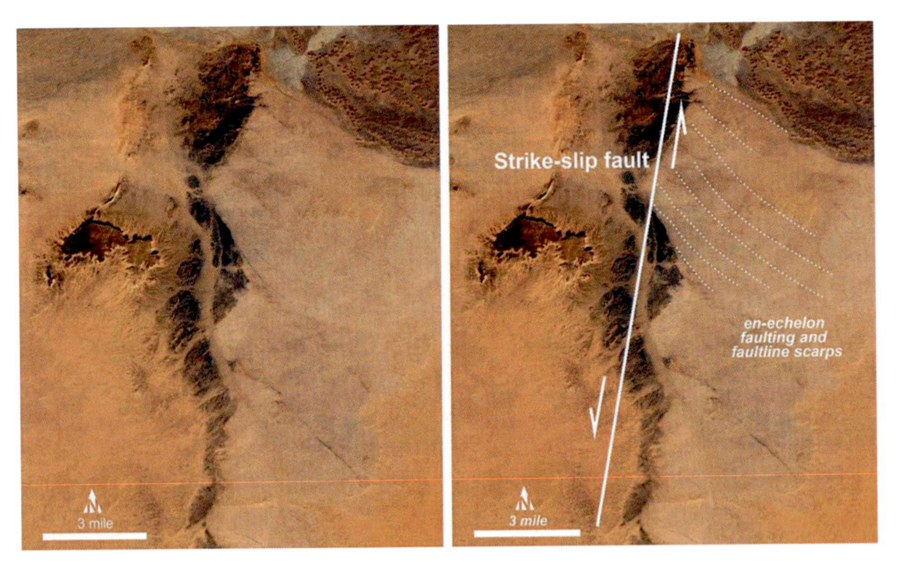

Fig. 24.6 Matched image showing (**a**) large strike-slip fault, (**b**) smaller flanking en-echelon faults (possible transform faulting), and (**c**) faultline scarps located in 26°26'N, 16°51′E southwest Haruj Volcanic Field (Google Earth, 2012a)

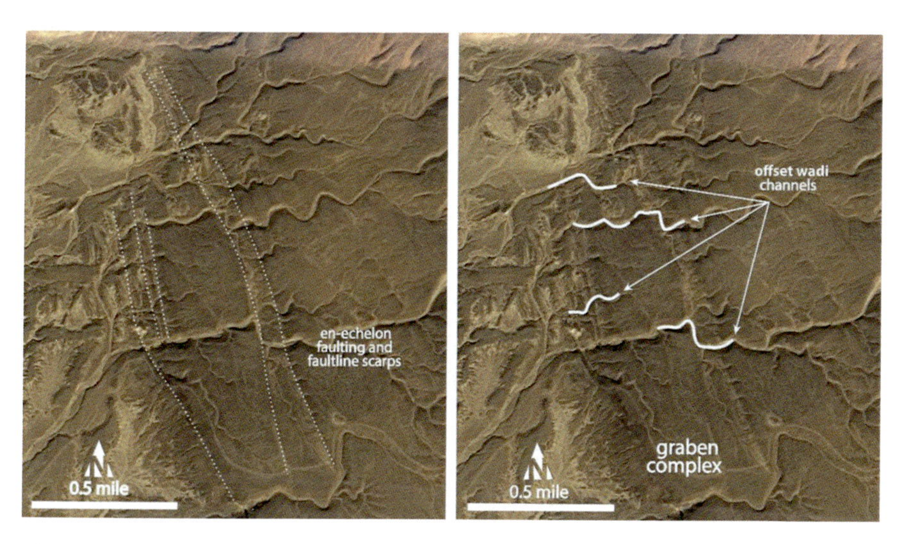

Fig. 24.7 Matched image pair displaying a (**a**) graben complex, (**b**) offset wadi channels, (**c**) clear en-echelon faulting, and (**d**) faultline scarps developed north of the Haruj Volcanic Field in 28°58'N, 17°48′E (Google Earth, 2012b)

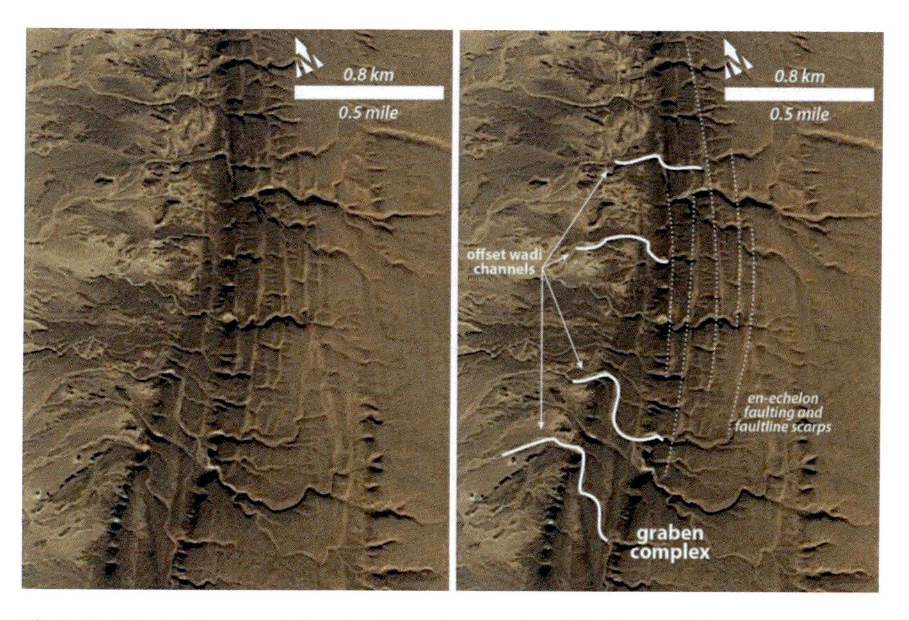

Fig. 24.8 Matched image pair showing (**a**) graben complex, (**b**) offset wadi channels, (**c**) distinctive en-echelon faulting, and (**d**) faultline scarps located in 28°56'N, 17°49′E north of the Haruj Volcanic Field (Google Earth, 2012b)

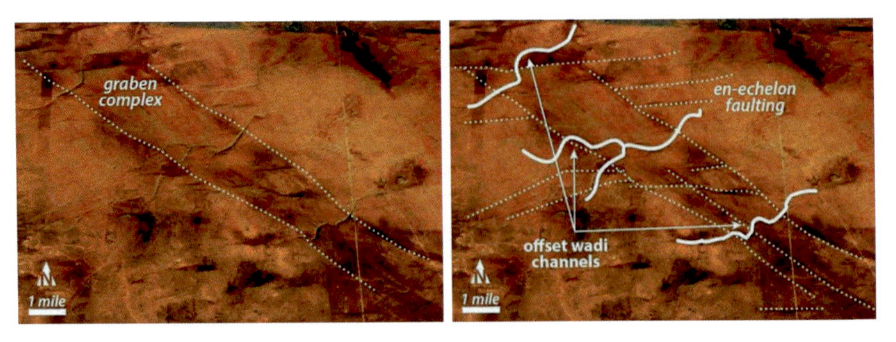

Fig. 24.9 Matched image pair illustrating (**a**) en-echelon faulting, (**b**) large graben feature trending SE-NW, and (**c**) offset wadi channels indicative of active fault displacement located in 30° 13'N, 18°28′E, northeastern Sirte, Libya (Google Earth, 2016)

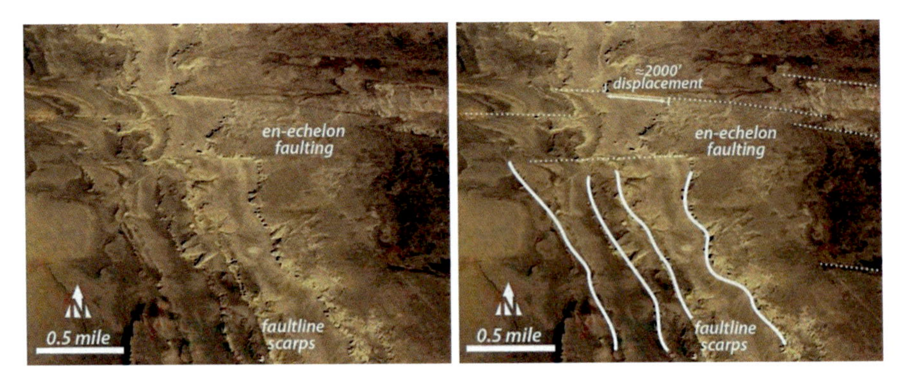

Fig. 24.10 Matched image pair showing (**a**) faultline scarps and (**b**) displacement of enechelon faulting in located 29°44'N,18°04′E eastern Sirte (Google Earth, 2015a)

faulting that is actively shifting wadi channel alignment, which a key indicator in identifying seismogenic geomorphology. Using remotely-sensed images is important in examining seismic geomorphologic features at various scales in various elevations.

In arid landscapes, such geomorphic features can be identified and classified through multi-scalar techniques and represent the first crucial process in identifying potentially hazards seismogenic features. It is through the remotely-sensing data that these seismogenic surface features may be identified first, then ground-truthed come second for further study, hazard assessment, and risk management. In a vast country like Libya, using ground observations to identify and assess seismic geomorphologic features would be too costly and time-consuming.

Using remote sensed imagery to identify, locate and classify similar features is less costly and ultimately cost-effective and efficient. An efficient plan to first use remote imagery to locate these features, and then undertake ground observation and feature assessment would be an ideal plan to locate, classify, and archive all potentially hazardous seismogenic features (and other related geomorphology) across Libya.

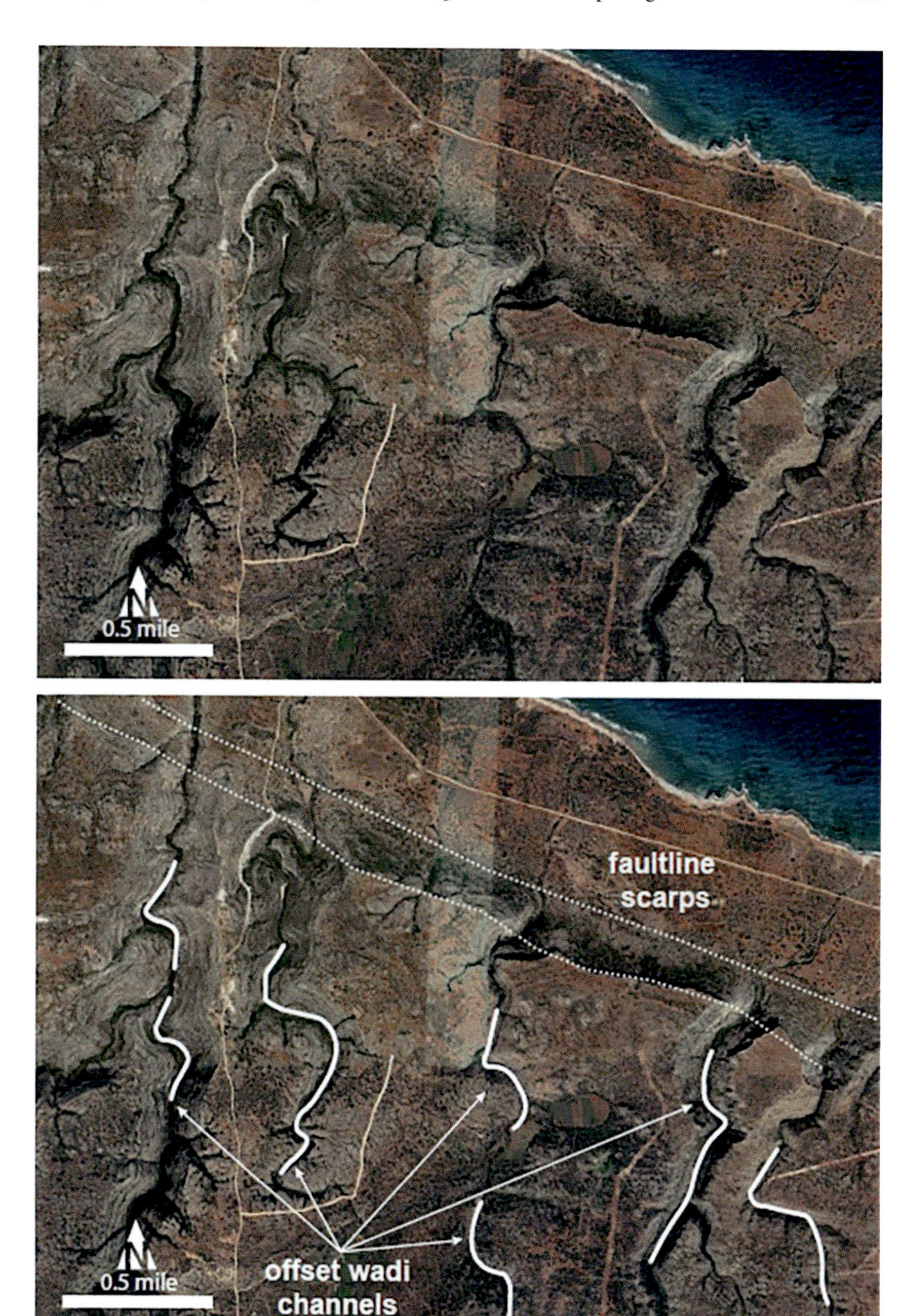

Fig. 24.11 Matched image illustrating (**a**) offset wadi channels and (**b**) faultline scarps between Al Hamamah and Susah in Al-Shuqluf, Al-Khirbet, and Amira Valleys (32°54'N, 21°45′E). Offset stream wadis and channels indicate the surface presence of faulting that is actively shifting wadi channel alignment – a key indicator in identifying seismogenic geomorphology (Google Earth, 2019b)

References

Abdunaser, K. M. (2015). Satellite imagery for structural geological interpretation in Western Sirt Basin, Libya: Implication for petroleum exploration. *Geosciences, 5*(1), 8–25. https://doi.org/10.5923/j.geo.20150501.02

Abdunaser, K. M., & McCaffrey, K. J. W. (2015). A new structural interpretation relating NW Libya to the Hun graben, western Sirt Basin based on a new paleostress inversion. *Journal of Earth System Science, 124*(8), 1745–1763. https://doi.org/10.1007/s12040-015-0631-4

Ambraseys, N. N. (1984). Material for the investigation of the seismicity of Tripolitania (Libya). In A. Brambati & D. Slejko (Eds.), *The OGS silver anniversary volume* (pp. 143–153).

Ambraseys, N. N., Melville, C. P., & Adams, R. D. (1995). *The seismicity of Egypt, Arabia and the Red Sea: A historical review*. Cambridge University Press.

Amelung, F., Jónsson, S., Zebker, H., & Segall, P. (2007). Widespread uplift and 'trapdoor'faulting on Galapagos volcanoes observed with radar interferometry. *Nature, 407*, 993–996.

Argus, D. F., Heflin, M. B., Peltzer, G., Crampé, F., & Webb, F. H. (2005). Interseismic strain accumulation and anthropogenic motion in metropolitan Los Angeles. *Journal of Geophysical Research: Solid Earth, 110*(4), 1–26. https://doi.org/10.1029/2003JB002934

Bellaoui, M., Hassini, A., & Bouchouicha, K. (2017). Remote sensed land surface temperature anomalies for earthquake prediction. *International Journal of Engineering Research in Africa, 31*, 120–134. https://doi.org/10.4028/www.scientific.net/JERA.31.120

Cakir, Z., de Chabalier, J. B., Armijo, R., Meyer, B., Barka, A., & Peltzer, G. (2003). Coseismic and early post-seismic slip associated with the 1999 Izmit earthquake (Turkey), from SAR interferometry and tectonic field observations. *Geophysical Journal International, 155*(1), 93–110. https://doi.org/10.1046/j.1365-246X.2003.02001.x

Campbell, A. S. (1968). *The Barce (A1 Marj) earthquake of 1963* (pp. 183–195). Geology and Archaeology of Northern Cyrenaica.

Chau, K. T., Sze, Y. L., Fung, M. K., Wong, W. Y., Fong, E. L., & Chan, L. C. P. (2004). Landslide hazard analysis for Hong Kong using landslide inventory and GIS. *Computers & Geosciences, 30*, 429–443. https://doi.org/10.1016/j.cageo.2003.08.013

Chen, K.-S., Serpico, S. B., & Smith, J. A. (2012). Remote sensing of natural disasters. *Proceedings of the IEEE, 100*(10), 2794–2797. https://doi.org/10.1109/JPROC.2012.2205835

Clark, M. R. (1978). Finding active faults using aerial photographs. *Earthquake Information Bulletin (USGS), 10*(5), 169–173.

Conant, L. C., & Goudarzi, G. H. (1967). Stratigraphic and tectonic framework of Libya. *AAPG Bulletin, 51*(5), 719–730. https://doi.org/10.1306/5d25c0c9-16c1-11d7-8645000102c1865d

Corti, G., Cuffaro, M., Doglioni, C., Innocenti, F., & Manetti, P. (2006). Coexisting geodynamic processes in the Sicily Channel. In Y. Dilek & S. Pavlides (Eds.), *Postcollisional tectonics and magmatism in the Mediterranean region and Asia* (Vol. 409, Issue 05, pp. 83–96). Geological Society of America Special Paper. https://doi.org/10.1130/2006.2409(05).

di Vita-Evrard, G. (1999). Libya: The lost cities of the *Roman Empire*. Konemann.

Dreger, D., & Kaverina, A. (2000). Seismic remote sensing for the earthquake source process and near-source strong shaking: A case study of the October 16, 1999 Hector mine earthquake. *Geophysical Research Letters, 27*(13), 1941–1944.

Fialko, Y., & Simons, M. (2001). Evidence for on-going inflation of the Socorro magma body, New Mexico, from interferometric synthetic aperture radar imaging. *Geophysical Research Letters, 28*(18), 3549–3552. https://doi.org/10.1029/2001GL013318

Froger, J. L., Fukushima, Y., Briole, P., Staudacher, T., Souriot, T., & Villeneuve, N. (2004). The deformation field of the August 2003 eruption at Piton de la Fournaise, Reunion Island, mapped by ASAR interferometry. *Geophysical Research Letters, 31*(14), 1–5. https://doi.org/10.1029/2004GL020479

Fu, B., Ninomiya, Y., Lei, X., Toda, S., & Awata, Y. (2004). Mapping active fault associated with the 2003 Mw 6.6 Bam (SE Iran) earthquake with ASTER 3D images. *Remote Sensing of Environment, 92*(2), 153–157. https://doi.org/10.1016/j.rse.2004.05.019

Fukushima, Y., Cayol, V., & Durand, P. (2005). Finding realistic dike models from interferometric synthetic aperture radar data: The February 2000 eruption at Piton de la Fournaise. *Journal of Geophysical Research: Solid Earth, 110*(3), 1–15. https://doi.org/10.1029/2004JB003268

Goodchild, R. G. (1968). *Graeco-Roman Cyrenaica, geology and archaeology of narthern Cyrenaica, Libya* (pp. 23–40). Geography and Archaeology of Northern Cyyrenaica.

Google Earth. (2012a, November 5). Southwest the Haruj Volcanic Field, Libya.26°26′10.94"N, 16°51′10.95″E, Maxar Technologies. Digital Globe 2019.

Google Earth. (2012b, September 3). North of the Haruj Volcanic Field, Libya. 28°58′20.77"N, 17°48′51.49″E, Eye alt 10.17 mil. Maxar Technologies. Digital Globe 2019.

Google Earth. (2015a, December 13). East Sirte, Libya. 29°44′25.52"N,18°04′02.16″E, Eye alt 40641 ft. SIO, NOAA, U.S. Navy, NGA, GEBCO. 2019.

Google Earth. (2015b, December 13). Libya 26°35′59.90"N, 18°13′51.67″E, Eye alt 1647.39 mil. SIO, NOAA, U.S. Navy, NGA, GEBCO. 2019.

Google Earth. (2016, December 25). Northeastern Sirte, Libya. 30°13′14.89"N, 18°28′18.89″E, Eye alt 31.28 mil. Maxar Technologies. Digital Globe 2019.

Google Earth. (2019a, January 26). Northeast Al Jabal Al Gharbi, Libya. 32°09′24.80"N, 12°32′41.26″E, Eye alt 12313 ft. Maxar Technologies. Digital Globe 2019.

Google Earth. (2019b, March 9). Al Jabal Al Akhdar, Libya. 32°54′00.98"N, 21°45′02.27″E, Eye alt 25595 ft. Maxar Technologies. Digital Globe 2019.*Google Earth. (November 5, 2012).*

Gordon, D. W., & Engdahl, E. R. (1963). An instrumental study of the Libyan earthquake of February 21, 1963. *Earthquake Notes, 34*(3–4), 50–56.

Gorny, V. I., Salman, A. G., Tronin, A. A., & Shilin, B. V. (2020). Terrestrial outgoing infrared radiation as an indicator of seismic activity. *arXiv preprint arXiv, 2001*, 11762.

Gupta, R. P., Saraf, A. K., Saxena, P., & Chander, R. (1994). IRS detection of surface effects of the Uttarkashi earthquake of 20 October 1991, Himalaya. *International Journal of Remote Sensing, 15*(11), 2153–2156.

Hooper, A., Prata, F., & Sigmundsson, F. (2012). Remote sensing of volcanic hazards and their precursors. *Proceedings of the IEEE, 100*(10), 1–50.

Johnston, A. C. (1989). The seismicity of "stable continental interiors." In *Earthquakes at North-Atlantic passive margins: Neotectonics and postglacial rebound* (pp. 299–327).

Jónsson, S., Zebker, H., Cervelli, P., Segall, P., Garbeil, H., Mouginis-Mark, P., & Rowland, S. (1999). A shallow-dipping dike fed the 1995 flank eruption at Fernandina Volcano, Galápagos, observed by satellite radar interferometry. *Geophysical Research Letters, 26*(8), 1077–1080.

Kebcasy, R. M. (1980). Seismicity and seismotectonics of Libya. *The Geology of Libya, 3*, 954–963.

Lewis, S. (2009). *Remote sensing for natural disasters: Facts and figures*. Science and Development Network.

Lu, Z., Wicks, C., Dzurisin, D., Thatcher, W., Freymueller, J. T., McNutt, S. R., & Mann, D. (2000). Aseismic inflation of Westdahl volcano, Alaska, revealed by satellite radar interferometry. *Geophysical Research Letters, 27*(11), 1567–1570. https://doi.org/10.1029/1999GL011283

Massonnet, D., & Feigl, K. (1995). Discrimination of geophysical phenomena in satellite radar interferograms. *Geophysical Research Letters, 22*(12), 1537–1540.

Massonnet, D., Rossi, M., Carmona, C., Adragna, F., Peltzer, G., Feigl, K., & Rabaute, T. (1993). The displacement field of the landers earthquake mapped by radar interferometry. *Nature, 364*(6433), 138–142. https://doi.org/10.1038/364138a0

Massonnet, D., Feigl, K., Rossi, M., & Adragna, F. (1994). Radar interferometric mapping of deformation in the year after the landers earthquake. *Nature, 369*(6477), 227–230. https://doi.org/10.1038/369227a0

Minami, J. K. (1965). Relocation and reconstruction of thr town of Barce, Cyrenaica, Libya, damaged by earthquake of 21 February 1963. In *Third World Conference on Earthquake Engineering* (pp. 96–110).

Mourabit, T., Abou Elenean, K. M., Ayadi, A., Benouar, D., Suleman, A. B., Bezzeghoud, M., ... Zuccolo, E. (2014). Neo-deterministic seismic hazard assessment in North Africa. *Journal of Seismology, 18*(2), 301–318.

Naranjo, M. F., Ebmeier, S. K., Vallejo, S., Ramón, P., Mothes, P., Biggs, J., & Herrera, F. (2016). Mapping and measuring lava volumes from 2002 to 2009 at El Reventador Volcano, Ecuador, from field measurements and satellite remote sensing. *Journal of Applied Volcanology, 5*(1), 1–11. https://doi.org/10.1186/s13617-016-0048-z

Norbert Brügge, G. (2015). The volcano Wau-an-Namus and further crater-like structures in context with the large flood basalt field of Al-Haruj (Libya). *International Journal of Geosciences.*

Ouzounov, D., & Freund, F. (2004). Mid-infrared emission prior to strong earthquakes analyzed by remote sensing data. *Advances in Space Research, 33*(3), 268–273. https://doi.org/10.1016/S0273-1177(03)00486-1

Pedersen, R., & Sigmundsson, F. (2006). Temporal development of the 1999 intrusive episode in the Eyjafjallajökull volcano, Iceland, derived from InSAR images. *Bulletin of Volcanology, 68*(4), 377–393. https://doi.org/10.1007/s00445-005-0020-y

Peltzer, G., Hudnut, K. W., & Feigl, K. L. (1994). Analysis of coseismic surface displacement gradients using radar interferometry: New insights into the landers earthquake. *Journal of Geophysical Research: Solid Earth, 99*(B11), 21971–21981. https://doi.org/10.1029/94JB01888

Price, E. J., & Sandwell, D. T. (1998). Small-scale deformations associated with the 1992 Landers, California, earthquake mapped by synthetic aperture radar interferometry phase gradients. *Journal of Geophysical Research: Solid Earth, 103*(B11), 27001–27016. https://doi.org/10.1029/98JB01821

Pritchard, M. E., & Simons, M. (2002). A satellite geodetic survey of large-scale deformation of volcanic centres in the central Andes. *Nature, 418*(6894), 167–171. https://doi.org/10.1038/nature00872

Reilinger, R., McClusky, S., Vernant, P., Lawrence, S., Ergintav, S., & Cakmak, ... Karam, G. (2006). GPS constraints on continental deformation in the Africa-Arabia-Eurasia continental collision zone and implications for the dynamics of plate interactions. *Journal of Geophysical Research: Solid Earth, 111*(5), 1–26. https://doi.org/10.1029/2005JB004051

Rui, J., Wang, C., Zhang, H., Jin, F., Zhang, Z., Liu, Z., ... Tang, Y. (2017). Landslide detection based on DEM matching. In *FRINGE Workshop.* Helsinki: European Space Agency.

Salamon, A., Hofstetter, A., Garfunkel, Z., & Ron, H. (2003). Seismotectonics of the Sinai subplate – The eastern mediterranean region. *Geophysical Journal International, 155*(1), 149–173. https://doi.org/10.1046/j.1365-246X.2003.02017.x

Saraf, A. (1998). Jabalpur earthquake of 22 May 1997: Assessing the damage using remote sensing and GIS techniques. In *Proceedings of the 11th Symposium on Earthquake Engineering* (Vol. 1, pp. 103–116).

Saraf, A. K. (2000). Cover: IRS-1C-PAN depicts Chamoli earthquake induced landslides in Garhwal Himalayas, India. *International Journal of Remote Sensing, 21*(12), 2345–2352.

Sivakumar, P. (2017). Which satellite takes the image of Google maps? *Quora.*

Skobelev, S. F., Hanon, M., Klerkx, J., Govorova, N. N., Lukina, N. V., & Kazmin, V. G. (2004). Active faults in Africa: A review. *Tectonophysics, 380*(3–4), 131–137. https://doi.org/10.1016/j.tecto.2003.10.016

Solano-Rojas, D., Wdowinski, S., Amelung, F., Cabral-Cano, E., Zhang, Y., & Walter, T. (2017). InSAR monitoring of the Popocatépetl volcano in Central Mexico. In *FRINGE workshop.* Helsinki: European Space Agency.

Suleiman, A. S., & Doser, D. I. (1995). The seismicity, seismotectonics and earthquake hazards of Libya, with detailed analysis of the 1935 April 19, M = 7.1 earthquake sequence. *Geophysical Journal International, 120*(2), 312–322.

Suwihli, S. S., & Paradise, T. R. (2020). Creating a Libyan earthquake archive: From classical times to the present. *Open Journal of Earthquake Research, 9*(4), 367–382.

Talebian, M., Fielding, E. J., Funning, G. J., Ghorashi, M., Jackson, J., & Nazari, ... Wright, T. J. (2004). The 2003 Bam (Iran) earthquake: Rupture of a blind strike-slip fault. *Geophysical Research Letters, 31*(11), 2–5. https://doi.org/10.1029/2004GL020058

Tarolli, P., Sofia, G., & Dalla Fontana, G. (2012). Geomorphic features extraction from high-resolution topography: Landslide crowns and bank erosion. *Natural Hazards, 61*(1), 65–83. https://doi.org/10.1007/s11069-010-9695-2

Tramutoli, V., Cuomo, V., Filizzola, C., Pergola, N., & Pietrapertosa, C. (2005). Assessing the potential of thermal infrared satellite surveys for monitoring seismically active areas: The case of Kocaeli (İzmit) earthquake, August 17, 1999. *Remote Sensing of Environment, 96*(3–4), 409–426. https://doi.org/10.1016/j.rse.2005.04.006

Tramutoli, V., Corrado, R., Filizzola, C., Genzano, N., Lisi, M., & Pergola, N. (2015). From visual comparison to robust satellite techniques: 30 years of thermal infrared satellite data analyses for the study of earthquake preparation phases. *Bollettino di Geofisica Teorica ed Applicata, 56*(2), 167–202. https://doi.org/10.4430/bgta0149

Trifonov, V. (1984). Application of space images for neotectonic studies. *Remote Sensing for Geological Mapping, 18*, 41–56.

Tronin, A. A. (2006). Remote sensing and earthquakes: A review. *Physics and Chemistry of the Earth, 31*(4–9), 138–142. https://doi.org/10.1016/j.pce.2006.02.024

Tronin, A. A., Hayakawa, M., & Molchanov, O. A. (2002). Thermal IR satellite data application for earthquake research in Japan and China. *Journal of Geodynamics, 33*(4–5), 519–534.

Tronin, A. A., Biagi, P. F., Molchanov, O. A., Khatkevich, Y. M., & Gordeev, E. I. (2004). Temperature variations related to earthquakes from simultaneous observation at the ground stations and by satellites in Kamchatka area. *Physics and Chemistry of the Earth, 29*(4–9), 501–506. https://doi.org/10.1016/j.pce.2003.09.024

U S Geological Survey (USGS). (2020). *New earthquake hazards program: Lists, maps, and statistics*. U.S. Geological Survey. https://doi.org/https://www.usgs.gov/natural-hazards/earthquake-hazards/lists-maps-and-statistics

Wicks, C. W., Thatcher, W., Dzurisin, D., & Svarc, J. (2006). Uplift, thermal unrest and magma intrusion at Yellowstone caldera. *Nature, 440*(7080), 72–75. https://doi.org/10.1038/nature04507

Yun, S. H., Zebker, H., Segall, P., Hooper, A., & Poland, M. (2007). Interferogram formation in the presence of complex and large deformation. *Geophysical Research Letters, 34*(12), 1–6. https://doi.org/10.1029/2007GL029745

Zebker, H. A., Rosen, P. A., Goldstein, R. M., Gabriel, A., & Werner, C. L. (1994). On the derivation of coseismic displacement fields using differential radar interferometry: The landers earthquake. *Journal of Geophysical Research: Solid Earth, 99*(B10), 19617–19634. https://doi.org/10.1029/94JB01179

Chapter 25
Seismological and Remote Sensing Studies in the Dead Sea Zone, Jordan 1987–2021

Najib Abou Karaki, Damien Closson, and Mustapha Meghraoui

Abstract The Dead Sea area is draining massive tourism and infrastructure investments. However, the area is prone to both induced anthropogenic and natural geological hazards, with indicators requiring innovative monitoring. Hazards are resulting from the zone's plate boundary tectonic setting and seismicity added to the generalized subsidence and sinkholes proliferation related to decades of accelerating water level lowering of this terminal lake.

The Jordan Dead Sea Transform Fault System (JDST) is an N-S trending and ~ 1000-km-long plate boundary that accommodates ~5 mm/year. left-lateral slip. We focus on the main research results concerning the whole spectrum of destructive seismicity components, i.e. Instrumental, historical, archaeo and paleoseismicity. Field investigations in earthquake geology and paleoseismology point out the identification of seismic gaps with long-term temporal quiescence reaching 851 years on the Jordan valley fault segment compared to 988 years for the Missyaf fault segment of the JDST further north in Syria (as per the year 2021). Destructive historical and instrumental seismicity were subjected to careful robust revision processes. The repetition of seismic events and related earthquake faulting parameters suggest a high level of seismic hazard and risk along the JDST.

From 1992 onwards, research based on space remote sensing techniques, Geographical information systems, and field data collection has been undertaken to develop a predictive model for salt karst hazards along the Dead Sea coast.

Radar and optical image processing produced images capable of being interpreted in a geographic information system (GIS).

N. Abou Karaki (✉)
Department of Geology, The University of Jordan, Amman, Jordan
e-mail: naja@ju.edu.jo

D. Closson
Department of Communication, Information, Systems and Sensors, Royal Military Academy, Zaventem, Brussels, Belgium

M. Meghraoui
University of Strasbourg, Strasbourg Cedex, France
e-mail: m.meghraoui@unistra.fr

M. M. Al Saud (ed.), *Applications of Space Techniques on the Natural Hazards in the MENA Region*, https://doi.org/10.1007/978-3-030-88874-9_25

The field observations were systematically georeferenced using a GPS and then imported into the GIS to be analysed with the processed satellite images. Each independent data source was used to establish an explanatory model for the prediction of areas at risk of collapse.

This approach has been improved over time due to the arrival of an ever greater number of optical and radar images. Image resolution has also increased, allowing inventories of sinkholes, especially in the most dangerous locations.

This Chapter contributes in filing the gap in the seismological, and remote sensing studies necessary to the sustainable preservation of this world class cultural heritage zone, the safe economic upgrading of the area and the safety of its inhabitants and visitors.

Keywords Dead Sea fault · Paleoseismicity · Radar interferometry · Seismicity · Sinkholes · Subsidence

25.1 Introduction

The Jordan Dead Sea Transform Fault System (JDST) is a north-south trending left lateral strike slip known as the Levant fault zone(Abou Karaki, 1987). It is a documented plate boundary of the transform type (Abou Karaki, 1987; Garfunkel et al., 1981; Meghraoui, 2015; Wilson, 1965). The JDST forms the limit between the Arabian and African (Sinai) plates accommodating the northward motion of Arabia to the east in relative to Africa to the west (Fig. 25.1a). Both plates show a northward movement toward Eurasia with different rates, the Arabian plate rate being about 18–25 mm/year. and the African plate having a slower movement rate of about 10 mm/year. (Reilinger et al., 2006). The JDST extends for about 1000 km passing through the Dead Sea. The Transform connects the Red Sea developing mid-oceanic ridge south of the entrance of the Gulf of Aqaba in the south to the triple junction area to the north in Turkey where it joins the East Anatolian Fault (EAF) and the Cyprus Arc (CA). In its north central part in Lebanon the JDST bends rightward and become oriented N30°E, and across Syria the northern segment trends N–S and bends towards NNE showing splays into several small fault segments (Fig. 25.1a) in southern Turkey (Westaway, 2004). A rate of left lateral displacement of 4 to 7 mm/year. has been estimated by several geological and geodetic studies along the JDST (Alchalbi et al., 2010; Ferry et al., 2007; Gomez et al., 2007; Le Béon et al., 2008; Meghraoui et al., 2003; Pe'eri et al., 2002; Wdowinski et al., 2004).

Geological and geomorphological studies show that the JDST is not acting as a tectonic structure with uniformly distributed deformation. We observe a difference in total motion between the south and the north segments of the transform: and also in the historical seismicity (Abou Karaki, 1987): a total cumulative sinistral slip of about 107 km has been documented along the southern part of the fault (Wadi Araba) and was accumulated 15–20 Ma since, when the JDST was initiated in the Middle Miocene (Freund et al., 1968; Quennell, 1958). In the northern part of the

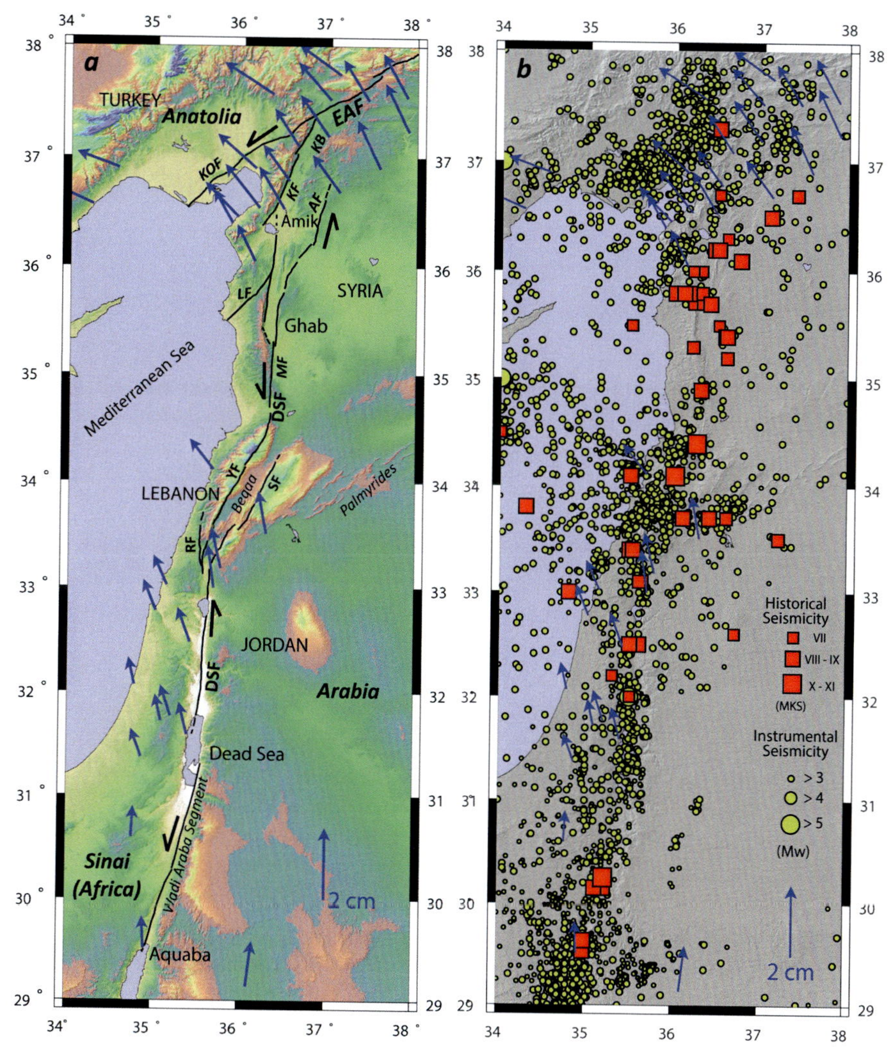

Fig. 25.1 (**a**) The Dead Sea fault segments, Fault mapping and slip rates from (Gomez et al., 2007; Khair et al., 2000; Meghraoui, 2015; Westaway, 2004) Abbreviations for some key tectonic features: *AGF* Al-Ghab fault, *AF* Afrin fault, *CA* Cyprus Arc, *KB* Karasu basin, *KF* Karasu fault, *KOF* Karatas-Osmaniye fault, *LF* Lattakia fault, *DSF* Dead Sea fault, *EAF* East Anatolian fault, *MF* Missyaf fault, *SF* Serghaya fault, *RF* Roum fault, *YF* Yammuneh fault. (**b**) Instrumental seismicity of the Dead Sea fault between 1964 and 2011, M > 3. Data are from: IRIS; Incorporated Research Institutions for Seismology (http://www.iris.edu/hq/), ISC; The International Seismological Center (http://www.isc.ac.uk/) and NEIC; The National Earthquake Information Center. The seismicity depending on international organizations lists for this area's earthquakes of magnitudes <4 is misleading. The clusters of yellow "epicenters" around (30°N, 36.4°E and 31.2°N,36.4°E) are most probably explosions these areas correspond to the "Phosphate Mines Co." locations of Al-Shidiah and Al-Hisa exploitation areas south of Jordan respectively where up to 10 Tons of explosives are detonated frequently on routine basis in the framework of the industrial exploitation of the mines. See Fig. 25.5b

system (Lebanon, AL-Ghab and Karasu basin; Fig. 25.1a) the total documented amount of slip is 70–80 km (Chaimov et al., 1990; Freund et al., 1970; Westaway, 1994; Westaway, 2004) which suggests that the difference in total motion between north and south along the JDST is at least ~25 km. This value was justified by the existence of the Palmyride fold belt and the Serghaya and Beqa'a faults (Ambraseys & Barazangi, 1989; Khair et al., 1997; Westaway, 1994).

25.2 General Seismotectonic Setting

The JDST consists in the north and the south zones (Fig. 25.1a) connected to each other by an active transpressive zone: the restraining Lebanese bend (Griffiths et al., 2000, Gomez et al., 2007). The principal issue is the ~25 km difference of accumulated slip rate between the northern and southern parts of the fault (Chaimov et al., 1990; Freund et al., 1970; Westaway, 1994). The JDST was divided into five major segments (Wadi Araba, Jordan Valley, Albeqa'a basin, Ghab Basin and Karasu Valley) (Khair et al., 2000); the segmentation was proposed due to the difference in geometry, geomorphology, geology and seismicity of each part (Fig. 25.1b). All these segments have N-S trending in general with small deviation to the NNE for the central and the northern segment of the fault.

The tectonic markers of Quaternary deformation are observable at all scales and on all sections along the JDST. Using the geological and geomorphological maps, the linear shape of the fault affecting late Pleistocene and Holocene deposits can be identified on Landsat, SPOT, and QuickBird satellite photographs as well as on aerial photographs. In the southern part the JDST zone crosses the western edge of the Gulf of Aqaba and extends further north for about 40 km crossing fluvial and alluvial terraces of Avrona Playa and the Wadi Araba Valley. Several fault branches outcrop showing normal geometry in this area, but they may result from the slip partitioning on branches of the main JDST (Niemi et al., 2001).

To the north, except for the accumulation of sand dunes that conceals the fault zone around latitude 30°N, the fault can be seen to affect successive stream channels and terraces. In the northern half of the Wadi Araba, the DSF fault exposes outstanding tectonic geomorphology with linear fault scarps showing offset alluvial fans and channels visible until the south eastern edge of the Dead Sea (Barjous & Mikbel, 1990). The southern end of the DSF meets the Gulf of Aqaba that shows NNE–SSW trending relay fault system with large pull-apart basins in the sea bottom that may act as a geometrical barrier that limits the Wadi Araba fault zone to the south. Indeed, the DSF including the Wadi Araba section is about 200-km long if we include its extension in the eastern edge of the Dead Sea area (Fig. 25.1a), it may constitute a single fault segment or two distinct segments if a major geometrical complexity (relay zone) is hidden below the mid-distance sand dunes.

25.2.1 *Dead Sea Area Fault Segments*

25.2.1.1 Wadi Araba

Wadi Araba is the southern part of DSF that starts from the Red Sea (Aqaba Gulf) at 29.5° N and extend for about 160 km till the basin of the Dead Sea at 31° N (Fig. 25.1a) (Klinger et al., 1997, 2000a). Along this segment, the fault have a sharp morphological discontinuity that can easily be traced across the Quaternary deposits and alluvium sediments, excluding where the fault is covered with sand dunes or cuts across very recent alluvial terraces. The principal fault is rather straight, striking N20° E, and showing limited structural discontinuities, with a simple geometry is reliable with basically pure strike slip motion (Meghraoui, 2015).

An estimation of the slip rate along Wadi Araba segment based on time varying geodetic study (GPS) of 4.9 ± 1.4 mm/year. was given by (Le Béon et al., 2008), this value rely on 6 years of time span and a locking depth of 12 km.

Regarding the historical seismicity catalogs, few seismic events are reported during the last 2000 years along the Wadi Araba. The largest reported events occurred in AD 1068, 1212, 1293 and 1458 (Abou Karaki, 1987; Abou Karaki et al., 1993; Ambraseys et al., 1994; Klinger et al., 2000b) These events seem to be smaller than the 1995 earthquake which struck the Gulf of Aqaba with a Mw ~ 7.3 (Abou Karaki, 1987, 2001a, b).

25.2.1.2 Jordan Valley

The N-S trending Jordan Valley extends for about 180 km between the Dead Sea pull-apart basin at 30.7° N and the Hula Basin in the north before connecting with the Lebanese restraining bend at 33.1° N. The northern end of this segment attests in the division of the fault into many fault branches trending toward the NNE, the Serghaya, Rashaya, Hasbaya, and Yammuneh faults (Fig. 25.1a). This segment is connecting the two pull-apart basins of the Dead Sea and the Tiberiade Lake. It was the object of paleoseismologic and geomorphologic studies which estimated its slip rate to be ranging from 2.5 to 10 mm/year (Galli, 1999; Marco et al., 1997).

The largest estimated magnitude for the paleoearthquakes in the Jordan Valley segment is Mw 7.3 (Ferry et al., 2007). Studies of macroseismic damage from historic events and archaeological evidence conclude that several large destructive earthquakes (Ms > 6) occurred in the northern Jordan Valley segment during the past 2000 year (Abou Karaki, 1987; Ambraseys et al., 1994; Guidoboni et al., 1994; Guidoboni & Comastri, 2005; Marco et al., 2003).

25.2.2 *Paleoseismicity*

The Levant area is one of the oldest inhabited regions of the world and by consequence it has among the richest written history, including on earthquakes. The Earthquakes as any other well-known natural events were recorded and described carefully. The old documents describe the earthquake effects on nature and man-made structures, such as faulting rupture, co-seismic deformation, landslide, springs appearing and disappearing, lives casualties, houses destruction, . . . etc. This allows us to have, nowadays, a very rich earthquake catalog for more than 3000 years in the region (Abou Karaki, 1987; Ambraseys, 2009; Guidoboni & Comastri, 2005; Sbeinati et al., 2005).

The Dead Sea area has the characteristics of a large pull-apart basin (100-km long, 17-km wide) limited by two north striking segments, the southern Wadi Araba fault zone and the northern Jordan valley fault segment. The latter segment follows the Jordan River, extends from the Dead Sea (right bank of the Jordan River) to the Tiberiade Lake and is made of a succession of 10–20-km-long subsegments (mostly on the left bank of the Jordan River) for a total length of 110 km. In the valley, small pull-apart basins and restraining bends limit the subsegments and show left-lateral slip of stream channels (Fig. 25.2a, b) (Ferry et al., 2007; Ferry et al., 2011).

25.2.3 *Archaeoseismicity*

The richness of exposed JDST and archaeological sites along the Jordan Valley constitutes an exceptional advantage for dating prehistorical earthquakes. There are about 120,000 recognized archaeological sites in Jordan, forming part of the cultural heritage of this open museum country. The number is expected to reach half a million sites after the introduction of new survey techniques. Most major Jordanian archaeological sites and other ancient sites in the neighbourhood of the JDST show evidences of earthquake related damage, often historically documented as well, some of these sites located in the immediate neighbourhood of the DSF have already been subjected to archaeoseismicity investigations (Galli, 1997, 1999; Galli & Galadini, 2001; Klinger et al., 2000a, b; Niemi et al., 2001; Niemi & Smith, 1999). More specifically. The areas of Aqaba (South of Wadi Araba), Petra (Middle of Wadi Araba), Dahl (northern Wadi Araba), Jerash, Amman (East of the Jordan valley), Umm Quais and Tabkat-Fahl (Northern Jordan Valley) represent a set of places singled out as a result of previous historical seismicity studies. All were affected by relatively large historical earthquakes of equivalent Mw 7.0 $\pm$ 0.5. during the last 2000 years. Field investigations revealed left-lateral faulting of the following archaeological sites: a total of 2.1-m offset walls of a Crusader castle during the 20 May 1202 and 30 October 1759 earthquakes (Ellenblum et al., 1998); the offset of walls in the Tiberiade area during the 18 January 749 earthquake (Marco et al., 2003); at Tell Saidiyeh and Ghor Katar in the Jordan Valley paleo-earthquakes have been identified in 759 B.C., 1150 B.C., 2300 B.C., and 2900 B.C. (Fig. 25.2a, b) in (Ferry et al., 2011), the 2.2-m offset wall of Qasr Tilah that can be related to an

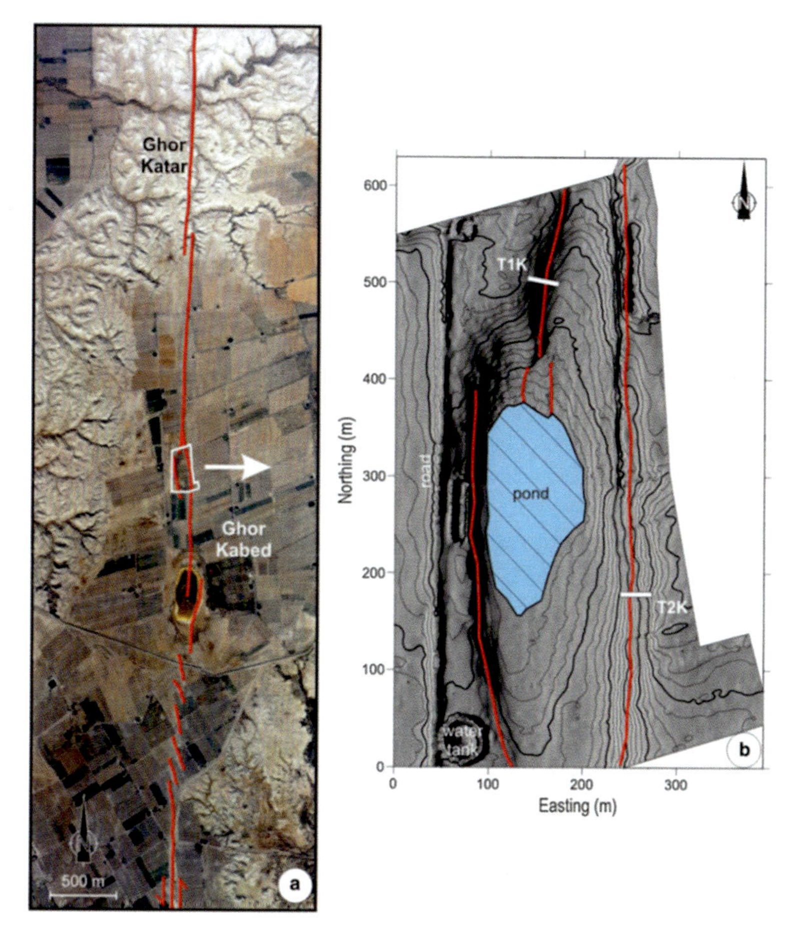

Fig. 25.2 (a) The Jordan valley fault (red lines) seen on an aerial photograph that exposes small pull-apart basins with left-lateral en echelon fault structures and stream channel offsets (at Ghor Katar), (**b**) Paleoseismic site with at Ghor Kabed pull-apart basin with high-resolution topography and fault branches affecting the Lisan lacustrine deposits (see also trench sites TK1 and TK2 in (Ferry et al., 2011))

earthquake event in 608–826 AD (Klinger et al., 2000a, b); the cumulative offset of the Roman–Byzantine–Islamic Qasr Tilah site of the northern Araba valley during the 634 or 659/660, 873, 1068, the minimum 1 m vertical slip of a qanat during the 1068 earthquake on the southern DSF near Eilat (Zilberman et al., 2005). These archaeoseismological studies of the JDST are complemented with tectonic geomorphology and paleoseismic investigations that constrain paleoearthquake faulting episodes and slip rates. In trenches, radiocarbon dating brackets the 1759 earthquake on the northern Jordan valley fault segment and the Serghaya fault branch through 3D trenching with 0.5 and 2.5-m left-lateral slip, respectively (Gomez et al., 2003; Marco & Agnon, 2005). Faulting of the 749 earthquake of the Jordan valley segment is resolved at Beyt Zayda (north of Tiberiade Lake) and Tell Saidiyeh (south Jordan valley) paleoseismic sites (Ferry et al., 2011; Marco & Agnon, 2005). The longest paleoseismic records on the JDST are determined in the Jordan valley with a sequence of 12 coseismic surface ruptures over the last 14 ka (Ferry et al., 2011). Although the correlation with individual fault segments is problematic, another source of earthquake records in the Dead Sea lake sediments and speleothems indicates a succession of seismites (El-Isa & Mustafa, 1986), synchronous to historical earthquakes in 1927, 1293, 1202/1212, 749, 551, 419, 33 AD and 31 BC and mid-second century BC (Kagan et al., 2011; Marco et al., 1997). From recent paleoseismic trenching (Klinger et al., 2015; Meghraoui, 2015) identified faulting events along the southern DSF in the Wadi Araba, which can be correlated with past earthquakes in 1458, 1212, 1068, and 363 AD, in addition of one faulting event bracketed between 806 AD and 1044 AD.

In conclusion, field investigations and studies in earthquake geology and paleoseismology conducted in the frame of the APAME Project (2003–2007) and other or subsequent studies point out the identification of seismic gaps with long-term temporal quiescence reaching 988 and 851 years (as per the year 2021) on the Missyaf fault segment in Syria and the Jordan valley segments of the JDST. The repetition of seismic events and related earthquake faulting parameters suggest a high level of seismic hazard and risk along the JDST including the Jordan Valley.

25.2.4 *Historical Seismicity*

The great importance of genuine historical seismicity catalogs, has been well recognized (Ambraseys, 1971). In our region, for historical and cultural reasons as the center of the ancient world civilizations and birthplace of writing, a wealth of Historical Seismicity material exists. This has been emphasized in many modern works (Abou Karaki, 1987, 1992, 1995a, 1996a, b, 1999; Al-Ghunaim, 2002; Amiran et al., 1994; Ben Menahem, 1991; Ghawanmeh, 1990; Guidoboni et al., 1994; Guidoboni & Comastri, 2005; Poirrier & Taher, 1980; Sbeinati et al., 2005). It is also necessary to recognize the importance and contributions of pioneering works of a great number of historians and scientists who compiled historical earthquake lists and catalogs, Impressive lists and references of these works is found in (Vered & Striem, 1977; Willis, 1928, 1933).

Nevertheless, rigorous analysis of the historical seismicity information in our area clearly demonstrated that most lists were contaminated by a variety of chronological and interpretation errors, some of them heavily, thus, to be fully useful historical seismicity lists need to be reassessed on detailed, objective and robust basis (Abou Karaki, 1987, 1995a), this was done as briefly introduced hereafter.

25.2.4.1 The General Mechanisms at the Origin of the Historical Seismicity Errors

Abou Karaki (1987, 1992, 1995a) defined the following, general, and in some cases, systematic causes and mechanisms of errors which led to a wide scale contamination of the historical seismicity catalogs depending mainly on Arabic manuscripts and documents in the Middle East and parts the wider Mediterranean and far East areas.

Timing and Chronological Calendar Systems Related Errors

The first step and most important factor for the correct identification of a historical earthquake is to accurately determine its time of occurrence. Any duplication of the timing may result in a direct duplication of the earthquake in future compilations and catalogs. The typical negative implications of this, are of some wide range consequences for seismic hazard evaluations and understanding of earthquake generating processes in the area. We agree that the greatest part of historical seismicity information related to this area, is provided by Arabic documents, manuscripts and chronicles, as was noticed by Ambraseys et al. (1994) p.7 "Arabic chronicles are the primary source for the history of the Middle east, and certainly for its earthquake history, from the ninth century at least until the end of the seventeenth centaury". However, the time window that looks like a restriction, is without any practical consequences because the core part of the area's available historical seismicity information is concentrated within the upper mentioned time interval specified in (Ambraseys et al., 1994) It is also well known that this area is characterized by the multiple aspects of its cultural heritage diversity. The area adopted a number of chronological calendar systems. Different calendars may have been simultaneously in use in a same place (Village, city, region, or stat). Hence a same and one earthquake may be reported from the same region associated with two or more dates in different calendar systems representing the same absolute point in time. Moreover, When a calendar is changed or modified at a certain point in time, usually following an important historical event, or simply, decision, in a given part of the area, chronicles did not uniformly follow the new style right away in all localities of the region. By far, the most important problem resulting in event duplications is the confusion between calendar systems in later compilations, and ambiguities related to this issue, a somewhat more complex case, is the more or less inaccurate careless date transformations between calendar systems. We identified six main mechanisms led to an equal number of error types in the historical seismicity of the area. These were defined by Abou Karaki (1987, 1992, 1995a) and further developed in this work, as follows:

Type I Errors

Instead of one genuine original earthquake, the date of which is eventually given in two different calendar systems, the earthquake is wrongly associated with two different dates given in one of the two calendar systems, e.g. the earthquake of the year 598 AH (After Hijra Muslim lunar Calendar) which is the same of 1202 AD, figuring in a catalog as two independent earthquakes associated respectively with the years 598 AD and 1202 AD, or, although less frequently, to 598 AH and 1202 AH)). Tens of such "earthquakes "that we call "false earthquake twins, or simply, *twins*" figures in Willis's list in his work of 1928, these errors were copied and passed on within the global seismicity lists of Seiberg in 1932 (Willis, 1933), such errors were corrected partially (rather inaccurately, and too late) in 1928 (Willis, 1933) then (Ambraseys, 1962).

Type II Errors

This type of errors refers to *twins* resulting from an inaccurate date transformation between calendar systems. This provides an explanation to the case of many major earthquakes with reasonably similar locations, effects, and descriptions occurring within ±2 years from each other in a given section of the fault, a clear example on this is found in the work of Taher (1979) (Taher, 1979). The earthquake of the month of Ramadan 130 AH is wrongly associated to the year 747 AD the correct year corresponding to the month of Ramadan in the year 130 or 131 AH should be 748 or 749 AD In some existing or future "complete" catalogs the years 746, 747, 748 and 749 AD would be listed as major earthquakes happening in almost the same geographic area. So the existence of two major earthquakes occurring in the same area and associated with the years 747 AD and 748 AD if given in a historical seismicity list would be the result of a type II error.

Type III Errors

This type of errors refers to a confusion between BC and AD dates. A full example illustrating this type of errors is shown in Table 25.1.

Type IV Errors

This is another source of timing errors potentially leading to multiple duplications in the historical seismicity catalogs. It refers to the difficulty to determine the exact year in the original manuscripts and historical seismicity primary sources, due to the frequent use of abbreviated numbers to represent the date in the original text. In many Arabic original manuscripts the full number representing the year of occurrence of a given event is in fact given partially. So a statement like "there was an earthquake in the year 37" might mean 37 AH or 137, 277,. . ., or 1437 AH and so on. Victims of such errors are typically those who depend upon translated short pieces of an original text. To overcome this problem, it was often necessary for us to perform the time consuming task of carefully reexamining a number of pages before

Table 25.1 An example illustrating the historical seismicity error type III, 2 independent earthquakes are listed instead of one

Date	Description	Io	ML	Type III EError in
525 BC	Of coast Sur, Sidon greatly damaged.	XI	7.5	Ben Menahem (1979)
525 AD	Of coast Sidon	IX-X	6.7	Ben Menahem (1979)

Notice the difference in the magnitude evaluations of Ben-Menahem (Ben Menahem, 1979), corresponding to this one and same earthquake and/or after a given purely pertinent description of the effects of an earthquake to find or clearly deduce its correct year of occurrence.

Manuscripts Reproduction Related Errors (Type V)

The only way to reproduce manuscripts was to recopy them manually, often by others than the original author. Different versions exist derived from an original work. The famous work of the Muslim tenth century paleographer Jalal-Eddine Al-Suyouti exists in twenty copies of variable qualities (Al-Sadani, 1971) p. XVII). One of them, namely that of the British Museum library, Ms. No. 5872, is known to be incomplete and to have some lack of accuracy problems (Al-Sadani, 1971) p. XIV). To illustrate this kind of difficulties by an example, the earthquake of 11 March 1068 AD is said to have destroyed the totality of Ramla except 2 *"houses"* (in Arabic *DARRAN*) in one manuscript or except 2 *"lanes"* (*DARBAN*) in another copy. The process of copying tend to produce, propagate and amplify errors. However this is not the only aspect of the problem, the ancient Arabic writing style is much more difficult to read and interpret. Recognizing the correct signification of a written word which could have a number of very different meanings is a matter of habit, training and context. Old style writing means more difficulties at least as far as the habit factor of the reader is concerned.

Erroneous Seismological Quantifications Based Upon Inaccurately Translated or Understood Texts (Type VI):-

This is best represented by the following case; Taher (1979) (Poirrier & Taher, 1980) presented "A full corpus of texts from Arabic sources and a summary French translation". This work is considered to be "By far the most valuable compilation of material on the seismicity of the region" Ambraseys et al. (Ambraseys et al., 1994, p.7). Although we agree with the general ideas implicated by the upper mentioned statement, it is necessary to say that some parts of Taher's translations were potentially very misleading from the seismological point of view. In his revision of the area's seismicity, Abou Karaki (1987) gave the following example concerning the earthquake of the year 425 AH = 1033 AD, Taher's French translation of part of the Arabic text concerning that earthquake began as follows *"Cette annèe un très violent tremblement de terre ravagea la Syrie et l'Egypt = This year a very violent earthquake ravaged Syria and Egypt"*(Taher, 1979) p. 35). A more accurate and faithful translation of the Arabic text would in fact be *"This year there was an increase (or multiplication) of earthquakes in Syria and Egypt"*(Abou Karaki, 1987), p.131). It is clear that the first seismologically erroneous translation would give a much higher magnitude for "the earthquake", if a magnitude calculation operation based on that translation and the "radius of perceptibility" concept is "committed". The work of Taher (1979) was the basis of the parametric-macroseismicity information catalog of Poirrier and Taher (1980). Despite the inaccuracies "It nevertheless remains the most authorative and reliable list of events in the region up to 1800, thanks to its reliance on primary sources" that was the opinion expressed by Ambraseys et al. (1994). However we think that primary sources when they exist are not quite enough, it should be mentioned here that "Taher's work is the starting point for (Ambraseys et al., 1994) retrieval and reassessment of historical information" (Ambraseys et al., 1994, p. 11). Our revision of Poirrier and Taher's work of 1980, and the application of Abou Karaki's algorithm for the detection of errors in the historical seismicity catalogs of the Arab region, (Abou Karaki, 1987, 1992, 1995a) allowed us to discover that 59 dates out of 240 ones associated with the historical earthquake's list of Poirrier and Taher (1980) are in fact erroneous. Some quantifications and interpretations are not less erroneous and misleading, and have already injected new ambiguities in the historical seismicity domain of this area. One example is provided by examining the following statement concerning, once more, the earthquake of March, 18th 1068 AD in Poirrier and Taher's remark "d" (Poirrier & Taher, 1980) p. 2199 (which should be "c" by the way) they wrote "*Al Djawzi reports that at Khaibar, the ground opened up and treasures were revealed. As Sayouti reports that at Tayma, the ground opened up. These features of ground deformation, usually restricted to the epicentral zone, plus the widespread destruction in Arabia suggest that this was an intraplate earthquake with its epicenter in Arabia despite the fact that there were sea waves on the Egyption and Israeli coasts. Perhaps we are dealing here with two close seisms*". Independently "?" of this statement Ambraseys and Melville (1989), and later Ambraseys et al. (1994), also enhanced this ambiguity by presenting a somewhat "artificially" strong case for the major 1068 AD earthquake along with two other less "well" documented ones (873 AD, 1588 AD) to be considered as intraplate earthquakes taking place inside northwestern Arabia, more than 150 to 250 km east of the nearest plate boundary zone in the area. This aspect along with other various sources of errors will be briefly discussed under our description and revision of the following case.

25.2.4.2 The "Ramla" Earthquake of 11 Jumada I, 460 AH, March 18, 1068 AD

We briefly show results of our revision processes and analysis as applied on the case of the Tuesday, 11 Jumada Al-Awla 460 AH 18 March 1068 AD well documented earthquake. This historical earthquake is reported in primary sources of Arabic documents and manuscripts, to have destroyed Ramla in Palestine (causing 15,000 to 25,000 causalities, mostly at Ramla as a result of a tsunami following the earthquake), and to have destroyed Ayla (modern Elat and Aqaba with the death of all but 12 of its inhabitants, Banyas to the North in southwestern Syria, suffered 100 casualties), and to have affected parts of Egypt, northwestern Arabia and Iraq. However the earthquake is referred to in most Arab manuscripts and primary sources as the earthquake of Ramla.

Our Analysis of some modern works revisiting the March, 18 1068 AD earthquake show important problems in interpreting the earthquake effects with some degree of information manipulation freely practiced on the historical data in other contributions as well. Some of the historical seismicity information has clearly been misinterpreted. We detected in a number of otherwise, usually "Authorative" historical seismicity works (Ambraseys et al., 1994; Ambraseys & Melville, 1989), a clear tendency and efforts to maximize the destructive effects of this earthquake in Arabia and to minimize these effects in Ramla and elsewhere. This led to shifting of the earthquake's epicenter hundreds of kilometers from the southern part of the JDST to northwestern Arabia near Tabuk about 250 km far away from the nearest recognized "axis" of the plate boundary. However, the simple plate tectonics and seismogenic sources basic principles are also certainly not in favor of the new location. The Ambraseys approach in dealing with historical texts was further criticized by a specialist "The Translations of Ambraseys were not complete, he altered the description of events in a way incompatible with the way ancient historical text should be treated" Al-Ghunaim (2002) p. 27. This led to very serious location and timing problems for this major earthquake that we examine hereafter.

Location Problems of the Earthquake of Ramla 18 March 1068 AD

Table 25.2 gives a synoptic summary allowing at this stage to compare between what we consider as the more realistic accounts concerning the Ramla earthquake affects and locations (Ambraseys, 1962) and the somewhat artificial radical evolution of ideas concerning this earthquake's description and location as revisited in (Ambraseys & Melville, 1989; Ambraseys et al., 1994).

Table 25.2 Showing the extent in the location evolution of the 1068 AD earthquake

Description after Ambraseys (1962) (Ambraseys & Melville, 1989)	Description after Ambraseys et al. (1994)
An earthquake in Palestine. Ramla was destroyed; it extended to the Hejaz. It reached also Wadi-el-Szafrh (Safra), Khaibar, Bedr (Badr), Yanbah(Yanb'u), Wadi Kora,Teima and Tabuk, and it extended as far as Kufa, **only two houses remained and 25,000 persons perished**	**A major Earthquake in the Hejaz and northern Arabia**...killed in all about 20,000 people. Aila was completely destroyed with all but 12 of its inhabitants. In Tabuk, three springs of water appeared and in Taima the ground was "split open." Near here a spring of water gushed out. The earthquake was felt at Khaibar, Medina (where the shock brought down two decorative crestings of the mosque of the prophet), Wadi al-Safra, Wadi al-Qura, Badr, and Yanbu'...**the earthquake was strong enough in Palestine to damage al-Ramla and ruin many houses with loss of life.** Damage was reported from Baniyas, where about 100 people were killed
In conclusion, it is clear from the comparison of the 2 texts (the **bold** parts in particular) that there was a shift of ideas implicating a serious shift of emphasis which led to a transfer of this earthquake's probable location from the DSF to a none probable one somewhere in Hejaz and northern Arabia. It should be mentioned that first hints implicating a new location of this earthquake to be in Arabia were suggested by (Ambraseys & Melville, 1989)	
This evolution is unjustified when closely examined after the original Arabic primary sources	

A possible hint and clue towards an explanation of the inherent cause of this transfer of location to Arabia may be in the following "It is a shame, in a sense, that there are not more earthquakes to record in Saudi Arabia itself, and that the focus of our attention has thus strayed inevitably to the surrounding regions." Ambraseys et al. (1994) page xix. Who also admitted the lake of robustness of the earthquake's location in Arabia later elsewhere "A major event in 1068 in the Hejaz in north-western Arabia is unusual not only because of its location but also because of the evidence admittedly slight, suggesting a surface rupture. The location of which must be sought in the region of Tabuk". Ambraseys and Jackson (1998) p. 399 point 9. It should be stressed that the reported "surface rupture" is not at all credible, the primary source this was based upon put it this way "Al Djawzi reports that at Khaibar, the ground opened up and **treasures were revealed**" (Poirrier & Taher, 1980*)* p. 2199. An extensive compilation of the Arabic sources of the earthquake of Tuesday 11 Jumada I 460 AH = 18 March 1068 AD is found in Al-Ghunaim (2002) pp.117–120.

Timing Problems and False Twins Related to 1068 AD Ramla Earthquake

After the application of Abou Karaki's algorithm for the detection of errors in the historical seismicity catalogs (Abou Karaki, 1987, 1992, 1995a), the following dates in various catalogs of the region must be regarded as very probably twins of false historical seismicity earthquakes 160 AH, 11 Jumada I 462 AH, 20 April 1067 AD, 20 April 1068 AD, 1069 AD, 2 Feb. 1070 AD, 25 Feb. 1070 AD, 26 Feb. 1070 AD.

These are all related to the Ramla 18 March 1068 AD earthquake, resulting from various possible errors populating a number of lists, references and catalogs, one or more of these or similar errors are still found and possibly propagating in the following works (Abou Karaki, 1987; Ambraseys, 1962, 2009; Ambraseys et al., 1994; Ambraseys & Melville, 1989; Amiran et al., 1994; Ben Menahem, 1979, 1991; Khair et al., 2000; Poirrier & Taher, 1980; Seiberg, 1932; Taher, 1979; Vered & Striem, 1977; Willis, 1928, 1933) a non-exhaustive list.

In conclusion, historical seismicity data was reassessed and subjected to an algorithm for the detection of errors in the historical seismicity catalogs (Abou Karaki, 1987, 1992, 1995a). Tens of false earthquakes were purged from the lists. The critical revision of the historical seismicity catalogs resulted in the following revised slim list Table 25.3.

25.2.5 *Instrumental Seismicity of the DSF, 1900–2021, M $\geq$ 6*

The first relatively nearby seismological station to the DSF was established very early in 1892 in Helwan Egypt, the second closer one started in Ksara in Lebanon in 1910, both along with 80 world stations recorded the Palestine (Jericho) earthquake of 11 July 1927 M 6 ¼ which caused more than 300 killed and widespread

Table 25.3 Historical seismicity list of the DSF for equivalent ~ M 6 or more based on a revised version of Abou Karaki (1987) and references therein

Date	φ°N	λ°E	Affected localities, intensity class, remarks	Io	Mag.	Fault segment (s) locations, remarks
31 BC	32	35.5	Qumran (2), Jerusalem(2), Jordan Valley	2	6.5	JVA (PS)
33 AD	32	35.5	Jerusalem	2	6.0	JVA
48	30	35.2	Jabal Rum, Petra, Tal El-Khalaifeh	2	6.5	WAR/GAQ
112	31.5	35.5	See Russel (1985).		6.5	WAR/JVA
19 05363	31	35.5	See Russel (1980), Guidoboni et al. (1994)	2	6.5	WAR/JVA
419	33	35.5	Safad (2), Aphek (2), Jerusalem (1)	2	6.0	JVA/BEQ
634/635	32.5	35.5	Palestine (30 days of Activity).		6.0	JVA
748/749 AD = Ramadan 130 AH. (or 131)	32	35.5	Jerusalem (2), Tiberias (2), Lod (2), Jerash (2), Jericho (2), Arad (2), Damascus (1), Syria (1), Egypt (1), Arabia (1), Mesopotamia (1)		7.0	JVA (PS) A minimum of 3 Shocks.
853/854	33	35.5	Tiberias	2	6.0	JVA
1033	32.5	35.5	Ramla, Tiberias, Jericho,Nablus Jerusalem, Akka, Gaza, Askalan	2 2	6.5	JVA
1047/48 439 AH	31	35.5	Ramla	2	6.0	Jordan Valley or Wadi Araba (JVA/WAR)
Tuesday 18 031068 AD 11 Jumada I 460 AH	31.5 30.0	35.5 35.0	Ramla (2), Banias (2), Jerusalem Aqaba (2), Tabuk (1), Tayma (1) Khaibar(1), Wadi Es Safra (1), Egypt (1)	2	7.0 6.5	WAR WAR or Gulf of Aqaba (GAQ)
24 121,105			Jerusalem	2	6.0	JVA
1156/59	31	35	Petra (2), Bethlehem (1), Egypt (1)	2	6.5	WAR
1160	32	35.5	St Jean Monastery	2	6.0	JVA
2 051212	30	35	Aqaba-Elat (2), Shobak (2), Karak (2), Cairo (2), Egypt (2)	2	6.5	WAR/GAQ
March/April 1260	32.5	35.5	Beisan (2), Galilee (2), Damascus (1)		6.5	JVA (Ambiguity) Abou Karaki (1987)
1261/1262	30	35	Karak, Shobak, Cairo		6.0	WAR
1269	32	35	Qualquilia (2), Imoisse (2), Sernour (2), Ballouta (2), Hajar al-Asher (2)	2	6.5	JVA, 8000 victims
January/ February 1293	31	35.5	Karak (2), Ramla (2), Gaza (1), Lod (1)	2	6.0	WAR
1458/1459	31	35.5	Karak (2) 100 victims in Karak	2	6.0	WAR

(continued)

Table 25.3 (continued)

Date	φ°N	λ°E	Affected localities, intensity class, remarks	Io	Mag.	Fault segment (s) locations, remarks
Jan 1546	32	35.5	Jerusalem (2), Hebron (2), Karak (2), Salt (2), Nablus (2), Jaffa (2), Gaza (2), Damascus (2)	2	6.5	JVA
5 01 1588	30	35	Aqaba (2), Cairo (2), Egypt, Sinai	2	6.5	WAR/GAQ
23 05 1834	32	35.5	East of the Dead Sea, Karak, Jerusalem, Gaza (2), Nablus (2), Askalane (2), Akka, Tiberias, Palestine, Syria	2	6.0	JVA
1 01 1837	33	35.5	Safad (2),Tiberias (2),Reineh (2), Ein Zeitoun (2),El-Jish (2), Sejera (2), Tyr (2),Sidon (2),Beirut (2), Damascus (2), Nazareth (1),Jerusalem (1),Hebron (1), Jericho (1), Tripoli (1).	2	6.5	JVA or north 5000 victims Asphalt Blocks in the Dead Sea.

Abou Karaki (1995a) 2 additional events (112 AD and 363 AD) were added from Russel (1980, 1985) and Guidoboni et al. (1994). Intensity classes 1 = Non-destructive [III, VI]. 2 = Destructive [VII-IX]. 3 = Very Destructive [X-XII]. Location errors of about ±1° are possible for the historical earthquakes, magnitude estimates for these earthquakes are generally minimum equivalent Ms. (Abou Karaki, 1987) The earthquakes listed in this catalog were individually discussed in Abou Karaki (1987) pp. 81–178. And 352–359. (PS) = C14 dated with additional Paleoseismicity evidences

destruction both sides of the DSF, the routine modern monitoring of the seismicity by a local network of seismological stations (Jordan Seismological Observatory, JSO) only started to operate East of the DSF in 1983. Just after the Institute of Petroleum Research and Geophysics (IPRG) based in Holon and operating west of the DSF. There were important location accuracy problems at the beginning of the JSO operations in particular (Abou Karaki, 1987). A number of swarms in 1983 near Haql at the gulf of Aqaba starting January 21, and in the Carmel Wadi El-Fara'a area in 1984 were analyzed and relocated (Abou Karaki et al., 1993; Abou Karaki, 1987, 1994, 1995a, b). Several earthquakes of these swarms were around ML = 5 the maximum magnitude did not exceed 5.2. Only 3 destructive or potentially destructive earthquakes occurred on the DSF during the instrumental period, their parameters are shown in Table 25.4.

To illustrate the revision processes aspect, we examine the results of our analysis as applied on the case of the earthquake of Palestine of 1927. The most destructive DSF earthquake during the instrumental seismicity era.

25.2.5.1 The Location of the Earthquake of Palestine 1927

The earthquake of Palestine (known also as the earthquake of Jericho), took place near the Damia area in the Jordan valley on the early afternoon of Monday 11-07-

Table 25.4 The M ≥ 6 class of instrumental earthquakes on the DSF

Date	φ°N	λ°E	Affected localities, intensity class, remarks	Io	Mag.	Fault segment (s) locations, remarks
11 07 1927	32	35.5	Nablus(2),Salt(2),Ramla(2),Jericho(2), Reineh(2). From Abou Karaki (1999)	2	6.2	JVA, Abou Karaki (1999)
03 08 1993	28.75	34.6	Aqaba- Elat 1, Haql 1, Cairo 1,	1	6.0	GAQ
22 11 1995	28.75	34.8	Nuweiba 2, Dahab(2), Haql (2), Aqaba-Elat (2), Tabuk (2) from (Abou Karaki, 2001a; Klinger et al., 1999).	2	7.3 *Mw*	GAQ (Abou Karaki 2001a)

The locations of the 1927 and 1995 earthquakes were fully discussed respectively in (Abou Karaki, 1999, 2001a). A brief reminder of the most destructive instrumental earthquake and related problems is reminded herein

1927, This Magnitude 6 ¼ earthquake, is still the most destructive earthquake since the earthquake of Safad back in 1837 AD on both sides of the river Jordan and the JDST. It caused more than 300 killed and widespread destruction. Figure 25.3 (Abou Karaki, 1987; 1999; Vered & Striem, 1977; Willis, 1928) and references therein. A reasonably accurate determination of this earthquake's epicenter is an important factor for a useful assessment of the seismic hazard in this part of the JDST. However, there were a number of new attempts to relocate this earthquake by applying rather modern location techniques using old data of questionable quality, recorded by that time seismological stations that were always suffering from known serious synchronization problems thus affecting the internal consistency of the different arrival times. A new calculated epicenter location resulted from those attempts (Avni et al., 2002; Shapira et al., 1993) clearly and anomalously shifting the location of the 1927 earthquake out of the main macroseismic area of damage. The quality of the new location (31.6°.0 N, 35°.4E) was analyzed, critically examined and finally assessed as being far less convincing than the old original ISS location (32°.0 N, 35°.5E) (Abou Karaki, 1995a)The main basis for the rejection of the new location could be seen on Fig. 25.3 and Table 25.4.

The (SAN) location Fig. 25.3 is a direct consequence of the lack of internal coherency in the data used as an input for the location computations. This is evidenced by examining the following elements:

The Earthquake was located assuming a normal depth and using the IASPEI travel time tables applied on data provided by 30 seismological stations. The calculated – observed residual values as derived from the data of Table 25.1 in (Shapira et al., 1993) were generally and unacceptably high. The nearest stations were (KSA) Ksara in Lebanon, situated at some 200 km north of the epicenter, has a calculated-observed residual of 4.6 s, and (HLW) in Helwan-Egypt at some 450 km south west of the epicenter has a residual of −2.5 s. 78% of the arrival times used in the computations yielded residual values (C-O) > |1.5 s| among these 22% were associated with (C-O) values > |4.0 s|. These elements show that the "new" epicenter

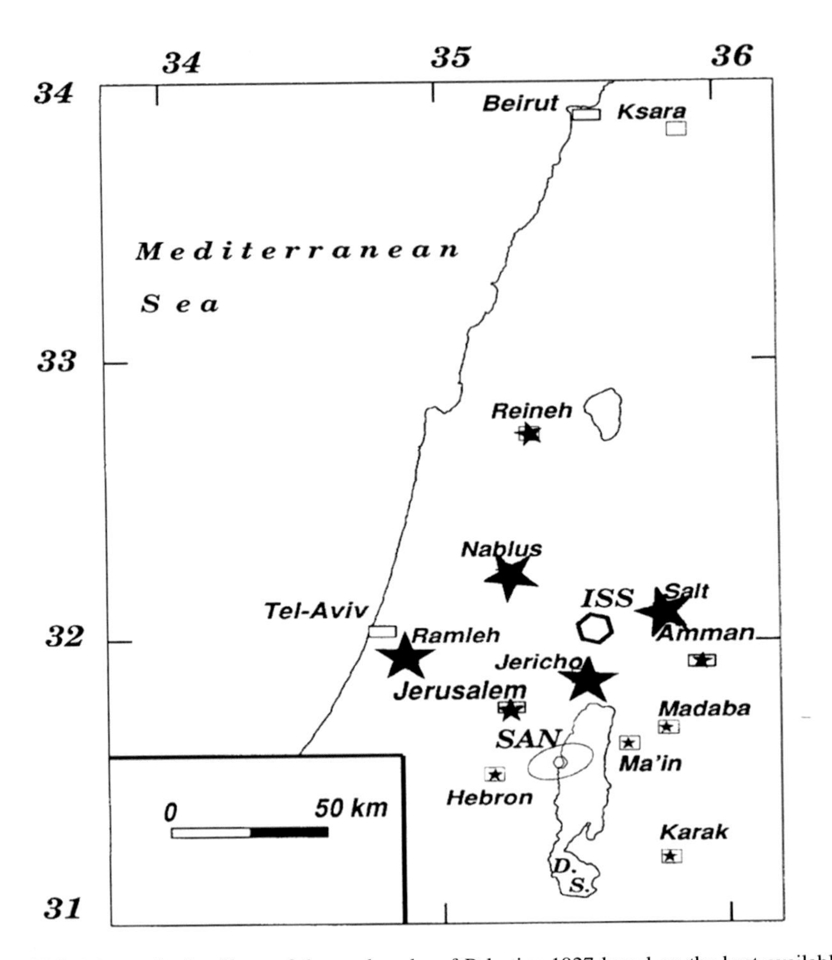

Fig. 25.3 Macroseismic effects of the earthquake of Palestine 1927 based on the best available primary sources (Abou Karaki, 1999). Damage proportional to the size of star, main destructive effects in Nablus, Salt, Ramla, and Jericho, minor damage in Hebron, Ma'in, Madaba. Original (ISS) epicenter location is shown. A new location, some 50 km to the south of the ISS location is also shown along with its "uncertainty ellipse" after (Avni et al., 2002; Shapira et al., 1993) (SAN). This location, clearly, does not fit the macroseismic data, it is based upon new computations using body waves arrival times as recorded by that time seismological stations which used to suffer from serious synchronization problems. The consequences of this are clearly shown in Table 25.4

is at best mathematically equivalent to the old one and the computational arguments of such nature and in such circumstances can't be used to justify or validate a new location of the epicenter. Furthermore the particular C-O values of KSA and HLW resulting from the new epicenter calculation mean that the epicenter should better be closer to KSA and more far away from HLW this is just the case of the ISS original epicenter (Table 25.5).

Yet it is noticed that the new location is already being uncritically followed and adopted by a number of more recent seismicity studies (Amiran et al., 1994; Jaber, 1994; Ken-Tor et al., 2001).

Table 25.5 Shows a representative part of the data used to calculate The (SAN) epicenter, those with epicenter-station distance <20°

Station	Phase	Arrival time	Calc-Obs (s)	Distance(deg)	Azimuth
KSA	P	13:04:43	4.6	1.8	1
HLW	P	13:05:10	−2.5	4.3	241
	S	13:06:07	2.9		
BAK	P	13:07:34	5.1	14.2	49
MKY	P	13:07:55	−3.0	16.4	5
	S	13:10:58	−4.2		
BEO	P	13:08:10	−1.6	17.5	321
	S	13:11:32	3.6		
BMP	P	13:12:13	4.2	19.2	303
NPL	P	13:08:37	4.1	19.3	303
LVV	P	13:12:23	0.7	19.8	337

Partially reproduced from Table 1 in (Shapira et al., 1993)

25.3 Conclusions About the JDST Seismicity – Identification of Seismic Gaps

It is important to notice some aspects of the seismicity of the Wadi Araba-Gulf of Aqaba segment(s) of the JDST. These segments, located south of the Dead Sea, are particularly interesting. From the historical seismicity data (Abou Karaki, 1987) the Gulf of Aqaba part of the transform passed from a clear state of seismic quiescence for the last few centuries, (since 1588 AD at least) (Abou Karaki, 1987), and from the status of the least seismically active segment of the transform during these centuries to the most active zone of the JDST since January, 1983 (Abou Karaki, 2001a), when it entered into a period of increased activity marked by a number of swarms, starting on January, 21, 1983 (Abou Karaki et al., 1993; Abou Karaki, 1987; Alamri et al., 1991; El-Isa et al., 1984; Russell, 1985) and culminating with the 7.3 Mw earthquake of November, 22 1995 (Abou Karaki, 2001a, b; Klinger et al., 1999). Paleoseimicity investigations Kanari et al. 2020 (Kanari et al., 2020) generally confirmed our conclusions from the historical seismicity results and catalogs (Abou Karaki, 1987, 2001a) and as shown in Table 25.3.

The Wadi Araba part (WAR) remained quiescent till now (Jan. 2022), this makes of the Wadi Araba-Gulf of Aqaba adjacent segments, the least and respectively most seismically active segments of (JDST), during the instrumental seismicity era.

In Conclusion, historical and instrumental seismicity of the JDSF along with relative seismic gap areas are shown on Fig. 25.4.

Finally Fig. 25.5a presents the JDSF instrumental seismicity 1900–2020 for $M \geq 4$ to avoid the noisy epicentres resulting from artificial events resulting from industrial mining explosions an example of which is given in Fig. 25.5b.

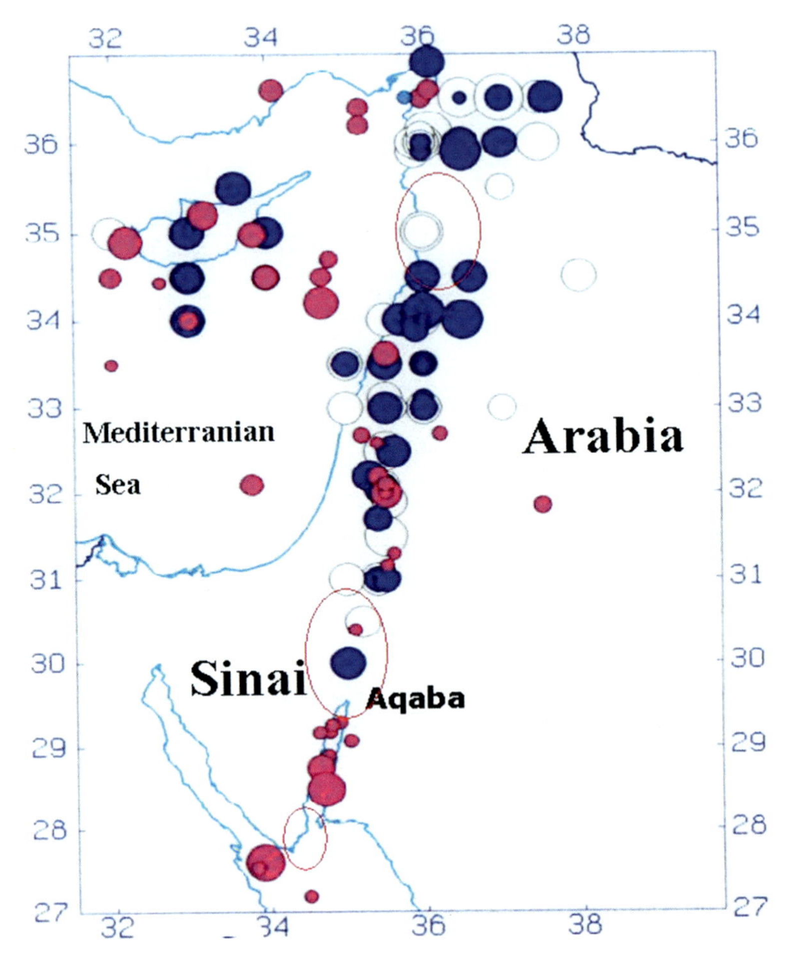

Fig. 25.4 Seismicity of the JDST from 31 BC up to 2020 AD. Open circles represent the historical earthquakes for the period from 31 BC–1200 AD. Bleu circles from 1200–1900 AD. With equivalent Ms. Magnitudes [6, 7.5] based on the revised catalog of Abou Karaki 1987 revised 2021. Instrumental seismicity epicentres from 1900 and on are represented in red solid circles for magnitudes from [5, 7.3] (Abou Karaki, 1994, 1995a, 2001a) and USGS, EMSC databases. Resulting locations of 3 relative seismic gap areas are represented in elliptical forms on this figure namely centred around the latitudes (28°N, 30°N and 35°N ± 0.5°)

25.4 Dead Sea Remote Sensing Studies

25.4.1 Problems Related to Subsidence in the Dead Sea Coastal Areas

Since the 1960s, the Dead Sea coastal areas are expanding because the Sea is undergoing what is an essentially induced, and extremely rapid decrease of its water level (Abou Karaki et al., 2007). The drop in the Dead Sea level caused the

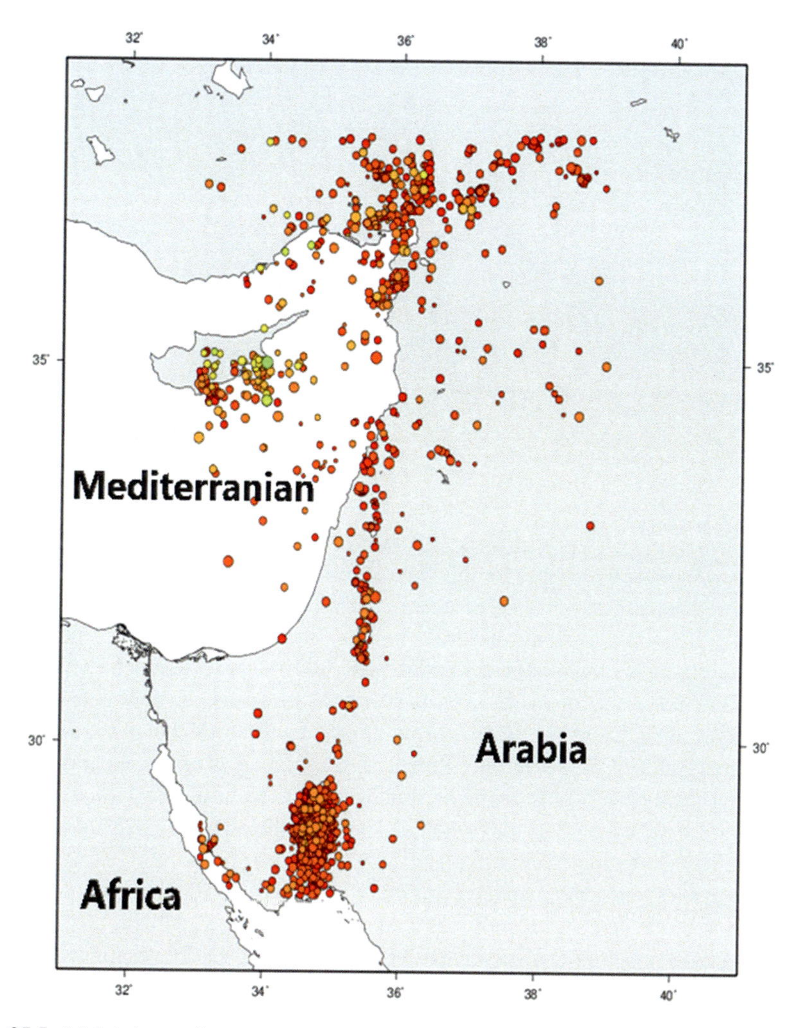

Fig. 25.5 (**a**) It is interesting to compare the locations of the relative seismic gaps represented in Fig. 25.4 with the instrumental seismicity map Fig. 25.5. For the magnitudes $\geq$4 It is still possible to identify the 2 main gaps which are comparable on the two figures ($\approx$ 30°N and 35°N. $\pm$ 0.5°). Epicentres are from the same sources of data as Fig. 25.1a In conclusion it is clear that limiting the represented magnitudes on the seismicity map of the JDST from $\geq$3 in Fig. 25.1a to $\geq$4 in Fig. 25.5 is more robust and useful for the tectonic significance of the epicentres distribution if the unchecked international seismicity monitoring organizations data is used. (**b**) Industrial detonations amounting to 10,000 kg of explosives are routinely used in mining operations of the Phosphate Mines Co. at the Phosphate extraction sites in Southern Jordan (El-Hisa and Al-Shidia Mines). This explains the "epicentral" non-tectonic clusters with low magnitudes around the coordinates (30°N, 36.4°E and 31.2°N, 36.4°E) in Fig. 25.1a

Fig. 25.5 (continued)

loss of more than a third of its water surface (Salameh & El-Naser, 1999, 2000). This is already inducing serious hazards in the area related to subsidence and hydrogeological problems (Abou Karaki et al., 2016, 2017; Salameh et al., 2019). The fundamental cause of these hazards is known to be the sharp imbalance in the water budget of the Dead Sea. This is provoking a steady shrinkage of the sea surface and the emergence of new unstable lands in the coastal areas of the sea coupled with the development and proliferation of rapid collapse features (Sinkholes) on both sides of the Dead Sea zone. If the Dead Sea level continues to decrease, these hazards are only expected to increase in intensity and extension.

Between 1992 and 2021, we have intensely studied the Jordanian part of the Dead Sea with remote sensing techniques (Abou Karaki et al., 2016, 2017, 2019; Al-Halbouni et al., 2017; Closson & Abou Karaki, 2009, 2014; Closson et al., 2003, 2005, 2010; Fiaschi et al., 2017; Watson et al., 2018) The nature of this work and the methods of image processing have evolved along with the discipline. We have tested many optical and radar sensors. One of our studies (Abou Karaki et al., 2016) was classified by (p.9, tables 4 and 5 in (Pricope et al., 2019)) as the study which used a maximum of satellite sources among 83 analyzed studies worldwide. Table 25.6 takes up the list of satellites used during the investigation period.

Table 25.6 provides an overview of the technologies' evolution over the past 60 years. In the 1990s, there were only a few optical satellites equipped with imaging systems capable of providing images with a resolution between 10 and 30 m (Spot, Landsat).

It was therefore impossible to visualize sinkholes with this data, except for the larger ones. Radar images provided an alternative but there was still no software capable of processing radar images in the manner of a freeware like SNAP.

Table 25.6 List of satellites used, investigation period, and image processing applied

Radar satellites	Periods of investigation	Image processing
ERS-1	1992–1999	Interferometry: InSAR (DSM); DInSAR; PS; SBAS
ERS-2	1997–2011	
Envisat	2003–2012	
ALOS-1	2008–2010	Interferometry: DInSAR
Cosmo-SkyMed	2011–2013	
Sentinel-1	2014–2021	Interferometry: DInSAR; PS; Sbas
Space Shuttle	1994	Visual interpretation of greyscale Intensity
Optical satellites	Periods of investigation	Image processing
Corona	1963–1971	Geocoding; visual interpretation of greyscale Intensity
Landsat 1 to 8	1972–2021	Geocoding improvement; textural filtering; band ratio and combination diachronic color composition; LULC classifications
Sentinel-2	2015–2021	
Spot	1992–1994, 2003	
Ikonos	2000, 2003, 2006	Geocoding improvement; textural filtering; band ratio and combination diachronic color composition; LULC classifications; detection of sinkholes,vegetation, and sel pillars in evaporation ponds
Aster	2000–2005	
Worldview	2010–2012	
Geoeye	2009	
Rapideye	2010	
Planet Lab	2018	
Bing; Google Map	2000–2021	Geocoding; visual interpretation

Soil moisture detection algorithms were used on optical images to detect wetlands (Closson & Abou Karaki, 2014) within land discovered because of the Dead Sea water level lowering (> 1 m/year). These moisture anomalies have been interpreted as the places where groundwater moves from the recharge zones to the ultimate base level (i.e. the Dead Sea), but also rivers eroding their beds in order to stay in balance with the base level (Abou Karaki et al., 2016).

We have observed that between the time of their emergence, in the 1980s, and the 2010s, these underground water flows gave rise to real karst networks capable of destroying agricultural land, destabilizing the installations and industrial dikes of the Arab Potash Company APC, as well as national roads bridges and hotels.

Figures 25.6 and 25.8 illustrate this point. Its principle can be generalized to the entire shores of the Dead Sea (Abou Karaki et al., 2019). These geomorphological maps were drawn from multiple observations made with optical Fig. 25.7 and radar images collected in 1992 to 2015 (Closson & Abou Karaki, 2009; Closson et al., 2003, 2005, 2010; Watson et al., 2018).

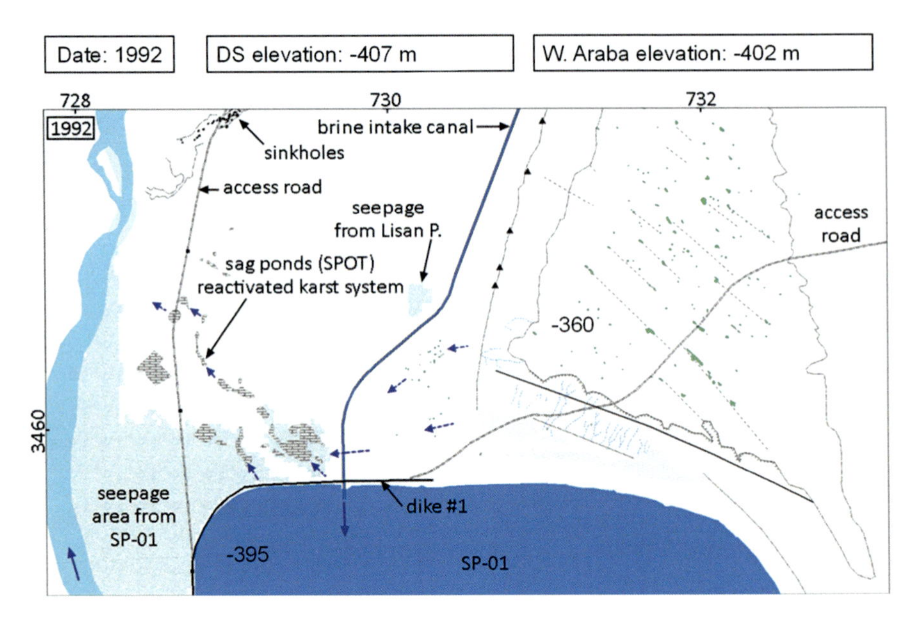

Fig. 25.6 Development of a karst system in the early 1990s with the Landsat and Spot satellites

Fig. 25.7 Satellite image (source: https://www.bing.com/maps) most likely acquired in 2015. This image is contemporary of Fig. 25.8

The region corresponds to Cape Molyneux (see the elevation of −360 m). The coordinates are expressed in UTM kilometers. Eastern side, a plateau bounded by a cliff dominates a gentle plain inclined to the North, towards the Dead Sea. This plain is crossed by the Wadi Araba (river to the west). In 1992, the Wadi Araba had an altitude of −402 m. It was therefore 42 m lower than the karst plateau of Cape Molyneux.

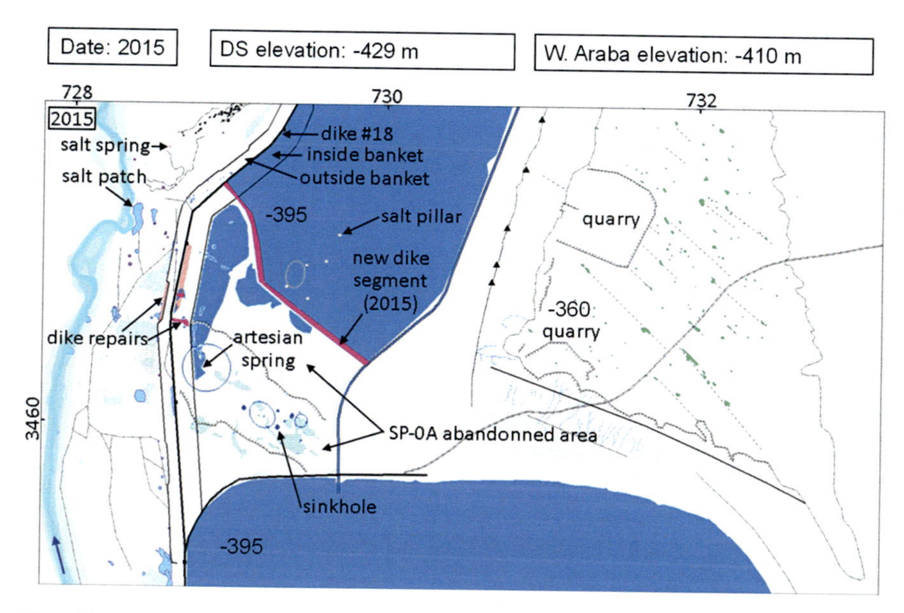

Fig. 25.8 Geomorphological map of the region west of Cape Molyneux in 2015. A levee (purple) was created to isolate a segment of the dam threatened with collapse. Many sinkholes attest to an intense flow of water below the basin

In winter, rainwater enters through the plateau because of hundreds of sinkholes developed along lineaments of tectonic origin (green spots and dashed lines). These lineaments and sinkholes are accessible via an access road that leads to the APC dikes' system.

The southern part of the cliff is of tectonic origin. A fault segment ESE-WNW (continuous black line) is represented. It is a small segment of a major transversal fault zone that connects the East and West fault systems of the Dead Sea Basin. This fault zone is kilometer-deep and drains some of the fresh water coming from the eastern part to the Wadi Araba. This fact is attested by the presence of fresh water sources in the Cape Molyneux area (oral communication of APC dike safety engineers, 2015).

South of Cape Molyneux is an evaporation basin (SP-01; −395 m). Given the difference of 7 m in water level between this basin and Wadi Araba (−402 m), the water stored in the reservoir naturally tends to escape to reach the local base level. Earthen dike #1 holds the water in the reservoir. Due to the significant fracturing of the subsoil, a portion of the brine manages to escape and reach the Wadi Araba.

This percolation of water greatly increases soil moisture. Very clear infrared signatures appear on the optical images where these percolations occur. We detected a karst system that was reactivated by water from the SP-01 basin and also fed by the transversal fault zone (underground flows giving rise to a development of shrub vegetation). The blue arrows indicate the flow direction subparalel to the transverse fault.

Optical and radar images have allowed us to study in great detail the impact of the gradual development of the karst system on the stability of the APC dikes built in the late 1990s (Fiaschi et al., 2017). So, a few years after the retreat of the Dead Sea, at the time when the karst system was being reactivated.

Data on land use land cover changes and soil nature have all been mapped: creation of new salt evaporation ponds; seepage areas contemporary to the feeding operations; partial draining and gradual drying of the ponds to increase the safety coefficient; various repairs; refilling sinkholes; widening of dikes; collapse of dikes; appearance and disappearance of springs; appearance and disappearance of vegetation in connection with an underground freshwater supply; faults and fractures; subsidence; landslides; soil collapse; structural lineaments; erosion resistance ...

With the increase in sensor resolution (Fig. 25.7), it has been possible to visualize sinkholes both inside and outside the evaporation basins. Combining optical observations and radar, at the end of 2012, it was possible to send warning signals to the APC. The observed elements have enabled security engineers to deploy considerable geophysical means. Prediction of collapses based on diachronic analysis of high- and very high-resolution radar and optical satellite images was one of the major culminations of two decades of research.

Figure 25.2 illustrates this point. Its interpretation is in (Fig. 25.8). It shows the amputated part of an evaporation basin. A levee (SE-NW and SSE-NNW) was created in early 2015 to completely isolate the most unstable part of the dike. 2.5 km were drained for repair. Detailed analysis of the bottom shows that it is perforated in multiple locations by sinkholes (Fig. 25.8). These are organized according to an underground karst network whose main axes were already visible in 1992 (Fig. 25.6), several years before the creation of the basin.

Cape Molyneux was surrounded by the waters of the Dead Sea until late 1970s. During the 1980s, major development work took place in the southern part of the terminal lake. The wadi Araba delta was transferred from the southern Dead Sea basin to the northern one via a 20 km long channel leading to Cape Molyneux. Naturally, this river has become the local base level for all surrounding water tables (including brine stored in evaporation basins). Wadi Araba, like all other rivers flowing into the terminal lake, is forced to rejuvenate its longitudinal profile to adapt to Dead Sea level. As a result, all water stored in basins and adjacent freshwaters are gradually being mobilized.

Large layers of water stored in the marls of the Lisan Peninsula were thus displaced by the force of gravity. Underground water flows appeared very early after the Dead Sea was withdrawn. These waters have taken all areas of tectonic weaknesses created either by the rise of the salt diapir or by the tectonic activity of the Dead Sea basin.

As early as the 1990s, radar images were able to detect millimeter movements of the ground in a relative way. This information has been very useful to us in improving knowledge in the field of neoctonics. The absence of geological maps, the study of tectonics on the border (military zone),etc. was more than compensated by the contribution of radar interferometry data. The detection of active faults in connection with the Lisan diapir allowed us to place sinkholes on a map of active

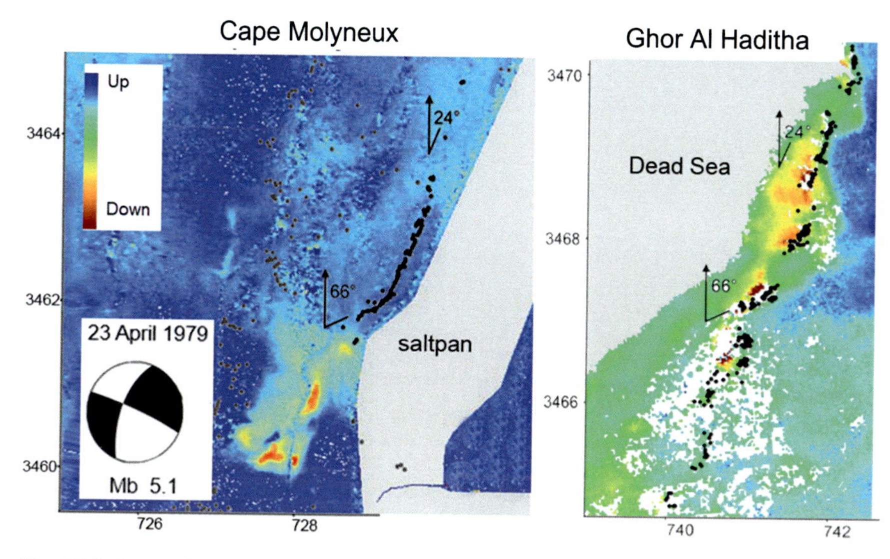

Fig. 25.9 Comparison between the distribution of sinkholes (black dots) of Cape Molyneux and Ghor Al Haditha. Two structural directions appear: N24E and N66E. N24E is in agreement with the EQ event of 23-04-1979 (Lisan). Focal mechanism after (Arieh et al., 1982). In Cape Molyneux region, the karst system is very active in the southern part of the lineament. The colors are those of an ALOS-2 differential interferogram. The bright colors correspond to the most intense subsidence during the observation period. Acquisition dates 01 April 2008–17 May 2008; T 46 days; perpendicular baseline 17 m; altitude of ambiguity 2705 m

faults and draw conclusions/models as to the genesis of sinkholes affecting the earthen dikes of the APC.

The main finding that can be attributed to the combination of image processing techniques with field observations is presented in Fig. 25.9.

Figure 25.9 is relevant in that it clearly shows the co-occurrence between elements of structural geology and ground collapses. The sinkholes that make up the two lineaments account for the majority of known collapses on the Jordanian side. Moreover, these two lineaments appeared at the same time, during the 1980s, a few years after the earthquake of 23-04-1979.

In this case, radar interferometry brings the dynamic dimension of the system. Optical sensors allow the counting and delimitation of sinkholes. Field observations provide information on the nature of the layers affected by collapses, as well as networks of cracks undetectable from space or even aerial photos. Differential radar interferometry shows the actual spatial extent of the phenomenon and its dynamics over time. Advanced techniques such as PS and SBAS have allowed us to establish a 25-year time series of data (Fiaschi et al., 2017).

Subsidence and sinkholes related problems and major expected resulting hazards in the sensitive Dead Sea area added to the recognized seismic hazards there (Abou Karaki, 1987; Bonnin et al., 1988) require mitigation, cooperation, funding and extensive multidisciplinary work (Abou Karaki & Closson, 2012).

Acknowledgments This work is supported by the Deanship of scientific research, The University of Jordan. www.ju.edu.jo

References

Abou Karaki, N. (1987). *Synthèse et carte sismotectonique des pays de la bordure orientale de la Méditerranée: sismicité du système de failles du Jourdain-Mer Morte*. Université Louis Pasteur de.

Abou Karaki, N. (1992). An algorithm for the detection of errors in the historical seismicity catalogs of the Arab region. In *Arabization: The Journal of the Arab Center for Arabization, Translation, Authorship and Publication. Damascus* (4), 139–153. (In Arabic).

Abou Karaki, N. (1993). *Testing the efficiency of seismological stations surrounding the Arabian Plate*. Abhath-al-Yarmouk, No.2, Vol. 2, (Pure and Eng. Series), 25–47.(In Arabic, English Abstract).

Abou Karaki, N. (1994). Analysis, relocation, and focal mechanism of the Carmel earthquake swarm of 1984. *Dirasat, 21B* (1), 281–291. (In English, Arabic Abstract).

Abou Karaki, N. (1995a). Re-evaluating the seismicity of the Jordan Dead Sea Transform System. In *Proceedings of the fifth Jordan geology conference and the third geology conference on the Middle East GEOCOME III*, Amman 3–5 October, 1994 (pp. 373–396).

Abou Karaki, N. (1995b). Testing routine locations of the Jordan Transform earthquakes. *Mu'tah Lil-Buhooth Wa Al-Dirasat, 10*(2), 9–26. (in Arabic, English abstract).

Abou Karaki, N. (1996a). Seismic hazard implications on the Jordanian population. In *Proceedings of the third Jordanian scientific week, Amman. The Higher Council for Science and Technology, environmental studies* (Vol. 5, pp. 279–303). (In Arabic)

Abou Karaki, N. (1996b). A comparative study of the seismicity of Jordan and the adjacent areas. *Al-Ta'ameen Al-Arabi, Journal of the General Arab Insurance Federation – Cairo, 14* (49), 36–56. (in Arabic,+ figure captions and Abstracts in English and in French).

Abou Karaki, N. (1999). Location of the earthquake of Palestine 11-7-1927. *A critical assessment, Abhath-al-Yarmouk, No.1, Vol. 8, pp. 9–34, (Basic Sciences and Engineering Series)*. (In Arabic, English Abstract).

Abou Karaki, N. (2001a). The Gulf of Aqaba earthquake activity of 1995: Geodaynamic context, analysis of the location of the main event (Mw =7.2) (in Arabic, English abstract). *Dirasat, 28*(1), 115–132.

Abou Karaki, N. (2001b). *Testing the efficiency of local and regional seismological stations involved in the monitoring of the Gulf of Aqaba activity (1995)* Al-Manara Vol. VII, No. 1 pp.25-60 In Arabic, English Abstract.

Abou Karaki, N., & Closson, D. (2012). *European Association of Geoscientists & Engineers-EAGE Workshop on Dead Sea Sinkholes, causes, effects & solutions*, Field Guidebook, 45 pages.

Abou Karaki, N., Closson, D., Salameh, E., de Schoutheete de Tervarent, M., & Barjous, M. (2007). Natural, induced and environmental hazards along the Dead Sea coast, Jordan. *Hydrogeologie und Umwelt, Würzburg, Heft, 33*(14), 1–25.

Abou Karaki, N., Fiaschi, S., & Closson, D. (2016). Sustainable development and anthropogenic induced geomorphic hazards in subsiding areas. *Earth Surface Processes and Landforms, 41*(15), 2282–2295.

Abou Karaki, N., Closson, D., Fiaschi, S., Galve, J. P., Alawabdeh, M., & Paenen, K. (2017). Can science save The Dead Sea ? Conference given at the World Science Forum (2017) Day 2, 8 November 2017, 16:30–18:00, Special Session. Available via https://www.youtube.com/watch?v=x15KeokjPfo. Accessed 28 Feb 2021.

Abou Karaki, N., Fiaschi, S., Paenen, K., Al-Awabdeh, M., & Closson, D. (2019). Exposure of tourism development to salt karst hazards along the Jordanian Dead Sea shore. *Hydrology and Earth System Sciences, 23*(4), 2111–2127.

Abou Karaki, N., Dorbath, L., & Haessler, H. (1993). La Crise sismique du golfe d'Aqaba de 1983: Implications tectoniques. *Comptes rendus de l'Académie des Sciences Paris*, t.317, Série II, 14111416.(In French + Abreged English version.

Alamri, A. M., Schult, F. R., & Bufe, C. G. (1991). Seismicity and aeromagnetic features of the. *Gulf of Aqaba (Elat) region Journal of Geophysical Research: Solid Earth, 96*(B12), 20179–20185.

Alchalbi, A., Daoud, M., Gomez, F., McClusky, S., Reilinger, R., Romeyeh, M. A., Alsouod, A., Yassminh, R., Ballani, B., & Darawcheh, R. (2010). Crustal deformation in northwestern Arabia from GPS measurements in Syria: Slow slip rate along the northern Dead Sea fault. *Geophysical Journal International, 180*, 125–135.

Al-Ghunaim, A. Y. (2002). *Earthquakes in Arab Records, their events and effects in Arab resources*. Geography Department, Kuwait University. (in Arabic).

Al-Halbouni, D., Holohan, E. P., Saberi, L., Alrshdan, H., Sawarieh, A., Closson, D., Walter, T. R., & Dahm, T. (2017). Sinkholes, subsidence and subrosion on the eastern shore of the Dead Sea as revealed by a close-range photogrammetric survey. *Geomorphology, 285*, 305–324.

Al-Sadani, A. (1971). *(Jalal-Eddine Al-Suyouti) Kasff Al-Salsala Wa Wasf Al-Zalzalah, in Arabic*. Rabat, Morocco.

Ambraseys, N. N. (1962). A note on the chronology of Willis's list of earthquakes in Palestine and Syria. *Bulletin of the Seismological Society of America, 52*(1), 77–80.

Ambraseys, N. N. (1971). Value of historical records of earthquakes. *Nature, 232*, 375–379.

Ambraseys, N. N. (2009). *Earthquakes in the Mediterranean and Middle East: A multidisciplinary study of seismicity up to 1900*, 947 Pages. Cambridge University Press.

Ambraseys, N. N., & Barazangi, M. (1989). The 1759 earthquake in the Bekaa Valley: Implications for earthquake hazard assessment in the Eastern Mediterranean region. *Journal of Geophysical Research, 94*, 4007–4013.

Ambraseys, N. N., & Jackson, J. A. (1998). Faulting associated with historical and recent earthquakes in the Eastern Mediterranean region. *Geophysical Journal International, 133*(2), 390–406. https://doi.org/10.1046/j.1365-246X.1998.00508.x

Ambraseys, N. N., & Melville, C. (1989). Evidence for intraplate earthquakes in northwest Arabia. *Bulletin of the Seismological Society of America, 79*, 1279–1281.

Ambraseys, N. N., Melville, C. P., & Adams, R. D. (1994). *The seismicity of Egypt, Arabia, and the Red Sea: A historical review*, edn, Vol., pp. Pages, Cambridge University Press.

Amiran, D. H. K., Arieh, E., & Turcotte, T. (1994). Earthquakes in Israel and adjacent areas: Macroseismic observations since 100 B.C.E. *Israel Exploration Journal*, 261–305.

Arieh, E., Rotstein, Y., & Peled, U. (1982). The Dead Sea earthquake of 23 April 1979. *Bulletin of the Seismological Society of Americ, 72*(5), 1627–1634.

Avni, R., Bowman, D., Shapira, A., & Nur, A. (2002). Erroneous interpretation of historical documents related to the epicenter of the 1927 Jericho earthquake in the Holly Land. *Journal of Seismology, 6*, 469–476.

Barjous, M., & Mikbel, S. (1990). Tectonic evolution of the Gulf of Aqaba – Dead Sea transform fault system. *Tectonophysics, 180*, 49–59.

Bazzari, M., Merghalani, H., & Badawi, H. (1990). *Seismicity of the Haql region, Gulf of Aqaba, Saudi Arabian Directorate General of Mineral Resources*. Open-File Reporte USGS-of-10-9, 39p.

Ben Menahem, A. (1979). Earthquake catalogue for Middle east (92 BC – 1980 AD). *Bollettino di geofisica teorica ed applicata, 21*, 245–310.

Ben Menahem, A. (1991). Four thousand years of seismicity along the Dead Sea rift. *JGR, 96*(B12), 20195–20216.

Ben-Menahem, A., Nur, A., & Vered, M. (1976). Tectonics, Seismicity and Structure of the Afro-Eurasian junction, the breaking of an incoherent plate. *Physics of the Earth and Planetary Interiors, 12*, 1–50.
Bonnin, J., Cara, M., Cisternas, A., & Fantechi, R. (1988). *Seismic hazard in mediterranean regions*. Springer.
Chaimov, T., Barazangi, M., Al-Saad, D., Sawaf, T., & Gebran, A. (1990). Crustal shortening in the Palmyride fold belt, Syria, and implications for movement along the Dead Sea fault system. *Tectonics, 9*, 1369–1386.
Closson, D., & Abou Karaki, N. (2009). Salt karst and tectonics: sinkholes development along tension cracks between parallel strike-slip faults, Dead Sea, Jordan. *Earth Surface Processes and Landforms, 34*(10), 1408–1421.
Closson, D., & Abou Karaki, N. (2014). Dikes stability monitoring versus sinkholes and subsidence, Dead Sea region, Jordan. In F. Holecz, P. Pasquali, N. Milisavljevic, & D. Closson (Eds.), *Land Applications of Radar Remote Sensing* (pp. 281–307). Intech.
Closson, D., Abou Karaki, N., Hansen, H., Derauw, D., Barbier, C., & Ozer, A. (2003). Space-borne radar interferometric mapping of precursory deformations of a dyke collapse, Dead Sea area, Jordan. *International Journal of Remote Sensing, 24*(4), 843–849.
Closson, D., Abou Karaki, N., Klinger, Y., & Hussein, M. J. (2005). Subsidence and sinkhole hazard assessment in the southern Dead Sea area, Jordan. *Pure and Applied Geophysics, 162*(2), 221–248.
Closson, D., Abou Karaki, N., Milisavljevic, N., Hallot, F., & Acheroy, M. (2010). Salt-dissolution-induced subsidence in the Dead Sea area detected by applying interferometric techniques to ALOS Palsar Synthetic Aperture Radar images. *Geodinamica Acta, 23*(1–3), 65–78.
El-Isa, Z., & Mustafa, H. (1986). Earthquake deformations in the Lisan deposits and seismotectonic implications. *Geophysical Journal of the Royal Astronomical Society, 86*, 413–424.
El-Isa Z., Merghelani H. M., Bazzari M. A (1984) The Gulf of Aqaba earthquake swarm of 1983 January–April, Geophysical Journal Research Astronomical Society,78, pp. 711–722.
Ellenblum, R., Marco, S., Agnon, A., Rockwell, T., & Boas, A. (1998). Crusader castle torn apart by earthquake at dawn, 20 May 1202. *Geology, 26*, 303–306.
Ferry, M., Meghraoui, M., Karaki, N. A., Al-Taj, M., Amoush, H., Al-Dhaisat, S., & Barjous, M. (2007). A 48-kyr-long slip rate history for the Jordan Valley segment of the Dead Sea fault. *Earth and Planetary Science Letters, 260*, 394–406.
Ferry, M., Meghraoui, M., Abou Karaki, N., Al-Taj, M., & Khalil, L. (2011). Episodic behavior of the Jordan Valley section of the Dead Sea fault from a 14-kyr-long integrated catalogue of large earthquakes. *Bulletin of the Seismological Society of America, 101*(1), 39–67. https://doi.org/10.1785/0120100097
Fiaschi, S., Closson, D., Abou Karaki, N., Pasquali, P., Riccardi, P., & Floris, M. (2017). The complex karst dynamics of the Lisan Peninsula revealed by 25 years of DInSAR observations. Dead Sea, Jordan. *Journal of Photogrammetry and Remote Sensing, 130*, 358–369.
Freund, R., Zak, I., & Garfunkel, Z. (1968). Age and rate of the sinistral movement along the Dead Sea rift. *Nature, 220*(5164), 253–255.
Freund, R., Garfunkel, Z., Zak, I., Goldberg, M., Weissbrod, T., & Derin, B. (1970). The shear along The Dead Sea rift. *Philosophical Transactions. Royal Society of London, 267A*, 107–130.
Galli, P. (1997). Archaeoseismological evidence of historical activity of the Wadi Araba – Jordan valley transform fault, Il Quaternario. *Italy Journal Quaternary Science, 10*, 399–404.
Galli, P. (1999). Active tectonics along the Wadi Araba – Jordan valley transform fault. *Journal of Geophysical Research, 104*(B2), 2777–2796.
Galli, P., & Galadini, F. (2001). Surface faulting of Archaeological relics. A review of case histories from the Dead Sea to the Alps. *Tectonophysics, 335*, 291–312.
Garfunkel, Z., Zak, I., & Freund, R. (1981). Active faulting in the Dead Sea rift. *Tectonophysics, 80*, 1–26.
Ghawanmeh, Y. (1990). *Earthquake effects on Belad El-Sham Settlements*. Dar El-Fikr, Amman-Jordan (in Arabic, English extended abstract).

Gomez, F., Meghraoui, M., Darkal, A. N., Hijazi, F., Mouty, M., Suleiman, Y., Sbeinati, R., Darawcheh, R., Al-Ghazzi, R., & Barazangi, M. (2003). Holocene faulting and earthquake recurrence along the Serghaya branch of the Dead Sea fault system in Syria and Lebanon. *Geophysical Journal International, 153*(3), 658–674.

Gomez, F., Karam, G., Khawlie, M., McClusky, S., Vernant, P., Reilinger, R., Jaafar, R., Tabet, C., Khair, K., & Barazangi, M. (2007). Global positioning system measurements of strain accumulation and slip transfer through the restraining bend along the Dead Sea fault system in Lebanon. *Geophysical Journal International, 168*, 1021–1028.

Guidoboni, E., & Comastri, A. (2005). *Catalogue of earthquakes and tsunamis in the Mediterranean area from the 11th to the 15th century*. Instituto National di Geofisica e vulcanologia.

Guidoboni, E., Comastri, A., Traina, G., Phillips, B., & Istituto nazionale di geofisica. (1994). *Catalogue of ancient earthquakes in the mediterranean area up to the 10th century*, edn, Vol., pp. Pages. Rome: Istituto nazionale di geofisica

Jaber, S. (1994). *Assessment and mitigation of earthquake hazard in the Greater Amman area, MsC Thesis*. Department of Geology, University of Jordan.

Kagan, E., SteinM, A. A., & Neumann, F. (2011). Intrabasin paleoearthquake and quiescence correlation of the late Holocene Dead Sea. *Journal of Geophysical Research, 116*(B4). https://doi.org/10.1029/2010JB007452

Kanari, M., Niemi, T. M., Ben-Avraham, Z., Frieslander, U., Tibor, G., Goodman-Tchernov, B. N., Wechsler, N., Abueladas, A., Al-Zoubi, A., Basson, U., & Marco, S. (2020). Seismic potential of the Dead Sea Fault in the northern Gulf of Aqaba-Elat: New evidence from liquefaction, seismic reflection, and paleoseismic data. *Tectonophysics, 793*. https://doi.org/10.1016/j.tecto.2020.228596

Ken-Tor, R., Agnon, A., Enzel, Y., Stein, M., Marco, S., & Negendank, J. (2001). High-resolution geological record of historic earthquakes in the Dead Sea basin. *JGR, 106*(B2), 2221–2234.

Khair, K., Tsokas, G. N., & Sawaf, T. (1997). Crustal structure of the northern Levant region: Multiple source Werner deconvolution estimates for Bouguer gravity anomalies. *Geophysical Journal International, 128*, 605–616.

Khair, K., Karakaisis, G. F., & Papadimitriou, E. (2000). Seismic zonation of the Dead Sea Transform fault area. *Annali di Geofisica, 43*(1), 61–79.

Klinger, Y., Avouac, J. P., & Abou Karaki, N. (1997). Seismotectonics of Wadi Araba Fault (Jordan). In *European Geophysical Society 22th General Assembly Vienna-Austria. Annales Geophysicae, Part I Society Symposia, Solid earth Geophysics & Natural hazards, Suppliment I to Vol. 15, Abstract C234.*

Klinger Y., Rivera L., Haessler H., Maurin J.C (1999) Active faulting in the Gulf of Aqaba: New knowledge from the Mw 7.3 Earthquake of 22 November 1995. Bulletin of the Seismological Society of America, 89, pp.1025–1036.

Klinger, Y., Avouac, J. P., Abou Karaki, N., Dorbath, L., Bourles, D., & Reyss, J. L. (2000a). Slip-rate on the Dead Sea transform fault in Northern Araba valley (Jordan). *Geophysical Journal International, 142*(3), 769–782.

Klinger, Y., Avouac, J. P., Dorbath, L., Abou Karaki, N., & Tisnerat, N. (2000b). Seismic behaviour of the Dead Sea fault along Araba valley. *Geophysical Journal International, 142*(3), 755–768.

Klinger, Y., Le Beon, M., & Al-Qaryouti, M. (2015). 5000 yr of paleoseismicity along the southern Dead Sea fault. *Geophysical Journal International, 202*, 313–327. https://doi.org/10.1093/gji/ggv134

Le Béon, M., Klinger, Y., Amrat, A. Q., Agnon, A., Dorbath, L., Baer, G., Ruegg, J. C., Charade, O., & Mayyas, O. (2008). Slip rate and locking depth from GPS profiles across the southern Dead Sea transform. *Journal of Geophysical Research, 113*, 1–19.

Marco, S., & Agnon, A. (2005). High-resolution stratigraphy reveals repeated earthquake faulting in the Masada Fault Zone, Dead Sea transform. *Tectonophysics, 408*(1–4), 101–112.

Marco, S., Stein, M., Agnon, A., & Ron, H. (1996). Long-term earthquake clustering: A 50,000-year paleoseismic record in the Dead Sea Graben. *Journal of Geophysical Research, 101*, 6179–6191.
Marco, S., Agnon, A., Ellenblum, R., Eidelman, A., Basson, U., & Boas, A. (1997). 817-year-old walls offset sinistrally 2.1 m by the Dead Sea Transform, Israel. *Journal of Geodynamics, 24*, 11–20.
Marco, S., Hartal, M., Hazan, N., Lev, L., & Stein, M. (2003). Archaeology, history, and geology of the 749 AD earthquake. *Dead Sea Transform Geology, 31*, 665–668. https://doi.org/10.1130G19516.1.
Marco, S., Rockwell, T. K., Heimann, A., Frieslander, U., & Agnon, A. (2005). Late Holocene activity of the Dead Sea transform revealed in 3D palaeoseismic trenches on the Jordan Gorge segment, Earth planet. *Science Letters, 234*, 189–205.
Meghraoui, M. (2015). Paleoseismic history of the Dead Sea Fault zone. In *Encyclopedia of Earthquake Engineering* (p. 20). Springer. https://doi.org/10.1007/978-3-642-36197-5_40-1.
Meghraoui, M., et al. (2003). Evidence for 830 years of seismic quiescence from paleoseismology, archaeoseismology and historical seismicity along the Dead Sea fault in Syria. *Earth and Planetary Science Letters, 210*, 35–52.
Niemi, T. M, Smith A. M (1999) Initial results of southern Wadi Araba, Jordan Geoarchaeological study: Implications for shifts in late quaternary aridity. Geochronology: An International Journal, Vol. 14, No. 8, 791–820.
Niemi, T. M., Zhang, H., Atallah, M., & Harrison, J. B. J. (2001). Late Pleistocene and Holocene slip rate of the northern Wadi Araba fault, Dead Sea transform, Jordan. *Journal of Seismology, 5*, 449–474.
Pe'eri, S., Wdowinski, S., Shtibelman, A., Bechor, N., Bock, Y., Nikolaidis, R., & van Domselaar, M. (2002). Current plate motion across the Dead Sea fault from three years of continuous GPS monitoring. *Geophysical Research Letters, 29*, 42–41.
Poirrier, J. P., & Taher, M. A. (1980). Historical seismicity in the near and middle east, North Africa and Spain from Arabic documents (VII–XVIII century). *Bulletin of the Seismological Society of America, 70*, 2185–2202.
Pricope, N. G., Mapes, K. L., & Woodward, K. D. (2019). Remote sensing of human–environment interactions in global change research: A review of advances, challenges and future directions. *Remote Sensing, 11*, 2783. https://doi.org/10.3390/rs11232783
Quennell, A. (1958). The structural and geomorphic evolution of the Dead Sea rift. *Quarterly Journal of the Geological Society, 114*, 1.
Reilinger, R., McClusky, S., Vernant, P., Lawrence, S., Ergintav, S., Cakmak, R., Ozener, H., Kadirov, F., Guliev, I., & Stepanyan, R. (2006). GPS constraints on continental deformation in the Africa-Arabia-Eurasia continental collision zone and implications for the dynamics of plate interactions. *Journal of Geophysical Research, 111*.
Russell, K. W. (1980). The Earthquake of Mai 19, AD 363. *Bulletin of the Seismological Society of America, 238*, 47–64.
Russell, K. W. (1985). The earthquake chronology of Palestine and Northwest Arabia fron the 2nd through the mid-8th century AD. *Bulletin of the Seismological Society of America, 260*, 37–59.
Salameh, E., & El-Naser, H. (1999). Does the actual drop in Dead Sea level reflect the development of water sources within its drainage basin? *Acta Hydrochimica et Hydrobiologica, 27*, 5–11.
Salameh, E., & El-Naser, H. (2000). Changes in the Dead Sea level and their impacts on the surrounding groundwater bodies. *Acta Hydrochimica et Hydrobiologica, 28*, 24–33.
Salameh, E., Alraggad, M., & Amaireh, M. (2019). Degradation processes along the new north-eastern shores of the Dead Sea. *Environmental Earth Sciences, 78*, 164. https://doi.org/10.1007/s12665-019-8155-x
Sbeinati, M. R., Darawcheh, R., & Mouty, M. (2005). The historical earthquakes of Syria: an analysis of large and moderate earthquakes from 1365 BC to 1900 AD. *Annals of Geophysics*.
Seiberg, A. (1932). Untersuchungen uber Erdbeben und Bruchollenbau in ostlichen Mittelmeerrgebiet. *Medizinish-Naturwissenschaftliche Gesellschaft, 18*, 159–273.

Shapira, A., Avni, R., & Nur, A. (1993). A new estimate for the epicenter of the Jericho earthquake of 11 July 1927. *Israel Journal of Earth Sciences, 42*, 93–96.

Taher, M. A. (1979). *Corpus des text Arabes relatifs aux tremblements de Terre at autres catastrophes naturelles de la conquete Arabe au XII H./XVIII J.C.*, LDD thesis University, Paris I, 675p.

Vered M., Striem H. L (1977) A macroseismic study and the implications of structural damage of two recent earthquakes in the Jordan rift, Bulletin of the Seismological Society of America, 67, 1607–1613.

Watson, R. A., Holohan, E. P., Al-Halbouni, D., Saberi, L., Sawarieh, A., Closson, D., Alrshdan, H., Abou Karaki, N., Siebert, C., Walter, T. R., & Dahm, T. (2018). Sinkholes, stream channels and base-level fall: A 50-year record of spatio-temporal development on the eastern shore of the Dead Sea. *Solid Earth Discussions, 10*, 1–43.

Wdowinski, S., Bock, Y., Baer, G., Prawirodirdjo, L., Bechor, N., Naaman, S., Knafo, R., Forrai, Y., & Melzer, Y. (2004). GPS measurements of current crustal movements along the Dead Sea fault. *Journal of Geophysical Research, 109.*

Westaway, R. (1994). Present-day kinematics of the Middle East and eastern Mediterranean. *Journal of Geophysical Research, 99*, 12071–12012. 12090.

Westaway, R. (2004). Kinematic consistency between the Dead Sea fault zone and the neogene and quaternary left-lateral faulting in SE Turkey. *Tectonophysics, 391*, 203–237.

Willis, B. (1928). Earthquakes in the holy land. *Bulletin of the Seismological Society of America, 18*, 72–105.

Willis, B. (1933). Earthquakes in the holy land – A correction. *Bulletin of the Seismological Society of America, 23*, 88–89.

Wilson, J. T. (1965). A new class of faults and their bearing on continental drift. *Nature, 207*, 343–347.

Zilberman, E., Amit, R., Porat, N., Enzel, Y., & Avner, U. (2005). Surface ruptures induced by the devastating 1068 AD earthquake in the southern Arava Valley. *Dead Sea rift, Israel, Tectonophysics, 408*, 79–99.

Index

M. M. Al Saud (ed.), *Applications of Space Techniques on the Natural Hazards in the MENA Region*, https://doi.org/10.1007/978-3-030-88874-9

D

E

G

H

I

J

K

L

S

Y

Z